ESSENTIALS OF ELECTRONIC TESTING
FOR DIGITAL, MEMORY AND
MIXED-SIGNAL VLSI CIRCUITS

FRONTIERS IN ELECTRONIC TESTING

Consulting Editor
Vishwani D. Agrawal

Books in the series:

Analog and Mixed-Signal Boundary-Scan: A Guide to the IEEE 1149.4 Test Standard
    A. Osseiran
    ISBN: 0-7923-8686-8
Design for At-Speed Test, Diagnosis and Measurement
    B. Nadeau-Dosti
    ISBN: 0-79-8669-8
Delay Fault Testing for VLSI Circuits
    A. Krstić, K-T. Cheng
    ISBN: 0-7923-8295-1
Research Perspectives and Case Studies in System Test and Diagnosis
    J.W. Sheppard, W.R. Simpson
    ISBN: 0-7923-8263-3
Formal Equivalence Checking and Design Debugging
    S.-Y. Huang, K.-T. Cheng
    ISBN: 0-7923-8184-X
On-Line Testing for VLSI
    M. Nicolaidis, Y. Zorian
    ISBN: 0-7923-8132-7
Defect Oriented Testing for CMOS Analog and Digital Circuits
    M. Sachdev
    ISBN: 0-7923-8083-5
Reasoning in Boolean Networks: Logic Synthesis and Verification Using Testing Techniques
    W. Kunz, D. Stoffel
    ISBN: 0-7923-9921-8
Introduction to I_{DDQ} Testing
    S. Chakravarty, P.J. Thadikaran
    ISBN: 0-7923-9945-5
Multi-Chip Module Test Strategies
    Y. Zorian
    ISBN: 0-7923-9920-X
Testing and Testable Design of High-Density Random-Access Memories
    P. Mazumder, K. Chakraborty
    ISBN: 0-7923-9782-7
From Contamination to Defects, Faults and Yield Loss
    J.B. Khare, W. Maly
    ISBN: 0-7923-9714-2

ESSENTIALS OF ELECTRONIC TESTING FOR DIGITAL, MEMORY AND MIXED-SIGNAL VLSI CIRCUITS

Michael L. Bushnell
Rutgers University

Vishwani D. Agrawal
Bell Labs, Lucent Technologies.

KLUWER ACADEMIC PUBLISHERS
Boston / Dordrecht / London

Distributors for North, Central and South America:
Kluwer Academic Publishers
101 Philip Drive, Assinippi Park
Norwell, Massachusetts 02061 USA
Telephone (781) 871-6600
Fax (781) 871-6528
E-Mail <kluwer@wkap.com>

Distributors for all other countries:
Kluwer Academic Publishers Group
Distribution Centre
Post Office Box 322
3300 AH Dordrecht, THE NETHERLANDS
Telephone 31 78 6392 392
Fax 31 78 6546 474
E-Mail <orderdept@wkap.nl>

Electronic Services <http://www.wkap.nl>

 Library of Congress Cataloging-in-Publication Data

Bushnell, Michael L. (Michael Lee), 1950-
   Essentials of electronic testing for digital, memory, and mixed-signal VLSI circuits / Michael L. Bushnell, Vishwani D. Agrawal.
         p. cm. -- (Frontiers in electronic testing)
   Includes bibliographical references and index.
   ISBN 0-7923-799-1-8 (alk. paper)
   1. Integrated circuits--Very large scale integration--Testing. 2. Digital integrated circuits--Testing. 3. Mixed signal circuits--Testing. 4. Semiconductor storage devices--Testing. I. Agrawal, Vishwani D., 1943- II. Title. III. Series.

TK7874.75.B87 2000
621.39'5--dc21   00-046212

Copyright © 2000 by Lucent Technologies and Michael L. Bushnell.

All rights reserved. No part of this publication may be reproduced, stored in a retrieval system or transmitted in any form or by any means, mechanical, photo-copying, recording, or otherwise, without the prior written permission of the publisher, Kluwer Academic Publishers, 101 Philip Drive, Assinippi Park, Norwell, Massachusetts 02061

This students edition for sale only in the Republic of China

Distributed by: SCI-TECH Publishing Co., Ltd.

To Margaret Kalvar for her patient understanding, love, and support, which makes my work possible — MLB.
To the women in my life, Premlata, Prathima, Victoria and Chitra, and to my son, Vikas — VDA.

TABLE OF CONTENTS

PREFACE   xv

ABOUT THE AUTHORS   xvii

I INTRODUCTION TO TESTING   1

1 INTRODUCTION   3
1.1 Testing Philosophy 4
1.2 Role of Testing 6
1.3 Digital and Analog VLSI Testing 7
1.4 VLSI Technology Trends Affecting Testing 9
1.5 Scope of this Book 15

2 VLSI TESTING PROCESS AND TEST EQUIPMENT   17
2.1 How to Test Chips? 18
    2.1.1 Types of Testing 18
2.2 Automatic Test Equipment 24
    2.2.1 Advantest Model T6682 ATE 24
    2.2.2 LTX Fusion ATE 28
    2.2.3 Multi-Site Testing 29
2.3 Electrical Parametric Testing 30
2.4 Summary 34

3 TEST ECONOMICS AND PRODUCT QUALITY   35
3.1 Test Economics 36
    3.1.1 Defining Costs 36
    3.1.2 Production 38
    3.1.3 Benefit-Cost Analysis 41
    3.1.4 Economics of Testable Design 42
    3.1.5 The Rule of Ten 44
3.2 Yield 44
3.3 Defect Level as a Quality Measure 47
    3.3.1 Test Data Analysis 48
    3.3.2 Defect Level Estimation 50

		3.4	Summary	53
4	**FAULT MODELING**			**57**
	4.1		Defects, Errors, and Faults	57
	4.2		Functional Versus Structural Testing	59
	4.3		Levels of Fault Models	60
	4.4		A Glossary of Fault Models	60
	4.5		Single Stuck-at Fault	70
		4.5.1	Fault Equivalence	72
		4.5.2	Equivalence of Single Stuck-at Faults	73
		4.5.3	Fault Collapsing	74
		4.5.4	Fault Dominance and Checkpoint Theorem	75
		4.5.5	Summary	78

II	**TEST METHODS**			**81**
5	**LOGIC AND FAULT SIMULATION**			**83**
	5.1		Simulation for Design Verification	83
	5.2		Simulation for Test Evaluation	88
	5.3		Modeling Circuits for Simulation	91
		5.3.1	Modeling Levels and Types of Simulators	91
		5.3.2	Hierarchical Connectivity Description	93
		5.3.3	Gate-level Modeling of MOS Networks	94
		5.3.4	Modeling Signal States	96
		5.3.5	Timing	98
	5.4		Algorithms for True-Value Simulation	101
		5.4.1	Compiled-Code Simulation	102
		5.4.2	Event-Driven Simulation	103
	5.5		Algorithms for Fault Simulation	105
		5.5.1	Serial Fault Simulation	106
		5.5.2	Parallel Fault Simulation	107
		5.5.3	Deductive Fault Simulation	109
		5.5.4	Concurrent Fault Simulation	113
		5.5.5	Roth's TEST-DETECT Algorithm	116
		5.5.6	Differential Fault Simulation	117
	5.6		Statistical Methods for Fault Simulation	120
		5.6.1	Fault Sampling	121
	5.7		Summary	125
6	**TESTABILITY MEASURES**			**129**
	6.1		SCOAP Controllability and Observability	131
		6.1.1	Combinational SCOAP Measures	132
		6.1.2	Combinational Circuit Example	134
		6.1.3	Sequential SCOAP Measures	140

TABLE OF CONTENTS

6.1.4 Sequential Circuit Example	142
6.2 High-Level Testability Measures	148
6.3 Summary	150

7 COMBINATIONAL CIRCUIT TEST GENERATION — 155

- 7.1 Algorithms and Representations 156
  - 7.1.1 Structural vs. Functional Test 156
  - 7.1.2 Definition of Automatic Test-Pattern Generator 157
  - 7.1.3 Search Space Abstractions 158
  - 7.1.4 Algorithm Completeness 159
  - 7.1.5 ATPG Algebras 159
  - 7.1.6 Algorithm Types 160
- 7.2 Redundancy Identification (RID) 168
- 7.3 Testing as a Global Problem 172
- 7.4 Definitions 172
- 7.5 Significant Combinational ATPG Algorithms 176
  - 7.5.1 D-Calculus and D-Algorithm (Roth) 176
  - 7.5.2 PODEM (Goel) 186
  - 7.5.3 FAN (Fujiwara and Shimino) 192
  - 7.5.4 Advanced Algorithms 197
- 7.6 Test Generation Systems 204
- 7.7 Test Compaction 205
- 7.8 Summary 206

8 SEQUENTIAL CIRCUIT TEST GENERATION — 211

- 8.1 ATPG for Single-Clock Synchronous Circuits 212
  - 8.1.1 A Simplified Problem 214
- 8.2 Time-Frame Expansion Method 214
  - 8.2.1 Use of Nine-Valued Logic 216
  - 8.2.2 Development of Time-Frame Expansion Methods 218
  - 8.2.3 Approximate Methods 222
  - 8.2.4 Implementation of Time-Frame Expansion Methods 222
  - 8.2.5 Complexity of Sequential ATPG 225
  - 8.2.6 Cycle-Free Circuits 225
  - 8.2.7 Cyclic Circuits 229
  - 8.2.8 Clock Faults and Multiple-Clock Circuits 231
  - 8.2.9 Asynchronous Circuits 232
- 8.3 Simulation-Based Sequential Circuit ATPG 238
  - 8.3.1 CONTEST Algorithm 239
  - 8.3.2 Genetic Algorithms 246
- 8.4 Summary 248

9 MEMORY TEST — 253
- 9.1 Memory Density and Defect Trends — 255
- 9.2 Notation — 258
- 9.3 Faults — 259
  - 9.3.1 Fault Manifestations — 259
  - 9.3.2 Failure Mechanisms — 260
- 9.4 Memory Test Levels — 261
- 9.5 March Test Notation — 262
- 9.6 Fault Modeling — 263
  - 9.6.1 Diagnosis Versus Testing Needs — 265
  - 9.6.2 Reduced Functional Faults — 266
  - 9.6.3 Relation Between Fault Models and Physical Defects — 276
  - 9.6.4 Multiple Fault Models — 278
  - 9.6.5 Frequency of Faults — 281
- 9.7 Memory Testing — 284
  - 9.7.1 Functional RAM Testing with March Tests — 284
  - 9.7.2 Testing RAM Neighborhood Pattern-Sensitive Faults — 286
  - 9.7.3 Testing RAM Technology and Layout-Related Faults — 294
  - 9.7.4 RAM Test Hierarchy — 295
  - 9.7.5 Cache RAM Chip Testing — 296
  - 9.7.6 Functional ROM Chip Testing — 300
  - 9.7.7 Electrical Parametric Testing — 301
- 9.8 Summary — 306

10 DSP-BASED ANALOG AND MIXED-SIGNAL TEST — 309
- 10.1 Analog and Mixed-Signal Circuit Trends — 309
- 10.2 Definitions — 314
- 10.3 Functional DSP-Based Testing — 317
  - 10.3.1 Concept — 317
  - 10.3.2 Mechanism of DSP-Based Testers — 319
  - 10.3.3 Waveform Synthesis — 320
  - 10.3.4 Waveform Sampling and Digitization — 322
- 10.4 Static ADC and DAC Testing Methods — 322
  - 10.4.1 Transmission vs. Intrinsic Parameters — 323
  - 10.4.2 Uncertainty and Distortion in Ideal ADCs — 325
  - 10.4.3 DAC Transfer Function Error — 325
  - 10.4.4 ADC Transfer Function Error — 326
  - 10.4.5 Flash ADC Testing Methods — 327
  - 10.4.6 DAC Testing Methods — 332
- 10.5 Realizing Emulated Instruments Using Fourier Transforms — 335
  - 10.5.1 Fourier Voltmeter — 345
  - 10.5.2 Testing of Analog Devices Using Non-Coherent Sampling — 350
  - 10.5.3 Coherent Multi-Tone Testing — 356
  - 10.5.4 ATE Vector Operations — 364

TABLE OF CONTENTS

- 10.6 CODEC Testing 366
  - 10.6.1 Considerations for CODEC Performance Tests 369
  - 10.6.2 CODEC Tests 372
- 10.7 Dynamic Flash ADC Testing FFT Technique 376
- 10.8 Advanced Topics 377
  - 10.8.1 Event Digitization 377
  - 10.8.2 Measuring Random Noise 380
- 10.9 Summary 382

11 MODEL-BASED ANALOG AND MIXED-SIGNAL TEST   385
- 11.1 Analog Testing Difficulties 386
- 11.2 Analog Fault Models 387
- 11.3 Levels of Abstraction 389
- 11.4 Types of Analog Testing 389
- 11.5 Analog Fault Simulation 390
  - 11.5.1 Motivation 391
  - 11.5.2 DC Fault Simulation of Nonlinear Circuits 391
  - 11.5.3 Linear Analog Circuit AC Fault Simulation 395
  - 11.5.4 Monte-Carlo Simulation 397
- 11.6 Analog Automatic Test-Pattern Generation 397
  - 11.6.1 ATPG Using Sensitivities 398
  - 11.6.2 ATPG Using Signal Flow Graphs 406
  - 11.6.3 Additional Methods 413
- 11.7 Summary 413

12 DELAY TEST   417
- 12.1 Delay Test Problem 417
- 12.2 Path-Delay Test 420
  - 12.2.1 Test Generation for Combinational Circuits ... 424
  - 12.2.2 Number of Paths in a Circuit 427
- 12.3 Transition Faults 428
- 12.4 Delay Test Methodologies 429
  - 12.4.1 Slow-Clock Combinational Test 429
  - 12.4.2 Enhanced-Scan Test 430
  - 12.4.3 Normal-Scan Sequential Test 431
  - 12.4.4 Variable-Clock Non-Scan Sequential Test 432
  - 12.4.5 Rated-Clock Non-Scan Sequential Test 434
- 12.5 Practical Considerations in Delay Testing 434
  - 12.5.1 At-Speed Testing 435
- 12.6 Summary 436

13 IDDQ TEST   439
- 13.1 Motivation 439
- 13.2 Faults Detected by I_{DDQ} Tests 441
- 13.3 I_{DDQ} Testing Methods 446

13.3.1 I_{DDQ} Fault Coverage Metrics 446
13.3.2 I_{DDQ} Test Vector Selection from Stuck-Fault Vector Sets . . 448
13.3.3 Instrumentation Problems 451
13.3.4 Current Limit Setting 452
13.4 Surveys of I_{DDQ} Testing Effectiveness 453
13.5 Limitations of I_{DDQ} Testing 455
13.6 Delta I_{DDQ} Testing 456
13.7 I_{DDQ} Built-In Current Testing 458
13.8 I_{DDQ} Design for Testability 460
13.9 Summary 460

III DESIGN FOR TESTABILITY 463

14 DIGITAL DFT AND SCAN DESIGN 465
14.1 Ad-Hoc DFT Methods 466
14.2 Scan Design 467
    14.2.1 Scan Design Rules 469
    14.2.2 Tests for Scan Circuits 471
    14.2.3 Multiple Scan Registers 474
    14.2.4 Overheads of Scan Design 474
    14.2.5 Design Automation 477
    14.2.6 Physical Design and Timing Verification of Scan 479
14.3 Partial-Scan Design 479
14.4 Variations of Scan 483
14.5 Summary 485

15 BUILT-IN SELF-TEST 489
15.1 The Economic Case for BIST 490
    15.1.1 Chip/Board Area Cost vs. Tester Cost 492
    15.1.2 Chip/Board Area Cost vs. System Downtime Cost . . 494
15.2 Random Logic BIST 495
    15.2.1 Definitions 495
    15.2.2 BIST Process 496
    15.2.3 BIST Pattern Generation 498
    15.2.4 BIST Response Compaction 512
    15.2.5 Built-In Logic Block Observers 519
    15.2.6 Test-Per-Clock BIST Systems 521
    15.2.7 Test-Per-Scan BIST Systems 521
    15.2.8 Circular Self-Test Path System 525
    15.2.9 Circuit Initialization 526
    15.2.10 Device Level BIST 526
    15.2.11 Test Point Insertion 528
15.3 Memory BIST 529
    15.3.1 Definitions 530

	15.3.2 March Test SRAM BIST	532
	15.3.3 SRAM BIST with MISR	534
	15.3.4 Neighborhood Pattern Sensitive Fault Test DRAM BIST	536
	15.3.5 Transparent Memory BIST Tests	539
	15.3.6 Complex Examples	539
15.4	Delay Fault BIST	540
15.5	Summary	543

16 BOUNDARY SCAN STANDARD — 549

16.1	Motivation	550
	16.1.1 Purpose of Standard	552
16.2	System Configuration with Boundary Scan	553
	16.2.1 TAP Controller and Port	553
	16.2.2 Boundary Scan Test Instructions	557
	16.2.3 Pin Constraints of the Standard	564
16.3	Boundary Scan Description Language	569
	16.3.1 BSDL Description Components	570
	16.3.2 Pin Descriptions	571
16.4	Summary	572

17 ANALOG TEST BUS STANDARD — 575

17.1	Analog Circuit Design for Testability	576
17.2	Analog Test Bus (ATB)	576
	17.2.1 Targeted Analog Faults	577
	17.2.2 Analog Test Access Port (ATAP)	579
	17.2.3 Test Bus Interface Circuit (TBIC)	580
	17.2.4 Analog Boundary Module (ABM)	583
	17.2.5 Instructions for 1149.4 Standard	585
	17.2.6 Other 1149.4 Standard Features	589
17.3	Summary	591

18 SYSTEM TEST AND CORE-BASED DESIGN — 595

18.1	System Test Problem Defined	596
18.2	Functional Test	597
	18.2.1 Microprocessor Test	598
18.3	Diagnostic Test	598
	18.3.1 Fault Dictionary	599
	18.3.2 Diagnostic Tree	600
	18.3.3 A System Test Example	602
18.4	Testable System Design	604
18.5	Core-Based Design and Test-Wrapper	606
18.6	A Test Architecture for System-on-a-Chip (SOC)	607
18.7	An Integrated Design and Test Approach	608
18.8	Summary	610

19 THE FUTURE OF TESTING 613

A CYCLIC REDUNDANCY CODE THEORY 615
A.1 Polynomial Multiplier 616
A.2 Polynomial Divider 617

B PRIMITIVE POLYNOMIALS OF DEGREE 1 TO 100 619

C BOOKS ON TESTING 621
C.1 General and Tutorial 621
C.2 Analog and Mixed-Signal Circuit Test 622
C.3 ATE, Test Programming, and Production Test 622
C.4 Board and MCM Test and Boundary Scan 623
C.5 Built-In Self-Test 624
C.6 Delay Fault Test 624
C.7 Design for Testability 624
C.8 Fault Modeling 625
C.9 Fault Tolerance and Diagnosis 625
C.10 Formal Verification 625
C.11 High-Level Test and Verification 626
C.12 I_{DDQ} Test 626
C.13 Memory Test 626
C.14 Microprocessor Verification and Test 627
C.15 Semiconductor Defect Mechanisms 627
C.16 System Test 627
C.17 Test Economics 627
C.18 Test Evaluation 628
C.19 Test Generation 628
C.20 Periodicals 628
C.21 Conferences and Workshops 629
C.22 Web Sites 629

BIBLIOGRAPHY 631

INDEX 671

PREFACE

The modern electronic testing has a *forty* year history. Test professionals hold some fairly large conferences and numerous workshops, have a journal, and there are over one hundred books on testing. Still, a full course on testing is offered only at a few universities, mostly by professors who have a research interest in this area. Apparently, most professors would not have taken a course on electronic testing when they were students.

Other than the computer engineering curriculum being too crowded, the major reason cited for the absence of a course on electronic testing is the lack of a suitable textbook. For VLSI the foundation was provided by semiconductor device technology, circuit design, and electronic testing. In a computer engineering curriculum, therefore, it is necessary that foundations should be taught before applications. The field of VLSI has expanded to systems-on-a-chip, which include digital, memory, and mixed-signal subsystems. To our knowledge this is the first textbook to cover all three types of electronic circuits.

We have written this textbook for an undergraduate "foundations" course on electronic testing. Obviously, it is too voluminous for a one-semester course and a teacher will have to select from the topics. We did not restrict such freedom because the selection may depend upon the individual expertise and interests. Besides, there is merit in having a larger book that will retain its usefulness for the owner even after the completion of the course.

With equal tenacity, we address the needs of three other groups of readers. The first group consists of engineers who, upon graduation, engage in any kind of electronic hardware design, testing, or manufacturing project. Parts I and III emphasize the needs of a design-oriented project and Parts I and II those of a test-oriented project. The second group consists of students of a VLSI design course who have not taken a course on testing. Parts I and III focus on their needs. The third group, consisting of post-graduate and research students, will find a complete coverage of topics with pointers to references where advanced material was omitted for a lack of space. Figure 1.6 shows several ways to read this book.

At the 1999 *International Test Conference* during a panel discussion titled, "Increasing Test Coverage in a VLSI Design Course," a panelist from the microelectronics industry gave the wish-list as: test economics, classical semiconductor defects, simple test pattern coverage, structured design for testability techniques (scan, boundary scan, BIST) for system-on-a-chip design, automatic test equipment (constraints and costs), and selected advanced topics (I_{DDQ} and delay faults.) We kept that list in mind while writing the book and we hope the teachers of VLSI design and electronic testing courses will too.

We are all too familiar with incompleteness of software debugging and hardware design verification. No "formal" method was used to verify the material in this book either. Despite all efforts to remove errors, we cannot guarantee that the readers will not find them. We will greatly appreciate the generosity of our readers if they inform us about any errors. We will make such findings available to all readers through our websites until the publisher gives us an opportunity to make corrections, with due acknowledgment to those who have pointed them out.

We have taught a course on testing at Rutgers University for the past ten years. Interaction with the students in the course and our master's and doctoral students had the greatest influence on our understanding of the subject. We would like to thank them. Special mention should be made of the class of Spring 2000, which used the draft and pointed out corrections and improvements. We are indebted to colleagues at Bell Labs and Rutgers for their advice and counsel. The enthusiasm and support of the world-wide test professionals was exceptional. A partial list of those we thank includes: Miron Abramovici, Mark Barber, Shawn Blanton, Amy Bushnell, Tapan Chakraborty, Srimat Chakradhar, Xinghao Chen, Dochan C. Choi, Rick Chruscial, Don Denburg, Jose de Sousa, David Fessler, Hideo Fujiwara, John Hayes, Michael Hsiao, James Jacob, Neil Kelly, Bill Kish, Kozo Kinoshita, Ken Lanier, Yuhai Ma, Pinaki Mazumder, Karen Panetta, Janusz Rajski, Elizabeth Rudnick, Manoj Sachdev, Kewal Saluja, and Sharad Seth. We thank our publisher Carl Harris for always encouraging us to proceed ahead and for being patient through schedule slips. We are thankful for the support of Al Aho, Dennis Ritchie, and Tom Szymanski, research managers at Bell Labs, and David Daut and Jim Flanagan of Rutgers University.

We also wish to thank the LTX Corporation, the Advantest Corporation, Samsung Electronics Company, Ltd., IBM, and Lucent Technologies for their cooperation in providing data for this book. In describing technical contributions we have tried our best to cite correctly. From those who find their work incorrectly cited, we beg forgiveness because such errors, caused by our ignorance, were unintentional.

Michael L. Bushnell
bushnell@caip.rutgers.edu
http://www-caip.rutgers.edu/
∼bushnell/rutgers.html

Vishwani D. Agrawal
va@research.bell-labs.com
http://cm.bell-labs.com/
cm/cs/who/va

ABOUT THE AUTHORS

Michael L. Bushnell is a Full Professor and a Board of Trustees Research Fellow in the Electrical and Computer Engineering Department at Rutgers University. He received his B.S. degree at Massachusetts Institute of Technology in 1975, and his M.S. degree in 1983 and his PhD degree in 1986, both from Carnegie Mellon University. He was selected in 1983 for the American Electronics Association Faculty Development Program, he received the Outstanding Graduate Student Teaching Award from Carnegie Mellon, and he was a *Presidential Young Investigator* of the National Science Foundation of the United States. His current VLSI CAD research interests are automatic digital, analog, and mixed-signal circuit test-pattern generation on serial and distributed computers, delay fault built-in self-testing, fault simulation, synthesis for testability, and low-power design. He coauthored 78 papers. His books on VLSI CAD are *Efficient Branch and Bound Search with Application to Computer Aided Design*, *Neural Models and Algorithms for Digital Testing*, and *Design Automation*. He holds 3 patents (2 more are pending) on BIST and test generation methods. He is an Editorial Board Member of the *Journal of Electronic Testing: Theory and Applications*. He served as the Program Co-Chair of India's annual *International Conference on VLSI Design* in 1995 and 1996.

Vishwani D. Agrawal is a Distinguished Member of Technical Staff in the Computing Sciences Research Center of Bell Labs (R&D arm of Lucent Technologies), Murray Hill, New Jersey, and a Visiting Professor of Electrical and Computer Engineering at Rutgers University, New Brunswick, New Jersey. He obtained his BSc degree from the University of Allahabad, Allahabad, India, in 1960, BE (honours) degree from the University of Roorkee, Roorkee, India, in 1964, ME degree from the Indian Institute of Science, Bangalore, India, in 1966, and PhD degree in electrical engineering from the University of Illinois, Urbana-Champaign, in 1971. His current interests are testing, synthesis for testability, and parallel algorithms. He has published over 200 papers and coauthored four books. He holds twelve U.S. patents on testing, design for testability, and low-power design. He is the Editor-in-Chief of the *Journal of Electronic Testing: Theory and Applications* and a past Editor-in-Chief of the *IEEE Design & Test of Computers* magazine. He is the Consulting Editor for the *Frontiers in Electronic Testing* book series of Kluwer Academic Publishers, Boston. In 1985, he co-founded India's annual *International Conference on VLSI Design*. From 1989 to 1990, he served on the Board of Governors of the IEEE

Computer Society, and in 1994, chaired the Fellow Selection Committee of that Society. He is a Fellow of the IEEE, a Fellow of the IETE (India), and a Member of the ACM. He has received five *Best Paper Awards*. In 1993, he received the *Distinguished Alumnus Award* of the University of Illinois at Urbana-Champaign. In 1998, he received the *Harry H. Goode Memorial Award* of the IEEE Computer Society for "innovative contributions to the field of electronic testing."

Part I
INTRODUCTION TO TESTING

Chapter 1

INTRODUCTION

> "... Of text-books, about which we hear so much, I never felt the want. I do not even remember having made much use of the books that were available. I did not find it at all necessary to load the boys with quantities of books. I have always felt that the true text-book for the pupil is his teacher. I remember very little that my teachers taught me from books, but I have even now a clear recollection of the things they taught me independently of books."
> — Mahatma Gandhi, in 1948 book, *An Autobiography or The Story of My Experiments With Truth* [238].

Now that you have picked up this book we want to convince you that it will help you. "To err is human," is an age-old excuse for sloppy work. Yet, while accepting imperfect products from others, we desire perfection. How can an imperfect human turn out a perfect product? Here is the formula:

> **Algorithm: Perfect**
> **Repeat until tested perfect:**
> {
>   **Redesign;**
>   **Remake;**
> }

Two points are relevant. First, the test should be designed to truly indicate the desired perfection, and second, the *repeat* loop in our algorithm will repeat indefinitely unless *redesign* and *remake* could respond to or correct the errors found in testing. Understanding the art and science of testing allows you to attain perfection in your work, otherwise an almost *superhuman* task. If we have convinced you that reading this book is worth your time, we caution you that this book covers only testing of analog and digital electronic systems with *very large scale integration* (VLSI.)

1.1 Testing Philosophy

If you are a student now, or were in the past, you are quite familiar with the word *test*, and probably hate it. Understanding the teacher's point of view will help. The teacher sets a *domain of knowledge* for testing, called the *course syllabus*. It may be the contents of a book, class notes, lectures, or some (arbitrary!) combination of all those. Next, comes the testing method. The teacher asks questions and analyzes the response, perhaps by matching answers to correct ones from the book. The quality of such a test system depends upon how well the test questions cover the syllabus. In VLSI testing also, one should know the *specification* (synonymous to the course syllabus) of the object being tested and then devise tests such that if the object produces the expected response then its conformance to the specification can be guaranteed.

Returning to our student analogy, since no one has infinite time, the number of questions must be limited, and they should be cleverly devised. The teacher now makes certain assumptions. Certain typical errors, ones that the student is likely to commit, are assumed. Questions are devised especially to uncover those errors and, if the student's answers are correct, the teacher grants the benefit of doubt, showing confidence in the implicit error model. Electronic testing also uses *fault modeling* and tests are generated for the assumed fault models. In testing, successful experience with a fault model gives it credibility, and eventually people expect reliability when a high percentage of the modeled faults is tested.

Finally, remember that, if you fail, you must repeat the course. This is similar to redesign and remake in our "Algorithm: Perfect." Of course, you can do better by asking your teacher, right at the beginning, about (1) the course syllabus and (2) error models (i.e., what you will be tested for), and then plan your studies to succeed. In VLSI, that is called *design for testability*.

Example 1.1 *Testing of students. In a course on xyzeeology, 70% of the students deserve to pass. We will call them "pass quality" students. Assuming that the number of students in the class is large, we will study the test process using a statistical basis. For a randomly selected student from the class, we define the following events:*

PQ: *student is pass quality* P: *student passes the test*
FQ: *student is fail quality* F: *student fails the test*

In our example, $Prob(PQ) = 0.7$. Assuming that only pass/fail grades are awarded, the remaining 30% students are of "fail quality," i.e., $Prob(FQ) = 0.3$. As we know, it is impossible to design a perfect test. However, our teacher does quite well and 95% of pass quality students actually pass the test. This is represented by conditional probabilities, $Prob(P|PQ) = 0.95$ and $Prob(F|PQ) = 0.05$. Similarly, the test correctly fails 95% of the fail quality students. A reader not familiar with basic concepts of the probability theory may wish to consult any basic text on the subject [677]. The diagram of Figure 1.1 illustrates the state transition caused by the test. The initial state, on the left, consists of all students in one group. The test separates them into two groups shown on the right as "passed" and "failed." Sizes

1.1 Testing Philosophy

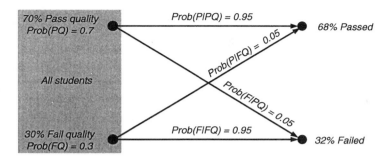

Figure 1.1: A pass/fail test.

of the passed and failed groups created by the test are given by the total probabilities of passing and failing, respectively. The total probability of passing is,

$$\begin{aligned} Prob(P) &= Prob(P|PQ) \times Prob(PQ) + Prob(P|FQ) \times Prob(FQ) \quad (1.1) \\ &= 0.95 \times 0.70 + 0.05 \times 0.30 = 0.68 \end{aligned}$$

Similarly, the total probability of failing is found to be $P(F) = 0.32$. Notice that the original group had 70% pass quality students, and the test has passed only 68%. Obviously, some pass quality students have been failed. But, are all passed students of pass quality?

We will examine the conditional probability $Prob(FQ|P)$ of a student belonging to the fail quality subgroup, given that he or she has passed. The joint probability of events FQ and P is given by:

$$Prob(FQ, P) = Prob(FQ|P)Prob(P) = Prob(P|FQ)Prob(FQ) \quad (1.2)$$

and, therefore:

$$Prob(FQ|P) = \frac{Prob(P|FQ)Prob(FQ)}{Prob(P)} \quad (1.3)$$

where $Prob(P)$ comes from Equation 1.1. Equation 1.3 is known as Bayes' rule [677] and is commonly used for drawing inferences from statistical data. We obtain $Prob(FQ|P) = 0.05 \times 0.30/0.68 = 0.022$. That is, 2.2% of passed students are of fail quality. We will call this the "teacher's risk." The teacher can reduce this risk by making the test more difficult, decreasing $P(P|FQ)$ closer to 0. However, that test can potentially fail a few more pass quality students. So, let us examine the "student's risk."

Applying the Bayes' rule, we obtain

$$Prob(PQ|F) = \frac{Prob(F|PQ)Prob(PQ)}{Prob(F)} = \frac{0.05 \times 0.7}{0.32} = 0.11 \quad (1.4)$$

This shows that 11% of failed students should have passed. We call this the "student's risk." Obviously, a pass quality student would not like to end up in the failed group.

To reduce the student's risk, the probability $P(F|PQ)$ will have to be reduced. This can be done by making the test easier. That will, however, also pass a few more fail quality students, worsening the quality of the passing batch. Thus, teacher's risk and student's risk are opposing criteria, requiring practical compromises. An ideal test, that minimizes both risks, should be so "tuned" that it fails no pass quality student and passes no fail quality student. Devising such a test is no mean task for our teacher.

Testing of electronic systems differs only slightly from the above scenario. A student may pass by correctly answering most, but not necessarily all, questions on the test. If a small number of answers is wrong, the teacher gives the student the benefit of doubt, for he or she may be having a bad day, or else could even learn correct answers in the future. There is no such benefit of doubt for a VLSI chip. Being inanimate, it cannot be having a bad day and certainly cannot learn. So, even a single incorrect test response will fail a VLSI chip. However, electronic tests are not perfect either. They may not cover certain faults and some bad chips will pass. We may also use some "nonfunctional tests" to prevent those bad chips from passing. Nonfunctional tests do not execute the specified function – an example is the quiescent current (I_{DDQ}) test discussed in Chapter 13. These tests can, in turn, fail some good chips. For VLSI, failing of good chips by tests is known as *yield loss*, which increases the cost of manufacturing.

In electronic testing, the teacher's risk is synonymous to the *consumer's risk*. It is related to bad chips being shipped to the consumer. The student's risk in the above example is similar to the *manufacturer's risk*, since failing of good devices increases the cost. We will examine these aspects of electronic testing in Chapter 3.

1.2 Role of Testing

If you design a product, fabricate and test it, and it fails the test, then there must be a cause for the failure. Either (1) the test was wrong, or (2) the fabrication process was faulty, or (3) the design was incorrect, or (4) the specification had a problem. Anything can go wrong. The role of *testing* is to detect whether something went wrong and the role of *diagnosis* is to determine exactly what went wrong, and where the process needs to be altered. Therefore, correctness and effectiveness of testing is most important for quality products (another name for perfect products.)

If the test procedure is good and the product fails, then we suspect the fabrication process, the design, or the specification. If all students in a class fail then it is often considered the teacher's failure. If only some fail, we assume that the teacher is competent, but some students are having difficulty. To select students likely to succeed, teachers may use prerequisites or admission tests for screening. Distributed testing along a product realization process catches the defect-producing causes as soon as they become active, and before they have done much damage. A well thought out test strategy is crucial to economical realization of products.

The benefits of testing are *quality* and *economy*. These two attributes are not

1.3 Digital and Analog VLSI Testing

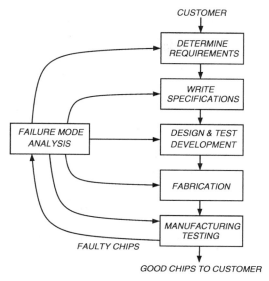

Figure 1.2: VLSI realization process (a naive version.)

independent and neither can be defined without the other. *Quality means satisfying the user's needs at a minimum cost.* A good test process can weed out all bad products before they reach the user. However, if too many bad items are being produced then the cost of those bad items will have to be recovered from the price charged for the few good items that are produced. It will be impossible for an engineer to design a quality product without a profound understanding of the physical principles underlying the processes of manufacturing and test.

1.3 Digital and Analog VLSI Testing

Before we shoot a picture we should examine the scenery. The scenery here is the process for realizing VLSI chips, shown crudely in Figure 1.2. *Requirements* are the user needs satisfied by the chip. They are often derived from the function of the particular application, for example controlling fuel injection in a car, controlling a robot arm, or processing pictures from a space shuttle.

One sets down the *specifications* of various types, which include function (input-output characteristics), operating characteristics (power, frequency, noise, etc.), physical characteristics (packaging, etc.), environmental characteristics (temperature, humidity, reliability, etc.), and other characteristics (volume, cost, price, availability, etc.).

The objective of *design* is to produce data necessary for the next steps of fabrication and testing. Design has several stages. The first, known as *architectural design*, produces a system-level structure of realizable blocks to implement the functional specification. The second, called *logic design*, further decomposes blocks into logic gates. Finally, the gates are implemented as physical devices (e.g., transistors) and a chip layout is produced during *physical design*. The physical layout is converted into photo masks that are directly used in the fabrication of silicon VLSI chips. Fabrication consists of processing silicon wafers through a series of steps involving photoresist, exposure through masks, etching, ion implantation, etc.

It is naive to think that every fabricated chip will be good. Impurities and defects in materials, equipment malfunctions, and human errors are some causes of defects. The likelihood and consequences of defects are the main reasons for testing!

Another very important function of testing is the process diagnosis. We must find what went wrong with each faulty chip, be it in fabrication, in design, or in testing. Or, we may have started with unrealizable specifications. The faulty chip analysis is called *failure mode analysis* (FMA.) FMA uses many different test types, including examination through optical and electron microscopes, to determine the failure cause and fix the process.

Examine Figure 1.2 now. The arrows out of the FMA block represent the corrective actions applied to the faulty steps of the realization process. Consider the process as a pipeline (or assembly line) with the flow direction from top to bottom, so the effort between the point where error occurred and the point of testing, where it was detected, is wasted. At the time the error is detected, the portion of the pipeline between these two points is filled with faulty product which will be either reworked or discarded. Wasted effort and material adds to the product cost. *Testing should, therefore, be placed closest to the point of error*. Many companies emphasize on *doing it right the first time*, or pursuing *the goal of zero defects*. This does not mean that humans, or even machines, cannot make mistakes. These goals are achievable in an error-prone environment, when errors are detected and corrected before damage occurs.

The VLSI realization process of Figure 1.3 has a distributed form of testing. The dotted lines (representing screening) show testing. Depending on the context, we give testing different names. Requirements and specifications are *audited*, design and tests are *verified*, and fabricated parts are *tested*. Each testing level performs two functions, and involves different technical personnel. The first function ascertains that the work still conforms to the objectives of previous levels and meets customer requirements. The second ascertains that things have been done according to the capabilities of the later process levels. For example, verification of design and test procedures should ensure that the design meets all functional and other specifications, and that it is also manufacturable, testable, and repairable.

Figure 1.3 also shows the level of involvement of various types of engineering personnel through the lifetime of a VLSI device. While this figure is typical for an *application specific integrated circuit* (ASIC), it applies to many other electronic devices as well. The process begins with a dialogue between the customer and the marketing engineer. As specifications are prepared, some involvement of those responsible for later activities (design, manufacture, and test) is advisable to ensure realizable specification. The systems engineer then begins by constructing an architectural block diagram. The architecture is verified by high-level simulation and each block is synthesized at the logic-level. The logic circuit is simulated for the same

1.4 VLSI Technology Trends Affecting Testing

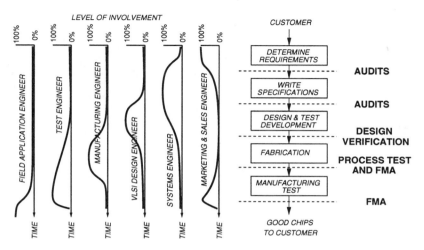

Figure 1.3: A realistic VLSI realization process.

stimuli (often produced by testbenches) as used for the high-level simulation. A *testbench* is hardware description language (HDL) code that, when executed, produces stimuli for the designed circuit [381] (see Sections 5.3 and 5.4.) Vectors generated by testbenches are compacted or augmented and run through a fault simulator to satisfy some specified fault coverage requirement. The VLSI design engineer generates a layout and verifies the timing against the specification. Manufacturing and test engineers then fabricate and test wafers, and package and test chips. All through this process, any failure modes are identified and process improvements are made to ensure a high yield of good devices. Finally, the sales and field application engineers interact with the customer. As "verification and test" related activities are distributed throughout the lifetime of the device, it is necessary that all engineering personnel have the knowledge of test principles.

1.4 VLSI Technology Trends Affecting Testing

The complexity of VLSI technology has reached the point where we are trying to put 100 million transistors on a single chip, and we are trying to increase the on-chip clock frequency to 1 GHz. Table 1.1 shows the proposed roadmap of the Semiconductor Industries Association (SIA.) These trends have a profound effect on the cost and difficulty of chip testing.

Rising Chip Clock Rates. Figure 1.4 shows microprocessor clock rate trends over the last 16 years. Microprocessors represent the leading edge in the VLSI technology trend of Table 1.1. The exponentially rising clock rate indicates several changes in testing over the next 10 years:

1. **At-Speed Testing.** It has been established that *stuck-fault* tests are more effective when applied at the circuit's rated clock speed [501], rather than at

Table 1.1: VLSI chips – present and future [297, 584].

Year	1997–2001	2003–2006	2009–2012
Feature size, μm	0.25–0.15	0.13–0.10	0.07–0.05
Millions of transistors/cm^2	4–10	18–39	84–180
Number of wiring layers	6–7	7–8	8–9
Die size, mm^2	50–385	60–520	70–750
Pin count	100–900	160–1475	260-2690
Clock rate, MHz	200–730	530–1100	840–1830
Voltage, V	1.2–2.5	0.9–1.5	0.5–0.9
Power, W	1.2–61	2–96	2.8–109

a lower speed. *Stuck-fault* testing covers all (or most) circuit signals assuming that a faulty signal may be permanently stuck-at logic 0 or 1. For a reliable high-speed test, the *automatic test equipment* (ATE) must operate as fast as, or faster than, the *circuit-under-test* (CUT.)

2. **ATE Cost.** At the time of writing (year 2000), a state of the art ATE can apply vectors at a clock rate of 1 GHz. The cost of such a tester rises roughly at the rate of $3,000 per pin. In addition, there is a fixed cost of function generators needed for mixed-signal circuits that can range between 0.5–1.2 million dollars [65]. Thus, devices with rated speed up to 1 GHz can be tested, though at a high cost. The semiconductor industry, however, faces two types of problems. First, the installed test capability in many factories around the world still allows only about a 100 MHz clock rate. By the time the present equipment is replaced by new systems, clock rates of chips are likely to go beyond 1 GHz. Second, as Figure 1.4 shows, the microprocessor clock rate in the year 2000 has already approached 1 GHz, exceeding the present state of the art of the ATE.

As the development of faster ATE continues, other test methods are also emerging. An *embedded-ATE* method [482], in which ATE functions such as high-speed vector generation and response analysis are added to the chip hardware, have been proposed. In another method, controllable delays are inserted in the chip hardware such that the critical path delay can be measured by a slow-speed tester [29]. A modified scan design (see Chapter 14) for at-speed test of Motorola's MPC7400 microprocessor has been described recently [654].

Example 1.2 *Testing cost. A state of the art ATE in the year 2000 applies test vectors at clock rates up to 500 MHz. It contains analog instruments (function generators, A/D converters and waveform analyzers.) The price of this tester for a 1,024 pin configuration is estimated as*

$$ATE \text{ purchase price} = \$1.2M + 1,024 \times \$3,000 = \$4.272M$$

1.4 VLSI Technology Trends Affecting Testing

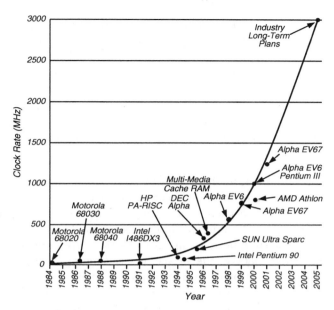

Figure 1.4: Microprocessor clock rates.

We compute the yearly running cost of the ATE by assuming a linear depreciation over five years, and an annual maintenance cost of 2% of the purchase price. The operating cost of the building, facilities, auxiliary equipment (wafer and chip handlers, fixtures, etc.), and personnel is estimated to be $0.5M. Thus:

$$\begin{aligned} Running\ cost &= Depreciation + Maintenance + Operating\ cost \\ &= \$0.854M + \$0.085M + \$0.5M = \$1.439M/year \end{aligned}$$

The tester is used in three eight-hour shifts per day and on all days of the year. Therefore:

$$Testing\ cost = \frac{\$1.439M}{365 \times 24 \times 3{,}600} = 4.5\ cents/second$$

The test time for a digital ASIC (application specific integrated circuit) is 6 seconds. That gives the test cost as 27 cents. Since the bad chips are not sold, their test cost must be recovered from the sale of good chips. If the yield is 65%, then the test component in the sale price of a good chip is 27/0.65 = 41.5 cents.

The test time of a chip depends on the types of tests conducted. These may include parametric tests (leakage, contact, voltage levels, etc.) applied at a slow speed, and vector tests (also called "functional tests" in the ATE environment) applied at high speed. The time of parametric tests is proportional to the number of pins since these tests must be applied to all active pins of the chip. The vector test time depends on the number of vectors and the clock rate. The total test time for digital chips ranges between 3 to 8 seconds. In general, mixed-signal or analog circuits have fewer pins than the digital chips. However, the tests are conducted at slower rates and test times can lie in the 3 to 6 seconds range. The handling of chips and probes, though mechanical, occurs at the speed of the machine. Such times are kept small by pipelining and parallelism used in handling. The time of handling is, however, included in the test time estimate.

3. **EMI.** A chip operating in the GHz frequency range must be tested for *electromagnetic interference* (EMI.) This is a problem because inductance in the wiring becomes active at these higher frequencies, whereas it could be ignored at lower frequencies. The inherent difficulties are: (1) Ringing in signal transitions along the wiring, because signal transitions are reflected from the ends of a bus and bounce back to the source, where they are reflected again; (2) Interference with signal propagation through the wiring caused by the *dielectric permeability* and the *dielectric permittivity* of the chip package; and (3) Delay testing of paths requires propagation of sharp signal transitions, resulting in high-frequency currents through interconnects, causing radiation coupling. *Delay testing* is necessary, because many factors may delay a signal propagating along a path. We must also test the chip interconnect carefully for radiation noise induced errors.

Increasing Transistor Density. Transistor feature sizes on a VLSI chip reduce roughly by 10.5% per year, resulting in a transistor density increase of roughly 22.1% every year. An almost equal amount of increase is provided by wafer and chip size increases and circuit design and process innovations [278]. This is evident in Figure 1.5, which shows a nearly 44% increase in transistors on microprocessor chips every year. This amounts to little over doubling every two years. The doubling of transistors on an integrated circuit every 18 to 24 months has been known as Moore's Law [475, 476, 477, 478] since the mid-1970s. Although many have predicted its end, it continues to hold, which leads to several results:

1. *Test complexity.* Testing difficulty increases as the transistor density increases. This occurs because the internal chip modules (particularly embedded memories) become increasingly difficult to access. Also, test patterns for sub-assemblies on the chip interfere with each other, due to the need to observe sub-assembly A through sub-assembly B while stimulating both sub-assemblies A and B from circuit inputs. Later chapters will show that test pattern generation computation time, in the worst case, rises exponentially with the number of chip *primary inputs* (PIs) and with the number of on-chip flip-flops.

Example 1.3 *Transistors versus pins. Consider a chip with a square area whose linear dimension on the side is d. The number of transistors, N_t, that*

1.4 VLSI Technology Trends Affecting Testing

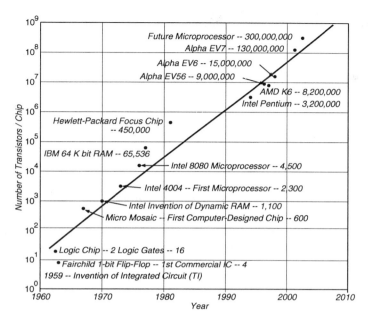

Figure 1.5: VLSI circuit transistor density.

can be placed on the chip is proportional to the chip area, d^2. The number of input/output (I/O) pins, N_p, is proportional to $4d$, since pins are placed on the periphery of the chip. We can thus express an approximate relation between N_p and N_t, as

$$N_p = K\sqrt{N_t} \tag{1.5}$$

where K is a constant. This simple relation was first observed empirically by Rent at IBM and is known as Rent's rule [680]. It has many applications and in Chapter 18, we will use a generalized form (Equation 18.3) to represent the number of terminal signals for a block of logic gates. As we shrink the feature size, for the same chip area, both N_p and N_t increase. But the number of transistors increases faster. Multilayer wiring allows more of the chip area to be utilized by transistors, but does not increase the number of pins, which must be placed at the chip boundary (an exception is the flip-chip technology, not considered here, in which the pins are placed on the chip area [54].) Since any test procedure must now access a larger number of devices (transistors or gates) and interconnects through a proportionately smaller number of pins, the test problem becomes more complex with the higher level of integration. Though it is not a very effective measure, the increase of test complexity is sometimes expressed as the ratio, N_t/N_p. For the data in the second column (1997-2001) of Table 1.1, this ratio for the largest chip is $10^7/900 = 1.1 \times 10^4$. For 2003-2006 and 2009-2012, we get 2.6×10^4 and 6.7×10^4, respectively. This shows the test complexity more than doubling every five or six years.

2. *Feature scaling and power dissipation.* The *power density* (power dissipation per unit area) of a CMOS chip is given by [185]:

$$Power\ density\ =\ C \times V_{DD}^2 \times f \qquad (1.6)$$

where C is the combined node capacitance per unit area that is switched per clock cycle, V_{DD} is the supply voltage, and f is the clock frequency. In general, C is proportional to the number of transistors per unit area and the average switching probability of signals. The basic objective of shrinking the device features is to increase the circuit speed and transistor density [185]. Suppose that the feature dimensions are divided by a constant $\alpha > 1.0$. The speed improves by a factor α because the individual node capacitance is reduced as $1/\alpha$ due to shorter wires and smaller devices. Reduced features increase the transistor density by factor α^2, resulting in an increase in C by factor α. Since increased electric field within transistors can degrade reliability, to keep the electric field unchanged the supply voltage is scaled down by factor $1/\alpha$. This scaling, known as the *constant electric-field* (CE) scaling, keeps the power density constant [185]. One can easily verify this fact by substituting $C \to \alpha C$, $V_{DD} \to V_{DD}/\alpha$, and $f \to \alpha f$ in Equation 1.6. In the smaller submicron region, the CE scaling is not practical because the threshold voltage of the transistor does not scale down with dimensions. As the supply voltage gets closer to the threshold voltage, the switching speed drops, defeating at least one purpose of scaling. In practice, therefore, V_{DD} is scaled by factor ϵ/α, where $\epsilon > 1.0$. Although the speed up of clock frequency is restored, this increases the power density by a factor ϵ^2 [185], causing a significant impact on testing:

  (a) Verification testing must check for power buses overloaded by excessive current. This causes a *brown-out* in the chip, just as overloading the electric power distribution network in a city causes a drop in supply voltage. This might cause the chip power bus lines to burn out due to metal migration, just as an old-fashioned fuse burns out in a fuse box.

  (b) Application of the test vectors may cause excessive power dissipation on the chip and burn it out, so the vectors must be adjusted to reduce power.

  (c) Shrinking features will eventually require the design of transistors with reduced threshold voltage. These devices have higher leakage current [652], which reduces the effectiveness of I_{DDQ} testing (discussed next.)

3. *Current testing.* A very successful recent approach to test chips is to check for elevated quiescent current. This method is called I_{DDQ} *testing* (see Chapter 13.) While switching, CMOS circuits exhibit an elevated current in the digital logic, which dies out quickly to a small quiescent current (I_{DDQ}) after the gate output settles to a steady state. Faults, such as transistors stuck-on, shorted wires, shorts from transistor gates to drains, etc., elevate the quiescent current. I_{DDQ} testing marks the chip as faulty if the measured quiescent current through ground busses of the chip exceeds a prespecified threshold.

Integration of Analog and Digital Devices onto One Chip. We seek this to reduce costs (i.e., one part is cheaper to manufacture and assemble into an electronic system than two separate parts, one analog and the other digital.) We increase speed with this approach, by eliminating chip-to-chip delay between an A/D converter and the *digital signal processor* (DSP) that processes the digitized data. When data goes between two chips, the driving chip inserts a delay to amplify and buffer the output signals, and the receiving chip inserts a delay to condition the signals and propagate them through a "lightening arrester" to eliminate voltage surges coming from people handling the chip. Integration onto one chip eliminates a significant delay, but brings new issues of testing mixed-signal circuits on one chip.

1.5 Scope of this Book

The main topic of this book is the "Test" half of the "Design and Test" block in Figure 1.3, which is almost in the middle of the picture. The figure also shows the involvement of engineers at various levels of VLSI development.

Part I of the book introduces you to testing. Chapter 2 covers requirements, specifications, testing, and fabrication. This background answers the *whys* before you start learning the *hows*. Testing is necessary, but how much? Answering this question requires experience, good judgement, and above all, *economic sense*. Chapter 3 gives basic engineering economics. Chapter 4 describes fault models.

Chapters in Part II cover test methods for digital and analog circuits. Tests are the circuit input signals that produce erroneous outputs in the presence of potential faults. Tests are derived from the specific circuit structure and the fault models. We discuss algorithms for *test generation* and *fault simulation* (for test evaluation.) Part III covers circuit *design for testability*.

Electronic design and test engineers of today have to deal with several types of subsystems, namely, analog, logic, and memory, which require different types of tests and design for testability methods. The book provides a rich selection of essential topics on all three types of circuits. The basics of combining these subsystems into a *system-on-a-chip* (SOC) are given in Chapter 18.

Teachers will find the book to be too large for a one-semester course. It is, however, easily adaptable for three types of courses:

- An introductory course on electronic testing.

- A comprehensive course on electronic testing (senior/graduate level).

- Test supplement (1/4 to 1/2) in a VLSI design course (senior/graduate level).

Teachers and students with either a *design* or a *test* orientation may wish to order the reading of the material differently. A suggested model is shown in Figure 1.6. Starting at Part I, a test-oriented reader will proceed clockwise, select more material in Part II, and somewhat less in Part III, depending on the time available. A design-oriented reader will also start at Part I, but will go anticlockwise, cover most of Part III, and depending on the available time sample Part II.

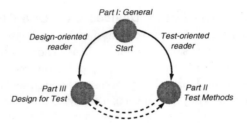

Figure 1.6: Ways to read this book.

Problems

1.1 *Chip testing.* Rework Example 1.1 for testing of VLSI chips whose true yield is 70% and the test probabilities are as shown in Figure 1.1. Obtain the *defect level* (DL), defined as the probability of finding a bad chip among those that passed the test.

1.2 *Chip testing.* Applying the test scenario of Example 1.1 to VLSI chip test, Prob($F|FQ$) is defined as the *defect coverage*. It is the probability with which the test detects any defect that occurs on the chip. Prob($P|FQ$) = 1 − Prob($F|FQ$) is known as the *escape probability*, which is the probability of a bad chip escaping (passing) the test. Keeping all other probabilities same as in Example 1.1, find the defect coverage necessary to reduce the defect level to 500 parts per million (ppm.)

1.3 *Test cost.* The analog part of a mixed-signal chip takes 1.5 s to test. The chip has 1,024 pins. There are 1,000 million vectors that must be applied to test the digital logic and the embedded memory block. The rated clock speed is 200 MHz. Assuming that the wafer yield is 70% and a 500 MHz ATE is used, calculate the test component that should be added in the price of a good chip.

1.4 *Test cost and self-test.* A mixed-signal VLSI chip has 1,024 digital pins. To reduce the testing cost, it is decided to design the digital part with *built-in self-test* (BIST) circuitry. Thus, most of the digital signals can be generated or analyzed on the chip and do not have to be supplied from/to the ATE. It can now be tested on a 256-pin ATE. Calculate the per second cost of testing.

1.5 *Test complexity.* Imagine a three-dimensional VLSI system in the form of a cube, where the volume contains transistors and interconnects, and pins are placed on the surface. We assume that the problems of power dissipation and cooling have been worked out to make such a device possible. Derive Rent's rule for this device. Is the test problem for the cube more, or less, complex than that for the flat chip?

Chapter 2

VLSI TESTING PROCESS AND TEST EQUIPMENT

> "... productivity increases as quality improves ... [but] inspection to improve quality is too late, ineffective, costly ... note that there are exceptions, circumstances in which mistakes and duds are inevitable but intolerable. An example is, I believe, manufacture of complicated integrated circuits. Separation of good ones from bad ones is the only way out ... It is important to carry out inspection at the right point for minimum total cost."
> — W. Edward Deming, in the book, *Out of the Crisis* [197].

VLSI chip testing is done in several different places by several different types of people. When a new chip is designed and fabricated for the first time, testing should verify correctness of design and the test procedure. This often requires the involvement of the design engineer and the testing may even take place in the design laboratory rather than in a factory. Based on the result, both the design and the test procedure may be changed. This is called *verification testing*.

Successful verification testing usually results in some good chips. These are the earliest chips and are normally used by the designers of systems that will use this design. A successful verification also signals the beginning of production. Production means large scale manufacturing. Fabricated chips are tested in the factory. This is called *manufacturing testing*.

Finally, when the manufactured chips are received by a customer, they may be again tested to ensure quality. This testing, known as *incoming inspection* (or *acceptance testing*), is conducted either by the user or for the user by some independent testing house.

This chapter gives a detailed view of the testing process, and describes *automatic test equipment* (ATE), which is an instrument used to test chips. Finally, the chapter concludes by describing parametric test.

2.1 How to Test Chips?

Figure 2.1 illustrates the basic principle of digital testing. Binary patterns (or *test vectors*) are applied to the inputs of the circuit. The response of the circuit is compared with the expected response. The circuit is considered good if the responses match. Obviously, the quality of the tested circuit will depend upon the thoroughness of the test vectors. Generation and evaluation of test vectors is one of the main subjects of this tutorial. In this chapter we describe the mechanics of VLSI chip testing, an important part of the manufacturing process.

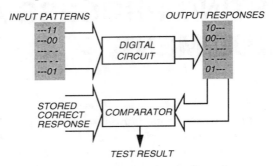

Figure 2.1: Principle of testing.

While Figure 2.1 illustrates the basics it does not tell the full story. VLSI devices are tested by *automatic test equipment* (ATE) that performs a variety of tests [642]. Modern ATE is a powerful computer operating under the control of a *test program* written in a high level language. Our discussion on test equipment is rather brief and introductory. It is included because we consider some basic familiarity with this material essential for engineers involved in design and testing of VLSI. Sources of more detailed information are provided in Appendix C.

2.1.1 Types of Testing

VLSI testing can be classified into four types depending upon the specific purpose it accomplishes [639].

Characterization

Also known as *design debug* or *verification testing*, this form of testing is performed on a new design before it is sent to production. The purpose is to verify that the design is correct and the device will meet all specifications. Functional tests are run and comprehensive AC and DC measurements are made. Probing of internal nodes of the chip, commonly not done in production testing, may also be required during characterization. Use of specialized tools such as *scanning electron microscopes* (SEM) and electron beam testers, and techniques such as *artificial intelligence* (AI) and *expert systems*, can be effective. A *characterization test* determines the exact limits of device operating values. We generally test for the worst case,

2.1 How to Test Chips?

because it is easier to evaluate than average cases and devices passing this test will work for any other conditions. We do this by selecting a test that results in a chip pass/fail decision. We then select a statistically significant sample of devices, and repeat the test for every combination of two or more environmental variables, and plot results. This essentially means repetitively applying functional tests and measuring various DC or AC parameters, as we vary different variables such as V_{CC} (the supply voltage.) This data is plotted as a *Shmoo plot* (see Figure 2.2) [442, 688], where '*' shows correct operation and '@' or nothing shows failure.

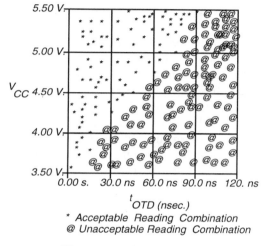

Figure 2.2: Shmoo plot.

We diagnose and correct design errors, measure chip characteristics for setting final specifications, and develop a production test program. Somewhat less comprehensive characterization testing is often continued throughout the production life of the device for possible improvements in the design and the process *yield*. Yield is the fraction (or percentage) of acceptable parts among all fabricated parts. Characterization testing may be done on the chips rejected during the production test or in the field.

Production

Every fabricated chip is subjected to production tests, which are less comprehensive than characterization tests yet they must enforce the quality requirements by determining whether the device meets specifications. The vectors may not cover all possible functions and data patterns but must have a high coverage of modeled faults. The main driver is cost, since every device must be tested. Test time (and therefore cost) must be absolutely minimized. Fault diagnosis (see Chapters 4 and 18) is not attempted and only a *go/no-go* decision is made. Production tests are typically short but verify all relevant specifications of the device. It is an outgoing inspection test of each device, and is not repetitive. We test whether some *device-under-test* (DUT) parameters are consistent with the device specifications under normal operating conditions. We test either at the speed required by the application of the device or at the speed guaranteed by the supplier.

Burn-in

All devices that pass production tests are not identical. When put to actual use, some will fail very quickly while others will function for a long time. *Burn-in* ensures reliability of tested devices by testing, either continuously or periodically, over a long period of time, and by causing the bad devices to actually fail. Correlation studies show that the occurrence of potential failures can be accelerated at elevated temperatures. For a detailed theory of the burn-in process, see Jensen and Petersen's book [340]. Briefly, two types of failures are isolated by burn-in: *Infant mortality failures*, often caused by a combination of sensitive design and process variation, may be screened out by a short-term burn-in (10-30 hours) in a normal or slightly accelerated working environment. *Freak failures*, i.e., the devices having the same failure mechanisms as the reliable devices, require long burn-in time (100-1,000 hours) in an accelerated environment. During *burn-in*, we subject the chips to a combination of production tests, high temperature, and over-voltage power supply. Refer to the *bathtub* component failure curve in Figure 3.3 in Chapter 3. In practice, a manufacturer must balance economic considerations against the device reliability. In any case, the elimination of infant mortality failures is considered essential.

Incoming Inspection

System manufacturers perform incoming inspection on the purchased devices before integrating them into the system. Depending upon the context, this testing can be either similar to production testing, or more comprehensive than production testing, or even tuned to the specific systems application. Also, the incoming inspection may be done for a random sample with the sample size depending on the device quality and the system requirement. The most important purpose of this testing is to avoid placing a defective device in a system assembly where the cost of diagnosis may far exceed the cost of incoming inspection.

Types of Tests. Actual test selection depends upon the manufacturing level (processing, wafer, or package) being tested. Although some testing is done during device fabrication to assess the integrity of the process itself, most device testing is performed after the wafers have been fabricated. The first test, known as *wafer sort* or *probe*, differentiates potentially good devices from defective ones [213, 639]. After this, the wafer is scribed and cut, and the potentially good devices are packaged.

Also, during wafer sort, a *test site characterization* is performed. Specially designed tests are applied to certain test sites containing specific test patterns. These are designed to characterize the processing technology through measurement of parameters such as gate threshold, polysilicon field threshold, bypass, metal field threshold, poly and metal sheet resistances, contact resistance, etc. [639].

2.1 How to Test Chips?

In general, each chip is subjected to two types of tests:

(1) *Parametric Tests.* DC parametric tests include shorts test, opens test, maximum current test, leakage test, output drive current test, and threshold levels test. AC parametric tests include propagation delay test, setup and hold test, functional speed test, access time test, refresh and pause time test, and rise and fall time test. These tests are usually technology-dependent. CMOS voltage output measurements are done with no load while TTL devices require current load.

(2) *Functional Tests.* These consist of the input vectors and the corresponding responses. They check for proper operation of a verified design by testing the internal chip nodes. Functional tests cover a very high percentage of modeled (e.g., stuck type) faults in logic circuits and their generation is the main topic of this tutorial. Often, functional vectors are understood as verification vectors, which are used to verify whether the hardware actually matches its specification. However, in the ATE world, any vectors applied are understood to be functional fault coverage vectors applied during manufacturing test. These two types of functional tests may or may not be the same.

Functional tests may be applied at an elevated temperature to guarantee specifications. For example, testing may be done at 85° C to guarantee 70° operation. This is called *guardbanding*. Another application is in *speed binning* to grade the chips according to performance. This may be done by applying the tests at several voltages and at varying timing conditions (e.g., clock frequency.)

The scenario described above represents a generality. The actual test plan for a VLSI device varies depending upon the specific application it is intended for, the manufacturer's test philosophy, available test equipment, and test economics.

While the general testing methodology is applicable to memory chips as well, there are some notable differences. Memory tests are *functional*; no stuck-type fault coverage is evaluated for these tests, which are designed to check functional attributes such as address uniqueness, address decoder speed, cell coupling, column and row coupling, data sensitivity, write recovery, and refresh. Elaborate testing may require long vector sequences. To increase throughput, memory testers sometimes have parallel testing capability. Specialized testers may even be able to repair redundant cells used in large memories (256 M bits) to enhance yield.

VLSI chip testing, in many ways, resembles the testing of other digital equipment such as circuit boards [509], but with differences. Circuit boards consist of previously-tested components. A primary objective of board testing is to check the printed wiring and the contacts between wires and components. It is possible to perform a bare-board testing of interconnections before the components are inserted. After the components (such as chips) are in place, *in-circuit testing* [69] (through a *bed-of-nails* fixture) is often used to verify the performance of individual components, although the bed-of-nails fixture is becoming obsolete. Finally, functional testing determines whether or not individual components, possibly designed with different technologies, function as a system and produce the expected response. The ATE employed for boards is different from that used for chips.

Test Specifications and Test Plan. The device specification document initiates the development activity, and contains the following information:

- Functional Characteristics – Algorithms to be implemented, I/O signal characteristics (timing waveforms, signal levels, etc.), data and control signal behavior, clock rate.

- Type of Device – Logic, microprocessor, memory, analog, etc.

- Physical Characteristics – Package, pin assignments, etc.

- Technology – CMOS (or gate array), custom, standard cell, etc.

- Environmental Characteristics – Operating temperature range, supply voltage, humidity, etc.

- Reliability – Acceptance quality level (defective parts per million), failure rate per 1,000 hours, noise characteristics, etc.

Test specifications, if not given explicitly, are derived from the above data. Based on these specifications, a *test plan* is generated. In the test plan the type of test equipment and the type of tests are specified. Selection of a tester depends on such parameters as throughput, clock rate, timing accuracy, test sequence length, tester availability, and cost. The types of test may include parametric, functional, burn-in, margin, speed sorting, etc. The fault coverage requirement should also be specified.

Testers. The basic purpose of a tester is to drive the inputs and to monitor the outputs of a device-under-test. Testers are popularly known as ATE (*automatic test equipment*.) Fast-changing VLSI technology has driven the development of modern ATE. Selection of ATE for a VLSI device must consider the specifications of the device. Major factors are speed (clock rate of the device), timing (strobe) accuracy, number of input/output pins, etc. Other considerations in selecting a tester are cost, reliability, serviceability, ease of programming, etc. Detailed description of testers may be found in Section 2.2 and in references cited in Appendix C.

Test Programming. Once the device-under-test has been mounted in the tester, three things are needed to conduct the test. These are the *test program*, the digital *test vectors*, and the analog test waveforms. Until recently, test programs were written manually. However, the use of CAD tools is now becoming widespread in this area.

As shown in Figure 2.3, the device specifications spawn several activities. The results of these activities are required for the test program. An automatic *test program generation* system (often called TPG) requires three types of inputs:

2.1 How to Test Chips?

(1) Tester specification and the information on the types of tests is obtained from the test plan.

(2) Physical data on the device (pin locations, wafer map, etc.) are obtained from the layout.

(3) Timing information on signals and test vectors (inputs and expected responses) are obtained from simulators.

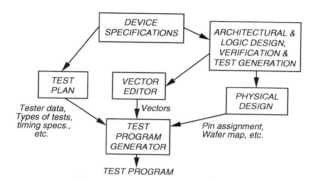

Figure 2.3: Test program generation.

The test program contains the sequence of instructions that a tester would follow to conduct testing. For example, a simple sequence of events will be apply power, apply clocks and vectors to input pins, strobe output pins, and compare output signals with stored expected response. Modern testers provide a choice of input signal waveforms, mask output signals, sense high impedance state, and have a variety of sophisticated capabilities. Since testers differ in their capabilities and programming languages, TPGs commonly generate a *tester-independent* program which can be customized to any specific ATE [298]. Also, since software simulators used in design verification differ from the ATE in their handling of signal format (logic values, timing, etc.), *vector editors* [47] are useful tools during test programming.

Test Data Analysis. The test data obtained from the ATE serves three purposes. First, it helps to accept or reject the device-under-test. Second, it provides useful information about the fabrication process. And third, it provides information about design weaknesses.

Failing tests quickly point to faulty devices. However, the devices that do not fail during test can be considered *good* only if the tests covered 100-percent of faults. Analysis of test data provides information on device quality [588, 589]. Due to the random variations in the fabrication process speed characteristics of devices vary. Test data analysis allows sorting of chips for higher than the nominal performance.

Failure mode analysis (FMA) of the failed devices provides further information for improving the VLSI processing. Failing devices often show patterns of repeated failures. The causes of these failures can point to weaknesses (sensitivity to process variations) in the design. Such information is useful for improving logic and layout design rules.

2.2 Automatic Test Equipment

The automatic test equipment is an instrument used to apply test patterns to a *device-under-test* (DUT), analyze the responses from the DUT, and mark the DUT as good or bad. The DUT is also sometimes called the *circuit-under-test* (CUT.) The ATE is controlled by a central UNIX work station or PC, and one or more additional CPUs is often built into it to provide housekeeping and data reduction capability. The tester has one or more *test heads*, which contain buffering electronics local to the DUT, but one mainframe with common instrumentation, power supplies, etc. The ATE is connected to external equipment that mechanically handles the wafers or IC packages being tested. Thus, while one chip or wafer is being tested in one test head (chip handler), another chip/wafer can be loaded into a second test head, so the tester overlaps mechanical handling of parts with electrical testing of parts. The chip movement mechanism is generally a pneumatic one, and many wafer probers also use compressed air to force the probes onto the wafer.

2.2.1 Advantest Model T6682 ATE

We illustrate typical ATE structure with the Advantest^{TM} Model T6682 tester [12], which is an extremely expensive, high-end ATE, shown in Figure 2.4. Figure 2.5 shows the block diagram of this ATE. The instrument electronics are made out of 0.35 μm VLSI chips. It has up to 1024 channels, so it can independently control and observe 1024 chip pins simultaneously. The test speed is either 250 MHz, 500 MHz, or 1 GHz. Its timing accuracy is ± 200 ps. The ATE can drive busses between -2.5 V to 6.0 V, and it can drive small amplitude 200 mV signals. The clock/strobe timing accuracy is ± 870 ps over the full timing range, and the overall timing accuracy is ± 200 ps. The clock setting resolution is 31.25 ps. The clock/strobe timing accuracy for AC measurements is 80 ps. This instrument is one of three available ones that can operate at 1 GHz frequencies. For clock rates exceeding 250 MHz, we use *pattern multiplexing* where we write two patterns in one tester cycle, which raises the operating rate to 500 MHz. We can also multiplex 2 pins to control 1 digital pin on the *device-under-test* (DUT), to achieve a data transfer rate of 1 GHz.

When testing wafers, a *probe card* mechanically interfaces the ATE's test head to a set of *probe needles* (or *probe membrane*), which actually contact the test pads on the wafer. The Advantest T6682 can support up to 1024 pins. Realize that for probing chips, inductance is a major problem, so each probe, even if digital, must have a careful balance between capacitance and inductance in the probe line to ensure accuracy. We visualize the probe as 1024 wires touching the chip in a very small area. Inductance in this system is proportional to the height above the ground plane, and the wires extend up vertically in the probe, so we essentially have

2.2 Automatic Test Equipment

Figure 2.4: Advantest T6682 test equipment with a single test head (foreground.)

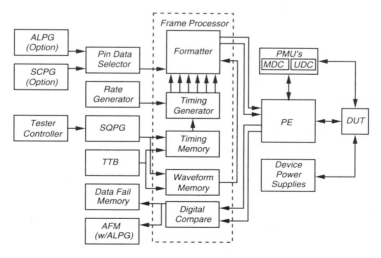

Figure 2.5: Block diagram of T6682 VLSI test system.

1024 tightly coupled transformer windings. Therefore, we will have severe crosstalk problems in this probe at 1 GHz frequency.

Pattern Generation: The standard *sequential pattern generator* (SQPG) stores patterns applied to the DUT, and contains 16 $MVectors$ of patterns, expandable to 64 $MVectors$. A vector is as wide as the number of pins supported for the DUT. The optional *algorithmic pattern generator* (ALPG) handles 32 independent address bits and 36 independent data bits. It can be assigned to any tester channel, and is useful for embedded memory testing, since it generates address and data lines in real time, up to 250 MHz speeds. It has an address descrambler. Pattern memory has 16 $M \times 3$ $bits$, and is expandable to 64 $M \times 3$ $bits$. It represents vectors with 3 $bits/pin$, so this allows six timing edges per pin, with a clock rate of 250 MHz. Two timing edges are used to generate transitions in the waveform applied to the pin, two more generate waveform or pin I/O control transitions, and the final two transitions are used for the comparison unit strobe. The *address failure memory* (AFM) module records addresses in the memory-under-test that failed. The optional *scan pattern generator* (SCPG) supports JTAG boundary scan (see Chapter 16.) It greatly reduces the test vector capacity needed for boundary scan, and can mask scan output data for up to 64 pins. It can have either 2 $GVector$ or 8 $GVector$ sizes, and operates at speeds up to 40 MHz. The *high-speed reload server* provides for rapid internal pattern transfer from a Host machine (via the ethernet), and essentially is a buffer. One must supply a test vector file and the good chip response for each vector to the ATE.

Response Checking: The concept of *matching*, or *pulse train matching*, is where the ATE matches patterns on one pin for up to 16 cycles. In *pattern matching* mode, the ATE matches a pattern on a number of DUT pins in one cycle. It determines whether the DUT Output matches an expected output, and the result is used to change the pattern generation sequence in real time. The output voltage resolution in the comparison units is 2 mV.

Frame Processor: This unit generates/compares DUT input/output waveforms at 250 MHz. It combines the DUT input waveform stimulus (from the pattern generators) with DUT output waveform comparison to form an event sequence. The notion of a *strobe time* is the interval after the pattern is applied when the outputs of the chip are to be sampled, to determine whether the chip responded correctly or incorrectly to the pattern.

Probing: The digital *pin electronics* (PE) is the local buffering circuitry, which is placed as close as possible to the DUT in order to provide maximum bandwidth and minimum parasitics. Figure 2.6 shows the configuration of the pin electronics. Each *pin* (*channel*) on the Advantest ATE has one comparator (DUT output-only) pin and a conventional driver/comparator pin. The pin electronics terminates in a *pogo pin* connector at the test head. The test head is then interfaced through a custom

2.2 Automatic Test Equipment

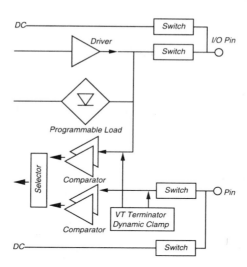

Figure 2.6: Pin electronics in Advantest T6682 ATE.

printed circuit board to a *wafer prober* for the testing of uncut and unpackaged wafers or to a *package handler* for the testing of packaged chips through a testing socket called a *contactor*. The ATE has between 128 and 1024 pins, which can be expanded in units of 128. The pin electronics uses liquid cooling to reduce electrical noise and improve its reliability, and supports CMOS, BiCMOS, ECL, and TTL digital testing. The ATE uses semiconductor relays for pin switching, rather than mechanical relays. In the pin electronics, one can independently set V_{IH}, V_{IL}, V_{OH}, V_{OL}, I_H (input high current), I_L (input low current), V_T (logic threshold voltage), and both dynamic clamp voltages for each channel. Each pin card has 64-channel multi-pins. Pin transition times are $0.55 \pm 0.12\ ns$, and signal edge setup has a resolution of $31.25\ ps$. The load on each channel is programmable. The Advantest pin electronics with the second comparator is unique in ATE. It is used for special *fly-by* wiring configurations where the DUT is placed at the midpoint of a transmission line between the driver-comparator pair and the stand-alone comparator.

The *parametric measurement unit* (PMU) applies voltages or currents at a pin and measures electrical responses (currents or voltages) at the same pin. In Figure 2.5, the *multi-DC unit* (MDC) and the *universal DC unit* (UDC) are provided for applying and measuring DC voltages and currents at pins. The PMU has six timing edges per channel. *Pin multiplexing* lets us combine the edges for two channels to double the operational speed for 1 pin, which is how the T6682 attains 1 GHz operation.

Mixed-Signal Test: The ATE has a *waveform generator*, a *digitizer*, *digital waveform capture memory*, a *sine wave generator*, an *audio front end*, and a *sampler*. The sampler converts analog signals into digital data using sampling. Mixed-signal test is discussed in Chapters 10 and 11.

Power Supplies: The ATE contains between 24 and 32 power supplies.

Software Control: The ATE runs Solaris UNIX on an UltraSPARC 167 MHz processor for non-real-time functions and runs a real-time OS on another Ultra SPARC 200 MHz processor for tester control. The ATE has disk storage, CD-ROMs, floppy disk, a monitor, a keyboard, a Hewlett-Packard GPIB* instrument interface, and an Ethernet interface. The ATE uses software called *Viewpoint* to debug, evaluate, and analyze VLSI chips

Test Description Language: The *test description language* (TDL) is a test programming language, used to describe a program that operates the ATE using the test vectors. The TDL provides this additional information for controlling the ATE: strobe times (for sampling the DUT outputs), voltage/current stimulus information, the clocking rate for vectors, the vector slew rate (rate at which waveforms rise or fall), and filtering information for sampling DUT signals. In TDL one can also set the resolution of the pin signal, which is useful for determining causes of failures in DUTs. Also, test vectors often must be edited, to adjust them to avoid burning out the chip by exceeding the maximum allowable power dissipation. Some chip signals must also be put into the high-impedance state to avoid burning out the DUT during testing.

Visualization Aids: ATE software can generate a fail bit map for testing a memory device. The software can also produce maps of wafers showing the passing chips, the failing chips, and the binning results for all chips. *Logic analyzer* software is useful for debugging the chip while it is on the tester. The *oscilloscope software* displays the DUT output analog waveforms after they have been sampled with high resolution. Software also assists in generating Shmoo plots for DUTs. Most testers offer a *virtual test* capability, where the test engineer can simulate the application of the vectors by the tester to the DUT on a work station. Virtual test reduces the actual tester time needed for test development, which is important given the high cost of testers.

2.2.2 LTX Fusion ATE

We discuss the LTX Fusion^{TM} tester (see Figure 2.7), because it is a modular series of ATE systems intended for *systems-on-a-chip* (SOC.) The models HF/HT/AC allow various trade-offs in integrating digital, analog, and memory testing using one tester, and span different price ranges. This is now required, since digital, analog, and memory devices are now combined into a SOC. The Fusion ATE is a single platform, which can be updated by adding/replacing modules to attach additional instruments and capabilities as needed. The LTX Fusion tester runs the enVision Operating System. The Fusion HF supports one or two test heads per tester with a maximum of 1024 digital pins and a maximum test rate of up to 1 GHz. It has a

---

*General Purpose Instrument Bus.

2.2 Automatic Test Equipment

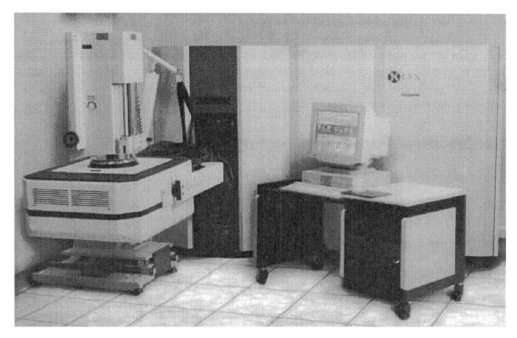

Figure 2.7: LTX Fusion ATE with a single test head.

memory depth of up to 64 $MVectors$. It also has specific pattern generation capability for serial scan requirements and memory test. It can hold two test programs in memory at once, but only one runs at a time.

The Fusion AC and HT models are intended for the testing of lower pin count DUTs and analog-intensive devices, and support up to 96 digital pins and a non-multiplexed testing rate up to 40 MHz. One million vectors can be stored in memory.

All of the Fusion configurations support multi-site testing and support a common set of analog instrumentation, including DSP-based synthesizers and digitizers, time measurement units, *smart* power testing instruments, and RF source and measurement capability up to 4.3 GHz.

2.2.3 Multi-Site Testing

In *multi-site testing*, one ATE tests several devices at the same time. This is frequently done both in probe and package test. Single test heads have been designed to handle multiple packages simultaneously. On this ATE, the DUT interface board has more than one socket on it. The motivation for this approach is that most of the cost is for the basic ATE, but adding a few additional instruments to enable multi-site testing is relatively inexpensive. Most ATE instruments can be replicated in the tester (e.g., more digital pins), and most ATE operating systems allow one to program one site, and then automatically expand the program to address more sites that use duplicate resources in the tester.

Digital and mixed-signal devices are frequently tested two or four at a time. One limitation is the number of instruments installed in the ATE to handle all of the required pins. Another limitation is the type of handling equipment available for a given chip package type. Memory devices are frequently tested 32 or 64 ICs at a time because test times are very long.

2.3 Electrical Parametric Testing

A typical test program proceeds as follows:

1. Probe test or wafer sort test (examine the die on the wafer before it is broken up into chips to weed out grossly defective devices.)

2. Contact test.

3. Functional and layout-related test.

4. DC parametric test.

5. AC parametric test.

Tests 1, 2, 4, and 5 are electrical tests, while test 3 is a functional test.

An *electrical fault* is observed at the chip pins, and affects the device input/output interface. It modifies the observed voltages/currents/delays at the pins. There are two kinds of electrical faults:

1. There may be a major deviation of voltage/current/delay from the part data book value.

2. There may be unacceptable limits of operation for the device.

Electrical testing is done without an understanding/analysis of the underlying fabrication process and chip layout. The tests are DC and AC parametric tests.

DC Parametric Tests. These tests measure steady-state electrical characteristics using Ohm's law: they force a terminal voltage and measure current, or vice versa. A *parametric measurement unit* (PMU) in the ATE does this, while forcing appropriate operating conditions.

Contact Test. This verifies that the chip pins have no opens or shorts [688]. We draw current out of the device and measure voltage at the input pin. Figure 2.8 [442] shows how this forward biases the protection diodes at the pin (which should have $\phi_0 = 0.7\ V$ in a $5\ V\ V_{CC}$ system.) Chip pins are protected by a series resistance connected to two Zener diodes. This protects the chip from α-particles and from electrostatic discharges inadvertently applied to its pins during handling, as the R absorbs a huge voltage drop and the two Zener diodes clamp the circuit node voltage

2.3 Electrical Parametric Testing

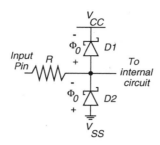

Figure 2.8: Protection circuits at chip inputs/outputs.

between supply $V_{CC}+0.7\ V$ and ground $V_{SS}-0.7\ V$. The tester contacts the device through R.

Method:
1. Set all inputs to $0\ V$.
2. Force a current I_{fb} out of the pin, which forward biases Diode D2. We expect I_{fb} to be 100 to 250 μA.
3. Measure pin voltage V_{pin}. R relates to V_{pin} by $V_{pin} = V_{SS} - 0.7 - I_{fb}R$
Possible test outcomes:
1. Contact short ($R \approx 0\ \Omega$) results in very low diode drop, so V_{pin} is $-0.075\ V$ (fails.)
2. No problem (passes.)
3. Pin is open circuited, so R is huge. This forces too much current through the diode, so I_{fb} is too large, and V_{pin} is therefore large ($-1.5\ V$) (fails.)

Power Consumption Test. This finds the worst case power consumption for static (steady input logic values) and dynamic (inputs changing dynamically during operation) situations.

Method:
1. Set ambient temperature to worst case conditions, and open circuit the device outputs.
2. Measure the maximum device current drawn from the power supply (I_{CC}) at the specified power supply voltage V_{IN}.
Possible test outcomes:
1. $I_{CC} > 70\ mA$ (fails.)
2. $40\ mA \leq I_{CC} \leq 70\ mA$ (passes.)

Output Short Current Test. This test verifies that the output current drive is sustained at high and low output voltages.

Method:
1. Cause the chip output to be a 1 with the appropriate input pattern.
2. Short the output pin to 0 V in the PMU.
3. Measure the short current (do not short the chip for very long, or the driver will burn out.)
Possible test outcomes:
1. Short current $> 40~\mu A$ (passes.)
2. Short current $\leq 40~\mu A$ (fails.)

Output Drive Current Test. For a specified output drive current, this test verifies that the output voltage is maintained.

Method:
1. Apply an input vector that forces the chip output to 0.
2. Simultaneously, force the output voltage to V_{OL} (0.4 V) and measure I_{OL}.
3. Repeat steps 1 and 2 with an output 1 pattern, forcing the output voltage to V_{OH} (2.4 V) and measure I_{OH}.
Possible test outcomes:
1. $I_{OL} < 2.1~mA$ (fails.)
2. $I_{OH} < -1~mA$ (fails.)

Threshold Test. This determines the V_{IL} and V_{IH} input voltages needed to cause the device output to switch from high to low (low to high.) $0 < V_{OL} < V_{IL} < V_{IH} < V_{OH} < V_{CC}$.

Method:
1. For each of the input pins, write a logic 0, along with a pattern that propagates the zero to an output. Read the expected output. Increase the input voltages in 0.1 V steps at the input and stop when an erroneous value occurs at the output.
2. For each of the input pins, write a logic 1, along with a pattern that propagates the one to an output. Read the expected output. Decrease the voltages in 0.1 V steps and stop when an erroneous value occurs at the output.
Possible test outcomes:
1. An erroneous output happens when the input 0 is above 0.8 V (passes.)
2. An erroneous output happens when the input 0 is below 0.8 V (fails.)
3. An erroneous output happens when the input 1 is below 2.0 V (passes.)
4. An erroneous output happens when the input 1 is above 2.0 V (fails.)

AC Parametric Tests. In AC parametric testing, we apply alternating voltages at some set of frequencies to the chip and measure the terminal impedance or dynamic resistance (reactance.) We select a DC bias level for these tests, which determine chip delays caused by input and output capacitances. However, the tests give no

2.3 Electrical Parametric Testing

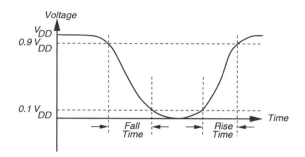

Figure 2.9: Determining rise and fall times.

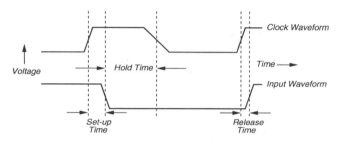

Figure 2.10: Determining setup, hold, and release times.

information on functional data or DC parameters.

Rise and Fall Time Tests.

Method:
1. Measure the time interval between two voltage levels of either a rising or falling edge (see Figure 2.9 [442].)

Tests for Set-Up, Hold, and Release Times. See Figure 2.10 [442] for the definitions of these times. The times are always measured from the input signal 50% point to the output signal 50% point.

Propagation Delay Tests. Propagation delay is the time between an input state change and a resulting output change, measured between 50% voltage levels, and specified with a minimum and maximum time pair.

Method:
1. Apply a standard output load (RC or RL.)
2. Apply an input pulse with a controlled rise and fall time.
3. At the tester, measure propagation delay between input and output changes.
Possible test outcomes:
1. Delay is between 5 ns and 40 ns (passes.)
2. Delays are outside this range (fails.)

2.4 Summary

Parametric tests are necessary to decide whether the chip pins meet various rise and fall times, setup and hold times, low and high voltage thresholds, and low and high current specifications. Functional tests determine whether the internal digital logic and analog sub-systems in the chip behave as intended. The major cost in testing is digital and analog functional tests. Parametric test is a tiny part of the cost, because it is very short, and test cost is proportional to the amount of tester time needed for the test. At present, the ATE cost is increasing, but the manufacturers claim that the actual test process cost is decreasing, because of throughput improvements to the ATE. The ATE itself becomes more expensive, because it must store more vectors than before and operate at much higher frequencies than before. The tester probe head is much more difficult to design than before, because of pin counts of 1024 pins/chip and higher clock rates. This leads to increased problems with inductance and electrical noise during digital test. For a more comprehensive study of the VLSI testing process, the reader may examine a recent survey article by Grochowski et al. [270] and the books listed in Section C.3.

Problems

2.1 *Test types.* You are making microprocessor chips, and need to have a very low product failure rate in the field to control your warranty costs. Describe the various types of tests that you would use on the microprocessor chips at their various stages of processing.

2.2 *Contact test.* In a contact test, we expect the input current limiting resistor R for an input pad to be 2000 Ω. If I_{fb} in this test ranges from 100 to 250 μA, what will be the expected pin voltage range V_{pin} during the contact test?

2.3 *Set-up time test.* Devise a parametric test for flip-flop set-up time of 360 *ps* for a 16-pin chip containing a common *clock* (for all 6 flip-flops), 6 *D* lines, 6 *Q* lines, a common *master-clear* line, and a common *master-set* line.

2.4 *Hold time test.* Devise a parametric test for flip-flop Hold time of 120 *ps* for a 16-pin chip containing a common *clock* (for all 6 flip-flops), 6 *D* lines, 6 *Q* lines, a common *master-clear* line, and a common *master-set* line.

2.5 *Threshold test.* Devise an improvement to the threshold parametric test procedure to speed up the test (and thus save money), while still correctly determining the input logic voltage thresholds.

Chapter 3

TEST ECONOMICS AND PRODUCT QUALITY

> "... Economics is the study of how men choose to use scarce or limited productive resources (land, labor, capital goods such as machinery, and technical knowledge) to produce various commodities (such as wheat, overcoats, roads, concerts, and yachts) and to distribute them to various members of society for their consumption..."
> — Paul Samuelson [560].

Engineering economics is the study of how engineers choose to optimize their designs and construction methods to produce objects and systems that will optimize their efficiency and hence the satisfaction of their clients. We discuss engineering economics concepts, such as the analysis of production and operational costs, and benefit versus cost analysis. These concepts, when applied to electronic test, lead to economic arguments that justify *design for testability* (DFT) [23].

Engineers are concerned with optimizing the *technological* efficiency. For example, in designing a heat engine, the prime consideration is to have as much heat energy converted into mechanical work as is possible. An economist, on the other hand, prefers to minimize the cost of obtaining mechanical work for the consumer. Apart from the technological (or energy conversion) efficiency, other factors such as fixed and variable costs of material, equipment, labor, insurance, etc. are important.

Lately, test economics has received noticeable attention. The relationship between testing cost and product quality is complex. For large electronic systems, testing accounts for 30% or more of the total cost. Still, it has been hard to justify the cost of DFT at the component level. We will examine the impact of DFT on the overall system cost rather than considering the component cost alone.

Study and practice of economic principles are as important for engineers as the laws of physics. We may find ourselves pursuing local goals such as *profitability, time to market,* and *beating the competition,* but the real goals are usefulness to society and preservation of the environment. Today, economics is a well developed science, divided into two main streams. *Microeconomics* is the study of economic laws at a small scale as affecting a company. *Macroeconomics* deals with the wealth of society at the national or international scale. We briefly introduce microeconomics, including costs, production, and benefit-cost analysis, and study the case for *built-in self-test* (BIST) in electronic products.

VLSI yield, product quality measured as defect level, and fault coverage are important concepts related to electronic production and testing. These are discussed in the later part of this chapter.

3.1 Test Economics

Testing is responsible for the quality of VLSI devices. Several tradeoffs are often necessary to obtain the required quality level at minimal cost. Costs include the cost of *automatic test equipment* (ATE) (initial and running costs), the cost of test development (CAD tools, test vector generation, test programming), and the cost of DFT [204, 522]. In the future, DFT will dominate test economics equations. The scan design technique can significantly reduce the cost of test generation and the BIST method can lower the complexity and cost of ATE. DFT techniques should, therefore, be included in the device specification and the test plan.

3.1.1 Defining Costs

Cost is a measurable quantity that plays a key role in economics. For example, we compare costs and benefits to select between alternatives, or minimize cost to optimize a design. As we define below, there are several types of costs.

Fixed Costs (FC.) These are the costs of things that are necessary but do not change with use. For example, if we wish to produce computers, we require a factory building and machinery which contribute to fixed costs. These costs do not change with the number of computers that are built, whether we build one or one thousand computers. Although fixed costs remain unchanged, the fixed costs per computer will reduce as the production is increased.

Variable Costs (VC.) These costs increase with the production output. The variable costs of producing one thousand computers will be thousand times greater than the variable costs of producing one computer. Variable costs generally consist of labor, energy, and raw materials.

Total Costs (TC.) Total costs are the sum of the fixed and variable costs, and increase with production output.

Average Costs (AC.) These are obtained by dividing the total costs by the number of units produced.

As an illustration, consider the cost analysis of car transportation for an individual. We take the purchase price of the car, say $25,000, as the *fixed cost*. We estimate

3.1 Test Economics

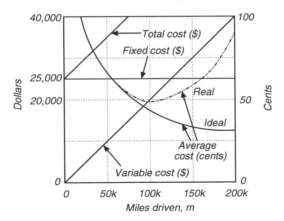

Figure 3.1: Costs of transportation.

the *variable costs* that account for gasoline, maintenance, repairs, and insurance, as 20 cents per mile. The *total cost* of traveling m miles is $25,000 + 0.2m$ dollars, so the average cost is:

$$\text{Average cost} = \frac{25,000}{m} + 0.2 \ dollars \ per \ mile \qquad (3.1)$$

This simple formula illustrates the principle of computing average cost. Depending upon the situation, various types of fixed and variable costs should be included. For example, suppose the owner of the car drives 10,000 miles per year for five years and sells it for $7,500. The fixed cost then reduces to $17,500 and the average cost will be (17,500/50,000)+0.2 dollars, or 55 cents per mile. If the car is used for ten years and then sold for $5,000, the average cost will be 40 cents per mile. When the car is used for twenty years and has no resale value, then the average cost will be 32.5 cents per mile. Two average cost curves are given in Figure 3.1. The monotonically decreasing average cost, shown as "ideal," is computed from the above formula. However, in reality, the repair costs may go up after the car has been driven over 100,000 miles. When the *aging* factor is taken into account, the average cost might be as shown by the rising curve (shown as real), called a *bathtub* curve. Thus, durability of equipment may be more important than the initial purchase price for realizing lower average operating cost.

Example 3.1 *John's economic selection of a car. John figures that his need for a car, which he plans to use for work, will be satisfied by a new mid-size car with 25 miles per gallon EPA (Environmental Protection Agency) rating. He assesses that the average cost of repair (other than the regular maintenance) will be $1,000 for each serious breakdown. This includes the actual cost of repair and the cost of his time off from work (he is a salesman and earns commission on the amount of sale.) Using the consumer's guide data on repairs of mid-size cars, he derives the* *following formula for the price (P) of a new car:*

$$P = 20{,}000 + \frac{20{,}000}{n} \quad dollars \tag{3.2}$$

where n is the number breakdowns per year. John drives about 15 thousand miles every year and assumes a linear depreciation to zero value over a period of ten years. The annual cost (C) of driving is:

$$C = 2{,}000 + \frac{2{,}000}{n} + K + 1{,}000\ n \quad dollars \tag{3.3}$$

where K is the gasoline and regular maintenance cost, which is assumed to be the same for all models. To minimize C, he sets the derivative $dC/dn = 0$, and obtains, $n = \sqrt{2} = 1.414$. From Equation 3.2, the price of the car he should select is determined as $34,144.

3.1.2 Production

Production is the process of making articles that society needs. Inputs to production are *labor, land, capital, energy,* and *enterprise* [472]. Enterprise means technical knowhow, organizational skills, etc. The inputs account for the cost of production. Although inputs vary widely, they can all be converted into dollar equivalents. Both fixed and variable costs may be included.

A *short-run* production means that some of the inputs are fixed. An example is the production in a factory over a period during which the size of the manufacturing facility remains fixed. In the short-run, output possibilities are limited. The company can hire more workers, order more raw material, perhaps add a shift, but that is about all. The *long-run* production is over a period during which the company can change all inputs, including the size of the manufacturing plant.

We first consider short-run production. *Production output, Q,* is a function of the inputs x, which presently accounts for only the variable costs of production:

$$Q = Q(x) \tag{3.4}$$

Technological Efficiency. Let us define:

$$\text{Average product} = \frac{Q}{x} \tag{3.5}$$

$$\text{Marginal product} = \frac{dQ}{dx} \tag{3.6}$$

The average product, or the product per unit of input, is called the *technological efficiency*. We maximize this efficiency by setting:

$$\frac{d}{dx}\frac{Q}{x} = 0 \quad \text{or} \quad \frac{1}{x}\frac{dQ}{dx} - \frac{Q}{x^2} = 0$$

3.1 Test Economics

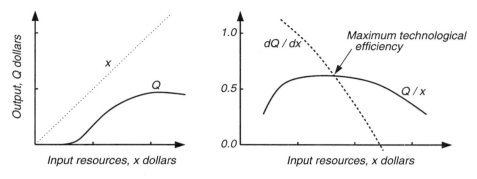

Figure 3.2: Maximizing technological efficiency.

which leads to the following relation:

$$\frac{Q}{x} = \frac{dQ}{dx} \qquad (3.7)$$

Figure 3.2 illustrates the above relation. The graph on the left shows a typical production output function, $Q(x)$. For very small input there is practically no output. We can explain the behavior of this function using the example of an electric motor. The input there is electrical energy and the output is mechanical work. Very small input will not be enough to overcome the inertia of the rotor. Initial start up is also slower because of the low speed friction of bearings. When the input exceeds certain value, the motor output increases with the input. It remains lower than the input because of the energy conversion losses (e.g., heat generated due to current flow.) As we increase the load, the motor draws more current. Eventually, the heat generated inside the wiring cannot all be radiated at the rate it is being generated, and the temperature rises. That in turn increases the wiring resistance and heat loss, reducing the output increment per unit input of electrical energy. This is shown by the reducing slope of Q. The behaviors of the average product, Q/x, and the marginal product or slope, dQ/dx, are shown in the graph on the right. According to Equation 3.7, the maximum technological efficiency occurs when the average product equals the marginal product.

A technologist maximizes this efficiency in the design by converting the variable cost into product as best as possible. This is like optimizing the energy conversion efficiency of the electric motor, which will be rated for optimum operation at the maximum technological efficiency point. As we shall see there are reasons for not operating the equipment at this point. We are told that cars are designed for a maximum fuel efficiency when driven at 55 miles per hour. Still many of us tend to drive faster (sometimes at the risk of getting a speeding ticket) because the cost of our time spent in transit may be greater (or so we think!)

Economic Efficiency. Engineers are good at optimizing the technological efficiency, but often ignore the total cost of the product. Economic efficiency is related to the *total cost* of production, which includes both fixed and variable costs. Thus,

Chapter 3. TEST ECONOMICS AND PRODUCT QUALITY

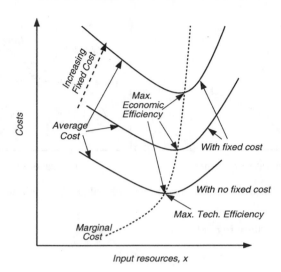

Figure 3.3: Maximum economic efficiency.

economic efficiency may also apply to long-run production. Suppose that X is the *total cost* (TC.) Then, the *average cost* (AC) is X/Q and *marginal cost* (MC) is dX/dQ. Marginal cost gives the cost of increasing the production level from the present average production level. The *economic efficiency* minimizes AC and is achieved when $MC = AC$. This is illustrated in Figure 3.3. Consider the point shown on this graph as the "Max. Tech. Efficiency." If we neglect the fixed cost then this gives the lowest product cost. However, it is impossible to produce any thing without fixed cost. For non-zero fixed cost, the average cost will be higher as illustrated by the other two average cost curves in Figure 3.3. If we were to produce at maximum technological efficiency, due to the actual non-zero fixed cost, the average cost of the product will be higher than the minimum cost given by the maximum economic efficiency.

The maximum economic efficiency may not occur at the same level of production that gives maximum technological efficiency. The difference is in whether we minimize the TC or the VC. More detailed considerations in long-run production take factors such as supply-demand and competition into account [46, 472].

The Law of Diminishing Returns. The shape of the production (Q/x) curve in Figure 3.2 is typical of all production environments. As inputs are increased, first the production rises and then decreases. This *fundamental* principle is known as the *law of diminishing returns*: *If one input of production is increased keeping other inputs constant, then the output may increase, eventually reaching a point beyond which increasing the input will cause progressively less increase in output.* In practice, various reasons cause diminishing returns. For example, when the input to a electric motor is increased, at some point its energy conversion efficiency begins to fall, due to greater heat loss. If more workers are put on the same job, they usually come in each other's way and the rate at which the work is done does not increase proportionately. Similarly, the phenomenon of diminishing speed up in multiprocessor computers is well known [373].

Increasing Returns to Scale. The case of *mass production* is worth considering. Production often increases faster than the increase of inputs, which is called *increasing returns to scale*. Some of the reasons are: (1) Technological factors and (2) Specialization. For example, a large company can afford to have a research and development unit to help utilize new technology in their production. They can also justify hiring specialists that a small company cannot afford. Another example is the *superlinear speed up* observed when multiprocessors are used to solve certain difficult search problems like test generation [603].

In the long run, however, the law of diminishing returns prevails. A company that has grown beyond control becomes inefficient due to reasons such as lack of communication among employees and counterproductive rivalries. There are recent examples of companies restructuring themselves into smaller business units, sometimes separate companies, for similar reasons.

Example 3.2 *Investing in equipment to reduce testing cost. In Example 1.2, we calculated the testing cost as 4.5 cents per second. Suppose that the production facility can locate (physically accommodate) five ATEs. The cost of the maintenance contract was 2% of the purchase price of an ATE. Suppose that additional ATEs can be included in the contract at half that price (bulk discount.) Similarly, the operating cost of additional ATEs is only half that of the first one, which was estimated as $ 0.5M in Example 1.2. This is because, the facility is shared and some tasks can be performed by the same person for multiple ATEs. Using the same cost assumptions as in Example 1.2, for five ATEs, we obtain:*

$$Running\ cost = \$\ 0.854M \times 5 + \$\ 0.585M \times 3 = \$\ 6.025M/year$$

Considering that all five testers are used around the clock, we obtain

$$Testing\ cost = \frac{6.025\ M\ dollars}{365 \times 24 \times 3,600 \times 5} = 3.82\ cents/second$$

This is an over 15% reduction in the testing cost provided by a single ATE. However, before purchasing the testers one should ascertain that the production volume of the wafer-processing plant and the demand of VLSI chips match the expanded capacity of the test shop. Any unused capacity will increase the testing cost by decreasing the denominator of the last equation.

3.1.3 Benefit-Cost Analysis

Benefits include income from sale of products or services, savings in cost and time, etc. *Costs* refer to the costs of labor, machinery, energy, finances, risks, etc.

All items are normally quantified and expressed in the same units (e.g., dollars.) We then define the *benefit-cost ratio* as follows:

$$\text{B/C ratio} = \frac{\text{annual benefits}}{\text{annual costs}} \tag{3.8}$$

Both benefits and costs should be estimated and averaged over the life time of the product. When there is a choice among products with similar utility, the *B/C ratio* provides a criterion for comparison. Product selection on the basis of just minimum cost or maximum benefits can lead to an imperfect result.

Example 3.3 *Weighing cost against benefit. A printed circuit board is repaired at an average cost of $350. This cost has several components. The average cost of diagnostic test is $300. The average price of a chip is $10 and any number of chips can be replaced at a cost of $40, which includes the final go/no-go type of system test. The repair shop's log on repaired boards shows that 95% of failed boards have a burnout failure of one of two specific chips. A test consultant advises that the shop should replace those two chips without running any diagnostic test. The cost of this repair is estimated to be $60 since two chips cost $10 a piece and chip replacement and system test cost $40. At this point it is expected that 5% of the boards will fail the system test. The shop must run the full diagnostic test on each failed board and repair it at an additional cost of $350. Thus, the average cost of repair is $(60 + 0.05 × 350) = $77.50, which is a substantial reduction over the original cost of repair.*

3.1.4 Economics of Testable Design

The changing trends of VLSI technology (see Chapter 1) have provided a new perspective to electronics manufacturing costs. Intel [607] recently reported that the combination of verification testing and manufacturing testing is its major capital cost, and not the $2 billion silicon fabrication lines. Many systems companies consider testing to be 50 to 60% of their equipment manufacturing cost. It may be cheaper to insert test hardware onto the chip rather than to provide testing using external ATE, if the test equipment cost becomes a concern. It has been shown [501] that at-speed testing is more effective than reduced-speed testing. However, the fastest-known ATE is always slower than the chips it will test. One possible solution to the high-speed test problem is the on-chip delay test capability. The added test circuitry will generate timing signals and capture the circuit's response (see Chapter 15.) Of course, then *Heisenberg's uncertainty principle* [289] of physics will apply: "Any attempt to observe a system will perturb the system behavior." This means that we will require a more careful consideration of the effect of the testability hardware on the normal system operation.

In the computer and communications industries, testing of electronic components is an important part of the business. Customers want reliable product at a reasonable cost. A manufacturer, with a poor design, may try to improve reliability by increased testing with the added cost passed on to the customer. However, in a

3.1 Test Economics

Table 3.1: Costs of built-in self-test (BIST.)

Level	Design and test	Fabri- cation	Production test	Maintenance test	Diagnosis and repair	Warranty repairs
CHIPS	+/−	+	−			
BOARDS	+/−	+	−		−	
SYSTEM	+/−	+	−	−	−	−

+ cost increase   − cost saving   +/− cost increase ≈ saving

competitive market economy, the consumer benefits by selecting the *best* product. To stay in business, the manufacturer must find means to provide the best product at the lowest cost. This requires production at the maximum economic efficiency.

The new advances in technology have given the capability to quickly design and manufacture very complex circuits at reasonable cost. However, as the cost of these products has reduced, the percentage of the total cost attributed to testing has increased [51, 192, 204]. In order to control testing costs, the designers must consider the test complexity. In electronic parts, DFT is important. Techniques such as *scan design*, BIST, and *boundary scan* simplify the test problem of electronic systems.

Consider the case of BIST. Test generation and response analysis circuitry is built into the VLSI chip or the printed wiring board. Tests can be run at the rated clock speed and the device can be efficiently tested even when it is embedded in a larger system. However, we must pay the price somewhere. The BIST circuitry will reduce the yield of the VLSI chip, thus increasing the cost. Unless this cost increase can be offset by cost reduction elsewhere, the use of BIST is not justified.

Systems planners and designers often face a decision of choosing from several alternatives. A typical question is: Should one use BIST? To answer the question, we must weigh costs against benefits. When such a trade-off is considered at the chip level, BIST offers some benefits in testing cost. However, if one considers the product life-cycle cost, because of the savings achieved, the decision is overwhelmingly in favor of BIST. Table 3.1 shows the impact of BIST on testing cost for chips, boards, and systems [36]. For simplicity, only qualitative impact (increase or decrease) on cost is shown in the table. Considering the cost of design and test development, we find that the additional cost of designing the BIST hardware somewhat balances the saving from test generation. Fabrication cost is increased at all levels due to the extra hardware. Production test cost is reduced due to more efficient tests and less expensive test equipment allowed by the self-testing design. Also, at board and system levels, BIST allows improved troubleshooting during assembly and integration. Of course, the maintenance cost has the greatest impact on the system operation. Diagnosis and repair costs are reduced at board and system levels. In alternative strategies, the lengthy or improper diagnosis is often responsible for much loss in revenues for the user due to service interruption and for the equipment manufacturer due to warranty repairs.

The main point of Table 3.1 is the significant benefit that BIST provides at the system level. A testability decision at one level cannot be made independent of other levels. This is because the chips are produced for use on boards, which are used in the system. The benefit of BIST may be small at the chip level, but the chips with BIST reduce the cost of testing at the board and system levels where benefits are greater. Even with relatively lower benefits at chip and board levels, BIST is still the best DFT alternative. Published work on economics of BIST [1, 48, 51, 192, 204] must be critically examined on this point.

3.1.5 The Rule of Ten

It is widely accepted in the electronics industry that chips must be tested before they are assembled onto printed circuit boards (PCBs), which, in turn, must be tested before they are assembled into systems. This is because experience has shown that the *rule of ten* holds [188]. If a chip fault is not caught by chip testing, then finding the fault costs 10 times as much at the PCB level as at the chip level. Similarly, if a board fault is not caught by PCB testing, then finding the fault costs 10 times as much at the system level as at the board level. Some claim that the rule of ten should be renamed the *rule of twenty*, because the chips, boards and systems are enormously more complex than when the empirical rule was first stated [188].

Example 3.4 *Spending more on test to save. A computer company provides a one-year free repair warranty on the CPU board. It is found that the average number of failures per board is three per year and the average cost of each repair is $200. The company sells 100,000 computers in one year and so the annual cost of warranty repair is 3 × 100,000 × $200 = $60M. The company is losing money on their product. To cut their loss on the warranty repair into half, they decide to reduce the failure rate of the CPU board to 0.5 failures per board per year. This will reduce the cost of warranty repair to $10M. The failure rate is brought down by applying a burn-in test to all CPU boards. In the burn-in test, also known as stress test or accelerated life test, the board is tested in an elevated temperature (e.g., 140°C) environment. The supply voltage may also be elevated to 1.5 × V_{DD}. In "static" burn-in, only the supply voltage is applied. In "dynamic" burn-in, the board is electrically exercised at the elevated temperature. Such tests are known to accelerate the failure mechanisms. Thus, most "weak" devices on the board will fail and will be replaced before the board is supplied to the customer. The duration of burn-in test can range from a few hours to several days, depending on the final failure rate desired [340]. The company must select a burn-in test duration suitable for reducing the failure rate to 0.5/year and can spend up to $200/board for test and repair, if a failure occurs. Thus, for 100,000 boards, an additional $20M will be spent. The manufacturer will be saving a cool $30M from the warranty expenditure. In addition, the customer will see only one breakdown in two years, in place of three per year.*

3.2 Yield

The *process yield* of a manufacturing process is defined as the fraction (or percentage) of acceptable parts among all parts that are fabricated. We will refer to

3.2 Yield

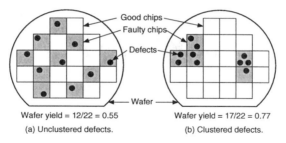

Wafer yield = 12/22 = 0.55
(a) Unclustered defects.

Wafer yield = 17/22 = 0.77
(b) Clustered defects.

Figure 3.4: Defect modeling for yield estimation.

the process yield simply as *yield*. Process variations, such as impurities in wafer material and chemicals, dust particles on masks or in the projection system, mask misalignment, incorrect temperature control, etc., can produce defects on wafers. The term *defect* generally refers to a physical imperfection in the processed wafer. The reader may refer to Section 4.1 for a formal definition. Typical defects are broken conductors, missing contacts, bridging between conductors, missing transistors, incorrect doping levels, and many other phenomena that can cause the circuit to fail. Some defects are observable through the optical or electron microscope. Others are not visible and can only be detected by electrical tests. To estimate the VLSI yield, defects are modeled as random phenomena.

The term *wafer yield* is sometimes used to refer to the average number of good chips produced per wafer. In the following discussion, we normalize the wafer yield by the number of chip sites on the wafer, and use it as the process yield. In general, though the process yield also accounts for the defects produced in other manufacturing steps such as scribing and packaging.

Figure 3.4 shows two possible scenarios of defect modeling. The defects are shown as dots and each faulty chip (shown in grey) contains one or more defects. In Figure 3.4(a), ten defects are randomly but uniformly distributed over the wafer and 12 of 22 chips are good, giving a wafer yield of $12/22 = 0.55$. The same number of defects, when distributed in random clusters as in Figure 3.4(b), produces a wafer yield of $17/22 = 0.77$. Higher yields are quite typical of the "clustered distribution" of defects. Fortunately, the actual chip manufacturing process is characterized by clustered defects. Random defects are characterized by two parameters:

- *defect density*, d, which is the average number of defects per unit area, and
- *clustering parameter*, α.

Thus, the average number of defects on a chip of area A is Ad. Suppose that we randomly pick a chip. Then, the number of defects, x, in it is an integer-valued random variable. The observed phenomenon of defect clustering is effectively modeled by assuming a *negative binomial* probability density function for x [546]:

$$\begin{aligned} p(x) &= \text{Prob(number of defects on chip} = x) \\ &= \frac{\Gamma(\alpha+x)}{x!\,\Gamma(\alpha)} \frac{(Ad/\alpha)^x}{(1+Ad/\alpha)^{\alpha+x}} \end{aligned} \qquad (3.9)$$

where $\Gamma(x)$ is the Gamma function, defined as [715]:

$$\Gamma(x) = \int_0^\infty e^{-x} x^{n-1} dx$$

The mean, $E(x)$, and variance, $\sigma^2(x)$, of x are given by:

$$E(x) = Ad \qquad (3.10)$$

$$\sigma^2(x) = Ad(1 + Ad/\alpha) \qquad (3.11)$$

The values of $E(x)$ and $\sigma^2(x)$ for the number of defects on a chip are obtained either from experimental measurements or by process simulation [167, 710]. Substitution in the above equations then leads to the determination of yield parameters, d and α. The yield is obtained as the probability, $p(0)$, of no defect on a chip. Thus, substituting $x = 0$ in Equation 3.9, we get:

$$Y = (1 + Ad/\alpha)^{-\alpha} \qquad (3.12)$$

When $\alpha \to \infty$, Equation 3.9 gives the *Poisson density function* with mean Ad, which corresponds to the unclustered distribution of defects (Figure 3.4(a)):

$$\text{For unclustered defects:} \quad p(x) = \frac{(Ad)^x e^{-Ad}}{x!} \qquad (3.13)$$

On substituting $x = 0$ this gives the yield, which can also be obtained by substituting $\alpha = \infty$ in Equation 3.12, as

$$Y_{\text{Poisson}} = e^{-Ad} \qquad (3.14)$$

Equation 3.14 gives very low yields. For example, if the average number of defects is, $Ad = 1.0$ and $\alpha = 0.5$, which are quite typical for a large VLSI chip, then $Y_{\text{Poisson}} = 1/e = 0.37$. A yield of 37% is considered to be too low for a profitable manufacturing. A more realistic yield, which characterizes a mature process, is obtained from Equation 3.12 as 58%.

When the processing of a newly designed chip is first started, the yield may be low. It may even conform to the prediction of Equation 3.14. The diagnosis of defects leads to process improvements to a stage where no further improvement is possible. The resulting process, called *matured*, has significantly higher yield as compared to the new process yield.

Example 3.5 *Cost of testability overhead. Consider a VLSI manufacturing process characterized by: defect density, $d = 1.25$ defects/cm^2, and clustering parameter, $\alpha = 0.5$. The area of a chip is $A = 8$ mm $\times$ 8 mm $= 0.64$ cm^2. Equation 3.12 gives the yield as:*

$$Y = \left(1 + \frac{0.64 \times 1.25}{0.5}\right)^{-0.5} = 0.62$$

3.3 Defect Level as a Quality Measure

Suppose that the process uses 8-inch wafers and the cost of processing a wafer prior to testing is $100. If a wafer has 400 chips, then the processing cost per chip is $100/(400 × 0.62) = 40 cents. Notice that the processing cost is distributed over the 62% good chips. Now suppose that the chip size is increased by 10% due to the addition of the hardware for design for testability (DFT) (see Chapters 14 and 15.) Then, the yield of the chips will be:

$$Y_{DFT} = \left(1 + \frac{0.64 \times 1.10 \times 1.25}{0.5}\right)^{-0.5} = 0.60$$

However, a wafer can now have only 400/1.1 ≈ 364 chips. Therefore, the processing cost per chip is $100/(364 × 0.60) = 46 cents, a 15% increase over the non-DFT design.

In the above example, we observe that the 2% yield reduction due to the DFT overhead is only one component of the cost increase. One must consider the entire economics of the manufacturing process to determine the impact of the overhead.

A good testing procedure can reject all (or most) defective parts. Testing, however, cannot improve the process yield. There are two ways of improving the process yield:

(1) *Diagnosis and Repair.* The parts that are found defective after test are diagnosed for specific failures which are then repaired. Although the yield is improved, this procedure increases the cost of manufacturing. The reason is that we first allow the process to make errors which are then corrected. A more economical procedure is to eliminate the sources of errors.

(2) *Process Diagnosis and Correction.* The defects found in the failed parts are traced to specific causes, which may be defective material, faulty machines, incorrect human procedures, etc. Once the cause is eliminated, the yield improves. Process diagnosis is the preferred method of yield improvement.

3.3 Defect Level as a Quality Measure

Quality of a product is a function of the user's satisfaction. The highest quality refers to the product meeting its requirements at lowest possible cost. In a manufacturing process, both criteria, *meeting the requirements* and *reducing cost*, determine the quality. Testing checks conformance to requirements. However, cost is reduced by enhancing the process yield.

A comprehensive test will thoroughly check all requirements. For complex VLSI components, such tests can be very expensive. Therefore, a cost tradeoff is often necessary. A test that can reduce the number of outgoing faulty parts to an acceptably small value can be considered a good test, especially if the test cost is also acceptable. For VLSI chips, the test quality is specified in terms of *defect level*, which is the fraction of faulty chips among the chips that pass the test, expressed as parts per million (*ppm.*) Other terms used for defect level are *reject ratio* and *field reject rate.*

The defect level is determined from the field return data. After VLSI chips leave the manufacturing facility, they may fail and be returned to the manufacturer (for a possible refund!) Some possible ways are listed below:

- *Failing acceptance test.* The customer (or user) conducts an *acceptance test* on the parts before they are mounted on printed circuit boards. If a part fails the test it is returned to the supplier.

- *Failing system test.* A board fails the *system test* but passes when one or more chips are replaced. The removed chips are returned to the chip supplier.

- *Failing maintenance test.* A *maintenance test* is conducted on a system operating in the field (usually on site) for both regular maintenance and diagnosis when an operational failure occurs. The faulty part, e.g., a board is located and replaced. While the system goes back to operation, the replaced board is returned to a repair shop, where faulty chips are found and replaced. The faulty chips are returned to the chip supplier.

The chips thus returned are examined by the manufacturer to determine the causes of failures. These causes may point to areas of improvement in specification, design, fabrication or test. Such improvements reduce the defect level. For VLSI chips, while a defect level of 500 *ppm* may be acceptable, 100 *ppm* or lower represents high quality.

3.3.1 Test Data Analysis

The above procedure of finding the defect level has several problems. First, not all failing parts are generally returned. Second, some returned parts are damaged due to mishandling or improper use and their cause of failure may be incorrectly diagnosed. Third, it takes a long time (usually a year or more) to collect sufficient data even on high-volume (million chips per year) parts. And fourth, as improvements are made the defect level reduces over time and, therefore, a time averaging of the number of returned parts gives an overly pessimistic defect level, especially for the early part of production and for low-volume production. The following technique of test data analysis provides an assessment of the defect level from the test data analysis of the manufacturing test. Although the field data is essential in the long run, the analysis helps the chip manufacturer to assess and improve the quality before the chips are supplied to the user.

Defects versus faults. We defined defects in the first paragraph of Section 3.2 as physical imperfections. The term *fault* is used to refer to electrical, Boolean, or functional malfunctions. In general, a physical defect in a chip can produce multiple faults. Thus, the spatial distribution of faults on a wafer is also clustered, sometimes even more so than the defects. We will rederive yield expressions for fault density. This will provide a relationship between the test process and yield and allow us to

3.3 Defect Level as a Quality Measure

measure the defect level. As we will find in Chapter 4, the "single stuck-at" fault model is the normal yardstick for VLSI tests. The present analysis also assumes this fault model, although the details can be worked out for other models as well.

Assume that a chip of area A on average has Af faults, where f is the average number of faults per unit area or *fault density* [191]. We follow the steps of Section 3.2 and characterize the random number (u) of faults by the fault density f and a clustering parameter β. Similar to Equation 3.9, we write a negative binomial probability density function for the random variable u, as:

$$\begin{aligned} q(u) &= \text{Prob(number of faults on chip} = u) \\ &= \frac{\Gamma(\beta+u)}{u!\,\Gamma(\beta)} \frac{(Af/\beta)^u}{(1+Af/\beta)^{\beta+u}} \end{aligned} \quad (3.15)$$

For discrete random variables, it is convenient to use the z-transform representation called the *probability generating function* (PGF). For random variable u, PGF is defined as [677]:

$$\begin{aligned} G(z) &= \sum_{u=0}^{\infty} q(u) z^u \\ &= [1+(1-z)Af/\beta]^{-\beta} \end{aligned} \quad (3.16)$$

where the last expression is obtained by substitution of Equation 3.15 for $q(u)$ and simplification [636].

Unlike Section 3.2, we will obtain the measured yield, $Y(T)$, when a test with fault coverage T is applied. We assume that $Y(1)$, i.e., the measured yield when all faults have been tested, is the true process yield Y of Section 3.2. One may argue that the last assumption is not a valid one since T is the coverage of stuck-at faults and many real faults are not the stuck-at type. Still, it is well recognized that the same tests detect many other faults, though perhaps with lesser efficiency. For example, a stuck-at fault is also a "delay fault," though the converse is not true. Thus, a circuit that passes all stuck-at fault tests has a good chance of passing delay fault tests also, but no guarantee can be given.

If the coverage of tests is T, then $1-T$ is the probability of a chip with one random fault passing the tests. If the chip has u faults, the probability of its passing the tests will be $(1-T)^u$. Since the probability of a chip having u faults is $q(u)$ as given by Equation 3.15, we can write the probability of a chip passing the tests, as:

$$Y(T) = q(0) + q(1)(1-T) + q(2)(1-T)^2 + \cdots = \sum_{u=0}^{\infty} q(u)(1-T)^u \quad (3.17)$$

The summation in the above equation is directly evaluated by substituting $1-T$ for z in Equation 3.16. Thus:

$$Y(T) = [1+TAf/\beta]^{-\beta} \quad (3.18)$$

The function $1 - Y(T)$ is called the *fallout rate*. It is an important relation that will be used in the next subsection for obtaining the yield parameters, f and β, from the chip test data. The true process yield is obtained by substituting $T = 1$, as

$$Y = Y(1) = [1 + Af/\beta]^{-\beta} \qquad (3.19)$$

When chip tests have a fault coverage T, the defect level is given by

$$DL(T) = \frac{Y(T) - Y(1)}{Y(T)} = 1 - \frac{Y(1)}{Y(T)} = 1 - \left(\frac{\beta + TAf}{\beta + Af}\right)^{\beta} \qquad (3.20)$$

This equation gives DL as a fraction that should be multiplied by 10^6 to obtain *parts per million* (*ppm*.) We can verify that for zero fault coverage, $DL(0) = 1 - Y(1)$, where $Y(1)$ is the process yield. For a 100% fault coverage, $DL(1) = 0$.

3.3.2 Defect Level Estimation

We will apply the preceding analysis to assess the defect level of a chip designed and tested at IBM under a SEMATECH experiment on test methods [501]. It is a bus interface controller ASIC chip containing 116,000 equivalent (two-input NAND) gates. It has 249 I/O signals and a 304-pin package. Some portions of the chip operate at a 40 MHz clock and others at 50 MHz. There are 5,280 scan latches (full-scan.) The chip operates with a 3.3 V supply. The die size is 9.4 $mm \times 8.8 \, mm$. It was fabricated using a 0.45μm CMOS process with three levels of metal.

Although four types of tests (stuck-at, functional, delay, and I_{DDQ}) were used to test the device, our analysis is based on stuck-at fault tests applied at the wafer level. The chip was designed in IBM's *level-sensitive scan design* (LSSD) style, which allows a scan flush test (see Chapter 14.) In this test, all flip-flops form a chain and both master and slave clocks are turned on, simultaneously. Signal transitions propagating through the long uninterrupted path test timing and many other faults. With additional scan tests, the total stuck-at fault coverage was 99.79% over a total of 375,142 faults. The fault coverage, as determined by a fault simulator, is shown in Figure 3.5. Here, the fault coverage is shown as a fraction instead of percentage.

Wafers were tested on an Advantest 3381 ATE, which applied vectors at 2.5 MHz. The cumulative chip fallout (fraction of chips failing up to a vector in the test set) is shown in Figure 3.6. This graph was obtained from the wafer level test of 18,466 chips. The fallout fraction rises to 0.2386 because the yield is near 76%. The two sets of data show similarity. As the fault coverage rises, more faulty chips fallout. That is all we can tell from these graphs. However, we would like to know how many bad chips are still in the "good" lot when the testing stops. That precisely is the question that the defect level answers.

A sample of the raw fault simulator and chip test data is shown in Table 3.2. The second and third columns of this table list the number of faults detected and the number of chips failing by the vector whose sequence number appears in the first column. The last two columns show the normalized cumulative data that are

3.3 Defect Level as a Quality Measure

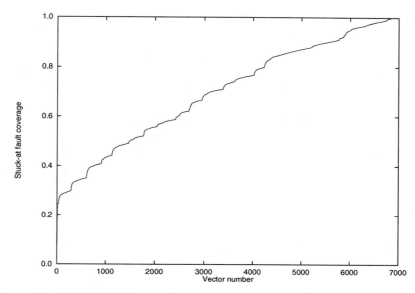

Figure 3.5: SEMATECH chip fault coverage obtained from fault simulator.

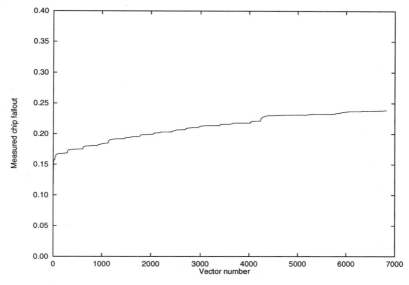

Figure 3.6: SEMATECH chip test data: chip fallout vs. test vector number.

Table 3.2: Fault simulator and ATE data for 18,466 SEMATECH chips.

Vector number	Measured incremental data		Normalized cumulative data	
	Faults det.	Chip fallout	Fault cov.	Chip fallout
1	26,587	1,673	0.071	0.0906
2	1,505	497	0.075	0.1175
3	2,923	5	0.083	0.1178
4	4,545	3	0.095	0.1180
5	21,841	2	0.153	0.1181
6	12,257	367	0.186	0.1379
7	959	159	0.188	0.1465
⋮	⋮	⋮	⋮	⋮
⋮	⋮	⋮	⋮	⋮
⋮	⋮	⋮	⋮	⋮
6,831	1	0	0.998	0.2386

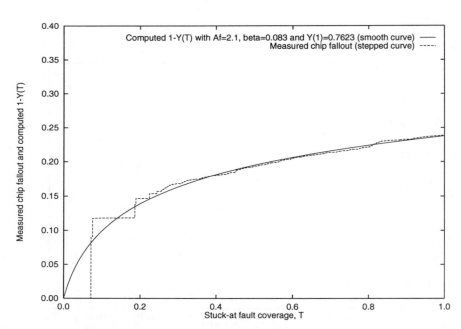

Figure 3.7: SEMATECH chip test data analysis: fallout rate vs. fault coverage.

plotted in Figures 3.5 and 3.6, respectively. In Figure 3.7 we have plotted the data from these two columns (stepped curve.) Using the nonlinear curve fitting in MATLAB [434], we fitted the fallout rate function $1 - Y(T) = 1 - [1 + TAf/\beta]^{-\beta}$ obtained from Equation 3.18. As shown in Figure 3.7, a close fit was obtained for $Af = 2.1$ and $\beta = 0.083$. We thus obtain, $f = Af/A = 2.1/(0.94 \times 0.88) = 2.54$ faults per sq. cm. When these values were substituted, Equation 3.19 gave a process yield $Y = 76.23$, and Equation 3.20 gave a defect level $DL = 168$ at fault coverage $T = 0.9979$.

The above analysis of data has several advantages. It allows us to determine the yield and fault distribution parameters. If those are not in the range expected for the manufacturing process being used, then we must diagnose and correct the process problems. If the defect level is too high, then the fault coverage of tests should be improved. The method has limitations too. The tests in the above example were run at slow speed. So, some delay and functional faults may have remained on chips that passed the tests. It is, therefore, necessary to tune the tests so that all possible defects have a non-zero probability of detection. Derivation of better tests has been a topic of research in recent years [112, 269].

Figure 3.8 shows the defect level for the SEMATECH chip as computed from Equation 3.20. For clarity in representing fault coverages close to 100%, we have used a reversed logarithmic scale as is sometimes used for probability graphs. For zero fault coverage, the defect level is $238,500$ *ppm*. This simply means that the lot of manufactured chips contains 76.23% good chips. A coverage of 99% reduces the defect level below $1,000$ *ppm*. Similarly, 99.9% and 99.99% coverages will drop defects levels below 100 and 10 *ppm*, respectively. While these fault coverages and defect levels are quite typical, we should remember the caution given in the previous paragraph. Thus, our conclusions are valid as long as the testing emulates real operating conditions. We might say that these will be the defect levels if the chips were to be used at a 2.5 *MHz* clock rate.

3.4 Summary

The basic idea behind economic theories is to allow improved usage of the available resources for the good of the society. In the design of a product the overall benefit/cost ratio for design, test, and manufacturing should be maximized. One must select the *most economical* design over the *cheapest design*. Our discussion on engineering economics has been adapted from the book by Mitchell [472]. Work specific to the economics of design for testability has been published by Ambler and coworkers [204]. Perhaps the earliest work on the yield of integrated circuits was published by Murphy [479]. Several models of yield analysis have been discussed by Glaser and Subak-Sharpe [254]. Recent work may be found in books by Ciciani [167] and Walker [710]. The analysis of defect level from the test data is based on statistical models of fault distribution. Commonly used models are due to Agrawal *et al.* [40], Seth and Agrawal [589], and Williams and Brown [725]. A comparative study of these models has been published by Das *et al.* [183]. In the last section, we

54 Chapter 3. TEST ECONOMICS AND PRODUCT QUALITY

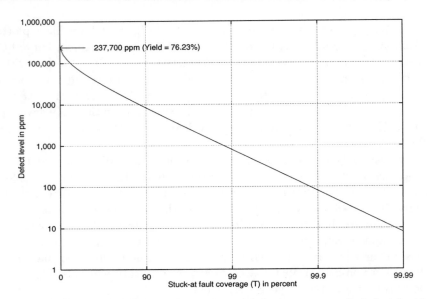

Figure 3.8: SEMATECH chip test data analysis: computed defect level.

have chosen an analysis due to de Sousa and Agrawal [191] because it incorporates the necessary fault clustering with the smallest number of parameters. The SEMATECH study is in progress and some parts are not yet published. Another study on VLSI test characteristics has been conducted by McCluskey and coworkers [226]. These and other unpublished studies in the industry show the inadequacy of the stuck-at fault model. As a result, there has been considerable interest on testing based on other fault models, usually dubbed as the *defect-oriented testing* [557].

Problems

3.1 *Economic decision.* The price of the type of car that Laura needs is given by Equation 3.2. She drives the car for personal use that amounts to about 5,000 miles per year. Assuming an average repair cost of $250 per breakdown and a linear depreciation to zero value over 20 years, determine the price of the most economical car for her.

3.2 *Economic decision.* An electronics manufacturing plant receives an order for supplying a large number of single-board systems. Components for the boards are purchased and stored in a warehouse whose rent is $10,000 per day. One technician assembles and tests one system each day. The cost of workspace (building, equipment, etc.) is estimated as $500n^2$ per day for n technicians.

(a) How many technicians should the plant employ on the job?

(b) If the cost of components for a system is $10,000 and the daily wages of a technician are $200, then what is the minimum cost of one system?

Problems

3.3 *Benefit-cost analysis.* Assume that the cost of conducting a quiescent current (I_{DDQ}) test, which measures the steady-state chip current for a few selected vectors (see Chapter 13), is 10% of the per chip cost of burn-in. I_{DDQ} test is non-functional and results in yield loss due to the rejection of some good chips. Suppose that your experimental data shows that when you lower the current threshold to a level where all burn-in failures are rejected by the the I_{DDQ} test, an additional 10% of good chips (that pass burn-in) also fail. Examine the following schemes:

(a) *Complete elimination of burn-in:* Show that this scheme is beneficial for chips whose total cost is less than 81% of the burn-in cost.

(b) *Apply burn-in test only to chips that fail I_{DDQ} test:* Show that this procedure will be beneficial as long as the burn-in yield is greater than 11.1%.

3.4 *Yield and cost.* Show that for a fractional increase Δ in the area A of a VLSI chip when hardware for design for testability is added, the cost increase is given by

$$\left[(1+\Delta)\left(1 + \frac{Ad\Delta}{\alpha + Ad}\right)^{\alpha} - 1\right] \times 100 \ \ percent$$

where d is the defect density and α is the defect clustering parameter. Calculate the percentage cost increase if the original chip area is 1 cm^2, $d = 1.25 \ defects/cm^2$, $\alpha = 0.5$, and the area overhead is 10%, i.e., $\Delta = 0.1$.

3.5 *Defect level and fault coverage.* Show that for a clustered fault distribution, if the required defect level is DL, then the fault coverage of tests should be

$$T = \frac{(\beta + Af)(1 - DL)^{1/\beta} - \beta}{Af} \times 100 \ \ percent$$

where f is fault density, β is fault clustering parameter, and A is chip area.

3.6 *Defect level and fault coverage.* Using test data analysis a chip production process has been characterized with fault density, $f = 1.45$ faults/sq. cm, and fault clustering parameter, $\beta = 0.11$. Given that the fault coverage of tests is 95%, calculate the defect level for a chip of 1 sq. cm area. What should the fault coverage be if the required defect level is: (a) $1,000 \ ppm$, and (b) $500 \ ppm$.

3.7 *Defect level.* Show that for an unclustered distribution of faults, as $\beta \to \infty$, the defect level is given by

$$DL(T) = 1 - Y^{1-T}$$

Note: This expression was first derived by Williams and Brown [725].

Chapter 4

FAULT MODELING

> " . . . *The extreme difficulty of obtaining solutions by conventional mathematical analysis has led in the past to the use of highly unrealistic models simply because they led to equations that could be solved. In fact, the applied mathematician has been engaged in a continual tussle with his conscience to decide how far he could go in the direction of distorting his model in order to make the equations tractable. The point is well made in the ancient jest about the examination question that began 'An elephant whose mass can be neglected . . .'.*"
> — M. V. Wilkes [720].

A doll house is a model of a household. So is the scaled model that an architect prepares for an airport. Both of these models have a visual similarity to the real entities they represent. We can use the doll house to determine the number of guests the living room can accommodate during a party. The model of the airport can, among other things, tell whether a passenger will find its appearance pleasing.

There are other models that have no resemblance to physical reality. For example, we model an electron as a sphere of certain charge and mass. This model can mathematically explain some (not all) behaviors displayed by the electron. However, this model may not have any resemblance to a real electron, which we have never seen.

In engineering, models bridge the gap between the physical reality and mathematical abstraction. They allow the development and application of analytical tools. They are thus essential in design. The most important models in testing are those of faults.

4.1 Defects, Errors, and Faults

Incorrectnesses in electronic systems are described in several ways. A reader may find that the terms *defect*, *error*, and *fault* are sometimes used in confusing ways in the literature on testing. In this book we will use them according to the following definitions.

Table 4.1: Typical printed circuit board (PCB) defects.

Defect type	Frequency of occurrence (%)
Shorts	51
Opens	1
Missing components	6
Wrong components	13
Reversed components	6
Bent leads	8
Wrong analog specifications	5
Defective digital logic	5
Performance defects	5

Definition 4.1 *Defect. A defect in an electronic system is the unintended difference between the implemented hardware and its intended design.*

Some typical defects in VLSI chips are [309]:

1. Process Defects – missing contact windows, parasitic transistors, oxide breakdown, etc.

2. Material Defects – bulk defects (cracks, crystal imperfections), surface impurities, etc.

3. Age Defects – dielectric breakdown, electromigration, etc.

4. Package Defects – contact degradation, seal leaks, etc.

Defects occur either during manufacture or during the use of devices. Repeated occurrence of the same defect indicates the need for improvements in the manufacturing process or the design of the device. Procedures for diagnosing defects and finding their causes are known as *failure mode analyses* (FMA) [52] and are beyond the scope of this book.

Since the manufacturing process of making *printed circuit boards* (PCBs) is different from that of VLSI chips, their defects are also different. Table 4.1 shows typical defects of PCBs and the frequency with which each defect has been observed [69].

Definition 4.2 *Error. A wrong output signal produced by a defective system is called an* error. *An error is an "effect" whose cause is some "defect."*

Definition 4.3 *Fault. A representation of a "defect" at the abstracted function level is called a* fault.

The difference between a defect and a fault is rather subtle. They are the imperfections in the hardware and function, respectively.

4.2 Functional Versus Structural Testing

Example 4.1 *Consider a digital system consisting of two inputs a and b, one output c, and one two-input AND gate. The system is assembled by connecting a wire between the terminal a and the first input of the AND gate. The output of the gate is connected to c. But the connection between b and the gate is incorrectly made – b is left unconnected and the second input of the gate is grounded. The functional output of this system, as implemented, is $c = 0$, instead of the correct output $c = ab$. For this system, we have:*

- *Defect: a short to ground.*

- *Fault: signal b stuck at logic 0.*

- *Error: $a = 1, b = 1$, output $c = 0$; correct output $c = 1$. Notice that the error is not permanent. As long as at least one input is 0, there is no error in the output.*

4.2 Functional Versus Structural Testing

Let us examine the testing of a ten-input AND function. Suppose that we apply an input pattern 0101010101, and observe a 0 output. This is a correct output, but what can we conclude: the gate under test is (A) an AND, (B) not a NAND, (C) not a NOR, or (D) not an OR function? Since the obtained output violates the truth tables of NAND and OR gates, only (B) and (D) are correct answers. We could use another pattern, 1111111111, to make sure that the gate is not a NOR. However, that does not guarantee that the given circuit will function correctly as an AND gate for all $2^{10} = 1024$ possible input patterns. Given ten inputs, it is possible to construct $2^{2^{10}}$ Boolean functions, and in the present situation our functional test must allow us to conclude that the function is AND and not one of the others. A complete functional test will check each entry of the truth table. Though possible with ten inputs, such a test will be too long and impossible to use with a real circuit with several hundred input lines.

Difficult as it is, the use of functional tests is often found necessary for verification of design. Methods of design verification lie outside the scope of this book, which focuses on hardware test. The purpose of *hardware test* (also referred to as *manufacturing test*) is to discover any faults caused due to manufacturing defects or errors. A basic assumption made is that the design being manufactured is correct. Back in 1959, Eldred derived tests that would observe the state of internal signals at primary outputs of a large digital system [215]. Such tests are called *structural* because they depend on the specific structure (gate types, interconnects, netlist) of the circuit.

One of the greatest advantages of structural testing is that it allows us to develop algorithms. Central to these algorithms are *fault models*. As we will see in later chapters, most test generation and test evaluation (fault simulation) algorithms are based on selected fault models.

4.3 Levels of Fault Models

Modeling of faults is closely related to the modeling of the circuit. In the design hierarchy, the *level* refers to the degree of abstraction. Thus, the *behavioral level* (sometimes referred to as *high level*) has fewer implementation details and fault models at this level may have no obvious correlation to manufacturing defects. High-level fault models play a greater role in the simulation-based design verification, than in testing. Exceptions are the functional fault models of semiconductor memories. Since the function of the memory is simple, exhaustive functional test is possible and is normally used in practice. We will study the fault models for memories in Chapter 9.

The *register-transfer level* (RTL) or logic level consists of a netlist of gates and the stuck-at faults at this level are the most popular fault models in digital testing. Other fault models at this level are *bridging faults* and *delay faults*. Bridging faults are discussed in various places in this book and Chapter 12 focuses on delay faults.

Transistor and other lower levels (referred to as *component levels*) include stuck-open types of faults that are also known as *technology-dependent* faults. Component-level faults are mainly modeled in analog circuit testing and we will study them in Chapters 10 and 11.

Finally, there are fault models that may not fit any of the design hierarchies. A typical example is the *quiescent current* (I_{DDQ}) defect discussed in Chapter 13. The usefulness of these models stems from the fact that they can represent some physical defects not represented by any other model. Therefore, sometimes these models have been called "realistic." Actually, there are far too many fault models that appear in the literature and a reader will find it convenient to refer to the following glossary.

4.4 A Glossary of Fault Models

Assertion Fault: An assertion expresses a property of a high-level function in the form: "*antecedent* ⇒ *consequent*," where *antecedent* and *consequent* can be simple predicates like "line L takes symbolic value v" or conjunctions of simple predicates. An *assertion fault* means that the corresponding property is not "true" for some input of the system. This fault model has been used for generating tests for a microprocessor [713].

Behavioral Faults: When the behavior of an electronic system is described in computer-readable form, it is generally written in a programming language (such as C) or some other hardware description language that resembles a programming language. At the behavioral level, also referred to as *functional* or *high level*, the variables or operations are not necessarily *electrical*, but correspond to the specific application domain. Behavioral faults refer to incorrect execution of the language constructs used in the description. Examples of behavioral faults are assertion faults, branch faults, and instruction faults. At the behavioral level, software test metrics [338], such as *statement coverage* and *branch coverage*, and an additional

4.4 A Glossary of Fault Models 61

toggle coverage, are also used although these do not conform to any specific fault model [656].

Branch Fault: This fault is modeled at the behavioral level where the circuit function is described in a programming language. A *branch fault* affects a branch statement and causes it to branch to an incorrect destination.

Bridging Fault: Usually modeled at the gate or transistor level, a *bridging fault* represents a short between a group of signals. The logic value of the shorted net may be modeled as 1-dominant (OR bridge), 0-dominant (AND bridge), or indeterminate, depending upon the technology in which the circuit is implemented. Non-feedback bridging faults are combinational and their coverage by stuck-at fault tests is normally very high. That is not always the case with feedback bridging faults that produce memory states in the otherwise combinational logic. Bridging faults are often used as examples of "defect-oriented faults" [557]. They are also modeled in programmable logic arrays (PLAs.) A reader interested in the analysis of bridging faults may refer to the book by Malaiya and Rajsuman [417].

Bus Fault: A *bus fault* specifies the status for each line in a bus as stuck-at-0, stuck-at-1, or fault-free. Thus, for an n-bit bus, there are $3^n - 1$ bus faults. A *total bus fault* assumes all lines of the bus to be stuck at the same 0 or 1 state [76, 77].

Cross-point Fault: These faults are modeled in *programmable logic arrays* (PLA.) In the layout of a PLA, input and output variable lines are laid out perpendicular to the product-lines. Crossing signal lines either form specific types of connections or remain unconnected at *cross-points*, depending on the function implemented. There are two types of cross-point faults. A *missing cross-point* means a missing connection at a crossing where a connection was intended. An *extra cross-point* means a faulty connection at a crossing where no connection was intended. Based on their influence on the logic function of the PLA, the cross-point faults are further classified as *shrinkage, growth, appearance*, and *disappearance* faults. Also see PLA faults.

Defect-Oriented Faults: Faults at the physical level that usually occur during manufacture are called *defects*. The electrical or logic-level faults that can be produced by physical defects are classified as *defect-oriented faults*. Examples of physical defects are broken (open) wires, bridges, improper semiconductor doping, and improperly formed devices. Some defect-oriented fault models are bridging faults, stuck-open faults, and increased I_{DDQ} faults [557]. PLA and some analog fault models are also classified as defect-oriented faults.

Delay Fault: These faults cause the combinational delay of a circuit to exceed the clock period [371]. Specific delay faults are *transition faults, gate-delay faults, line-delay faults, segment-delay faults*, and *path-delay faults*. See Chapter 12.

Functional Faults: See behavioral faults.

Gate-Delay Fault: The fault increases the input to output delay of a single logic gate, while all other gates retain some nominal values of delay [371]. The increase in the delay of the faulty gate is called the *size* of the gate-delay fault. The reader may refer to Iyengar *et al.* [327] for details of gate-delay faults.

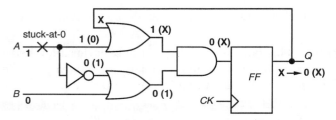

Figure 4.1: An initialization fault.

Hyperactive Fault: A hyperactive fault causes a large number of signals in the circuit to differ from their correct values. The fault thus produces very high fault-related activity in the circuit. If not readily detected, fault simulators usually remove hyperactive faults for later consideration to save CPU time and memory [142].

Initialization Fault: Circuits with memory elements (e.g., flip-flops) are designed so that they can be initialized by applying suitable input signals. Faults that interfere with such an initialization procedure are called *initialization faults*. A typical example of such a fault is the clock line of a flip-flop being stuck in the inactive state. Initialization faults are sometimes detected as potentially detectable faults. See Chapter 8.

Example 4.2 *Initialization Fault. In the circuit of Figure 4.1, consider a fault that permanently grounds the signal A. Such a fault is called a stuck-at-0 fault and will be discussed in detail in Section 4.5. In our notation, where the faulty value differs from the correct value, the faulty value is shown in parentheses. We assume that the initial state of the circuit (i.e., output Q of the flip-flop FF) is unknown, denoted as X. To set Q to 0, we apply $A = 1$ and $B = 0$. After the application of the clock CK, the fault-free circuit output is initialized to 0, but the faulty circuit remains in the unknown state. Such a fault that prevents the circuit from being initialized is called an initialization fault.*

Instruction Fault: Usually modeled in programmable systems like microprocessors or digital signal processors, an *instruction fault* causes an intended instruction to be incorrectly executed. Commonly considered faulty cases involve a target instruction producing a wrong result or the execution of an unintended instruction [660].

Intermittent Fault: A fault that appears and disappears as a function of time is called an *intermittent fault*. A fracture in an interconnect may produce an intermittent open for some time before it becomes a permanent fault. Intermittent faults can be of any type, e.g., stuck-at fault or a bridging fault, with its presence in time described probabilistically [90].

Line-Delay Fault: This fault models rising and falling delays of a given signal line. In contrast with the transition fault where the transition can be propagated through any path, a test for a *line-delay fault* must propagate the transition through the longest sensitizable path. A single fault assumption limits the number of faults

4.4 A Glossary of Fault Models

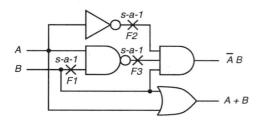

Multiple stuck-at fault	Output at $\overline{A}\,B$	Test
F1, F2	$\overline{A}\,B$	Redundant
F1, F3	$\overline{A}\,B$	Redundant
F2, F3	B	11
F1, F2, F3	B	11

Figure 4.2: An example of multiple stuck-at faults.

to twice the number of lines. Tests can detect all spot defects and many (not all) distributed delay defects. This model was recently proposed by Majhi et al. [413, 414].

Logical Faults: These faults affect the state of logic signals. Normally, the state may be modeled as $\{0, 1, X \text{ (unknown)}, Z \text{ (high impedance)}\}$, and a fault can transform the correct value to any other value. Several types of faults can be modeled at the logic level. However, the term *logical faults* often implies stuck-at faults.

Memory Faults: Faults modeled in memory blocks are *single cell stuck-at-*[0,1] faults, *pattern sensitive faults*, *cell coupling faults*, and single stuck-at faults in the address decoder logic [442, 688]. See Chapter 9.

Multiple Fault: A *multiple fault* represents a condition caused by the simultaneous presence of a group of single faults. Frequently considered multiple faults consist of the same type of single faults. For example, multiple stuck-at faults [85], or multiply-testable path-delay faults [245]. Multiple stuck-at faults are usually not considered in practice because of two reasons:

1. The number of multiple stuck-at faults in a circuit with n single fault sites is $3^n - 1$, which is too large a number even for circuits of moderate size.

2. Tests for single stuck-at faults are known to cover a very high percentage (greater than 99.6%) of multiple stuck-at faults [13, 315, 331, 341] when the circuit is large and has several outputs.

There are situations, however, where consideration of multiple stuck-at faults is important. First, diagnostic or fault location procedures, developed with the single fault assumption, do not work well when a multiple fault is present [94]. Second, a circuit with redundant single stuck-at faults can malfunction in the presence of a multiple fault even when the faulty circuit passes the test [528]. The following example illustrates the phenomenon.

Example 4.3 *Multiple stuck-at faults. The circuit of Figure 4.2 has three redundant stuck-at faults: shown as F1, F2, and F3. A set of three vectors, 00, 01, and 10, detects all other single stuck-at faults. A set of multiple faults contains all possible combinations of single faults. If any one of the single fault components of a multiple fault is detectable then there is a good chance that the multiple fault will be detected.*

Following Gharaybeh et al. [245], we will define a "multiply-testable stuck-at" fault as one where all single fault components are redundant. This circuit has four such faults that are shown in the table in Figure 4.2. Two of these, (F1,F2) and (F1,F3), produce the same output functions as the original circuit and are redundant. The other two faults, (F2,F3) and (F1,F2,F3), change the output function and are detectable by a vector, 11. However, this vector was not in the original single fault test set. So, the circuit, though faulty, will pass the tests.

The problem of multiple faults, as exposed in the above example, can be remedied either by enhancing the tests to cover multiply-testable faults, or by making the circuit fully single-fault testable via removal of redundant faults. Test generation for multiple faults [6, 85] is possible but is not commonly attempted due to high complexity. Instead, a designer resorts to removing the redundant faults and making the circuit fully single-fault testable.

With tests that cover 100% of single stuck-at faults, one can expect to detect almost all multiple stuck-at faults with a few rare exceptions. Such exceptions arise due to *fault masking* [7]. If a test t_i for a fault f_i fails to produce the fault effect at an observable output in the presence of another fault f_j, then f_j masks f_i. This phenomenon is also known as the *test invalidation problem*. In this situation f_i will remain undetected if both faults were present. However, f_j may still be detected by its single-fault test because fault masking is often unidirectional. Cases of *circular fault masking*, where two or more faults can mask each other, are possible but extremely rare. Invalidation of stuck-at fault tests by the presence of delay faults has also been reported [118].

Non-classical Fault: Although a *non-classical fault*, in general, refers to a fault other than a stuck-at fault, the term has been used for the stuck-open and stuck-short faults of MOS technologies [39].

Oscillation Fault: These faults cause oscillating signals in the faulty circuit when the fault-free circuit remains stable. Such a condition can occur due to certain single stuck-at faults in sequential circuits that contain combinational feedback. Oscillations can also occur in a purely combinational circuit if a bridging fault produces feedback [417]. Oscillation faults are also referred to as *star-faults* [142].

Parametric Fault: Such a fault changes the values of electrical parameters of active or passive devices from their nominal or expected values. Examples are the threshold voltage of a transistor (active device) and values of resistors and capacitors (passive devices.) See Chapters 2, 10, and 11.

Path-Delay Fault: This fault causes the cumulative propagation delay of a combinational path to increase beyond some specified time duration. The combinational path begins at a primary input or a clocked flip-flop, contains a connected chain of logic gates, and ends at a primary output or a clocked flip-flop. The specified time duration can be the duration of the clock period (or phase), or the vector period. Propagation delay is defined for the propagation of a signal transition through the path. Thus, for each combinational path there are two path-delay faults, which correspond to the rising and falling transitions, respectively. See Chapter 12.

4.4 A Glossary of Fault Models

Pattern Sensitive Fault: This fault causes an incorrect behavior in a certain part of the circuit only when a specific state occurs in some other part. Usually modeled in memories, a typical example is a fault condition that prevents writing a 1 in a memory cell when its physical neighbors have 0s stored in them. See Chapter 9.

Permanent Fault: Any faulty behavior that does not change with time is called a *permanent fault*. Faults that are not permanent and affect the circuit only at certain times (often at random instants) are called intermittent faults.

Physical Faults: These faults cause physical changes in the circuit. Examples of physical faults are broken wires, bridges (shorts) between conductors carrying unconnected signals, shorted or open transistors, etc. These faults also sometimes referred to as "defect-oriented faults."

Pin Fault: When a circuit is modeled as an interconnect of modules, the terminals of those modules are referred to as *pins*. This term is adopted from the technology of *printed circuit boards* (PCBs), which contain interconnecting wiring between the pins of the mounted chips. Pin faults are the stuck-at faults on the signal pins (not power and ground pins) of all modules in the circuit. Pin faults are commonly modeled in high-level designs where the internal gate-level structure of modules may not be known.

PLA Faults: A *programmable logic array* (PLA) is a physical implementation of two-level AND-OR combinational logic. The design consists of three sets of parallel wires: *inputs*, *product-terms*, and *outputs*. Three types of faults are modeled in a PLA:

1. Stuck-at faults on inputs and outputs.

2. Cross-point faults – These occur at the points where product lines cross input or output lines. An extra cross-point fault means a contact between crossing lines where no contact was intended. A missing cross-point fault is a missing contact between crossing lines. A PLA implements two-level AND-OR (or equivalent) logic. Based upon their effect on the product term cubes (as represented on a Karnaugh map), a missing cross-point in the AND plane is called a *growth fault*. An extra cross-point in the AND plane is called a *shrinkage fault*. Similarly, for the OR plane such faults are termed *disappearance* and *appearance* faults. The growth and disappearance faults are equivalent to stuck-at faults in the equivalent two-level logic circuit.

3. Bridging faults – Shorting between any set of lines that is not covered by the extra cross-point faults is classified as a bridging fault. Meaningful bridging faults correspond to the shorting of lines that are physically close to each other on the layout of the PLA.

PLA faults have been extensively studied in the literature and a reader will find concise information in the book by Agrawal and Seth [39] or details in the book by Fujiwara [230].

Potentially Detectable Fault: When a test is applied to a sequential circuit, certain faults produce an unknown state at the output when a deterministic output is expected in the fault-free circuit. This condition is known as potential (or probabilistic) detection. In contrast, deterministic detection requires that both faulty and fault-free outputs be different and definite (0 or 1.) Generally, stuck-at faults that can only be detected potentially are called *potentially detectable faults*. These faults form a subset of the initialization faults. For example, the fault A stuck-at-0 in the circuit of Figure 4.1 is a potentially detectable fault (detected only if the flip-flop happens to power up in the "1" state.) See Chapter 8.

Quiescent Current (I_{DDQ}) Fault: These faults are relevant to the CMOS (complementary metal oxide semiconductor) technology. In the steady state (i.e., when the gate is not switching) the CMOS logic gate provides no conducting path between the power supply and ground. Thus, the steady state current, also known as the *leakage* or *quiescent* (I_{DDQ}) current, of a CMOS gate is on the order of only a few microamperes. Under various fault conditions in the gate, this current can rise by several orders of magnitude thus allowing fault detection via current measurement. Faults detectable by this method are called I_{DDQ} faults.

Historically, quiescent current testing for CMOS circuits was proposed in 1981 [393]. Later developments may be found in a recent book [134]. We discuss the essentials of this technique in Chapter 13.

Race Fault: Stuck-at faults that cause a race condition in the circuit are called *race faults*. For a certain initial state and input, the final state of an asynchronous sequential circuit can vary depending on specific delays of its logic gates. Such a condition is known as a *race*. In the absence of the exact delay values, a logic simulator usually assumes a unit delay for each gate. When a race condition occurs, the simulator will place unknown signal values. Thus, a race fault may appear similar to an initialization fault. Sometimes, race and oscillation faults are grouped together as *star-faults* [142].

Redundant Fault: Consider a combinational circuit. Any fault that does not modify the input-output function of the circuit is called a *redundant fault*. A redundant fault cannot be detected by any test. Such faults can be removed from the circuit without changing its output function. Removal of redundant stuck-at faults is often used for circuit optimization.

The above definition applies to sequential circuits as well. However, identification and removal of redundant faults is a more complex process, which we will not discuss here. In general, the faults in sequential circuits for which no test can be found are classified as *untestable faults*. Redundant faults form a subset among untestable faults. For combinational circuits, the two types are the same.

Example 4.4 *Redundant stuck-at fault. Consider the circuit shown in Figure 4.3(a). Its output function is $\bar{A}B$. Now suppose that the B input of the NAND gate has a s-a-1 fault. Then the output of the NAND gate will be $\bar{A}$. This will produce a function $\bar{A}B$ at the output of the circuit, which is the same as the fault-free function. This fault is, therefore, redundant. The reader can verify that no values of A and B will produce different outputs from the faulty and fault-free circuits.*

4.4 A Glossary of Fault Models

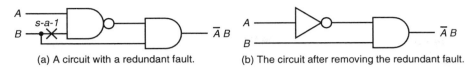

(a) A circuit with a redundant fault.   (b) The circuit after removing the redundant fault.

Figure 4.3: An example of a redundant stuck-at fault.

Since the circuit is combinational, this fault can be removed by a simple procedure: The fanout of B on which the fault lies is deleted and the corresponding input of the NAND gate is set to 1. The output of the NAND gate is now $\bar{A}$ and it can be replaced by a NOT gate. The circuit after removal of the redundant fault is shown in Figure 4.3(b).

Notice that in the circuit of Figure 4.3(a) signal B fans out and then reconverges at the AND gate. This structure, known as *reconvergent fanout*, is a necessary (though not sufficient) condition for a redundant single stuck-at fault. The circuit of Figure 4.3(b) has no reconvergent fanout. Its structure is known as a *tree* (rooted at the output.) A combinational circuit with a tree structure cannot have any redundant single stuck-at fault. For details of redundant stuck-at faults in combinational circuits the reader may refer to Chapters 5 and 7.

Segment-Delay Fault: A segment of length L is a chain of L combinational gates. Such a segment can be contained in one or more input to output paths. A *segment-delay fault* increases the delay of a segment such that all paths containing the segment will have a path-delay fault. If L is taken as the maximum combinational depth of the circuit, then segment-delay faults become the same as path-delay faults. For $L = 1$, segment-delay faults become identical to transition faults. Two faults, corresponding to two types (rising and falling) of transitions are modeled for each segment. The segment-delay fault model has been proposed very recently [292, 371].

Structural Faults: The structure of a circuit may refer to its topology or to physical geometry. However, the term *structural faults* is commonly used not for faults modeled in the layout, but rather in gate-level interconnects. Examples of structural faults are single stuck-at faults and bridging faults.

Stuck-at Fault: This fault is modeled by assigning a fixed (0 or 1) value to a signal line in the circuit. A signal line is an input or an output of a logic gate or a flip-flop. The most popular forms are the single stuck-at faults, i.e., two faults per line, *stuck-at-1* (s-a-1 or sa1) and *stuck-at-0* (s-a-0 or sa0.) See Section 4.5 for details of stuck-at faults.

Stuck-Open and Stuck-Short Faults: As we will discuss in Chapter 5, the modeling and simulation of MOS devices requires special considerations. For understanding the operation of purely digital MOS circuits, the simple "ideal switch" model of the transistor is useful. Considering a MOS transistor as an ideal switch, a defect is modeled as the switch being permanently in either the open or the shorted state. In general, a MOS logic gate consists of more than one transistor. This fault model assumes just one transistor to be *stuck-open* or *stuck-short*. Stuck-short is also

Chapter 4. FAULT MODELING

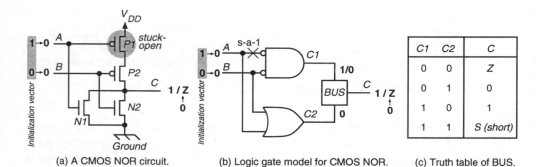

Figure 4.4: A CMOS NOR gate.

referred to as *stuck-on*. It is found that the input-output behavior of a faulty MOS logic circuit cannot be exactly represented by the stuck-at fault model. The models we discuss were proposed in 1976 by Case [115] and later analyzed by Wadsack [701]. The following examples illustrate the model.

Example 4.5 *Stuck-open fault in a CMOS gate.* Figure 4.4(a) shows a NOR gate implemented in CMOS technology. P1 and P2 are pMOS transistors that will be shorted when their gate-inputs, A and B, are 0. The same values on the gate-inputs of nMOS transistors, N1 and N2, make them open. Thus $A = B = 0$ will connect the output C to V_{DD} while isolating it from ground. Either $A = 1$ or $B = 1$ will connect C to ground while isolating it from V_{DD}. Consider the fault P1 stuck-open. If we apply $A = B = 0$, then P1 and P2 are shorted in the fault-free circuit but only P2 is shorted in the faulty circuit. N1 and N2 are open in both circuits. Thus, the output C is 1 in good circuit but is "floating" (not connected either to V_{DD} or to ground) in the faulty circuit. We denote this state as Z and $1/Z$ denotes the good/faulty states in Figure 4.4(a). In an actual CMOS circuit, the node C will have some parasitic capacitance with left-over charge from the previous operation of the circuit. To detect the fault we must ensure that Z assumes a value 0. This is done by preceding the $A = B = 0$ input vector by an initialization vector, $A = 1, B = 0$, which sets output C to 0 in the faulty circuit by discharging the node capacitance to the ground potential. The complete test consists of two vectors, $10 \rightarrow 00$, which produce an output $0 \rightarrow 1$ in the good circuit and $0 \rightarrow 0$ in the faulty circuit.

Switch-level test generation algorithms [14, 538] allow automation of the above procedure. However, it is often convenient to use a gate-level model [335] as the next example illustrates.

Example 4.6 *Gate-level model for stuck-open fault.* Figure 4.4(b) shows a gate-level model of the CMOS NOR circuit [335]. Here every series interconnection of transistors between a supply node (V_{DD} or ground) to output is replaced by an AND gate. Similarly, a parallel interconnection is replaced by an OR gate. The control inputs of pMOS transistors feed these gates via an inversion and those of nMOS transistors feed directly. The output is produced by a BUS network whose truth table

4.4 A Glossary of Fault Models

is shown in Figure 4.4(c). When the two inputs of BUS are different, the output assumes the upper value (C1.) For a 00 input the BUS produces a high impedance (or memory) state shown as Z. For a 11 input the BUS produces a logic state shown as S. Often interpreted as unknown (X) in three-state simulation (see Chapter 5), the S state in this model means a short circuit between the supply nodes. Note that 00 and 11 inputs of BUS cannot occur in the circuit without a fault because the two logic functions feeding the BUS are complementary.

A stuck-open fault of a pMOS transistor is modeled as a stuck-at-1 fault at the corresponding input signal and that of an nMOS transistor as a stuck-at-0 fault. Test generation for P1 stuck-open is illustrated in Figure 4.4(b). First, an input vector 00 produces an output 1/Z. Then, Z is set to 0 and is justified by input 10, the initialization vector.

Example 4.7 *Stuck-short fault. In the gate-level model of Figure 4.4(b), a stuck-short fault of a pMOS transistor is represented as a stuck-at-0 fault at the corresponding input to a logic gate. Similarly, a stuck-short fault of an nMOS transistor is represented as a stuck-at-1 fault of a gate input. For detecting the fault P1 stuck-short, we will analyze a stuck-at-0 fault at the site where stuck-at-1 is shown in the figure. We see that an input vector 10 produces a 0/S output at C. Since S represents a short circuit between the supply nodes, this test will produce a high current in the faulty circuit even when it reaches the steady state. This current is usually several orders of magnitude larger than the normal quiescent current of a CMOS circuit. Thus, a measurement of the device current will detect the fault (see Chapter 13.)*

Modeling of stuck-open and stuck-short faults has been extensively studied in the literature. The CMOS design for testability is discussed in a book by Jha and Kundu [342]. Among switch-level fault simulators that model the entire circuit at one level are MOSSIM described by Schuster and Bryant [578], and SLS by Barzilai *et al.* [68]. MOTIS, a mixed-level fault simulator that combines switch and gate levels, was developed by Lo *et al.* [404]. The SATISFAULT program of Meyer and Camposano [462] combines register-transfer level (RTL) models with switch-level models. Among the programs developed for switch-level test generation are those described by Leet *et al.* [389], Einspahr and Seth [212], and Lee *et al.* [386].

In spite of much research and development since 1976, these faults have not been assimilated into the VLSI test methodology. It is generally known that the coverage of stuck-open faults lags by about 10 to 15% behind that of stuck-at faults, mainly because the tests are generated for the latter type of faults [39]. The fundamental characteristics of these faults can be summarized as:

- The effect of a *stuck-open* fault is to produce a floating state at the output of the faulty logic gate. It can be detected by a test that detects a stuck-at fault on that output provided it was appropriately initialized by the previous vector. Thus, detection of these faults requires "two-vector tests." Also, any possible hazard (timing-related variation) at the output node can change the initialization and invalidate the test. For a guaranteed detection of a stuck-open fault, the two vectors should not produce any hazard at the output of the logic gate containing the faulty transistor [335, 540].

- The effect of a *stuck-short* fault is to produce a power supply to ground conducting path. When the fault is activated, the logic state at the output of the gate containing the faulty transistor depends upon the relative impedances of transistors producing the conducting path and the switching threshold of other devices. However, a definite detection of the fault is possible by quiescent current (I_{DDQ}) measurement.

Transistor Faults: Stuck-open and stuck-short faults are generally referred to as *transistor faults*.

Transition Fault: It is assumed that in the fault-free circuit all gates have some nominal delays and that the delay of a single gate has changed. The gate delay, usually an increase over the nominal value, is assumed to be large enough to prevent a passing transition from reaching any output within the clock period, even when the transition propagates through the shortest path. Possible *transition faults* of a gate are *slow-to-rise* and *slow-to-fall* types and hence the total number of transition faults is twice the number of gates. Transition faults model spot defects and are also called the *gross-delay faults* [371]. Also see Chapter 12.

Also see Chapter 12.

Untestable Fault: A fault for which no test can be found is called an *untestable fault*. There are two classes of untestable faults:

1. Faults that are redundant, i.e., whose presence does not change the input-output behavior of the circuit.

2. Faults that change the input-output behavior of the circuit but no test can be found by a given method of testing or test generation. Initialization faults of sequential circuits belong to this class (see Chapter 8.)

4.5 Single Stuck-at Fault

We assume that the circuit is modeled as an interconnection (called a *netlist*) of Boolean gates. A stuck-at fault is assumed to affect only the interconnection between gates. Each connecting line can have two types of faults: stuck-at-1 and stuck-at-0 (commonly written as s-a-1 and s-a-0.) Thus, a line with a stuck-at-1 fault will always have a logic state 1 irrespective of the correct logic output of the gate driving it.

In general, several stuck-at faults can be simultaneously present in the circuit. A circuit with n lines can have $3^n - 1$ possible stuck line combinations. This is because each line can be in one of the three states: s-a-1, s-a-0, or fault-free. All combinations except one having all lines in fault-free states are counted as faults. Clearly, even a moderate value of n will give an enormously large number of multiple stuck-at faults. It is a common practice, therefore, to model only single stuck-at

4.5 Single Stuck-at Fault

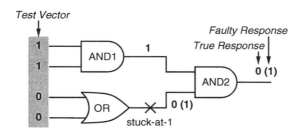

Figure 4.5: An example of a single stuck-at fault.

faults. An n-line circuit can have at most $2n$ single stuck-at faults. This number is further reduced by a technique known as *fault collapsing*, as discussed later.

Definition 4.4 *Single stuck-at fault. Three properties (or assumptions) characterize a single stuck-at fault:*

1. *Only one line is faulty.*

2. *The faulty line is permanently set to either 0 or 1.*

3. *The fault can be at an input or output of a gate.*

Example 4.8 *Consider the circuit of Figure 4.5. A stuck-at-1 fault as marked at the output of the OR gate means that the faulty signal remains 1 irrespective of the input state of the OR gate. If the normal output of the OR gate is 1, as it would be if the inputs were 01, 10, or 11, then this fault will not affect any signal in the circuit. However, a 00 input to the OR gate will produce a 0 output in the normal circuit. The faulty circuit will have a 1 there. Figure 4.5 shows the normal(faulty) value as 0(1), which is applied to the AND2 gate at the output. This signal state must be propagated to the output of the AND2 gate, which is an observable output of this circuit. This is done by setting the other input of AND2 as 1, which is justified by setting the inputs of AND1 as 11. Now we have the input vector 1100 as a test for the s-a-1 fault since for this vector the normal output (true response) and the faulty output differ.*

This example illustrates the basic features of a *single stuck-at* fault. Notice that gates are assumed to function correctly. Only the signal interconnects are considered to be faulty. The circuit of Figure 4.5 has seven lines, which are the potential sites for single stuck-at faults. The number of possible faults is 14. As the next example shows, fanouts give rise to additional fault sites.

Example 4.9 *Consider the exclusive-OR function implemented by the circuit of Figure 4.6. As shown, the single fault h s-a-0 is detectable by a 10 input. Signal lines g, h, and i, commonly known as a signal net, carry the same signal value. The 10 input also activates single s-a-0 faults on g and i. But, only g s-a-0 is detectable*

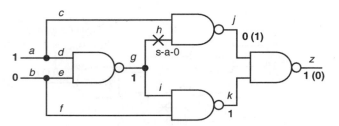

Figure 4.6: Single stuck-at fault sites.

by this input. The effect of fault i s-a-0 is blocked from propagating to the primary output z by $f = 0$, which uniquely sets $k = 1$. We notice that the faults on the fanout branches of a net are not identical. In a logic circuit, a net contains a stem or source (g in this circuit) and fanout branches (h and i.) The stem is the output of some gate and fanout branches are inputs of some other gates. To consider all possible faults, we model single stuck-at faults on the stem and all fanout branches of the net. Considering all nets in the circuit, this is equivalent to modeling faults on inputs and outputs of all gates (see Property 3 of Definition 4.4.)

The reader should verify that the exclusive-OR circuit of Figure 4.6 has 12 fault sites and hence we would model 24 single stuck-at faults in this circuit. To reduce this number, we will use the concepts of fault equivalence and fault dominance, discussed next.

4.5.1 Fault Equivalence

Let us consider a single-output combinational circuit with n input variables. We will denote its output function as $f_0(V)$, where V is an n-bit Boolean vector. To consider two faults, designated as fault 1 and fault 2, let the output function in the presence of fault 1 be $f_1(V)$ and that in the presence of fault 2 be $f_2(V)$.

Any test V for fault 1 must produce different values for $f_0(V)$ and $f_1(V)$. This condition can be expressed as:

$$f_0(V) \oplus f_1(V) = 1 \tag{4.1}$$

Similarly, a test for fault 2 must satisfy

$$f_0(V) \oplus f_2(V) = 1 \tag{4.2}$$

When fault 1 and fault 2 have exactly the same tests, i.e., all vectors that satisfy Equation 4.1 also satisfy Equation 4.2, and vice-versa, then the Boolean functions on the left hand sides of the two equations are identical. That is,

$$[f_0(V) \oplus f_1(V)] \oplus [f_0(V) \oplus f_2(V)] = 0 \tag{4.3}$$

With some manipulation this leads to

$$f_1(V) \oplus f_2(V) = 0 \tag{4.4}$$

4.5 Single Stuck-at Fault

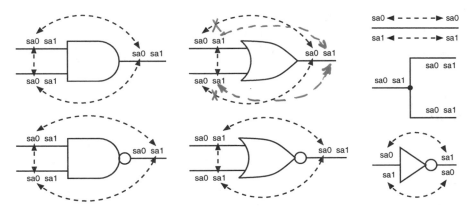

Figure 4.7: Equivalent fault collapsing for Boolean gates, wires, and fanouts.

Equation 4.4, known as the *indistinguishability condition*, shows that the two faulty functions are identical when the faults have the same set of tests.

Definition 4.5 *Fault equivalence. Two faults of a Boolean circuit are called equivalent iff they transform the circuit such that the two faulty circuits have identical output functions. Equivalent faults are also called indistinguishable and have exactly the same set of tests.*

4.5.2 Equivalence of Single Stuck-at Faults

Equation 4.4 can determine the equivalence of any pair of faults. For an n-line circuit that has $2n$ single stuck-at faults, we will have to apply this equation to $2n^2 - n$ pairs of faults to determine all equivalences. Besides, manipulations of large Boolean functions can be cumbersome. Alternatively, if we determine equivalences among faults of simple Boolean gates, then those results can be applied to arbitrarily large circuits.

Consider a k input AND gate. It has $k+1$ s-a-0 faults on its input and output lines. Each s-a-0 fault transforms the AND gate to a constant 0 output function. Thus all s-a-0 faults are equivalent. No such equivalence relation is found among the $k+1$ s-a-1 faults. Similar analysis of other Boolean gates leads to the equivalences shown in Figure 4.7, which also includes connecting wire and fanout that are necessary components for a circuit. A two-way arrow shows pair-wise equivalence. Note that the NOT gate is identical to a single-input NAND or NOR gate. Similarly, the wire (or a non-inverting buffer) is identical to a single-input AND or OR gate.

The case of the fanout shown in Figure 4.7 needs explanation [569]. The fact that no fault equivalences are possible seems to violate the conclusions of the last subsection. To understand this, we need to expand the notion of a test for a multi-output circuit. Let us denote the stem signal as a and the two fanout branches as y and z. We specify its function as $(y = a, z = a.)$ Consider the three s-a-0 faults all of which are tested by an input 1. The output functions and tests for these faults

Table 4.2: Stuck-at-0 faults of a fanout structure.

Fault	Test, $input \to outputs$	Faulty function
stem a s-a-0	$1 \to 1/0, 1/0$	$(y = 0,\ z = 0)$
branch y s-a-0	$1 \to 1/0, 1$	$(y = 0,\ z = a)$
branch z s-a-0	$1 \to 1, 1/0$	$(y = a,\ z = 0)$
branches y s-a-0 and z s-a-0	$1 \to 1/0, 1/0$	$(y = 0,\ z = 0)$

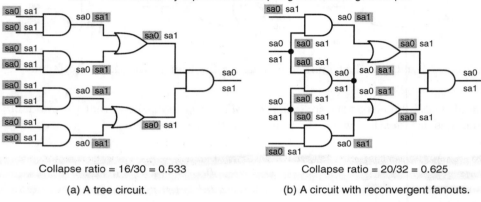

Figure 4.8: Examples of equivalence fault collapsing.

are summarized in Table 4.2, which also includes a multiple s-a-0 fault of the two branches. The test outputs containing fault effects are shown as 1/0, meaning a fault-free state 1 and faulty state 0. For a multi-output function, all outputs should be considered while applying the Definition 4.5. It is evident from Table 4.2 that the three single s-a-0 faults of the fanout have neither the same test nor the same faulty function. However, the stem fault a s-a-0 is equivalent to the multiple s-a-0 fault of the two branches. Similar arguments apply to the s-a-1 faults in the fanout.

4.5.3 Fault Collapsing

The set of all faults in a circuit can be partitioned into *equivalence sets*, such that faults in an equivalent set are equivalent to each other. Equivalence sets divide faults into disjoint sets because if a fault exists in two equivalence sets then those sets can be merged together as one equivalence set.

The process of selecting one fault from each equivalence set is called *fault collapsing*. The set of selected faults is known as the *equivalence collapsed set*. The relative size of the equivalence collapsed set with respect to the set of all faults is the *collapse ratio*:

$$\text{Collapse ratio} = \frac{|\text{Set of collapsed faults}|}{|\text{Set of all faults}|} \qquad (4.5)$$

4.5 Single Stuck-at Fault

The process of generating the equivalence fault set, known as *equivalence fault collapsing*, is illustrated for single stuck-at faults of combinational circuits by the following example.

Example 4.10 *Equivalence fault collapsing. Figure 4.8 shows two circuits, one without a fanout and the other with fanouts. Fault collapsing is performed in a level-by-level pass from inputs to output using local gate fault equivalences shown in Figure 4.7. The procedure begins at primary inputs and a gate is not processed until all gates feeding its inputs have been processed. At a gate, first input faults are examined. Only one among the equivalent faults is retained. Then any input faults that are equivalent to some output fault are deleted. Note that the collapse ratio is around 50 or 60% and that more reduction occurs in the absence of fanouts.*

The input to output pass produces a unique result since a Boolean gate has a single output and no collapsing is possible through fanouts. An output to input pass can also be used, but one has to choose an input from the multiple inputs of a gate where the equivalent fault is retained. Depending on the choice made, the collapse ratio can be higher than or the same as that obtained by the input to output collapsing. Fault collapsing within and across sequential elements is more complex [140, 149] and is usually not attempted. So the faults are collapsed for the combinational logic alone. The collapsing process of the above example is quite typical and is often used in practice. Results on fault collapsing on several benchmark circuits [99, 100] processed by Bell Labs' GENTEST system [160] are given in Table 4.3. Larger circuits allow more fault collapsing as indicated by smaller collapse ratios. Circuits with large fanins also reduce the collapse ratio. An example is the s9234 circuit. The circuit c499 contains a large number of exclusive-OR modules. These are not primitive Boolean gates and do not allow fault collapsing among input and output faults (see the next example.) As a result, the collapse ratio (0.76) is high. The Boolean gate implementation of c499 is the circuit c1355, which allows greater collapsing.

Example 4.11 *Fault collapsing by functional equivalence. Figure 4.9 shows a gate-level implementation of an exclusive-OR (XOR) circuit in which faults have been collapsed by the procedure of the last example. Faulty output functions corresponding to four s-a-1 faults, F1, F2, F3, and F4, are also shown. We find that F1 and F4 are equivalent and F2 and F3 are equivalent. Such equivalences, where fault sites have no directed path in between, can be found by using Equation 4.4. There are algorithms [7, 449] for finding these but they are difficult to apply to large circuits. In spite of the additional fault collapsing, we find that no equivalence between the input and output faults of the XOR function can be established. As shown in Figure 4.9, when the XOR function is modeled as a module, all six faults on the inputs and outputs should be included. This reduces the amount of fault collapsing in the circuit as we observed in the case of circuit c499 in Table 4.3.*

4.5.4 Fault Dominance and Checkpoint Theorem

According to Figure 4.7, fault equivalence requires us to consider $n + 2$ stuck-at faults for an n-input gate. Figure 4.10 shows a three-input AND gate for which we

Table 4.3: Equivalence fault collapsing for stuck-at faults in benchmark circuits.

Circuit name	No. of gates	No. of inputs	No. of outputs	Number of faults		
				All	Collapsed	Collapse ratio
c432	160	36	7	864	524	0.61
c499	202	41	32	998	758	0.76
c880	383	60	26	1,760	968	0.55
c1355	546	41	32	2,710	1,606	0.59
c1908	880	33	25	3,816	2,041	0.54
c2670	1,193	233	140	5,340	2,943	0.55
c3540	1,669	50	22	7,080	3,651	0.52
c5315	2,307	178	123	10,630	5,663	0.53
c7552	3,513	207	108	15,104	8,084	0.54
s27	10	4	1	52	32	0.62
s9234	5,597	19	22	10,572	3,862	0.37
s38584	19,257	12	278	78,854	36,303	0.47

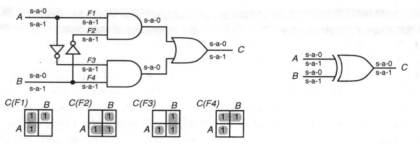

Figure 4.9: Functional equivalence fault collapsing in an exclusive-OR circuit.

will analyze two single s-a-1 faults shown as $F1$ and $F2$. Suppose that $T(F1)$ is the set of all tests for $F1$ and $T(F2)$ is the set of all tests for $F2$. $T(F1)$ contains one vector and $T(F2)$ has seven vectors. As shown in Figure 4.10, $T(F2)$ is larger and completely contains $T(F1)$. According to the following definition, fault $F2$ dominates fault $F1$.

Definition 4.6 *Fault dominance. If all tests of fault F1 detect another fault F2, then F2 is said to dominate F1. The two faults are also called "conditionally" equivalent with respect to the test set of F1. When two faults F1 and F2 dominate each other, then they are equivalent.*

In an alternative form of fault collapsing, known as *dominance fault collapsing*, we further eliminate the dominating faults from the equivalence collapsed set. For the AND gate shown in Figure 4.10, we will thus eliminate the s-a-1 fault from the output. The figure also shows the three-input AND gate with four faults left after dominance fault collapsing. Since the output s-a-0 is equivalent to any input s-a-0

4.5 Single Stuck-at Fault

Figure 4.10: An example where fault $F2$ dominates fault $F1$.

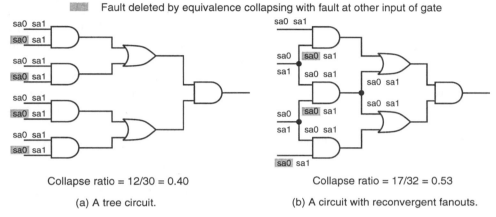

Figure 4.11: Examples of dominance fault collapsing.

(see Figure 4.7), we have moved all faults to inputs. Similar collapsing is possible for all other Boolean gates. Thus we can summarize dominance fault collapsing as:

1. An n-input Boolean gate requires $n + 1$ single stuck-at faults to be modeled.

2. To collapse faults of a gate, all faults from the output can be eliminated retaining one type (s-a-1 for AND and NAND; s-a-0 for OR and NOR) of fault on each input and the other type (s-a-0 for AND and NAND; s-a-1 for OR and NOR) on any one of the inputs.

3. The output faults of the NOT gate, the non-inverting buffer, and the wire can be removed as long as both faults on the input are retained. No collapsing is possible for fanout.

Example 4.12 *Dominance fault collapsing. Figure 4.11 shows the application of the above rules of dominance fault collapsing to the circuits without and with fanouts. Collapsing is done in an output to input pass. A comparison with Figure 4.8 shows the lower collapse ratio for dominance fault collapsing. For the fanout-free circuit, the collapsed fault set only contains input faults. This is an important result. For the circuit with fanouts, the collapsed set contains faults located at "checkpoints" defined below.*

Theorem 4.1 *Fault detection in fanout-free circuit. A test set that detects all single stuck-at faults on all primary inputs of a fanout-free circuit must detect all single stuck-at faults in that circuit.*

Definition 4.7 *Checkpoints. Primary inputs and fanout branches of a combinational circuit are called the* checkpoints.

Theorem 4.2 *Checkpoint theorem. A test set that detects all single stuck-at faults of the checkpoints of a combinational circuit detects all single stuck-at faults in that circuit.*

Proofs of these theorems can be constructed from the notions of fault equivalence and dominance. Checkpoints provide a starting set for dominance fault collapsing in which further reduction is possible with the three rules specified above. When tests are specially generated for detecting multiple faults using a procedure given by Bossen and Hong [85], the number of checkpoints can be reduced by eliminating primary input stems that fanout. This will eliminate the second and third primary inputs in the circuit of Figure 4.11(b).

Abramovici et al. [7] point out that the dominance relations in a combinational circuit may not remain valid when that circuit is embedded in a sequential circuit. This is because a fault can dominate several faults. For example, the fault $F2$ in Figure 4.10 dominates all input s-a-1 faults. Hence, in the collapsed set $F2$ is represented by three faults, which are not equivalent to each other. In a sequential circuit, considering multiple time-frames (see Chapter 8) $F2$ can have many more activations than each of dominated fault, thus providing a greater chance of fault effect cancellation.

A "dominated" fault can become redundant due to the circuit structure. If the fault set used for test generation is obtained by dominance fault collapsing, then no test may be obtained for the dominating fault even when it is detectable. In fact, the checkpoint theorem (Theorem 4.2) gives no guarantees if the test set has a less than 100% coverage of the checkpoint faults, some of which may be redundant [11].

In spite of the greater reduction of the fault list (smaller collapse ratio) that dominance fault collapsing provides, the two reasons cited above limit its use. In practice, therefore, equivalence fault collapsing is more popular and is often recommended.

4.5.5 Summary

Many workers believe that Eldred's 1959 paper [215] laid the foundation for the stuck-at fault model. Eldred's main contribution was to break away from functional testing and demonstrate the practicality of testing at the hardware level. His paper did not mention the stuck-at fault. The term "stuck-at fault" appeared in the 1961 paper by Galey, Norby, and Roth [237]. It is possible that other researchers of that time may have used it as well. Many people think that the "stuck-open" fault model was first mentioned in Wadsack's 1978 paper [701]. Actually, it was proposed by Case in 1976 [115], a fact brought to our notice by Don Ross.

It is essential to grasp the ideas behind the stuck-at fault model, which is most fundamental to digital testing. Chapters 5 through 8 develop algorithms based on these. In addition, one must gain working knowledge of models used in testing of memory (Chapter 9) and analog circuits (Chapters 10 and 11.) Fault models most likely to gain significance in the near future are the delay fault models discussed in Chapter 12.

Problems

4.1 *Boolean functions.* Show that there are 2^{2^n} distinctly different Boolean functions with n inputs.

4.2 *Initialization faults.* Show that the output of the circuit in Figure 4.1 cannot be initialized as $Q = 1$ when the fault A stuck-at-0 is present. Can you find another fault in that circuit that will prevent initialization?

4.3 *Fault counting.* Determine the total number of stuck-at (single and multiple) faults for a logic circuit with n lines.

4.4 *Fault counting.* Compute the total number of stuck-at (single and multiple) faults for the exclusive-OR circuit of Figure 4.6.

4.5 *CMOS faults.* For a two-input CMOS NAND circuit:

   (a) Find a two-pattern test for each single-transistor stuck-open fault.

   (b) Rearrange the eight vectors in a compact set and show that this set can be constructed from the single stuck-at fault tests for the NAND gate.

   (c) For each stuck-at fault of the NAND gate, find an equivalent transistor (stuck-open, stuck-short, or a combination) fault.

4.6 *Fault models.* Define the following fault models using examples where possible:

   (a) Bridging fault
   (b) Cross-point fault
   (c) Hyperactive stuck-at fault
   (d) I_{DDQ} fault
   (e) Multiple stuck-at fault
   (f) Path-delay fault
   (g) Potentially-detectable stuck-at fault
   (h) Stuck-open fault
   (i) Stuck-short fault
   (j) Transition fault

4.7 *Fault indistinguishability.* Derive Equation 4.4 from Equation 4.3.

4.8 *Functional equivalence.* Show that the two faults c s-a-0 and f s-a-1 are equivalent in the circuit of Figure 4.12.

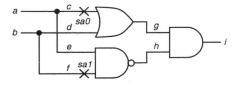

Figure 4.12: Circuit for Problem 4.8.

4.9 *Functional equivalence.* Show that in the exclusive-OR circuit of Figure 4.6, faults c s-a-1 and f s-a-1 are equivalent.

4.10 *Fault collapsing for test generation.* List the equivalence collapsed set of single stuck-at faults for the circuit of Figure 4.9. Find a minimal set of vectors to detect all faults.

4.11 *Equivalence and dominance fault collapsing.* For the circuit of Figure 4.13:

(a) What is the number of all potential fault sites?
(b) Derive the equivalence collapsed set. What is the collapse ratio?
(c) Derive the dominance collapsed set. What is the collapse ratio?

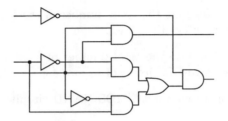

Figure 4.13: Circuit for Problem 4.11.

4.12 *Dominance fault collapsing.* For the circuit of Figure 4.14:

(a) Enumerate the minimal set of single stuck-at faults that must be tested according to the Checkpoint Theorem.
(b) Show that the two faults d s-a-0 and g s-a-1 are equivalent.

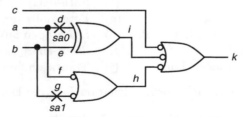

Figure 4.14: Circuit for Problem 4.12.

Part II

TEST METHODS

Chapter 5

LOGIC AND FAULT SIMULATION

> sim·u·la·tion (sĭm′yə-lā′shən) *n* **1** : the act or process of simulating : FEIGNING **2** : a sham object : COUNTERFEIT **3 a** : the imitative representation of the functioning of one system or process by means of the functioning of another ⟨a computer ∼ of an industrial process⟩ **b** : examination of a problem often not subject to direct experimentation by means of a simulating device
>
> sim·u·la·tor (sĭm′yə-lā′tər) *n* : one that simulates; *esp* : a device that enables the operator to reproduce or represent under test conditions phenomena likely to occur in actual performance
>
> — *Webster's New Collegiate Dictionary*, The G. & C. Merriam Company, Springfield, Massachusetts, U.S.A.; 1981, pp. 1074-1075.

Assume for a moment that you are a dress designer. You have designed an exquisite garment. Before you cut the cloth, you would like to make sure that the design is correct and the pieces will fit together. Once the cloth is cut, it will be too late to discover any errors in the design. Therefore, a dress designer often uses paper cuttings to verify the design. That is a form of simulation. The idea of simulation is to predict the result of building the desired object without actually building it.

There are many other examples of simulation. Scaled models of bridges and aircrafts are made and tested to verify designs. In earlier days, "breadboards" were constructed to verify the design of VLSI chips. These were printed circuit (or wire-wrap) boards with discrete components that implemented the chip's function [302]. Today, much of the scaled models and breadboards have been replaced by computer simulations. Thus, a *simulator* generally means a computer program.

5.1 Simulation for Design Verification

Simulation serves two distinct purposes in electronic design. First, it is used to verify the correctness of the design and second, it verifies the tests. The first form of simulation is illustrated in Figure 5.1. The process of realizing an electronic system

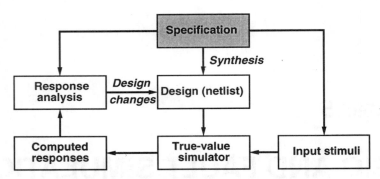

Figure 5.1: Simulation for design verification.

begins with its *specification*, which describes the input/output electrical behavior (logical, analog, and timing) and other characteristics (physical, environmental, etc.) The specification, shown as a shaded block in the figure, is the starting point for the design activity. The process of synthesis produces an interconnect of components (called a *netlist*.) The design is verified by a *true-value* simulator. True-value means that the simulator will compute the response for given input stimuli without injecting any faults in the design. The input stimuli are also based on the specification. Typically, these stimuli correspond to those input and output specifications that are either critical or considered *risky* by the synthesis procedures. A frequently used strategy is to exercise all functions with only *critical* data patterns. This is because the simulation of the exhaustive set of data patterns can be too expensive. However, the definition of "critical" often depends on designer's heuristics.

The true-value simulator in Figure 5.1 computes the responses that a circuit (if built using the netlist) would have produced if the given input stimuli were applied. In a typical design verification scenario, the computed responses are analyzed (either automatically, or interactively, or manually) to verify that the designed netlist performs according to the specification. If errors are found, suitable changes are made, until responses to all stimuli match the specification.

This simulation-based design verification method has strengths and weaknesses. Its strength lies in the details of the circuit behavior that can be simulated. For example, logic, timing, and analog behaviors can be simulated. Another advantage is in the use of hierarchy. For example, a design can be first simulated at a higher behavior level. Instead of a netlist, the design may be described in a programming language such as C. Such a description is compact, but does not contain the detailed timing information. Neither is the electrical behavior (or possible malfunction) of components considered at this level. Once this design is verified, higher-level blocks (or C language subroutines) are replaced by logic-level netlists. At this point, a logic simulator is used for verification. The process may be repeated by replacing some or all portions by transistor-level or circuit-level implementations. Simulation is used in this way for verifying very large electronic systems.

The weakness of this method is its dependence on the designer's heuristics used in generating the input stimuli. To contain the complexity, these stimuli are non-

5.1 Simulation for Design Verification

exhaustive and, therefore, a guarantee of conformance to specification is impossible. Such a guarantee is possible with a *formal verification method* [380], which mathematically proves the correctness of the design. A restricted form of formal verification, known as *model checking* [168, 453], verifies finite state concurrent systems by an exhaustive search of the state-space. It verifies whether a given specification is true. According to Clarke et al. [168], an efficiently implemented model checking procedure will always terminate with a yes/no answer and can be run on moderate-sized machines, though not on an average desktop computer. Thus, the high complexity of formal methods allows their use only at the higher behavior level. In spite of the incompleteness, simulation provides a better check on the manufacturability of the design. An ideal system of design verification should combine the behavior-level formal verification with the logic and circuit-level simulation.

Example 5.1 *Simulating an adder circuit: Consider a combinational logic circuit designed to add two 32-bit binary integers. This circuit has 64 binary inputs and 33 outputs. To completely verify the correctness of the implemented logic, we must simulate 2^{64} input vectors and check that each produces the correct sum output. This circuit may have about 200 gates and a fast logic simulator may require $1\mu s$ to simulate one vector. The time required to complete the simulation is:*

$$\frac{2^{64} \times 10^{-6}}{3600 \times 24 \times 365} \approx 584,942 \ years$$

This is clearly impractical. So, the designer must simulate some selected vectors. For example, one may add pairs of integers where both are non-zero, one is zero, and both are zero. Then add a large number (say, 10^6) randomly generated integer-pairs. Such heuristics, though they seem arbitrary, can effectively find many possible errors in the designed logic. The next example illustrates a rather simple heuristic.

Example 5.2 *Design verification heuristic for a ripple-carry adder: Figure 5.2 shows the logic design of an adder circuit. The basic building block in this design is a full-adder that adds two data bits, A_n and B_n, and one carry bit, C_n, to produce sum and carry outputs, S_n and C_{n+1}, respectively. For logic verification, one possible strategy is to select a set of vectors that will apply all possible inputs to each full-adder block. For example, if we set $A_0 = B_0 = 0$ and apply all four combinations (00, 01, 10, 11) to A_1 and B_1, and then set $A_0 = B_0 = 1$ and again apply the four combinations to A_1 and B_1, these eight vectors include all eight inputs for the $FA1$ block. Logic simulation of these vectors will thoroughly check $FA1$, since the states of all three outputs S_1 and C_1 are provided by the simulator. One significant advantage of simulation is that all internal signals of the circuit can be examined. This reduces the complexity of verification. Table 5.1 gives a set of eight vectors for verifying a 4-bit ripple-carry adder. Interestingly, the regular pattern in each vector allows us to expand the width of the vector, without increasing the number of vectors, for applying this heuristic to an adder of arbitrarily large size.*

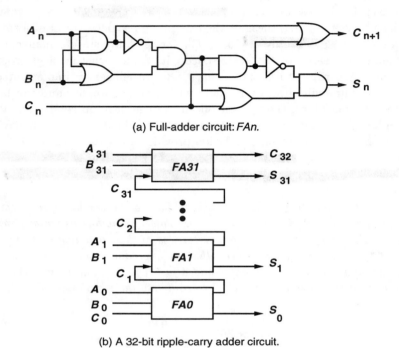

Figure 5.2: Logic design of a 32-bit ripple-carry adder.

Table 5.1: Design verification vectors for a 4-bit ripple-carry adder circuit.

Vector no.	Bits: $C_0 A_0 B_0 A_1 B_1 A_2 B_2 A_3 B_3$	Input $C_n A_n B_n$ to FAn
1	000000000	000 applied to all FAs
2	001010101	001 applied to all FAs
3	010101010	010 applied to all FAs
4	011001100	011 applied to FA0, FA2 & 100 applied to FA1, FA3
5	100110011	100 applied to FA0, FA2 & 011 applied to FA1, FA3
6	101010101	101 applied to all FAs
7	110101010	110 applied to all FAs
8	111111111	111 applied to all FAs

5.1 Simulation for Design Verification

The vectors of Table 5.1 have been derived without any regard for testing the stuck-at faults we discussed in the previous chapter. These vectors aim at verifying the functional correctness of the design via true-value simulation. Once the circuit is verified, manufacturing tests are derived. These can be based on the verification vectors. The stuck-at fault coverage of verification vectors may or may not be always high. In this example, the first seven vectors cover all stuck-at faults. One may, therefore, use only the first seven vectors in the manufacturing test.

In general, one must select an adequate set of vectors from the verification vectors using a fault simulator. Verification vectors can be *compacted* [273] by removing vectors that do not increase fault coverage. If the coverage is found to be inadequate (stuck-at fault coverage of 95 to 100% is desirable) then fault coverage vectors are added using test generation algorithms discussed in later chapters. Still the importance of verification vectors in testing should not be overlooked. In the above examples, we only considered the verification of the logic function. However, a complete verification must also consider the timing behavior of the design. The following example illustrates how we can verify a design, or test a device, for timing.

Example 5.3 *Timing verification of the ripple-carry adder: In an ideal sense the logical correctness means that the circuit will produce the correct result if signals were applied at a slow rate. In practice, however, hardware elements (e.g., gates and interconnects) have delays. Special tests are necessary to verify or test the delay of a designed or manufactured circuit. A delay test must propagate a signal change through the longest combinational path. For the four-bit ripple-carry adder, the longest path begins at C_0 and ends at C_4. It is tested by rippling a $0 \rightarrow 1$ transition through this path. One possible test consists of two vectors of Table 5.1: vector number 2, followed by vector number 6. Another vector-pair with a similar property is vector number 3, followed by vector number 7. For a timing verification, at least one of these vector-pairs must be simulated using a true-value simulator, which also models delays. These vector-pairs, though desirable for manufacturing test also, may not be included if one simply selects vectors for a 100% stuck-at fault coverage.*

Generation of verification vectors for sequential circuits is a complex problem for which numerous heuristics are used. Designers often select critical functions from the system specification. However, a suitable criterion for the completeness is difficult to find. One simple heuristic, suggested in the literature [366, 492], requires the verification vector sequence to execute all transitions of the state diagram. Although the worst-case length of such a sequence (also called a *transition tour*) can be very large, in practice, sequences of average length $n \cdot \log n$ can be generated for state machines with n states. One difficulty with this method is that the specification of a system often gives only an incomplete transition graph. The fault coverage of the verification vector sequence in such cases turns out to be low for specific implementations in which the unspecified (or *don't care*) transitions have all been specified. It then becomes necessary to augment the verification vectors [348].

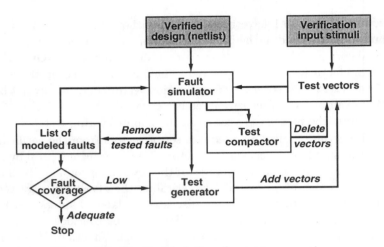

Figure 5.3: Fault simulation for test generation.

5.2 Simulation for Test Evaluation

Another type of simulator, known as a *fault simulator*, is used for the development of manufacturing tests. A fault simulator incorporates many of the details of a true-value simulator. In addition, it also has the capability to simulate the circuit response in the presence of faults.

A typical application of a fault simulator in the digital design process is illustrated in Figure 5.3. Fault simulation is normally done after the design has been verified and the verified circuit netlist and verification vectors are available. These two inputs to the fault simulator are shown as the shaded boxes in Figure 5.3. The fault simulator performs two functions:

1. It determines the coverage of a given set of input stimuli (vectors) for a given fault model or a given fault list.

2. With the help of other programs (a test generator or a vector compacter [273]), it can produce a set of vectors with a given fault coverage for manufacturing test.

If no fault list is supplied, the fault simulator generates one for the specified fault model. For example, it may make a list of all single stuck-at faults by using the equivalence fault collapsing as discussed in Chapter 4. It then simulates the initially supplied verification vectors. Simulation produces lists of faults, detected by each vector. The detected faults are deleted from the fault list. *Fault coverage*, defined as the ratio of the number of detected faults to that of faults in the initial fault list, is computed. Once an adequately high fault coverage (98-100%) is reached, the simulation may be stopped. If there are unsimulated vectors, they can be removed to reduce the vector set. On the other hand, if the vectors do not produce an adequate fault coverage, some of the undetected faults from the fault list can be

5.2 Simulation for Test Evaluation

given to a test generator program (see Chapter 7 on combinational and Chapter 8 on sequential ATPG) to produce new vectors.

It is also possible to use the fault simulation result to remove those vectors that do not detect any new faults not already covered by previously-simulated vectors. Thus, a set of vectors can be *compacted* without reducing the fault coverage. Such deletion of unproductive vectors is easily done for combinational circuits and is often used effectively for *random test generation* as described in Chapter 7 on combinational ATPG. For sequential circuits, however, deletion of a vector will modify the state of the circuit and thereby change the fault detection behavior of subsequent vectors. There are sophisticated test compaction algorithms that can achieve as much as 50% or greater reduction of the vector set without reducing the fault coverage [273]. Test compacters are discussed briefly in Chapter 7.

Example 5.4 *Fault simulation: Consider a four-bit ripple-carry adder circuit of the type shown in Figure 5.2. This circuit contains four full-adder blocks of Figure 5.2(a.) It has 36 logic gates, 9 primary inputs, and 5 primary outputs. A fault simulator [160] makes a list of 186 single stuck-at faults. Equivalence fault collapsing (see Chapter 4) option in the simulator reduces it to 114 faults. The result of fault simulation with vectors of Table 5.1 is shown in Table 5.2. We make several observations:*

- *The last vector does not detect any new fault. We can use the first six vectors for testing the circuit without any reduction in the fault coverage. Even though the last two vectors were necessary for the design verification heuristic of the last section, they are found to be redundant for testing. For large circuits, it is not unusual to have several million vectors in the verification vector set. A fault simulator can significantly reduce the size of the vector set, thus reducing the time required for the manufacturing test.*

- *The collapsed fault set is about 40% smaller than the set of all faults. A 40-50% reduction in the fault list is quite typical.*

- *Although the coverages of collapsed and uncollapsed fault lists differ, they are quite close. The CPU time for simulation of all faults on a SUN Sparc 2 computer was 33 ms, and reduced to 16 ms for the collapsed fault set. For large circuits, the saving in the CPU time may be significant.*

In general, when fault detection *is* desired, a collapsed fault list is used for the coverage analysis and for test generation. The only exception is made when fault diagnosis is the objective (see Chapter 18.) By definition, detection means finding the presence of any set of faults. So, if the collapsed list is used, the presence of a fault means that any of the equivalent faults could be present. The purpose of diagnosis, on the other hand, is to identify the specific fault that has occurred.

A possibility, not illustrated by the above example, is that the coverage of verification vectors can also be lower than 100%. For very large sequential circuits, coverages in the 70-75% range are not uncommon. The vectors can still be compacted

Table 5.2: Result of fault simulation of four-bit adder with Table 5.1 vectors.

Vector number	186 Uncollapsed faults		114 Collapsed faults	
	Detected	Coverage	Detected	Coverage
1	61	33%	37	32%
2	113	61%	65	57%
3	125	67%	77	68%
4	143	77%	89	78%
5	162	87%	102	89%
6	186	100%	114	100%
7	186	100%	114	100%
8	186	100%	114	100%

Table 5.3: Reverse-order simulation of vectors 1-6 of Table 5.2.

Vector number	114 Collapsed faults	
	Detected	Coverage
6	45	40%
5	73	64%
4	98	86%
3	114	100%
2	114	100%
1	114	100%

either by a simple truncation or by other techniques [273]. However, the 25-30% undetected faults identified by the fault simulator are used to generate more vectors to enhance the fault coverage.

Example 5.5 *Reverse-order simulation:* *In the previous example, we found that the last two vectors (below the line after vector 6 in Table 5.2) can be dropped without reducing the fault coverage. We thus compacted the vector set from 8 to 6. Reverse-order simulation is a simple technique for further compacting the vector set of a combinational circuit. Note that vector 6 is essential since it detects some faults that could not be detected by all 5 previous vectors. It is, however, possible that vector 6 could detect faults covered by some previous vectors, making those vectors unnecessary. This could not be determined earlier because of "fault dropping" by the simulator. That is, a fault detected by a vector is immediately dropped from consideration and is not simulated for the following vectors. So, we will simulate vector 6 first. Similarly, we can argue that vector 5 could also detect faults covered by some earlier vectors. A simple way to find that out is to simulate the vectors in the reverse order. The result is shown in Table 5.3. We can now use four vectors above the line in this table for a 100% coverage. Unfortunately, the reverse-order simulation cannot be used for sequential circuits.*

Table 5.4: A summary of modeling levels and simulators.

Modeling level	Circuit description	Signal values	Timing resolution	Application
Function, behavior, or RTL	Programming language-like HDL	0, 1	Clock boundaries	Architectural and functional verification
Logic	Connectivity of Boolean gates, flip-flops and transistors	0, 1, X and Z	Zero-delay, unit-delay or multiple-delay	Logic verification and test
Switch	Transistor connectivity and sizes, and node capacitances	0, 1, and X	Zero-delay	Logic verification
Timing	Transistor connectivity and tech. data; node capacitances	Analog voltage	Fine-grain time	Timing verification
Circuit	Connectivity of active and passive components; tech. data	Analog voltage and current	Continuous time	Digital timing and analog circuit verification

5.3 Modeling Circuits for Simulation

The main inputs to a simulation program are: (1) a circuit description, (2) stimuli, and (3) user commands. In this section, we will discuss the first of these. The circuit is described using a *hardware description language* (HDL.) Popular HDL's are Verilog [667] and VHDL [171]. The readers are advised to familiarize themselves with these or any other HDL used in their work environment. The discussion in this section is kept independent of any specific language.

5.3.1 Modeling Levels and Types of Simulators

The *level* of a simulator is related to the detail of modeling. Typical levels are *function* (or *behavior*), *logic*, *switch*, *timing*, and *circuit*. A summary of various levels, their modeling characteristics, and applications is given in Table 5.4. We will define these levels in this subsection, but our discussion in this chapter will mainly focus on the logic level. This is because logic and switch levels are the only ones where fault analysis has been done. Today's fault simulators work predominantly at the logic-level, although the circuit model may be a mixture of function, logic and switch level modules.

Function or Behavior Level. This is the highest level. Only the architecture (an interconnect of functional modules such as adders, multiplexers, decoders, multipliers, registers, etc.) is considered. The functions of the modules are expressed in programming language constructs. No implementation details are assumed to be known. Although no processing or propagation delays of signals are considered, signal changes are normally assumed to be synchronized with respect to some clock signal. This makes the simulation of a module similar to an input to output data transfer, even though there may not be explicit registers at the ports of the module. Therefore, this level is sometimes referred to as the *register-transfer level* (RTL.) A significant advantage of this level is that stimuli generation and response checking functions can also be blended with the circuit model as additional HDL code. The code that generates stimuli is referred to as *testbench* [381].

Logic Level. The circuit is modeled as an interconnect of Boolean gates. In addition to the standard gates (AND, OR, NAND, NOR, and NOT), sometimes complex gates such as XOR are also considered building blocks. Sequential elements, such as flip-flops, can be modeled as gate-level interconnects, or as stand-alone functional blocks. In order to correctly simulate MOS circuits, most logic simulators also accept transmission gates (or pass transistors) and MOS buses as valid elements.

Switch Level. Switch-level simulation was invented by Bryant [103]. Here, the entire circuit is modeled as an interconnect of MOS transistors, which are treated as ideal switches. Any circuit node that has a conducting path to the supply voltage, V_{DD}, assumes a logic 1 state. Similarly, any node with a conducting path to *ground* assumes a logic 0 state. The states of a set of floating nodes connected among themselves, but having no connection either to V_{DD} or to *ground*, are determined based on charge sharing among the node capacitances. The determination of the state of a node that has conducting paths to both V_{DD} and *ground* is more tricky. Actual sizes (channel lengths and widths) of transistors are used to determine the state of such nodes. Thus, the circuit data must also contain transistor sizes and node capacitances. Switch-level simulation can accurately simulate the logic behavior of many MOS circuits that have no exact logic gate structure. However, no timing information is considered at the switch-level. These simulators have been used for logic verification of high-performance full-custom microprocessor and *digital signal processor* (DSP) chips. Switch-level techniques have also been used to simulate stuck-at and stuck-open faults [578]. However, the fault-effect propagation procedures become too complex at this level and their use in the industry has been limited. Most logic-level fault simulators, contain a limited capability of simulating switch-level components, as we pointed out above [404].

Timing Level. This level uses a circuit description similar to that used for the switch-level simulation. That is, the connectivity of transistors, their sizes and types, and node capacitances are needed. In addition, technology data specifying the transistor voltage-current characteristics are also used to compute charging or discharging currents for the nodes. Thus, a timing simulator computes node voltages as a function of time (or, in practice, at closely spaced discrete times) [145]. Also, to speed up the simulation, the circuit can be partitioned into *channel-connected*

5.3 Modeling Circuits for Simulation

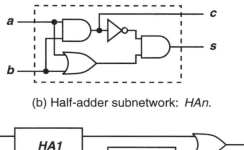

(b) Half-adder subnetwork: *HAn*.

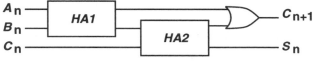

(a) Full-adder subnetwork: *FAn*.

Figure 5.4: Hierarchical description of a full-adder circuit.

components (see Example 5.6.)

Circuit Level. This is the lowest level and represents the ultimate in accuracy for the simulation of electronic systems. The circuit is assumed to be composed of electrical elements such as resistors, capacitors, inductors, and transistors. Equations relating branch or loop currents and node voltages are developed and solved by numerical methods [446]. These simulators are vital tools for the design of analog integrated circuits. For digital circuits, they are used where MOS structures cannot be accurately simulated by a logic or switch-level simulator. One area where a circuit simulator is indispensable is where high timing accuracy is desired. So, the critical path of a large digital chip may be simulated by a circuit simulator. Library cells used in the design of standard cell ASICs (*application specific integrated circuits*) are generally characterized by a circuit level simulator. The SPICE simulator [485], originally developed at the University of California, Berkeley, and its numerous derivatives have been used in the industry for the last 25 years.

5.3.2 Hierarchical Connectivity Description

The description of a circuit consists of the names of input and output signals, and the function of the circuit. The function is generally described as an interconnection of simpler building blocks or subcircuits. The subcircuits serve a similar purpose as *subroutines* in a program. These are again described by their input and output signal lists, and the function. The function of a subnetwork may be described as an interconnection of simpler subnetworks. In this hierarchy, the lowest level consists of Boolean gates, flip-flops, tristate gates (transmission gates or pass transistors), and buses.

Consider the example of the 32-bit adder circuit of Figure 5.2. The circuit description will contain lists of 64 (or 65 if C_0 is an input) input and 33 output signals. It will also contain a connectivity description for 32 modules of full adder (*FAn*) type. The *FAn* may be described via a gate-level connectivity as shown in

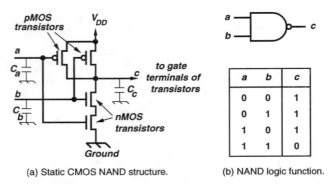

Figure 5.5: CMOS implementation of a NAND logic gate.

Figure 5.2(a), or it may use further hierarchy. The latter case is shown in Figure 5.4, where the *FAn* is built with two half adders or *HAn*. The subnetwork *HAn* is shown in Figure 5.4(a.) Lower level subnetworks, which are frequently reused, are generally kept in *standard cell libraries*. Most HDLs [171, 667] used to describe circuits support hierarchy. Circuit connectivity description without hierarchy is usually referred to as the *flat description*. A flat description of the 32-bit ripple-carry adder of Figure 5.2(b) contains 288 logic gates.

5.3.3 Gate-level Modeling of MOS Networks

In today's semiconductor technology, digital logic functions are realized by MOS (*metal-oxide semiconductor*) transistors [718]. A MOS transistor is a three-terminal device. Two terminals, *source* and *drain*, form a semiconductor *channel*. The third terminal, *gate*, controls the conductivity of the channel. The gate is insulated from the channel and represents a capacitive load to the controlling signal source. There are two types of MOS transistors. For an nMOS device, the high state (usually, V_{DD}) of the gate makes the channel conductive, causing almost a short between source and drain. In this state, the *on-resistance* of the channel depends on the dimensions or size (length and width.) For the low state (usually, *ground*), the channel acts as an open-circuit. The pMOS device works in a complementary manner.

Figure 5.5 shows a NAND gate in the *complementary-MOS* (CMOS) design style. Signals a and b feed into high impedances since they are connected to the insulated gates of transistors. The sources of signals a and b only see equivalent capacitive loads, C_a and C_b, respectively. We assume that the output signal c feeds only to the gate terminals of some other transistors. This is modeled by the total capacitance C_c of node c. As usual, we denote the voltage level V_{DD} as logic 1 and the ground voltage as logic 0. When $a = b = 1$, the output c is grounded through the two nMOS devices. This offers a ground voltage to all gates connected to c. When either one among a or b is 0, c is isolated from ground and is connected to V_{DD} through one or both pMOS devices. Thus, the capacitor C_c is charged to V_{DD}. In either state of c, only a transient charging current flows. Once the study-state is reached, no current

5.3 Modeling Circuits for Simulation

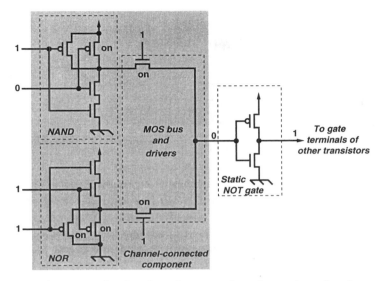

Figure 5.6: An example showing the necessity of transistor-level modeling.

is drawn from the supply (neglecting the quiescent or leakage current, I_{DDQ}.) In this specific configuration, the transistor implementation, called a static CMOS gate, can be modeled as an ideal Boolean gate of NAND type. Similar static designs of NOR and NOT gates are also used. The following example shows that not every CMOS structure can be mapped onto a Boolean gate.

Example 5.6 *Necessity of transistor-level models: Consider the MOS circuit of Figure 5.6. The outputs of the NAND and NOR gates are connected to a bus via two nMOS driver transistors. In a normal operation of the circuit, it is usually intended that only one driver will be turned on. However, due to signal delays, it is possible that both drivers are momentarily turned on. They may also be turned on due to design errors. These errors can be detected by simulation provided the modeling is correct. For the signal values shown in Figure 5.6, there is a conducting path formed by transistors shown as "on." If we assume that all devices have the same on-resistance, then because of the two parallel transistors in the NOR gate the output to ground resistance will be lower than the V_{DD} to output resistance. Depending on the threshold voltage of transistors in the NOT gate, the output of the MOS bus may be interpreted as a logic 0.*

Correct simulation requires partitioning of the circuit into "channel-connected components" [27, 103, 145]. A channel-connected component is an arbitrary interconnection of MOS transistors with three types of external connections: (1) inputs feeding only into gate-terminals of transistors in the component, (2) outputs feeding only into gate-terminals of transistors in other components, and (3) connections to V_{DD} or ground. States of nodes inside a component are determined by analyzing the node capacitances and conducting paths between V_{DD} and ground formed by the connected channels of "on" transistors. Thus, for given states of input signals, the

96 Chapter 5. LOGIC AND FAULT SIMULATION

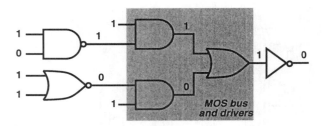

Figure 5.7: An erroneous gate-level model of an MOS bus.

outputs of a channel-connected component can be determined without considering the rest of the circuit. The circuit of Figure 5.6 has two channel-connected components, one in the shaded area and the other, the NOT gate. Whenever a Boolean gate equivalent MOS structure forms a channel-connected component, it can be safely modeled as a logic gate. The NOT gate satisfies that condition. However, that is not the case with NAND and NOR gates, and the MOS bus. The three elements are, therefore, modeled as one channel-connected component.

Figure 5.7 shows an often-used logic model for the MOS bus. Since the bus signal is produced by an OR gate, it is also called a wired-OR *structure. If both control inputs are turned on, as is shown in Figure 5.7, the 1 input will dominate. This models the situation in which the* pull-up *path through the MOS circuit has lower impedance than the* pull-down *path. Similarly, a wired AND or 0-dominating bus can be modeled. In some typical MOS circuits, the dominance can dynamically change. The logic model always works correctly if the bus block is exactly like a two-to-one (many-to-one, in general) multiplexer, i.e., only one control input is turned on at a time. That is why, for the values shown in the figure, the output is different from that obtained in Figure 5.6.*

The preceding example shows that in order to correctly simulate MOS circuits, a logic simulator should have some switch-level modeling capability. Most modern simulators are mixed-mode types, so that large VLSI circuits can be economically modeled [27]. Typically, large embedded blocks of memory are modeled as functional modules. Random logic is commonly modeled at the logic gate-level using hierarchy. Within random logic some portions involving buses and tristate drivers would be modeled at the switch-level. Some blocks, such as adders, multipliers, multiplexers, registers, etc., for which automatic synthesis programs are available, may be originally modeled at the function-level and then, as the design progresses, replaced by their lower-level implementations.

5.3.4 Modeling Signal States

Purely combinational logic can be modeled with two states, usually denoted as [0,1] or [*low, high*]. When the circuit is sequential, it can have internal states, which are usually assumed to be unknown at the beginning of the simulation. This assumption is also a good representation of the reality, because when a circuit is

5.3 Modeling Circuits for Simulation

Table 5.5: Truth tables for three-state logic.

Inputs		Output		
a	b	AND (ab)	OR $(a+b)$	NOT $(\bar{a})$
0	0	0	0	1
0	1	0	1	1
0	X	0	X	1
1	0	0	1	0
1	1	1	1	0
1	X	X	1	0
X	0	0	X	X
X	1	X	1	X
X	X	X	X	X

powered up, the flip-flops can be in any of the two possible states. In fact, the purpose of the initial vectors is to bring the circuit to a known state. Thus, a simulator starting with all flip-flops in the unknown state also verifies whether the applied vectors can bring the circuit to a known state.

There are additional reasons, as we will discuss in the next subsection, which can set a flip-flop into an ambiguous state even after it had been initialized. Therefore, for simulation of sequential circuits it is essential that we represent signals by three states: 0, 1, and *unknown*, usually denoted as X. Table 5.5 gives the three-state truth tables for two-input AND, OR, and NOT gates. All other gates (NAND, NOR, XOR, etc.), or gates with larger numbers of inputs, can be implemented using these three types and thus their truth tables can be derived.

The three-state logic is pessimistic. In some cases, it cannot determine a definite 0 or 1 value for a signal, even though that value can be easily obtained. For example, in the multiplexer circuit of Figure 5.8(a), the three-state logic of Table 5.5 determines the output to be unknown. We notice, however, that both data inputs are 1 and so the state of the control input should not matter. The output should be 1. This can be determined by the *symbolic simulation* of Figure 5.8(b.) Here, X is treated as a symbolic representation of the signal. Thus, when inverted by the NOT gate, it appears as $\overline{X}$. Similarly, the OR gate is symbolically evaluated as $X+\overline{X}=1$. While, symbolic simulation can be effective if performed locally in a circuit, so that it involves a small number of signals, it is impractical for large circuits. In general, many signals may have to be symbolically represented and the expressions can become large. In general, symbolic simulation is expensive in memory and CPU time. Therefore, three-state simulation is most common for digital logic circuits.

There are some circuits for which three logic states are not sufficient. These are the circuits that cannot be correctly modeled with Boolean gates alone. We saw an example in Figure 5.6. In such circuits, signal values 0 and 1 are applicable only to "driven" nodes. A node in a CMOS circuit is called *driven* if it has a conducting path either to V_{DD} setting the node to 1, or to *ground* setting it to 0. In steady-state, a correctly functioning CMOS gate will not have simultaneous conducting paths to

Chapter 5. LOGIC AND FAULT SIMULATION

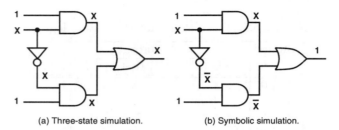

Figure 5.8: Pessimism in three-state logic and symbolic simulation.

V_{DD} and *ground*. If such simultaneous paths do occur, as is illustrated in Figure 5.6, and the state of the node cannot be determined by the simulator, then it is set to an unknown (X) state.

Now consider the case when both nMOS transistors (drivers) of the MOS bus in Figure 5.6 have been turned off by applying 0s to their gate terminals. The bus output is not in a driven state. This is defined as the *floating state*. The voltage of this node depends upon the charge stored on its parasitic capacitance. The logic value of a floating node is specified as *high impedance state* and is denoted by the symbol Z. Unlike X, which is interpreted as unknown logic state, Z is interpreted dynamically as follows:

- In CMOS circuits the leakage resistance of floating nodes is so high that they are assumed to hold their charge indefinitely. The *leakage resistance* refers to the resistance from the supply (V_{DD}) to *ground*. It consists of the resistances of transistors that are in the cutoff state. In other words, a floating node has an infinite memory. Its logic value is considered the same as its driven value before it became floating.

- If two or more floating nodes get connected due to the switching of some transistors, then the total charge on them gets shared according to their relative capacitances. All connected floating nodes assume the same voltage. Based on the switching thresholds of transistors whose gate terminals are connected to the floating nodes, the high impedance (Z) state may be interpreted as 0 or 1. If there is uncertainty about relative values of the node voltage and the switching thresholds, then Z is interpreted as X.

- If a floating node gets connected to V_{DD} or *ground* due to the switching of its drivers, then its state no longer remains Z and is changed to 1 or 0.

Most MOS circuits containing dynamic logic require four-state logic (0,1,X,Z) representation of signals.

5.3.5 Timing

In an electronic circuit, electrical parameters such as voltage and current are used to encode the information being processed. Thus, in a digital system, a high

5.3 Modeling Circuits for Simulation

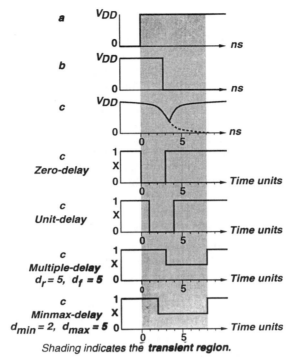

Figure 5.9: Timing models of the NAND gate of Figure 5.5.

level of voltage (i.e., close to V_{DD}) represents the logic 1 state and a low level of voltage (i.e., close to *ground*) represents the logic 0 state. In a real circuit, however, the voltage is an analog quantity, i.e., it is a continuous function of time. Therefore, it takes a finite amount of delay before a digital signal can be changed from 0 to 1, or vice-versa.

Signals experience two types of delays. The time interval between an input change (cause) and the output change (effect) of a gate is called the *inertial delay* or *switching delay*. The time interval between the generation of a signal transition at a gate output (source) and its arrival at the input of a fanout gate (destination) is known as the *propagation delay* or *transport delay*.

Let us first examine the switching delay. Figure 5.9 shows the output of the CMOS NAND circuit of Figure 5.5 for a given set of inputs (a and b), as would be obtained by various timing models. The third waveform is a simulated analog response as obtained by a circuit simulator [446] or a timing simulator [145]. We have set the time reference (somewhat arbitrarily) at the beginning of the *transient region* (shaded), which begins when the signal a rises. This creates a discharge path for C_c through the nMOS transistors, both of which are now shorted. The output signal would have dropped to almost 0 in about 5 ns as shown by the dotted curve. The exact shape and duration of this discharge waveform depends upon active (transistor V-I characteristics) and passive (C_c and other parasitics) component parameters of the circuit. The discharging of C_c is, however, interrupted by the input b dropping at 3 ns. This cuts the discharge path off and sets up a charging path through one of the pMOS transistors. As a result, the output goes up, back to V_{DD}.

Digital simulators produce quite different responses in the transient region, depending upon how delays are modeled. The last four waveforms in Figure 5.9 correspond to popular delay models:

- *Zero-delay:* All gates and interconnects are assumed to have no delay. Any signal change is entirely determined by the logic function of gates. This model is particularly useful for logic simulation of combinational circuits, i.e., circuits with no feedback and no memory states.

- *Unit-delay:* Each gate is assumed to have one unit of delay. No delay is assumed for interconnects. The unit of delay is entirely *fictitious* and has no real relationship to actual delays of the circuit. However, it serves an important purpose of maintaining the proper sequencing of signal changes (events.) Thus, circuits with feedback can also be simulated with the unit-delay model.

- *Multiple-delay:* All delays are modeled as multiples of some time unit. In Figure 5.9, this unit is 1 ns. It can be much smaller for a very high speed circuit or if greater timing accuracy is desired. Each gate is assigned a rise-delay (d_r) and a fall-delay (d_f) [27]. These are the delays with which the gate output will rise or fall, after the input change that caused the output to change. Determined from circuit parameters, the rise and fall delays of a gate can have two different values. In Figure 5.9, $d_r = d_f = 5$. So the output c should fall at time unit 5 after the rise of a. However, before that happens, i.e., at time 3, b falls causing c to rise again. Since $d_r = 5$, this rise can only occur 5 units after the fall of b. Thus, c definitely attains a logic 1 state at that time. A logic simulator will usually evaluate c at least for a portion of the transient region to be in the unknown state (X.) In Figure 5.9, this is displayed as a level half way between 0 and 1. Notice that the multiple-delay waveform approximates the behavior of the analog waveform. A more detailed multiple-delay model would specify separate rise and fall delays for each pair of input-output port of a gate or a multioutput module.

- *Minmax-delay:* Process variations in the manufacture of VLSI circuits make the exact characterization of switching delays difficult. To account for such variations, delays are often characterized statistically. Thus, the switching delay of a gate can be specified with two limits, d_{min} and d_{max}. It is also possible to give the min-max range for both rise and fall delays [79]. In Figure 5.9 (see last waveform), we assume equal rise and fall delays with $d_{min} = 2$ and $d_{max} = 5$. The rise of a at time 0 causes an ambiguity interval $(2, 5)$ in the output. This means that the output signal at c could fall any time in this interval. Similarly, the fall of b at time 3 causes an ambiguity interval $(5, 8)$. The actual ambiguity interval, $(2, 8)$, is the "union" of these two intervals.

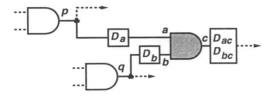

Figure 5.10: Propagation and switching delays for a gate.

The output appears as unknown (X) during this interval. The minmax-delay model, though pessimistic, allows effective debugging of large VLSI circuits.

Interconnects of an electronic circuit are like transmission lines. A signal transition at the output of a gate would have to travel through fanout branches of varying lengths before arriving at the inputs of destination gates. Thus *propagation delays* are usually specified for each gate-output and gate-input pair. Rise and fall delays are often considered the same, although either the multiple-delay or the minmax-delay model can be used. It is convenient to specify the propagation delay as *input delays*, separately for each gate input.

Figure 5.10 shows the complete set of delays specified for a two-input gate with inputs a and b, and output c. The signal for a is generated at p and that for b, at q. Four delays for this gate are:

1. Input delay D_a is the propagation delay for the interconnect $p \to a$.
2. Input delay D_b is the propagation delay for the interconnect $q \to b$.
3. Output delay D_{ac} is the switching delay for an output change caused by a change at a.
4. Output delay D_{bc} is the switching delay for an output change caused by a change at b.

In general, each of the four delays can have separate rise and fall components, thus requiring up to eight numbers to specify delays for a gate.

An alternative approach is to consider *fanout nets* as circuit elements just like gates. A fanout net simply transmits the signal from the stem to each branch with some delay. Thus, a net with k branches will have k delays, one for each fanout. In this case, gates will only have the switching delays.

5.4 Algorithms for True-Value Simulation

Simulation of an electronic circuit is the process of computing its signals as a function of time. Time is a continuous parameter and that is the way it is treated for analog signals. For most digital circuits, only certain discrete values of signals are meaningful. Therefore, digital circuit simulators tend to either ignore the fine-grain variations (transients) between those meaningful values or model the transients

- Step 1: Levelize circuit and produce compiled code
- Step 2: Initialize data variables (flip-flops and other memory)
- Step 3: For each input vector
    Set primary input variables
    Repeat until steady-state or maximum iteration-count reached
        Execute compiled-code
        Report or save variable values

Figure 5.11: Pseudocode for a compiled-code simulator.

approximately. Thus, the time is assumed to advance in discrete "jumps" and signals acquire values from a set of meaningful values. The change of a signal from one value to another is defined as an *event*. Once the time is discretized, the simulator basically computes the events occurring in the circuit as a result of the applied primary input signals (source events.) For this reason, digital simulation is also known as the *discrete event simulation*.

5.4.1 Compiled-Code Simulation

In this method of simulation the circuit is described in a language that can be compiled and executed on a computer. The circuit description, normally in an HDL such as VHDL [171] or Verilog [667], is levelized and converted into a programming language. *Levelization* ascertains that the evaluation of a signal is preceded by the evaluations of its sources. In circuits with feedback, global levelization may not be possible. However, portions can be locally levelized. Levelization is described in Chapter 6, Algorithm 6.1.

Signals are treated as variables in the code and can be typed (Boolean, integer, etc.) suitably. Functions such as AND, OR, etc., are directly converted into program statements. High level functions, such as memory blocks, adders, multipliers, etc., are modeled as subroutines. Flip-flops are modeled with internal states as data variables.

For every input vector, the code is repeatedly executed until all variables have attained steady values. Such iterations are necessary for circuits that have feedbacks. Figure 5.11 gives the pseudo-code for a compiled-code simulator.

Compiled-code simulators are very effective where two-state (0,1) simulation suffices. Otherwise, the larger number of signal states makes the code complex. The two-state simulation is useful for combinational logic and for sequential logic which is already initialized. Another scenario is the high-level design verification where the circuit is described using functional modules and stimuli are generated by testbenches. Because the code execution can be very fast on a computer, such simulators are capable of high speed.

Large gate-level circuits present several problems. First, timing problems –

5.4 Algorithms for True-Value Simulation

glitches, race conditions, etc. – are not modeled in a compiled-code simulator. The simulator does detect oscillations when the iterations do not converge, but it is difficult to deal with the situation unless the unknown (X) state is available. The second problem is that of the inefficiency incurred by the evaluation of the entire code when only a few signals may be changing. In digital circuits, generally only 1-10% of signals are found to change at any time. For this reason, event-driven simulators (discussed in Subsection 5.4.2) run much faster at the gate-level. Finally, any detailed timing (such as multiple-delay or minmax-delay) is almost impossible to simulate in the compiled-code. For these reasons, the use of compiled-code simulators is usually limited to high-level design verification.

5.4.2 Event-Driven Simulation

Event-driven simulation [683, 684] is a very effective procedure for discrete-event simulation. It is based on the recognition that any signal change (event) must have a *cause*, which is also an event. Thus, an event causes new events, which in turn may cause more events. An event-driven simulator follows the path of events. Consider a circuit at the gate-level. Suppose, all signals are in steady-state when a new vector is applied to primary inputs. Some inputs change, causing events on those input signals. Gates whose inputs now have events are called *active* and are placed in an *activity list*. The simulation proceeds by removing a gate from the activity list and evaluating it to determine whether its output has an event. A changing output makes all fanout gates active, which are then added to the activity list. The process of evaluation stops when the activity list becomes empty.

An event-driven simulator only does the necessary amount of work. For logic circuits, in which typically very few signals change at a time, this can result in significant savings of computing effort. However, the biggest advantage of this technique is in its ability to simulate any arbitrary delays. This is done by a procedure known as *event scheduling*.

Suppose the evaluation of an active gate generates an event at its output. If the gate has a switching delay of Δ units, then the event should take effect Δ time units later. For correctly considering the effects of delays, the simulator distributes the activity list in time. *Event scheduling* is the procedure of distributing the activity caused by events over time according to the specified delays. The following example illustrates event-driven simulation with circuit delays.

Example 5.7 *Event-driven simulation: Figure 5.12 shows a combinational circuit with an input vector $a = b = c = 1$. The gates and their output signals are given the same name. For single-output gates this does not cause any confusion. All gates are assumed to have fixed delays whose value are specified on the gates. Gates have the same rise and fall delays and interconnects have no additional delays. The signal values shown in the figure are the steady-state values for the initial vector. We will denote the simulated time by variable t, which can take discrete values of 0, 1, 2, etc. These values are assigned to the slots of a time stack. Attached to each time*

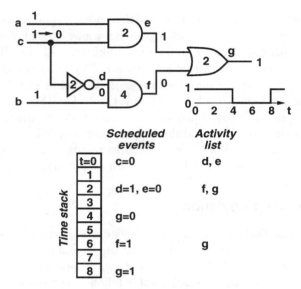

Figure 5.12: An example of event-driven simulation.

slot there is an event list of scheduled events and an activity list. We will perform the simulation for an input event $1 \to 0$ on c occurring at $t = 0$:

- $t = 0$. Add $c = 0$ to the event list. Place fanouts d and e in the activity list. Remove d from activity list and evaluate it. Since the delay of the NOT gate is 2 units, add the new event $d = 1$ to the event list of time slot $t = 2$. Similarly, e is removed from the activity list and its evaluation produces an event $e = 0$ added to the event list of time slot $t = 2$, since the delay of gate e is also 2 units. This leaves the activity list empty, so we advance time.

- $t = 1$. No events. Advance time.

- $t = 2$. Scheduled events are $d = 1$ and $e = 0$. Place fanouts of d and e, i.e., gates f and g, in the activity list. Evaluations of f and g, using the currently scheduled values of d and e, produce two new events, namely, $f = 1$ scheduled at $t = 6$, and $g = 0$ scheduled at $t = 4$. Since there are no more gates in the activity list, we advance time.

- $t = 3$. No events. Advance time.

- $t = 4$. Scheduled event $g = 0$ (primary output) has no fanout. So the activity list remains empty and we advance time.

- $t = 5$. No events. Advance time.

- $t = 6$. Scheduled event $f = 1$ adds g to the activity list. Its evaluation produces event $g = 1$, which is scheduled two time units later at $t = 8$. Since there are no more gates in the activity list, we advance time.

5.5 Algorithms for Fault Simulation

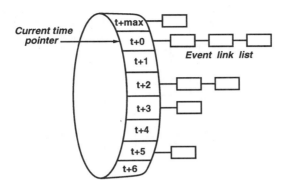

Figure 5.13: A circular stack (time wheel) for event-driven simulation.

- $t = 7$. *No events. Advance time.*

- $t = 8$. *Scheduled event $g = 1$ (primary output) produces no active gates, so we can advance time. But, there are no events scheduled for any future time and simulation stops.*

This simulation produced events for all signals in the circuit. Also, these events are arranged according to the simulated time t. Suppose that we wish to observe the output signal g, and we find that it has two events, namely, $g = 0$ at $t = 4$ and $g = 1$ at $t = 8$. Knowing its initial value of 1, we can plot the signal as a function of time. This is shown in Figure 5.12 near the output of the circuit.

In the above example we notice that the time stack would continue to grow as simulation progresses. This can be avoided. We make two key observations: For any "current" time, all new events are scheduled only at future times. Also, the difference between the current time and the farthest time at which we can schedule an event is limited by the maximum delay a gate has in the entire circuit. Thus, in an efficient implementation all event scheduling occurs relative to the current time. We can view the time array as a circular stack, often called the *time wheel* [683, 684]. This is shown in Figure 5.13. The number of slots in the time wheel equals $max + 1$, where max units is the largest delay experienced by any event (gate + interconnect) in the circuit. For advancing the time, the current time pointer is moved to the next time slot, which then becomes $t + 0$ and the times of all other slots advance by one unit. All immediate past events attached to the last slot the time pointer just moved from are saved in the result file before designating that slot as time $t + max$.

5.5 Algorithms for Fault Simulation

As outlined in Section 5.2, a fault simulator must classify the given target faults in a circuit as *detected* or *undetected* by the given stimuli. The basic problem is illustrated in Figure 5.14. The block C( ) is the fault-free circuit and blocks C(f1) through C(fn) are copies of the same circuit with faults f1 through fn permanently

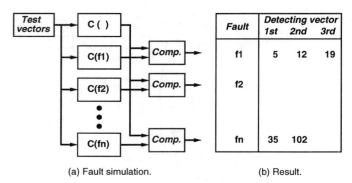

(a) Fault simulation.   (b) Result.

Figure 5.14: Fault simulation problem.

inserted. The same vectors are applied to all blocks and the outputs of the faulty circuits are compared in the comparators shown as *Comp*. All signal lines in Figure 5.14(a) are buses that provide parallel access to all primary inputs and outputs. If any comparator shows a mismatch, the corresponding fault is noted as detected by the vector being simulated. Thus, the simulator records the numbers of vectors that detect a fault. For example, in Figure 5.14(b), fault f1 is detected by vectors 5, 12, and 19. Fault fn is detected by vectors 35 and 102, and f2 is not detected by any vector. In addition, the simulator may also record the specific output at which a fault is detected. While the multiple detection of a fault, as shown in Figure 5.14(b) is desirable for fault diagnosis (discussed in Chapter 18), it is expensive in computation time. Therefore, when fault fn is detected for the first time by vector 35, the simulation of block C(fn) is suspended beyond that vector. This procedure, known as *fault dropping*, considerably speeds up the fault simulation process.

We notice that, except for fault dropping, the effort of simulating n faults is equivalent to either simulating a circuit that is n times larger, or repeating the original true-value simulation n times. The algorithms in the following subsections attempt to reduce this effort.

5.5.1 Serial Fault Simulation

This is the simplest algorithm for simulating faults. The circuit is first simulated in the true-value mode for all vectors and primary output values are saved in a file. Next, faulty circuits are simulated one by one. This is done by modifying the circuit description for a target fault and then using the true-value simulator. As the simulation proceeds, the output values of the faulty circuit are *dynamically* compared with the saved true responses. The simulation of a faulty circuit is stopped as soon as the comparison indicates detection of the target fault. All faults are simulated serially in this way.

A serial fault simulator repeatedly uses a true-value simulator. Its implementation is, therefore, simple. It can simulate any fault that can be introduced in the circuit description. Thus, apart from the stuck-at and stuck-open types, bridging, delay, and analog faults can also be simulated. For this reason, analog circuit faults are usually simulated by this method. Also, a serial fault simulator can easily simulate all types of fault conditions, such as fault-induced races, hazards, loss of initialization, etc., in sequential circuits, which present difficulties to other types of simulators.

For n faults, the CPU time of a serial simulator can be almost n times that of a true-value simulator. When fault-dropping is used, the CPU time can be significantly lower, especially if many faults are detected by a few earlier vectors in the set. Its memory requirement basically amounts to that of the true-value simulator. It has been successfully used with hardware simulators and in multiprocessing environments. However, as we will discuss next, there are more *intelligent* algorithms to reduce the effort of fault simulation.

5.5.2 Parallel Fault Simulation

Let us assume that the circuit consists of only logic gates and we wish to simulate stuck-at faults. Also, signals are assumed to take only binary, i.e., 0 and 1, values. Also, all gates are assumed to have the same delay. It is under these conditions that parallel fault simulation is most effective. So, each circuit in Figure 5.14 is almost identical to the fault-free circuit, except for one line that contains a stuck-at fault. In fact, the connectivity and individual gate functions are identical in all circuits. The idea of parallel fault simulation is to use the bit-parallelism of logical operations in a digital computer. For a 32-bit machine word, an integer consists of a 32-bit binary vector. A logical AND or OR operation involving two words performs simultaneous AND or OR operations on all respective pairs of bits. This allows a simultaneous simulation of 32 circuits with identical connectivity, but with possibly different signal values.

For a large number of faults, a parallel fault simulator will process $w - 1$ faults in one pass, where w bits is the machine word size. This is because one bit of the word is used for the signal value of the fault-free circuit. Thus, $w - 1$ faults are simulated in the same CPU time as that taken by the true-value simulation. If no fault-dropping is used, a parallel fault simulator will run about $w - 1$ times faster than a serial fault simulator. Both simulators will gain speed by fault dropping. In serial fault simulator, the pass is terminated as soon as the single target fault is detected. In parallel fault simulator, all $w - 1$ faults must be detected before a pass can be terminated. Therefore, the serial fault simulator gains more by fault dropping.

Example 5.8 *Parallel fault simulation: Figure 5.15 shows a circuit that is being simulated for two faults: c stuck-at-0 and f stuck-at-1. The computer has a three-bit word. To simulate the fault-free and two faulty circuits in parallel, the signal on each line is expressed as one word. The state of the left-most bit (bit 0) represents the signal value in the fault-free circuit, the middle bit (bit 1) that in the circuit with c s-a-0, and the right-most bit (bit 2) that in the circuit with f s-a-1. We apply a vector $a = 1$, $b = 1$. Since the line a and the stem b are not affected by any fault, these have the same values for all three circuits. The fault on c is present only in the*

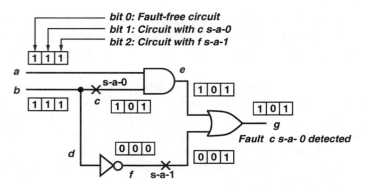

Figure 5.15: An example of parallel fault simulation.

first faulty circuit and so it affects the middle bit. When the words on a and c are ANDed, we get the three-bit word for line e. Line f is the inversion of d, hence its values, where the right bit (bit 2) is affected by the s-a-1 fault, are given by the word 001. The output g is obtained by a bit-by-bit OR of e = 101 and f = 001. Thus, g = 101. We notice that only the output of the circuit with c s-a-0 (middle bit) differs from that of the fault-free circuit (left bit.) Hence, that fault is detected. The other fault, which produces the same output as the fault-free circuit, is not detected.

A parallel fault simulator lacks the capability to model accurate rise and fall delays of signals. This is because all signal changes corresponding to several circuits must be computed together. In general, a signal may rise in one circuit while it falls in another. In a parallel simulator, combinational logic is modeled with either zero-delay or unit-delay. Sequential logic is modeled with unit-delay. Because of the idealized model of delay, the simulator must contain special algorithms to deal with race and oscillation faults [668].

Parallel fault simulators are implemented both as compiled-code or event-driven simulators. The algorithm can be extended for multiple-valued logic simulation. For example, four states (0, 1, X, and Z) can be encoded in two bits. It is necessary to use an encoding scheme so that all logic gates can be modeled with the Boolean operations available in the computer. Some of these restrictions limit the use of such simulators for VLSI circuits containing bidirectional buses and tristate gates.

A true-value logic simulator can be used as a parallel fault simulator by a simple fault injection method. The following example illustrates the method.

Example 5.9 *Fault injection for parallel fault simulation: The two faults simulated in the last example are modeled in Figure 5.16 by inserting logic gates in the circuit. A stuck-at-0 fault is modeled by inserting a two-input AND gate at the fault site (see the shaded AND gate.) The other input of this gate is permanently set to 1, except for the bit position corresponding to the fault it models. In this case, the middle bit (bit 1) is set to 0 to model the c s-a-0 fault. Similarly, the faulty circuit with f s-a-1 is modeled by inserting an OR gate in the signal f. The other input of this OR gate is set to 0, except for the right bit position (bit 2), which represents*

5.5 Algorithms for Fault Simulation

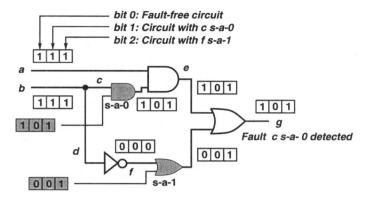

Figure 5.16: Fault injection for parallel fault simulation.

the modeled fault. Once these gates are inserted, simulation does not require any further consideration of faults, except the examination of the primary output bits for fault detection. This procedure can be extended to model a multiple fault.

The idea of parallel fault simulation was proposed by Seshu and Freeman [586, 587] and parallel fault simulators were widely used in the 1960s and 70s. Programs such as HILO [224] and TEGAS [668] were used by many companies. However, more efficient fault simulation algorithms soon emerged.

5.5.3 Deductive Fault Simulation

The circuit model used for this simulator is similar to that we discussed in the previous subsection. That is, the circuit is modeled with Boolean gates assuming either zero or unit delay for each gate, and signal states are binary (0,1) variables. In the *deductive* method [56], only the fault-free circuit (C( ) in Figure 5.14) is simulated. All signal values in each faulty circuit are deduced from the fault-free circuit values and the circuit structure. Since the circuit structure is the same for all faulty circuits, all deductions are carried out simultaneously. Thus, a deductive fault simulator processes all faults in a single pass of true-value simulation augmented with the deductive procedures. This gives the deductive simulators a tremendous speed, but only when the modeling conditions can be satisfied. Variable (separate rise and fall) delays, multiple signal states, and transistor or functional models, though possible [458], require major changes in the implementation of a deductive fault simulator and slow down the execution.

The simulation proceeds by simulating a vector in the true-value mode, which can be done by either a compiled-code or an event-driven mechanism. Before simulating the next vector, a deductive procedure is applied to all lines in a level-order (for combinational logic) from inputs to outputs. In this process, fault lists are generated for each signal. The fault list of a signal is derived from the fault lists at the inputs of the gate producing that signal and any faults associated with that gate. In a

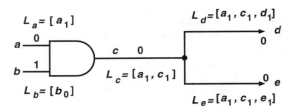

Figure 5.17: Fault lists in a deductive fault simulator.

Table 5.6: Fault list propagation in a deductive fault simulator.

Gate type	Inputs a	b	Output c	Output fault list L_c
AND	0	0	0	$[L_a \cap L_b] \cup c_1$
	0	1	0	$[L_a \cap \overline{L_b}] \cup c_1$
	1	0	0	$[\overline{L_a} \cap L_b] \cup c_1$
	1	1	1	$[L_a \cup L_b] \cup c_0$
OR	0	0	0	$[L_a \cup L_b] \cup c_1$
	0	1	1	$[\overline{L_a} \cap L_b] \cup c_0$
	1	0	1	$[L_a \cap \overline{L_b}] \cup c_0$
	1	1	1	$[L_a \cap L_b] \cup c_0$
NOT	0	-	1	$L_a \cup c_0$
	1	-	0	$L_a \cup c_1$

circuit with feedbacks, the fault lists of a signal may change several times. Only after the fault lists become stable, the simulator proceeds with the next vector.

The fault list of a signal at any time during simulation contains the *names* of all faults in the circuit that can change the state of that line. Thus a fault originally appears in the fault list of the output signal of the gate which the fault is physically associated with. An example is shown in Figure 5.17. Here a fault a_k means signal a stuck-at-k, with $k = 0$ or 1. For the AND gate with primary inputs $a = 0$ and $b = 1$, the output is $c = 0$. Since a and b are primary inputs, their fault lists simply contain their own faults activated by the present signal values. We will denote their fault lists as sets, $L_a = [a_1]$ and $L_b = [b_0]$. Since $b = 1$, the path from a to c is sensitized, but the path from b to c is not sensitized. Therefore, the fault list of c contains a_1. In addition, it also contains the fault c s-a-1 as the current state of c is 0. The fanouts d and e simply adopt the fault list from their source stem c and add their own respective faults.

Table 5.6 gives the rules for fault list propagation. The output fault list of a gate is generated by set operations such as union ($\cup$), intersection ($\cap$) and complementation (-), among fault lists. For example, when both inputs of a two-input AND gate are 0, to propagate through, the effect of a fault must be present on both inputs. This is achieved by the intersection, $L_a \cap L_b$, of the two input fault lists. In addition, the fault c_1 is included. Table 5.6 gives the rules for two-input primitive gates. For

5.5 Algorithms for Fault Simulation

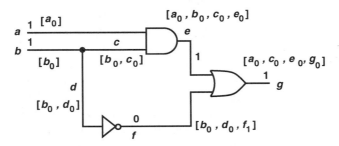

Figure 5.18: An example of deductive fault simulation.

gates with more inputs and for other gates such as the exclusive-OR, propagation rules can be derived by expanding them in terms of two-input primitive gates. This technique is quite general and has been used for deductive fault simulation of larger functions [458].

Example 5.10 *Deductive fault simulation: Consider the vector (1,1) applied to the circuit of Figure 5.18. We will simulate the s-a-0 and s-a-1 faults on all lines a through g for this vector. First, true-value simulation is carried out to determine all signal values, which are shown in the figure. This requires a single pass from inputs to the output. If the circuit had feedbacks, more than one pass might be required in the true-value simulation. Next, we conduct a second input to output pass for fault list generation and propagation. The lists of primary inputs just contain the respective s-a-0 faults that are active there. Thus:*

$$L_a = [a_0] \quad \text{and} \quad L_b = [b_0]$$

Fault lists for fanouts c and d are obtained by adding their locally active faults to the fault list L_b of the stem (Figure 5.17.) The fault list for e is obtained by the propagation rules of Table 5.6, as:

$$L_e = [L_a \cup L_c] \cup e_0 = [a_0, b_0, c_0, e_0]$$

The fault list of d simply propagates through the NOT gate. Thus:

$$L_f = L_d \cup f_1 = [b_0, d_0, f_1]$$

Using the propagation rule for OR gate, we get:

$$L_g = [L_e \cap \overline{L_f}] \cup g_0 = [a_0, c_0, e_0, g_0]$$

Notice that the fault b_0 is present at both inputs of the OR gate. That is, this fault will invert both inputs and the output of the faulty circuit will remain 1, the same as the fault-free output. That is why b_0 is absent from L_g. This completes the fault propagation, which tells us that four faults, a s-a-0, c s-a-0, e s-a-0, and g s-a-0, are detected by the (1,1) input vector.

Similar to the parallel fault simulation method, the deductive method also leads to difficulties in fault simulation of sequential circuits. First, fault list propagation rules should be extended to deal with the three-state (0, 1, X) logic. This is possible, but adds greater complexity to the algorithm. Nevertheless, it is necessary to represent the initial state of the circuit. Second, we should work out a method to deal with memory elements. If flip-flops are implemented with feedback among logic gates, the normal fault propagation rules of Table 5.6 can be used. However, any fault-related race conditions should be analyzed via complex procedures [56]. When the algorithm is correctly implemented, the simulator identifies *race faults* (faults causing race conditions) and assumes an unknown (X) state for the memory element in the faulty circuit. The fault then may be detected as a *potentially detectable fault* if the unknown state of the faulty circuit propagates to a primary output.

A technique, often used to simplify the simulation of logic circuits with asynchronous feedback, is to break all feedbacks by inserting ideal delay elements. These elements are assumed to have delay that is larger than the total delay of the combinational logic. They propagate the logic values and fault lists from their inputs to outputs with that delay. Thus the problem of sequential circuit simulation is transformed into iteratively repeated simulation of a combinational circuit. All delay elements start in the unknown (X) state. Combinational logic is simulated and fault lists are propagated to primary outputs or the inputs of the delay elements. This completes one iteration. Then the fault-free signal values and fault lists at the inputs of the delay elements are transferred to their respective outputs. If this causes any changes at the inputs of the combinational logic, then another iteration of true-value simulation and fault list propagation is done. Iterations stop when all signals and fault lists stabilize, i.e., remain unchanged from the previous iteration. If any oscillatory behavior is detected, either the oscillating signal is set to the unknown (X) state, or the fault causing the problem is dropped from consideration.

This process becomes much faster for synchronous circuits where all memory is implemented as clocked flip-flops. If clock faults are to be simulated, then the simulation will be similar to that for asynchronous circuits described above. However, when the clock faults are not simulated (also no faults internal to flip-flops are simulated) then the flip-flops can be assumed to work as ideal memory elements. The clock is often treated as an implicit signal that allows data transfer through the memory element only once per vector (see Example 5.13.)

The LAMP (*logic analyzer for maintenance planning*) system [142], developed and used at Bell Labs during the 70s, is an example of a deductive fault simulator. Besides logic gates, LAMP could also simulate functional memory blocks in the fault simulation mode [458]. This mode, which is quite common in the mixed-mode of circuit modeling, does not model faults in the storage cell array of the memory but propagates the effects of external faults through the memory. With the advent of the MOS technology, however, strictly logic-level simulation became deficient and LAMP was replaced by concurrent fault simulators, LAMP2 [8] and MOTIS [404].

5.5 Algorithms for Fault Simulation

5.5.4 Concurrent Fault Simulation

The concurrent fault simulation algorithm [684, 685] is the most general method of fault simulation. It can handle various types of circuit models, faults, signal states, and timing models. It basically extends the event-driven simulation method to the simulation of faults in the most efficient way.

Let us again consider the fault simulation problem as depicted in Figure 5.14. A concurrent fault simulator models the problem as follows:

1. *Events.* Consider a line (signal) in the fault-free circuit C( ). An event (signal change) on this line is called a *good-event*. Three attributes specify a good-event: line designation (or signal name), type of change (e.g., 0-to-1, or any change among permissible signal states), and the time of change. The same line also exists in faulty circuits, C(f1) through C(fn), and events on it in those circuits, only when they differ from the good-event, are called *fault-events*. A fault event is specified by the same three attributes required for a good-event and an additional designation of the fault (site and type.)

2. *Structure.* The circuit C( ) is modeled in the same way as for the true-value simulation (see Section 5.3.) Any hierarchy is flattened and modules (though referred to as gates in this discussion) can be at any level or abstraction. In general, a module can have internal state variables and multiple outputs. The structure contains the connectivity information (fanouts and fanins) and gate (or module) functions. Each gate is called a *good-gate*. A *fault-list*, usually in the form of a linked-list, is associated with each good-gate. Elements of this list are called *bad-gates*. A bad-gate is not faulty itself but is affected by some fault. At least one of its signals at input or output terminals or internal states differs in value from the corresponding good-gate. The bad-gate derives its name from the fault that causes the signal difference. A bad-gate has the same input-output function as the good-gate since faults are modeled in signals and not in the function. Let us denote faulty circuit gates corresponding to gate g in C( ) as g_i in C(fi), where fi ($i = 1, 2, \ldots n$) are faults being simulated. Suppose, only gates g_6 and g_{18} have any signal differences with g at some time during the simulation, then at that time the fault-list of g will have two bad-gates specified as: (fault f6; terminal values of g_6) and (fault f18; terminal values of g_{18}.) The circuit structure of good-gates with associated fault-lists is a very compact and complete representation of the $n + 1$ circuits shown in Figure 5.14.

3. *Faults.* Faults are assumed to be permanent and affect signal values at terminals or internal states (if any) of modules. The fault information (fault site and type) is stored with the good-gate connected to the fault site. Faults at primary inputs and primary outputs can be modeled by attaching simple buffer gates. Whenever the signal values of a good-gate make a fault active, a bad-gate is inserted in the fault-list of that good-gate.

Chapter 5. LOGIC AND FAULT SIMULATION

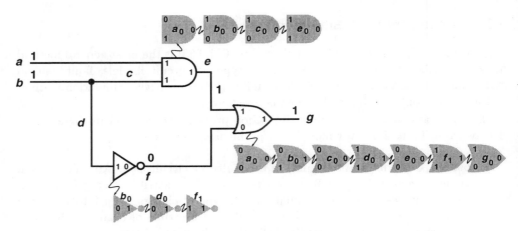

Figure 5.19: Fault-lists (bad-gates) in concurrent fault simulation.

The simulation proceeds exactly in the same manner as the event-driven simulation described in the Subsection 5.4.2. The circuit activity is simulated by event scheduling and processing. Usually a unit-delay model is assumed, though any other form of delays can also be simulated. Whenever the signal value at a fault site differs from the faulty state, the fault becomes active. The good-gate attached to that site instantly insert a bad-gate with the fault name and differing signal values to its fault-list.

All good-events and fault-events make good-gates active for evaluation. Good-events also make bad-gates active for evaluation. However, a fault-event with a certain fault name only activates the bad gates with the same fault name. All active gates, both good and bad, are evaluated and can potentially generate new events. Good-events can be generated only when the activation of a good-gate is caused by a good-event. All events either caused by fault-events or caused by good-events activating bad-gates are fault-events. Upon evaluation, any bad-gate whose signals become identical to the corresponding good-gate is removed from the fault-list. This bad-gate is said to "converge" to the good-gate and the process is called *convergence*. The process of creating new bad-gates, either because of activation of a fault site or because of a good-gate evaluation due to a fault-event, is called *divergence*. The following example illustrates several features of the concurrent fault simulation.

Example 5.11 *Concurrent fault simulation: Figure 5.19 shows a three-gate combinational circuit being simulated for an input vector 11. All single stuck-at faults are concurrently simulated. We will use the subscript notation for faults. Thus, fault a s-a-0 will be called a_0. Faults are modeled on all gate inputs, primary input b, and primary output g. Good-gates are shown as gates drawn in solid lines and without shading. Bad-gates are shown in grey shade with the corresponding fault name written in the center. Signal values at inputs and output of each gate are written inside the gate. Figure 5.19 shows the steady-state after the input vector 11*

5.5 Algorithms for Fault Simulation

was applied. Thus, there are no changing signals (events) shown. To each good-gate, a number of bad-gates are attached in a linked-list structure (shown as tiny wiggly arrows.) Attached to the good AND gate are four bad-gates, a_0, b_0, c_0, and e_0. Notice that at least one value around a bad-gate differs from the good-gate and the difference is caused by the corresponding fault. Fault lists for the other two gates are also shown in the figure. At the primary output g, any bad-gate whose output differs from that of the good-gate indicates fault detection. Thus, faults a_0, c_0, e_0, and g_0 are detected. With fault-dropping, we would have removed these faults from further consideration. In this exercise, however, we will not drop faults.

Notice that our fault detection result agrees with that obtained by deductive fault simulation in Figure 5.18. But the fault lists there were shorter. In the deductive simulator the fault-list is for a signal and contains only the faults that affect (or are detected at) that signal. In the concurrent simulator the fault-list is for a gate and even faults that affect the inputs of that gate are included in the list. Fault-lists in a concurrent simulator are, therefore, comparatively longer. The advantage, though not as clear in logic simulation, is significant when more complex functional modules (memories and RTL or behavioral models) are simulated.

In Figure 5.20, we simulate a 1 to 0 ($1 \rightarrow 0$) good-event at a. Examine the changes shown in the AND good-gate and its fault-list (associated bad-gates.) The top input of all except the a_0 bad-gate change. Only the good-gate output changes, producing a $1 \rightarrow 0$ event on signal e. After these evaluations, bad-gates a_0 and e_0 have identical signal values as the good-gate and hence they converge. They are removed from the fault-list. At this point, one good-event $1 \rightarrow 0$ on e and no bad-events have been generated. The OR gate is evaluated and produces a $1 \rightarrow 0$ good-event on g. Bad-gates are also evaluated but none generates any bad-event. After evaluation, bad-gates a_0, c_0, e_0 and g_0 converge to the good OR gate. These are removed from the fault-list. However, the processing of the ($1 \rightarrow 0$) good-event at a is not complete.

As Figure 5.21 shows, changing of signal a activates fault a_1. Thus, a diverging bad-gate labeled a_1 is inserted in the fault-list of the AND gate. Newly diverging gates are shown with lighter grey shading in Figure 5.21. Similarly, another bad-gate e_1 is also added. These two generate bad-events, which when processed at the OR gate produce further divergence of bad-gates a_1 and e_1 there. The change caused by the good-event at the OR gate, discussed above, results in the divergence of another bad-gate g_0. This completes the simulation.

All bad-gates at the output g have a different output value than that of the good-gate. Therefore, detected faults are b_0, d_0, f_1, a_1, e_1, and g_1.

Our example illustrates only some features of the concurrent fault simulation algorithm. Useful techniques such as *multi-list traversal* (MLT) allow simulation of multiple output functions that may also have internal states. An interested reader should study the book by Ulrich et al. [684] to learn about the complete capabilities of this algorithm. Its significant advantages are efficiency (elimination of redundant computation) and modeling flexibility (fault simulation for anything that can be simulated.) Some notable concurrent fault simulators are the MARS

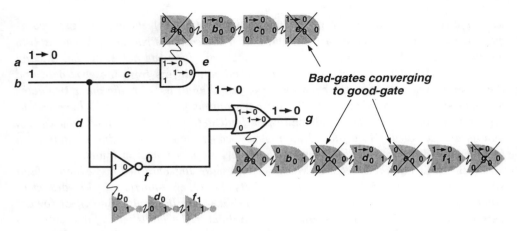

Figure 5.20: Event processing and convergence in concurrent fault simulation.

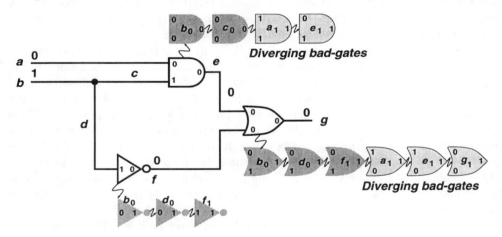

Figure 5.21: Bad-gate divergence in concurrent fault simulation.

hardware accelerator [19, 82], CREATOR [235], MOZART [236], MOTIS [404], and FMOSSIM [578]. With the exception of FMOSSIM, which models the entire circuit at the switch-level, all use mixed-level (transistor, logic, and functional) modeling.

5.5.5 Roth's TEST-DETECT Algorithm

Roth [551] devised an easy-to-program algorithm for simulating faults in combinational circuits. All gates are assumed to have zero delay. The circuit is simulated for a vector in the true-value mode. This determines the states of all lines. Next, faults are analyzed one at a time to determine which faults are detected by the presently simulated vector. The analysis is based on Roth's D-calculus that allows a composite representation of a signal in the fault-free and faulty circuits. Assuming a two-value simulation, the signal state (fault-free value, faulty value) can take four

5.5 Algorithms for Fault Simulation

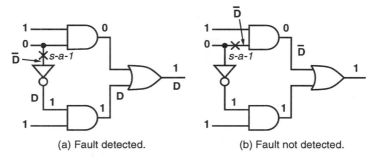

(a) Fault detected. (b) Fault not detected.

Figure 5.22: An example of fault simulation by the TEST-DETECT algorithm.

possible assignments. These are symbolically denoted as: $0 = (0,0)$, $1 = (1,1)$, $D = (1,0)$, and $\overline{D} = (0,1)$. Details of the D-calculus can be found in Chapter 7 on combinational ATPG.

To simulate a fault, we start at the fault site. If the fault is activated, a D or $\overline{D}$ is placed at the fault site. These values are propagated through fanouts and gates, using the true-values or the newly placed symbolic states. If a D or $\overline{D}$ reaches a primary output, the fault is considered detected at that output. The signal values are restored to their true-values before analyzing the next fault. When all faults have been analyzed, we proceed with the simulation of the next vector. The following example illustrates the TEST-DETECT procedure.

Example 5.12 *TEST-DETECT: Consider the multiplexer circuit with input (1,0,1) as shown in Figure 5.22. The binary logic values shown are a result of the true-value simulation. In Figure 5.22(a), we consider the s-a-1 fault on the lower fanout of the second primary input. The fault is active and the input to the inverter is changed to $\overline{D}$. We will now perform an even-driven simulation by evaluating the active gates. The $0 \to \overline{D}$ event makes the NOT gate active. Its evaluation puts a $1 \to D$ at its output, making the lower AND gate active. Evaluation of AND causes an output event $1 \to D$ and the activation of the OR gate. Evaluation of OR produces a $1 \to D$ at the output, indicating the detection of the fault. Next, we restore the signals to the original fault-free values and apply the procedure to the next fault, which is not detected. The reader can easily follow the steps in Figure 5.22(b).*

5.5.6 Differential Fault Simulation

Cheng and Yu [161] made two improvements in the TEST-DETECT algorithm. First, they eliminated the use of the D-calculus by relying on logic events alone. Second, they eliminated the explicit restoration to true-value needed in TEST-DETECT before analyzing the next fault. Their algorithm, known as the *differential fault simulation*, works as follows. We first consider the simulation of combinational circuits:

1. Given a vector set and a fault list, start with the first vector as the *current vector*.

2. Simulate the current vector in the true-value mode and store the primary output values in an *output file.*

3. Activate a fault such that the true-value at its site is opposite to the faulty value by placing an event at the fault site. For example, if the true value of the site is 0 and it is a s-a-1 fault, then a $0 \rightarrow 1$ event is placed.

4. Continue with event-driven simulation until no gate remains active. If any newly simulated output differs from the saved values in the *output file*, then the fault is detected. Remove the fault from the list if fault dropping is desired.

5. If all faults have been analyzed for the *current vector* and there are unsimulated vectors, then advance to the next vector and go to Step 2 *(stop if vectors are exhausted.)* Otherwise, proceed to the next step.

6. Activate the next fault by placing two events in the circuit, whose signal values still correspond to the previous fault: (a) an event at the site of the previous fault to restore that signal to its true-value, and (b) an event at the site of the fault now being simulated to change the signal there to the faulty state. With these two events, proceed to Step 4.

The main idea that makes this algorithm different from TEST-DETECT is in Step 6. Reworking the example of Figure 5.22, which is left to the reader, will make several advantages clear. The absence of the D-calculus makes its implementation in a true-value simulation program rather simple. The simultaneous restoration from the last fault and the injection of the next fault roughly reduces the number of simulation passes to half of those required for TEST-DETECT.

Both TEST-DETECT and the differential algorithms, and for that matter any combinational circuit fault simulation algorithm, can be extended to simulate synchronous sequential circuits. Such fault simulation assumes that the circuit has no timing problems. That is, the maximum combinational circuit delay is smaller than the period of the clock controlling the flip-flops. Thus, the combinational logic can be simulated with zero delay and flip-flops are assumed to be ideal storage devices with an implicit clock that is synchronized with the application of input vectors. Only the faults in the combinational logic are simulated.

The extension of combinational algorithms simply requires that every flip-flop output has a *current fault list* and every flip-flop input has a *next fault list*. The *current fault list* contains those faults whose effects are present at the output of the flip-flop. Similarly, the *next fault list* contains those faults whose effects have reached the combinational output feeding the flip-flop. In the same way as the effects of faults in the combinational logic are propagated, the fault effects in the *current fault list* are also propagated. For an input vector, a fault can produce several fault effects. All of these are propagated along with any effects of the same fault if present in the *current fault list*. If any fault effect reaches a flip-flop input, it is added to the *next fault list*. When a new input vector is applied, all true signal values are transferred from flip-flop inputs to outputs. Similarly, all *current fault*

5.5 Algorithms for Fault Simulation

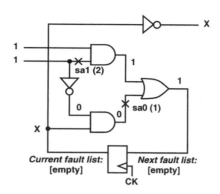

Figure 5.23: Differential fault simulation example: first vector (1,1).

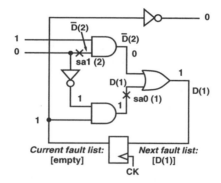

Figure 5.24: Differential fault simulation example: second vector (1,0).

lists are replaced by the corresponding *next fault lists*. As usual, only the faults whose effects reach primary outputs are treated as detected. The following example illustrates this simulation procedure.

Example 5.13 *Fault simulation of a synchronous sequential circuit:* Consider the simulation of two faults shown as (1) and (2) in Figure 5.23. The vector set contains three vectors, the first of which is simulated in Figure 5.23. The initial state of the flip-flop is assumed to be unknown and is denoted as X. Both current and next fault lists *are empty. The fault-free circuit states indicate that none of the faults is activated. After simulating the second vector (1,0) in Figure 5.24, we find that both faults are activated. The effects of fault (1) are denoted as* $D(1)$ *and those of fault (2), as* $\overline{D}(2)$. *Only* $D(1)$ *reaches the flip-flop input and is added to the* next fault list. *No fault is detected. Figure 5.25 shows the simulation of the third vector (0,1.) The* current fault list *is updated with* $D(1)$, *which now propagates to the primary output. Thus, fault (1) is detected. No new fault effect is produced by this vector. Since no fault effect is propagated to the flip-flop, the* next fault list *is now empty. At this point, fault (1) can be dropped, and subsequent vectors will only simulate fault (2), until that is detected too.*

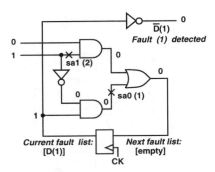

Figure 5.25: Differential fault simulation example: third vector (0,1).

Notice that we used TEST-DETECT to work out the above example. We also added the flavor of concurrency by introducing multiple (but non-interacting) fault effects. A reader can easily work it out with the differential algorithm. In general, both TEST-DETECT and *differential fault simulation* algorithms are well-suited for synchronous sequential circuits. A parallel implementation of the differential algorithm, called PROOFS [496], has proved to be very successful in the industry. Here fault lists are propagated through the combinational logic using the bit parallelism of the computer.

5.6 Statistical Methods for Fault Simulation

The complexity of true-value simulation per vector is $O(G)$, where G is the number of gates in the circuit. This is exactly so for combinational logic because for each input vector every gate is evaluated no more than once. For event-driven simulation, the number of active gates per vector is still proportional to the total number of gates. This estimate of complexity is also applicable to synchronous sequential circuits, since basically the simulation involves evaluation of combinational logic. Only asynchronous circuits, if they have the problem of oscillations, can take more simulation time.

If we assume that the number of vectors needed to verify or test a circuit might increase in proportion to the number of gates, then the total complexity for all vectors will be $O(G^2)$. Since stuck-at faults are associated with the inputs and outputs of gates, the number of such faults is also proportional to the number of gates. For fault simulation, in the worst case, we repeat true-value simulation for each fault (see Figure 5.14.) Thus, the worst case complexity of fault simulation is $O(G^3)$. Fault dropping in serial fault simulation, or the use of clever methods such as parallel, deductive, concurrent, or differential, can save time but cannot do better than the true-value simulation. Therefore, the complexity of fault simulation of a G gate VLSI circuit is between $O(G^2)$ and $O(G^3)$. This has been experimentally observed [257].

5.6 Statistical Methods for Fault Simulation

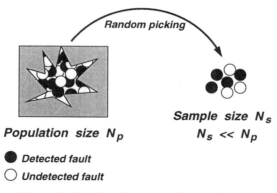

Figure 5.26: Random sampling of faults.

5.6.1 Fault Sampling

Fault sampling is a popular technique used to reduce the effort of fault simulation. In this technique, a subset of faults is randomly picked from the set of all faults. This subset, usually a small fraction of the complete fault set, is known as a *fault sample* or simply a *sample*. The faults in the sample are simulated and the *sample coverage*, i.e., the ratio of detected faults to all faults in the sample, is used as an estimate of the fault coverage in the complete fault set. The accuracy (or error bound) of the estimated coverage depends on the absolute number of faults in the sample. This number is known as the *sample size*. The error bound of the estimate can be reduced by increasing the sample size. Thus, we determine the sample size from the required accuracy with which we wish to estimate the coverage. This sample size is independent of the total number of faults in the circuit, which is assumed to be very large as compared to the sample size.

To determine the sample size that we should use, let us consider the statistician's model of Figure 5.26. A box is filled with black balls (detected faults) and white balls (undetected faults.) Given a circuit and the vector set the balls have already been colored. It is just that we do not know the fraction C of black balls in the box. This fraction is the true fault coverage. The total number, N_p, of balls (faults) is called the *population size*. We shuffle the balls in the box and randomly collect a sample of N_s balls (sample size = N_s.)

In a method, known as *sampling with replacement*, we pick one ball, note its color and then put it back in the box. So when the next ball is picked, the proportion of black and white balls remains the same as before. Although this scheme is simpler to analyze, it has the disadvantage of picking the same ball again. Therefore, fault sampling is done *without replacement*. We will develop an analysis of fault sampling using following variables:

N_p Total number of faults in the circuit for which coverage is to be determined.

C Unknown but true fault coverage of given vectors, $0 \leq C \leq 1$. *This is the quantity being estimated.*

CN_p Actual (but unknown) number of faults detectable by the given vectors.

N_s Number of randomly sampled faults from the set of N_p faults. N_s is known. Normally, $N_s \ll N_p$.

c Sample coverage, a random variable with range, $0 \leq c \leq 1$.

x Value of c determined from sample fault simulation, $0 \leq x \leq 1$.

xN_s Number of sampled faults detected by given vectors. This is a known quantity that is determined by the fault simulator.

The number of ways in which we can pick N_s faults from a set of N_p faults is:

$$\text{Ways of obtaining sample of size } N_s = \binom{N_p}{N_s} = \frac{N_p!}{N_s!(N_p - N_s)!} \quad (5.1)$$

where ( ) is a notation for a *binomial coefficient*, frequently used in combinatorial analysis [363], which gives the number of combinations of N_p objects taken N_s at a time. We simulate N_s faults and find that exactly xN_s faults are detected by the given vectors. This gives the measured *sample coverage*, x. These xN_s detected faults must have been obtained from the total of CN_p detectable faults. Similarly, the sample has $(1-x)N_s$ undetected faults obtained from the total of $(1-C)N_p$ undetectable faults. The number of ways in which a sample with coverage x is possible is the product of ways of sampling the detectable and undetectable fault subpopulations. Thus:

$$\text{Ways of obtaining sample coverage } x = \binom{CN_s}{xN_s}\binom{(1-C)N_p}{(1-x)N_s} \quad (5.2)$$

The probability of a fault sample giving a value x for the random variable c is obtained upon dividing Equation 5.2 by 5.1:

$$p(x) = \text{Prob(sample coverage, } c = x) = \frac{\binom{CN_s}{xN_s}\binom{(1-C)N_p}{(1-x)N_s}}{\binom{N_p}{N_s}} \quad (5.3)$$

This is known as the *hypergeometric probability density function* of a discrete-valued random variable [677]. The random variable c can only take discrete values, 0, $1/N_s$, $2/N_s$, ..., 1. When N_s is large, c can be treated as a continuous variable and Equation 5.3 is conveniently approximated by a *Gaussian probability density function* (also called *normal probability density* [677]) with mean, $E(c) = C$, and variance σ^2 [21]:

$$p(x) = \text{Prob}(x \leq c \leq x + dx) = \frac{1}{\sigma\sqrt{2\pi}} e^{-\frac{(x-C)^2}{2\sigma^2}} \quad (5.4)$$

5.6 Statistical Methods for Fault Simulation

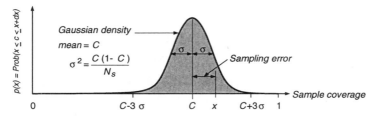

Figure 5.27: Gaussian probability density of sample fault coverage c.

We have already substituted C, the true coverage, as the mean of c. This can be derived by using Equation 5.3 and the definition of statistical mean. It shows that the measured sample coverage c is an *unbiased* estimate of the true fault coverage C [196]. The variance of c is given by [21]:

$$\sigma^2 = \frac{C(1-C)}{N_s}\left(1 - \frac{N_s}{N_p}\right) \approx \frac{C(1-C)}{N_s} \qquad (5.5)$$

The factor $(1 - N_s/N_p)$ accounts for sampling without replacement. It reduces the variance exactly to 0 when $N_s = N_p$, i.e., when all faults are simulated. For most practical cases, $N_s \ll N_p$. So, the approximation in Equation 5.5 remains valid and actually gives the same variance for coverage as if the sample was taken with replacement. This means that when the circuit is large and the sample size is small, fault samples taken either with or without replacement give almost the same result.

Variance is a measure of the amount by which the measured values of c deviate from C. According to Equation 5.5, the parameter mainly responsible for reducing this deviation is the sample size N_s. Thus, the sample size should be determined from the desired accuracy of estimation and not from the size of the circuit – a point about sampling that is usually misunderstood. If a circuit has 10,000 faults and we find that a sample of 1,000 faults (10% of population) can give us the desired accuracy, then for another circuit with 100,000 faults we still need to sample 1,000 faults for the same accuracy. This amounts to a 1% sample. As the circuit size increases, the sample fraction drops, increasing the advantage of the fault sampling method over simulation of all faults.

The probability density function of Equation 5.4 is shown in Figure 5.27. Most values of x lie in the shaded region where $p(x)$ is non-zero. It is a practice to assume that the range for x is from $C - 3\sigma$ to $C + 3\sigma$. This is known as the *three-sigma range* [196]. The actual probability of x being within the three-sigma range is the shaded area that evaluates to 0.997 [219]. We define the sampling error as $|x - C|$ and determine its high confidence (0.997 probability) limits by setting it to three times the standard deviation (σ):

$$|x - C| = 3\sigma \quad \text{or} \quad |x - C| = 3\sqrt{\frac{C(1-C)}{N_s}} \qquad (5.6)$$

where the expression for σ has been substituted from Equation 5.5. In general, this equation can be written for any confidence probability and was derived by Millot in

1923 [196] as a quadratic equation:

$$(x - C)^2 = \lambda^2 \frac{C(1-C)}{N_s} \tag{5.7}$$

where $\lambda = 3$ for a three-sigma confidence-level. Here x is the measured sample fault coverage and N_s is the known sample size. However, the true fault coverage C is unknown. So we solve the equation for C. The two roots give the upper and lower bounds for the coverage [21]. A simple approximation can be obtained for large sample size, typically $N_s \geq 1,000$, and provides the coverage estimate as follows [35]:

$$3\sigma \text{ coverage estimate} = x \pm \frac{4.5}{N_s}\sqrt{1 + 0.44 N_s x (1-x)} \tag{5.8}$$

Experimental results [35] show that Equation 5.8 provides good estimates.

Example 5.14 *Fault coverage by fault sampling: The ISCAS'89 [99] benchmark circuit s35932 has 39,096 collapsed faults. To determine the coverage of a set of 372 vectors, $N_s = 1,000$ faults were randomly selected. Bell Labs' simulator [160] required 11s of CPU time on a SUN Ultrasparc to complete the simulation and report a sample fault coverage, $x = 0.887$ or 88.7%. Equation 5.8 gives a 3σ estimate of 0.887 ± 0.030. Simulation of all faults required 94 s and produced a coverage of 0.871.*

Example 5.15 *Finding sample size: Suppose we want the 3σ sampling error not to exceed $\pm\Delta$. Then, from Equation 5.8 we can write:*

$$\Delta^2 = \frac{4.5^2}{N_s^2}(1 + 0.44 N_s x (1-x)) \approx \frac{4.5^2}{N_s} 0.44 x (1-x) \tag{5.9}$$

where sample size N_s is assumed to be large. Also, since the maximum value of $x(1-x)$ is 0.25, which occurs when $x = 0.5$, we get:

$$N_s = \frac{4.5^2 \times 0.44 \times 0.25}{\Delta^2} = \frac{2.2275}{\Delta^2} \tag{5.10}$$

For $\Delta = 0.02$, we obtain $N_s = 5,569$. Notice that this sample size is calculated for the worst case. If $x = 0.90$, according to Equation 5.8 the 3σ range will be 0.90 ± 0.012.

Fault sampling has been used for many years in the industry. As circuits have grown larger, its use has become more widespread. Sampling estimates are approximate but are adequate for assessing product quality from test data (Chapter 3.) Sampling methods are considered inadequate for fault diagnosis (Chapter 18) because they provide no data on unsampled faults.

Equations 5.8 and 5.10 are convenient ways of determining the fault coverage estimate and sample size. There are other sampling methods that can provide narrower estimates [177, 454]. A multi-pass sampling procedure, known as *sequential sampling* [21, 115, 709], requires simulation of small sets of faults such that each pass revises the coverage estimate closer to the correct value. Another technique, called *stratified sampling*, is useful when the fault population is unevenly distributed over the circuit. This has been used for finding fault coverage at the register-transfer level [655, 657] and for defect-oriented faults [265]. In this method, the circuit is partitioned and separate samples are drawn from each partition. The partition coverages are combined according to the sizes of partitions. Although we have only discussed coverage estimation, fault sampling can also be used for test generation. In that case, the sample size is related to the testability of the circuit [591].

5.7 Summary

Simulators are essential tools for digital design verification and testing. In the future, we might see greater use of formal verification techniques [168, 380, 453]. That is not likely to diminish the importance of simulators because their ability to model lower-level detail will remain critical for manufacturability of designs. Among the applications of fault simulators are VLSI product quality evaluation discussed in Chapter 3 and fault diagnosis discussed in Chapter 18. They also help the test generation process studied in Chapters 7 and 8.

Both logic and fault simulator programs consume enormous computing resources. When the design schedule is impacted, speed up mechanisms are sought. Popular techniques involve parallel and distributed processing [49, 50, 63].

To limit the size of this chapter, we had to omit several techniques that are worth mentioning. The *parallel value list* (PVL) method [474, 625] combines the parallel and concurrent techniques and was implemented in a commercial simulator, HILO3. The *critical path tracing* technique [7, 10, 459, 460] is a fault-independent method that traces sensitized paths from output to inputs on the basis of signal values obtained from true-value simulation. The *statistical fault analysis* (STAFAN) [336, 698] is another fault-independent technique that statistically determines controllabilities, observabilities, detection probabilities, and fault coverage from true-value simulation. *Parallel-pattern single-fault propagation* (PPSFP) [704] is a serial fault simulator for combinational circuits that simulates several vectors in parallel using the bit parallelism of a computer word. In a *parallel iterative simulator* (PARIS), the parallel vector simulation is applied to clocked sequential circuits [267, 375]. In the beginning all but the first vector must assume flip-flops to be in unknown states. The simulation for a block of parallel vectors is iterated until signal values have stabilized.

Problems

5.1 Rearrange the eight vectors of Table 5.1 into a minimal-length sequence that applies all possible inputs to each full-adder block of a four-bit ripple-carry adder and also activates the longest path.

Chapter 5. LOGIC AND FAULT SIMULATION

5.2 Draw a state diagram for a two-bit shift register. A simulation-based verification strategy checks for the existence of all four states and verifies the execution of all transitions. Starting with a sequence to initialize the circuit in the 00 state, derive a complete minimal-length input sequence for verification. *Hint: The minimal-sequence traces an Eulerian path in the state diagram, i.e., executes each directed edge exactly once. An Eulerian path in a directed graph is defined as a continuous path that traces each edge exactly once [363].*

5.3 Add a *clear* input to the two-bit shift register to synchronously initialize the circuit in the 00 state. Draw the state diagram with two primary inputs, *shift* and *clear*. Is it possible to find an Eulerian path in the state diagram? If not, consult the literature on graph theory [363] to find the condition for the existence of a minimal-length verification sequence.

5.4 Using the fault simulation result of Table 5.2 find a minimal-length vector sequence that has 100% fault coverage and also activates the longest path.

5.5 Explain why the reverse-order fault simulation is not a practical test compaction technique for sequential circuits.

5.6 Which type of simulator will you use for: (a) verifying the architecture of a digital system, (b) checking the design of an analog circuit, (c) verifying the logic of an MOS bus, (d) simulating the critical timing path of a large digital circuit, and (e) logic verification of a large digital circuit implemented with static CMOS gates?

5.7 Suppose that the bus in the circuit of Figure 5.6 is initially in the driven state of 1 when control inputs of both nMOS transmission gates are changed to 0. What will be the state of the bus? How will the logic model of Figure 5.7 simulate this situation?

5.8 Sketch a circuit for a 0-dominating (*wired AND*) bus with two drivers using Boolean gates.

5.9 Derive a logic model for the circuit of Figure 5.6 that represents the memory state of the MOS bus.

5.10 Will three-state logic simulation initialize the flip-flop (FF) in the circuit of Figure 5.28 when clock CK is applied? If not, then devise an input sequence that will initialize FF to a 0 output and will be correctly simulated using the three-state logic.

5.11 A two-input OR gate has rise and fall switching delays of 3 and 5 units, respectively. Consider the case when one input is fixed at 0 and the other input has a rising rectangular pulse of width 6 units. (a) What is the width of the pulse at the output? (b) If the rise and fall switching delays are changed to 5 and 3 units, respectively, then what will be the width of the output pulse produced by the same inputs?

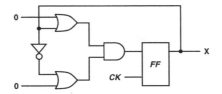

Figure 5.28: Circuit for Problem 5.10.

5.12 Reexamine the cases (a) and (b) of the previous problem for an input pulse of 1 unit width. *Note that the gate filters out a pulse narrower than its own delay – a phenomenon known as spike suppression.*

5.13 Explain what action an event-driven true-value simulator will take when it evaluates a zero-delay gate.

5.14 For implementing an event-driven simulator for unit-delay simulation, what is the minimum number time slots needed?

5.15 Suppose that the fault coverage increases linearly from 0 to 100% as a large number of vectors is applied. Show that a serial fault simulator will take about half the CPU time with fault dropping than without. Assume that in fault dropping any fault is dropped the first time it is detected.

5.16 Show that when n faults are simulated without fault dropping, a parallel fault simulator on a w-bit word computer will run $(n+1)(w-1)/n$ times faster than a serial fault simulator.

5.17 Assuming a four-bit machine word, demonstrate parallel fault simulation of vector (1,0,1) for the three single stuck-at-1 faults on the second primary input and its two fanouts, respectively, in the circuit of Figure 5.22.

5.18 Derive the output fault list L_c for an exclusive-OR gate, $c = a \oplus b$, in terms of the input fault lists, L_a and L_b.

5.19 Explain why a concurrent fault simulator requires more memory than a deductive fault simulator.

5.20 A VLSI design center is equipped to design digital circuits with embedded memory blocks. The circuits are modeled at the logic level but the simulator must have a four-state (0,1,X,Z) signal representation. Only single stuck-at faults are to be simulated. In order to select a fault simulator, you must answer the following questions:

(a) Which of the simulators among serial, parallel, deductive, and concurrent, can simulate the circuits?

(b) Which is the best choice among these four simulators?

5.21 What can you say about the detectability of a multiple stuck-at fault (f1,f2), if concurrent simulation of single stuck-at faults, f1 and f2, produces the following result?

(a) Both faults are detected.

(b) Only f1 is detected.

(c) None is detected.

5.22 Show that the computing effort of a concurrent fault simulator is equivalent to the true-value simulation of a circuit that is $1 + \alpha N/2$ times larger, where N is the number of faults simulated and α is the average fraction of gates affected by a fault. Assume that half of the faults are active at any time and simulation is done without fault dropping.

5.23 Use the TEST-DETECT algorithm to determine whether the stuck-at-1 fault on the second primary input of the multiplexer in Figure 5.22 is detected by the vector (1,0,1.)

5.24 Simulate the two faults in Figure 5.22(a) and (b) using the differential fault simulation algorithm.

5.25 *Fault sampling.* A circuit contains 1,000,000 collapsed stuck type faults. A set of 4,000 randomly-selected faults is simulated and 3,900 faults are found to be detectable by the test vectors. What is the estimated range of fault coverage?

5.26 *Fault sampling.* Determine the sample size for fault simulation to achieve a ±2% accuracy in the coverage estimate when the actual coverage is known to be above 70%.

Chapter 6

TESTABILITY MEASURES

> " . . . the perception of testability is highly dependent on one's involvement with testing. For example, a design engineer might consider the complexity of test vector generation as a measure of circuit testability. To a test engineer, testability may mean compatibility of design with the test equipment. Quality engineers often relate testability to fault coverage. . . . Testability is the property of a circuit that makes it easy (and sometimes possible!) to test."
> — Agrawal and Seth, in their 1988 book [39].

The notions of controllability and observability of signals in a circuit originated in automatic control theory. *Controllability* for a digital circuit is defined as the difficulty of setting a particular logic signal to a 0 or a 1. *Observability* for a digital circuit is defined as the difficulty of observing the state of a logic signal. These measures are important for circuit testing, because while there are methods of observing the internal signals of a circuit, they are prohibitively expensive. Electron beam testing, for example, can actually scan the VLSI chip-under-test and produce a picture of the chip layout. The signals at logic 0 will appear in one color in the image, and those charged to logic 1 appear as another color. However, this testing method is used only for specialized purposes, because of its very high cost. Therefore, we must instead set internal signals by setting signals at *primary inputs* (PIs), and we must observe internal signals by arranging to propagate their values to *primary outputs* (POs.) The controllability and observability measures are useful because they approximately quantify how hard it is to set and observe internal signals of a circuit. *Testability analysis* usually has two significant attributes:

- It involves circuit topological analysis, but no test vectors. It is a static type of analysis.

- It has linear complexity, because otherwise testability analysis is pointless and one might as well use *automatic test-pattern generation* (ATPG) or fault simulation.

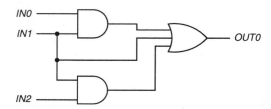

Figure 6.1: Reconverging fanouts causing correlation among signals.

Nearly all of the testability analysis methods in this chapter, and most of those used in industry, have these attributes.

Goldstein developed the SCOAP testability measures [262], and contributed a linear complexity algorithm to compute them. He defined separate zero and 1-controllabilities for each signal as measures of the *effort* (or *difficulty*) of setting the line to a logic 0 or 1 value. However, SCOAP has significant inaccuracies due to the assumption that signals at *reconvergent fanout* stems are treated as independent, when they frequently are not. Figure 6.1 shows a circuit in which the reconvergent fanout stem signal *IN1* fans out to three different gates, and then all of these diverging paths *reconverge* at the OR gate producing the signal *OUT0*. The assumption of signal independence, which provides the speed of computation to SCOAP, severely reduces the accuracy of the measure. SCOAP is poor at predicting which individual faults will remain undetected and which will be detected. However, it is highly effective in predicting relative coverage levels of the entire circuit fault sets [38]. The numerical values of the SCOAP measures range between zero and infinity. Higher values point to testability problems.

Another type of testability measures are probability-based. These overcome some limitations of SCOAP. Being related to signal probabilities, they are easier to interpret and their values always range between 0 and 1. The *1-controllability* ($C1$) is the probability of a signal value on line l being set to 1 by a random vector. The *0-controllability* ($C0$) is the probability of a signal value on line l being set to 0 by a random vector. Parker and McCluskey [512] give an algebraic procedure for computing line controllabilities. Seth and Agrawal developed PREDICT, a numerical technique for this purpose [590]. They break the circuit into partitions, known as *supergates*, which completely include reconvergent fanouts. However, their algorithm has exponential computation cost in the number of reconvergent fanouts in each supergate. In the worst case, the entire circuit may be one supergate. While these techniques can compute exact probabilities, one must recognize that the computational complexity of the problem is exponential in the size of the circuit. Seth and Agrawal suggest several heuristics for approximation. Savir *et al.* [568] developed another procedure known as the *cutting algorithm*. They cut selected fanout lines to make the circuit a tree structure and initialize the cut lines to a controllability range $[0, 1]$. The modified network has no reconvergent fanout and controllability ranges are readily computed for all lines.

The fault detection probability can be viewed as the *1-controllability* of a signal that is the XOR of the good and faulty circuit outputs. Thus the methods for computing probabilistic controllability can be used for computing fault detection probabilities. The disadvantage of this approach is that it almost doubles the size of the circuit for which controllabilities must be computed. An alternative approach is to define the observability $OB(l)$ of a line l as the probability of sensitizing a path from l to a PO. An obvious, though incorrect, definition of fault detection probability leads to a simple product formula. For example, the probability of detecting the *stuck-at-0* (sa0) fault on line l by a random input can be written as $C1(l) \times OB(l)$. Brglez [98] developed COP, a probabilistic testability algorithm, which derives computational efficiency by neglecting signal correlations. The error in the product formula for detection probability was observed by Savir [563]. Since the control and observation of a line are not independent events, their probabilities cannot be multiplied. To correct this error, Jain and Agrawal defined *0-observability* and *1-observability* of a line l as probabilities of l being observed – given that it assumed the appropriate value of 0 and 1 [336]. Since their observabilities are *conditional* probabilities, they can be multiplied to appropriate controllabilities, without error, to obtain the probabilities of fault detection. The PREDICT algorithm of Seth and Agrawal [590, 592] uses these definitions of observabilities to obtain exact detection probabilities. Ratiu also developed a fast testability analysis program [536]. Recently, Lee et al. [388] have developed useful high-level testability measures.

6.1 SCOAP Controllability and Observability

Goldstein [262] invented an algorithm to determine the difficulty of controlling (called *controllability*) and observing (called *observability*) signals in digital circuits. Thigpen and Goldstein implemented a computer program to compute controllabilities and observabilities [263]. These notions originated in control theory, and earlier work was done by Rutman [555]. However, Goldstein was the first to propose a systematic, efficient algorithm to compute these measures, which he called SCOAP*. It is still widely used.

SCOAP consists of six numerical measures for each signal (l) in the circuit:

1. Combinational 0-controllability, $CC0(l)$

2. Combinational 1-controllability, $CC1(l)$

3. Combinational observability, $CO(l)$

4. Sequential 0-controllability, $SC0(l)$

5. Sequential 1-controllability, $SC1(l)$

6. Sequential observability, $SO(l)$

---

*Acronym for *Sandia Controllability/Observability Analysis Program*.

Roughly speaking, the three combinational measures are related to the number of signals that may be manipulated to control or observe l. The three sequential measures are related to the number of time-frames (or clock cycles) needed to control or observe. The controllabilities range between 1 and ∞, and observabilities lie between 0 and ∞. The higher the measures for a line, the more difficult it will be to control or observe.

6.1.1 Combinational SCOAP Measures

In Goldstein's [262] method of calculating controllabilities, the first step is to set the difficulty of controlling each *primary input* (PI) to 0 (called $CC0$) to the value 1 and the difficulty of controlling each PI to 1 (called $CC1$) to the value 1. We progress through the circuit in a forward pass, in *level order*. The *level* of a logic gate is the maximum of the distances (in logic gates) of its various inputs from the PIs. Thus if we calculate controllabilities of logic gates in order of increasing level number, then we will only process logic gates whose input signal controllabilities ($CC0$ and $CC1$) have already been determined.

For each logic gate that we traverse, we add 1 to the controllability. This accounts for the logic depth. If a logic gate output is produced by setting only one input to a controlling value, then:

$$\text{output controllability} = min \text{ (input controllabilities)} + 1$$

If a logic gate output can only be produced by setting all inputs to a non-controlling value, then:

$$\text{output controllability} = \sum \text{(input controllabilities)} + 1$$

If an output can be controlled by multiple input sets (e.g., a two-input XOR gate where "01" and "10" input sets will both cause a 1 output), then:

$$\text{output controllability} = min \text{ (controllabilities of input sets)} + 1$$

Figure 6.2 shows the output controllability calculation for all of the basic digital logic gates. The error arises in the controllability calculation due to reconvergent fanout (see Figure 6.1) where the reconverging signals may correlate, and therefore the controllability becomes inaccurate at the reconvergence point. Goldstein's procedure may overestimate or underestimate the controllability difficulty by assuming that reconverging signals are independent.

After all controllabilities are established, we compute observabilities in a reverse pass starting from *primary outputs* (POs) and moving backwards to the PIs. We first set the output observability difficulty (called CO) to 0, making no distinction between logic 0 and 1 in observabilities. For a logic gate with an input signal that needs to be observed, the difficulty of observing that input equals the observability of the output plus the difficulty of setting all other inputs to non-controlling values, plus 1 to account for the logic depth. For instance, the difficulty of observing the

6.1 SCOAP Controllability and Observability

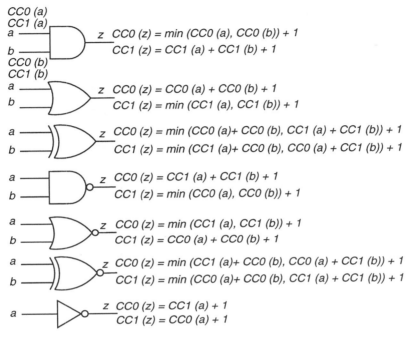

Figure 6.2: SCOAP controllability calculation.

top input a of the AND gate in Figure 6.3 becomes $CO(z) + CC1(b) + 1$. Figure 6.3 shows the observability calculation for all of the basic logic gates.

The accuracy problem arises when we compute the observability of a fanout stem with n branches. A *fanout stem* is merely a signal (a logic gate output or a PI) that branches to many different places, each of which is called a *fanout branch*. The necessary conditions for controlling, observing, and testing faults on fanout branches frequently differ from the corresponding conditions for the stem. One might be inclined to bound this stem observability between min (all fanout branch observabilities) and max (all fanout branch observabilities). In the former case, a min function is used because we assume that the events of observing a signal through each fanout stem branch are independent, and therefore we observe through the branch with the lowest observability difficulty. In the latter case, a max function is used because we assume that the events of observing a signal through each fanout stem branch are totally dependent, and therefore the branch that is hardest to observe gives the correct observability. These two scenarios ignore the reality that a signal may need to be simultaneously propagated through some or all fanout branches to make it observable. If that were the case, then the stem observability becomes:

$$\sum \text{ or } min \text{ (some or all fanout branch observabilities)}$$

An even rarer worst case is that all branches must be able to propagate the fault effect in order for it to be observable, but the propagation conditions for the branches

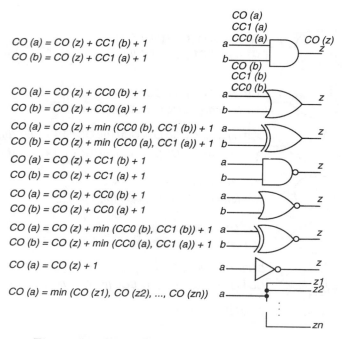

$CO(a) = CO(z) + CC1(b) + 1$
$CO(b) = CO(z) + CC1(a) + 1$

$CO(a) = CO(z) + CC0(b) + 1$
$CO(b) = CO(z) + CC0(a) + 1$

$CO(a) = CO(z) + min(CC0(b), CC1(b)) + 1$
$CO(b) = CO(z) + min(CC0(a), CC1(a)) + 1$

$CO(a) = CO(z) + CC1(b) + 1$
$CO(b) = CO(z) + CC1(a) + 1$

$CO(a) = CO(z) + CC0(b) + 1$
$CO(b) = CO(z) + CC0(a) + 1$

$CO(a) = CO(z) + min(CC0(b), CC1(b)) + 1$
$CO(b) = CO(z) + min(CC0(a), CC1(a)) + 1$

$CO(a) = CO(z) + 1$

$CO(a) = min(CO(z1), CO(z2), ..., CO(zn))$

Figure 6.3: SCOAP observability calculation.

are not independent events. The observation condition for one branch may disable propagation for another, so the stem observability is ∞, even though the individual branches have low observability difficulties. These pathological cases occasionally happen. The computational complexity of exact controllability and observability computation is similar to that of the *automatic test-pattern generation* (ATPG) computational complexity [231], so Goldstein uses the following approximation:

$$CO(\text{stem}) = min(CO(\text{branches}))$$

Observability calculation errors often occur, so we should expect that ATPG algorithms guided by controllability/observability measures may be misled. With Goldstein's approximation, controllability/observability calculation has only $O(2 \times n)$, or $O(n)$ complexity, and is quick. Note that the controllability and observability numbers approximate the number of circuit lines that must be set to control or observe a given circuit line. An astonishing fact is that in combinational circuits of 10 to 30,000 logic gates, with hard-to-test internal signals, $CC0$, $CC1$, and CO difficulty measures may overflow 16-bit integers and require 32-bit integers to be represented.

6.1.2 Combinational Circuit Example

Consider the sequential circuit in Figure 6.4. We will assume that the flip-flops in this circuit have special testing hardware attached to them so that it is possible

6.1 SCOAP Controllability and Observability

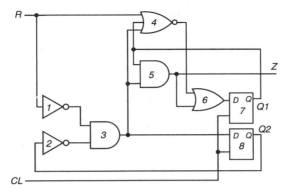

Figure 6.4: A sequential circuit example.

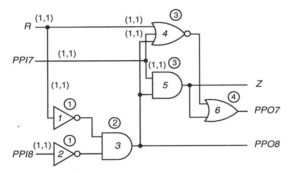

Figure 6.5: Combinational circuit with controllabilities at level 0.

to read out the present state of the flip-flops, and also it is possible to set the present state of the flip-flops. Then, for testing purposes, the two flip-flops can be modeled as a pair of primary input, primary output pairs. The D line is referred to as a *pseudo-primary output*, (PPO), since it can be observed by the flip-flop testing hardware, and the Q line is referred to as a *pseudo-primary input* (PPI), since it can be set by the flip-flop testing hardware. Figure 6.5 shows the circuit of Figure 6.4 with the flip-flops replaced by PPI-PPO pairs. Flip-flop *7* in Figure 6.4 has been replaced by a PPI for the $Q1$ line (called *PPI7*) and a PPO for the D line (called *PPO7*) in Figure 6.5. Flip-flop *8* in Figure 6.4 has been replaced by a PPI for the $Q2$ line (called *PPI8*) and a PPO for the D line (called *PPO8*) in Figure 6.5.

We first label the gates in *level order*, which means that each gate encountered along any path from a PI to a PO is labeled with a level number giving its maximum distance (in gates) from a PI. In Figure 6.5, the level numbers are shown above each gate in circles. In this example, PIs R, *PPI7*, and *PPI8* have distance 0. Next, the level number algorithm looks at where the fanouts of these PIs go, and labels each of the fanouts with level number 1. Thus, in Figure 6.5, the top two fanins of gate *4*, the top fanin of gate *3*, and the inputs to inverters *1* and *2* are labeled with level 0. When all of the inputs to a logic gate are labeled, we label the output of the gate with the maximum level number of its inputs plus 1. In this case, INVERTERS *1* and *2* are immediately labeled with the level number of their inputs plus 1, which is 1. Because of that, we label the inverter fanouts with level number 1. Since both inputs to AND gate *3* are now labeled with level 1, we label the output of AND gate *3* as being at level number 2, and we label the bottom inputs of logic gates *4* and *5* with level number 2. Now, since all inputs to NOR gate *4* are labeled with level numbers, we can label the output of the NOR gate with level 3, and therefore the top input of OR gate *6* as level number 3. Also, since both inputs to AND gate *5* are labeled with level numbers, we label AND gate *5* as being at level number 3, the maximum of the level numbers of its inputs plus 1. Then, the bottom input of OR gate *6* is labeled with level number 3, and therefore its output is labeled as being at level number 4, the maximum level number of its inputs plus 1. The following algorithm does the level ordering:

Algorithm 6.1 *Levelization algorithm. Level number labeling from PIs to POs.*

1. *Assign level number 0 to all primary inputs.*

2. *For each PI fanout:*

    (a) *Label that circuit line with the level number of the PI, and*

    (b) *Queue the logic gate driven by that fanout line.*

3. *While the queue is not empty:*

    (a) *Dequeue the next logic gate in the queue.*

    (b) *If all of the gate fanins are labeled with level numbers, then label the logic gate and its fanouts with the maximum of its input level numbers + 1 and queue all fanouts of the logic gate. Otherwise, requeue the logic gate.*

Depending on the application, we can adjust this algorithm to label the POs of the circuit with one more level number than the logic gates driving them, or we can label the POs with the level number of the logic gates that drive them.

Figure 6.5 shows the operation of Goldstein's controllability and observability calculation algorithm. Each line is labeled with a $(CC0, CC1)$ pair to show its controllabilities. We first assign $(1,1)$ to all PIs: R, $PPI7$, and $PPI8$. Next, these ordered pairs are propagated to all fanout branches of the PIs, leading to the labeling of the top two inputs of gate *4*, the top input of gate *5*, and the inputs to INVERTERS *1* and *2* with $(1,1)$. Referring to Figure 6.2, we see that for the inverter, $CC1(OUTPUT) = CC0(INPUT) + 1$ and $CC0(OUTPUT) = CC1(INPUT) + 1$. This leads to the outputs of INVERTERS *1* and *2* being labeled with $(2,2)$, and both inputs of logic gate *3* being labeled with $(2,2)$, as shown in Figure 6.6. The algorithm processes each logic gate in level number order. This guarantees that when a logic gate is processed, all of its input controllabilities will be known, so the output controllabilities can be computed. So far, we have processed all of the PIs at

6.1 SCOAP Controllability and Observability

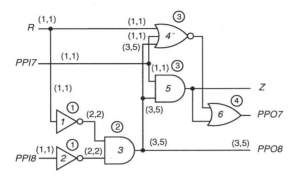

Figure 6.6: Combinational circuit with controllabilities through level 2.

level 0, and the two INVERTERS at level 1. Next, we process AND gate 3. From Figure 6.2, we see that for the AND gate:

$$CC0(3) = min(CC0(1), CC0(2)) + 1 = min(2, 2) + 1 = 3$$
$$CC1(3) = CC1(1) + CC1(2) + 1 = 2 + 2 + 1 = 5$$

This causes the AND gate 3 output, the bottom inputs to gates 4 and 5, and $PPO8$ to be labeled with $(3, 5)$ in Figure 6.6. Next, the algorithm processes logic gates at level number 3. For NOR gate 4:

$$CC0(4) = min(CC1(R), CC1(PPI7), CC1(3)) + 1$$
$$= min(1, 1, 5) + 1 = 2$$
$$CC1(4) = CC0(R) + CC0(PPI7) + CC0(3) + 1 = 1 + 1 + 3 + 1 = 6$$

Hence, logic gate 4 and the top input of logic gate 6 in Figure 6.7 are labeled with $(2, 6)$. For AND gate 5:

$$CC0(5) = min(CC0(PPI7), CC0(3)) + 1 = min(1, 3) + 1 = 2$$
$$CC1(5) = CC1(PPI7) + CC1(3) + 1 = 1 + 5 + 1 = 7$$

Therefore, logic gate 5, the bottom input of logic gate 6, and PO Z in Figure 6.7 are labeled with $(2, 7)$. Now we process OR gate 6 at level 4. For OR gate 6:

$$CC0(6) = CC0(4) + CC0(5) + 1 = 2 + 2 + 1 = 5$$
$$CC1(6) = min(CC1(4) + CC1(5)) + 1 = min(6, 7) + 1 = 7$$

This causes OR gate 6 and $PPO7$ in Figure 6.7 to be labeled with $(5, 7)$. The circuit is now completely labeled with the final controllabilities.

Next, we renumber the circuit level numbers from POs backwards to PIs, so that each gate is labeled with the maximum distance in logic gates of any of its fanouts from a PO. The procedure is similar to Algorithm 6.1, except that we now start at POs and work backwards. Figure 6.8 shows all gate level numbers in square boxes.

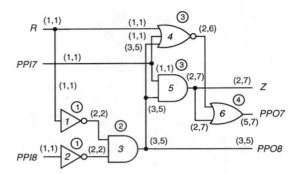

Figure 6.7: Final combinational circuit controllabilities.

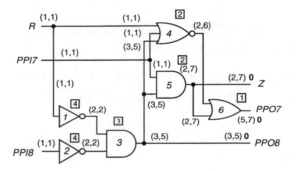

Figure 6.8: Combinational circuit with observabilities through level 1.

We now compute observabilities, which are shown in **boldface** in the figures after the ($CC0, CC1$) ordered pairs. We process logic gates in level order from POs backwards to PIs. In Figure 6.8, we assign 0 to PO Z and PPOs *PPO7* and *PPO8*. This causes observability 0 to be assigned to gate *6*, as well. We cannot yet assign observabilities to gates *5* and *3*, because they are fanout stems, and not all fanout branch observabilities are known. So, we process OR gate *6* at level 1 in Figure 6.9. For each input, according to Figure 6.3, we have:

$$CO(4) = CC0(5) + CO(6) + 1 = 2 + 0 + 1 = 3$$
$$CO(5) = CC0(4) + CO(6) + 1 = 2 + 0 + 1 = 3$$

This is shown in Figure 6.9. Next, we process NOR gate *4* at level 2 in Figure 6.9. According to Figure 6.3, for a three-input OR gate:

$$CO(\text{1st input}) = CC0(\text{2nd input}) + CC0(\text{3rd input}) + CO(\text{output}) + 1$$

Therefore,

$$CO(R) = CC0(PPI7) + CC0(3) + CO(4) + 1 = 1 + 3 + 3 + 1 = 8$$
$$CO(PPI7) = CC0(R) + CC0(3) + CO(4) + 1 = 1 + 3 + 3 + 1 = 8$$
$$CO(3) = CC0(R) + CC0(PPI7) + CO(4) + 1 = 1 + 1 + 3 + 1 = 6$$

6.1 SCOAP Controllability and Observability

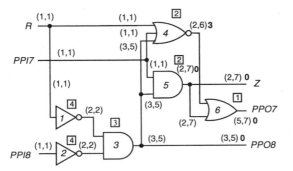

Figure 6.9: Combinational circuit with observabilities through level 2.

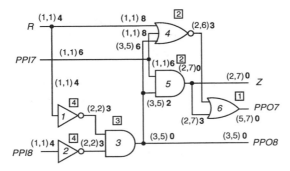

Figure 6.10: Final combinational circuit observabilities.

This is shown in Figure 6.10. Next, we process AND gate 5 at level 2 in Figure 6.9. All of its fanout branch observabilities are now known. Since its stem observability is the minimum of all branch observabilities, $CO(5) = min(0,3) = 0$. According to Figure 6.3, for a two-input AND gate:

$$CO(\text{any input}) = CC1(\text{other input}) + CO(\text{output}) + 1$$

Therefore,

$$CO(PPI7) = CC1(3) + CO(5) + 1 = 5 + 0 + 1 = 6$$
$$CO(3) = CC1(PPI7) + CO(5) + 1 = 1 + 0 + 1 = 2$$

This is shown in Figure 6.10.

Next, we process AND gate 3 at level 3 in Figure 6.10, according to Figure 6.3. $CO(3) = min(6, 2, 0) = 0$. Thus,

$$CO(1) = CC1(2) + CO(3) + 1 = 2 + 0 + 1 = 3$$
$$CO(2) = CC1(1) + CO(3) + 1 = 2 + 0 + 1 = 3$$

These are shown in Figure 6.10. Now we process the two INVERTERS at level 4. From Figure 6.3, $CO(input) = CO(output) + 1$. For INVERTER 1, we get

Table 6.1: Sequential SCOAP measures for two-input gates.

The sequential controllability measures $SC0$ and $SC1$ for combinational gates are given by Figure 6.2, but without adding the 1 shown in the equations.	The sequential observability measure SO for a combinational gate is given by Figure 6.3, but again without adding the 1 shown in the equations.

$CO(R) = 3 + 1 = 4$ and for INVERTER 2, we get $CO(PPI8) = 3 + 1 = 4$. For PI R, which is a fanout stem, the observability is the minimum of all branch observabilities, so $CO(R) = min(8,4) = 4$. For PPI $PPI7$, $CO(PPI7) = min(6,8) = 6$. For PPI $PPI8$, the observability is 4. In Figure 6.10, all controllabilities and observabilities are now known. Note that the computational complexity of this algorithm is $O(2n) = O(n)$, where n is the number of gates. This is because we process all inputs and logic gates once in the forward pass to compute controllabilities, and then processes all logic gates and inputs again in the backward pass to compute observabilities. We see that the fanout branches from PPI $PPI7$ to gate 4 and from R to gate 4 are the hardest signals to observe. The hardest signals to control to logic 1 are $PPO7$ and all fanouts from gate 5. The hardest signal to control to logic 0 is $PPO7$.

6.1.3 Sequential SCOAP Measures

There are two main differences in the sequential measures from the combinational controllability and observability measures:

1. One increments the sequential measure by 1 only when signals propagate from flip-flop inputs to Q or $\overline{Q}$ outputs, or from flip-flop outputs backwards to D, C (clock), SET, or $RESET$ inputs.

2. One must iterate in calculating controllability numbers in sequential circuits because of feedback loops involving flip-flops.

Sequential controllabilities $SC0$ and $SC1$ roughly measure the number of times various flip-flops must be clocked to control a signal. Thus, if a given line l can only be set to 1 by clocking flip-flop a twice and flip-flop b three times, then we would expect $SC1(l) = 5$. *Sequential observability SO* measures the number of times various flip-flops must be clocked to observe a signal. In a sequential circuit, the combinational controllabilities and observabilities roughly measure the number of lines that must be set, over all of the required clock periods, in order to control or observe a combinational signal. The sequential controllability and observability equations for basic logic gates differ from the equations for combinational gates only in that a 1 is not added as we move from one level of logic to another, but rather a 1 is added when a signal passes through a flip-flop. The procedure to convert the combinational measure formulas of Figures 6.2 and 6.3 to those of sequential measures is given in Table 6.1.

6.1 SCOAP Controllability and Observability

Figure 6.11: Resettable, negative-edge-triggered D flip-flop.

Figure 6.11 shows a resettable negative-edge-triggered D flip-flop. In order to control the Q line to 1, one must set D to 1, cause a falling clock (C) edge (first a 1 and then a 0), and control the $RESET$ line to 0 to avoid clearing Q. The combinational and sequential difficulties of controlling Q to a 1 are:

$$CC1(Q) = CC1(D) + CC1(C) + CC0(C) + CC0(RESET) \quad (6.1)$$
$$SC1(Q) = SC1(D) + SC1(C) + SC0(C) + SC0(RESET) + 1$$

$CC1$ measures how many *lines* in the circuit must be set to make Q as 1, whereas $SC1$ measures how many *flip-flops* in the circuit must be clocked to set Q to 1. There are two ways to set Q to a 0. We can either use the $RESET$ line while holding clock C at 0, or clock a 0 into Q through the D line. Thus,

$$\begin{aligned} CC0(Q) = min \quad & (CC1(RESET) + CC0(C), \\ & CC0(D) + CC1(C) + CC0(C) + CC0(RESET)) \end{aligned} \quad (6.2)$$
$$\begin{aligned} SC0(Q) = min \quad & (SC1(RESET) + SC0(C), \\ & SC0(D) + SC1(C) + SC0(C) + SC0(RESET)) + 1 \end{aligned}$$

The D line can be observed at Q by holding $RESET$ low and generating a falling edge on the clock line C:

$$CO(D) = CO(Q) + CC1(C) + CC0(C) + CC0(RESET) \quad (6.3)$$
$$SO(D) = SO(Q) + SC1(C) + SC0(C) + SC0(RESET) + 1$$

$RESET$ can be observed by setting Q to a 1 and using $RESET$:

$$CO(RESET) = CO(Q) + CC1(Q) + CC0(C) + CC1(RESET) \quad (6.4)$$
$$SO(RESET) = SO(Q) + SC1(Q) + SC0(C) + SC1(RESET) + 1$$

There are two ways to indirectly observe the clock line C:

1. Set Q to 1 and clock in a 0 from D, or
2. Reset the flip-flop and clock in a 1 from D.

$$\begin{aligned} CO(C) = min \quad [&CO(Q) + CC0(RESET) + CC1(C) + \\ & CC0(C) + CC0(D) + CC1(Q), \\ & CO(Q) + CC1(RESET) + CC1(C) + \\ & CC0(C) + CC1(D)] \end{aligned} \quad (6.5)$$

$$\begin{aligned}SO(C) = min \quad & [SO(Q) + SC0(RESET) + SC1(C) + \\ & SC0(C) + SC0(D) + SC1(Q), \\ & SO(Q) + SC1(RESET) + SC1(C) + \\ & SC0(C) + SC1(D)] + 1\end{aligned}$$

Here is the algorithm to set both combinational and sequential measures:

Algorithm 6.2 *Compute combinational and sequential measures [262]:*

1. *For all PIs I, set $CC0(I) = CC1(I) = 1$ and $SC0(I) = SC1(I) = 0$.*

2. *For all other nodes N, set $CC0(N) = CC1(N) = \infty$ and $SC0(N) = SC1(N) = \infty$.*

3. *Working from PIs to POs, use the $CC0$, $CC1$, $SC0$, and $SC1$ equations to map logic gate and flip-flop input controllabilities into output controllabilities. Iterate until the controllability numbers stabilize in feedback loops.*

4. *For all POs U, set $CO(U) = SO(U) = 0$.*

5. *For all other nodes N set $CO(N) = SO(N) = \infty$.*

6. *Working from POs to PIs, use the CO and SO equations and the pre-computed controllabilities to map output node controllabilities of gates and flip-flops into input controllabilities. For fanout stems Z with branches Z1, ..., ZN, $SO(Z) = min(SO(Z1), ..., SO(ZN))$ and $CO(Z) = min(CO(Z1), ..., CO(ZN))$.*

7. *If any node remains with $CC0(SC0) = \infty$, then that node is 0-uncontrollable. If any node remains with $CC1(SC1) = \infty$, then that node is 1-uncontrollable. If any node remains with $CO = \infty$ or $SO = \infty$, then that node is unobservable. These are sufficient but not necessary conditions.*

Convergence of the iterations is guaranteed because the controllability numbers are monotonically non-increasing between iterations. Usually, the algorithm converges in a few (two to three) iterations. The equations for various latches and flip-flops are analogous to those for the D flip-flop. Controllability cost functions can be adjusted with *correction terms* to account for the additional difficulty of setting circuit lines due to reconvergent fanout [530, 9, 92]. The SCOAP measures can be enhanced to reflect this by adding in the term $f_l - 1$ to every controllability calculation for $CC0$, $CC1$, $SC0$, and $SC1$, where f_l is the number of fanouts of a line l.

6.1.4 Sequential Circuit Example

On circuit lines in this example, $CC0$, $CC1$, and CO are shown as $(CC0, CC1)$**CO** and $SC0$, $SC1$, and SO are shown as $[SC0, SC1]$**SO** below the combinational measures. The observabilities are always shown in **bold** to avoid

6.1 SCOAP Controllability and Observability

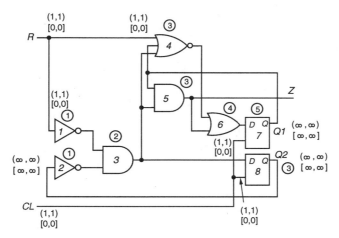

Figure 6.12: Initialization for computing sequential SCOAP measures.

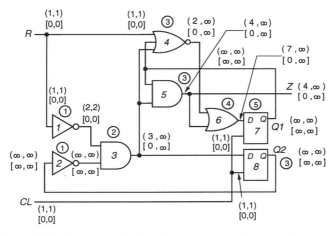

Figure 6.13: Sequential circuit measures after one iteration.

confusion. The forward level numbers for logic gates and flip-flops are shown circled over the gates. For backward level numbers during observability calculation, we just use the forward level numbers in decreasing order from 5 down to 1. For level numbering flip-flops, we treat them like ordinary logic gates, and ignore their feedback loops.

Figure 6.12 shows the initial measures for the example of Figure 6.4. Signals R and CL, and all of their fanouts, are set to $(CC0, CC1) = (1,1)$ and $[SC0, SC1] = [0,0]$. All other nodes, particularly $Q1$ and $Q2$, are set to $(CC0, CC1) = (\infty, \infty)$ and $[SC0, SC1] = [\infty, \infty]$.

In Figure 6.13, we start the input to output computation of controllabilities. Thus the output of INVERTER *1* is set to $(CC0, CC1) = (2,2)$ and $[SC0, SC1] = [0,0]$ according to Figure 6.3 and Table 6.1. INVERTER *2* is more interesting,

144   **Chapter 6. TESTABILITY MEASURES**

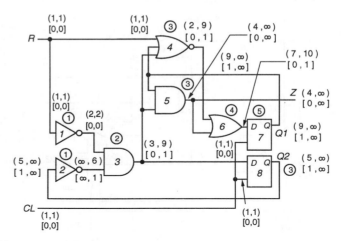

Figure 6.14: Sequential circuit measures after two iterations.

because the feedback loop means that its measures must remain at ∞. For AND gate *3*, since 0 is a controlling value, we can determine $CC0$ and $SC0$ measures of its output from Figure 6.3 and Table 6.1, as shown in Figure 6.13. However, its $CC1$ and $SC1$ measures remain at ∞. A similar situation arises at AND gate *5* in Figure 6.13. For NOR gate *4*, $CC1(4) = min(CC0(R), CC0(Z), CC0(3)) + 1 = min(1, \infty, 3) + 1 = 2$. Similarly, $SC1(4) = 0$. An analogous situation occurs at OR gate *6*. At this point, 0-controllabilities are defined on both D inputs to the flip-flops, but 1-controllabilities remain ∞.

Figure 6.14 shows the situation after two iterations. By Equations 6.2, $CC0(7) = CC0(6) + CC1(CL) + CC0(CL) = 7 + 1 + 1 = 9$. The $CC1$, $SC0$, and $SC1$ measures for both flip-flops follow from Equations 6.2 and 6.1. Now, output controllabilities of INVERTER *2* change to $(\infty, 6)$ and $[\infty, 1]$. This allows AND gate *3*'s output $CC1$ measure to change to 9 and its output $SC1$ measure to change to 1. For NOR gate *4*, the output 1-controllabilities are now defined, because all of its input 0-controllabilities have been defined, so gate *4* has $CC1 = 9$ and $SC1 = 1$. AND gate *5* cannot have its output measures updated, because the 1-controllabilities of its inputs are still undefined. The 1-controllabilities of OR gate *6* can now be defined, because its topmost input now has defined 1-controllabilities. So, $CC1(6) = 10$ and $SC1(6) = 1$. At this point, Figure 6.14 shows the situation where the controllabilities of both D inputs to the flip-flops are now completely defined.

Figure 6.15 shows the situation after three iterations. Equations 6.2 and 6.1 cause flip-flop *7*'s output controllabilities to become $(9, 12)$ and $[1, 2]$. Flip-flop *8*'s controllabilities become $(5, 11)$ and $[1, 2]$. INVERTER *2*'s output controllabilities become $(12, 6)$ and $[2, 1]$. However, this causes no change in AND gate *3*'s output controllabilities. NOR gate *4*'s $CC1$ increases to 14, and AND gate *5*'s $CC1$ is now defined as 22 and its $SC1$ is now defined as 3. The changes at gates *4* and *5* now cause a change at gate *6*: $CC1$ increases to 15. Figure 6.15 illustrates the situation at this point.

6.1 SCOAP Controllability and Observability

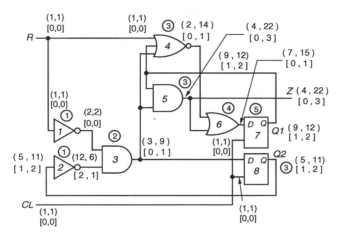

Figure 6.15: Sequential circuit measures after three iterations.

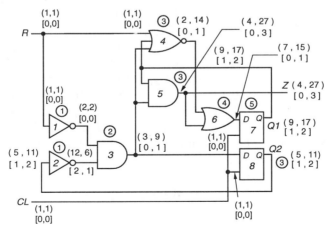

Figure 6.16: Sequential circuit measures after stabilization.

Chapter 6. TESTABILITY MEASURES

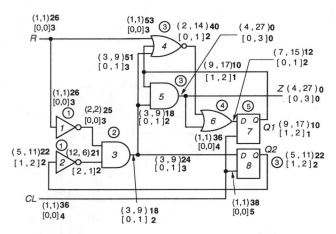

Figure 6.17: Sequential circuit final observability measures.

Figure 6.16 illustrates the results of the final iteration, in which the circuit stabilizes. $CC1$ of flip-flop 7 increases to 17, but all other flip-flop controllability measures do not change. This change has no effect on NOR gate 4, but it causes $CC1$ of AND gate 5 to increase to 27. Thus, $CC1$ of PO Z increases to 27. This change has no effect on OR gate 6, so the circuit controllabilities are now completely stabilized.

Rather than presenting the calculation of observabilities in running text, we list in Table 6.2 all of the observabilities in their order of computation, and give the equation or table that allows us to compute the observability.

The only difficult step is computing CO from line CL to 7 which is $CO = CO(Q) + CC1(CL) + CC0(CL) + CC0(D) + CC1(Q) = 10+1+1+7+17 = 36$ by Equation 6.5. For the same line, $SO = SO(Q) + SC1(CL) + SC0(CL) + SC0(D) + SC1(Q) + 1 = 1+0+0+0+2+1 = 4$, also by Equation 6.5. For the CO of the line from CL to line 8, $CO = CO(Q) + CC1(CL) + CC0(CL) + CC0(D) + CC1(Q) = 22+1+1+3+11 = 38$ by Equation 6.5. Similarly, $SO = SO(Q) + SC1(CL) + SC0(CL) + SC0(D) + SC1(Q) + 1 = 2+0+0+0+2+1 = 5$, also by Equation 6.5. Therefore, $CO(CL) = min(36, 38) = 36$ and $SO(CL) = min(4, 5) = 4$. Observabilities are now completely computed in Figure 6.17. Note that no iterations are required to compute observabilities. The clock line CL is the hardest signal to observe, both combinationally and sequentially.

SCOAP may also be used to predict the length of the test vector set for a circuit. The testabilities of the stuck-at faults at node x are defined as:

$$T(x \text{ stuck-at-0}) = CC1(x) + CO(x) \quad (6.6)$$
$$T(x \text{ stuck-at-1}) = CC0(x) + CO(x)$$
$$\text{Testability index} = log \sum_{all\ f_i} T(f_i)$$

In order to detect a fault at x, one must set x to the opposite value from the fault

6.1 SCOAP Controllability and Observability 147

Table 6.2: Observability calculation steps for the circuit of Figure 6.17.

Signal	Observability measure	Observability value	Equation or Table
Z	CO	0	Figure 6.3
Z	SO	0	Table 6.1
5	CO	0	Figure 6.3
5	SO	0	Table 6.1
7	CO	10	Figure 6.3
7	SO	1	Table 6.1
3	CO	18	Figure 6.3
3	SO	2	Table 6.1
6	CO	12	Equations 6.3
6	SO	2	Equations 6.3
5 to 6	CO	14	Figure 6.3
5 to 6	SO	2	Table 6.1
Confirm: $CO(5) = 0$ and $SO(5) = 0$			
4	CO	40	Figure 6.3
4	SO	2	Table 6.1
R to 4	CO	53	Figure 6.3
R to 4	SO	3	Table 6.1
Confirm: $CO(7) = 10$ and $SO(7) = 1$			
3 to 4	CO	51	Figure 6.3
3 to 4	SO	3	Table 6.1
Confirm: $CO(3) = 18$ and $SO(3) = 2$			
1	CO	25	Figure 6.3
1	SO	3	Table 6.1
2	CO	21	Figure 6.3
2	SO	2	Table 6.1
R to 1	CO	26	Figure 6.3
R to 1	SO	3	Table 6.1
Set: $CO(R) = 26$ and $SO(R) = 3$			
8	CO	22	Figure 6.3
8	SO	2	Table 6.1
3 to 8	CO	24	Equations 6.3
3 to 8	SO	3	Equations 6.3
Confirm: $CO(3) = 18$ and $SO(3) = 2$			

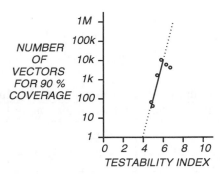

Figure 6.18: Test vectors versus testability index.

and observe x at a PO. Figure 6.18 [39] shows a linear relationship between the number of vectors needed for 90% fault coverage and the *Testability index* for a number of circuits.

6.2 High-Level Testability Measures

Lee, Wolf, Jha, and Acken have extensively researched behavioral synthesis for testability of VLSI circuits [388]. *Behavioral synthesis* is the process of automatically translating a *behavioral* level circuit description into an optimized *logic* level circuit description. From this research has emerged the first viable measures for testability analysis of finite state machines at the *behavioral* level. They also produced the first behavioral-level synthesis program, PHITS, to improve design testability while optimizing the chip design for minimal chip area.

Improved testability at the behavioral level means designing a register-transfer structure to which one can easily apply a test sequence to excite a fault from the PIs (*easy controllability*) and to propagate the fault effect to the POs (*easy observability*.) They analyzed testability via the *data flow graph* (DFG), where each *node* represents a register and each *arc* represents a combinational circuit of a functional module connecting two registers, or an input to a register, or a register to an output port. An example of an arc would be a full adder taking inputs from two input registers and supplying the sum of the two input registers to a third register. The three registers would be nodes, and there would be arcs from each of the two input registers to the third register to represent the full adder. The *length of a sequential path* between two registers is the number of arcs along the path. The *sequential depth* between a register pair is the length of the shortest sequential path between them. The sequential depth can be viewed as a high-level testability measure.

This system succeeds by following two data path synthesis rules:

1. Rule *SR1*. Improve observability and controllability of registers. Whenever possible, allocate a register to at least one PI or PO variable.

2. Rule *SR2*. Reduce the sequential depth between registers. Reduce the sequen-

6.2 High-Level Testability Measures

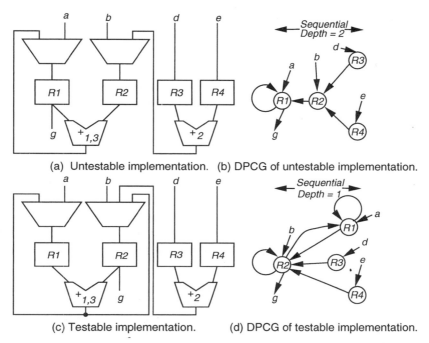

Figure 6.19: Example of high-level testability enhancement.

tial depth from a controllable register to an observable register.

Application of these rules depends only on the *register-transfer level* (RTL) design structure.

Example 6.1 *Consider the data path of Figure 6.19(a) [388]. Registers R1 and R2 and adder $+_{1,3}$ are easy-to-test because of relatively easy controllability of a and b (inputs to R1 and R2) and easy observability of g. However, registers R3 and R4 and adder $+_2$ have problems. Although d and e are easy to control, R3, R4, and adder $+_2$ are hard to observe from g. The fault must propagate through an additional register R2 before observation of it from g is possible. Note that a register self-loop has sequential depth 0. The DPCG in Figure 6.19(b) shows the sequential depths for this implementation:*

$R1 \to R1$: 0 $e \to g$: 4
$R2 \to R1$: 1 $d \to g$: 4
$R3 \to R1$: 2 $a \to g$: 2
$R4 \to R1$: 2 $a \to R1$: 1
                    $b \to g$: 3

One can reduce the sequential depths using very little chip area. Figure 6.19(c) shows an alternate implementation, produced by using a different hardware sharing (binding) of the required hardware operators to the actual hardware during various time steps. Figure 6.19(d) shows the DPCG of this alternate implementation. It produced these sequential depths, some of which are reduced:

$R3 \to R1$: 2

$a \to R1$: 1

$d \to R3$: 1

$e \to R4$: 1

$e \to g$: 3

$b \to g$: 2

$a \to g$: 3

$d \to g$: 3

$e \to R2$: 2

$d \to R2$: 2

$R4 \to R1$: 2

They used STEED [247] to generate tests for both versions of this 4-bit circuit. The fault coverage for the Figure 6.19(a) implementation was only 89.28% (obtained in 24.5 CPU s), but for the Figure 6.19(c) implementation it was 100% (obtained in 10.7 CPUs.) Their prediction turned out to be true: For the implementation of Figure 6.19(a), most undetected faults were aborted faults at R3, R4, and adder $+_2$.

They obtained additional results of the SR1 and SR2 rules on the *DiffEq* benchmark, where they achieved 99.78% fault coverage in 709.0 CPU s, with a maximum sequential depth of 1. On the 4-bit *Tseng* benchmark, they achieved 99.21% fault coverage in 426.8 CPU s, with a maximum sequential depth of 1. These results confirm the value of their testability measures.

6.3 Summary

What do controllability and observability tell us? [38, 324] They are incredibly useful in guiding ATPG algorithms during *backtracing* [258] (see Chapter 7), since they give the backwards path of least resistance in justifying a desired internal circuit signal by setting signals at PIs. The backtracing method uses $CC0$ measures, if it is tracing backwards in logic to justify a logic 0, and it uses $CC1$ measures for backtracing to justify logic 1. Observability measures tell us the path of least resistance in forward propagating fault effects to POs, although we have to be cautious since controllabilities have errors from reconvergent fanout and observabilities have errors introduced by fanout stems. This guidance may reduce ATPG time by orders of magnitude. Also, the measures tell the VLSI designer which parts of his circuit are extremely hard-to-test, and indicate where redesign or special-purpose test hardware is mandatory to achieve high (100%) fault coverage. Finally, the probability versions of controllabilities and observabilities are incredibly useful for estimating fault coverage [293, 294, 295, 296, 334] and test vector set length. Fault coverage estimation for a given vector set can reduce CPU time by orders of magnitude over ordinary fault simulation, and generally has a 3 to 5% error. It is mainly appropriate for early stages of design, where a check is needed to see whether the circuit will be testable. The controllability/observability measures will be used to guide many of the automatic test-pattern generators discussed in Chapters 7 and 8.

Problems

6.1 *SCOAP.* For the circuit of Figure 6.1, compute the combinational SCOAP testability measures (both controllability and observability.)

6.2 *SCOAP.* For the circuit of Figure 6.20, compute the combinational SCOAP testability measures (both controllability and observability.)

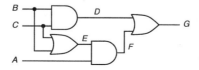

Figure 6.20: Circuit realizing $G = A(B \vee C) \vee BC$.

6.3 *SCOAP.* For the circuit of Figure 6.21, compute the combinational SCOAP testability measures (both controllability and observability.)

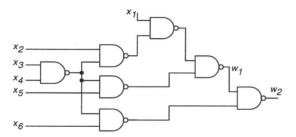

Figure 6.21: Circuit for Problem 6.3.

6.4 *SCOAP.* For the circuit of Figure 6.22, compute the combinational SCOAP testability measures (both controllability and observability.)

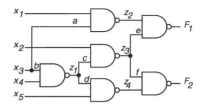

Figure 6.22: Circuit for Problem 6.4.

6.5 *SCOAP.* For the circuit of Figure 6.23, compute the combinational SCOAP testability measures (both controllability and observability.)

6.6 *High-level testability.* For the circuit of Figure 6.24 [249], compute the high-level sequential depth testability measure for all input-output pairs of signals. Also compute the measure for all register-transfer paths.

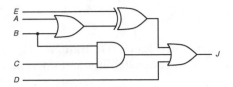

Figure 6.23: Circuit for Problem 6.5.

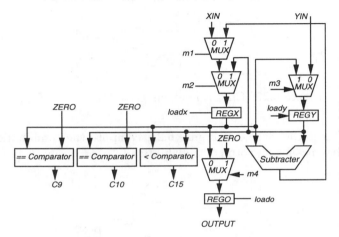

Figure 6.24: Circuit for Problem 6.6.

6.7 *SCOAP.* For the circuit of Figure 6.25, compute the combinational and sequential SCOAP testability measures (both controllability and observability) (including those for CK and $\overline{RESET}$.)

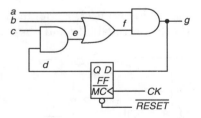

Figure 6.25: Circuit for Problem 6.7.

6.8 *SCOAP.* For the circuit of Figure 6.26, compute the combinational and sequential SCOAP testability measures (both controllability and observability, and including the $CLOCK$ and $\overline{RESET}$ signals.)

6.9 *SCOAP.* For the circuit of Figure 6.27, compute the combinational and sequential SCOAP testability measures (both controllability and observability, and including the $\overline{RESET}$ and $CLOCK$ signals.)

Problems

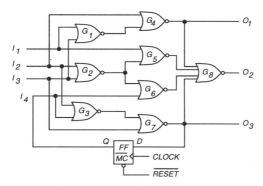

Figure 6.26: Circuit for Problem 6.8.

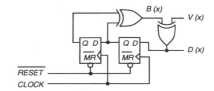

Figure 6.27: Circuit for Problem 6.9.

6.10 *SCOAP*. For the circuit of Figure 6.28, compute the combinational and sequential SCOAP testability measures (both controllability and observability, and including the $CLOCK$ and $\overline{RESET}$ signals.)

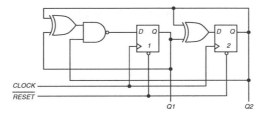

Figure 6.28: Circuit for Problem 6.10.

6.11 *SCOAP*. For the circuit of Figure 6.29, compute the combinational and sequential SCOAP testability measures (both controllability and observability.)

6.12 *SCOAP*. For the circuit of Figure 6.30, compute the combinational and sequential SCOAP testability measures (both controllability and observability, and including the $CLOCK$ and $\overline{RESET}$ signals.)

6.13 *SCOAP*. For the circuit of Figure 6.31, compute the combinational and sequential SCOAP testability measures (both controllability and observability, and including signals $CLOCK$ and $\overline{RESET}$.)

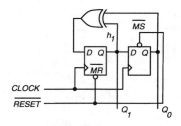

Figure 6.29: Circuit for Problem 6.11.

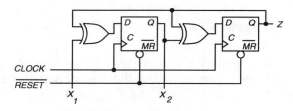

Figure 6.30: Circuit for Problem 6.12.

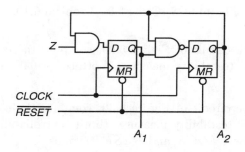

Figure 6.31: Circuit for Problem 6.13.

Chapter 7

COMBINATIONAL CIRCUIT TEST GENERATION

> *"In order for the . . . operation of a test . . . to guarantee that a computing system has no faulty components, the test conditions . . . should be devised at the level of the components themselves, rather than at the level of programmed orders . . . This is the only way in which all conditions of operation of each logical function can be uniquely . . . defined and all logical components within each logical function can be made to perform the task to which they are assigned . . . thereby producing a minimum program which tests and detects failure . . ."*
> — Richard D. Eldred, in 1959 paper,
> "Test Routines Based on Symbolic Logical Statements" [215].

With the above words, Eldred began the era of structural logic circuit testing at Datamatic. Roth's work at IBM marked the true beginning of systematic generation of tests for hardware faults in digital computers and laid the mathematical basis for test-pattern generation. *Automatic test-pattern generation* (ATPG) is the process of generating patterns to test a circuit, which is described strictly with a logic-level *netlist* (schematic.) These algorithms usually operate with a fault generator program, which creates the minimal collapsed fault list (see Chapter 4) so that the designer need not be concerned with fault generation. In a certain sense, ATPG algorithms are multi-purpose, in that they can generate circuit test-patterns, they can find *redundant* or unnecessary circuit logic, and they can prove whether one circuit implementation matches another circuit implementation [409, 410, 717].

We first describe algorithms and representations needed by ATPG. We then introduce *redundancy identification* (RID), a very important benefit of ATPG algorithms. Controllability and observability testability measures (see Chapter 6) are used in all major ATPG algorithms. Finally, we present several key combinational ATPG algorithms, and show their behavior with examples.

Chapter 7. COMBINATIONAL CIRCUIT TEST GENERATION

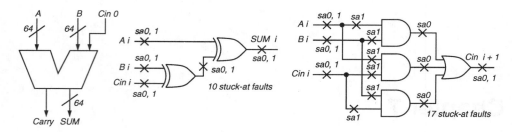

Figure 7.1: 64-bit adder: functional test vs. structural stuck-at fault test.

7.1 Algorithms and Representations

We now discuss the broad categories of ATPG algorithms and the search space representations used by these algorithms.

7.1.1 Structural vs. Functional Test

First, we explain the difference between *structural* and *functional* test. Eldred [215] is credited with switching this field from functional to structural test generation. However, the first publication of the *stuck-at logic* 0 (sa0) or 1 (sa1) fault for test generation was by Galey, Norby, and Roth in 1961 [237]. Later, Seshu and Freeman mentioned the stuck-at fault model for parallel fault simulation [587]. In 1963, Poage presented a theoretical analysis of stuck-at faults [523]. The first structural test method was used to test the Honeywell *Datamatic 1000*, a vacuum tube/diode second-generation mainframe computer.

Functional ATPG programs generate a complete set of test-patterns to completely exercise the circuit function. Figure 7.1 shows a 64-bit ripple-carry adder, and gives a very naive (and inefficient) logic design for one bit slice of the adder, which is sufficient to make our point. From a *functional* point of view, the adder has 129 inputs and 65 outputs. Therefore, to completely exercise its function, we need 2^{129} = 680,564,733,841,876,926,926,749, 214,863,536,422,912 input patterns, producing 2^{65} = 36, 893, 488, 147, 419, 103, 232 output responses. The fastest *automatic test equipment* (ATE), at present, operates at 1 GHz. This ATE would take $2.1580566142 \times 10^{22}$ years to apply all of these patterns to the *circuit-under-test* (CUT), assuming that the tester and circuit can operate at 1 GHz. Thus, we see that an exhaustive functional test is impractical, except for small circuits, and today most circuits tend to be huge.

Structural test, on the other hand, only exercises the minimal set of stuck-at faults on each line of the circuit, after discarding equivalent faults. If we use *fault equivalence* (see Chapter 4), then each bit-slice in the adder would only have 27 faults (see Figure 7.1), ignoring fault equivalence along the carry lines. This adder has no redundant hardware and the total structural fault list will have no more than $64 \times 27 = 1,728$ faults. So we need, at most, 1,728 test-patterns. The 1 GHz ATE would apply these patterns in 0.000001728 s, and since this test-pattern set covers all possible structural stuck-at faults in the adder, it achieves exactly the same fault

7.1 Algorithms and Representations

coverage as the intractable functional test-pattern set described above. Frequently, the circuit designer will provide a limited subset of the functional test-patterns for the circuit, but those typically cover only 70 to 75% of the total number of faults. Testing for only 75% of the modeled failures is of limited value – it will catch only the most severe defects. Thus, we see the importance of ATPG algorithms. The vectors they produce supplement the functional test vectors from the designer to raise the stuck-at fault coverage to 98% or higher levels.

7.1.2 Definition of Automatic Test-Pattern Generator

ATPG algorithms inject a fault into a circuit, and then use a variety of mechanisms to activate the fault and cause its effect to propagate through the hardware and manifest itself at a circuit output. The output signal changes from the value expected for the fault-free circuit, and this causes the fault to be detected. Fault effects are propagated from an AND/NAND gate input to its output by setting other inputs to 1, a *non-controlling* value for AND/NAND. Fault effects are propagated from an OR/NOR gate input to its output by setting other inputs to 0, a *non-controlling* value for OR/NOR. Fault effects are propagated from an XOR/XNOR gate input to its output by setting all other inputs to 0 or 1 as is convenient.

E-beam testing [672] allows observation of internal circuit signals by "developing" a picture of the circuit that shows the internal nodes charged to logic 0 in one color and those charged to logic 1 in a different color. This eliminates the need to propagate fault effects to *primary outputs* (POs.) However, this method is impractically expensive, is only used for very specialized applications, and in a sense converts an intractable testing problem into another intractable image processing problem, since some mechanism must now look at a VLSI chip image and determine whether all signals are "colored" correctly. ATPG algorithms are extremely valuable, in that they propagate an abnormal voltage reading from the internals of the circuit to a PO, where an ATE can examine the voltage and determine whether it is correct.

Scan-Design for Microprocessor Testing. At present, the preferred method for testing at least parts of Intel Pentium^{TM} and AMD K6^{TM} microprocessors uses combinational ATPG. A *scan-chain* inserter adds special-purpose MUX and clocking hardware to every circuit flip-flop for testing, so that in *scan* mode, the flip-flops are converted into a giant shift register, and the entire state of the microprocessor can be shifted out through a special test-mode port called *scan_out* (see Chapter 14). Similarly, a desired initial flip-flop state can be serially shifted into flip-flops through a special test-mode port called *scan_in*. This approach converts a difficult sequential circuit ATPG problem into a more tractable combinational circuit ATPG problem, at the expense of:

1. Using 5 to 20% of the chip area for the scan chain hardware in large chips.

2. Slowing down all flip-flops because of the added scan chain MUX delays.

3. Reserving one or more additional pins for scan chain control.

4. Lengthening the test-pattern sequence. This occurs because to set the machine having n flip-flops to any desired initial state requires n clocks of the scan chain. The application of the desired test-pattern requires 1 additional clock, followed by n additional clocks to read out the flip-flop state (in the event that the fault effect is captured in a flip-flop, rather than propagated to the circuit output.) For multiple tests, these can be overlapped (see Chapter 14.)

However, scan design coupled with combinational ATPG is the most popular test method with microprocessor and other VLSI chip designers, because it is very likely to generate a test set with close to 100% fault coverage. Besides, the test development time is predictable and can be accounted for in the new product introduction schedule. The state-of-the-art of sequential ATPG frequently causes major design delays, due to algorithm problems and untestable hardware, and can delay a product introduction. In Chapters 14 and 16 we will cover scan design in greater detail, but we see that, at least for now, the combinational ATPG programs are extremely important.

7.1.3 Search Space Abstractions

All ATPG programs need a data structure describing the search space for test patterns.

Binary Search Trees. Consider the *binary tree* in Figure 7.2(b) for the circuit *primary inputs* (PIs) in Figure 7.2(a). Goel [256, 258] first used these trees in the combinational ATPG literature. The tree represents all eight choices for circuit input patterns. At the topmost node, if the left branch is selected, then signal A is set to 0 (the $\overline{A}$ branch), but if the right one is selected, then A is 1. At the second and third levels in the tree, subsequent values are selected for other circuit inputs, first for B and then for C. This covers all possible input patterns. The leaf nodes of the tree are labeled with the good machine output that the corresponding input values will cause. All ATPG algorithms *implicitly* search this tree to find test-patterns, and in the worst case, must examine the entire tree to prove that a fault is *untestable*. We wish to avoid a complete examination, because the number of tree leaves is:

$$2^{number\_primary\_inputs}$$

and rises exponentially.

Binary Decision Diagrams. Any switching function can be completely described by the *binary decision diagram* (BDD), which was invented by Lee in 1959 [385]. The present discussion is based on the work of Akers who applied BDDs to solve the problem of testing [43, 44]. Figure 7.2(c) shows the BDD for the circuit of Figure 7.2(a). In order to read the diagram, we start at the topmost *root* node, and follow a path from that node to one of the two bottommost nodes, 0 or 1, which gives the circuit output value. The product of the Boolean literals along the path

7.1 Algorithms and Representations

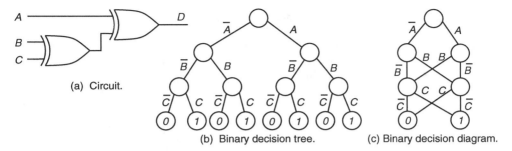

(a) Circuit. (b) Binary decision tree. (c) Binary decision diagram.

Figure 7.2: Different representations of a circuit.

gives a circuit maxterm or minterm, depending on whether we end up at the 0 or 1 output node. For example, the leftmost path in the BDD is $\overline{A}\,\overline{B}\,\overline{C}$, which produces the circuit output 0, and this is consistent with the circuit function. The rightmost path in the BDD is $A\,B\,\overline{C}$, which produces the circuit output 1, also consistent with the circuit function. We can verify that all BDD paths are consistent with the logic function. BDDs have been used for ATPG [234, 635], but suffer from the problems of computational intractability, particularly for multiplier circuits. One observes vast changes in compute time, depending on the order in which circuit PIs are expanded in the BDD [104].

7.1.4 Algorithm Completeness

The notion of ATPG algorithm *completeness* means that in order to generate a test-pattern, the algorithm must ultimately be able to search the entire *binary decision tree*, if necessary, to generate a test-pattern even for a hard-to-test fault. If the fault is *untestable*, then after searching the entire tree no test is found. This means that the circuit behaves correctly even in the presence of that fault. It is important for an ATPG algorithm to be complete, or it may not attain the required fault coverage.

7.1.5 ATPG Algebras

The ATPG *algebra* is a higher-order Boolean set notation with the purpose of representing both the "good" and the "failing" circuit (or machine) values simultaneously. This has the advantage of requiring only *one* pass of ATPG to determine signal values for both machines. Since a test vector requires that a difference be maintained between the two machines, it is computationally fastest to represent both machines in the algebra, rather than maintaining them separately. Roth [551] showed how multiple-path sensitization, required to test certain combinational circuits, could be done with his five-valued algebra given in Table 7.1.

Later, Muth [481] showed that in order to test finite state machines, the X symbol must be expanded to cover the cases where one among the good or failing machine values may be known, but the other machine value is unknown. Table 7.1 shows Muth's nine-valued algebra, which frequently benefits combinational circuit

Table 7.1: Roth's five-valued and Muth's nine-valued algebras.

Symbol	Meaning	Roth's 5-valued algebra		Muth's 9-valued algebra	
		Good machine	Failing machine	Good machine	Failing machine
D	(1/0)	1	0	1	0
$\overline{D}$	(0/1)	0	1	0	1
0	(0/0)	0	0	0	0
1	(1/1)	1	1	1	1
X	(X/X)	X	X	X	X
$G0$	$(0/X)$	—	—	0	X
$G1$	$(1/X)$	—	—	1	X
$F0$	$(X/0)$	—	—	X	0
$F1$	$(X/1)$	—	—	X	1

ATPG, as well [116]. For computing the response of a logic gate to input symbols of the algebra, we expand the symbols into the good machine/bad machine values given in column 2 of Table 7.1. We then independently compute the logic gate response for both machines and combine the output values back into the algebra.

7.1.6 Algorithm Types

We classify various types of ATPG algorithms, and present their complexity.

Exhaustive. In this approach, for an n-input circuit, we generate all 2^n input patterns. For the reasons discussed above, this is infeasible unless the circuit is partitioned into cones of logic, each with 15 or fewer inputs. We can then perform exhaustive test-pattern generation for each cone. However, those faults that require multiple cones to be activated in a synergistic way during testing may not be tested.

Random – Used With Algorithmic Methods. In 1972 at the University of Illinois, Agrawal and Agrawal [26] suggested the use of *random pattern generation* (RPG) for testing. An essential part of the RPG scheme is a fault simulator that selects useful patterns (Figure 7.3.) While generating tests for the boards of the ILLIAC IV parallel computer they reported that the coverage of random patterns would often saturate between 60-80% and that switching to a D-algorithm based program at that point proved beneficial. This limitation of RPG, which relates to the testability of the circuit, was observed by Eichelberger *et al.* [210] for *programmable logic arrays* (PLAs.) It has been realized that patterns with equally likely 0s and 1s, as is used to start the RPG in Figure 7.3, may not be the best choice [15, 16, 20]. When the probabilities of 0s and 1s are different from 0.5, the patterns are called *weighted random patterns* (WRP.) Such patterns have been used by Schnurmann *et al.* [571], Waicukauski and Lindbloom [705], and several others. A method for finding an

7.1 Algorithms and Representations   161

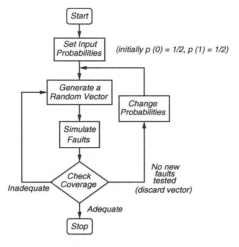

Figure 7.3: Random pattern generation method.

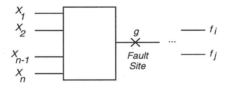

Figure 7.4: Boolean difference.

optimum set of probabilities for inputs is given by Wunderlich [739]. For application of RPG to sequential circuits, see the discussion of the simulation-based methods in Chapter 8. David's recent book [187] comprehensively covers this subject.

Symbolic – Boolean Difference. Sellers *et al.* [582, 583] use *Shannon's Expansion Theorem* to characterize Boolean circuits. For example, an arbitrary Boolean function $F(X_1, X_2, \ldots, X_n)$ can be expanded about any variable, say X_2, as:

$$F(X_1, X_2, \ldots, X_n) = X_2 \cdot F(X_1, 1, \ldots, X_n) + \overline{X_2} \cdot F(X_1, 0, \ldots, X_n)$$

Assuming a logic function: $g = G(X_1, X_2, \ldots, X_n)$ for the fault site shown in Figure 7.4, we express the outputs as:

$$f_j = F_j(g, X_1, X_2, \ldots, X_n) \quad ; \quad 1 \leq j \leq m \tag{7.1}$$
$$X_i = 0 \text{ or } 1 \text{ for } 1 \leq i \leq n$$

Notice that these output equations express the circuit outputs in terms of the *primary inputs* X_i and the fault site signal g. Sellers *et al.* define the *Boolean difference*, or *Boolean partial derivative*, of a circuit as:

$$\frac{\partial F_j}{\partial g} = F_j(1, X_1, X_2, \ldots, X_n) \oplus F_j(0, X_1, \ldots, X_n) \tag{7.2}$$

They express the fault detection requirements for g s-a-0 at output f_j as:

$$G(X_1, X_2, \ldots, X_n) = 1 \qquad (7.3)$$

$$\frac{\partial F_j}{\partial g} = F_j(1, X_1, X_2, \ldots, X_n) \oplus F_j(0, X_1, \ldots, X_n) = 1 \qquad (7.4)$$

Equation 7.3 says that to test a stuck-at-0 fault at g, the logic gate G must sensitize the fault site by driving it to logic 1. Equation 7.4 says that in order to detect the fault, the Boolean difference of some output with respect to the fault site g must be 1 (i.e., the output must change its signal value when the fault site signal switches from 1 to 0.) Unfortunately, due to high complexity the *Boolean difference* is not an efficient way to compute test patterns for large circuits.

Path Sensitization Methods. Path sensitization at the logic gate level of representation is currently the preferred ATPG method. The approach consists of three steps [550]:

1. *Fault sensitization*, in which a stuck-at fault is activated by forcing the signal driving it to an opposite value from the fault value. This is necessary to ensure a behavioral difference between the good circuit and the faulty circuit. Fault sensitization is also known as *fault activation* or *fault excitation*.

2. *Fault propagation*, in which the fault effect is propagated through one or more paths to a PO of the circuit. For some faults, it is necessary to simultaneously propagate the fault effect over multiple paths to test it. In general, the number of paths may rise exponentially in the number of logic gates in the circuit. Fault propagation is also known as *path sensitization*.

3. *Line justification*, in which the internal signal assignments previously made to sensitize a fault or propagate its effect are justified by setting PIs of the circuit.

In the second and third steps, we may find a *conflict*, where a necessary signal assignment contradicts some previously-made assignment. This forces the ATPG algorithm to *backtrack* or *backup*, i.e., discard a previously-made signal assignment and make an *alternative assignment*.

Consider the example in Figure 7.5 [550]. In all examples in this chapter, we will label PIs and POs with capital letters, and every other signal line in the circuit with a lower-case letter. Note that PI B (a *fanout stem*) fans out to two AND gates, whose outputs are h and i. The *fanout branches* from B to the inputs of the two AND gates are labeled f and g. It frequently happens that tests for faults on B are different from tests for faults on f, which are also different from tests for faults on g. That is why we must label every distinct *line* or *wire* of a signal *net*. We generate a test for B stuck-at-0. For *fault sensitization*, we set B to 1, and this leads to the signal assignments $f = D$ and $g = D$ (see Table 7.1.)

Fault propagation requires us to select among three scenarios:

7.1 Algorithms and Representations

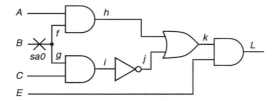

Figure 7.5: A combinational circuit example for path sensitization.

1. Propagation along the path $f - h - k - L$, or

2. Propagation along the path $g - i - j - k - L$, or

3. Simultaneous propagation along both paths $f - h - k - L$ and $g - i - j - k - L$.

We choose path $f - h - k - L$ for propagation. This means that for every AND gate along the path, the off-path inputs should be set to non-controlling values (1), and similarly for every OR gate along the path the off-path inputs should be set to 0. This results in the signal assignments $A = 1$, $j = 0$, and $E = 1$. *Line justification* now requires us to justify any internal signals where we assumed a value assignment. In this case, the only one is j. We can assign $i = 1$ to justify $j = 0$ by backwards logic simulation of inverter j. However, AND gate i then needs to have an output of 1, but it already has input $g = D$. Backwards logic simulation, using the 5-valued algebra (see Table 7.1), reveals that there is no way to get $i = 1$ when an input already was set to D. Therefore, we *backtrack* and retract the assignment $j = 0$ and try the alternative assignment $j = 1$. However, this immediately blocks fault propagation along path $f - h - k - L$. We conclude that the only viable options may be scenarios 2 and 3 listed above.

We choose scenario 3. We change our *fault propagation* approach, and now make the assignments $A = 1$, $C = 1$, and $E = 1$ to ensure fault propagation. Forward logic simulation from these assignments yields $i = D$, $h = D$, $j = \overline{D}$, and $k = 1$, since D OR $\overline{D} = 1$. This is obtained by computing 1/0 OR 0/1 = 1/1. The *D-frontier* is the cut-set separating the circuit portion labeled with Xs from the portion labeled with D or $\overline{D}$, where we include only the D or $\overline{D}$ signals closest to the outputs in the frontier. Finally, $L = 1$ and the *D-frontier* has disappeared at OR gate k, which means that the fault is untestable via these multiple paths. It only remains to return and try fault propagation along path $g - i - j - k - L$. We set $C = 1$, $h = 0$, and $E = 1$ to propagate the fault. Forward logic simulation gives $i = D$, $j = \overline{D}$, $k = \overline{D}$, and $L = \overline{D}$. It remains to *justify* $h = 0$. This is achieved by *backwards logic simulation* for AND gate h, with input $f = D$, by setting input $A = 0$. The only test for B stuck-at-0 is $ABCE = 0111$, and this produces the output $L = 0$ in the good machine, and $L = 1$ in the failing machine.

This ATPG procedure is only correct for acyclic combinational circuits. We will discuss procedures for sequential circuits in Chapter 8. In particular, any circuit with feedback paths, flip-flops, or implicit latches expressed as combinational logic will frequently force this procedure into an infinite loop. Fault propagation and line justification during ATPG consist of an intermixing of signal assignment operations, forward logic gate simulation, backwards logic gate simulation, and backtracks.

Boolean Satisfiability and Implication Graph Methods. The *Boolean satisfiability* problem means satisfying a Boolean expression or equation. An n-bit Boolean vector consists of a set of n binary variables, x_i, $i = 1, 2, \ldots, n$. Symbolically, a variable x_i or its complement $\bar{x}_i$ is referred to as a *literal*. The *two-satisfiability* (2-SAT) problem refers to finding a set of values for x_i's that will satisfy an equation of the type:

$$\sum \alpha_k \beta_k = 0 \quad (\text{non} - \text{tautology}) \quad \text{or} \quad \prod (\alpha_k + \beta_k) = 1 \quad (\text{satisfiability}) \qquad (7.5)$$

where α_k and β_k are any two literals, and summation and product are Boolean OR and AND operations, respectively. A term or a factor in Equation 7.5 is called a *Boolean clause* or simply a *clause*. The 2-SAT problem is characterized by each clause having just two literals. The 2-SAT problem is solvable in polynomial time [190]. When the clauses in the Boolean expression contain three literals, the problem is known as the *three-satisfiability* (3-SAT) problem. The solution of 3-SAT has exponential time complexity.

Chakradhar *et al.* [121, 124, 130] and Larrabee [383, 384] have derived Boolean satisfiability formulations for the ATPG problem. Given a target fault, one derives an energy function of a neural network or a Boolean product of sums expression in terms of signal variables of the circuit such that any test for the target fault will minimize the energy function or satisfy the Boolean expression. The energy minimization is shown to be equivalent to a Boolean sum of products expression. The rest depends on finding efficient ways for solving the satisfiability problem.

These methods have been extended by others [291, 605, 648], and are now the fastest known ATPG algorithms for huge circuits. In these methods, the Boolean function of every logic gate is captured in equations that relate the input and output signals of the gate. Consider the signal relationships of the AND gate in Figure 7.6:

$$\begin{aligned} &\text{If } a = 0, \text{ then } z = 0 \\ &\text{If } b = 0, \text{ then } z = 0 \\ &\text{If } z = 1, \text{ then } a = 1 \text{ AND } b = 1 \\ &\text{If } a = 1 \text{ AND } b = 1, \text{ then } z = 1 \end{aligned} \qquad (7.6)$$

For each constraint a cube is designed so that if signals are consistently labeled, that cube will become 0. If any signal value around the logic gate is inconsistent with the gate function, then some cube will become 1. We simply sum up the cubes to obtain the following Boolean equation, where cubes are shown in order of the above conditions:

$$\bar{a}\,z + \bar{b}\,z + z\,\overline{ab} + a\,b\,\bar{z} = 0 \qquad (7.7)$$

which simplifies to:

$$\bar{a}z + \bar{b}z + ab\bar{z} = 0 \qquad (7.8)$$

7.1 Algorithms and Representations

Figure 7.6: A two-input AND gate.

This equation is satisfied only when a, b, and c assume values that are consistent with the function of the AND gate. The first two terms are 2-SAT terms and the third is a 3-SAT term. An alternative representation is called the *pseudo-Boolean equation* [124]. We convert a Boolean expression into a pseudo-Boolean form by replacing the Boolean OR and AND operators by arithmetic addition and multiplication, and treating signals as "real" variables that can assume values 0.0 and 1.0. Complementation $\bar{x}$ is replaced by $1.0 - x$. For the AND gate, this form is:

$$F_{pseudo-Bool} = 2z + ab - az - bz - abz = 0.0 \qquad (7.9)$$

derived from Equation 7.8. A third representation is the *energy function* [126], obtained by letting the variables assume any real value in the range (0,1).

An alternative and easier way to derive the Boolean equation representation is by the *Boolean false expression* [125] defined as:

$$f_{AND}(a, b, z) = z \oplus (ab) = \bar{a}z + \bar{b}z + ab\bar{z} \qquad (7.10)$$

The expression on the right hand side in Equation 7.10 evaluates to logic 0 only when a, b, and z take values that are consistent with the AND function. The complement of f_{AND} is called the *truth expression* or the *satisfiability expression*, which evaluates to logic 1 only when values of a, b, and z are consistent with the AND function. These expressions can be derived for any complex Boolean function using the exclusive-OR definition of Equation 7.10. The Boolean false expression can be directly converted into the energy function of a *neural network* by replacing the Boolean operators with arithmetic operators. The variables then represent the states of neurons, which can either assume continuous values or discrete 0 and 1 values as in the Hopfield model [304]. Applications of the Hopfield model of neural networks to testing problems have been described in a book [127].

The non-intuitive aspect of these formulations is that logic gate outputs, as well as inputs, appear in the expressions. The advantage of this approach is that we can write an energy function for every logic gate (or Boolean function module) in a circuit, and then sum all of those functions into a single energy function for the entire circuit. If the function value is 0, then all signals are consistently labeled; otherwise, they are not.

A really efficient way to minimize the energy function or find satisfying variable assignments for the false or truth functions is the *implication graph*. This graph has a node for every literal. Thus, a Boolean variable x is represented by two nodes, x and $\bar{x}$. A node can be "true" or "false." For $x = 1$, the x node assumes the true state. For $x = 0$, the $\bar{x}$ node becomes true. A two-variable "if ... then" clause is represented as a directed edge from the literal representing the "if" condition to the literal representing the "then" clause. The graph can then be transformed into

Chapter 7. COMBINATIONAL CIRCUIT TEST GENERATION

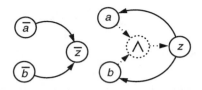

Figure 7.7: Implication graph (solid) and 3-SAT term (dotted) for 2-input AND.

a *transitive closure* [122] graph so that when a node is set to true, all reachable nodes are also set to true. This allows very efficient analysis of signal implications, because the transitive closure can determine more global signal relationships in the graph than other branch-and-bound search methods. Figure 7.7 (solid lines) shows the implication graph for the "if ... then" clauses of Equation 7.6. Note that only binary implications (those involving two literals) can be represented by an edge. The node with $\wedge$ (dotted lines) denotes an ANDing operator, and represents the 3-SAT term of Equation 7.8 (last clause of Equation 7.6) [291]. We cover these algorithms briefly in advanced topics (see Section 7.5.4).

Computational Complexity. Ibarra and Sahni [317] analyzed the computational complexity of ATPG. They found that it is an *NP-Complete* problem, which means that no polynomial expression for the compute time function was found, and the problem is presumed to have exponential complexity. We will informally discuss how this arises. In the worst case, with no_pi inputs, there are 2^{no_pi} different input combinations to try in depth-first fashion in the binary decision tree. When no_ff flip-flops are present in the circuit, there are potentially 4^{no_ff} different initial flip-flop states for ATPG to consider. This is because a flip-flop can be in either 0 or 1 state in the fault-free circuit and also in 0 or 1 state in the faulty circuit. Thus, the state-space of a flip-flop contains four elements (also see Section 8.2.) Finally, the work to forward simulate or reverse simulate all logic gates, as appropriate, rises proportionately to n, the number of logic gates. In the worst case, this work has to be done for all potential combinations of PIs and initial flip-flop states. The complete expression for the worst-case ATPG computational complexity is:

$$O(n \times 2^{no\_pi} \times 4^{no\_ff})$$

The above proof considers ATPG to be mathematically equivalent to the problem of *Boolean satisfiability*.

The entire history of ATPG algorithms has been a process of improving heuristic algorithms and procedures to (1) find all necessary signal assignments for a test as early as possible, and (2) search as little of the above decision space as possible. The worst-case decision space is $2^{no\_pi} \times 4^{no\_ff}$. For logic simulation, the computational complexity is $O(n)$. For combinational fault simulation, the complexity os $O(n^2)$[277], and for sequential fault simulation, the complexity is estimated to be between $O(n^2)$ and $O(n^3)$, based on empirical measurements. This means that, whenever possible, we will use fault simulation to avoid ATPG computations. For

7.1 Algorithms and Representations

Table 7.2: History of algorithm speedups.

Algorithm	Estimated speedup over D-Algorithm (normalized to D-ALG CPU time)	Year
D-ALG [551]	1	1966
PODEM [258]	7	1981
FAN [229, 232, 233]	23	1983
TOPS [360]	292	1987
SOCRATES [576]	1574† ATPG System	1988
Waicukauski et al. [708]	2189† ATPG System	1990
EST [110, 253]	8765† ATPG System	1991
TRAN [122]	3005† ATPG System	1993
Recursive learning [376]	485	1995
Tafertshofer et al. [648]	25057	1997

instance, we use RPG and fault simulation to get tests. When that fails, we use ATPG for hard-to-test faults. If we find a pattern for a fault, then we simulate that pattern against all remaining undetected faults, in the hope that we will "accidentally" test additional faults.

VLSI designers have become accustomed to analog circuit simulators (e.g., SPICE [163, 484, 486]), which let them model the actual defects in circuit behavior and see analog signal aberrations. The problems of analog modeling for ATPG are:

1. The huge number of different faults possible in large circuits.

2. The exponential complexity of the algorithm (i.e., for a sequential circuit with only 20 flip-flops, sequential ATPG may take days of computing.) Since sequential ATPG is slow at the Boolean level of representation, it will be even slower at the analog level of representation.

3. ATPG for transistor structures must model bidirectional and tri-state behavior (see Chapter 4.) Fault models and ATPG algorithms exist but are more complex than their logic gate counterparts [105, 287, 538]. Though there are test generators that can operate at the transistor level [173, 212, 250, 386, 389], the prevailing test methodology continues to rely upon gate-level stuck-at faults.

Table 7.2 shows the history of accelerating combinational ATPG. The speedups in the table are very approximate, and should be treated as order-of-magnitude estimates, due to the difficulty in normalizing the CPU times of the older CPUs on which the earlier algorithm experiments were run, and due to implementation differences. The earliest and last two table entries are not ATPG systems (see Section 7.6), but the middle entries are. The ATPG system, using random-pattern generation, has an unfair advantage over the pure algorithm execution experiments. Also, the TRAN algorithm and the algorithm of Tafertshofer et al. perform much better than the other systems on particularly huge circuits, which is not reflected

Table 7.3: Test vectors for all circuit faults.

Fault	Test	Response (good/failing)
a sa0	$A = 0$	D
a sa1	$A = 1$	$\overline{D}$
b sa0	$A = 1$	$\overline{D}$
b sa1	$A = 0$	D

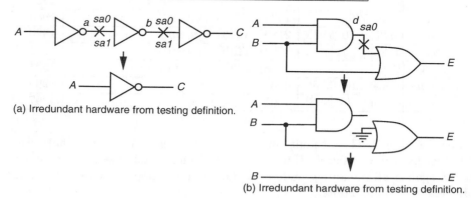

Figure 7.8: Redundancy definition for testing purposes.

by these benchmark results. We see that improvements come slowly, and that they have barely kept pace with *Moore's Law* [477, 478].

7.2 Redundancy Identification (RID)

Combinational ATPG algorithms provide a major side benefit. *They can determine when the circuit has unnecessary, or redundant, hardware.* In testing, one can remove redundant hardware and the circuit will still function exactly the same way as before. However, there are still circuits with unnecessary hardware that are fully testable, so we do not consider that hardware (in testing) to be redundant. We only consider hardware that is untestable for stuck-faults to be redundant. In Figure 7.8(a), the first two inverters of the three inverters are unnecessary, unless they are part of a chain of cascaded amplifiers needed to amplify A to drive C. In most cases, however, they are unnecessary. However, notice that the stuck-at-0 and stuck-at-1 faults on the first two inverters are fully testable by the vectors shown in Table 7.3, so in the testing sense, these inverters are NOT redundant.

Now consider Figure 7.8(b). In order to excite the stuck-at-0 fault on line d, we must set both AND gate inputs, A and B, to 1. However, this causes an input of the OR gate, E, to be 1, and blocks the fault effect from propagating from line d to output E. In this case, the conditions for sensitizing the fault also block its effect propagation, and this means that the circuit behaves identically whether or not the fault is present. This is an *untestable* fault, and in combinational circuits

7.2 Redundancy Identification (RID)

untestable faults indicate redundant hardware. We remove the redundant hardware by noticing that since the fault does not change the circuit behavior, we can assume that line d is always at logic 0, and so we permanently ground it. Then, we can (1) remove the OR gate and replace it with a wire (since it no longer serves any useful function), and (2) discard the AND gate and input A for the same reason. The *irredundant* circuit degenerates into a wire from B to E.

We now discuss the benefits of redundant hardware removal. In the 1970s, logic gates were bulky and expensive, so it was highly important to eliminate redundant hardware to control costs. This is no longer true – at present a 2-input AND gate in *ultra large scale integrated* (ULSI) circuit technology costs roughly 0.012 cents [584], so there is little cost reduction from removing the redundant AND gate. However, notice that our irredundant logic circuit is faster. Instead of requiring two logic gate delays and the wiring propagation delay, it only requires the wiring propagation delay. Also, notice that the circuit consumes less power, since two logic gates were eliminated. Both of these motivations are critical at present, since the stress is on speed and controlling power dissipation.

There is an even more critical *reliability* reason for redundancy removal. Figure 7.9 shows a pernicious *fault masking* problem. Figure 7.9(a) shows a conflict indicating that q sa1 is a redundant fault. However, Figure 7.9(b) shows that f sa0 is a testable stuck-at-0 fault, either through path $f - e - l - q - Z$ or path $f - g - m - q - Z$. In this case, the ATPG algorithm chose path $f - e - l - q - Z$ through the redundant fault site q sa1. In Figure 7.9(b), the redundant fault q sa1 is not present, and the test for f sa0 correctly tests for fault f sa0. However, in Figure 7.9(c) the redundant fault q sa1 is present, and it blocks the propagation for fault f sa0, which is also present, and causes the test for f sa0 to assume the good machine value at the output. *For multiple stuck-faults, if one of the multiple faults is redundant, the presence of the redundant fault may mask (hide) the presence of other, testable faults.* This is a serious compromise of circuit reliability. At present, ULSI technology is increasingly used in systems critical to human safety or wealth, such as automobile engine controls, jet aircraft control systems, jet engine controls, high-speed bullet train controls, telephone exchange controls, and stock selection programs (running on a PC.) Therefore, it is essential for *reliability* that all redundant hardware be removed from the circuit.

Figure 7.10(a) shows Boolean function $OUT0 = A \cdot B + \overline{A} \cdot C$ as a Karnaugh map. In the map we see that the implicants for $A \cdot B$ and $\overline{A} \cdot C$ do not overlap. Therefore, some designers also include the implicant $B \cdot C$, which redundantly covers the transition of A switching between 0 and 1 while both B and C are 1. This redundant implicant eliminates hazards in this switching situation for the function at the output $OUT0$. The implicant in the Karnaugh map and the corresponding additional gate in the implementation are shown in grey shade in Figure 7.10. Notice that there is a redundant stuck-at-0 fault in this circuit on line e. The presence of this fault has no effect on the function, but if it is present, it will mask any testable fault whose test pattern uses the path going through line e to propagate the fault to the $OUT0$ line. In addition to the redundant fault problem caused by this hardware,

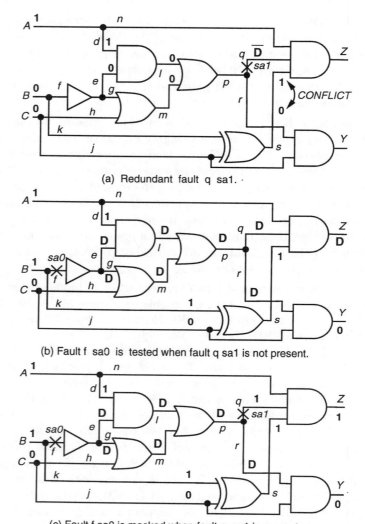

Figure 7.9: Fault masking problem.

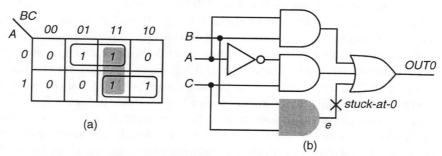

Figure 7.10: Redundant implicant to eliminate hazards.

7.2 Redundancy Identification (RID)

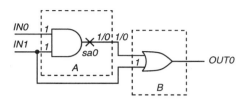

Figure 7.11: Testing as a global circuit problem.

there is additional chip area and power required by this redundant logic.

In Section 7.3, we will show how fully-testable sub-assemblies with no redundant logic may be combined into an assembly that has redundant hardware, and therefore, is slower, wastes power, and may be unreliable if multiple stuck-faults are present. For the design of a new microprocessor, a typical design team may have 12 to 30 ULSI circuit designers, and the microprocessor might contain 12.5 million logic gates. It is impossible for the designers to manually determine whether redundant hardware exists in the design. The commercially available Synopsys^{TM} logic synthesis system [75, 381] will synthesize moderately-sized designs into irredundant hardware, but huge designs must be partitioned before they can be synthesized. The act of partitioning introduces the potential for redundant hardware, and ATPG algorithms are one of the best methods for finding this redundant hardware. An accompanying *redundancy removal* tool will examine the redundant fault list produced by ATPG and remove the corresponding unwanted hardware. When removing redundant hardware, the removal of one redundant fault may make other formerly redundant faults testable, so we use the procedure below. In a real sense, the ATPG algorithm and redundancy remover complete the logic optimization that was incompletely done by a logic synthesis algorithm.

Algorithm 7.1 *Redundancy removal.*
Repeat until there are no more redundant faults:
{
 Use ATPG to find all redundant faults;
 **Remove all redundant faults with non-overlapping fault
  effect areas;**
}

Implication-based RID algorithms that do not require ATPG have been described by Iyer [328], and by Iyer and Abramovici [329]. These algorithms analyze the circuit without targeting any specific faults and can find many, but not all, redundant faults. Similar algorithms are implemented using the implication graph and transitive closure [28, 399].

Elimination of logic redundancy can reduce chip area, delay, power consumption, and improve testability. In some typical instances redundancy may not be avoidable. When an irredundant circuit is mapped onto a standard cell library it may become necessary to have extra gates or gate inputs to minimize the overall area. Keutzer *et al.* [352] show that redundancy is not necessary for the speed of logic, but point out

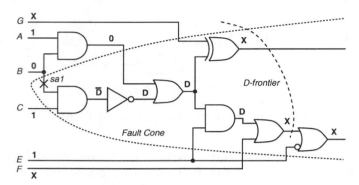

Figure 7.12: Fault cone.

that one may have to use extra logic to have an optimum-speed irredundant circuit. Gharaybeh *et al.* [244] report that they had to add as many as 41% more gates to obtain an irredundant delay-optimized design of the benchmark circuit s1238.

7.3 Testing as a Global Problem

Figure 7.11 shows two sub-assemblies A and B. By itself each sub-assembly is fully testable for all input and output signals stuck-at logic 0 or 1 (sa0 or sa1.) However, a local test-pattern of A for a sa0 fault on its output has both inputs at logic 1. When we apply this pattern to the composite circuit, both inputs of A and the bottom input of B are set to logic 1. The top input of B receives the fault effect 1/0, which means that if the fault is not present, the input is 1, and if it is present, the input is 0. However, the bottom input of 1 for B forces the output of B ($OUT0$) to be logic 1, and blocks observation of the fault effect coming from A. This simple example illustrates that testing is a global problem: Combinations of fully-testable modules in a logic circuit are not necessarily fully-testable, and may not be testable by the same patterns that would test the modules individually. In this case, the output of A sa0 is untestable.

7.4 Definitions

Definitions presented here are common to all ATPG algorithms.

Definition 7.1 *The* fault cone *is the portion of a circuit whose signals are reachable by a forward trace of the circuit topology starting at the fault site.*

Figure 7.12 shows an example circuit, a fault (on the lower fanout of B), the fault cone, the circuit labeling in the five-valued algebra, and the *D-frontier*.

Definition 7.2 *A* forward implication *results when the inputs to a logic gate are significantly labeled so that the output can be uniquely determined.*

7.4 Definitions

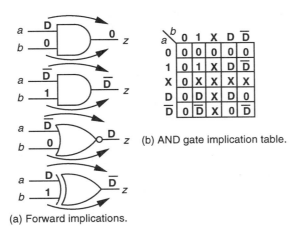

(b) AND gate implication table.

(a) Forward implications.

Figure 7.13: Forward implication examples (a) and implication table (b).

Arrows in Figure 7.13(a) show the forward implications for various logic gates. The arrows point from known signals to signals inferred by the forward implication. One of the best ways to implement a forward implication routine is in Figure 7.13(b), where a two-dimensional table gives the labeling for the AND gate output c in terms of the inputs a and b. However, a forward implication routine can also be written procedurally, using *if-then-else* statements. Algorithm 7.2 shows how to handle a NAND gate with an arbitrary number of inputs, using the AND gate and INVERTER implication tables. The methods for other logic gates are similar.

Algorithm 7.2 *NAND gate forward implication.*
Copy the first input labeling to the gate output.
For each additional input:
 {
  Combine the next input with the present gate output using AND gate forward implication table.
  Store the table value at the present gate output.
 }
Use INVERTER gate forward implication table to invert the present gate output.
Store the table value at the present gate output.

Definition 7.3 Backward implication *is the unique determination of all inputs of a gate for given output and possibly some of the inputs.*

Using tables to implement backward implications is cumbersome for gates with more than 2 inputs. Therefore, backward implications are usually implemented procedurally. Figure 7.14 shows examples of backwards implications.

Definition 7.4 *The* implication stack *[550] is a push-down stack that records each signal set in the circuit by the ATPG algorithm, and whether the alternate signal*

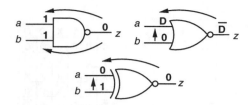

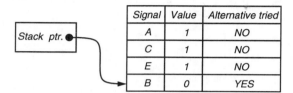

Figure 7.14: Backward implication examples.

Signal	Value	Alternative tried
A	1	NO
C	1	NO
E	1	NO
B	0	YES

Figure 7.15: Implication stack.

value has already been tried for that signal. This is an efficient way of representing which portion of the circuit's binary decision tree has already been traversed by the ATPG algorithm, so that search of that tree portion is not repeated.

In the circuit of Figure 7.12, assume that signals were set in the order A, C, E, and B. Furthermore, assume that the ATPG algorithm first tried setting signal B to 1, and that this failed, so B has been set, instead, to 0. Figure 7.15 shows the implication stack.

Definition 7.5 *The* D-frontier *[550] is the set of all gates with D or $\overline{D}$ at inputs and X at the output. The D-frontier divides the circuit into two parts, one with fault effects (D and $\overline{D}$) and the other without.*

Figure 7.12 shows the D-frontier for the circuit, given the implication stack in Figure 7.15, after A, C, E, and B are set but while F and G are unassigned.

Definition 7.6 *The ATPG algorithm may decide to* backtrack *(or back-track) [550], when it finds one of the following conditions:*

- *The D-frontier becomes empty, meaning that there is no possibility of propagating a fault effect through the circuit.*

- *A signal must be simultaneously set to both 0 and 1 in order to satisfy the testing conditions for the test vector, but this is obviously impossible.*

When a backtrack occurs, the ATPG algorithm removes one or more signal assignments from the implication stack, and then selects the alternate assignment for a signal already on the stack. This causes the binary decision tree to be explored in depth-first fashion.

7.4 Definitions

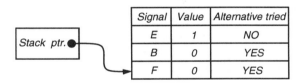

Figure 7.16: Backup of ATPG algorithm.

Figure 7.16 shows the stack after a backtrack caused by the algorithm determining that a 1 on F will block the fault propagation, and therefore the alternative assignment of 0 for F is being tried.

Definition 7.7 *Objectives [258] are determined by the ATPG algorithm, and serve as desired goals to be achieved during ATPG. Frequently, many intermediate signal assignments are needed in order to achieve an objective. Naturally, in the course of these intermediate assignments, backtracks may occur before the objective can be achieved, or it may even be discovered that the objective is unachievable, in which case a better objective must be constructed. Objectives are useful because they guide the ATPG search through the binary decision tree, to avoid areas of infeasible or difficult solutions. Most of the improvements to ATPG algorithms over the years have come from better selection of objectives, coupled with a reduction in backtracks. A typical objective would be to "set signal A to 0."*

Definition 7.8 *Backtrace [258] is an operation used to determine which PI should be set to achieve an objective. It is most frequently implemented using Goldstein's [263] combinational controllability and observability measures (see Chapter 6), although other measures can be used (and sometimes more successfully.)*

Figure 7.17 shows the objective $J = 1$. For OR gate J, we wish to trace backwards through it and find the easiest way to justify $J = 1$ at the PIs. We need only set one input of OR gate J to 1, so we should pick the easiest one to set (i.e., the one with lowest $CC1$), which is the bottom one. That means that *backtrace* determines that $D = 1$ satisfies the objective $J = 1$ with the least effort. If the *objective* were $J = 0$, then matters would be different. To justify that objective, we would set all three inputs to OR gate J to 0. We should begin by picking the *hardest* sub-objective first, which will be setting the top input to 0, since it has the highest controllability difficulty for logic 0 ($CC0$.) We achieve a 0 on the top input to OR gate J with the signal assignments $A = 1$ and $B = 1$ by tracing the hardest backwards path first, followed by setting $E = 0$. This allows us to proceed to justifying the middle input of J to 0, which we achieve with the assignment $C = 0$. Finally, we assign $D = 0$ to justify $J = 0$.

Backtrace [258] is this entire process of determining the easiest set of PI assignments that justify an objective. Its advantage is that we delete all of the intermediate nodes between the *objective* and the PIs from the ATPG algorithm search space. Since backtrace is frequently done, it must be highly efficient.

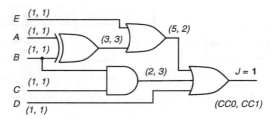

Figure 7.17: Backtracing of ATPG algorithm.

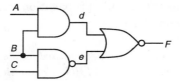

Figure 7.18: Singular cover examples for logic gates.

Definition 7.9 Branch-and-bound search *is used to efficiently search the binary decision tree for suitable test-patterns. At each level of the tree, the* branching operation *determines which input variable will be set to what* value *(0 or 1.) The* bounding operation *avoids exploring large portions of this tree by restricting the search decision choices, since complete exploration is impractical. ATPG algorithms require us to make decisions about exploring this tree, frequently with only limited information. We use* heuristics *or even* heuristic algorithms *to bound the tree search.*

7.5 Significant Combinational ATPG Algorithms

Combinational ATPG algorithms are best learned by starting at the beginning of the field, and studying the algorithms in the order of increasing sophistication. The seminal and first algorithm was Roth's D-Algorithm (D-ALG) [550], which established the calculus and algorithms for ATPG using D-cubes. The next development was Goel's [258] PODEM algorithm. He efficiently used path propagation constraints to limit the ATPG algorithm search space, and introduced the notion of *backtrace*. The third significant development was Fujiwara and Shimono's FAN algorithm [229, 232, 233]. They efficiently constrained the backtrace to speed up search, and took advantage of signal information to limit the search space. There are several later and more advanced algorithms that we will discuss only briefly.

7.5.1 D-Calculus and D-Algorithm (Roth)

We discuss Roth's D-Algorithm [550] in depth, since it is the basis of all subsequent ATPG algorithms. Armstrong [55] also did early work on ATPG.

7.5 Significant Combinational ATPG Algorithms

Table 7.4: Singular cover of AND and NOR gates.

Gate type	Inputs		Output	Gate type	Inputs		Output
AND	a	b	d	NOR	d	e	F
1	0	X	0	4	1	X	0
2	X	0	0	5	X	1	0
3	1	1	1	6	0	0	1

Definitions.

1. The *singular cover* of a logic gate is the minimal set of input signal assignments needed to represent *essential prime implicants* in the Karnaugh map of that logic gate, for both output cases of 0 and 1. Figure 7.18 shows an example. Table 7.4 gives the singular cover for the AND and NOR gates in this circuit.

2. A *D-cube* is a collapsed truth table entry that can be used to characterize an arbitrary logic block. Combine rows 3 and 1 of the AND gate singular cover, express it in Roth's five-valued algebra (with row 3 as the good machine and row 1 as the bad machine), and you get the AND gate *propagation* D-cube "D 1 D." Interchange the role of the two inputs and you get the additional cube "1 D D." AND these two D-cubes together and you get a third cube "D D D." Realize that you can replace D in these cubes by $\overline{D}$ and they are still valid, and you now have the complete set of D-cubes describing how fault effects can be propagated through the AND gate. However, if one cube signal is changed from D to $\overline{D}$, then the sense of all D and $\overline{D}$ signals in the cube must be inverted. You can verify that the cube set for the NOR gate is "D 0 $\overline{D}$," "0 D $\overline{D}$," and "D D $\overline{D}$."

3. The *D-intersection* operation [550] is defined as the set of circumstances under which different cube labelings for different logic gates can coexist in the circuit. The rule is that if one cube assigns a specific signal value, then other cubes must assign either that same signal value or X to the signal. Equations 7.11 give the simpler rules of cube intersection.

$$0 \cap 0 = 0 \cap X = X \cap 0 = 0$$
$$1 \cap 1 = 1 \cap X = X \cap 1 = 1 \quad (7.11)$$
$$X \cap X = X$$

For example, "0 X X" ∩ "1 X X" = ϕ (the empty cube.) This means that the cubes are incompatible and cannot be used together. There are two kinds of D-cubes: (1) Those where only one input coordinate is D or $\overline{D}$, and (2) Those where multiple input coordinates are D or $\overline{D}$. Table 7.5 gives the complete D-intersection operation. If ψ or ϕ occur during D-intersection, then the cubes are incompatible. If both μ and λ occur, then the cubes are incompatible. If

Table 7.5: Complete D-intersection operation.

∩	0	1	X	D	$\overline{D}$
0	0	ϕ	0	ψ	ψ
1	ϕ	1	1	ψ	ψ
X	0	1	X	D	$\overline{D}$
D	ψ	ψ	D	μ	λ
$\overline{D}$	ψ	ψ	$\overline{D}$	λ	μ

ϕ – empty
ψ – undefined (same effect as ϕ)
μ or λ – requires inversion of D and $\overline{D}$

only μ occurs, then $D \cap D = D$ and $\overline{D} \cap \overline{D} = \overline{D}$. If only λ occurs, transform the second cube as follows: $D \Rightarrow \overline{D}$ and $\overline{D} \Rightarrow D$. Then, apply the μ intersection rules. D-intersection is commutative and associative.

4. Cube A *D-contains* cube B if the set of A cube vertices contains (is a superset of) the B cube vertices.

5. The *primitive D-cubes of failure* (PDF) model faults in a logic circuit, and can model any (1) stuck-at-0 fault, (2) stuck-at-1 fault, (3) bridging fault (short circuit), or (4) arbitrary change in logic gate function (e.g., from AND to OR.) For the AND gate in Table 7.4, the *primitive D-cube of failure* of the AND gate output stuck-at-0 is "1 1 D," since in the *good* machine both inputs must be set to 1 to cause the gate output to have a 1, but the stuck-fault causes the output to be 0 in the *bad* machine. The two *primitive D-cubes of failure* for the AND gate output stuck-at-1 are "0 X $\overline{D}$" and "X 0 $\overline{D}$." The *primitive D-cubes of failure* will be distinctly different from the *propagation D-cubes*, since the former model a failure at that gate, while the latter model the conditions under which fault effects can propagate through the gate. A simple wire x stuck-at-0 (1) would be represented as a single-element cube D ($\overline{D}$) for the signal x. Each collapsed circuit fault has a unique set of *primitive D-cubes of failure*.

6. The *implication procedure* consists of modeling the fault with the appropriate *primitive D-cube of failure* (PDF), selecting *propagation D-cubes* to propagate the fault-effect to a circuit output (called the *D-drive* procedure), and selecting singular cover cubes to justify internal circuit signals (called the *consistency* procedure.) Unfortunately, the D-Algorithm selected cubes and singular covers very arbitrarily during test generation. When cube intersection failed, the procedure backed up to the last decision point (at which a cube was selected) and selected an alternate cube, instead.

Generation of Complex Fault Models. It is frequently necessary to generate fault models for bridging faults and complex logic gate function changes. The following

7.5 Significant Combinational ATPG Algorithms

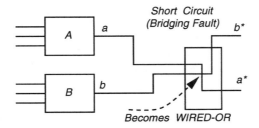

Figure 7.19: Wired-OR bridging fault short-circuiting wires a and b.

Table 7.6: Cube intersection for short circuit (bridging fault.)

Cube-set	a	b	$a*$	$b*$	Cube-set	a	b	$a*$	$b*$
$\alpha 0$	0	X	0	X					
	X	0	X	0	Primitive D-cubes				
$\alpha 1$	1	X	1	X	of failure	1	0	1	$\overline{D}$
	X	1	X	1	for bridging	0	1	$\overline{D}$	1
$\beta 0$	0	0	0	0	fault				
$\beta 1$	X	1	1	1					
	1	X	1	1					

procedure constructs the *primitive D-cubes of failure* for these complex faults.

1. Construct a cube set $\alpha 1$ when the good machine output is 1 and the set $\alpha 0$ when the good machine output is 0.

2. Construct a cube set $\beta 1$ when the failing machine output is 1 and the set $\beta 0$ when the failing machine output is 0.

3. Change the output of all $\alpha 1$ cubes to 0 and D-intersect each such cube with every $\beta 0$ cube. If any cubes result, change their output coordinate to D to indicate that these cubes excite the condition where the good machine output should be 1, but it instead became the bad machine output 0.

4. Change the output of all $\alpha 0$ cubes to 1 and D-intersect them with the $\beta 1$ cubes. If any cubes result, change their output coordinate to $\overline{D}$ to indicate that these cubes excite the condition where the good machine output should be 0, but it instead became the bad machine output 1.

Bridging Fault Example. Figure 7.19 shows how the output wires from logic gates a and b are short circuited, and due to the nature of the transistor circuitry became a wired-OR logic function. This could happen due to a missing spot of insulator that allowed wires on two different metal layers to short on a chip. According to the above procedure for generating cubes to model this bridging fault, Step 3 produces no cubes, because the altered $\alpha 1$ cubes are incompatible with the $\beta 0$ cube.

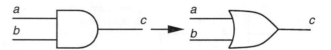

Figure 7.20: Fault transforming an AND gate into an OR gate.

Table 7.7: Cube intersection for gate function change fault.

Cube-set	a	b	c	Cube-set	a	b	c
$\alpha 0$	0	X	0	Primitive			
	X	0	0	D-cubes	0	1	$\overline{D}$
$\alpha 1$	1	1	1	of failure			
$\beta 0$	0	0	0	for AND	1	0	$\overline{D}$
$\beta 1$	1	X	1	becoming OR			
	X	1	1				

However, you can verify that Step 4 produces the *primitive D-cubes of failure* listed in Table 7.6.

Logic Gate Function Change Example. Figure 7.20 shows how a fault transforms an AND gate into an OR gate. This rare defect could happen when a very large blob of polysilicon landed on multiple transistors in a logic gate, merging them into a single transistor. By the above procedure, Step 3 produces no cubes, because the altered $\alpha 1$ cube is incompatible with the $\beta 0$ cube. However, Step 4 produces the *primitive D-cubes of failure* listed in Table 7.7. This is an extremely powerful fault modeling method.

D-Algorithm. The *test cube* refers to the set of all PI, intermediate, and PO circuit signals that are set to get a test for the fault.

Algorithm 7.3 *D-ALG.*
   Number all circuit lines in increasing level order from PIs to POs;
   Select a primitive D-cube *of the fault to be the* test cube *and put the*
      logic gate output(s) with inputs labeled as D ($\overline{D}$) onto the D-frontier;
   D-drive ();
   Consistency ();
   return ();
D-drive () /* Create a D-chain of propagation cubes from the fault to a PO */
{
    while (untried fault effects exist on the D-frontier)
    {
       Select next untried D-frontier *gate for fault propagation forward;*
       while (untried fault effect fanouts still exist)
       {
          Select the next untried fanout of the fault effect to be propagated;

7.5 Significant Combinational ATPG Algorithms 181

      Generate the next untried propagation D-cube, *for the next*
       fanout branch of the fault effect, to drive the fault effect to a PO;
      D-intersect *the selected* propagation D-cube *with current test cube;*
      if (intersection fails or is undefined) continue (go to end of loop);
      if (all possible propagation D-cubes *were tried and failed) break;*
      if (intersection succeeded)
      {
        Add *the* propagation D-cube *to the test cube and recreate the*
         D-frontier *with all propagating fault effects closest to POs;*
        Save D-frontier, *algorithm state, test cube, fanout branches*
         tried, fanout number, and fault effect number on the stack;
        break;
      }
      else if (intersection fails with D and $\overline{D}$ in test cube) Backtrack ();
      else if (intersection fails) break (go to outer loop);
    }
    if (all fault effects were tried and could not be propagated) Backtrack ();
  }
  return;
}
Consistency () /* *Justifies all unjustified internal circuit signals* */
{
  g = *coordinates of the test cube with only 1s and 0s;*
  if (g consists only of primary inputs) declare the fault testable and stop;
  for (each unjustified signal in g)
  {
    Select highest-numbered unjustified 0 or 1 signal z in g (not a PI);
    if (inputs to logic gate driving z are labeled as both D and $\overline{D}$) break;
    while (there are untried singular covers of the gate driving z)
    {
      Select the next untried singular cover of the gate driving z;
      if (there are no more untried singular covers)
      {
        if (no untried choices in imp. stack) stop: fault untestable;
        else if (untried alternatives exist in consistency mode)
          then pop implication stack and try alternative assignment;
        else
        {
         Backtrack ();
         D-drive ();
        }
      }
      if (the singular cover D-intersects with z) delete z from g, add
       the inputs to the singular cover to g and break;

Table 7.8: Truth table for simple circuit.

Inputs			Output
a	b	c	F
0	0	0	0
0	0	1	0
0	1	0	0
0	1	1	1
1	0	0	0
1	0	1	0
1	1	0	0
1	1	1	0

```
                if (intersection fails) mark this singular cover as failing;
            }
        }
        return;
    }
    Backtrack ()
    {
        if (a PO exists with the fault effect) Consistency ();
        else pop prior implication stack setting to try an alternative assignment;
        if (no untried choices in implication stack) stop: fault untestable;
        else return;
    }
```

Automatic Test-Pattern Generation Examples. In the following examples, for clarity we assume that D-ALG has all single and multiple input *propagation D-cubes* with single or multiple input D ($\overline{D}$) values tabulated for all types of logic gates. In reality, Roth's procedure dynamically generated all *propagation D-cubes*, whether they had single or multiple D ($\overline{D}$) inputs (double D cubes), as they were needed.

Example 7.1 Roth's First Example. *Consider the circuit of Figure 7.18. Table 7.8 shows its truth table. Table 7.9 shows all of the singular cover and D-cubes for this circuit in a tabular form. The D-cubes give the conditions under which a difference between the good and failing machines can be propagated through a gate. The cube interpretation is that if the Ds in the cube are set to 0 (1), then all $\overline{D}$s in the cube must be set to 1 (0). Table 7.10 shows the actual order in which D-ALG might select cubes for testing the fault d sa0. The algorithm selected the next cube, to intersect with the other previously-selected cubes in order to test the fault, according to the gate level ordering.*

7.5 Significant Combinational ATPG Algorithms

Table 7.9: Singular cover and D-cubes for Example 7.1.

A	B	C	d	e	F	
\multicolumn{7}{c}{Singular cover}						
1	1		1			
0			0			
	0		0			Used
	1	1		0		for
	0			1		justifying
		0		1		lines
				1	0	
			1		0	
			0	0	1	
\multicolumn{7}{c}{Propagation D-cubes}						
D	1		D			
1	D		D			Conditions
D	D		D			under which
		D	1		$\overline{D}$	difference
		1	D		$\overline{D}$	between
		D	D		$\overline{D}$	good and failing
			D	0	$\overline{D}$	machines can
			0	D	$\overline{D}$	propagate through gate
			D	D	$\overline{D}$	

Table 7.10: Steps of Example 7.1 for fault d sa0.

Step	A	B	C	d	e	F	Type of cube
1	1	1		D			Primitive D-cube of failure for AND gate
2				D	0	$\overline{D}$	Propagation D-cube for NOR gate
3		1	1		0		Singular cover of NAND gate

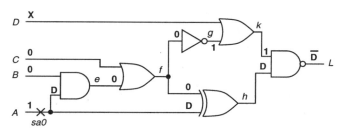

Figure 7.21: Circuit for Example 7.2.

Table 7.11: Steps of Example 7.2 for line A sa0 fault.

	Step	A	B	C	D	e	f	g	h		h	k	L	
D-drive	1	D												
	2	D					0		D					
	3	D					0		D	$\cap$	D	1	$\overline{D}$	Ds driven to POs
	=	D					0		D			1	$\overline{D}$	
consistency	4 or							1			1			
	5 not						0	1						
	6 or			0		0	0							
	7 and		0				0							
	tc^7	D	0	0		0	0	1	D			1	$\overline{D}$	
						D-chain dies								

tc^n is the Test Cube after step n.

Example 7.2 Roth's Second Example. *Figure 7.21 [550] shows Roth's second circuit example. We wish to test line A stuck-at-0. Table 7.11 shows the selection of the primitive D-cube of failure for the fault site (Step 1), and the subsequent selection of two cubes to propagate that fault effect to a circuit output. However, the final phase of the D-Algorithm uses the singular cover to justify internal signals selected by these cubes. This means that the singular cover must contain the prime implicants of all of the logic gates for this operation to work. Table 7.11 shows the remaining algorithm cube selection steps to generate a complete test for the fault.*

Example 7.3 Complex D-Algorithm Example. *Figure 7.22 shows the circuit for this example. Table 7.12 gives various decision steps for the fault s sa1 and Table 7.13 provides propagation D-cubes and singular cover. There are no backtracks. The second part of Table 7.12 shows the outcome for the fault u sa1. If the singular cover cubes are selected so that inputs are first set to 0, rather then 1, the test generation for this fault results in three backtracks before D-Algorithm corrects its initial error of setting $B = 0$. In this example, D-Algorithm works very well. However, its serious limitations are that it insists on propagating the D-frontier strictly in the order of increasing circuit signal numbers listed in the D-frontier, and it insists on justifying unjustified internal signals in the order of decreasing signal numbers in the active test cube. These methods are often counterproductive, although here they work well, because better heuristics are needed for deciding which signals are best to manipulate at each stage of the computation. Also, the algorithm normally selects a singular cover to set an input to 1 first, rather than to set it to 0 first. A better heuristic is needed for selecting the value of the assignment. Also, D-ALG insists on propagating all D-frontier fault effects to POs, whenever possible. This is unnecessary, since only one fault effect propagated to a PO is sufficient for a test.*

7.5 Significant Combinational ATPG Algorithms

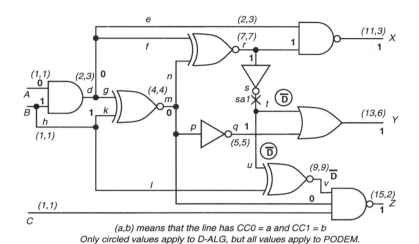

Figure 7.22: An example of D-ALG and need for improvements.

Table 7.12: Steps of Figure 7.22 for s sa1 and u sa1.

Step	A	B	C	d	m	q	r	s	v	X	Y	Z	Type of cube
Steps of Example 7.22 for s sa1.													
1								1	$\overline{D}$				PDF
2						0	1	$\overline{D}$			$\overline{D}$		Prop. D-cube Y
3	1					0	1	$\overline{D}$	D		$\overline{D}$		Prop. D-cube v
4	1	1			1	0	1	$\overline{D}$	D		$\overline{D}$	D	Prop. D-cube Z
5	1	1	1	1	0	1	$\overline{D}$	D		$\overline{D}$	D		Sing. cover m
6	1	1	1	1	1	0	1	$\overline{D}$	D		$\overline{D}$	D	Sing. cover d
Steps of Example 7.22 for u sa1.													
	A	B	C	d	m	q	r	u	v	X	Y	Z	
1								1	$\overline{D}$				PDF
2	0							1	$\overline{D}$	D				Prop. D-cube v
3	0	1				1		1	$\overline{D}$	D		$\overline{D}$		Prop. D-cube Z
4	0	1	1	1				1	$\overline{D}$	D		$\overline{D}$		Sing. cover r
Backtrack – $d = 1$ and $B = 0$ cannot justify $m = 1$													
	0	1			1			1	$\overline{D}$	D		$\overline{D}$		
Backtrack – no alternative singular cover to justify $r = 1$													
	0							1	$\overline{D}$	D				
Backtrack – need to try alternate propagation D-cube for v													
								1	$\overline{D}$				
5	1							1	$\overline{D}$	$\overline{D}$				Prop. D-cube for v
6	1	1			1			1	$\overline{D}$	$\overline{D}$	D			Prop. D-cube for Z
7	1	1	1	1				1	$\overline{D}$	$\overline{D}$	D			Sing. cover for r
8	1	1	1	1	1			1	$\overline{D}$	$\overline{D}$	D			Sing. cover for A

Table 7.13: Propagation D-cubes and singular cover for Example 7.3.

A	B	C	d	m	q	r	s	v	X	Y	Z	A	B	C	d	m	q	r	s	v	X	Y	Z
D	1		D									1	1		1								
1	D		D									0	X		0								
D	D		D									X	0		0								
		0		D	$\overline{D}$									0		0	1						
		1		D	D									0		1	0						
		D		0	$\overline{D}$									1		0	0							
		D		1	D									1		1	1							
				0	D		$\overline{D}$									0	0		1				
				1	D		D									0	1		0				
				D	0		$\overline{D}$									1	0		0	1			
				D	1		D									1	1		1				
																	0	1					
				D	$\overline{D}$												1	0					
				1			D		$\overline{D}$							1		1			0		
				D			1		$\overline{D}$							0		X			1		
				D			D		$\overline{D}$							X		0			1		
						0	D			D								0	0			0	
						D	0			D								1	X			1	
						D	D			D								X	1			1	
0							D	$\overline{D}$				0							0	1			
1							D	D				0							1	0			
D							0	$\overline{D}$				1							0	0			
D							1	D				1							1	1			
		1		1				D			$\overline{D}$			1		1				1			0
		1		D				1			$\overline{D}$			0		X				X			1
		1		1				D			$\overline{D}$			X		0				X			1
		D		D				D			$\overline{D}$			X		X				0			1

7.5.2 PODEM (Goel)

In the late 1970s, semiconductor DRAM memory was introduced into IBM mainframe computers, and this memory required *error correction and translation* (ECAT) circuits to raise its reliability to acceptable levels. Unfortunately, D-ALG failed to generate tests for these circuits, because its search was too undirected. ECAT circuits consists of XOR logic gate trees, where all external gate inputs have to be set before an output signal is defined. Goel [258] introduced the PODEM* algorithm to deal with these circuits. Many standard ATPG concepts (described earlier) came from PODEM:

- PODEM expanded the binary decision tree only around the PI variables, and

---
*Acronym for *Path Oriented DEcision Making*.

7.5 Significant Combinational ATPG Algorithms

not around all circuit signals. This greatly reduced the size of the tree from 2^n to 2^{no_pis}, where n is the number of logic gates and no_pis is the number of primary inputs. This greatly accelerated ATPG.

- D-ALG tended to continue intersecting D-cubes, even when the D-frontier disappeared. Goel introduced a subroutine (X-PATH-CHECK) to test whether the D-frontier still existed. If not, PODEM would backtrack immediately, and this also greatly accelerated the search.

- Goel introduced *objectives* into ATPG algorithms. The *initial objective* was selected to bring the ATPG algorithm closer to propagating a D or $\overline{D}$ to a PO.

- Goel recognized that choosing primary inputs to set was important in efficiently realizing objectives. He invented *backtracing* (defined earlier in Section 7.4.) Backtracing obtained a PI assignment given the initial objective. PODEM used the length of paths between its initial objective and POs to measure the difficulty of sensitizing a path. This was simply given by the logic *level* of the logic gate at the start of the path, which was the minimum number of logic gates between the start of the path and any PO. Objectives were selected by level to pick the *easiest* objective to achieve. After objectives were selected, backtracing determined PI assignments to justify these objectives. This was determined using controllability measures. In fact, Goel was the first to use controllability/observability measures effectively to guide the ATPG algorithm, although Rutman [555] had experimented with this earlier. Goel used a logic gate level distance to measure testability. Here we will use SCOAP controllability ($CC0$ and $CC1$) (described earlier in Chapter 6) and level distance observability measures to illustrate PODEM. However, any other controllability and observability measures may be used with PODEM.

Figure 7.23 shows the high-level flow of the PODEM algorithm. For selecting an objective, PODEM found the logic gate with D or $\overline{D}$ on its inputs that was closest to a PO. If X-PATH-CHECK indicated that a path existed with unassigned signals from that gate to a PO, it would set the gate objective of obtaining a 1 on that gate output if it was an AND or an OR gate and a 0 if it was a NAND or a NOR gate to propagate the D forward. Opposite objectives were used to propagate $\overline{D}$.

Next, backtracing occurred by tracing backwards from the objective. If all logic gate inputs had to be set to achieve the objective, PODEM backtraced through the hardest-to-control input first. That way, if controlling that input failed, it wasted no time trying easier-to-control inputs. If only one logic gate input needed to be set to achieve the objective, PODEM backtraced through the easiest-to-control input. The selected input during backtracing became the next objective to satisfy.

Let us reconsider the circuit of Figure 7.22. Table 7.14 shows the steps that PODEM follows during ATPG for the fault s sa1, and Table 7.15 shows PODEM's implication stack and objectives as ATPG proceeds, based on gate SCOAP controllability measures.

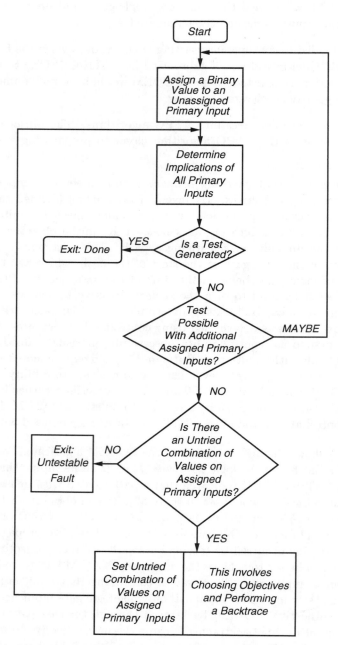

Figure 7.23: PODEM algorithm high-level flow chart.

7.5 Significant Combinational ATPG Algorithms

Table 7.14: PODEM ATPG decision steps for example in Figure 7.22.

Step	Action
1.	Select path $s - Y$
2.	Initial objective: Set r to 1
3.	Backtrace from r
4.	Implications in stack: $A = 0$
5.	Forward implications: $d = 0$, $X = 1$
6.	Initial objective: Set r to 1
7.	Backtrace from r again
8.	Implications in stack: $A = 0$, $B = 1$
9.	Forward implications: $k = 1$, $m = 0$, $r = 1$, $q = 1$, $Y = 1$, $s = \overline{D}$, $u = \overline{D}$, $v = \overline{D}$, $Z = 1$.
10.	X-PATH-CHECK shows paths $s - Y$ and $s - u - v - Z$ are blocked (the D-frontier disappeared), so backtrack.
11.	Set $B = 0$ (alternate assignment)
12.	Forward implications: $d = 0$, $X = 1$, $m = 1$, $r = 0$. Conflict – fault not sensitized. Backtrack.
13.	Set $A = 1$ (alternate assignment)
14.	Backtrace from r again.
15.	Set $B = 0$.
16.	Forward implications: $d = 0$, $X = 1$, $m = 1$, $r = 0$. Conflict – fault not sensitized. Backtrack.
17.	Set $B = 1$ (alternate assignment)
18.	Forward implications: $d = 1$, $m = 1$, $r = 1$, $q = 0$, $s = \overline{D}$, $X = 1$, $Y = \overline{D}$, and the fault is tested.

In the example, since path $s - Y$ has level distance 1 from a PO, and path $s - u - v - Z$ has level distance 2 from a PO, PODEM chooses path $s - Y$ as the fault propagation path. X-PATH-CHECK determines that there is an X path along path $s - Y$ to PO Y, so the *initial objective* is to set r (the driver of s) to a 0 to sensitize the fault. Since both inputs of XOR gate r must be controlled to define r, PODEM picks the hardest-to-control input n with an objective of setting it to 0. This leads to various intermediate objectives listed in Table 7.15, and results in the implication $A = 0$. A forward implication from that results in $d = 0$, and $X = 1$. However, this still does not define r, so backtracing from r occurs again. This leads to the primary input assignment $B = 1$. Forward implications from that are $k = 1$, $m = 0$, $r = 0$, $q = 1$, $Y = 1$, $s = \overline{D}$, $u = \overline{D}$, $v = \overline{D}$, and $Z = 1$.

At this point, X-PATH-CHECK indicates that the original propagation path, $s - Y$, no longer has an X path to PO Y, and that path $s - u - v - Z$ also cannot propagate a fault effect. The *D-frontier* is { }. The algorithm backtracks and chooses the alternate assignment for B, $B = 0$. Forward implications from this yield $d = 0$, $X = 1$, $m = 1$, $r = 1$. There is a conflict at s because the fault is not sensitized, so

Table 7.15: PODEM's objectives/implication stack for Figure 7.22 example.

Objective	Implication stack
Set r to 1 (initial)	$A = 0$
Set n to 0	
Set m to 0	
Set g to 0	
Set r to 1 (initial)	$A = 0$
Set n to 0	$B = 1$
Set m to 0	
Set k to 1	
	$A = 0$
	$B = 0$
	$A = 1$
Set r to 1 (initial)	$A = 1$
Set n to 0	$B = 0$
Set m to 0	
Set k to 0	
	$A = 1$
	$B = 1$

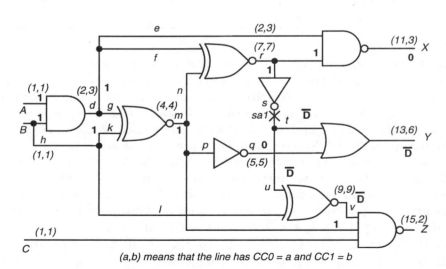

(a,b) means that the line has CC0 = a and CC1 = b

Figure 7.24: Example of successful PODEM ATPG.

7.5 Significant Combinational ATPG Algorithms

the algorithm again backtracks. It removes all assignments for B and sets $A = 1$. No forward implications are possible. PODEM backtraces from r, and goes through the intermediate objectives listed in Table 7.15. It sets $B = 0$. Forward implications yield $d = 0$, $X = 1$, $m = 1$, and $r = 1$. These is again a conflict at s because the fault is not sensitized, so PODEM again backtracks. It now sets $B = 1$, the alternate assignment for B. This yields the forward implications $d = 1$, $m = 1$, $r = 0$, $q = 0$, $s = \overline{D}$, $X = 1$, and $Y = \overline{D}$ so the fault is tested. Figure 7.24 shows the final circuit labeling where the test-pattern for the fault s sa0 is $ABC = $"$11X$" and the response $XYZ = $"$1\overline{D}X$."

If one repeats the above example with no controllability measures, where backtracing always traces through the first logic gate input and first tries to achieve the objectives by setting that input to 0, then the exact same sequence of decisions will be followed. This does not mean that the controllability measure does not help, but instead indicates that this circuit example is pathological. In general, controllability measures greatly accelerate ATPG.

Algorithm 7.4 shows pseudo-code for the backtracing operation, Algorithm 7.5 shows objective setting, and Algorithm 7.6 shows PODEM pseudo-code.

Algorithm 7.4 *Backtrace* (s, v_s). *Translate objective into PI assignment.*
```
{
    v = v_s;
    while (s is a gate output)
    {
        if (s is NAND or INVERTER or NOR) v = v̄;
        if (objective requires setting all inputs)
            Select unassigned input a of gate s with hardest controllability to value v;
        else
            Select unassigned input a of gate s with easiest controllability to value v;
        s = a;
    }
    /* s is now a primary input */
    return (s, v) /* Gate and value to be assigned */
}
```

Algorithm 7.5 *Objective* (g, v)
```
{ /*Target fault gate g stuck-at v */
    if (gate g is unassigned) return (g, v̄);
    Select a gate P from the D-frontier;
    Select an unassigned input l of P;
    if (gate G has controlling values) c = controlling input value of gate G;
    else if (0 value easier to get at input of XOR/EQUIVALENCE gate) c = 1;
```

      else $c = 0$;
      return $(l, \bar{c})$;
}

Algorithm 7.6 *PODEM.*
Podem (fault, vfault)
{
  while (no fault effect at POs)
    {
      if (xpathcheck (D-frontier)) /* Returns true if test possible because
                                    X-path exists from D-frontier to POs */
        {
          (l, v_l) = Objective (fault, vfault);
          (pi, v_{pi}) = Backtrace (l, v_l); /* Find the PI to set */
          Imply (pi, v_{pi});   /* Assign the value to the PI
                                and Compute all forward implications */
          if (Podem (fault, vfault) = SUCCESS) then return (SUCCESS);
          /* We need to backtrack */
          (pi, v_{pi}) = Backtrack (); /* Return alternate assignment to try */
          Imply (pi, v_{pi});
          if (Podem (fault, vfault) = SUCCESS) then return (SUCCESS);
          Imply (pi, "X"); /* Mark pi as unknown */
          return (FAILURE);
        }
      else if (implication stack exhausted) /* no more possibilities */
          return (FAILURE);
      else Backtrack ();
    }
  return (SUCCESS);
}

The basic idea of PODEM is to limit the search space to primary inputs without compromising the completeness. That is done by using the backtrace.

7.5.3 FAN (Fujiwara and Shimino)

Fujiwara and Shimono [232] introduced several novel concepts to further limit the ATPG search space and to accelerate backtracing in their FAN[†] algorithm.

Immediate Implications. In a number of situations, PODEM misses opportunities to immediately assign values that are uniquely determined to signals. Fujiwara

---
[†]Acronym for *Fanout-oriented test generation*.

7.5 Significant Combinational ATPG Algorithms

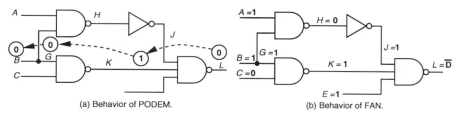

Figure 7.25: FAN's early determination of unique signals.

and Shimono [232] developed the concept of *immediate assignment of uniquely-determined signals*. Figure 7.25(a) [232] shows how PODEM would backtrace from the output signal L through the circuit, would pick the NAND gate input to control, signal K, and justify $K = 1$ by backtracing and setting $G = 0$, which forces $B = 0$. This is an unfortunate assignment, because $B = 0$ forces the assignments $H = 1$ and $J = 0$, which prevents achievement of the objective $L = 0$. In Figure 7.25(b), FAN instead sets $J = 1$, $K = 1$, and $E = 1$, because they all eventually have to be set to 1 to justify $L = \overline{D}$, so FAN does this immediately. But $J = 1$ uniquely determines $H = 0$, which uniquely determines $A = 1$ and $B = 1$. The only way left to obtain $K = 1$ is to set $C = 0$, which FAN does. The benefit of all of this is to eliminate all search operations from this ATPG problem, whereas PODEM must waste time making an untenable assignment $B = 0$ and then later retracting it during search.

Unique Sensitization. Fujiwara and Shimono [232] observed an additional opportunity to find signals that are uniquely determined. In Figure 7.26(b), we see that in portions of this circuit, there is only one path over which the fault can propagate. We might as well set the off-path inputs along these path segments to allow signal propagation immediately, rather than discovering these signal assignments during search. For the fault E stuck-at-0, the fault must propagate through lines F, H, K, and M. Therefore, the off-path inputs to the NAND gates driving those lines must be set to 1 to allow the fault effect to propagate. This immediately implies the *unique sensitization* assignments $C = 1$, $G = 1$, $J = 1$, $A = 1$, and $L = 1$. Now, backward logic simulation of NAND gate C shows that the only way to justify $C = 1$, when $A = 1$ already, is to set $B = 0$. By contrast, in Figure 7.26(a), we see that PODEM would backtrace from F whose good machine value is 0, and set an objective $C = 1$. Further backtracing leads PODEM to erroneously conclude $A = 0$, which implies $J = 0$, which cuts off fault-effect propagation. Again, FAN uniquely determines signals while PODEM wastes time searching for a conflict and ultimately retracts the unfortunate initial signal assignments.

Headlines. Fujiwara and Shimono [232] developed the notion of *headlines*, which are points where the circuit can be partitioned such that a cone of logic driven by PIs can be isolated from the rest of the circuit by cutting a single line, called the *headline*. This means that either a logic 0 or a logic 1 can be justified from the headline back to the circuit PIs. In Figure 7.27(a) [232], H and J are headlines.

194 **Chapter 7. COMBINATIONAL CIRCUIT TEST GENERATION**

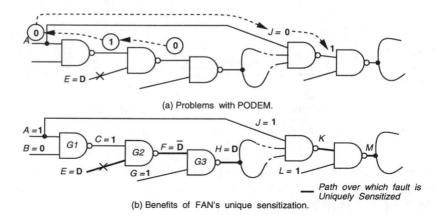

Figure 7.26: FAN's unique sensitization.

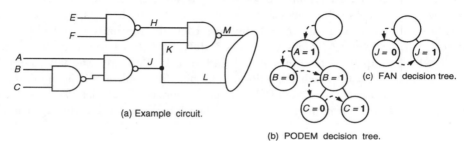

Figure 7.27: FAN's use of headlines.

The advantage of *headlines* is that one can remove the decision tree representing a family of PIs and replace it with one choice for the headline value during ATPG. Signal assignments to the removed PIs are deferred until the ATPG algorithm knows that it has a viable assignment for the headline. This eliminates the unnecessary work of searching the decision tree shown in Figure 7.27(b) that PODEM does, and allows FAN to search the much simpler decision tree shown in Figure 7.27(c). The initial *headline objective* is erroneously set to $J = 0$. PODEM justifies this with a backtrace, and sets $A = 1$, $B = 1$, and $C = 0$. When PODEM discovers later that we really need $J = 1$ to propagate the fault effect, PODEM revisits the decision tree, retracts $C = 0$ and sets $C = 1$. FAN, however, defers backwards search from the headline J until it is certain of the final headline objective $J = 1$. This limits its search space to the tree shown in Figure 7.27(c). FAN can then justify $J = 1$ with $A = 0$ much more efficiently than PODEM can.

Multiple Backtrace. The final contribution [232] from FAN is the breadth-first *multiple backtrace* procedure. We illustrate the inefficiency of depth-first backtracing in PODEM. In Figure 7.28 [232], we see that PODEM will make six backtraces to justify $C = 0$. The first backtrace sets objectives of $B = 1$ and $A = 0$. It finally assigns $PI1 = 1$. This process is laboriously repeated five more times until we

7.5 Significant Combinational ATPG Algorithms

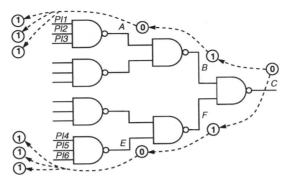

Figure 7.28: FAN's multiple backtrace.

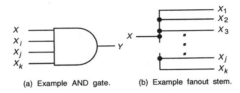

(a) Example AND gate. (b) Example fanout stem.

Figure 7.29: Examples of multiple backtrace.

have $PI2 = 1$, $PI3 = 1$, $PI4 = 1$, $PI5 = 1$, and $PI6 = 1$. This is happening because PODEM backtraces in a *depth-first* fashion, so FAN instead backtraces in a *breadth-first* fashion. For each *fanout stem*, FAN computes the ordered triple $(s, n_0(s), n_1(s))$, where s is the fanout stem signal objective number, $n_0(s)$ is the number of times a 0 is needed on that signal, and $n_1(s)$ is the number of times a 1 is needed on that signal. The counts of 0s and 1s for fanout stems are determined during breadth-first backtracing from objectives to headlines. Consider the AND gate in Figure 7.29(a) [232]. When AND gate input X is the easiest-to-control input during backtracing, $n_0(X) = n_0(Y)$ and $n_1(X) = n_1(Y)$. For all other inputs X_i, $n_0(X_i) = 0$ and $n_1(X_i) = n_1(Y)$. These relations are used, because in order to justify a 0 on Y, it is easiest to justify it on the easiest-to-control input. However, in order to justify a 1 on Y, then it must be justified on all inputs.

For the fanout stem in Figure 7.29(b) [232], FAN's backtracing relationships are:

$$n_0(X) = \sum_{i=1}^{k} n_0(X_i) \quad \text{and} \quad n_1(X) = \sum_{i=1}^{k} n_1(X_i)$$

So, we say that the number of votes for 0s and 1s at the stem is the tally of all votes from the branches. During the multiple backtrace, FAN searches backwards in breadth-first fashion, propagating the votes for 0s and 1s backwards. At a fanout stem p that is not reachable from the fault site, if $n_0(p) > 0$ and $n_1(p) > 0$, there is a conflict. FAN sets p to 0 if $n_0(p) > n_1(p)$ and otherwise sets $p = 1$. If there is no conflict, then FAN continues backtracing in breadth-first fashion. But, with a conflict, after the fanout stem signal value is voted on, FAN immediately does a forward implication from the fanout stem signal on p. This will find a signal conflict much sooner than the single backtrace procedure, which means that FAN will rapidly find a prior search decision that causes conflict and reverse it much sooner than PODEM. Algorithm 7.7 shows the multiple backtrace pseudo-code.

Algorithm 7.7 *Multiple backtrace.*

*multiple_backtrace(*current_objectives*)*
  {
    repeat
      {
        remove entry (s, v_s) *from* current_objectives;
        if (s is a head_objective*) add* (s, v_s) *to* head_objectives;
        else if (s is not a fanout stem and s is not a primary input)
          {
            mark the inputs driving the gate with the objective s as follows:
              all inputs i needing non-controlling values to achieve v_s *get* $n_0(s)$
                or $n_1(s)$ *votes as appropriate for the non-controlling value* v_i;
              easiest-to-control input i needing a controlling value to achieve v_s
                gets $n_0(s)$ *or* $n_1(s)$ *votes as appropriate for controlling value* v_i;
            if (i is a fanout branch)
              {
                $j = fanout\_stem(i);$
                $n_0(j)+ = n_0(i);$
                $n_1(j)+ = n_1(i);$
                add j to stem_objectives;
              }
            else add i to current_objectives;
          }
      }
    *until (*current_objectives $= \phi$*);*
    *if (*stem_objectives $!= \phi$*)*
      {
        $(k, n_0(k), n_1(k)) =$ *highest level stem from* stem_objectives;
        if $(n_0(k) > n_1(k))$ $v_k = 0;$
        else $v_k = 1;$
        if $((n_0(k)\ != 0)$ *&&* $(n_1(k)\ != 0)$ /* *k has contradictory requirements* */
          && (k is not reachable from fault)) return $(k, v_k);$
        add (k, v_k) *to* current_objectives;
        *return (multiple_backtrace (*current_objectives*));*
      }
    remove one objective (k, v_k) *from* head_objectives;
    return $(k, v_k);$
  }

7.5 Significant Combinational ATPG Algorithms

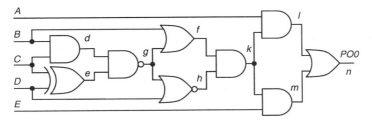

Figure 7.30: Example of dominators.

The main contribution of FAN is to introduce the immediate implications of signal assignments, unique sensitization, headlines, and multiple backtrace.

7.5.4 Advanced Algorithms

We briefly summarize additional ATPG algorithms that have been developed.

Dominator ATPG Programs. Kirkland and Mercer [360] developed TOPS, which found even more immediate signal assignments than FAN using *dominators*. A *dominator* is a circuit signal through which the fault effect *has* to pass in order to be detected a particular PO. An *absolute dominator* is a dominator through which the fault effect *has* to pass to be detected at any PO. In Figure 7.30, l and n are absolute dominators of A, k and n are absolute dominators of B, and g, k, and n are absolute dominators of C. k and n are absolute dominators of D and m and n are absolute dominators of E. For absolute dominators of a fault, we must set the inputs to those absolute dominators that are not in the fault effect cone to non-controlling values to propagate the fault effect through the absolute dominator. These are *mandatory assignments*, meaning that they can be determined by topological circuit analysis rather than by search. This would sometimes cause faults to be proved redundant without any search. During search, whenever any dominators of a fault become the constant values 0 or 1, fault propagation has been cut off, and the algorithm can back-up immediately.

Learning ATPG Programs.

SOCRATES (Schulz et al.) Schulz *et al.* introduced SOCRATES [574, 576, 577], an ATPG program that does *static* and *dynamic learning*. The *learning procedure* systematically sets all circuit signals to 0 and 1, and discovers what other signal values are implied. These implications are saved in the circuit netlist, in the form of *implication* arcs, and are used during search when they cause additional signals to be assigned. SOCRATES also uses *dynamic learning*, which invokes the learning procedure between search steps. This is costly but effective, because additional signal relationships can be learned when signals are already partially set in the circuit. Not all implications are worth learning. Consider Figure 7.31. The implication

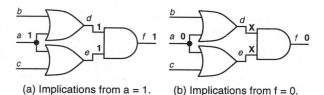

(a) Implications from a = 1.   (b) Implications from f = 0.

Figure 7.31: Implications from $a = 1$ and $f = 0$.

$(d = 0) \Rightarrow (f = 0)$ is not worth learning, because the normal AND gate logic implication procedure will discover it, anyway. Schulz was the first to use the *Boolean Contrapositive* relationship (*Modus Ponens*), which is $(p \Rightarrow q) \Leftrightarrow (\neg q \Rightarrow \neg p)$. This means that if we learn a forward implication triggered by signal p being 1, then we have also learned a backwards implication triggered by signal q being 0. In Figure 7.31(a), setting $a = 1$ also sets signals $d = 1$, $e = 1$, and $f = 1$. The *learning criterion* will cause SOCRATES to learn $(f = 0) \Rightarrow (a = 0)$ from the implication $(a = 1) \Rightarrow (f = 1)$ by applying Modus Ponens as in Figure 7.31(b), because these conditions are met:

1. $f = 1$ requires all inputs of AND gate f to be at non-controlling values, and

2. A forward implication contributes to the assignment $f = 1$.

This learning criterion is sufficient, but necessary, to establish that an implication cannot be performed by the normal logic implication procedure. The learning procedure is applied statically to the circuit before search begins, and frequently discovers redundant faults. The procedure can also be applied dynamically, at each step of the search algorithm for a test for a specific fault, but this usage is quite costly.

Figure 7.32(a) shows the improved unique sensitization procedure of SOCRATES [577]. Here, the D-frontier has only one signal and all paths from the D-frontier to POs pass through distinct gates. *Instruction 1* of SOCRATES says that when x is the only D-frontier signal, and gates $G = \{g_1, g_2, ..., g_n\}$ dominate x, then for all gates $g \in G$, we apply the non-controlling value to all inputs of g that cannot be reached from x. In Figure 7.32(a), Instruction 1 causes signal cc to be assigned 1. For multiple D-frontier signals, *Instruction 2* is illustrated in Figure 7.32(b) [577]. Let x be the only signal on the D-frontier or a dominator of the only signal on the D-frontier. Signal x branches to the gates $g_1, g_2, ..., g_n$, all of which require the same non-controlling value. This non-controlling value is assigned to all signals $y \neq x$ that branch out to all gates $g_1, g_2, ..., g_n$. In Figure 7.32(b), this causes the assignment $b = 1$.

Later versions of SOCRATES [576] also used the *Constructive Dilemma*, which is $[(a = 0) \Rightarrow (i = 0)] \wedge [(a = 1) \Rightarrow (i = 0)] \Rightarrow (i = 0)$. This means that if both assignments of 0 and 1 to a cause $i = 0$, then $i = 0$ is deduced independently of the setting of a. Figure 7.33 illustrates how the assignment $i = 0$ is determined to be independent of setting signal a. Also, *Modus Tollens* is used to determine when backtracking is necessary. Modus Tollens is $(f = 1) \wedge [(a = 0) \Rightarrow (f = 0)] \Rightarrow$

7.5 Significant Combinational ATPG Algorithms

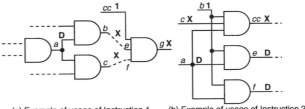

(a) Example of usage of Instruction 1. (b) Example of usage of Instruction 2.

Figure 7.32: Usage of Instructions 1 and 2.

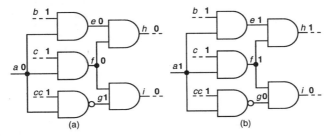

Figure 7.33: Example of Constructive Dilemma.

($a = 1$). Finally, SOCRATES dynamically computed dominators and dynamically learned implications, given the current settings of circuit signals. This identified more opportunities for implying signals, at the significant expense of running the learning and dominator algorithms at every decision step.

EST (Giraldi and Bushnell.) Giraldi and Bushnell [110, 252, 253] introduced the *evaluation frontier (E-frontier)*, which is a partial circuit functional decomposition that is equivalent to a node in the binary decision diagram describing the search space. The E-frontier is the cut-set between the part of the circuit with known labelings and the part with unknown (X) labelings, and includes the D-frontier. A new E-frontier is generated by every decision step of ATPG. EST[‡] tested for search space equivalence by hashing E-frontiers (circuit decompositions) into a hash table and seeing if they matched prior learned E-frontiers. When a match occurred, it meant that the search for the current fault had encountered the exact same functional decomposition learned during a prior fault. In most cases, EST could terminate search and either backtrack immediately, or substitute signals in the prior fault test-pattern (and the output response) for unknown signals in the current fault test-pattern and output response. This was a form of *dynamic programming*, which was first used in chess game playing programs. EST was the first and only ATPG algorithm to learn tests for the current fault from tests for prior faults. Figure 7.34(a) shows ATPG using EST for the fault B sa1, during which five E-frontiers (labeled 1 to 5) are generated. After the fault is tested, Figure 7.34(b) shows EST generating a test for a different fault, h sa1. E-frontiers 1 through 3 are generated, but then

---

[‡]Acronym for *Equivalent STate hashing*.

200 Chapter 7. COMBINATIONAL CIRCUIT TEST GENERATION

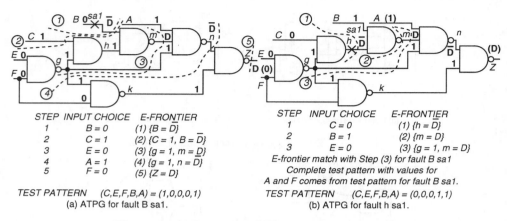

Figure 7.34: Example of EST using E-frontiers.

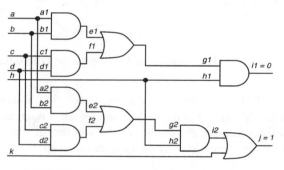

Figure 7.35: Circuit demonstrating recursive learning.

the hash table indicates that E-frontier 3 matches E-frontier 3 from the prior fault, B sa1. At this point, EST terminates search, and fills in the PO value and the missing PI values (these are all shown in parentheses) from information stored with the matched E-frontier. The test for B sa1 will not test the fault h sa1, and yet the search space information for the former helps to test the latter.

EST also contributed the *complete (parallel) multiple backtrace*, which speeded up backtracing by doing all signal backtraces in parallel. Finally, EST contributed a calculus of redundant faults, which discovered additional redundant faults by drawing implications from previously-proven redundant faults.

Recursive Learning (Kunz and Pradhan.) Kunz and Pradhan [376] pioneered the notion of *recursive learning*, in order to accelerate the FAN ATPG algorithm. They applied SOCRATES-style learning recursively to determine even more circuit signals and learn them as circuit implications. Figure 7.35 illustrates recursive learning. Lines $i1$ and j are unjustified. They find the necessary assignments, not by using an implication stack and decision making, but by recursive learning which learns information about the current value assignments in the circuit.

7.5 Significant Combinational ATPG Algorithms

The main advance over SOCRATES-style learning is that the learning procedure is called recursively, and the maximum recursion depth determines how much is learned about the circuit. Time complexity is exponential in r_{max}, the maximum recursion depth, but memory grows linearly with r_{max}. Table 7.16 shows the different recursive learning activities for the circuit of Figure 7.35, where column 0 reflects value assignments before and after recursive learning is done. The recursion causes learning to happen at column 1, which in turn recursively causes learning to occur in column 2. The outcome of this is that the necessary assignment $k = 1$ is learned.

Algorithm 7.8 *Demo_recursive_learning.*
```
{
    for each unjustified line
    {
        for each input: justification
        {
            assign controlling value;
            make implications and set up new list of resulting unjustified lines;
            if (consistent) Demo_recursive_learning ();
        }
        if (there are one or several signals f in circuit, so that f assumes
            same logic value V for all consistent justifications) then learn f = V,
            make implications for all learned signal values;
        if (all justifications are inconsistent) learn that the current
            situation of value assignments is consistent;
    }
}
```

A reader interested in finding the details of the recursive learning technique and its applications should reexamine the papers by Kunz and Pradhan [376, 377] and a recent book by Kunz and Stoffel [378].

Legal Assignment Test Generators. Rajski and Cox [531] maintained a *set* of legal signals on each circuit line (a power-set) for implications rather than using just a single signal for implications.

Implication Graph ATPG Algorithms. Chakradhar *et al.* developed the NNATPG [121, 127][§] algorithm family. These ATPG algorithms model logic gate behavior using implication graphs (see Section 7.1.6), which also enable redundancy identification [28, 399] and the modeling of transistor-level faults for ATPG [173, 250]. Chakradhar *et al.* also developed TRAN [122, 128], using a graph *transitive closure* algorithm to perform ATPG for huge circuits very rapidly.

[§]Acronym for *Neural Net ATPG*.

Table 7.16: Example of *Demo_recursive_learning*.

0. Learning level	1. Learning level	2. Learning level
(valid signal values)		unjust. line $e1 = 0$;
$i1 = 0$ (unjust.)	unjust. line $i1 = 0$;	1. justify: $a1 = 0$
$j = 1$ (unjust.)		$\Rightarrow a2 = 0$
	1. justify: $g1 = 0$	$\Rightarrow \mathbf{e2 = 0}$
enter	$\Rightarrow e1 = 0$ (unjust.)	
learning $\rightarrow$	$\Rightarrow f1 = 0$ (unjust.)	2. justify: $b1 = 0$
	enter next	$\Rightarrow b2 = 0$
	recursion $\rightarrow$	$\Rightarrow \mathbf{e2 = 0}$
	$e2 = 0$	$\Leftarrow=========$
		unjust. line $f1 = 0$;
		1. justify: $c1 = 0$
		$\Rightarrow c2 = 0$
		$\Rightarrow \mathbf{f2 = 0}$
		2. justify: $d1 = 0$
		$\Rightarrow d2 = 0$
		$\Rightarrow \mathbf{f2 = 0}$
	$f2 = 0$	$\Leftarrow=========$
	$\Rightarrow g2 = 0$	
	$\Rightarrow i2 = 0$	
	$\Rightarrow \mathbf{k = 1}$	
	2. justify: $h1 = 0$	
	$\Rightarrow h2 = 0$	
	$\Rightarrow i2 = 0$	
	$\Rightarrow \mathbf{k = 1}$	
$k = 1$	$\Leftarrow=========$	

The transitive closure of an implication graph is found by computing all global signals implied by a new signal assignment by tracing paths through the implication graph. The value of this approach was that TRAN made better signal assignment decisions using transitive closure, and backed up far less often than the other ATPG methods.

Figure 7.36 illustrates how the transitive closure algorithm effectively works. Figure 7.37(a) shows the implication graph, which represents all literals in the circuit. Signal d is set to 0, by adding the edge $(d, \overline{d})$ to the implication graph. This means that if d is set to 1, then d implies $\overline{d}$, which is not possible, so d is never set to 1. Figure 7.37(b) shows the resulting implication graph with the edge $(d, \overline{d})$ added (pointing from d to $\overline{d}$.) There are now paths $\overline{a} - d - \overline{d} - a$, $\overline{F} - d - \overline{d} - F$, and $\overline{b} - d - \overline{d} - b$ in the implication graph. The transitive closure algorithm discovers these paths, and concludes that $\overline{a} \Rightarrow a$, $\overline{F} \Rightarrow F$, and $\overline{b} \Rightarrow b$. Since these conclusions mean that when the literal is set to 0, it must also be set to 1, then the literal can never be set to 0. The transitive closure algorithm thus deduces that $a = 1$, $b = 1$,

7.5 Significant Combinational ATPG Algorithms

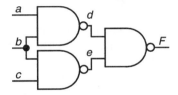

Figure 7.36: Example of computation of necessary assignments.

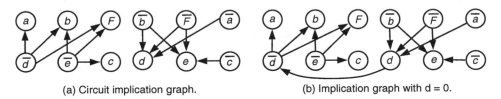

(a) Circuit implication graph. (b) Implication graph with d = 0.

Figure 7.37: Implication graphs for the example of Figure 7.36.

and $F = 1$. Other ATPG algorithms would be unable to draw these conclusions. Now consider the case when the circuit in Figure 7.36 has $F = 1$. The transitive closure for this graph is in Figure 7.38. When $F = 1$, the term deF in the Boolean false function of gate F reduces to de. By including $\overline{F} = 0$ and $de = 0$ in the graph, we cause the edges $(\overline{F}, F)$, $(d, \overline{e})$, and $(e, \overline{d})$ to be added to the graph. From this graph, the algorithm will also conclude that $\overline{b} \Rightarrow b$. Therefore, b must be set to 1 as a consequence of setting F to 1.

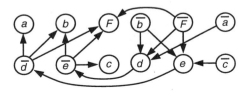

Figure 7.38: Implication graph for Figure 7.36 as a consequence of $F = 1$.

Larrabee [383, 384], starting with the Boolean difference and Chakradhar [121], with the neural network model, arrived at the satisfiability formulation of the ATPG problem. Both solved the problem with the help of implication graphs. The use of *path variables* by Larrabee and transitive closure by Chakradhar significantly added to the efficiency of the solution.

Henftling et al. [291] used the transitive closure method, and added an *ANDing node* to the implication graph (see Figure 7.7 in Section 7.1.6.) This allowed them to represent 3-SAT relationships directly in the graph very effectively. Whenever ATPG binds a signal, which converts a 3-SAT relation to a 2-SAT relation, some *ANDing* node will automatically collapse into an implication arc. Several implementations of satisfiability-based ATPG programs have been reported in the literature: GRASP [605], NEMESIS [384], TEGUS [638], TRAN [128], and a program by Tafertshofer et al. [648].

BDD-Based ATPG Algorithms. The methods based on *binary decision diagrams* (BDDs) deserve to be mentioned, because they are interesting, although not effective for test generation for multipliers, where the BDD becomes infinite. Gaede *et al.* [234] developed CATAPULT, an early BDD ATPG algorithm. Stanion and Bhattacharya developed TSUNAMI [635], which propagated partial BDDs along the fault propagation path. For circuits without the extremely reconvergent fanout of multipliers, TSUNAMI was extraordinarily effective.

7.6 Test Generation Systems

Early work of Agrawal and Agrawal [26] shows that an efficient test generation system may contain a random pattern generator, a fault simulator, and an algorithmic pattern generator. Considering today's large circuits, a test compacter should also be added to the list of tools. Some performance criteria are fault coverage, fault efficiency, vector set size and CPU time. The coverage and efficiency are defined as:

$$\text{Fault coverage} = \frac{\text{number of detected faults}}{\text{total number of faults}} \qquad (7.12)$$

$$\text{Fault efficiency} = \frac{\text{number of detected faults}}{\text{total number of faults} - \text{number of undetectable faults}} \qquad (7.13)$$

where both quantities can be expressed as fractions or percentages. Undetectable (or untestable) faults, which include redundant faults, consume large amounts of CPU time (see Chapter 4.) In combinational circuits, often the logic modeling of MOS structures (bus, tristate and bidirectionals, etc.) gives rise to undetectable faults [131, 690].

SOCRATES [576] is a typical test generation system. It begins with random test generation, optionally with weighted pattern probabilities, combined with concurrent fault simulation and fault dropping. So, 32 random patterns can be generated in parallel, and then one concurrent fault simulation can process all patterns. Typically, this process is continued until 64 random patterns have been tried, and no more faults are detected. After that, we perform several passes of automatic test-pattern generation. The first pass is done usually with only 10 backtracks allowed per fault, and then each generated pattern is fault simulated against all remaining faults, and detected faults are dropped. Later passes of ATPG increase the allowed number of backtracks to 50, then 100, and finally 10,000. Among the things that affect the ATPG performance are testability measures [402]. Various decisions within the ATPG are guided by testability measures (see Chapter 6.) The approximate nature of these measures is well known [38]. Because of different heuristics used in various measures, each provides incorrect information about a different part of the circuit. An experimental study by Chandra and Patel [137] suggests that successively using several testability measures can improve the performance of an ATPG program. Inputs to the ATPG system are the circuit netlist (often with standard cell libraries) and a list of target faults (which could be generated by a fault collapsing program), and outputs from the system are a test vector file, a summary file, a list of undetected faults, a list of redundant faults, a list of aborted faults, and a backtrack distribution file. An implementation of SOCRATES by Waicukauski *et al.* [708] is commercially available. Examples of other combinational ATPG systems are TRAN [128], used at NEC, and IBM's TestBench.

7.7 Test Compaction

At present, many combinational ATPG methods use RPG to inexpensively get to about 60% fault coverage, and then use the ATPG algorithm to generate tests for the remaining faults. However, randomly-generated patterns may not be very good, in the sense that a more complex pattern may exist that activates a higher percentage of the hardware, and in one test determines whether a number of faults are present, rather than just one or two. Therefore, at the end of ATPG, all testpatterns are fault simulated in the reverse order of their generation, so patterns from the test generator are simulated first, and patterns from RPG are simulated last. When the fault coverage reaches 100%, all remaining RPG patterns are discarded. This method greatly shortens the test-pattern sequence, since patterns from the ATPG algorithm are likely to activate a higher percentage of the circuit than RPG patterns, and therefore test more faults. This compaction method saves significant test time (and cost) by shortening the pattern sequence, and is described further in Section 5.2.

An additional *static compaction* method is suitable only for patterns generated by an ATPG program, where the unassigned inputs are left as X. Test patterns are compatible if they do not specify conflicting values for any PI. Two compatible tests t_a and t_b are combined into one test $t_{ab} = t_a \cap t_b$ using the D-intersection operator of Table 7.5. The detected faults will be the union of faults detected by t_a and t_b [7]. The compacted test set depends on the order in which vectors are compacted. Consider the following test set [7]:

$$t_1 = 01X \quad t_2 = 0X1 \quad t_3 = 0X0 \quad t_4 = X01$$

By first combining t_1 and t_3, and then t_2 and t_4, we obtain the compacted test set:

$$t_{13} = 010 \quad t_{24} = 001$$

Optimal static compaction algorithms are impractical, so heuristic algorithms are used. In *dynamic compaction* [259] every partially-complete vector from ATPG is processed immediately after it is generated, by assigning 0 or 1 to PIs with X values to extend the vector to detect additional faults. The vectors are selected for this only if the percentage of PIs with X values exceeds a user-specified percentage. Goel and Rosales [260] compare an arbitrary selection method for secondary target faults with a method that analyzes the internal circuit values produced by a partially-specified vector to select a secondary target fault for which ATPG is more likely to succeed. They showed that dynamic compaction produces smaller test sets than static compaction, and is worthwhile. More advanced dynamic compaction was done by Abramovici *et al.* [9], where fault simulation by *critical path tracing* was used to select a secondary target fault already activated by the partially-specified test pattern. Ayari and Kaminska [60] experimentally show that for large combinational circuits, dynamic compaction can reduce the test set by 50%. Pomeranz *et al.* [524] describe a PODEM-based program, COMPACTEST, that targets independent faults to produce very compact test sets. Akers *et al.* [45] define an *independent fault set* in which no two faults can be detected by the same test. Thus, the tests for independent faults can be considered "essential." Since finding a maximum independent fault set is a difficult problem, one must use heuristics.

7.8 Summary

We have introduced the terminology and search mechanisms of the most successful combinational ATPG algorithms. An understanding of at least the D-Algorithm, PODEM, and FAN is essential. This information will carry over into Chapter 8, where we will employ it to explain sequential automatic test-pattern generation. Another critical part in this chapter that digital designers must know is the redundancy identification and removal algorithm. In today's design environment, it is necessary that ATPG handles the MOS logic [93, 322, 323, 690]. In MOS circuits, tri-state devices, buses, and bidirectional components are also referred to as *non-Boolean primitives*. A good example to study is NEC's TRAN system that not only models non-Boolean primitives but can also identify redundancies in such structures, which may be removed [131].

Another important area is *hierarchical test generation*, which has been studied by several authors. For ATPG to handle modules, such as adders, comparators, multiplexers, etc., one would compute the *propagation D-cubes* for these modules and their singular cover [95]. Then, one can propagate fault effects and justify signals through these modules. For accurate testing of internal module faults, it would be necessary to replace the module primitive with the logic-level circuit for the module. Min *et al.* [467] summarize the work on hierarchical test generation and show a nine times speed up over the flat gate-level test generation for a 24 × 24 multiplier circuit.

Combinational circuit ATPG is considered to be a matured science by many. Forty years of development has produced three types of algorithms: (1) path sensitization methods, (2) simulation-based methods, and (3) Boolean satisfiability and neural network methods. While incremental improvements of these methods will continue, there is always a place for something radically different.

Problems

For each circuit, and for each decision step in the specified test-pattern generation algorithm, label a copy of the circuit with all circuit signal values, identify the D-

frontier, *show the implication stack contents, clearly state when the algorithm backs up, and count the number of backtracks. For problems involving the PODEM and FAN algorithms, compute all controllabilities and observabilities for the circuit using SCOAP and show them with your work. For problems using the SOCRATES algorithm, compute all controllabilities and observabilities for the circuit using SCOAP and show them with your work. Identify all signal relationships that are learned due to static learning. Describe the full dynamic learning algorithm for the circuit, and discuss whether it is worth its tremendous cost. Identify each application of the* Constructive Dilemma *and of* Modus Tollens *[574, 576, 577]*.

7.1 *Cubes.* For a two-input AND gate and a two-input Exclusive-OR gate, develop the singular cover of the gates, the propagation D-cubes, and primitive D-cubes of failure for a sa1 fault on one of the gate inputs.

7.2 *Stuck-at fault testing.*

   (a) Find only three tests that together test all single stuck-at faults in a two-input OR gate.

   (b) If the OR gate is replaced by another Boolean gate (AND, NAND, or NOR) with the same number of inputs and an output, then can the three tests determine that the given gate is not an OR?

   (c) Are the three tests still sufficient for the above decision if the OR gate was replaced by an exclusive-OR (XOR) gate? If not, why?

7.3 *D-ALG.* Use Roth's D-ALG to perform ATPG for the sa1 fault on the fanout branch h in the modified Schneider's example [570] circuit shown in Figure 7.39.

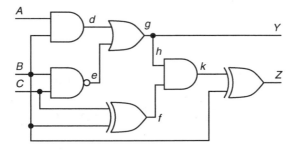

Figure 7.39: Modified Schneider's circuit.

7.4 *D-ALG.* Use Roth's D-ALG to perform ATPG for the sa0 fault on the fanout branch gate g to h for the circuit in Figure 7.39.

7.5 *PODEM.* Perform ATPG for the fault line h sa1 in the circuit of Figure 7.39 using PODEM and SCOAP measures.

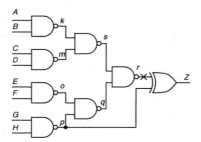

Figure 7.40: Circuit for Problem 7.6.

7.6 *PODEM.* Perform ATPG for the line r sa0 in the circuit of Figure 7.40 using PODEM and SCOAP measures.

7.7 *PODEM and FAN.* For the circuit of Figure 7.41 and the fault line r sa1, perform ATPG using both PODEM and FAN with SCOAP measures, and using the multiple backtrace with FAN.

   (a) For FAN, identify the following: *initial objectives, final objectives, fanout objectives, headlines,* and *head objectives*.

   (b) For each gate in Figure 7.41, identify its absolute dominators.

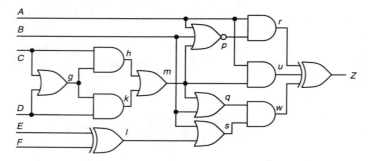

Figure 7.41: Circuit for Problems 7.7 through 7.12.

7.8 *PODEM.* Generate a test with the PODEM ATPG algorithm for the fault g sa1 in Figure 7.41.

7.9 *PODEM.* Generate a test with the PODEM ATPG algorithm for the fanout branch fault $C - h$ sa1 in Figure 7.41.

7.10 *FAN.* Generate a test with the FAN ATPG algorithm for the fault r sa0 in Figure 7.41.

7.11 *SOCRATES.* Generate a test with the SOCRATES algorithm for the fanout branch fault $m - p$ sa0 in Figure 7.41.

7.12 *SOCRATES.* For the circuit of Figure 7.41, perform SOCRATES ATPG using SCOAP measures for the fault line g sa0.

7.13 *D-ALG.* Perform ATPG on the circuit in Figure 7.35 using D-ALG to test the fault $h1$ sa1.

7.14 *PODEM.* Perform ATPG on the circuit in Figure 7.35 using PODEM and SCOAP measures to test the fault $h1$ sa1.

7.15 *FAN.* Generate a test with the FAN ATPG algorithm for the fault k sa1 in Figure 7.24.

7.16 *FAN.* Perform ATPG on the circuit in Figure 7.35 using FAN and SCOAP measures to test the fault $h1$ sa1.

7.17 *SOCRATES.* Generate a test with the SOCRATES algorithm for the fault n sa1 in Figure 7.24.

7.18 *SOCRATES.* Perform ATPG on the circuit in Figure 7.35 using SOCRATES and SCOAP measures to test the fault $h1$ sa1.

7.19 *Redundancy proofs.* Figure 7.42 shows a circuit. Find tests or redundancy proofs for: (1) The single fault d sa0 and (2) The single fault m sa0. Show the test pattern and the fault effect at q if the fault is testable. If any of these faults is redundant, remove the redundancy and reexamine the other fault. What is the Boolean function of the circuit and how could you obtain its minimal implementation?

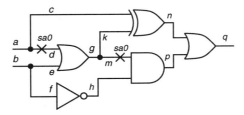

Figure 7.42: Circuit for Problem 7.19.

7.20 *PODEM.* Generate a test with the PODEM ATPG algorithm for the fault Z sa1 in Figure 7.24.

7.21 *SOCRATES.* Generate a test with the SOCRATES algorithm for the fault C sa1 in Figure 7.24.

7.22 *FAN.* Generate a test with the FAN ATPG algorithm for the fault d sa1 in Figure 7.30.

7.23 *SOCRATES.* Generate a test with the SOCRATES algorithm for the fault f sa1 in Figure 7.30.

7.24 *PODEM.* Generate a test with the PODEM ATPG algorithm for the fault e sa1 in Figure 7.30.

7.25 *PODEM.* Generate a test with the PODEM ATPG algorithm for the fanout branch fault $B - d$ sa1 in Figure 7.39.

7.26 *Static compaction.* Statically compact the following test patterns, first in forward order, and then in reverse order. Which compaction order is superior?

$$
\begin{array}{ccccc}
t_1 & X & 0 & 1 & 0 \\
t_2 & 1 & X & X & 0 \\
t_3 & 0 & X & X & 0 \\
t_4 & X & 1 & 0 & 0 \\
t_5 & 1 & 1 & 0 & 0
\end{array}
$$

Chapter 8

SEQUENTIAL CIRCUIT TEST GENERATION

> *"If you have mastered solving puzzles in two and three dimensions, you are then ready for puzzles in four dimensions."*

Most of us can add or multiply single-digit decimal numbers with ease. For calculating with numbers involving more digits, many people require paper and pencil. Notice that we always use the same basic methods, but with larger numbers we need to write down and reuse intermediate results. A similar paradigm is used in digital circuits that perform large computations. These circuits involve combinational logic and memory elements (usually in the form of flip-flops.) The combinational part produces an instantaneous result that is stored in the memory elements. The stored data may be reprocessed by the same combinational logic. Thus, a long sequence of computations can be economically performed by a small amount of hardware. The memory allows spreading of the computation over time.

The influence of sequential computing on testing can be profound. Going back to the example of human computing, suppose we restrict ourselves to single-digit numbers, so that no memory operation (paper and pencil or the memory in the brain) is involved. It seems easy to test the ability of a person to correctly calculate by asking a few questions. The test procedure here is similar to that of testing combinational logic. However, we can easily verify that such a test is not sufficient for checking the person's ability to deal with large numbers. For example, our simple test does not involve the manipulation of carries in addition. If we were allowed to also observe the scribbling on the paper then that could be some help. Still there can be far too many variations that must be tested. Besides, many people prefer to use their own memory rather than paper. That will restrict our ability to check the intermediate results, adding further complication to the test procedure.

Almost all digital systems of any significant size are realized as sequential circuits. These circuits contain combinational logic and flip-flops. Their testing is more complex than that of the combinational logic, discussed in the last chapter, for two reasons:

1. *Internal memory states.* The circuit contains internal memory whose state is not known at the beginning of the test. The test must, therefore, initialize the circuit to a known state. After test inputs are applied, the final state of the internal memories must be inferred only indirectly from primary outputs. Only in special cases can the internal memory be made controllable and observable for testing, sometimes at the cost of extra hardware.

2. *Long test sequences.* A test for a fault in sequential logic essentially contains three parts: (a) initialization of the internal memory, (b) a combinational test to activate the fault and bring its effects to the boundary of the combinational logic, and (c) if the fault has affected one or more memory elements, then observation of the state of one of the affected elements at a primary output. Thus, the test of a fault may be a sequence of several vectors that must be applied in the specified order. In comparison, any fault in a combinational circuit can be detected by a single vector.

In this chapter, we will first consider the simplest form of sequential circuit, i.e., a single-clock synchronous circuit, for studying various methods of test generation. Later, we will outline applications of those methods to multiple-clock synchronous and asynchronous circuits.

8.1 ATPG for Single-Clock Synchronous Circuits

A synchronous circuit consists of combinational logic and flip-flops, and is often represented in the form shown in Figure 8.1. The circuit in the large block is purely combinational. Some outputs of this block feed a set of flip-flops, which control some inputs of the block. Bold lines in the figure show groups of signals. The combinational logic has two types of inputs. Those at the top are external inputs known as *primary inputs* (PI). The inputs on the left side, called *pseudo-primary inputs* (PPIs) or *present state* (PS), are supplied by the flip-flops. Similarly, the combinational logic has two types of outputs. Those at the bottom are externally observable and are known as *primary outputs* (PO). The outputs on the right are called *pseudo-primary outputs* (PPO) or *next state* (NS) and feed into flip-flops. Input vectors are applied to PIs and observable outputs are produced at POs. It is the lack of a direct contact with the PPI and PPO that makes the detection of a fault in the combinational logic difficult.

We also assume that the flip-flops in Figure 8.1 are ideal memory elements that function under the control of a *clock* signal. The frequently-used data (or D) flip-flop, shown in Figure 8.2, is a one-bit memory element. It consists of two latches, *master* (M) and *slave* (S), each implemented with cross-coupled gates. The data to be stored in latches is applied in true and complemented forms only when the clock opens the latch. For example, when clock is high (1), the master latch will be in the "closed" state. It stores the previous state and is unaffected by the signal D. The slave latch (S) is, however, "open" and the state of M is transferred to S with an

8.1 ATPG for Single-Clock Synchronous Circuits

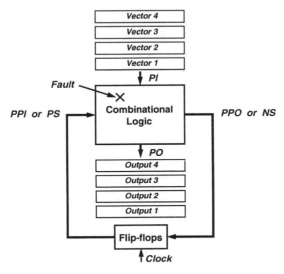

Figure 8.1: A popular representation of a synchronous sequential circuit.

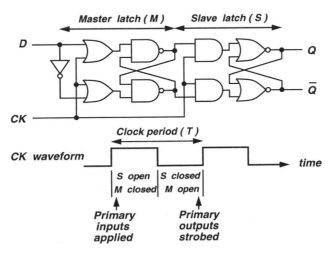

Figure 8.2: A master-slave flip-flop.

inversion. When the clock is low (0), M opens and allows the signal D to be loaded with an inversion, but S is closed and stores the previous state of M.

The *clock period*, T, contains one high and one low state. For the circuit of Figure 8.1, *Clock* is a free-running signal that repeats its high/low waveform periodically. Vectors at PIs are synchronized with the clock. A new vector is applied just after the clock goes from low to high ($0 \rightarrow 1$.) This avoids any simultaneous change of the data and clock signals at a flip-flop, a condition that can cause a *race*. The outputs are assumed to reach their final steady-state just before the next rising edge of the clock. Thus, one vector is applied and one output produced per clock period.

This operation of a sequential circuit is called "synchronous." It works perfectly as long as the time of signal propagation through the combinational logic does not exceed the clock period. With this condition satisfied, the flip-flops can be treated as ideal memory elements with an implicit clock. That is, for each PI vector and PPI state, the resulting PO and PPO state are produced, and the PPO state becomes the PPI state for the next vector. In general, the memory elements can be realized in many different ways and the number and wave shapes of clock signals can differ as well, but the synchronous model of Figure 8.1 is still used with minor adaptations.

8.1.1 A Simplified Problem

The combinational part of the circuit is modeled at the Boolean gate-level and all single stuck-at faults are considered in it. Flip-flops are treated as ideal memory elements, whose clock signal is not explicitly represented. Thus, no faults in the clock signal are modeled. Similarly, internal faults of flip-flops are not modeled; their output and input faults are modeled as faults on input and output signals of the combinational logic. The methods for test generation can be classified into two categories:

1. *Time-frame expansion.* In this method a model of the circuit is created such that tests can be generated by a combinational ATPG method. This procedure is very efficient for circuits described at the Boolean gate-level. Its efficiency degrades significantly with cyclic structure, multiple-clocks, or asynchronous circuitry.

2. *Simulation-based methods.* In these methods a fault simulator and a vector generator are used to derive tests. In general, tests can be generated for any circuit that can be simulated. Thus, a simulator that takes delays into account may be able to produce race-free tests for asynchronous circuits. Also, circuits modeled at other levels (register-transfer, transistor, etc.) can be treated.

8.2 Time-Frame Expansion Method

In order to apply a combinational ATPG algorithm to the sequential circuit of Figure 8.1, we must understand two basic differences between combinational and

8.2 Time-Frame Expansion Method

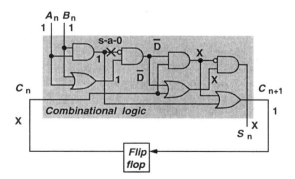

Figure 8.3: A serial adder circuit.

sequential circuits. The first difference is that a test for a fault in a sequential circuit may consist of several vectors. A combinational ATPG, on the other hand, is capable of generating only a single vector for a target fault. The second difference is due to the uninitialized (assumed to be unknown) memory states of the sequential circuit. A combinational ATPG can deal with unknown (X) signal states, but we find that the five-value logic system (Table 7.1), usually effective for combinational circuits, is insufficient for sequential circuits. The following examples illustrate these aspects of sequential circuits.

Example 8.1 *A serial adder. Figure 8.3 shows a serial version of the adder circuit of Figure 5.2. When the nth bits, A_n and B_n, are applied to this circuit, it produces the sum, S_n, and loads the carry, C_{n+1}, in the flip-flop. Thus, two 32-bit binary integers are added as follows:*

- *Initialization – A 00 input is applied to initialize the flip-flop in the 0 state. This corresponds to $C_0 = 0$. The output at this time is ignored.*

- *Serial addition – Inputs (A_0, B_0), (A_1, B_1), . . . , (A_{31}, B_{31}), are applied as successive two-bit vectors while sum bits, S_0, S_1, . . . , S_{31}, appear at the output. Then, a 00 input is applied to force the final carry bit at the output as S_{32}, while initializing the flip-flop to 0 for the next addition.*

Notice that the adder of Figure 8.3 is realized with one full-adder and one flip-flop. In comparison, the adder of Figure 5.2 uses 32 full-adders. In the serial adder, the hardware complexity is reduced by reusing the full-adder repeatedly. The price, however, is paid in the longer time to complete the computation. This scenario is quite typical of sequential circuits, especially those used in data path circuitry.

This example shows an equivalence between sequential and combinational circuits. To apply the combinational ATPG procedures we can thus "unroll" the sequential circuit into a larger combinational circuit. This unrolling is called "time-frame expansion." The following example illustrates test generation.

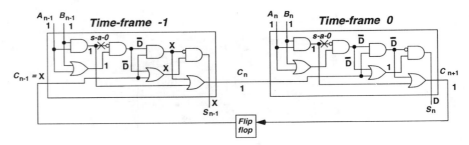

Figure 8.4: Test generation by time-frame expansion method.

Example 8.2 Time-frame expansion. *Consider the s-a-0 fault shown in Figure 8.3. We assume that the state of the flip-flop is unknown. So we set $C_n = X$ and apply the D-Algorithm procedure to the combinational logic. The fault is activated by $A_n = B_n = 1$, but cannot be propagated to the output S_n, which assumes the unknown value. We can easily see that we should set C_n to 1 to propagate a $\overline{D}$ to S_n. We should, therefore, precede this vector by one or more initialization vectors to set the circuit in the required state. Alternatively, $C_n = 0$ could also have propagated the fault effect to S_n, though $C_n = X$ will not work!*

For generating a two-vector test we use the expanded circuit model of Figure 8.4. Here, the combinational logic is repeated twice. The right block (time-frame 0) receives its state input from the left block (time-frame −1) in the same way as the carry signal was routed in the combinational adder of Figure 5.2. In fact, this circuit performs the same function as that of Figure 8.3. However, it takes only half the number of clock cycles (though the clock period may have to be doubled) since two inputs are applied simultaneously. Similarly, this test generation model of the circuit can be expanded to any desired length. As we shall later see, the maximum length of this expansion has a limit.

In Figure 8.4, we refer to the right-most block as time-frame 0. This contains the primary output where the fault is actually detected. The left-most block (time-frame −1 in this case) has its state inputs in the unknown (X) state. In general, they can be in any other given known state. The fault is present in all frames. So, we must generate a combinational test that will detect the multiple fault. In Figure 8.4, this test consists of four input bits all of which are 1. The sequential circuit test is an initialization vector 11 followed by another 11 vector that produces a D at the output. Notice that the fault is activated in both time-frames, although that is not always necessary. For example, another test 00 followed by 11 (had we used the alternative choice of $C_n = 0$) activates the fault only in time-frame 0.

8.2.1 Use of Nine-Valued Logic

A signal can assume different values in the fault-free and faulty circuits. It is, therefore, convenient to denote both values as a composite state. In Section 7.1.5, we discussed two logic systems used in test generation. The five-valued logic system,

8.2 Time-Frame Expansion Method

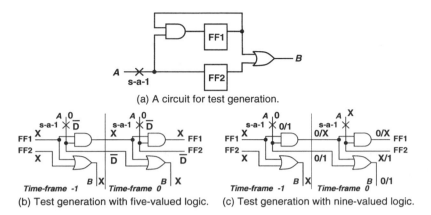

Figure 8.5: Test generation with five and nine-valued logic systems.

also known as the D-calculus [550], is used in the above example. In this system, D and $\overline{D}$ represent the signals that differ in the two circuits. While this system is simple and quite effective, there are situations that are not adequately handled by it. It was shown by Muth [481] that for sequential circuits, especially where initialization may be affected by the fault, a nine-valued logic system is required. In this system, a signal can assume any combination of the three values (0,1,X) in the two circuits. Thus a differing signal may be 0 in the fault-free circuit and X in the faulty circuit. We will designate it as 0/X (Subsection 7.1.5.) The following examples illustrate the advantage of the nine-valued system.

Example 8.3 *The necessity of the nine-valued logic. Figure 8.5(a) shows a circuit with two flip-flops, FF1 and FF2. Initially, the flip-flop states are unknown. FF2 is initialized after any input at A. FF1 is initialized to 0 any time the $A = 0$ input occurs and remains in that state thereafter. After the initialization of FF1, B tracks A with one clock delay. We consider the s-a-1 fault on input A. In the faulty circuit the output B will be a constant 1 and the fault is thus detectable. Figure 8.5(b) is a two time-frame construction of the test generation process with the five-valued logic, showing that the fault cannot be detected. This is because the five-valued D-calculus assumes that if the fault-free signal is in a known state then the faulty signal can differ from it but should also be in a known state. In this example, we notice that the fault does not allow initialization of FF1 and produces signal state 0/X, which is regarded as X in the five-valued system. Figure 8.5(c) repeats the process with the nine-value system and finds that an input $A = 0$ after the application of one clock detects the fault at B as 0/1. Most implementations of sequential circuit test generators use the nine-valued system.*

In the above example, the use of the nine-valued system was essential to find a test. Notice that the test output was 0/1. That means the faulty circuit produces a different but known output. There exist other situations, also related to the initialization, where such a test is not possible. The first is the case of an *initialization or*

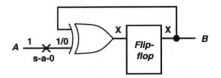

Figure 8.6: An example of a multiple-observation test.

potentially detectable fault discussed in Section 4.4. Here the test propagates a 0 or 1 value to a primary output of the fault-free circuit and produces an X at the faulty circuit output. In actual application of such a test to the hardware, the detection depends on the "power-up" state of the affected flip-flops. Derivation of tests for potentially detectable faults is only possible when the nine-valued logic is used.

Another type of initialization fault is detected by a *multiple-observation test*. According to the generally accepted definition of a test, the outputs of the fault-free and faulty circuits must differ at a specific observation time. In some cases, such a test is not possible. However, by observing an output for more than one vector, the fault can be detected. This test is called a *multiple-observation test* [525]. The following example illustrates such a test.

Example 8.4 Multiple-observation test. *The circuit of Figure 8.6 produces a toggling $(0 \rightarrow 1 \rightarrow 0 \cdot \cdot \cdot)$ output when $A = 1$. When $A = 0$, the output is either steady 0 or steady 1. The circuit has no initialization hardware and can "power-up" in any state. Although used in practice, such a circuit is difficult to analyze by a three-state simulator. Hence, it is difficult to verify a digital design that is not initializable. A similar difficulty is encountered in test generation. Consider the s-a-0 fault on the input A. Assuming the flip-flop to be in the unknown (X) state, the nine-valued logic system fails to produce a test. This is not a potentially detectable fault because the fault-free circuit, which must produce a deterministic output, is not initializable. We notice, however, that a definite test is possible if we observe the output B for at least two vectors, while A is held at 1. The fault-free circuit will produce either a $0 \rightarrow 1$ or a $1 \rightarrow 0$ output. The output of the faulty circuit will be either $0 \rightarrow 0$ or $1 \rightarrow 1$.*

In actual testing of circuits, the unknown outputs are masked by the tester. That is, the unknown bits are not compared with the expected true response. In the multiple-observation test, some unknown outputs should be strobed and saved for analysis. Such test procedures, though useful for testing uninitializable circuits, are more expensive.

8.2.2 Development of Time-Frame Expansion Methods

Soon after the publication of the D-Algorithm by Roth [551] in 1967, Kubo [372] and Putzolu and Roth [530] presented the idea of time-frame expansion using the five-valued logic system. Then, in his 1976 paper Muth [481] explained that a nine-valued logic was necessary, especially in cases where the fault affected the

8.2 Time-Frame Expansion Method

initialization of the circuit. Just a year later, Snethen's paper on SOFTG [613] advanced the idea of *backtrace*. Until then the D-Algorithm's strategy of one-level justification followed by forward implication was universally used. In that procedure, one sets the inputs of a gate to justify its outputs and immediately performs all forward implications. The inputs, thus set, are justified next. Typically, this process may be carried out through many steps before a conflict is found, making backtracks rather costly. Snethen's idea of *backtrace* meant that the justification of any signal value must be carried out only at primary inputs. Goel [258] used backtrace to formulate an elegant algorithm (PODEM) for combinational circuits. Thereafter, combinational circuit ATPG became quite practical.

In the PODEM algorithm, we advance the fault effect gate-by-gate toward a primary output. Any signal value needed is backtraced to primary inputs, followed by a forward implication (simulation.) If the simulation indicates a conflict with the basic objective of propagating the fault effect to a primary output, we must backtrack to some previously-set primary input, change the value there, and resimulate. When applied to sequential circuit time-frames, the complexity of this procedure increases very rapidly. There are two reasons. First, as the number of time-frames increases, the number of primary inputs, or decision points for a PODEM-like algorithm, increases. Second, the fault effect may have to be propagated through several time-frames before reaching a primary output. Thus, the number of steps in fault effect propagation increases rapidly as more time-frames are required.

Marlett proposed [432, 433] a solution for the problems mentioned above. He selects a primary output for fault detection and then selects a path between that output and the fault site. Test generation begins in the "final" time-frame in which the fault is supposed to have been detected. A fault effect (0/1 or 1/0 depending on the fault type and the number of inversions on the selected path) is placed at the output in that time-frame. The reader will immediately recognize this as time-frame 0 in all our examples. The rest of test generation consists of justifications via backtraces to primary inputs and state variables. His *extended backtrace* (EBT) algorithm breaks the procedure down into a series of current and previous vectors. For time-frame i ($i > 0$), the current vector C_i contains the next states. The previous vector P_i contains the present state and primary input values. For time-frame 0, C_0 contains the fault effect for the selected output and don't cares for all other states and primary outputs. P_0 is constructed by backtraces through this time-frame. An attempt is made to justify C_0 with the largest number of don't cares in the present state portion of P_0. Upon justification of C_0, if P_0 contains any present state variable in the specified state, then time-frame -1 is analyzed. A C_{-1} vector is constructed using the state portion of P_0. The justification of specified values in C_{-1} determines P_{-1}. This process continues through to some time-frame $-n$ in which C_{-n} is justified with only the primary input portion of P_{-n}. Sequential ATPG programs using EBT were developed in the mid-80s by Marlett [433], and Mallela and Wu [419]. These programs performed adequately, but would give up on certain large circuits.

In 1988, Cheng published further innovations. In his BACK algorithm,

Cheng [159] selects a primary output for fault detection, but unlike Marlett, does not restrict propagation through any one path from the fault site to that output. The process of selecting an output and activation of paths is based on a testability measure called *drivability*. The drivability of a line is a measure of the effort of driving the fault effect from the fault site to that line. Since the fault effect can traverse through several paths and may arrive as 0/1 or 1/0 at the line, two drivabilities are calculated. Drivabilities are fault-specific and are calculated from the SCOAP controllability measures (see Example 8.5.) Prior to computing drivability a controllability calculation for all lines in the fault-free and faulty circuits is performed. For test generation, the output with "easiest" drivability is first selected. A faulty state is placed in time-frame 0. Then, justification backtrace is guided by both drivability and controllability measures. If the justification is for a line affected by the fault, then the backtrace is guided by drivabilities. Otherwise, it is guided by controllabilities. Drivability, being a fault-specific measure, provides a much improved guidance. Also, taking all paths simultaneously is a better heuristic since some fault effects can only be propagated if multiple paths are sensitized. As a result, the BACK algorithm significantly reduces the number of backtracks made in the search for a test sequence.

Example 8.5 Drivability. *Drivability calculation involves some minor modifications in the SCOAP measures [117, 159]. Only one set of controllability measures is used. These are similar to the 0 and 1 combinational controllabilities (CC0 and CC1) of SCOAP (see Chapter 6.) We remind the reader that these measures represent "effort" and hence larger values indicate greater effort needed. For the implicit clock model of flip-flop, it is assumed that no effort is needed to control the clock. Thus, the output controllabilities for a flip-flop are obtained by adding the depth (1) to its input controllabilities. Figure 8.7 shows a part of a circuit with two primary outputs. We assume that controllabilities have been calculated and are as shown in parentheses. Consider the s-a-1 fault on the upper input of the AND gate A. Setting this line to 0 produces a 0/1 state on it. Therefore, its 0/1 drivability, $d(0/1)$, equals the 0-controllability, which is 4. It is impossible to have a 1/0 at the site of this fault, i.e., $d(1/0) = \infty$. In actual implementation, a very large integer like 99999999 might represent ∞.*

In order to drive the fault effect to the output of A, its fault-free input must be 1. Hence, its output drivabilities are obtained by adding the CC1 of the other input and the depth factor 1 to input line drivabilities. That gives the drivabilities for the output line of A as $d(0/1) = 9$ and $d(1/0) = \infty$. For an OR gate, we will simply use CC0 in a similar fashion. Any inversion, such as in the NOR gate B, will interchange the values of the two drivabilities. Drivabilities are carried forward through the flip-flop by adding a large depth, 100 in this example, to account for the fact the lines on the two sides belong to neighboring but different time-frames. Alternatively, the overall depth of the combinational logic can be used as the depth for a flip-flop.

Multiple paths converge at gate D. For the output of this gate, the drivability is computed three ways and the minimum is taken as its drivability. For $d(0/1)$, the

8.2 Time-Frame Expansion Method

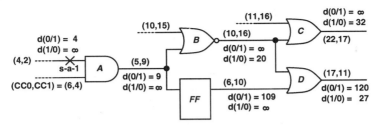

Figure 8.7: Drivabilities for guiding sequential ATPG.

three ways lead to ∞ from the upper input, 120 from the lower input, and ∞ when both inputs are drivable. When both inputs are drivable, the drivabilities are added to find the drivability of the output. Similarly, for $d(1/0)$ of the output of D, the three values are 27, ∞, and ∞, respectively, and the minimum is 27. Drivabilities for the output of C are $d(0/1) = \infty$ and $d(1/0) = 32$, respectively.

These drivabilities indicate that the output of D is a better choice for detecting the fault. Although this output is drivable to both 0/1 and 1/0, the preferred state is 1/0 since lower drivability means reduced effort. Drivabilities of the two inputs of D are 20 and ∞, hence a preferred justification is 1/0 for the upper input and X/0 for the lower input.

Cheng's second contribution is a data structure called SPLIT [158], especially designed for the nine-valued logic. In this data structure, the fault-free and faulty circuits are modeled as two separate copies of the circuit joined at primary inputs. A signal can have any of the three values (0,1,X) independently in the two copies. However, an additional logical variable with states (equivalence, difference) is maintained for each signal in the fault-free circuit. *Equivalence* means that the signal is not affected by the fault and *difference* implies a definite or potential fault effect. This data structure allows signal justifications to be performed rather simply. A justification may be performed in one copy at a time via backtraces. Since the two copies are joined at primary inputs, the implied values are also set in the other copy. The procedure allows the least restrictive justification values and hence reduces the possibilities of backtracks. Besides, the data structure is memory-efficient and simple to program. Cheng [158] gives examples to illustrate its effectiveness.

Several implementations of the time-frame expansion method have been reported. Some of these are ESSENTIAL [575], GENTEST [72, 160], HITEC [497], and SEST [150]. While most of these work with "backward time" or with a combination of "backward" and "forward" time, a program exclusively with forward time has also been reported. This program, called FASTEST [351], uses a SCOAP-like testability analysis (see Chapter 6) to determine the length of a time-frame array such that the appropriate controllability measure (e.g., CC0 for s-a-1 fault) at the fault site in the right-most time-frame is below some threshold. The combinational ATPG algorithm PODEM (Chapter 7) is then applied to the array to detect the fault. If the fault can only be detected at a next state output, then forward time-frames are added until it is detected at a primary output.

8.2.3 Approximate Methods

The presence of the fault in all time-frames adds tremendous complexity to the test generation process. The circuit states in all time-frames, previous to the time-frame in which the fault is detected, have to be justified for both fault-free and faulty circuits. To simplify the procedure an approximate method can be used. Here, only the time-frame of fault *activation* is assumed to have the fault. Thus, a combinational test is generated in a single time-frame. If the fault can only be detected at a flip-flop input, then "future" time-frames are added without the fault and the fault effect is propagated through them to a primary output. Similarly, the states of the activation time-frame are justified through fault-free "previous" time-frames. The SCIRTSS system of Hill and Huey [300], and the STALLION program of Ma *et al.* [408], use the *state transition graph* (STG) of the fault-free circuit to derive test sequences. The tests so derived must be verified by a fault simulator. In practice, these tests do work for most faults. However, it is easy to find examples where the multiple instances of the fault will invalidate the approximate test. Another problem with the use of the STG is its size. For a circuit with N_{ff} flip-flops, the STG will have $2^{N_{ff}}$ states and a large number of edges. So, for more than 10 or so flip-flops one must work with a partial STG, adding further approximation to the technique. A fundamental assumption used in most approximate methods is that a *universal reset* primary input signal initializes all flip-flops correctly in both fault-free and faulty circuits.

Another approximate test generator, STEED [248], takes a different approach. For the combinational logic, on-sets and off-sets are first generated for each primary output and next state variable. An *on-set* of an output of a combinational circuit is an input vector that sets that output to 1. Similarly, an *off-set* of that output is an input vector that sets the output to 0. Intersections of these on-sets and off-sets produce vectors for fault effect propagation and state justification. Once again, as the circuits get larger, the complexity of generating, storing, and manipulating the on-sets and off-sets increases. Besides, approximate tests must be validated by a fault simulator.

8.2.4 Implementation of Time-Frame Expansion Methods

Commonly used algorithms for sequential circuit test generation are *backward state justification* procedures shown in Figure 8.8. The construction of a test sequence for a given target fault begins with the last vector, which detects the fault at a primary output. In Figure 8.8, this is accomplished in time-frame 0 (called the fault detection time-frame) by input vector v_0 and state vector s_0. The vector v_0 is binary-valued and s_0 is a nine-valued vector. These are determined by a combinational ATPG algorithm like PODEM or FAN, as discussed in the last chapter. Next, the state vector s_0 is justified in time-frame -1 by input v_{-1} and state s_{-1}, again by the combinational ATPG algorithm. This process continues until some time-frame $-k$ such that s_{-k} matches a given initial state. The test vector sequence then contains $k + 1$ vectors applied in the order, $v_{-k}, v_{-k+1}, \ldots, v_0$. All time-frames -1

8.2 Time-Frame Expansion Method

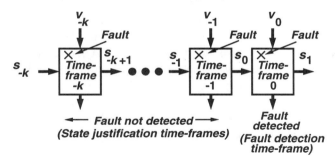

Figure 8.8: Test generation by backward state justification.

through $-k$ are called justification time-frames. The fault is assumed to be present in all time-frames and can be activated any number of times before detection. This procedure is called "backward time" because the vectors are actually derived in the reverse order. The initial state s_{-k} is determined in one of three different ways, depending upon the specific implementation of the algorithm:

1. *Start with uninitialized state.* The initial state s_{-k} is set to all Xs. It is assumed that both fault-free and faulty circuits are in some unknown states.

2. *Start with the final state of the previous tests.* This is the case where we already have some vectors that could not detect the target fault. In a typical scenario, these may be verification or initialization vectors provided by the circuit designer, or may be test vectors generated for previously-targeted faults. The nine-valued state vector s_{-k} is determined by simulating the presently targeted fault. The fault simulator starts with both fault-free and faulty circuits in unknown states. The new test sequence is, of course, concatenated to the previous vector sequence.

3. *Start with a known initial state.* Some circuits are designed to "power-up" in a known state [73, 721]. It is usually not possible to simulate the often analog or asynchronous power-up initialization circuitry by a logic or fault simulator. Thus, we would assume that both fault-free and faulty circuits start with a known binary-state s_{-k}. The vectors so produced are valid tests as long as the fault does not interfere with the initialization circuitry.

A test generator GENTEST [72, 160], developed at Bell Labs, implements the time-frame expansion algorithm. It uses the BACK principle and the SPLIT data structure mentioned in the last subsection. Also, built into the test generator is a parallel version of a differential fault simulator (see Chapter 5.) GENTEST works in multiple passes. In the first pass, based on the circuit size, a reasonable but small per-fault time limit is selected. The test generator prepares a list of collapsed (or uncollapsed at the user's option) fault list. Faults can be randomly sampled if so desired (see Chapter 5.)

GENTEST starts with one fault from the fault list. If a test is found, all faults in the list are simulated and detected faults are removed from the fault list and placed in a detected fault list. Potentially detected faults are marked as such but are not removed, unless the user exercises the "potential detection option." At this point the fault simulation is suspended and the test generator restarts with a new fault target from the list. However, the fault-free and faulty circuit states are saved for the restart of simulation when more vectors become available.

If the target fault is not detected within the per-fault CPU time limit, it is abandoned, but not removed from the list, and another fault is targeted. Once all faults have been processed and the fault list still contains more than a prespecified number, the per-fault CPU time limit is increased and a new test generation pass begins. The entire process ends when either a required fault coverage is achieved or some maximum per-fault CPU time limit has been exceeded.

A typical GENTEST run for the ISCAS '89 benchmark circuit s35932 [99] produced the following result on a SUN Ultra II (200 MHz) workstation:

```
Primary inputs = 36
Primary outputs = 320
State elements = 1728
Total Faults = 39096
test generation time = 4251 s
fault simulation time = 97 s
total vectors = 327
detected faults = 33630
untestable faults = 2
undetected faults = 5464
0 untestable faults were potentially detected
68 undetected faults were potentially detected
faults tried = 6116
time limit per fault = 833 ms
fault coverage = 86.019%
fault efficiency = 86.024%
Note: GENTEST did not target certain faults in the last cycle.
The dynamically adjusted maximum effort per fault is too
small for those faults.  Use the -t or -k option to
increase the cpu effort per fault.
```

As the test generator's note indicates, this fault coverage can be enhanced by increasing the per-fault time limit. GENTEST also contains an I_{DDQ} option, which assumes a complete observability for any fault that is activated. When this option is used, GENTEST models 22,493 faults in s35932. It generates 93 vectors to cover all of these faults. The CPU time is only 313 s. Strangely, these vectors have a stuck-at fault coverage of 87.02%, which is higher than that obtained in the above run. Though this result cannot be generalized, it needs investigation.

The HITEC program [497], developed at the University of Illinois, contains several unique features. It combines forward and reverse time processing. Test sequence generation for a target fault is begun in a time-frame where the fault is activated.

8.2 Time-Frame Expansion Method

This is designated as time-frame 0. First, in a forward time phase, the fault is activated by setting PIs and state inputs of time-frame 0. Also, propagation of the fault effect to a PO of time-frame 0 is attempted. When that is not possible, the fault effect is propagated to some flip-flop. and forward time-frames (numbered greater than 0) are added to propagate it until it reaches a PO. All necessary signal assignments in the forward time-frames are backtraced to PIs of time-frames numbered ≥ 0 and the present state flip-flops of time-frame 0. If this phase is successful, then the reverse time justification phase begins. The flip-flop states assigned in time-frame 0 are now justified through one or more negative-numbered time-frames (< 0), until all flip-flops in the "most negative" time-frame have don't care states. A conflict can occur any time during this process and, when that happens, the test generator backtracks to the last decision point and makes an alternative choice.

The HITEC test generation system incorporates a parallel fault simulator [496] and can achieve high fault coverages and efficiencies for large sequential circuits. It has been in use by researchers at many universities and its algorithms have been implemented in a test generation program that is commercially available from Synopsys.

8.2.5 Complexity of Sequential ATPG

It is possible to bound the number of time-frames needed. It would be fruitless to repeat the state during justification for the following reason. If a subsequence of states has the same first and last state, then that subsequence can be replaced by just one vector (either the first or the last vector in the subsequence) without affecting the success of the justification process. The size of the state vector equals the number of flip-flops in the circuit, N_{ff}. Hence the total number of distinct state vectors is $9^{N_{ff}}$, when the nine-valued logic is used. If the state is not justified with $k_{max} = 9^{N_{ff}}$ time-frames, then the fault may be considered untestable. The maximum number of justification time-frames is reduced if we assume that the fault-free circuit remains initialized. Then a flip-flop in the fault-free circuit will only assume two (0,1) values, while the faulty circuit flip-flop can have any of the three (0,1,X) states. Thus $k_{max} = 6^{N_{ff}}$. Furthermore, if the faulty circuit also remains initialized, then $k_{max} = 4^{N_{ff}}$ [7]. This shows the importance of initialization in testing and the difficulty of detecting the faults that affect initialization. Next, we will examine two classes of sequential circuits and their ATPG complexities.

8.2.6 Cycle-Free Circuits

This classification is based on the connectivity structure of the circuit. We assume that the circuit consists of combinational gates and flip-flops. The state of a flip-flop Fi is considered dependent on another flip-flop Fj if there exists a combinational path from Fj to Fi. Such dependence can be expressed by a directed graph, usually called the *s-graph* [154, 379]. Figure 8.9 shows a circuit and its s-graph. This is a *cycle-free circuit* since there is no feedback among flip-flops. As

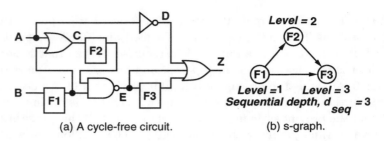

(a) A cycle-free circuit.   (b) s-graph.

Figure 8.9: A cycle-free circuit and its s-graph.

a result, the s-graph is acyclic. A cycle-free circuit is also called a *feedback-free* or *pipeline* circuit.

In a *directed acyclic graph* (DAG) there must be at least one vertex with in-degree 0. The in-degree of a vertex is the number of incoming edges [41, 363]. Similarly, the out-degree of a vertex is the number of outgoing edges. In the s-graph, the node with 0 in-degree corresponds to a flip-flop whose state depends exclusively on primary inputs. There can be more than one such vertices. We will levelize the s-graph by assigning a level number 1 to all 0 in-degree vertices. In Figure 8.9(b), F1 has level 1. The rest of the graph is then levelized such that the level number of a vertex is one greater than the maximum level feeding into it. Thus, the level of F2 is 2 and that of F3 is 3.

We define the *sequential depth*, d_{seq}, of the circuit as the maximum level number. Clearly, the sequential depth is the number of flip-flops on the longest directed path in the s-graph. The following is a useful result:

Theorem 8.1 (Initialization) *Given a cycle-free structure, the fault-free circuit and the faulty circuit with a non-flip-flop fault are always initializable [152].*

Proof: Examining the levelization procedure, we find that flip-flops in level 1 are initialized directly by a single vector at primary inputs. Once that is done, using the level 1 flip-flops and primary inputs, flip-flops in level 2 are initialized by a second vector. Carrying this process on level by level, all flip-flops will be initialized with the application of d_{seq} input vectors. Since a single stuck-at fault in the combinational logic changes neither the acyclic nature of the s-graph, nor the level numbers of flip-flops, the same arguments apply to the faulty circuit. ∎

According to Theorem 8.1, both fault-free and faulty circuits are initializable to known states, which may or may not be identical. Thus, after the application of d_{seq} vectors, there will be no unknown state left either in the fault-free circuit or in the faulty circuit. If the primary outputs differ any time after the application of d_{seq} vectors, the detection will be definite and not potential. If the two circuits produce identical outputs beyond the first d_{seq} vectors, then the fault may only be detected as a potential detect in the pre-initialization phase. The latter type is called a *partially detectable fault* [527].

Theorem 8.2 Test complexity of a cycle-free circuit. *A test for a non-flip-flop*

8.2 Time-Frame Expansion Method

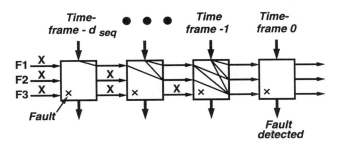

Figure 8.10: Time-frame expansion of a sequential depth 3 cycle-free circuit.

fault in a cycle-free circuit can always be found with at most $d_{seq} + 1$ time-frames, unless the fault is untestable.

Proof: Consider the circuit of Figure 8.9 with a sequential depth, $d_{seq} = 3$. Similar arguments will apply to any cycle-free circuit. Suppose we derive an input vector and a state vector in time-frame 0 that detect the target fault at a primary output. According to Theorem 8.1, both fault-free and faulty circuits are initializable. So, we assume that the state vector in time-frame 0 contains no unknowns (X.) From the arguments following the proof of Theorem 8.1, all flip-flops are initialized by at most d_{seq} vectors. Thus, we add d_{seq} time-frames prior to the time-frame 0, as shown in Figure 8.10, which illustrates the circuit with three flip-flops. In Figure 8.10, the top line connecting the time-frames carries the state of F1, the middle line that of F2, and the bottom line that of F3. On the left hand side all three flip-flops are in the unknown state. The lines inside combinational logic blocks show the signals that control a flip-flop. For example, in time-frame −1, F3 is controlled by primary inputs, F1, and F2. All combinations of F1, F2, and F3 states that are possible in this circuit can be obtained by primary inputs of these three time-frames. Thus, the four (or $d_{seq} + 1$, in general) time-frame combinational logic must provide a test if one is possible. Since there are no unknown state variables at the input of time-frame 0, the test in that time-frame will be deterministic. The tests in earlier time-frames can be of the potential detection type, since some flip-flops there are still in the unknown state. ∎

Example 8.6 *We can generate a test for any fault (all are testable) in the circuit of Figure 8.9 using at most four time-frames. Figure 8.11 illustrates the test generation for "input A of OR gate C stuck-at-0." The test generator takes the following steps:*

1. *A simple analysis of the circuit topology in Figure 8.9 tells us that the fault effect can be propagated to the output Z through two paths, $C - F2 - E - Z$ and $C - F2 - E - F3 - Z$. The first path, being shorter (fewer flip-flops) is chosen as the primary path. Real test generators may use more complex analyses for grading the propagation paths (see Example 8.5.)*

2. *In time-frame 0, we place a 0/1 at Z and justify it by $A = 1$, $F1 = 1$, $F2 = 1/0$, and $F3 = 0/X$. Some values shown in Figure 8.11 differ, since*

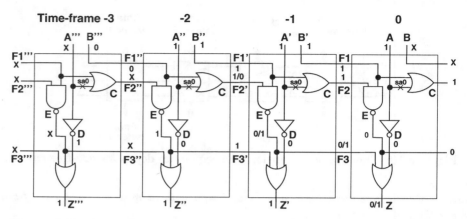

Figure 8.11: A test for input A of C s-a-0 fault in the circuit of Figure 8.9.

only the final test is shown. Notice three things. First, fault effect propagation is attempted through the shortest path. Second, F3 is set to 0/X and not to 0 because all paths are being simultaneously activated. This condition is enforced by the BACK algorithm [159]. Third, F3 is set to 0/X and not to 0/1, which is the least restrictive condition allowed by the nine-valued logic. The last two choices result in significant saving in the number of backtracks.

3. *We move to time-frame -1 to justify $F1 = 1$, $F2 = 1/0$, and $F3 = 0/X$. Setting $A' = 1$, $B' = 1$, $F1' = 0$, $F2' = X$, and $F3' = X$ as inputs to time-frame -1 meets the first two requirements but causes a conflict on $F3$. So, we backtrack to time-frame 0.*

4. *The two path objectives are reversed. $Z = 0/1$ is obtained by $F2 = 1/X$ and $F3 = 0/1$. All signals for time-frame 0 are shown in Figure 8.11.*

5. *In time-frame -1, we justify $F1 = 1$, $F2 = 1/X$, and $F3 = 0/1$, by $A' = 1$, $B' = 1$, $F1' = 1$, $F2' = 1/0$, and $F3' = 1$. This changes $F2$ to 1 but that is not a conflict with $1/X$.*

6. *In time-frame -2, we justify $F1' = 1$, $F2' = 1/0$, and $F3' = 1$, by $A'' = 1$, $B'' = 1$, $F1'' = 0$, $F2'' = X$, $F3'' = X$.*

7. *In time-frame -3, we justify $F1'' = 0$, $F2'' = F3'' = X$, by $A''' = X$, $B''' = 0$. Thus, a four-vector test is found as $(A, B) = (X, 0), (1, 1), (1, 1), (1, X)$, which produces a 0/1 output on the last vector.*

A time-frame expansion type of test generation program, GENTEST [72], produced 17 vectors for detecting all 21 single stuck-at faults in the combinational logic of this circuit.

8.2 Time-Frame Expansion Method

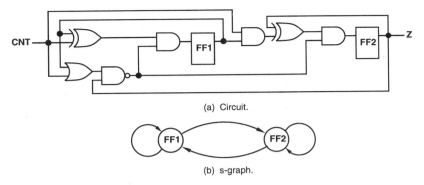

(a) Circuit.

(b) s-graph.

Figure 8.12: A modulo-3 counter without initialization input.

8.2.7 Cyclic Circuits

A circuit whose s-graph contains cycles is called *cyclic*. The modulo-3 counter circuit in Figure 8.12(a) is a cyclic circuit as is indicated by its s-graph in Figure 8.12(b). When $CNT = 1$, it increments its state (FF2,FF1), every time the flip-flops are clocked. Thus, if FF2=0 and FF1=1, the state 01 will change to $10 \rightarrow 00 \rightarrow 01 \rightarrow 10 \rightarrow \cdots$ on successive clocks. For $CNT = 0$, the circuit holds its state. In some applications where synchronization of the pulse at Z, appearing at every third clock, is not required this circuit will correctly perform the function.

The circuit of Figure 8.12 has the same type of test problem as the circuit of Figure 8.6. A test generation program [72] finds that all of its 32 faults are untestable. This is because the state of the fault-free circuit cannot be determined by the test generator. Consider the fault Z s-a-0. For any input, the output will be X/0. However, when the fault-free circuit is in the unknown state, fault detection is meaningless. To remedy this situation, a designer must provide an initialization input. In Figure 8.13, the input $CLR = 1$ will set the circuit in FF1=FF2=0 state. Since the state is set on the application of the clock after CLR becomes 1, this operation is called *synchronous initialization*.

The circuit of Figure 8.13 has the same s-graph as shown in Figure 8.12. However, the test generation program [72] generated 9 vectors to detect all but 4 of the 36 faults. Four faults shown in Figure 8.13 were found to be untestable. Of these, the three s-a-1 faults of the $\overline{CLR}$ signal were potentially detectable. The remaining untestable s-a-0 fault is a partially detectable fault. It cannot be detected after the fault-free circuit has been initialized. However, it can be detected prior to initialization if the circuit powers-up in the 11 state (FF1=FF2=1.) Notice that this is not a valid state in this implementation of the modulo-3 counter. The length of the longest vector sequence produced by the test generator was 5, showing that a maximum of five time-frames were needed. Even though a cyclic circuit can be made initializable by adding hardware, some faults in the added circuitry can only be potentially tested. Those are the faults that interfere with the initialization process.

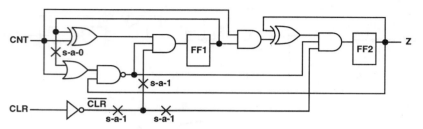

Figure 8.13: A modulo-3 counter with synchronous clear input.

Not all cyclic circuits are uninitializable. For example, the circuits of Figures 8.3 and 8.5(a) are also cyclic, but a test generator can produce tests for them. However, the cyclic structure adds to the complexity of test generation. It is also a prerequisite for any possible uninitializability problem. In practice, if the test generator produces very low fault coverage then the designer should suspect the problems related to the initialization of the circuit.

Besides the synchronous clear of Figure 8.13, *asynchronous clear* and *preset* signals are also effective. Irrespective of the type of initialization signal used, it is important that it should be controllable from a primary input. For example, if the modulo-3 counter is embedded in a larger circuit and the CLR signal is internally generated, then the designer should provide a primary input that will override the internal CLR.

The s-graph of a cyclic circuit cannot be levelized. Thus, no sequential depth is defined. For a cycle-free circuit with N_{ff} flip-flops, sequential depth $d_{seq} \leq N_{ff}$. Thus, the length of time-frame expansion is bounded by N_{ff}, which is a substantial reduction over the worst-cases of $9^{N_{ff}}$ or $4^{N_{ff}}$ we calculated earlier. Those higher bounds are still valid for circuits with cyclic s-graphs.

Table 8.1 shows ATPG results for four ISCAS89 [99] benchmark circuits of similar sizes. These results were obtained by Bell Labs' test generation program, GEN-TEST [72]. CPU times in the table are for a SUN Sparc 2 workstation. The first two circuits, s1196 and s1238, are cycle-free and have a sequential depth of 4. The test generator is able to either detect or determine untestability of each fault in these circuits. The other two circuits, s1488 and s1494, have cycles and hence their sequential depths are indeterminate. Here, the test generator abandoned many faults because of the per-fault CPU time limit (100 s and 81 s, respectively) used in these runs. Fault coverages are lower than those for the cycle-free circuits, and fault efficiencies are also lower. Furthermore, cyclic circuits have some potentially detected faults, which are treated as undetected faults in this study. Fault coverages and fault efficiencies are calculated according to Equations 7.12 and 7.13, respectively. In Table 8.1, the maximum sequence length is the longest test sequence for any fault. Though smaller than the upper-bounds, these are comparatively longer for cyclic circuits. In fact, many more time-frames may be required if the abandoned faults were to be detected or found untestable. Such an exercise will require increasing the per-fault CPU time limit. CPU times for cyclic circuits are already larger by three

8.2 Time-Frame Expansion Method

Table 8.1: Test generation for cycle-free and cyclic circuits.

Circuit name	s1196	s1238	s1488	s1494
Primary inputs	14	14	8	8
Primary outputs	14	14	19	19
Flip-flops	18	18	6	6
Gates (with 2 or more inputs)	388	428	550	558
Inverters	141	80	103	89
Structure	Cycle-free	Cycle-free	Cyclic	Cyclic
Sequential depth	4	4	–	–
Total faults	1242	1355	1486	1506
Detected faults	1239	1283	1384	1379
Potentially detected faults	0	0	2	2
Untestable faults	3	72	26	30
Abandoned faults	0	0	76	97
Fault coverage (%)	99.76	94.69	93.14	91.57
Fault efficiency (%)	100.00	100.00	94.79	93.43
Maximum sequence length	3	3	24	28
Total vectors	313	308	525	559
CPU s	10	15	19941	19183

orders of magnitude. In Chapter 14, we will discuss a technique, known as partial scan, to convert a cyclic circuit into a cycle-free circuit for testing.

8.2.8 Clock Faults and Multiple-Clock Circuits

Our discussion so far has focused on single-clock circuits. All flip-flops were controlled by one clock, which was a primary input to the circuit. For test generation this clock was modeled only implicitly. That is why many of our circuit diagrams show flip-flops without clock signals. It was assumed that one input vector is applied per clock cycle (see the clock waveform in Figure 8.2.) This approach provides simplicity to test generation. However, there is a loss of generality.

First, faults on the clock line are not modeled and hence no tests were obtained for them. Second, many circuits that contain multiple clocks cannot be handled as such. Third, some memory elements in certain circuits may not be controlled by the clock signal.

In a synchronous circuit, all signals are synchronized with respect to the clock signal. The clock signal provides the time reference. A fault in the clock line disturbs the synchronization and in some ways makes the faulty circuit asynchronous. In this subsection, we present a synchronous method for modeling flip-flops with explicit clock and other signals that may be either synchronous or asynchronous. These models are frequently used for simulation and test generation of sequential circuits and allow faults to be modeled for flip-flop inputs. These do not allow modeling of faults inside flip-flops, which will be discussed in the next subsection.

An explicit clock model of a flip-flop is shown in Figure 8.14. Here, the *ideal flip-flop* is the same flip-flop we have been using for the implicit single-clock case.

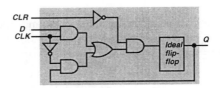

Figure 8.14: An explicitly clocked flip-flop with asynchronous clear.

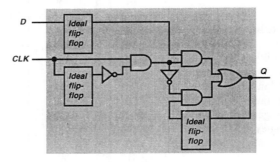

Figure 8.15: A positive edge-triggered flip-flop with explicit clock.

It is a basic delay element that provides a delay of one vector period. The logic in the shaded region in Figure 8.14 is used for modeling the function of the flip-flop. Faults inside this logic are usually not modeled. This flip-flop is shown with an asynchronous clear (CLR) signal. That is, when $CLR = 1$, the flip-flop state changes to 0 irrespective of the state of the clock signal CLK. Only when $CLR = 0$, clock $CLK = 1$ will read the data into the flip-flop. For $CLK = 0$, D will not be stored but the flip-flop will keep its previous state. Note that the CLR signal in Figure 8.13 was a synchronous clear. There, the clock was implicit and the flip-flop was cleared only after the application of the clock. Various types of flip-flops can be modeled using the ideal flip-flop.

Figure 8.15 shows a positive edge-triggered flip-flop. Here, only when a rising edge occurs at CLK, D is propagated to Q. For all other states of CLK, the output Q retains its state. When explicit clock models of flip-flops are used, synchronous and asynchronous clear and preset signals can be modeled. In general, any number of clocks can be present in a circuit. Clocks can even be gated by other signals or clock inputs of flip-flops can be fed by external or internal combinational signals. If the operation of the flip-flop is accurately modeled, then the test generator will correctly deal with faults on clock, clear, and preset signals.

8.2.9 Asynchronous Circuits

A digital circuit is defined as *combinational* if its steady-state output is completely determined by the present inputs. Thus, a combinational circuit does not store any data for the future time. In actual implementation, these circuits contain logic gates *without feedback signals*. A sequential circuit, on the other hand,

8.2 Time-Frame Expansion Method

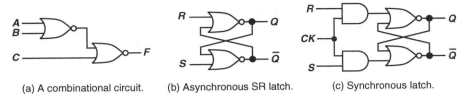

(a) A combinational circuit. (b) Asynchronous SR latch. (c) Synchronous latch.

Figure 8.16: Types of logic circuits.

has combinational feedback which can store signal states. Its output, therefore, depends on both primary inputs and internal states. The internal states, in turn, may depend upon previous primary inputs.

Figure 8.16 gives examples of different types of circuits. Figure 8.16(a) is a combinational circuit as characterized by the feedback-free structure. Its output F is determined uniquely by inputs, A, B, and C. Figure 8.16(b) shows an asynchronous SR (set-reset) latch. An input $S = 0, R = 1$ produces $Q = 0, \overline{Q} = 1$. Input $S = 1, R = 0$ reverses the outputs. For these two input conditions, the latch acts like a combinational circuit. Following any of these inputs, if the input changes to $S = R = 0$, then no new output is produced and Q and $\overline{Q}$ retain their previous values. This is the *storing state* of the latch. Notice that the inputs in the storing state produce a sensitized path around the feedback. Such a feedback path is also called a *combinational loop* or *asynchronous loop* or simply *loop*. The two operations of injecting signals into the loop and storing them are quite typical of asynchronous logic. However, these are only the "good" operations.

Some bad operations are also typical of asynchronous logic. Consider the input $S = R = 1$ applied to the SR latch of Figure 8.16(b). The output $Q = \overline{Q} = 0$ is instantly produced. However, trouble arises if we next go to the storing state by applying $S = R = 0$. Let us examine two cases:

1. *The two NOR gates have equal delays,* Δ. Assuming that $S = R = 1$ is changed to $S = R = 0$ at time $t = 0$, at $t = \Delta$ the output changes to $Q = \overline{Q} = 1$. At $t = 2\Delta$, the output changes again to $Q = \overline{Q} = 0$. With a period of Δ, $(Q, \overline{Q})$ oscillate as $(0,0) \rightarrow (1,1) \rightarrow (0,0) \cdots$

2. *The upper NOR gate is much slower than the lower NOR gate.* The input change $S = R = 1 \rightarrow S = R = 0$ causes $\overline{Q}$ to change to 1, while Q is still 0. The new input 01 of the upper NOR gate is now consistent with $Q = 0$, which stays as such. Thus, the latch attains the state $Q = 0, \overline{Q} = 1$. This state would have been reversed had the lower NOR gate been slower.

The malfunction we described is known as a *race*. As the loop is sensitized, the signal at the output of the "slow" gate literally races to its own input, thus forcing the output to remain stable. Also, notice that the number of inversions around the loop must be even for stability. If no gate is too slow, then the loop will experience oscillating values, which will finally stabilize to some state favoring the slower of the gates. For exactly equal delays of gates, the oscillations may continue indefinitely.

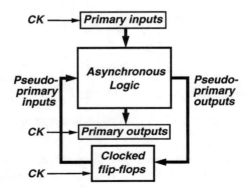

Figure 8.17: A general asynchronous circuit.

The operation we just described is termed *asynchronous* because:

1. Signals can change at any time.

2. Some signals may depend on the past inputs.

3. Steady state signal values may depend on circuit delays.

The design, verification, and testing of asynchronous circuits present complex problems. Therefore, digital circuits of any appreciable size use *synchronous* operation. This is accomplished by a synchronizing signal or clock. Figure 8.16(c) shows a synchronous version of the SR latch. The synchronizing clock signal $CK = 0$ puts the latch in the "store" state, isolating it from the inputs S and R. A correct logic design of the circuit must ensure that $S = R = 1$ never occurs (note the NOT gate at the input in Figure 8.2.)

Many simulators can accurately model delays and hence simulate asynchronous circuits. Thus, simulation-based test generators are the best choice for large asynchronous circuits. These are discussed in the next section. Since time-frame expansion methods rely on path sensitization procedures like D-Algorithm or PODEM, which ignore the timing behavior of the circuit, they take a "conservative" approach.

A general sequential circuit is represented by the schematic of Figure 8.17. Comparing this to the synchronous circuit of Figure 8.1, we find that the combinational logic is here replaced by asynchronous logic, which contains feedback among gates. In spite of the asynchronous logic, which is unclocked, most digital systems have a clock which synchronizes the flip-flops, primary inputs, and primary outputs. This clock is shown as the signal CK in Figure 8.17.

To isolate the combinational logic, we split the asynchronous logic into two parts: (1) feedback-free combinational logic and (2) a set of delay elements synchronized with a *fast model clock*, $FMCK$. $FMCK$ runs much faster than the system clock CK. Its purpose is to repeatedly evaluate the combinational logic and stabilize asynchronous signals before CK clocks the flip-flops, applies new PIs, or observes POs.

8.2 Time-Frame Expansion Method

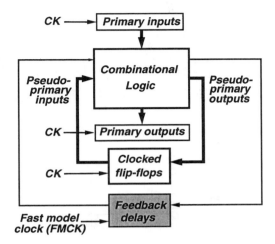

Figure 8.18: A test generation model for asynchronous circuits.

To create the model of Figure 8.18, all loops in the asynchronous logic are identified. A set of signals, called the *feedback set*, is found such that if the corresponding interconnects are cut then the circuit will have no feedback. This set is the same as the feedback edge set defined for a directed graph [41]. The graph in this case contains gates as vertices and signals as directed edges. The minimum number of signals that must be cut is known as the *feedback index* [686]. The problem of finding the absolute minimum feedback set is very complex, but heuristics can quickly find near minimal solutions. Generally used heuristics are similar to those employed for solving the *minimum feedback vertex set* (MFVS) problem discussed in Chapter 14.

Fictitious delay elements are inserted at the signals in the feedback set. Since all combinational gates are assumed to have zero delay, these delay elements hold the signal until all combinational signals have been evaluated. In Figure 8.18, these delays are taken out in a separate block (shown shaded.) Thus, we have created a two-clock model for test generation. The normal system clock CK controls all synchronous memory elements (flip-flops.) The modeling clock $FMCK$ controls the asynchronous delays. The $FMCK$ clock is necessary because we have modeled the logic with zero delays. A time-frame expansion type of test generator deals with this circuit in two phases (Figure 8.19):

1. *System clock (CK) phase.* The operation of the circuit is synchronous with respect to CK. Thus, one input vector is generated for each period of CK. Similarly, the primary output is observed only once in the period. In Figure 8.19, the blocks C are time-frames of the combinational logic. The unshaded blocks correspond to the CK phases. The inputs to this block are the input vector, previous flip-flop states, or *pseudo-primary inputs* (PPIs), and the states of the feedback delays. It produces a *primary output* (PO), and the next states of flip-flops (*pseudo-primary outputs*, PPOs.)

Chapter 8. SEQUENTIAL CIRCUIT TEST GENERATION

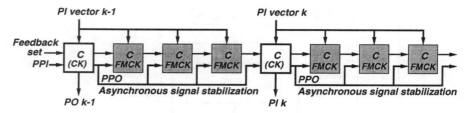

Figure 8.19: Two-phase time-frame expansion for asynchronous circuits.

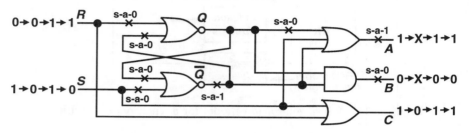

Figure 8.20: An asynchronous circuit for test generation.

2. *Fast modeling clock (FMCK) phase.* Following the system clock phase, which provides new inputs to the combinational logic, a series of "fast" time-frames exercise the logic until signals become stable. For practical reasons, a small fixed number of time-frames is used (three in Figure 8.19.) If some signal does not become stable, then an oscillation can be assumed and that signal is set to the unknown state. During this phase, the primary inputs and the clocked flip-flop states are held without change, and no primary output is examined for fault detection.

Thus, each input vector produces a new set of PPO and asynchronous states (feedback signal states.) In a simplified version of asynchronous circuit test generation, we ignore the feedback states. Each input vector then begins with feedback signals in the unknown states. The test generator favors those primary input values that uniquely determine the feedback signals. For example, for the NOR latch of Figure 8.16(b), inputs 01 and 10 will be favored. Also, the loops are preprocessed and a *sensitization value set*, i.e., a set of signal values that sensitize a path around the loop, is generated for each loop. For example, the sensitization set for the NOR latch of Figure 8.16(b) is $S = R = 0$. Then, whenever the signal states coincide with the sensitization set, outputs of all gates in the loop are set to X. This approach avoids any oscillatory evaluation of signals. However, the conservatism prevents the generation of tests where asynchronous states are essential. This approach is used in GENTEST [72], as the next example illustrates.

Example 8.7 *Consider the asynchronous circuit shown in Figure 8.20. We will discuss test generation by the memory-less model. The feedback set is $(Q, \overline{Q})$. The loop sensitization condition is $R = S = 0$. Thus, whenever this signal combination*

8.2 Time-Frame Expansion Method

occurs, the test generator sets the feedback set to the unknown state, $Q = \overline{Q} = X$. Gentest [72] produced the following result on a SUN Sparc 2 workstation:

```
Primary inputs = 2
Primary outputs = 3
State elements = 0
Total Faults = 23
test generation time = 33 ms
fault simulation time = 16 ms
total vectors = 4
detected faults = 15
untestable faults = 8
undetected faults = 0
0 untestable faults were potentially detected
0 undetected faults were potentially detected
faults tried = 12
time limit per fault = 0.8 ms
fault coverage = 65.2%
```

The four test vectors, corresponding outputs, and eight untestable faults are shown in Figure 8.20. We make several observations:

- Not all faults identified as untestable are really untestable. They are really untestable by a single-vector test, which is a limitation of the combinational model. For example, the s-a-0 fault on the Q input of the OR gate A is testable by two vectors, $(S,R) = (1,0), (0,0)$. Still, there are several faults that are either not detectable even by multiple vector tests, or can only be detected potentially or as race faults. A generally low fault coverage is quite typical of asynchronous circuits.

- Fortunately, the generated test sequence does not cause a race condition in the fault-free circuit, which is a requirement for useful tests but is not imposed by the test generator. For example, if we generate a test, the fault list is ordered as "C s-a-0" followed by "s-a-0 on S input of A," then the two tests $S = R = 0$ and $S = R = 1$, applied in that order, will produce a race in the NOR latch in the fault-free circuit. If the asynchronous logic is embedded in a sequential circuit, the ordering of vectors cannot be arbitrarily changed. Such race conditions should be found by a simulator and the vectors causing them should be discarded or modified. Alternatively, the test generator should recognize the race producing sequences and generate alternative tests.

Asynchronous circuits continue to be difficult to test. Tools and techniques are only adequate for small circuits. The typical situation often encountered involves large synchronous circuits with a small amount of asynchronous circuitry embedded in the combinational logic. In addition, tests for faults in the clock circuitry require asynchronous techniques. The major difficulty of finding good tests for asynchronous circuits arises due to the inadequate treatment of delays. Analysis of races and hazards can improve the tests, but requires additional computation [141, 96]. Determining the steady-state without the complete delay information can be troublesome, too. A recent method gives specific attention to the time-frames used for signal stabilization [64]. Finally, when it comes to handling of delays, logic simulators are more advanced than test generators and the early proposal of Seshu and Freeman [587] for simulation-based test generation is still attractive.

8.3 Simulation-Based Sequential Circuit ATPG

The application of a fault simulator for test generation was suggested by Seshu and Freeman in the early 1960s [587]. They used a compiled-code simulator and the faults were *serially* injected. Random vectors were used in the 1970s with fault simulation to select only those vectors that increased the fault coverage [26]. While this strategy was quite successful with some combinational circuits, for hard to test circuits it had to be backed up with algorithmic (non-random) vectors. Breuer [89] devised a simulation-based method for sequential circuits. In his method, several randomly generated vectors were simulated for some "present state" of the circuit and the best vector (according to specified criteria) was included in the test sequence. The circuit state was then advanced before simulating a new set or random vectors. Schuler *et al.* [573] were the first to use a *concurrent fault simulator* (CFS) for test generation. They simulated a set of random vectors. Each vector was simulated for the same given starting state of the circuit. The vector that detected the largest number of faults was selected. The states of the good and all faulty circuits were changed corresponding to the selected vector. The test generator then advances to the selection of the next vector from a new set of random vectors. Parker [508] reported an adaptive method of making the random vector source circuit-specific.

One of the greatest advantages of these methods is that before a vector is selected as a test, it is simulated. As an event-driven simulator analyzes both logic and timing behavior of the circuit, the selected vector is guaranteed to be free from harmful races or hazards. Many other test generators completely neglect timing information and produce hazardous tests.

Several observations were made by Schuler *et al.* [573]. They experienced a serious shortage of available memory required to simulate a large number of faults. They suggested using a small subset of faults. However, the problem of finding a proper subset had no existing solution. They also reported that for a given computing time, the fault coverage remained somewhat low unless extra observation points were inserted in the circuit. Their circuits contained up to 1,000 gates and were small by today's standard.

The problem of low fault coverage when no extra observation points are inserted has been reported by other workers as well [343, 361]. These authors did not use CFS. However, the difficulty lies not in the simulation algorithm, but in the way vectors are selected. The vector that detects either the target fault or the largest number of faults at primary outputs is the natural choice. When the faults are very difficult to detect, none of the trial vectors may detect anything. Selection of

8.3 Simulation-Based Sequential Circuit ATPG

a test vector from a reasonable number of random vectors is very inefficient in this situation. Also, when the circuit is sequential, not every test vector may produce a fault effect at primary outputs. In general, several vectors are required to bring the circuit to a state such that the fault can be activated. Again, several vectors may be needed to propagate the effect of the fault to a primary output. Thus, a test for a fault may consist of a sequence of vectors where only the last vector produces a fault effect at a primary output. Generation of such a sequence is highly improbable by a process that only considers single vectors and relies on fault detection information at primary outputs. Output observation can produce better results if vector sequences, like those produced by a genetic algorithm, are used instead of single vectors (see Subsection 8.3.2.)

The above observations motivated further exploration. A fault simulator computes the fault activity at all internal nodes of the circuit. This information is frequently ignored since we use the simulator only to gather fault detection data. Takamatsu and Kinoshita [649] have used CFS for generating tests for combinational circuits. They use the test generation algorithm, PODEM (see Chapter 7), to generate a test for some target fault. PODEM involves a series of backtrace and forward implication operations. The backtrace determines a value for some primary input to accomplish some objective like fault activation or fault propagation. Forward implication ascertains that the input value determined by the backtrace does not contradict the objective. In case of a contradiction, the input value must be changed via backtracks. For an algorithm like PODEM, which enumerates primary input values until a test is found, the CPU time of test generation largely depends upon the number of backtracks. Takamatsu and Kinoshita find that a backtrack can often be avoided by changing the target fault. In their CONT-2 algorithm, the forward implication is similar to CFS. Thus, fault activity information about all undetected faults is available. When a contradiction occurs, CONT-2 will abandon the current target fault and select some other target fault that has a greater chance of being detected. The new target can be a fault that is already active and whose effect is present at some signal close to a primary output. PODEM and CONT-2, which is based on PODEM, are test generation algorithms applicable only to combinational circuits.

The CONTEST algorithm, devised by Agrawal, Cheng, and Agrawal [31, 153], uses a directed-search approach for generating tests for sequential circuits. This algorithm works in a closely knit fashion with CFS. The basic idea is to obtain test vectors by successive modification of primary input bits based upon cost functions that are computed by the simulator. More advanced test generators use genetic algorithms [445].

8.3.1 CONTEST Algorithm

The test generation process can be subdivided into three phases. In Phase 1 initialization vectors are generated. The purpose of these vectors is to bring flip-flops in the circuit to known states irrespective of their starting state. Phase 2 begins with vectors that are either supplied by the designer or generated in Phase

1. A fault list is generated in the conventional manner. For example, this list may contain all single stuck faults or a subset of such faults. These faults are simulated using a fault simulator. If the coverage is adequate, the test generation would stop. Otherwise, tests are generated with all undetected faults as targets. In the initial stages of test generation, the fault list is usually long and the objective of this phase is to generate tests by concurrently targeting all undetected faults. At the end of Phase 2, if the fault coverage has not reached the required level then Phase 3 is initiated. In this phase, test vectors are generated for single faults targeted one at a time.

Phase 1: Initialization. Here, the cost is defined simply as the number of flip-flops that are in the unknown state. Initially, the cost may be equal to the number of flip-flops in the circuit. The goal in the initialization phase is to reduce this cost to 0. This cost function is derived only from good circuit simulation and is not related to the faulty circuit behavior. If the circuit is hard to initialize, one may relax the criterion for exiting to the next phase by allowing a small number of flip-flops, say 10%, to remain uninitialized.

Using this cost function, the circuit is driven to the easiest initialized state instead of any specified state. All flip-flops are assumed to be in the unknown state at the beginning and the cost function is equal to the number of flip-flops in the circuit. Before applying a trial vector, signal states are saved for restoration in case the trial vector is not accepted. To start the process, any trial vector (a randomly generated or user-supplied vector) can be used. This is called the "current vector." Subsequent trial vectors are generated by changing the bits of the current vector. The integer n is used to denote the bit position in the current vector that is changed. The clock bits (only in the synchronous mode) are treated separately. The user specifies the clock sequence. During simulation, the input data bits are kept fixed whenever the given clock sequence is applied. In combinational or asynchronous sequential circuits, all input bits are treated as data.

After simulation of a trial vector, the "trial cost" is computed as the number of flip-flops that are in the unknown state. If the trial cost is lower than the current cost, then the trial vector is saved. If the trial cost is zero, then the initialization phase is complete. Otherwise, the current cost is updated, signal states are saved, the accepted trial vector becomes the current vector, n is set to 1, a new trial vector is generated by changing the nth data bit, and the process of simulation, cost calculation, etc., is repeated.

A trial vector is not accepted as an initialization vector if the corresponding trial cost is not lower than the current cost. In that case, the bit number n is advanced and the process is repeated with a new trial vector. When all bits of a current vector have been changed without lowering of the cost, this process will stop, indicating that initialization is impossible with this scheme, i.e., a local cost minimum is reached. One possible strategy is to restart with a new randomly-selected current vector.

Phase 2: Concurrent fault detection. The initialization vectors may already have detected some faults. Some others may have been activated but not detected. As a result, effects of active faults will be present at internal nodes of the circuit. For

8.3 Simulation-Based Sequential Circuit ATPG 241

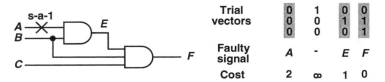

Figure 8.21: An example of the distance cost function used in Phase 2.

an active fault, a suitable cost function is the shortest distance to a primary output from any fault effect caused by the fault. The distance here is simply the number of logic gates on the path. The smaller this cost, the closer the fault is to being detected. When a fault is detected, its cost will be zero. The objective in test generation is to reduce the cost by propagating the fault effect forward, gate by gate, until it reaches a primary output. If the fault is not activated, i.e., no fault effect is present anywhere in the circuit, then the cost is defined as infinite.

Figure 8.21 gives a simple example to illustrate how the *distance* cost function works. The given fault is signal A stuck-at-1 and the initial vector is 000. The fault effect appears at signal A; thus, initial cost is 2. After simulating three trial vectors, the search terminates and a test is found.

When there are several undetected faults, cost C_i is computed for each fault i for some input vector and internal state. Similarly, the cost C'_i is obtained for a candidate trial vector. A comparison of C_i and C'_i determines whether to accept the candidate vector or reject it. Since there can be several undetected faults, there are two *lists* of cost functions instead of just two numbers. The search for tests should be guided by a group of faults instead of a single target fault. One can devise simple rules to determine the acceptance of a vector. For example, if the combined cost of 10% of the lowest-cost undetected faults is found to decrease, then the new vector may be accepted. Experience has shown that for many circuits, the test vectors for all stuck-at faults are usually clustered instead of being evenly distributed in the input vector space. Figure 8.22 shows the input vector space with dots representing tests for undetected faults from the fault list. In the beginning there are many undetected faults and the vector space may have large clusters of tests as shown in Figure 8.22(a). Starting at any initial vector A, the cost function will steer the search toward large clusters. When only a few faults are left, their tests will be a few isolated vectors. In Figure 8.22(b), test generation in Phase 2 has followed the path from A to B. At B, the combined cost provides very little "direction." Hence, a single target fault strategy may be needed.

If flip-flops are modeled as functional primitives, they may be treated differently from individual gates such as AND or OR. Propagating a fault through a gate only needs setting appropriate values at the inputs of the gate. In contrast, propagation through a flip-flop requires first setting the appropriate value at its data input and then activating the clock signal. In cost computation, therefore, a large constant, say 100, is assigned to a flip-flop as its distance contribution.

Phase 2 begins with fault simulation of initialization vectors. The faults thus

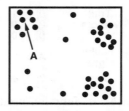 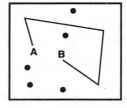

(a) Many undetected faults.   (b) Few undetected faults.

Figure 8.22: Directed search in Phase 2 and the need for Phase 3.

detected are eliminated from consideration. Generation of trial vectors is performed in a manner similar to the initialization phase. Costs for trial vectors are obtained from the result of CFS. Prior to simulation, gates are levelized starting from primary outputs. Thus the level of a gate directly gives its distance from primary outputs. The cost of a trial vector is easily computed using the levels of the faulty circuit gates. The concurrent phase stops when all single bit changes in a current vector produce no cost reduction. This will normally happen when the number of faults left in the target set is small. The test vectors for these faults sparsely populate the vector space and, therefore, the collective cost function does not provide any significant guidance.

Phase 3: Single fault detection. The cost function in this phase is based on a SCOAP-like testability measure. In SCOAP (see Chapter 6), each signal is assigned three measures, 1-controllability, 0-controllability, and observability, respectively. All measures are integer-valued and a higher value of the measure for a signal indicates that it might be difficult to control or observe. In Phase 3, testability measures are dynamically computed. Their values depend upon the circuit structure as well as on input vectors. The dynamic nature is essential in this application since the measures are used to compare the suitability of vectors for detecting a target fault.

$DC1(i)$ and $DC0(i)$, are defined as dynamic 1 and 0 controllabilities for node i. These are related to the *minimum* number of primary inputs that must be *changed* and the *minimum* number of additional vectors needed to control the value of node i to 1 or 0. The number of inputs required to be changed is further defined as the *dynamic combinational controllability* (DCC) and the number of required vectors is called the *dynamic sequential controllability* (DSC.) In order to keep the test sequence short, DSC is weighted heavier than DCC. For example, $DC1(i)$ and $DC0(i)$ could be the weighted sums of DCC and DSC, say, DSC times 100 plus DCC. If the current logic value of node i is 1, then $DC1(i)$ is defined as:

$$DC1(i) \mid_{V(i) = 1} = 0 \qquad (8.1)$$

where $V(i)$ is the logic value on node i. Similarly, if the current logic value on node i is 0,

$$DC0(i) \mid_{V(i) = 0} = 0 \qquad (8.2)$$

This definition follows from the fact that no input change is needed to justify a 1(0) on node i if the value is already 1 (0.) Under other conditions, $DC1(i)$ and $DC0(i)$

8.3 Simulation-Based Sequential Circuit ATPG

will assume nonzero values. For example, for the output line i of an AND gate with m inputs, $DC1(i)$ and $DC0(i)$ are computed as:

$$DC1(i) \mid_{V(i) = 0 \text{ or } X} = \sum_{j=1}^{m} DC1(k_j) \qquad (8.3)$$

$$DC0(i) \mid_{V(i) = 1 \text{ or } X} = \min_{1 \leq j \leq m} DC0(k_j) \qquad (8.4)$$

where k_j is the jth input line of the gate. Here, min means the minimum of m quantities and its use is similar to that in the SCOAP testability measures. Primary input controllabilities are set to 0 or 1 depending on their current state. As explained, the controllability of sequential elements is weighted heavier. Dynamic controllabilities for a flip-flop output i are defined as:

$$DC1(i) \mid_{V(i) = 0 \text{ or } X} = DC1(d) + K \qquad (8.5)$$

$$DC0(i) \mid_{V(i) = 1 \text{ or } X} = DC0(d) + K \qquad (8.6)$$

where d is the input data signal of the flip-flop and K is a large constant, say, 100.

In order to detect a stuck fault, the test generator must first find a sequence of vectors to activate the fault, i.e., set the appropriate value (opposite of the faulty state) at the fault site, and then find another sequence to sensitize a path to propagate the fault effect to a primary output. Thus, the cost function should reflect the effort needed for activating and propagating the fault. The activation cost, $AC(i_{s-a-j})$, of node i stuck-at-j fault is defined as:

$$AC(i_{s-a-j}) = DCv(i) \qquad (8.7)$$

where v is 0 if node i is stuck-at-1 and v is 1 if node i is stuck-at-0. This follows from the consideration that the cost of activating a stuck-at-0 (stuck-at-1) fault is the cost of setting up a 1 (0) at the fault site. The propagation cost is basically a dynamic observability measure. For a fanout stem i with n fanout branches, it is:

$$PC(i) = \min_{1 \leq j \leq n} PC(i_j) \qquad (8.8)$$

where i_j is the jth fanout branch of i. For an input signal i_a of an m-input AND gate whose output signal is i, we have

$$PC(i_a) = PC(i) + \sum_{\substack{1 \leq k \leq m \\ k \neq a}} DC1(i_k) \qquad (8.9)$$

where i_k is the kth input line of the gate. Similar formulas are easily derived for other types of gates. The cost function for test generation for a single target fault is derived from the activation cost and the propagation cost defined above. For an undetected fault F, line i stuck-at-v, that is not activated, the cost is defined as:

$$Cost(F)\vert_{F\ not\ activated} = K1 \times AC(i_{s-a-v}) + PC(i) \qquad (8.10)$$

where $K1$ is a large constant that determines the relative weighting of the two costs. If the fault has been activated, and N_F is the set of the nodes where the fault effect appears, then:

$$Cost(F)\vert_{F\ activated} = \min_{i\ \in\ N_F} PC(i) \qquad (8.11)$$

Notice that in this cost function reconvergent fanouts are ignored. This approximation provides computational simplicity but may occasionally result in failure to detect a fault. Once activation and propagation costs are computed for each fault in the list, the lowest-cost fault is targeted for detection. New input vectors are created to further lower the cost of the target fault until it is detected or the search is abandoned due to a local cost minimum. As new vectors are added to the sequence, CFS eliminates any other detected faults from the list. This phase ends when either an adequate coverage is achieved or all faults that were left undetected at the end of Phase 2 have been processed.

A program implementing the CONTEST algorithm (CONcurrent TEst generator for Sequential circuit Testing) [31, 153] accepts a logic-level circuit description in a hardware description language. The test generator works in two modes: *synchronous* and *asynchronous*. In the synchronous mode, clock signals and their transition sequence within a period must be specified. The test generator follows each change in primary inputs by a clock sequence. In the asynchronous mode, no clock signal is identified and the test generator treats all primary inputs alike. For circuits that are largely synchronous with a limited amount of asynchronous circuitry, test generation in the synchronous mode works better initially. If the coverage by this mode is inadequate, then, for the remaining faults, asynchronous mode can be used. This is because the speed of test generation depends upon the number of primary inputs that must be manipulated. In the synchronous mode, clock signals are prespecified and are not manipulated.

An optional fault list is an input to the test generator. If this is not given, the system generates a list of collapsed single stuck-at faults. CONTEST contains an event-driven CFS. Race analysis in feedback structures is automatic and is performed through special modeling features. By default, potentially detectable faults (that produce an unknown faulty output) are considered detected. This option can be turned off by the user. If the number of changes in a signal for the same input vector exceeds a prespecified number, then the simulator assumes oscillation and sets the signal to the unknown state.

In Phase 1, the user can specify the acceptable percentage of uninitialized flip-flops. The default is 10 percent. Also, Phase 2, which normally follows Phase 1, can be independently run if the user supplies functional vectors or initialization vectors. Experience has shown that Phase 2 can achieve a coverage of 65 to 85 percent. Phase 3 can also be independently run if the size of the given fault list is small.

8.3 Simulation-Based Sequential Circuit ATPG

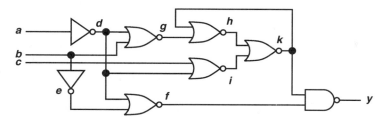

Figure 8.23: An asynchronous circuit for test generation by CONTEST.

Table 8.2: CONTEST results (CON: CONTEST, GEN: GENTEST.)

Circuit data				Fault coverage (%)		Fault eff. (%)	Number of test vectors		VAX 8650 CPU s	
Name	Gates	FFs	Faults	CON	GEN	GEN	CON	GEN	CON	GEN
MANNY	26	7	67	100.00	83.95	83.95	32	219	5	NA
SSE	207	6	454	83.26	83.20	99.50	561	676	291	1134
MULT4	382	15	540	97.04	92.78	93.15	364	148	838	1490
TLC	355	21	772	94.69	94.64	94.64	1256	5340	3312	32590
PLANET	690	6	1582	95.13	57.71	61.00	1439	132	3120	19388
MI	779	18	1629	94.53	NA	NA	1358	NA	1261	NA
CHIP-A	1112	39	1643	93.73	84.11	88.48	1031	384	98432	NA
CHIP-B	1539	73	2533	91.28	NA	NA	1034	NA	77904	NA

Example 8.8 *Asynchronous circuit. Consider the circuit of Figure 8.23. This is an asynchronous sequential circuit. Muth [481] used this example to illustrate the necessity of a nine-value path sensitization algorithm when a test for the fault "d stuck-at-1" is attempted. CONTEST produced four vectors: 000, 100, 101, 111. The last two vectors are the same as given in Muth's paper. Two extra vectors are generated because CONTEST starts with an arbitrary 000 vector and then brings the circuit to the appropriate state.*

Some results are shown in Table 8.2 [31, 153]. The circuit MANNY is asynchronous. All others are synchronous. SSE and PLANET are finite state machines with combinational logic implemented as a programmable logic array. MI is another finite state machine with random logic implementation. MULT4 is a four bit Booth multiplier circuit synthesized by an automatic synthesis system and TLC is a traffic light controller circuit. CHIP-A and CHIP-B are CMOS custom chips. CHIP-B contains one asynchronous flip-flop. For comparison, the results of a time-frame type of sequential circuit test generator (GENTEST [72]) are included in Table 8.2, CONTEST consistently produced better fault coverage and required less CPU time as compared to GENTEST. In some cases, due to circuit model incompatibilities, GENTEST aborted. The unavailable data are shown as "NA" in the table. GENTEST was ineffective for several circuits because of their complex sequential structure and asynchronous behavior. In other cases, GENTEST could identify redundant faults that were used to obtain fault efficiencies. A simulation-based technique like CONTEST cannot identify redundancies and hence does not provide the fault efficiency.

Consider the TLC Circuit to examine the run time performance of the directed-search method. Even though there are only 21 flip-flops, it is a highly sequential circuit with an internal counter and the ratio of logic to primary inputs is high (only 4 primary inputs.) For some faults, more than 200 vectors are needed to initialize the circuit to appropriate states to activate and propagate the fault effects to primary outputs. That means that more than 200 copies of the combinational portion are created in the iterative-array model. This is one of the reasons why GENTEST required significantly more CPU time than CONTEST for the TLC circuit.

As evident from Table 8.2, CONTEST frequently generated more vectors than the GENTEST program. This is a consequence of the one-bit change heuristic. In general, neighboring vectors in a CONTEST-generated sequence have fewer bit changes than in sequences generated by GENTEST. As a result, more vectors are needed to take the circuit to a desired state, starting from some given state. In some cases, this may be desirable because too many simultaneous input changes can produce hazards in the logic or produce power supply fluctuation due to current surge. The single-bit change strategy has also been used by other workers [587]. In general, the vector sequence length directly affects the test time and long sequences may have to be compacted to reduce the testing cost.

When the final fault coverage is lower than the desired goal, two options are possible. The first option is to start with a different (randomly selected) vector and attempt generation of tests for the undetected faults. The second option is to expand the one-bit change heuristic to include two-bit, three-bit, ... changes. One should, however, expect a rapid increase in the amount of computations. Lioy *et al.* have taken a different trial vector approach [403]. They implemented the directed search approach in the MOZART concurrent fault simulator [236]. When a single vector with one-bit change does not provide cost reduction, their program examines the cost with multiple vectors. This strategy has the advantage of being able to get out of some local minima. They were able to obtain excellent results for some ISCAS '89 sequential benchmark circuits [99]. However, their conclusion was that the simulation-based technique was not very efficient for circuits with a complex feedback structure. Such structures are, in fact, known to be troublesome for other test generation algorithms also (see Subsection 8.2.6.) Among other possible strategies for trial vectors is the method based on genetic algorithms.

8.3.2 Genetic Algorithms

The process of test generation in CONTEST is *evolutionary*, in the sense that a test sequence is evolved by accepting and rejecting vectors according to their fault detection characteristics. Improved results are possible if trial vectors are generated by some "learning" process. For example, we can probabilistically favor the generation of the type of vectors that were more successful in the past. That is the basic idea of *genetic algorithms*, introduced by Holland [303]. An interested reader will also find the book by Goldberg to be useful [261]. A recent book by

8.3 Simulation-Based Sequential Circuit ATPG

Mazumder and Rudnick [445] discusses VLSI design and test applications of genetic algorithms.

Test generators based on genetic algorithms resemble CONTEST in several ways [445]. One uses the three-phase process. The cost function is replaced by a *fitness function*, which is now maximized instead of being minimized. Various types of fitness functions are computed via true-value or fault simulation. The basic difference, however, lies in the method of generating trial vectors. The procedure works with a set of vector sequences, called the population, which is improved iteratively. Each iteration is called a new generation. Vectors of a generation are produced from those of the previous generation, using operations known as *crossover*, *mutation*, and *selection*. In crossover bits from two vectors of the old generation are combined to construct two vectors for the new generation. In mutation, bits of a vector from the old generation are manipulated to create a vector for the new generation. In selection two individuals are selected, with selection biased toward more highly fit individuals. The fitness of the new generation is evaluated by simulation of the required characteristics such as initialization or fault detection. Creation of vectors in later generations is biased toward higher fitness.

One of the earliest simulation-based programs that used *genetic algorithms* (GA) was CRIS [556]. A simple fitness function, evaluated from true-value simulation, was used. It would favor those vector sequences that increased the signal activity in the circuit. This program had only limited success, perhaps because increased signal activity only improves controllability but does not necessarily increase observability. A later version of the program included fault simulation. Another program used *adaptive GA* [631]. Here, crossover and mutation probabilities were dynamically reduced for more fit individuals. The authors showed advantages of the adaptive scheme though their implementation was only for combinational circuits.

Compared to the early versions, significantly improved results were achieved by the GATEST program developed by Rudnick *et al.* [553]. In that program accurate fault simulation data was used for evaluating the fitness function. In a GA framework, new populations via crossover, mutation, and selection can be very quickly generated. However, fault simulation of large populations consumes enormous computing resources. GATEST simulates 100 to 300 randomly sampled faults to compute the fitness. These strategies worked well and the results compared favorably with the time-frame expansion program HITEC. GATEST used less CPU time than HITEC, produced shorter test sequences and sometimes (though not always) obtained higher fault coverages. More importantly, the results established the practicality of GA-based test generation and showed the necessity of fault simulation for fitness assessment.

Another recent program, GATTO [174], targets one single fault at a time. The circuit is assumed to be already initialized and a target fault is chosen from among those already active. The GA then generates vectors to propagate the fault effect toward POs.

Several strategies of the previous programs are combined in the STRATEGATE program developed by Hsiao *et al.* [310, 311]. This program uses GA in multiple

Table 8.3: Sequential ATPG by STRATEGATE [311].

Circuit name	Number of faults total	Number of faults detected	Fault coverage	Number of vectors	HP J200 (256MB) CPU time
s1423	1,515	1,414	93.3%	3,943	1.3 hours
s5378	4,603	3,639	79.1%	11,571	37.8 hours
s35932	39,094	35,100	89.8%	257	10.2 hours

phases to activate the fault in the combinational circuit (time-frame 0), justify the required state through previous time-frames, and propagate the fault effect through later time-frames. Table 8.3 shows a sample of results obtained for three benchmark circuits by STRATEGATE on a HP J200 computer with 256MB RAM. The test generator starts with the circuit in a completely unknown state. Only coverages are given since the simulation-based ATPG cannot identify redundant faults. Although CPU times are large, the fault coverages are significantly higher than those reported for time-frame expansion programs such as GENTEST [72, 160] and HITEC [497]. The reader may compare the result for the circuit s35932 in Table 8.3 with that obtained by GENTEST in Section 8.2.4.

Years of research on genetic algorithms for sequential ATPG has produced some highly improved programs. In the true sense of the word, this evolution will continue and more improvements may be forthcoming. Similarly, the quest for improved implementations of time-frame expansion algorithms also continues.

8.4 Summary

Sequential ATPG is practical for arbitrarily large circuits with adequate testability properties such as good initializability and cycle-free or limited-cycle structure. The University of Illinois program, HITEC [497], has been the basis for a commercial ATPG system, available for several years. GENTEST [72, 160] has been used within Lucent Technologies. At least two other sequential ATPG systems have been in use at IBM and NEC, respectively. IBM's program incorporates sophisticated branch and bound techniques originally developed at Rutgers University [150]. It has been used to generate tests for their partial scan microprocessor [151]. It is also commercially available to the users of IBM's TestBench system. NEC's program, SATURN [120], is based on various neural network and graph theoretic algorithms [127], and has been used within the company to generate tests for VLSI chip sets containing several million transistors. These programs employ the time-frame expansion technique.

In view of the large CPU times required for sequential ATPG, multi-processing has been explored. A popular technique is to distribute the fault list over a network of workstations that independently generate tests. This procedure is known as *fault-parallelism*. The inter-processor communication is minimized by only sharing the generated tests [63]. Significant speedups have been reported for the GEN-

TEST [603] and ESSENTIAL [369] programs. Interestingly, in cases where the detection of hard-to-detect faults is strongly influenced by the circuit state, even *superlinear speedup* is possible [18]. Superlinear speed up refers to the speed up of the program by a factor greater than the number of processors used. Sienicki [603] has analyzed the conditions for such speed up and has given adaptive techniques to obtain the best advantage of parallelization. Krauss *et al.* [370], using a distributed system of 100 workstations, first divide the fault list among workstations. When only the hard-to-detect faults are left over, they use several processors to cooperatively explore the vector space for tests targeting one fault at a time. This procedure is known as *search-space parallelism*. They observed speed ups between 42 and 92 for various circuits. They also reported that parallelization of test generation produced more vectors, which had to be compacted.

Simulation-based methods of test generation derive their efficiency from fault simulators such as a concurrent fault simulator (CFS.) Since the selection of a test vector depends upon the cost comparison, several trial vectors have to be simulated before a decision is made. A CFS implementation simulates the trial vectors in series. A more efficient implementation will be to use MDCCS (*multi-domain concurrent and comparative simulation*) [684] such that costs for many trial vectors are concurrently evaluated. The simulation-based method is applicable to all types of circuits, combinational or sequential. Its best advantage is in sequential, particularly asynchronous, circuits. In such circuits, timing of signals cannot be neglected and, therefore, the time-frame expansion methods run into difficulties (see Subsection 8.2.9.) With the simulation-based method, any circuit that can be simulated, can be tested. Among simulation-based techniques genetic algorithms have produced the best results.

Problems

8.1 *Race condition.* Suppose that all gates in the flip-flop circuit of Figure 8.2 have one unit of delay. Analyze all signal waveforms when a falling transition at D and a rising transition at CK occur, simultaneously. What timing condition should data (D) and clock (CK) signals satisfy for race-free operation?

8.2 Show that any fault on the primary output of the serial adder circuit of Figure 8.3 can be obtained with at most two time-frames. *Hint:* Note that the primary output fault does not interfere with the initialization for the flip-flop which can be set in either 0 or 1 state by a single vector.

8.3 Show that any single stuck-at fault on primary inputs of the circuit in Figure 8.3 can be detected by two vectors when the initial state of the flip-flop is unknown.

8.4 Determine a test sequence for the s-a-0 fault on the output line of the flip-flop in the circuit of Figure 8.3.

8.5 Show that a test for the fault A s-a-0 in the circuit of Figure 8.24 cannot be obtained using the five-valued logic of the D-calculus. Obtain a test for this fault using the nine-valued logic.

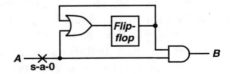

Figure 8.24: Circuit for Problem 8.5.

8.6 *Initialization fault.* Derive a test for the A s-a-1 fault in the circuit of Figure 8.25. Does the test provide a definite or a potential detection?

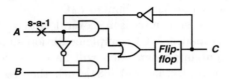

Figure 8.25: Circuit for Problems 8.6 and 8.7.

8.7 Devise a multiple observation test for the fault shown in Figure 8.25. Is a multiple observation test still possible if the inverter in the feedback path was shorted?

8.8 Compute drivabilities for all lines in the circuit of Figure 8.9 for the fault B s-a-0.

8.9 *Approximate test.* The single clock synchronous sequential circuit in Figure 8.26(a) has two inputs CLR and A. $CLR = 1$ initializes the flip-flop to 0. Using only the combinational part shown in Figure 8.26(b), derive a test vector (CLR, A, PS) to detect the A s-a-0 fault at the output Z. Find a justification sequence, assuming the combinational logic to be fault-free in previous time-frames. Verify whether this test sequence will work when the fault is present in all time-frames. If the test does not work, then derive an alternative test assuming the fault to be present in all time-frames.

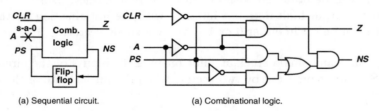

Figure 8.26: Circuit for Problem 8.9.

Problems

8.10 Prove that a fault that is untestable in the stand-alone combinational logic is also untestable in the sequential circuit.

8.11 Prove that a fault in the combinational logic of a synchronous sequential circuit is untestable if no combinational test vector can be justified using fault-free time-frames. *Hint:* See the paper by Agrawal and Chakradhar [30].

8.12 *Pseudo-combinational circuit.* Derive a combinational circuit by replacing all flip-flops by shorting wires in the circuit of Figure 8.9. This is known as the *pseudo-combinational* transformation, which can be applied to any cycle-free clocked sequential circuit [463]. Derive a test for the fault D s-a-0 in the pseudo-combinational circuit. Verify that the vector sequence obtained by repeating this vector three times will detect the D s-a-0 fault in the original sequential circuit. Note that the number of repetitions equals the sequential depth.

8.13 Prove that if a combinational test vector can be obtained for a fault in the pseudo-combinational circuit, then that vector repeated as many times as the sequential depth will always detect the corresponding fault in the sequential circuit. *Hint:* See the paper by Min and Rogers [468]

8.14 Prove that a synchronous sequential circuit that is not initializable, must be cyclic.

8.15 *Cyclic circuits.* Redefine the s-graph by including PIs and POs as additional vertices. Levelize the graph starting from PI vertices using the minimum distance rule. Draw the new types of levelized s-graphs for circuits of Figures 8.9 and 8.13. What do the depths of these graphs represent in terms of the length of test sequences?

8.16 *Race fault in asynchronous circuit.* Derive a test for the s-a-1 fault at the output of the NOT gate in the circuit of Figure 8.27. Is this a race fault?

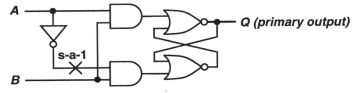

Figure 8.27: Circuit for Problem 8.16.

8.17 *Oscillation fault.* The asynchronous circuit of Figure 8.28 is designed to have no memory state. Derive a test for the s-a-1 fault on the C input of the NAND gate and show that it is an oscillation fault. Redesign the fault-free function as a combinational circuit.

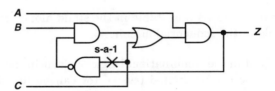

Figure 8.28: Circuit for Problem 8.17.

8.18 *Simulation-based initialization.* Using the cost function as the number of flip-flops in the unknown state and unit Hamming distance trial vectors from an initial vector 00, generate initialization sequences for the circuits of Figures 8.9 and 8.13. Can this procedure initialize the circuit of Figure 8.12?

8.19 *CONTEST.* Starting with an initial vector 00, use the Phase 1 and Phase 2 procedures of CONTEST to derive a test sequence for the serial adder circuit to detect the s-a-0 fault shown in Figure 8.3. Even though there is a single target fault, use the distance cost function.

8.20 *CONTEST.* Given an initialization vector 00, use the Phase 3 (single fault detection) procedure of CONTEST to obtain a test sequence for the s-a-0 fault of the serial adder circuit in Figure 8.3.

Chapter 9

MEMORY TEST

> "... Advances in semiconductor memories have been very impressive. Their density ... is ever increasing. ... algorithms with a test time of ... order $O(n^2)$, where n is the number of bits in the memory chip, are no longer acceptable for testing current ... multi-Mega-bit memory chips." — Ad J. van de Goor [688].

An engineer joining the electronics industry in 1970 would have witnessed the invention of the first *Dynamic RAM* (DRAM) memory chip, the Intel 1103 DRAM, with 1024 bits. In 1993 the first papers on the 256 Mega-bit DRAM were published, and in 1997 the first samples of the 256 Mega-bit DRAM were introduced. In this time frame, the cost of computer memory has declined from 1 cent/bit to 120×10^{-6} cents/bit. The industry is anticipating 275 Giga-bit (Gb) DRAMS in 2021, at a cost of 0.66×10^{-6} cents/bit [584], although such extrapolations of historical trends are speculative. This extraordinary technological advance and cost reduction has enabled the widespread use of virtual memory in personal computers, a revolution in computer graphics, and rapid retrieval of volumes of data unimaginable in the 1960s.

Ever since Kilburn's invention of virtual memory for the Manchester University Ferranti Atlas computer, we have used a hierarchy of memory to give the appearance of a huge, high-speed memory attached to every computer, when in fact this illusion is achieved at low cost by a hierarchy of memories ranging from a high-speed, expensive, small, random access memory to a low-speed, inexpensive, and large memory. In the Ferranti Atlas, the fastest memory was magnetic core memory, medium speed memory was a drum drive, and low-speed memory was magnetic tape. At present, the fastest memory in the hierarchy is the *static RAM* in the microprocessor cache, *dynamic RAM* is the off-chip main memory, and Winchester moving head disk drive technology is at the bottom of the hierarchy. The widespread deployment of virtual memory, and the increase in typical computer memory sizes to several Mega-bytes has only been possible because of increasing memory chip density and cost-effective memory testing. Virtual memory has freed tens of millions of programmers from the need to code software in assembly language, and from the need to carefully construct software overlays. We have used the dramatic cost reduction achieved by memory technology to reduce the labor content of writing software. Also, lower-cost memory enabled the invention of the personal computer, with its major innovation of incorporating the display memory into the processor main memory in order to speed up image drawing and simplify graphics programming. The compound annual growth rate in semiconductor memory world sales has been averaging at 7% [442]. Semiconductor memories are about 35% of the entire semiconductor market. In 1975, DRAM cost $ 8,000/Mbyte, but in 1985 it cost $ 500/Mbyte. It now costs between $ 1.56/Mbyte and $ 1.88/Mbyte in 2000.

The significant types of semiconductor memory are listed below:

1. *Dynamic Random Access Memory* (DRAM) has the highest possible density but a slow access time of 20 ns. Bits are stored as charge on a single capacitor, but the memory must be refreshed, typically every 2, 4, or 6 ms, if information is not to be lost.

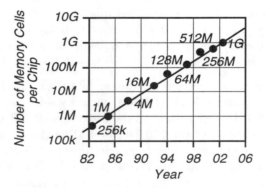

Figure 9.1: Number of bits per DRAM chip in volume production.

2. *Static Random Access Memory* (SRAM) has the fastest possible speed, with a 2 ns access time. Bits are stored in cross-coupled latches, and the memory need not be refreshed.

3. *Cache DRAM* (CDRAM) combines both SRAM and DRAM on the same chip, in order to accelerate block transfer between the SRAM cache and the slow DRAM.

4. *Read-Only Memories* (ROMs) have every bit content programmed by the presence or absence of a transistor at manufacturing time, and do not lose information when power is shut off.

5. *Erasable, Programmable Read-Only Memories* (EPROMs) are ROMs that can be programmed in the field. The entire contents is erased by applying ultraviolet light, and then the EPROM can be reprogrammed.

6. *Electrically Erasable, Programmable ROMs* (EEPROMs) are programmable in the field, and words in them can be selectively erased by electrical means.

9.1 Memory Density and Defect Trends

Most of the information in this chapter came from two outstanding memory testing texts by van de Goor [688] and Mazumder and Chakraborty [442].

9.1 Memory Density and Defect Trends

Table 9.1: DRAM family characteristics over time.

Mode	Bandwidth (Mb/s)	Volt.	Chip sizes	Time period
FP	25	5 V	256 kb/1 Mb/4 Mb/16 Mb/64 Mb	1986-97
EDO	33 – 60	3.3 / 5 V	4 Mb/16 Mb/64 Mb/128 Mb	1990-99
SDRAM	66 – 133	3.3 V	16 Mb/64 Mb/128 Mb/256 Mb	1995-
DDR	200 – 300	2.5 V	128 Mb/256 Mb/512 Mb/1 Gb	1999-
DRDRAM	600 – 800	2.5 V	128 Mb/144 Mb/256 Mb/ 288 Mb/512 Mb/596 Mb/1 Gb	1999-

In memory technology, the number of bits/chip quadruples roughly every 3.1 (or π) years. The bit count per chip continues to increase exponentially, and this causes the memory price to decrease exponentially. Figure 9.1 shows the density rate of increase. Exponential density increase implies exponential decrease in area per memory cell, which implies an exponential decrease in the size of the capacitor used in a DRAM to remember a single bit. Table 9.1 [164] shows how the families of DRAMs evolved over the years. Various access methods include *fast page mode* (FP), *extended data output* (EDO), *synchronous DRAM* (SDRAM), *double data rate* (DDR), and *direct Rambus DRAM* (DRDRAM.) Table 9.2 [395, 164] quantifies these trends. Manufacturers are now experimenting with new high dielectric constant, ferroelectric materials [442] such as *barium strontium titanate* (BST.) These

Table 9.2: DRAM characteristics.

Memory size (bits)	Feature width (μm)	Chip area (mm^2)	Clock rate (MHz)	Supply voltage (V)
16 Mb	0.35	25	66 – 100	2.5 or 3.0 or 3.3 or 5
64 Mb	0.30	40	100 – 133	2.5 or 3.3 or 5
128 Mb	0.23	55	100 – 800	2.5 or 3.3
256 Mb	0.17	120	100 – 400	2.5 or 3.3
512 Mb	0.13	200	100 – 400	2.5 or 3.3
1 Gb*	0.11	500†	133**	1.8

* Initial samples available now, production scheduled for late 2002.
** Clock rate is the input clock rate.
† Estimated size.

materials allow a greater capacitance to be maintained in the same physical space as before, leading to a density improvement. Present day DRAM capacitors made from *oxynitride oxide* have a dielectric constant of 5.0, but BST has a dielectric constant of > 500. Present-day DRAM designs now have *stacked capacitors*, where the storage capacitor is manufactured on top of a transistor, and are now made with deep sub-micron feature sizes (less than 0.5 μm.) These technological changes are forcing DRAM testing to be more oriented towards the physical DRAM layout in order to achieve acceptable fault coverage. For the future, one possible technology is *silicon-on-insulator* (SOI), where individual DRAM transistors are electrically insulated from each other by a dielectric layer, rather than being insulated from each other by reverse-biased *p-n* junctions.

In order to keep memory prices economical, the test cost per memory chip (which is directly related to the test time) cannot increase significantly. The number of bits per chip continues to increase exponentially and fault sensitivity increases, so faults become more complex. Tests for detecting the coupling faults between adjacent memory cells are more complicated and take longer than individual cell tests. For these reasons, we will not consider the early memory testing methods here. These methods had test times proportional to $O(n \times log_2(n))$ or even $O(n^2)$ (n is the number of memory bits.) Table 9.3 explains why these methods are now prohibitively expensive, so we will only consider $O(n)$ complexity memory tests in this chapter. The table assumes a memory cycle time of 60 ns.

Table 9.3: Test time (s) in terms of memory size n.

	Number of test algorithm operations			
n	n	$n \times log_2 n$	$n^{3/2}$	n^2
$1Mb$	0.063	1.26	64.5	18.33 hr
$4Mb$	0.252	5.54	515.4	293.2 hr
$16Mb$	1.01	24.16	1.15 hr	4691.3 hr
$64Mb$	4.03	104.7	9.17 hr	75060 hr
$256Mb$	16.11	451.0	73.30 hr	1200959.9 hr
1 Gb	64.43	1932.8	586.41 hr	19215358.4 hr
2 Gb	128.9	3994.4	1658.61 hr	76861433.7 hr

Market Share Trends. Table 9.4 documents the shifting memory market as it moves away from EEPROMs and EPROMs and towards DRAMs. Because of this shift, we will only cover DRAM, SRAM, and ROM testing here. Interested readers can consult van de Goor [688] for EEPROM and EPROM testing methods. Table 9.2 gives DRAM cell parameters for recent DRAM chip generations.

Defect Trends. Since we now have exponentially less charge stored per memory cell, as well as much closer proximity of memory cells to each other, cell coupling faults are now common and should be routinely tested for. Also, cells are becoming more vulnerable to manufacturing process disturbances. We should always assume that multiple memory faults will be present, and test appropriately. In addition to

9.1 Memory Density and Defect Trends

Table 9.4: Memory market share (%.)

	1988	1990	1994
DRAMs	56	54	58
SRAMs	17	22	21
ROMs	8	8	6
EPROMs	17	14	12
EEPROMs*	2	2	3

*Includes flash EEPROMs.

the cell coupling faults, *linked* cell coupling faults (defined below) must now also be tested for. The yield of memory chips would be nearly 0%, since every chip has defects, if not for the practice of including redundant rows and columns in the memory array. During initial manufacturing test, the manufacturer develops a map of the faulty rows and columns in the array. The column and row address decoders are then rewired using either a laser or by blowing fuses to reroute these defective rows and columns to working spare rows and columns included in the memory.

Test Time Complexity. Memory tests should deliver the best fault coverage possible given a certain test time. Also, we must consider the distinction between memory manufacturing test and memory end-user test. Since manufacturing test will end with the defective memory rows and columns being rewired to spare rows and columns, manufacturing test must also include *diagnosis*, which not only indicates that the memory chip is defective, but also shows *which* location is defective. The manufacturer may also use a variety of tests, such as initial production *characterization* tests, to determine which failures are really occurring, versus a variety of production tests for high-volume, high-yield production. End users might wish to forgo DRAM testing, except that the new ULSI *system-on-a-chip* (SOC) technology will include DRAM on a custom-designed chip having several microprocessors as well. Therefore, DRAM testing also is the ASIC designer's problem. End users who do not include DRAM in their systems-on-a-chip will still have to perform incoming parts test, board level tests, and system tests on their purchased DRAM components. Therefore, most of the tests covered here remain relevant to end users.

The major change in memory testing is that tests are now based on *fault models*, and tests are now proven to have complete coverage for particular fault models. A fault model is an abstraction of the error caused by a particular physical fault(s.) The purpose of the fault model is to simplify the testing procedure and reduce its cost, while still retaining the capability of detecting the presence of the modeled fault. Thus, the fault model need not accurately model the physical fault, as long as it still indicates the presence of the physical fault. Also, tests are now proven to be minimal length for the given set of fault models covered by the tests. In addition, it has been shown that memory tests with high fault coverage do not necessarily test a high percentage of the defects (*defect coverage*) occurring in production [442]. This happens when the fault model does not correctly model the actual defects that the particular fabrication line and DRAM geometry are susceptible to. With deep sub-micron chip feature sizes, DRAM chips are increasingly subject to peculiar, layout-specific failures. Therefore, we now use *inductive fault analysis* [596] to analyze the chip layout, and determine which fault models correctly model the actual physical defects that occur. Additional layout and fabrication line specific tests (I_{DDQ} and I_{DDT} tests) have been added to boost the DRAM defect coverages over those achievable by the functional tests.

9.2 Notation

nFET	MOS enhancement mode transistor
pFET	MOS enhancement mode transistor
0	A cell is in logical state 0
1	A cell is in logical state 1
x	A cell is in logical state 0 or 1
A	A memory address
ABF	AND Bridging fault
AF	Address decoder fault
ANPSF	Active Neighborhood Pattern Sensitive Fault
APNPSF	Active and Passive Neighborhood Pattern Sensitive Fault
B	Memory width, the number of bits in a memory word
BF	Bridging Fault
C	A Memory Cell
CF	Coupling Fault
CFdyn	Dynamic Coupling Fault
CFid	Idempotent Coupling Fault
CFin	Inversion Coupling Fault
DRF	RAM Data Retention Fault
k	Size of a neighborhood
M	Set of memory cells, words, or addresses
n	Number of memory bits: $n = 2^N$
N	Number of address bits
neighborhood	For pattern sensitive faults, this is the immediate cluster of cells whose pattern provokes a fault in the base cell
NPSF	Neighborhood Pattern Sensitive Fault
OBF	OR Bridging fault
PNPSF	Passive Neighborhood Pattern Sensitive Fault
SAF	Stuck-at Fault
SCF	State Coupling Fault
SNPSF	Static Neighborhood Pattern Sensitive Fault
SOAF	Stuck-Open Address Decoder Fault
TF	Transition Fault

9.3 Faults

Definition 9.1 Failures, errors, and faults. *A system is defined here either as a purely electronic system or a mixed electronic, electromechanical, chemical, and photonic device system. Such systems are coming into common use through* micro electro-mechanical system *(MEMS) technology that combines all of the above-listed devices on a single chip. A common MEMS system is the air bag controller in modern automobiles. A system* failure *occurs when system behavior is incorrect or interrupted. Failures are caused by* errors*, which are manifestations of* faults *in the system. A fault is present in the system when there is a physical difference between good and incorrect (failing or bad) system behavior, but some time may elapse before a fault causes a detectable system error.*

9.3.1 Fault Manifestations

Faults can either be permanent or non-permanent.

Permanent. These faults are caused by the following mechanisms, and can be modeled with a fault model, since they will exist indefinitely.

- Bad Electrical Connections (missing or added)
- Broken Components (this could be an IC mask defect or a silicon-to-metal or a metal-to-package connection problem)
- Burnt-Out Chip Wire
- Corroded Connection Between Chip and Package
- Chip Logic Error

Non-Permanent. Non-permanent faults are present only part of the time, and occur randomly. They have no well-defined fault model. These faults particularly afflict memory integrated circuits, but memory system designers handle these problems by including *information redundancy*, or redundant error-correcting codes, in each system memory location. Therefore, it is now customary to detect three-bit errors in memory systems, and to automatically correct up to two simultaneous bit errors in a memory system. Non-permanent faults are further subdivided into *transient* and *intermittent* faults. Transient faults are caused by environmental conditions, while intermittent faults are cause by non-environmental conditions.

Here are the common causes of transient faults:

- Cosmic Rays
- α-Particles (ionized Helium atoms)
- Air Pollution (causes temporary wire short or open)

- Humidity (causes temporary wire short)
- Temperature (causes temporary logic malfunction)
- Pressure (causes temporary open or short circuit in wiring)
- Vibrations (causes temporary open circuit in wiring)
- Power Supply Fluctuations (causes temporary logic error due to low voltage)
- Electromagnetic Interference (causes signal coupling between wires)
- Static Electrical Discharges (causes memory cells to change state)
- Ground Loops (causes misinterpretation of logic values between chips)

Here are the common causes of intermittent faults:

- Loose Connections
- Aging Components (logic gate delays change and relative signal arrival times therefore change)
- Hazards and Races in Critical Timing Paths (from bad design)
- Resistors, Capacitors, and Inductors Vary (causing timing faults)
- Physical Irregularities (a narrow wire, which causes a high resistance connection)
- Electrical Noise (causes memory cells to change state)

Intermittent faults can be modeled by the permanent fault models, but the test for the intermittent fault may have to be continuously repeated until the fault is detected.

9.3.2 Failure Mechanisms

Definition 9.2 Corrosion. *This electromechanical failure is caused by the presence of a voltage drop and various ions such as Cl^- and Na^+, which function as catalysts for the corrosion. Higher-quality package sealing deters corrosion, but CMOS is more susceptible to it due to lower voltages than TTL or ECL.*

Definition 9.3 Electromigration *occurs in Al wiring on chips, because the electrons flowing in the wires collide with Al grains of materials. The grains are dislocated in the direction of the electron current, which eventually burns the wire out in the same way that an old-fashioned fuse would.*

Definition 9.4 Bonding Deterioration. *This happens at the spot welds of the Au (gold) wires of the package to the Aluminum chip pads. Interdiffusion of Au-Al causes the bond to open circuit.*

Definition 9.5 Ionic Contamination. *Mobile ions in the environment, particularly Na^+, diffuse into the semiconductor package and diffuse into the gate oxide of a FET transistor. Here, they change the threshold (turn-on) voltage of the device, and cause a malfunction.*

Definition 9.6 Alloying. *This is caused by Al atom migration from the printed wiring on top of the Silicon chip into the Si. This failure turns into a shorted pn junction or an open contact between metal and the silicon.*

Definition 9.7 Radiation and Cosmic Rays. *Trace radioactive elements in the chip package emit α-particles, which have energies of up to 8 MeV. The α-particle collides with the Silicon chip lattice, and generates electron-hole pairs, which destroy information in RAM cells and cause information loss. These are the major cause of soft memory errors. Cosmic rays bombard the earth from outer space, particularly at higher altitudes because the atmosphere is thinner, and particularly afflict memory chips, because their transistor sizes are smaller than in logic chips. Computers in Mexico City or Denver are much more likely to suffer from cosmic rays than computers in New York City.*

Definition 9.8 *The activation energy E_a of a failure mechanism describes the temperature dependency of the mechanism. A higher E_a means that the failure is more likely to occur as temperature rises than failure mechanisms with lower E_a. Van de Goor [688] gives the activation energies for the more important failure mechanisms in Table 9.5.*

Table 9.5: Activation energies of major failure mechanisms.

Failure mechanism	Activation energy E_a
Corrosion of metallization	$0.3 - 0.6$ eV
Electrolytic corrosion	$0.8 - 1.0$ eV
Electromigration	$0.4 - 0.8$ eV
Bonding (purple plague)	$1.0 - 2.2$ eV
Ionic contamination	$0.5 - 1.0$ eV
Alloying (contact migration)	$1.7 - 1.8$ eV

9.4 Memory Test Levels

Figure 9.2 [688] shows the chip, array, and board levels of functional memory testing. Chip testing must be done with a memory fault model to make it economical. Memory array testing must also test the chip select and control logic. Memory board testing must test the memory array, the refresh logic, the error detection and correction logic, the board selector hardware, and the memory board controller. Electrical parametric tests are also important for memory systems, and are covered

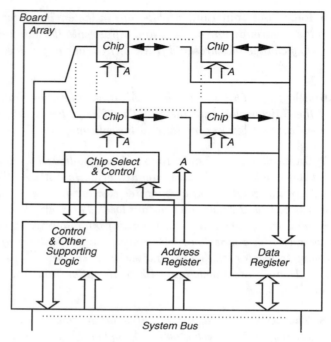

Figure 9.2: Memory system with chip, array, and board levels.

later. Since we assume a knowledge of common memory SRAM, DRAM, and ROM circuits, please refer to Weste and Eshraghian [718] for standard memory circuit designs.

9.5 March Test Notation

Notation. The following notation will be used throughout the rest of this chapter.

r	Memory action: A read operation
w	Memory action: A write operation
r0	Memory action: Read a 0 from the memory location
r1	Memory action: Read a 1 from the memory location
w0	Memory action: Write a 0 to the memory location
w1	Memory action: Write a 1 to the memory location
↑	Write a 1 to a cell containing 0 or the cell has a rising transition
↓	Write a 0 to a cell containing 1 or the cell has a falling transition
↕	Complement the cell contents
⇑	Increasing memory addressing order
⇓	Decreasing memory addressing order
⇕	Addressing order can be either increasing or decreasing=index⇕
→	Write a 0 to a cell containing a 0
→	Write a 1 to a cell containing a 1

9.6 Fault Modeling

$\Rightarrow$	Write value x to a cell already containing x
$\forall$	Denotes any memory write operation: $\forall \in \{\uparrow, \downarrow, \updownarrow, \rightarrow, \overset{\rightarrow}{\ }, \Rightarrow\}$
$< \ldots >$	Denotes a particular fault, described by ...
$< I/F >$	I is the fault sensitizing condition $\forall \in \{\uparrow, \downarrow, \updownarrow, \rightarrow, \overset{\rightarrow}{\ }, \Rightarrow\}$
	F is the faulty cell value $F \in \{0, 1, \uparrow, \downarrow, \updownarrow\}$
$< I1, \ldots, In-1; In/F >$	
	Denotes a fault involving n cells
	$I1, \ldots, In-1$ give fault sensitization conditions for cell n.
	In gives the conditions to sensitize the fault in cell n.
	If In is empty, then In/F is written as F.

Suk and Reddy [646] described the *march tests*, which consist of a finite operation sequence. Abadir and Reghbati [4, 689] and van de Goor [688] precisely defined and described these tests. The march test is applied to each cell in memory before proceeding to the next cell, which means that if a specific pattern is applied to one cell, then it must be applied to all cells. This is either done in increasing memory address order ($\Uparrow$, or from 0 to $n-1$) or decreasing order ($\Downarrow$.) The address order may be irrelevant ($\Updownarrow$.) Note, however, that the only real requirement for the march tests is that the address orders $\Uparrow$ and $\Downarrow$ must be the inverses of each other. An *operation* can be r0, r1, w0, or w1, which were defined earlier. Consider the MATS+ [362, 490, 4]* march test, written as $\{\ \Updownarrow (w0);\ \Uparrow (r0, w1);\ \Downarrow (r1, w0)\ \}$. It has the three march elements M0: $\Updownarrow (w0)$, M1: $\Uparrow (r0, w1)$, and M2: $\Downarrow (r1, w0)$. These are written with commas or semicolons separating them, and the entire march sequence is enclosed in braces. All operations of a march element are done before proceeding to the next address.

Figure 9.3 presents the interpretation of this notation for MATS+ as a testing algorithm. Interpretation of all subsequent march tests is left up to the reader. For all march tests, one may interchange the $\Uparrow$ and $\Downarrow$ address ordering notation throughout the test, and the test will still remain valid.

The march tests are a preferred method for RAM array testing, whether by means of an external tester or through *built-in self-testing* (covered later in Chapter 15.) Their $O(n)$ complexity, regularity, and symmetry are the reasons for this preference. However, tests for *neighborhood pattern sensitive faults* (NPSFs) cannot be performed by march tests, because one must treat the *base cell* differently from other cells in the neighborhood [442]. Note that a *complete* test achieves 100% fault coverage with respect to its model. An *irredundant* test is complete and has the property that removal of any test operation results in an incomplete test.

9.6 Fault Modeling

Physical examination of memory is not possible. The only other possible testing mechanism is to compare logical behavior of faulty memory against good memory. This requires modeling of the *physical* faults as *logical* faults. We model the memory system as an interconnected set of functional blocks. Testing must detect and/or

---

*Acronym for *Modified Algorithmic Test Sequence*.

M0: { March element $\updownarrow$ $(w0)$ }
  for $cell := 0$ to $n-1$ (or any other order) do
  begin
    write 0 to A [$cell$];
  end;

M1: { March element $\Uparrow$ $(r0, w1)$ }
  for $cell := 0$ to $n-1$ do
  begin
    read A [$cell$]; { Expected value = 0 }
    write 1 to A [$cell$];
  end;

M2: { March element $\Downarrow$ $(r1, w0)$ }
  for $cell := n-1$ down to 0 do
  begin
    read A [$cell$]; { Expected value = 1 }
    write 0 to A [$cell$];
  end;

Figure 9.3: MATS+ march test algorithm.

locate physical faults. The actual physical faults depend heavily on the circuit technology (CMOS, nMOS, ECL, TTL, etc.) and on the transistor schematic. Modeling faults as logical faults makes the testing approach more independent of the technology and the manufacturing process, and therefore more general. One disadvantage of logical fault modeling is that it may not be possible to relate a failure detected by a test to the actual physical defect, because of the high level of fault modeling.

The *behavioral* or *black-box model* of a memory system models the memory as a state machine, which has states for all possible combinations of the memory contents. This model is not used because of its excessive number of states, and because it does not adequately model coupling faults. We will model memory faults using the *functional*, or *gray-box model*. Figure 9.4 [688] shows a detailed functional memory model, while Figure 9.5 [688] shows a simplified functional model, consisting of an address decoder, a memory cell array, and read/write logic. The advantage of functional models is that they have enough detail of data paths and adjacent wiring runs in the memory to adequately model the coupling faults, which must be tested. We will not use the *logic gate model* of memories, because memory cells are designed using transistors and capacitors, and not logic gates, so the logic gate model does not correspond to physical reality. One may also model a memory with an *electrical model*, which allows detailed fault localization, at the expense of greatly increased testing cost. We will cover electrical models later. The *geometrical model* of a memory implies complete knowledge of the chip layout. This allows *inductive fault analysis* [596] to be used during initial memory chip production when we are

9.6 Fault Modeling

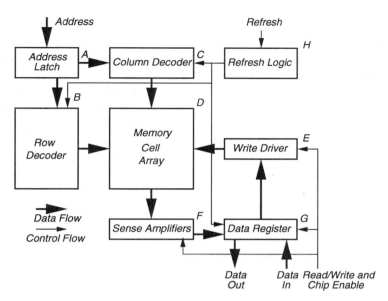

Figure 9.4: Functional memory model.

trying to redesign the memory layout to increase the memory chip yield. Inductive fault analysis allows us to populate the layout with defects and, using Monte-Carlo analysis, we can conclude which shorts, opens, bridging faults, and missing layers are actually possible, given a typical defect size. The penalty of using the geometrical model is a great increase in the fault model computation time.

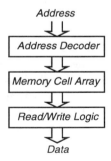

Figure 9.5: Simplified functional memory model.

9.6.1 Diagnosis Versus Testing Needs

During initial memory chip production, we would want very detailed testing and geometrical modeling. We would perform inductive fault analysis, and *diagnosis*, or fault localization, would be highly critical. A result of this analysis would be a redesigned set of geometric masks for manufacturing the memory chip. In the redesigned masks, wires might be separated if they were prone to bridging or coupling faults in the original design. Electrical drive strengths of transistors would be increased in parts of the chip where signals are consistently arriving too slowly. Finally, detailed electrical element values of the basic DRAM cell (the capacitance C, the word-line cell width/length ratio W_n/L_n, and the strength ratios of the bit line precharging transistor K_p relative to the word-line transistor K_n) would be adjusted to improve electrical performance, and reduce the likelihood of failure. The result of this would be improved masks, and greatly improved memory chip yield. At this point, the level of memory testing would be raised to the functional level, to reduce testing cost, although some electrical testing would be continued. For end users who purchase memory chips, diagnosis is not interesting, since once a fault is detected by functional testing of a memory chip, the only option is to discard the faulty chip and replace it with another one.

9.6.2 Reduced Functional Faults

Early work on functional RAM fault models was done by Thatte and Abraham [659]. They proposed fault models consisting of SAFs, TFs, and CFs. We will use van de Goor's [689] reduced functional DRAM chip model in Figure 9.4. This model can also be used for ROM, EPROM, or EEPROM testing by discarding some of the blocks in the model. It can also be used for SRAM chip testing by discarding the refresh logic block. The memory array containing n bits can be organized as n/B addresses each containing a word of B bits, or as n addresses each containing 1 bit. In many DRAM chips, an access occurs when the Address Latch contents change and the chip enable line is activated. The low-order address bits operate the Column Decoder, which selects the appropriate columns of the Memory Cell Array, while the high-order address bits operate the Row Decoder, which selects a row in the Memory Cell Array.

During a read, the Bit Lines of the columns to be read are first precharged to a logic 1 voltage, and then the cells in the rows to be read are activated. The memory will then couple its capacitors representing the data in the selected row and columns onto the bit lines for those columns. If the capacitors contain a low voltage, the voltage on the bit lines will drop. If the capacitors contain a high voltage, the bit line voltages hold steady. Then, the Sense Amplifiers compare the voltages on the selected columns with reference voltages, and if the bit line voltage for a given column is less than the reference voltage, the column is interpreted as a 0, otherwise the column is interpreted as a 1. During a write operation, the Sense Amplifiers are not used, but instead the Write Driver drives the columns to be written to the desired values. When the row to be written is activated, the capacitors representing the memory locations for those columns are coupled to the driven bit lines, and this causes the memory capacitors to change to the voltage state represented by the bit lines, since the bit lines are much stronger than individual memory capacitors. The benefit of this organization is that each memory row requires only 1 word line for addressing it, and each column requires only 1 bit line to read or write all of the cells in that column. These bidirectional bit lines save enormous chip area and also

9.6 Fault Modeling

reduce the number of pins on the memory chip, since the external data bus is also bidirectional.

During memory refresh, the Refresh signal is asserted, the Data Register is enabled, and the column decoder selects all columns. The Row Decoder selects the row required by the Address Latch contents, and all bits in that row are read and simultaneously refreshed. The refresh logic disables the Data Register. Please refer to Weste and Eshraghian [718] or Dillinger [202] for detailed descriptions of memory circuits, including Basic Cells, Decoders, Write Circuits, Sense Amplifiers, and Refresh Logic.

Table 9.6 lists some of the functional faults that can occur in a memory. A *cell* refers either to a memory cell or a data register. A *line* is any wiring connection in the memory. When we are not interested in fault diagnosis (localization), we will model the memory using the reduced or simplified functional model of Figure 9.5 [659, 491]. When this is done, the faults of Table 9.6 can be mapped into the *reduced functional faults* of Table 9.7. These faults are sufficient [491, 659] for functional memory testing.

Table 9.6: Subset of functional memory faults.

	Functional fault		Functional fault
a	Cell stuck	j	Open circuit in address line
b	Driver stuck	k	Shorts between address lines
c	Read/write line stuck	l	Open circuit in decoder
d	Chip-select line stuck	m	Wrong address access
e	Data line stuck	n	Multiple simultaneous address access
f	Open circuit in data line	o	Cell can be set to 0 but not to 1 (or vice versa)
g	Short circuit between data lines		
h	Crosstalk between data lines	p	Pattern sensitive cell interaction
i	Address line stuck		

Stuck-At Faults. The *stuck-at-fault* (SAF) is one in which the logic value of a cell or line is always 0 (SA0) or always 1 (SA1). The cell/line is always in the faulty state and cannot be changed. Stuck-at 0 (stuck-at 1) faults are described by the notation $< \forall/0 >$ ($< \forall/1 >$), meaning that for all actions affecting the memory location, its response is the given value 0 (or 1.)

Table 9.7: Reduced functional faults.

Notation	Fault
SAF	Stuck-at fault
TF	Transition fault
CF	Coupling fault
NPSF	Neighborhood pattern sensitive fault

Van de Goor [688] gives the necessary condition for a test to *detect* and *locate* all stuck-at faults:

From each cell, a 0 and a 1 must be read.

Figure 9.6(a) [688] shows the *state transition diagram* for a good memory cell. S_0 is the state where the cell contains logic 0, while S_1 is the state where it contains logic 1. A SA0 fault is sensitized by a w1 operation, while a SA1 fault is sensitized by a w0 operation. After sensitization, a SA0 fault is detected by a r1 operation, while a SA1 fault is detected by a r0 operation.

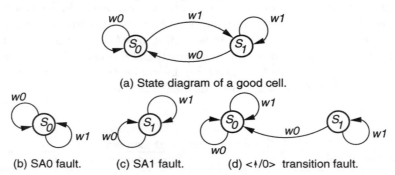

Figure 9.6: State transition diagram model for stuck and transition faults.

Transition Faults. The *transition fault* (TF) is a special case of the SAF, in which a cell fails to make a $0 \rightarrow 1$ (up) transition or a $1 \rightarrow 0$ (down) transition when it is written. An *up transition fault* is denoted as $<\uparrow /0>$, while a *down transition fault* is denoted as $<\downarrow /1>$. Figure 9.7 shows van de Goor's [688] flip-flop model for a memory cell. When a cell with a $<\uparrow /0>$ transition fault can power up in the 0 state, it may appear to be a SA0 fault. However, a coupling fault with another cell can cause this faulty cell to revert to the 1 state, so the stuck-at fault cannot model a transition fault. According to van de Goor [688], a test that must detect and locate all transition faults has this necessary condition:

Each cell must undergo a $\uparrow$ transition and a $\downarrow$ transition, and be read after each, before undergoing any further transitions.

Figure 9.6(d) [688] shows the state diagram for a $<\uparrow /0>$ transition fault. This fault is sensitized by the the w0 operation, followed by a w1 operation. A subsequent r1 operation will detect the fault. Shorting of adjacent bit and word lines or of adjacent cells in DRAM memory is so common that TF tests are essential for correct memory testing.

Coupling Faults. A *coupling fault* (CF) means that a transition in memory bit j causes an unwanted change in memory bit i. The *2-coupling fault* is a coupling fault involving two cells [429, 491, 646, 688]. A write operation that generates an $\uparrow$ or $\downarrow$

9.6 Fault Modeling

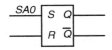

Figure 9.7: Flip-flop model of a transition fault.

transition in cell j changes the contents of cell i. The 2-coupling fault is a special case of the *k-coupling fault*, which has the 2-coupling fault behavior with respect to cells i and j, except that faulty behavior occurs only when another $k-2$ cells are in a particular state. The k cells must be restricted, as is done in the *neighborhood pattern sensitive fault*, to make the k-coupling fault model practical [284, 286, 491, 506]. We will treat the *inversion CF* and *idempotent CF*, which are specialized 2-coupling faults. *Bridging* and *state coupling* faults may involve any number of cells and lines, and are caused by a *logic* level, not a transition. Another coupling fault is the *dynamic coupling fault* (CFdyn), in which a read or write operation on cell j forces cell i to 0 or 1.

Inversion Coupling Faults. An *inversion coupling fault* (CFin) means that a ↑ or ↓ transition in cell j inverts the contents of cell i. Cell i is said to be *coupled* to cell j, which is the *coupling cell*. We use the notation $<\uparrow;\updownarrow>$ for C_i and C_j, where the $\updownarrow$ means that cell C_i contents were inverted. The two possible CFin types are $<\uparrow;\updownarrow>$ and $<\downarrow;\updownarrow>$. A test for all CFins must satisfy this necessary condition [688]:

> For all cells that are coupled, each should be read after a series of possible CFins may have occurred (due to writing into the coupling cells), and the number of coupled cell transitions must be odd (to prevent the CFins from masking each other.)

In Figure 9.8(a) we present the state diagram for a pair of good cells [688]. Nodes labeled S_{ij} represent the states of cells i and j. Figure 9.8(b) shows the modified state diagram to represent the inversion coupling fault $<\uparrow;\updownarrow>$. We sensitize the fault with a ↑ transition write to cell j (operation w1/j.) Fault detection occurs when cell i is read.

Theorem 9.1: *Not all linked CFins can be detected by march tests* [442].
Proof: Consider three cells i, j, and k (with address relationships Address(i) < Address(j) < Address(k).) Cell k is $<\uparrow;\updownarrow>$ coupled to cell i and to cell j, and both i and j are visited either before or after k is visited by a march element. Then any march element or combination of march elements can traverse these three cells in either of these two ways:

1. Using the sequence $i \to j \to k$ and/or its reverse. In this case: (i) The two CFins will mask each other for any march element marching 'up', and (ii) Neither will be triggered for an element marching 'down'. Therefore, the march test fails to detect the linked CFins.

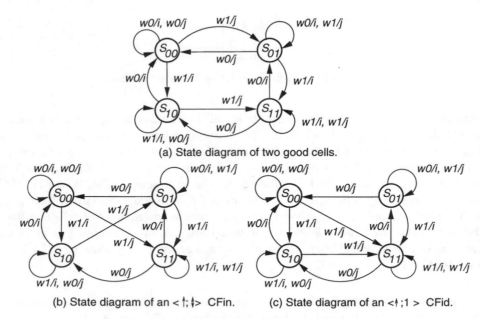

Figure 9.8: State diagrams for CFin and CFid.

2. Using the sequence $k \to i \to j$ and/or its reverse, in which case any 'up' or 'down' march would mask the two CFins. ∎

Idempotent Coupling Faults. An *idempotent coupling fault* (CFid) is where an ↑ or ↓ transition in cell C_j sets cell C_i to 0 or 1. This is denoted as $<\uparrow; 0>$ or $<\uparrow; 1>$, depending on whether cell i is set to 0 or 1, for a rising transition for cell j. The other two idempotent coupling faults are $<\downarrow; 0>$ and $<\downarrow; 1>$. A test to *detect* all CFids has this necessary condition:

> *For all coupled cells, each should be read after a series of possible CFids may have happened (by writing into the coupling cells), such that the sensitized CFids do not mask each other (the coupled cells are read while their state is opposite from the good machine state.)*

A CF is *asymmetric* [506] when it causes the coupled cell to undergo only a ↑ or a ↓ transition. The CF is *symmetric* when the coupled cell experiences both transitions due to the fault. The inversion coupling fault CFin $<\uparrow; \updownarrow>$ is a symmetric fault. The CF is *one-way* if it is sensitized only by a rising or falling transition of the coupling cell, and *two-way* if either transition sensitizes it. Figure 9.8(c) shows the state transition diagram for the CFid fault $<\uparrow; 1>$.

Dynamic Coupling Faults. A *dynamic coupling fault* (CFdyn) occurs between cells in different words. A read or write operation on one cell forces the contents of the second cell either to 0 or 1. This is a more general case of the CFid, because a

9.6 Fault Modeling

CFdyn can be sensitized by any read or write operation, where as a CFid can only be sensitized by a writing a change (transition write operation) to the coupling cell. We denote a CFdyn as $<r0|w0;0>$ where | denotes the *or* of the read and write operations, which must be done to the coupling cell [688]. There are four CFdyn faults: $<r0|w0;0>$, $<r0|w0;1>$, $<r1|w1;0>$, and $<r1|w1;1>$.

Bridging Faults. A *bridging fault* (BF) is a short circuit between two or more cells or lines. It is a bidirectional fault, so either cell/line can affect the other cell/line. A 0 or 1 state of the coupling cell causes the fault, rather than a coupling cell transition. With the *AND bridging fault* (ABF), the logical bridge value is the AND of the shorted cells/lines. The four possible ABFs are $<0,0/0,0>$, $<0,1/0,0>$, $<1,0/0,0>$, and $<1,1/1,1>$. The notation is the good machine values for cells i and j, followed (after the slash) by their bad machine values. With the *OR bridging fault* (OBF), the logical bridge value is the OR of the shorted cells/lines. The four possible OBFs are $<0,0/0,0>$, $<0,1/1,1>$, $<1,0/1,1>$, and $<1,1/1,1>$.

State Coupling Faults. The *state coupling fault* (SCF) [194] is where the coupling cell/line j is in a given state y that forces the coupled cell/line i into state x. The four SCFs are $<0;0>$, $<0;1>$, $<1;0>$, and $<1;1>$. Figure 9.9 [442] shows a Mealy machine model of the state coupling fault, along with a more complete model of the transition fault [106, 107, 169].

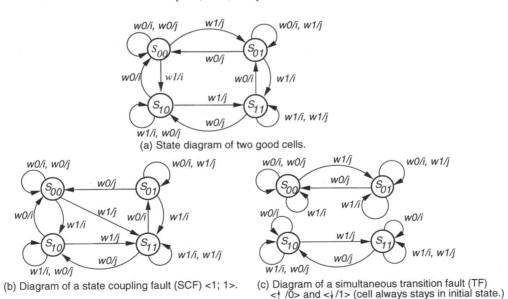

Figure 9.9: State diagram model for state coupling and transition faults.

Neighborhood Pattern Sensitive Coupling Faults. In a *pattern sensitive fault* (PSF), the content of cell i (or the ability of cell i to change) is influenced by the contents of all other memory cells, which may be either a pattern of 0s and 1s or a pattern of transitions. The PSF is the most general *k-coupling fault*, where $k = n$ (all of the memory.) The *neighborhood* is the total number of cells involved in this fault, where the *base cell* is the cell-under-test, and the *deleted neighborhood* is the *neighborhood* without the *base cell*. In the PSF model, the neighborhood could be anywhere in the memory array, whereas in a *neighborhood pattern sensitive fault* (NPSF), the neighborhood must be in a single position surrounding the base cell. All known algorithms are for NPSFs, which must now be tested for because of reduced memory cell capacitance in high-density DRAMs.

We consider three types of NPSFs:

- With the *active NPSF* (ANPSF) [645] (also called *dynamic* [558]), the base cell changes due to a change in the pattern of the deleted neighborhood. One deleted neighborhood cell has a transition, while the rest of the neighborhood (including the base cell) has a given pattern. An ANPSF test has this necessary condition [688]:

  Each base cell must be read in state 0 and state 1, for all possible deleted neighborhood pattern changes.

  A Type-1 neighborhood [644] (see Figure 9.10) has four deleted neighborhood cells. We describe a fault as $C_{i,j} < d_0, d_1, d_3, d_4; b >$. $C_{i,j}$ is the base cell location, d_0, d_1, d_3, and d_4 are the deleted neighborhood patterns, and b is the fault effect in the base cell. $C_{i,j} < 0, \downarrow, 1, 1; 1 >$ denotes an ANPSF fault where the base cell $C_{i,j}$ is initially 0, d_1 experiences a $\downarrow$ transition, while d_0, d_3, and d_4 contain 011. The fault effect is to switch the base cell to 1. When the base cell becomes 0, we write this as $C_{i,j} < 0, \downarrow, 1, 1; 0 >$ and when the base cell is inverted, we write this as $C_{i,j} < 0, \downarrow, 1, 1; \updownarrow >$.

Figure 9.10: Type-1 neighborhood definition.

A Type-2 neighborhood [644] (see Figure 9.11) has eight deleted neighborhood cells, and is a more complex fault model than a Type-1 neighborhood. The Type-1 neighborhood assumes either that diagonal cell couplings are insignificant or that if there is a diagonal cell coupling, then the fault will also cause a vertical or horizontal cell coupling, which the Type-1 neighborhood can model. A Type-2 neighborhood is needed when diagonal couplings are significant, and do not necessarily cause a vertical/horizontal coupling.

- A *passive NPSF* (PNPSF) [645] means that a certain neighborhood pattern

9.6 Fault Modeling

Figure 9.11: Type-2 neighborhood definition.

prevents the base cell from changing. The necessary condition to detect and locate a PNPSF fault is [688]:

Each base cell must be written and read in state 0 and in state 1, for all deleted neighborhood pattern permutations.

We introduce the notation $\uparrow/0$ ($\downarrow/1$) for the base cell fault effect to indicate that the base cell cannot be changed from 0 (or 1.) The fault preventing the base cell $C_{i,j}$ from changing from 0 is denoted as $C_{i,j} < 0,0,1,1; \uparrow/0>$. If it cannot change from 1, we write this as $C_{i,j} < 0,0,1,1; \downarrow/1>$. If it cannot change regardless of content, we write this as $C_{i,j} < 0,0,1,1; \updownarrow/x>$.

- With a *static NPSF* (SNPSF) [558], the base cell is forced into a particular state when the deleted neighborhood contains a particular pattern. This differs from the ANPSF in subtle way, because there need not be a transition in the deleted neighborhood to sensitize an SNPSF fault. The necessary condition is [688]:

Each base cell must be read in state 0 and in state 1, for all deleted neighborhood pattern permutations.

A SNPSF in a Type-1 neighborhood where $C_{i,j}$ is forced to 0 when d_0 through d_4 contain 0101 is $C_{i,j} < 0,1,0,1; -/0>$. When $C_{i,j}$ is forced to 1, we write $C_{i,j} < 0,1,0,1; -/1>$.

Fault 1	Fault 2	Fault 3	Fault 4
No Cell Accessed for A_x	No Address to Access cell C_x	Multiple Cells Accessed with A_y	Multiple Addresses for Cell C_x

Figure 9.12: Address decoder faults.

Address Decoder Faults. An *address decoder fault* (AF) represents an address decoding error, in which we assume that the decoder logic does not become sequential [659, 491]. We also assume that the fault is the same during both read and write operations. We discuss only bit-oriented memory, in which each word contains only one bit. Consult van de Goor [688] for the multi-bit word case. Van de Goor [688] classifies these faults into four cases: Fault 1: No cell is accessed for a certain address, Fault 2: No address can access a certain cell, Fault 3: With a particular address, multiple cells are simultaneously accessed, and Fault 4: A particular cell can be accessed with multiple addresses. Figure 9.12 [688] illustrates these faults. Finally, all address decoders built out of CMOS logic gates (most are at present) can also exhibit CMOS stuck-open faults [463]. Consult van de Goor [688] for this particular test. Figure 9.13 [688] shows how multiple numbers of address faults are combined (and these must be tested for.) Because there are as many addresses as there are cells, none of the address decoder faults can stand alone.

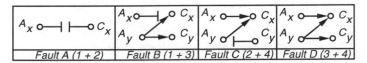

Figure 9.13: Combinations of AFs that must be tested.

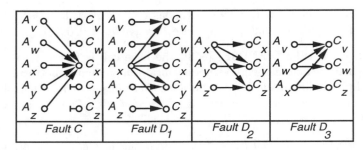

Figure 9.14: Illustration of conditions for proof.

Table 9.8: Conditions for address decoder fault detection.

Condition	March element	Condition	March element
1	$\Uparrow (rx, \ldots, w\bar{x})$	2	$\Downarrow (r\bar{x}, \ldots, wx)$

Theorem 9.2: *A march test satisfying Conditions 1 and 2 in Table 9.8 detects all address decoder faults* [688].

1. The ... in a march element shows any number of read or write operations.

2. Before the Condition 1 element, there must be a wx march element.

3. The value of x can be chosen freely as 0 or 1, but it must be the same throughout the test.

Proof (by enumerating all cases) [688]:

- *Sufficiency.*

9.6 Fault Modeling

- *Faults A and B*: Faults A and B (see Figure 9.14) are detected by every test that detects SAFs in the memory cell array. When address A_x is written and read, cell C_x will appear either SA0 or SA1. Therefore, either Condition 1 or Condition 2 will detect the fault.
- *Fault C*: Fault C is detected by first initializing the entire memory to an *expected* value h (which may be x or $\bar{x}$.) Any subsequent march element operation that reads the expected value h and ends by writing $\bar{h}$ detects fault C:
    * Elements marching $\Uparrow$ will write $\bar{h}$ to A_v. Reading A_w detects the fault (see Figure 9.14.)
    * Elements marching $\Downarrow$ will write $\bar{h}$ to A_z. Reading A_y detects the fault (see Figure 9.14.)

  Therefore, any one of Conditions 1 or 2 will detect fault C.
- *Fault D*: The memory may return a random result when a fault causes multiple cells to be read simultaneously for a given address (due to noise.) For fault D to be detected in this scenario, we assume that the result of reading A_x is unpredictable, since addressing location x causes multiple simultaneous reads. The fault must therefore be generated when A_x is written, and detected when either A_w or A_y is read. This can be done with either $\Uparrow$ or $\Downarrow$ marches.
    * Condition 1: $\Uparrow$ (rx,...,w$\bar{x}$) detects cases D_1 and D_2. When A_x is written with $\bar{x}$, cells $C_y \ldots C_z$ are also written with $\bar{x}$. This is caught when C_y is read, because $\bar{x}$ will be read when x is expected.
    * Condition 2: $\Downarrow$ (r$\bar{x}$,...,wx) detects cases D_1 and D_3. When A_x is written with x, cells $C_w \ldots C_v$ are also written with x. This is detected when C_w is read, because x is read while $\bar{x}$ is expected.

  The sufficient case is now proven.

- *Necessity.* We remove one operation at a time from the two conditions, and prove that any operation removal invalidates the test.
    - Leaving out rx from Condition 1 leaves a test that is unable to detect faults A or B when they always return the value $\bar{x}$.
    - Leaving out r$\bar{x}$ from Condition 2 gives a test that is unable to detect faults A or B when they always return x.
    - Leaving out rx or w$\bar{x}$ from Condition 1 gives a test that is unable to detect fault D_2.
    - Leaving out the r$\bar{x}$ or wx from Condition 2 results in a test that is unable to detect fault D_3.
    - Leaving out both write operations results in a test than cannot detect faults C and D_1.

Q.E.D.

Read/Write Logic Faults. Van de Goor and others [491, 659, 688] give a proof that SAFs, TFs, CFs, and NPSFs in the read/write logic block of the memory can be detected by tests for SAFs, TFs, CFs, and NPSFs in the memory array. Therefore, we will drop the Read/Write Logic Block from our Reduced Functional Memory System model.

9.6.3 Relation Between Fault Models and Physical Defects

Reduced Functional Fault Modeling. We reprint van de Goor's [688] discussion of the mapping of physical defects into the reduced functional faults. Table 9.9 shows the reduced functional fault in the left column and the covered functional fault in the right column. *Address decoder faults* (AFs) will be covered later. A special case occurs when the read/write chip control line is stuck at 'read'. The memory cells cannot be written, so they all appear to be stuck. If the line is stuck at 'write', the memory can be written, but during a read either all 0s or all 1s will be read, depending on the technology. So, fault c is covered by the SAF model.

Table 9.9: Mapping of functional faults onto reduced functional faults.

Reduced functional fault		Functional fault
SAF	a	Cell stuck
SAF	b	Driver stuck
SAF	c	Read/write line stuck
SAF	d	Chip-select line stuck
SAF	e	Data line stuck
SAF	f	Open circuit in data line
CF	g	Short circuit between data lines
CF	h	Crosstalk between data lines
AF	i	Address line stuck
AF	j	Open circuit in address line
AF	k	Shorts between address lines
AF	l	Open circuit in decoder
AF	m	Wrong address access
AF	n	Multiple simultaneous address access
TF	o	Cell can be set to 0 but not to 1 (or vice versa)
NPSF	p	Pattern sensitive cell interaction

Examples of Fault Modeling. We reprint van de Goor's [688] excellent discussion. Figure 9.15 [688] shows some SRAM cell defects. Defect q (inverse node shorted to V_{DD}) causes a SA0 fault. Defect r (true node shorted to V_{SS}) causes a SA0 fault. Defect s (open true node gate) causes a SA0 fault. Defect t (an open *word line* (WL)) causes all cells after the WL fault to be inaccessible (AF.) When such cells are read, the BL and $\overline{BL}$ signals are not driven (since the open WL prevents activation of

9.6 Fault Modeling

Q_5 and Q_6.) If a single-input read circuit is used, this maps into a SAF fault for the inaccessible cells. For a differential amplifier read circuit, this may also be a SAF fault (depending on the design.)

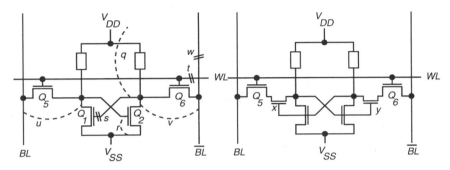

Figure 9.15: Static RAM cell with physical defects.

Defect u (a short between the true node and BL) will pull BL down if the cell contains a 0, but will not affect BL if the cell contains 1. Cells along BL will enter state 0 when the cell with defect u contains 0. This is the state coupling fault $<0;0>$. Defect v (short between inverse node and $\overline{BL}$) is similar, as is the state coupling fault $<1;1>$. Defect w (open $\overline{BL}$) prevents cells after the open defect from passing a logic 0 value on $\overline{BL}$. Cells after this defect containing 0 will be correctly read. When they contain 1, the result depends on the type of read circuit (if it is single-input, this is a SAF.)

With defects x and y, a new transistor is created due to polysilicon defects [194]. With defect y and logic 0 in the cell, the cell is disconnected from $\overline{BL}$, and this is a $<\uparrow/0>$ transition fault. Defect x is a $<\downarrow/1>$ transition fault.

From van de Goor [688], another example of DRAM defects is presented in the single-device DRAM cell array of Figure 9.16. Defect aa is a shorted capacitor, which is a SA0 fault. Defect bb is a capacitor-WL short, which is a SA1 fault. Defect cc, a short between two WLs, is an AND bridging fault between cell pairs located in the same column for the two shorted WLs. Defect dd is a state coupling fault. Defect ee is an AND bridging fault with all cells in the same column. Defect ff, a short between two bit lines, is an AND bridging fault between pairs of cells on the same word line and on the shorted bit lines.

A short between a WL and a BL causes the cell at the crosspoint to be SA1 [688], as follows. During a read, the WL is driven high (for cell selection), so BL becomes 1. A w1 operation successfully writes a 1 in the crosspoint cell, while a w0 operation has no effect, since the fault changes the BL data to 1. The crosspoint appears to be $<1;1>$ coupled with all cells in the same column, as well as coupled with all cells in the same row.

The present, extremely high density of 256 Mbit DRAM chips makes all of these above defects quite likely. Therefore, correct DRAM testing requires testing for ALL of the reduced functional faults.

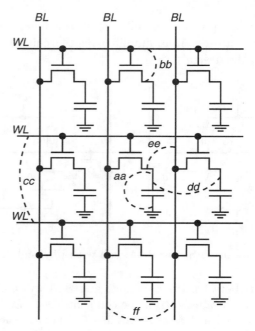

Figure 9.16: DRAM cell array with physical defects.

9.6.4 Multiple Fault Models

When coupling faults are used, we also assume that any number of the various different faults can occur *simultaneously*, which frequently happens in chip manufacturing.

Linkage. Faults may also be *linked* [506], meaning that a fault may influence the behavior of other faults. *Unlinked* faults do not influence the behavior of other faults.

Fault Masking Example. Figure 9.17(a) [688] shows two coupling faults. The first fault is cell i $<\uparrow; 1>$ coupled to j, and the second is k $<\uparrow; 0>$ coupled to l. The march test { $\updownarrow (w0); \Uparrow (r0, w1); \updownarrow (w0, w1); \updownarrow (r1)$ } will detect these two faults, provided that $i \neq k$. However, the test fails to detect the combination of the faults when $i = k$ (see Figure 9.17(b).) Linkage causes the test to pronounce the circuit as good. This is called *fault masking*. However, the linked faults will be detected by van de Goor's [688] march test { $M0 : \updownarrow (w0); M1 : \Uparrow (r0, w1); M2 : \Downarrow (w0, w1); M3 : \Uparrow (r1, w0, w1)$ }. The $<\uparrow; 1>$ CF is detected by the r0 operation of march element M1. The $<\uparrow; 0>$ CF is detected by the r1 operation of march element M3. The linked fault is detected by the r1 operation of M3 when it operates on cell i.

Figure 9.18 shows the fault hierarchy. Fault m pointing to fault n means that a test for fault m also detects fault n. It is now necessary to test for linked faults of

9.6 Fault Modeling

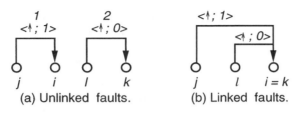

Figure 9.17: Example illustrating masking of coupling faults.

different types in the current manufacturing environment.

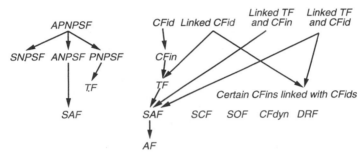

Figure 9.18: Fault hierarchy.

Fault Combinations.

Linked, Similar Type Faults. Since the SAF involves only one cell, and only one SAF can happen at a time in a single cell, then SAFs cannot be linked with other SAFs. By a similar argument, linked transition faults cannot occur in any one cell.

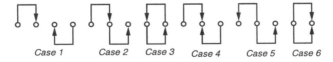

Figure 9.19: Definition of linked coupling faults.

Coupling faults require careful analysis, as shown in Figure 9.19. The notation is that an arrow from fault i to fault j means that fault i affects fault j (j is the *coupled cell*, while i is the *coupling cell*.) Cases 1, 2, 3, and 5 represented unlinked faults, because each of the coupled cells is only coupled to a single other cell, and there is only one type of coupling. Case 4 is a linked fault because a cell is coupled to multiple cells, and Case 6 is a linked fault because there are multiple coupling mechanisms.

Generally, linked CFins cannot be detected by march tests [688]. However, tests for *idempotent CFs* (CFids) will detect *inversion CFs* (CFins) [688]. A CFid fault may have the same effect as a CFin fault. Consider when memory contains all 0s, and then a $<\uparrow;1>$ CFid fault has the same effect as a $<\uparrow;\updownarrow>$ CFin fault. As a result, not all CFids linked with CFins are detectable by march tests. Figure 9.20(a) shows a non-maskable linked pair of faults, while Figure 9.20(b) shows a maskable linked pair of faults. CFids linked with CFins are not detectable by march tests when an odd number of CFin coupling cells are present (in increasing address order) between the coupling cell of the CFid and the coupled cell of the CFid, and the CFin coupling cells sensitize faults in the same sense ($\uparrow$ or $\downarrow$) as the CFid. In Figure 9.20(b), there is one CFin cell, and both the CFin and CFid faults are sensitized by the same $\uparrow$ transition.

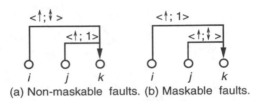

(a) Non-maskable faults. (b) Maskable faults.

Figure 9.20: Fault masking through linkage.

Neighborhood Pattern Sensitive Faults are only caused by patterns in a cell's deleted neighborhood, and during their tests, the base cell is read after each unique deleted neighborhood pattern is applied. Therefore, NPSFs are *unlinked*. The combination of tests for active and passive NPSFs (APNPSFs) covers all NPSFs (including SNPSFs.) The base cell is forced to a particular value by a SNPSF. This fault can be detected by a test for an ANPSF (when the base cell is changed to the particular value) and a test for a PNPSF (the base cell already has its forced value, and as a result cannot be changed.)

Linked, Different Type Faults. In the fault hierarchy, if a test for a fault higher in the hierarchy is run, it will cover faults at a lower hierarchical level, so they need not be tested for separately. Tests for TFs and CFs read the 0 or 1 state of every cell in the memory, so they also satisfy the condition for testing the SAFs. This is true even when TFs and CFs are linked with the SAF. Since tests for all NPSFs always read every cell in state 0 or 1, they also detect SAFs. A SAF in the base cell masks a NPSF, but a NPSF never masks a SAF in the base cell. So, tests for ANPSFs, PNPSFs, or SNPSFs always detect SAFs, even when the SAFs are linked to the NPSFs. However, TFs linked with CFs require a new test. This is because the TF may mask a CF, while the CF masks the TF. A TF linked with a NPSF cannot always be detected by the test for the NPSF. However, TFs can be detected when linked with PNPSFs by a test either for PNPSFs or APNPSFs.

Linked Address Decoder Faults. Table 9.10 gives the conditions for detecting AFs linked with other faults.

9.6 Fault Modeling

Table 9.10: Tests for linked AFs.

Fault	Test required
AFs linked with SAFs	Any test for AFs
AFs linked with TFs	Any AF test that reads every cell after writing it to 0, and reads every cell after writing it to 1
AFs linked with CFs	Any CFid and/or CFin test that can find AFs in the absence of CFs
AFs linked with NPSFs	AFs may mask the NPSFs. Test first for NPSFs, and then apply a separate test for AFs

9.6.5 Frequency of Faults

Table 9.11 [170] lists common faults found in DRAMs and SRAMs. Note that the possible faults differ between these two types of memory. In a DRAM, a read operation is followed by a write-back refresh cycle. Both reads and transition writes are both likely to trigger DRAM faults. It was once assumed that SRAM read operations were fault-free [442]. However, an examination of Table 9.11 shows many cases where an SRAM will be faulty during a read operation. Since a DRAM stores charge for a bit on a capacitor, while an SRAM uses a cross-coupled pair of inverters, the DRAM is far more susceptible than the SRAM to various charge leakage fault mechanisms caused by leakage currents.

Veenstra [697] presents several tables of fault frequency data in SRAM chips. Figure 9.21 [697] shows the relationship between the test and the achieved fault coverage. The fault model included SAFs, CFids, and SNPSFs. The poor performance of the MATS and MARCHING 1/0 tests shows that 100% stuck fault coverage still leads to many failures. The CF tests (MARCH A, MARCH B, and MARCH C) are best because they detect a large number of physical faults. The TL-SNPSF1G SNPSF test performs poorly, because pattern sensitive faults are unlikely in low-density SRAMs.

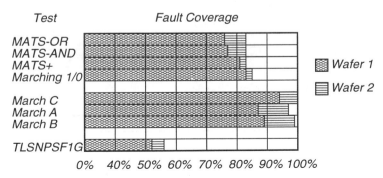

Figure 9.21: Influence of test on fault coverage in SRAMs.

Veenstra [697] also investigated the influence of addressing order on detecting dynamic faults such as sense amplifier recovery and write recovery faults. Fig-

Table 9.11: SRAM and DRAM faults.

DRAM or SRAM faults	Fault model
Shorts and opens in the memory cell array	SAF, SCF
Shorts and opens in the address decoder	AF
Access time failures in the address decoder	Functional Fault
Stray coupling capacitances between adjacent cells	CF
Bit line shorted to word line (causes excessive I_{DDQ} current) [456, 455, 596]	
Transistor gate shorted to channel (causes excessive I_{DDQ} current) [456, 455, 596]	
Transistor stuck-open fault (causes excessive I_{DDQ} current) [456, 455, 596]	SOF

Pattern sensitive faults (causes excessive I_{DD} power supply current) [643]:
    Diode-connected transistor short between 2 cells
    Open transistor drain
    Gate oxide short
    Bridging Fault
	PSF

Faults found only in SRAM	**Fault model**
Open-circuited pull-up device (causes data loss) [442]	DRF
Excessive coupling capacitance between bit lines [357]	CF
Faults found only in DRAM	**Fault model**
Data retention fault (*sleeping sickness*)	DRF
Refresh line stuck-at fault	SAF
Bit-line voltage imbalance fault [442] due to weak inversion currents	PSF
Coupling between Word and Bit line	CF
Single-ended bit-line voltage shift	PSF
Precharge and decoder clock overlap	AF

9.6 Fault Modeling

ure 9.22 [697] shows his results. The various MATS+i tests generate sequential addresses along the same word line, along the same bit line (column), with increments different from 1 along the same word or bit line, and by accessing a different word and bit line on every address. From the superior performance of the march tests, one can conclude that coupling fault tests are best for testing the recovery faults. DRFs are detected when the addressing order follows the word lines, but write recovery faults are detected when the order follows the bit lines.

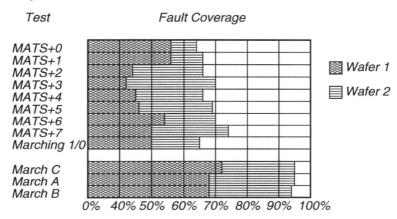

Figure 9.22: Influence of addressing order on fault coverage in SRAMs.

Dekker [193] shows data on faults caused for various spot defect sizes. The *critical area* is the chip area where the spot defect may damage the active part of the layout (see Table 9.12.) The *critical path length* is the length of parallel wires, which are separated by a dimension equal to the spot defect size. The probability of a short is determined by the length of the wires. From his result, one can conclude that CFids and TFs are a problem only on long wires.

Table 9.12: Critical path length as a function of fault class.

Fault class	Spot defect size (μm)						
	< 2	< 3	< 5	< 7	< 9	< 2	< 9
Stuck-at	78	213	227	269	269	51.3%	49.8%
Stuck-open	32	64	64	64	64	21.0%	11.9%
Transition	0	36	38	38	38	0%	7.0%
State coupling	15	15	51	71	71	9.9%	13.2%
Idemp. coupling	0	0	0	0	18	0%	3.3%
Data retention	27	29	80	80	80	17.8%	14.8%
Total	152	357	460	522	540	100%	100%

Validation of Dekker's fault model with a scanning electron microscope is shown in Table 9.13 [195]. Observe that transition faults rarely occurred, while idempotent coupling faults occurred frequently (due to certain address decoder faults.) Note that 10% of the faults could not be classified.

Table 9.13: Validation of fault model.

Cluster	Number of devices	Fault class
0	714	Stuck-at and Total failure
1	169	Stuck-open
2	18	Idemp. coupling
3	9	State coupling
4	8	?
5	5	?
6	26	Data retention
–	–	?
14	2	?

9.7 Memory Testing

9.7.1 Functional RAM Testing with March Tests

Nomenclature. This subsection is only concerned with layout-independent RAM testing. We establish a defect model for the RAM, based on the most likely layout and design defects that we predict. We abstract this defect model into a functional *fault model*, and this leads to the reduced functional faults. We refer to unlinked inversion coupling faults as *CFins*, unlinked idempotent coupling faults as *CFids*, and linked idempotent coupling faults as *linked CFids*.

Mapping AFs into Functional Memory Array Tests. March tests are capable of detecting AFs. However, NPSF tests are not capable of detecting AFs, and therefore must be augmented with an Address Decoder test. A march test can detect all address decoder faults if it satisfies these conditions:

- It must read the value x (0 or 1) from cell 0, write $\bar{x}$ to cell 0, read the value x from cell 1, write $\bar{x}$ to cell 1, and so on for the entire memory.

- It must read the value $\bar{x}$ (0 or 1) from cell $n-1$, write x to cell $n-1$, read the value $\bar{x}$ from cell $n-2$, write x to cell $n-2$, and so on for the entire memory.

These can be summarized by saying that the march test must contain these two march elements: $\Uparrow$ (rx, ..., w$\bar{x}$) and $\Downarrow$ (r$\bar{x}$, ..., wx). The basic principle is that as the memory writing and examination operation moves through memory, any address decoder fault that causes unexpected accesses of memory locations will cause those locations to be written to an unexpected value. As the march test proceeds, it will discover those locations and report a fault.

Tests. Table 9.14 summarizes which faults and linkages are covered by all of the march tests. Table 9.15 presents the actual march tests. As mentioned previously, the addressing orders can be interchanged. Note that in a march test that detects all faults of a particular type it is permissible to complement the data values for all

9.7 Memory Testing

of the read and write operations *provided that* when a march test has to detect AFs, a read operation accessing multiple cells on the physical memory produces a random result. If, instead, the multiple read operation produces either the AND or OR of read data values, then the data values of march test read and write operations cannot be complemented without reducing address fault coverage. Van de Goor [688] has provided detailed proofs that these march algorithms detect the listed faults, and that the march algorithms are of minimal length.

The MATS march test can be used for complete AF coverage when a fault causing multiple memory cells to be read simultaneously returns the OR of the memory word values. If, instead, the memory technology returns the AND of the multiple memory cells during a faulty read, then all 0s and 1s in the MATS algorithm should be interchanged to obtain complete AF fault coverage. If the faulty memory behavior is unknown during multiple simultaneous reads, use the MATS+ march test instead.

Table 9.14: Irredundant march test summary.

Algorithm	Fault coverage								Operation count
	SAF	AF	TF	CF in	CF id	CF dyn	SCF	Linked faults	
MATS	All	Some							$4 \cdot n$
MATS+	All	All							$5 \cdot n$
MATS++	All	All	All						$6 \cdot n$
MARCH X	All	All	All	All					$6 \cdot n$
MARCH C−	All	All	All	All	All	All	All		$10 \cdot n$
MARCH A	All	All	All	All				All linked CFids, Some CFins linked with CFids	$15 \cdot n$
MARCH Y	All	All	All	All				All TFs linked with CFins	$8 \cdot n$
MARCH B	All	All	All	All				All linked CFids, All TFs linked with CFids or CFins, Some CFins linked with CFids	$17 \cdot n$

Figure 9.23 shows the behavior of the MATS+ march test when cell (2, 1) has a SA0 fault. MATS+ has march elements $M0$, $M1$, and $M2$. The fault is detected by march element M2 as it moves from the highest memory address downward and expects to read a 1 in cell (2, 1), but instead gets a 0. Figure 9.24 shows how MATS+ detects the fault cell (2, 1) SA1. The fault is detected by march element M1 as it moves from the lowest memory address upward and expects to read a 0 in cell (2, 1), but instead gets a 1. Figure 9.25 shows how MATS+ detects the multiple address decoder fault, where cell (2, 1) is unaddressable, and address (2, 1) maps instead to an access of cell (3, 1). This is multiple fault type C, which is a combination of address decoder faults 1 and 3. Since all writes to cell (2, 1) have no effect, and any read of cell (2, 1) produces a random result, the defective cell will be

Table 9.15: Irredundant march test algorithms.

Algorithm	Description	Ref.
MATS	{ $\updownarrow (w0); \updownarrow (r0, w1); \updownarrow (r1)$ }	[362, 490]
MATS+	{ $\updownarrow (w0); \Uparrow (r0, w1); \Downarrow (r1, w0)$ }	[4, 736]
MATS++	{ $\updownarrow (w0); \Uparrow (r0, w1); \Downarrow (r1, w0, r0)$ }	[688]
MARCH X	{ $\updownarrow (w0); \Uparrow (r0, w1); \Downarrow (r1, w0); \updownarrow (r0)$ }	[688]
MARCH C−	{ $\updownarrow (w0); \Uparrow (r0, w1); \Uparrow (r1, w0);$ $\Downarrow (r0, w1); \Downarrow (r1, w0); \updownarrow (r0)$ }	[429]
MARCH A	{ $\updownarrow (w0); \Uparrow (r0, w1, w0, w1); \Uparrow (r1, w0, w1);$ $\Downarrow (r1, w0, w1, w0); \Downarrow (r0, w1, w0)$ }	[646]
MARCH Y	{ $\updownarrow (w0); \Uparrow (r0, w1, r1); \Downarrow (r1, w0, r0); \updownarrow (r0)$ }	[688]
MARCH B	{ $\updownarrow (w0); \Uparrow (r0, w1, r1, w0, r0, w1);$ $\Uparrow (r1, w0, w1); \Downarrow (r1, w0, w1, w0); \Downarrow (r0, w1, w0)$ }	[646]

detected either by march element M1 when it reads cell (2, 1) (if the read returns a 1 when a 0 was expected), or by march element M2 when it reads cell (2, 1) (if the read returns a 0 when a 1 was expected.) In Figure 9.25(e), march element M1 writes a 1 to cell (2, 1), but that instead has the effect of writing cell (3, 1). This is detected when element M1 operates on cell (3, 1), because it first reads a 0 (but gets an unexpected 1), and then it writes a 1 to the cell. If, instead, the address of cell (3, 1) mapped into an access of cell (2, 1), then march element M2 would detect this error as it descended from highest to lowest addresses in memory. It would expect to read a 1 from cell (2, 1), but would get a 0 instead.

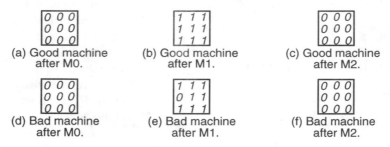

Figure 9.23: MATS+ detection of cell (2, 1) SA0 fault.

9.7.2 Testing RAM Neighborhood Pattern-Sensitive Faults

Assumptions and Testing Requirements. We always assume that read operations of memory cells are fault-free in the NPSF testing algorithms in order to make them practical. Note, in particular, that this is not necessarily true for present-day DRAM technology. Here is the necessary condition to detect and locate an *active neighborhood pattern sensitive fault* (ANPSF): *Each base cell must be read in state 0 and state 1, for all possible transitions in the deleted neighborhood pat-*

9.7 Memory Testing

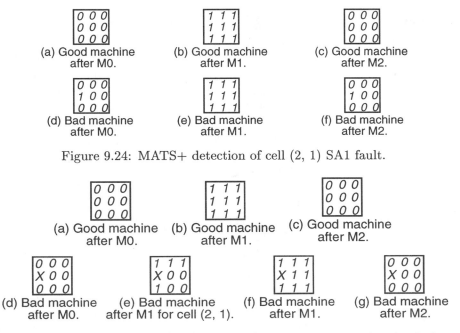

Figure 9.24: MATS+ detection of cell (2, 1) SA1 fault.

Figure 9.25: MATS+ detection of cell (2, 1) multiple address decoder faults.

tern [688]. There are two different possible values for the base cell (0 and 1), $k-1$ ways of choosing the deleted neighborhood cell which must undergo one of two possible transitions ($\uparrow$ or $\downarrow$), and 2^{k-2} possibilities for the remaining neighborhood cell contents. The total number of *active neighborhood patterns* (ANPs) is $2 \times (k-1) \times 2 \times 2^{k-2} = (k-1) \times 2^k$. Table 9.16 [688] shows all possible patterns for a Type-1 neighborhood.

Here is the necessary condition to detect and locate a *passive neighborhood pattern sensitive fault* (PNPSF): For each of the 2^{k-1} deleted neighborhood patterns, the two possible transitions $\uparrow$ and $\downarrow$ must be verified. Therefore, the total number of PNPSFs is $2 \times 2^{k-1} = 2^k$ (see Table 9.16 [688].) The total pattern count for APNPSFs is therefore $(k-1) \times 2^k + 2^k = k \times 2^k$.

Here is the necessary condition to detect and locate a *static neighborhood pattern sensitive fault* (SNPSF): We must apply the 2^k combinations of 0s and 1s to the k-cell neighborhood, and verify by reading each cell that each pattern can be stored. Hayes [286] states that transition writes, non-transition writes, and read operations can cause PSFs.

Optimal Write Sequences. It is essential to minimize the number of *writes* during NPSF testing, in order to obtain the shortest possible test.

Eulerian and Hamiltonian Sequences. A *Hamiltonian sequence* is used for writing during SNPSF tests. Patterns in a k-bit Hamiltonian sequence differ by only

Table 9.16: *Active* (ANP) and *passive* (PNP) neighborhood patterns.

b	0000000000000000111111111111111	0000000000000000111111111111111
0	↑↑↑↑↑↑↑↑↓↓↓↓↓↓↓↓↑↑↑↑↑↑↑↑↓↓↓↓↓↓↓↓	0000111100001111000011110000111
1	0000111100001111000011110000111	0011001100110011001100110011001
3	0011001100110011001100110011001	0101010101010101010101010101010
4	0101010101010101010101010101010	↑↑↑↑↑↑↑↑↓↓↓↓↓↓↓↓↑↑↑↑↑↑↑↑↓↓↓↓↓↓↓↓
b	0000000000000000111111111111111	↑↑↑↑↑↑↑↑↑↑↑↑↑↑↑↑↓↓↓↓↓↓↓↓↓↓↓↓↓↓↓
0	0000111100001111000011110000111	0000000011111111000000001111111
1	↑↑↑↑↑↑↑↑↓↓↓↓↓↓↓↓↑↑↑↑↑↑↑↑↓↓↓↓↓↓↓↓	0000111100001111000011110000111
3	0011001100110011001100110011001	0011001100110011001100110011001
4	0101010101010101010101010101010	0101010101010101010101010101010
b	0000000000000000111111111111111	
0	0000111100001111000011110000111	
1	0011001100110011001100110011001	
3	↑↑↑↑↑↑↑↑↓↓↓↓↓↓↓↓↑↑↑↑↑↑↑↑↓↓↓↓↓↓↓↓	
4	0101010101010101010101010101010	

1 bit from their preceding pattern, as this minimizes the number of writes needed to generate the patterns. The *Gray code* is a Hamiltonian sequence. ANPSFs, PNPSFs, and APNPSFs are tested with an *Eulerian sequence*. An *Eulerian graph* has a node for each k-bit pattern of 0s and 1s and there is an arc between two nodes, if and only if they differ by exactly one bit [688]. When two nodes are connected, they are connected by only two arcs (see Figure 9.26 [688].) The arcs in this graph correspond to *active neighborhood patterns* (ANPs), *passive neighborhood patterns* (PNPs), and *active and passive neighborhood patterns* (APNPs) of a k-bit neighborhood. An *Eulerian sequence* traverses each arc in the graph exactly once, while a *Hamiltonian sequence* traverses each node in the graph exactly once. We use these sequences to obtain the minimal number of pattern writes.

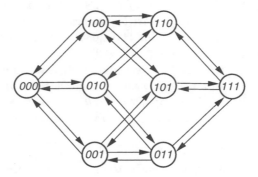

Figure 9.26: Eulerian graph.

A Hamiltonian sequence lets one apply all 2^k SNPs to a neighborhood with k initial writes for the first pattern, followed by $2^k - 1$ writes for the remaining $2^k - 1$

9.7 Memory Testing

patterns. An Eulerian sequence lets one apply all $k \times 2^k$ APNPs to a neighborhood with k writes to initialize the neighborhood, and 1 additional write for each of the $k \times 2^k$ APNPs.

Testing Neighborhoods Simultaneously. When a cell is written, we change k different neighborhoods (Type-1 or Type-2.) We wish to test the neighborhoods simultaneously, using the *tiling* and *two-group* methods.

Tiling Method. The *tiling method* totally covers memory with non-overlapping neighborhoods. This is known as a *tiling group*, and the set of all neighborhoods in the group is called the *tiling neighborhoods* [688]. Figure 9.10 [688] depicts this for a 5-element Type-1 neighborhood. Cell 2 is always the base cell, and the deleted neighborhood cells are numbered as shown. Figure 9.11 [688] shows the tiling for a 9-element Type-2 neighborhood, where cell 4 is the base cell. In a Type-1 neighborhood, when we tessellate the neighborhoods, we have $\frac{n}{5}$ base cells. However, while we are applying all of the test patterns to the $\frac{n}{5}$ base cells 2, it turns out that we also apply the appropriate patterns to the memory when cells 0 are base cells, cells 1 are base cells, cells 3 are base cells, and cells 4 are base cells. This reduces the pattern length from $n \times 2^k$ patterns to $\frac{n}{k} \times 2^k$ patterns. Under this method, each cell is simultaneously a base cell and a deleted neighborhood cell for other base cells (see Figure 9.27 [688].) A similar argument and reduction holds for Type-2 neighborhoods. This argument breaks down when we test ANPSFs and PNPSFs separately, because with those patterns the k cells in the neighborhood are not treated identically.

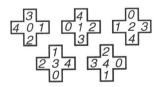

Figure 9.27: Base cell 0, 1, 2, 3, and 4 Type-1 tiling neighborhoods.

Two-Group Method. For the *two-group method*, a cell is simultaneously a base cell in one group and a deleted cell in the other group, and vice versa (see Figure 9.28 [688].) With this *duality* property, cells are divided into two groups, *group-1* and *group-2*, in a checkerboard pattern. Base cells of group-1 are deleted neighborhood cells of group-2, and vice versa. Each group has $n/2$ base cells b and $n/2$ deleted neighborhood cells formed by 4 subgroups A, B, C, and D. This only works for Type-1 neighborhoods, because in Type-2, 9 cell neighborhoods, there are both middle cells 1, 3, 5, and 7 and corner cells 0, 2, 6, and 8. Unfortunately, duality (a cell is either a base cell or a deleted neighborhood cell) does not hold here.

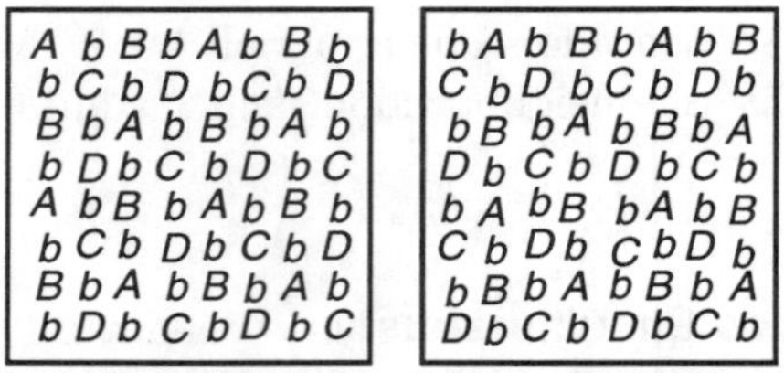

```
A b B b A b B b      b A b B b A b B
b C b D b C b D      C b D b C b D b
B b A b B b A b      b B b A b B b A
b D b C b D b C      D b C b D b C b
A b B b A b B b      b A b B b A b B
b C b D b C b D      C b D b C b D b
B b A b B b A b      b B b A b B b A
b D b C b D b C      D b C b D b C b
```

(a) Labels of cells of   (b) Labels of cells of
    group-1.   group-2.

Figure 9.28: Cell labels in the two-group method.

Basic Algorithms. Figure 9.29 [688] presents the basic NPSF fault location algorithm and Figure 9.30 [688] presents the basic fault detection algorithm. Table 9.17 [688] gives algorithm performance metrics and Table 9.18 [688] summarizes the detailed NPSF algorithms presented here.

Step 1: write *base-cells* with 0;
Step 2: loop
        apply *a pattern*; { it could change the *base-cell* from
                0 to 1. }
        read *base-cell*;
        endloop;
Step 3: write *base-cells* with 1;
Step 4: loop
        apply *a pattern*; {it could change the *base-cell* from
                1 to 0. }
        read *base-cell*;
        endloop;

Figure 9.29: NPSF location algorithm.

APNPSF Testing Algorithms. We will discuss the TLAPNPSF1G algorithm, which uses the two-group method and writes onto cells in group-1 or group-2 with the same label. Table 9.19 [688] shows the length 64 optimal 4-bit Eulerian graph sequence used by this algorithm. This table gives the *deleted active neighborhood patterns* (DANPs), which are the *active neighborhood patterns* (ANPs) without the base cell values. The algorithm has two passes: the first initializes base cells of groups 1 (2) with 1s (0s) in a checkerboard fashion. The second pass initializes base cells with all 0s (see Figure 9.31 [688].)

Step (1) initializes the memory for the first pass, and then reads all cells to verify the initialization. In Step (2), we apply two different Eulerian sequences, one starting with all 0s and the other starting with all 1s. This is needed because all cells in group-1 are initialized to 1 and those in group-2 are initialized to 0. The statement $write(i,j), j \in \{1,2\}$ changes the contents of the deleted neighborhood of all cells in group-j. The ↑ or ↓ symbol in entry i of Table 9.19 determines which

Step 1: write *base-cells* with 0;
Step 2: loop
　　　　apply *a pattern*; { it could change the *base-cell* from 0 to 1. }
　　　endloop;
　　　read *base-cell*;
Step 3: write *base-cells* with 1;
Step 4: loop
　　　　apply *a pattern*; {it could change the *base-cell* from 1 to 0. }
　　　endloop;
　　　read *base-cell*;

Figure 9.30: NPSF detection algorithm.

Table 9.17: NPSF testing algorithm summary.

Algorithm	Fault location?	Fault coverage					Operation count
		SAF	TF	\multicolumn{3}{c	}{NPSF}	
				A	P	S	
TDANPSF1G	No	L		D			$163.5 \cdot n$
TLAPNPSF1G	Yes	L	L	L	L	L	$195.5 \cdot n$
TLAPNPSF2T	Yes	L	L	L	L		$5122 \cdot n$
TLAPNPSF1T	Yes	L	L	L	L		$194 \cdot n$
TLSNPSF1G	Yes	L				L	$43.5 \cdot n$
TLSNPSF1T	Yes	L				L	$39.2 \cdot n$
TLSNPSF2T	Yes	L				L	$569\frac{7}{9} \cdot n$
TDSNPSF1G	No	L				D	$36.125 \cdot n$

L: Locates faults.　　　　　　　　　　D: Detects faults.

Table 9.18: NPSF testing algorithms.

Algorithm	Neighborhood	Method	Fault location?	Figure	k	Ref.
TDANPSF1G	Type-1	2 Group	No	None	5	[646]
TLAPNPSF1G	Type-1	2 Group	Yes	9.31	5	[688]
TLAPNPSF2T	Type-2	Tiling	Yes	None	9	[688]
TLAPNPSF1T	Type-1	Tiling	Yes	None	5	[688]
TLSNPSF1G	Type-1	2 Group	Yes	None	5	[688]
TLSNPSF1T	Type-1	Tiling	Yes	9.32	5	[189]
TLSNPSF2T	Type-2	Tiling	Yes	None	9	[688]
TDSNPSF1G	Type-1	2 Group	No	None	5	[688]

cell A_i is changed. The inverse of $write(i,j)$ for cells of the other group $(3-j)$ is $write((i + 52 + 33p + 58q) \mod 65, 3 - j)$, where $p = 0, q = 0$ for $1 \leq i \leq 12$; $p = 1, q = 0$ for $13 \leq i \leq 32$; $p = 0, q = 1$ for $33 \leq i \leq 52$; and $p = 1, q = 1$ for $53 \leq i \leq 64$. The Step (2) read operations check various NPSFs. The one at label L2 checks for PNPSFs of group-2. The one at L3 tests the base cells of group-1 for ANPSFs. Checks are made for PNPSFs (ANPSFs) at label L4 (L5) for group-1 (group-2.)

At Step (3), the group-1 base cells are initialized again to 0s and read along with base cells in group-2 (which are already 0.) At Step (4), a single Eulerian sequence is traced by $write(i, 1)$, followed by 4 successive read operations to test PNPSFs, ANPSFs, PNPSF, and ANPSFs for group-2, group-1, group-1, and group-2, respectively. Therefore, all possible APNPSFs with an initial base cell value of 1 in group-1 cells are tested during Steps (1) and (2). All possible APNPSFs with an initial base cell value of 0 in group-1 cells are tested during Steps (3) and (4). The strength of this algorithm is that it generates all ANPs and PNPs for each neighborhood, and tests any cell whose contents change or that is affected by a change in another cell. Tests are done before and after the changes are made.

Table 9.19: Optimal write sequence for TLAPNPSF1G.

Oper	DANP 0134	Oper	DANP 0134	Oper	DANP 0134	Oper	DANP 0134
1	000↑	17	0↑00	33	00↑0	49	10↓0
2	00↑1	18	↑100	34	001↑	50	1↑00
3	0↑11	19	110↑	35	↑011	51	11↑0
4	01↓1	20	11↑1	36	1↑11	52	111↑
5	010↓	21	↓111	37	1↓11	53	111↓
6	01↑0	22	011↓	38	↓011	54	11↓0
7	↑110	23	0↓10	39	001↓	55	1↓00
8	1↓10	24	↑010	40	00↓0	56	10↑0
9	↓010	25	1↑10	41	↑000	57	101↑
10	0↑10	26	↓110	42	100↑	58	10↓1
11	011↑	27	01↓0	43	1↑01	59	↓001
12	↑111	28	010↑	44	↓101	60	0↑01
13	11↓1	29	01↑1	45	0↓01	61	↑101
14	110↓	30	0↓11	46	↑001	62	1↓01
15	↓100	31	00↓1	47	10↑1	63	100↓
16	0↓00	32	000↓	48	101↓	64	↓000

SNPSF Testing Algorithms. TLSNPSF1T [189] is an algorithm that locates SNPSFs, using the tiling method and a 5-bit Hamiltonian sequence ($k = 5$.) It writes the seed pattern with n writes (see Figure 9.32), and then uses a Hamiltonian sequence in each of $\frac{n}{2}$ tiling neighborhoods to write the remaining $2^k - 1$ successive

9.7 Memory Testing

(1) initialize *all cells-1 with 1 and cells-2 with 0*;
  L1: read *them out*;
(2) $p := 0; q := 0$;
  for $i := 1$ to 64 do
  begin
    if $i = 13$ then $p := 1, q := 0$; { All cells-2 are 0, $j = 64$ }
    if $i = 33$ then $p := 0, q := 1$; { All cells-2 are 1, $j = 52$ }
    if $i = 53$ then $p := 1, q := 1$; { All cells-2 are 0, $j = 32$ }
    write $(i, 1)$; { Write deleted neighborhood cells of group-1 }
  L2: read *all cells written by this write operation*;
      { Test group-2 for PNPSFs }
  L3: read *all cells-1*; { Test group-1 for ANPSFs }
    $j = (i + 52 + p * 33 + q * 58) \bmod 65$;
    write $(j, 2)$; { This is the inverse of write $(i, 1)$ }
  L4: read *all cells written by this write operation*;
      { Test group-1 for PNPSFs }
  L5: read *all cells-2*; { Test group-2 for ANPSFs }
  end;
(3) write 0 *in all cells-1*;
  read *all cells-1 and all cells-2*;
(4) { NPSF test with initial value of base cells: cells-1 = 0 and cells-2 = 0 }
  for $i := 1$ to 64 do
  begin
    write $(i, 1)$; { Write deleted neighborhood cells of group-1 }
  L6: read *all cells written by this write operation*;
      { Test group-2 for PNPSFs }
  L7: read *all cells-1*; { Test group-1 for ANPSFs }
    write $(i, 2)$; { This is the inverse of write $(i, 1)$ }
  L8: read *all cells written by this write operation*;
      { Test group-1 for PNPSFs }
  L9: read *all cells-2*; { Test group-2 for ANPSFs }
  end;

Figure 9.31: TLAPNPSF1G algorithm.

patterns in $(2^k - 1) \times \frac{n}{k}$ cell writes. The total complexity is $39.2n$. This algorithm can also be used on a Type-2 neighborhood.

(1) initialize *all cells with* 0; *read* 0 from all cells;
(2) for $j := 1$ to $2^k - 1$ do
    begin
        write *pattern no. j of the chosen Hamiltonian sequence*;
        { This is done by flipping a bit (i.e., cell) of pattern no. $j - 1$ }
        read *all cells*;
    end;

Figure 9.32: TLSNPSF1T algorithm.

9.7.3 Testing RAM Technology and Layout-Related Faults

The problems with the prior memory testing approaches are that DRAMs may be repaired or may have their address lines deliberately scrambled. As a result, consecutive addresses may not be adjacent, so the previously described coupling fault tests will not be effective. Also, the Gigabit DRAMS have new kinds of defects, which the march or NPSF tests may not cover. It is now common to use *inductive fault analysis* (IFA) [420, 596] to model a single defect per memory cell. Here is the procedure:

1. Generate defect sizes, locations, and layers based on what might actually happen in the real fabrication line.

2. Place the defects on a model of the layout.

3. Extract the schematic and electrical parameters for the defective cell.

4. Evaluate the results of testing the defective cell, possibly performing Monte-Carlo analysis using a tool such as VLASIC [711].

Dekker performed this analysis [193, 194, 195] and found faults caused by actual defects modeled as broken wires, shorts between wires, missing contacts, extra contacts, and newly-created parasitic transistors. He mapped these defects into the following functional faults:

1. SAF in a memory cell.

2. A *stuck-open fault* (SOF) in a memory cell.

3. A TF in a memory cell.

4. A *state coupling fault* (SCF) between two memory cells.

5. A CFid between two cells.

9.7 Memory Testing

6. A *data retention fault* (DRF), caused by a broken pull-up device, in which the cell loses its contents over time [193, 688].

He proposed the augmented march test IFA-9 [193] to test for these faults in combinational read/write logic. The new march element, *Delay*, means wait for 100 *ms* before continuing, in order to test for data retention faults (*sleeping sickness*). The extension to sequential read/write logic is IFA-13 [193], which also tests for SOFs. These tests are summarized in Table 9.20 and described in Table 9.21. The fault model was validated with a scanning electron microscope.

Table 9.20: IFA augmented march test summary.

Algorithm	Actual physical defect fault coverage							Operation count
	SAF	TF	AF	SOF	SCF	CFid	DRF	
IFA-9	All	All	All		All	All	All	$12 \cdot n +$ Delays
IFA-13	All	All	All	All	All	All	All	$16 \cdot n +$ Delays
MARCH G	All	All	All		All linked CFids All TFs linked with CFids or CFins Some CFins linked with CFids		All	$23 \cdot n +$ Delays

Table 9.21: IFA augmented march test algorithms.

Algorithm	Description	Ref.
IFA-9	$\{ \Uparrow (w0); \Uparrow (r0, w1); \Uparrow (r1, w0); \Downarrow (r0, w1);$ $\Downarrow (r1, w0); Delay; \Uparrow (r0, w1); Delay; \Uparrow (r1) \}$	[193]
IFA-13	$\{ \Uparrow (w0); \Uparrow (r0, w1, r1); \Uparrow (r1, w0, r0); \Downarrow (r0, w1, r1);$ $\Downarrow (r1, w0, r0); Delay; \Uparrow (r0, w1); Delay; \Uparrow (r1) \}$	[193]
MARCH G	$\{ \Updownarrow (w0); \Uparrow (r0, w1, r1, w0, r0, w1);$ $\Uparrow (r1, w0, w1); \Downarrow (r1, w0, w1, w0); \Downarrow (r0, w1, w0);$ $Delay; \Updownarrow (r0, w1, r1); Delay; \Updownarrow (r1, w0, r0) \}$	[532]

Dekker [195] proposed these new fault models, and Meershoek [456] validated Dekker's fault models of Table 9.13. He compared the IFA march test performance against conventional march tests in Table 9.22. A higher score in this table signifies a greater fault coverage of the actual layout and fabrication process faults that occur. Meershoek has proposed a shorter IFA-6 test, and it performed very slightly worse than the IFA-13 test in this table, in that it failed 3792 defective devices, while IFA-13 failed 3795 defective devices. Analysis showed that IFA-13 can detect the single CFid faults $<\uparrow; 0>$, $<\uparrow; 1>$, $<\downarrow; 0>$, and $<\downarrow; 1>$ whereas IFA-6 can only detect linked coupling faults, in which both cells influence each other.

9.7.4 RAM Test Hierarchy

Multiple faults are most likely, as is fault linkage, during DRAM manufacturing. Table 9.23 summarizes which tests to use when a given set of faults is expected during

Table 9.22: Validation of IFA tests.

Test	Score	Test time
MATS+	7	$5 \cdot n$
MATS+ and Delay	18	$8 \cdot n + 2 \cdot Delay$
MARCH C	61	$11 \cdot n$
MARCH C and Delay	89	$14 \cdot n + 2 \cdot Delay$
IFA-9 and Delay	91	$12 \cdot n + 2 \cdot Delay$
IFA-13	80	$13 \cdot n$
IFA-13 and Delay	92	$16 \cdot n + 2 \cdot Delay$

memory production.

9.7.5 Cache RAM Chip Testing

Cache DRAM (CDRAM) [205, 742] is a promising high-speed memory that combines static RAM and dynamic RAM onto the same chip. This essentially combines the cache memory and the slower main memory of a virtual memory system on a single chip, which lessens the bus data transfer bottleneck when a block must be transferred between main memory and the cache. Since the bus is now an internal chip bus, rather than a chip-to-chip bus, it runs many times faster. There are additional applications of cache DRAM, such as multi-media or mass storage replacement [634], as well.

Figure 9.33 [365] shows the cache DRAM functional block diagram. Its organization is 256K words × 16 bits of DRAM core, 8 words × 16 bits of *read data buffer* (RB), 8 words × 16 bits of *write data buffer* (WB), and 1k words × 16 bits of SRAM. The DRAM and SRAM parts can operate concurrently and independently, if desired, because address pins for the SRAM and the DRAM are completely separated (As0-9 versus Ad0-9) in order to improve system performance. Also, SRAM and DRAM control pins are also separated (CC0#, CC1#, WE#, CMs#, RAS#, CAS#, DTD#, CMd#.)

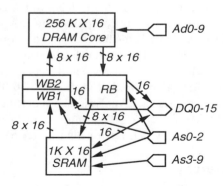

Figure 9.33: Block diagram of 256K × 16 b Cache DRAM.

9.7 Memory Testing

Table 9.23: Test hierarchy.

Faults in memory	Unlinked fault test	Linked fault test
SAF	MATS, MATS+	MATS, MATS+
TF	MATS++	MATS++
CFin	MARCH X	†
CFid	MARCH C−	MARCH A
CFdyn	MARCH C−	
SCF	MARCH C−	
SOF	IFA-13	
DRF	IFA-13	
ANPSF	TLAPNPSF1G	
PNPSF	TLAPNPSF1G	
SNPSF	TLAPNPSF1G or TLSNPSF1T	
AF	MATS, MATS+, MATS++, IFA-13 MARCH X, C−, A, B, or Y	
AF + SAF + TF	MATS++	MATS++
AF + SAF + CFin	MARCH X	MARCH X †
AF + SAF + CFid	MARCH C−	MARCH A
AF + SAF + CFin + CFid	MARCH C−	MARCH A ‡
AF + SAF + TF + CFin	MARCH X	MARCH Y †
AF + SAF + TF + CFid	MARCH C−	MARCH B
AF + SAF + TF + CFin + CFid	MARCH C−	MARCH B ‡
AF + SAF + TF + CFin + CFid + CFdyn + SCF	MARCH C−	
SAF + ANPSF	TLAPNPSF1G	
SAF + TF + PNPSF	TLAPNPSF1G	
SAF + SNPSF	TLAPNPSF1G or TLSNPSF1T	
SAF + TF + APNPSF + SNPSF	TLAPNPSF1G	
AF + SAF + SOF + TF + SCF + CFid + DRF	IFA-13	

† Linked CFins cannot be tested by march tests.
‡ We assume that CFins are not linked to other CFins.

Tests Performed.

1. *DRAM Functional Test.* Konishi [365] connects the data input/output lines to the SRAM, the WB, and the RB, and not to the DRAM core. As a result, data to and from the DRAM moves as a 16 bit × 8 word block, via the WB and the RB. When DRAM core memory cells are tested, the RB and the WB are also automatically tested. SRAM address bits As0-2 select 1 of 8 words in the data buffers as the DRAM column address. An additional signal WE# was provided to control the WB and the RB. In other respects, testing the DRAM core proceeds similarly to testing a 256K × 16 synchronous DRAM with a burst length of 8. Stalnaker [634] recommends the use of standard pattern libraries, such as march and scan tests.

2. *SRAM Functional Test* takes negligible time, and is equivalent to testing a 1K × 16 synchronous SRAM.

3. *Data Transfer Test.* The data transfer between the DRAM and the RB or WB is tested, and also the data transfer between the SRAM and the RB or WB is tested.

4. *High-Speed Operation Test* is performed at frequencies of up to 100 MHz. The dual *pattern generators* (PGs) and the pin multiplexing features of the ATE are used. In order to operate a driver at 100 MHz with a 50 MHz clock in the ATE, two clock pin data lines are multiplexed together as one, with one ATE PG dedicated to each. This is called *pin multiplexing*. This allows 100 MHz clock rate testing with a 50 MHz ATE, which is a significant cost saving.

5. *Concurrent Operation Test.* Before this test, every chip component was verified. All that remains is to test the data transfer from the DRAM to the RB while the SRAM is being accessed (this is also known as a *contention test*.) The dual ATE PGs are used, one to generate the SRAM address and control signals, and the other to generate the corresponding DRAM signals.

6. *Cache Miss Test* [634]. During a cache (SRAM) read cycle, the hardware compares the row address to the cache directory. If the addresses match, we have a cache "HIT," and the data will be read from the cache at its fast access time. For a cache write cycle, data can be read from the cache (to free up a cache block) and both the DRAM array and the SRAM cache are written when the WE# signal is pulsed. When the row address is not in the cache directory, we have a cache "MISS," and this operation becomes similar to a normal DRAM cycle. For a read operation, data from the new row is loaded into the cache and the cache directory is updated. For a write cycle, the new data is written into the DRAM, but not into the cache. Testing requires that simple patterns be written to verify each of these "HIT" or "MISS" conditions. It is only necessary to maintain the cache directory with one address register while using the other address register for DRAM memory writing.

9.7 Memory Testing

Testing Extremely Fast DRAMS. The Rambus^{TM} DRAMs now have the fastest pin bandwidth of any available DRAM, since they combine a large DRAM with an extremely fast bus. The problem of testing these devices is to utilize conventional page-mode memory testers to avoid a massive investment in extremely expensive ATE [240].

Their approach was to utilize existing ATE for die-sort, burn-in, final test, and failure analysis. Rambus utilizes high-speed ATE only for the final test of the high-speed interface logic. This approach requires *design-for-testability* (DFT) hardware to be put on the Rambus^{TM} DRAM. A special test mode allows direct memory core access from the pins, which bypasses the high-speed bus and allows existing ATE to be used for memory testing. Secondly, a *phase-locked loop* (PLL) bypass mechanism is added to allow testing of the DRAM protocol logic by low-speed, conventional memory testers.

For final test, the Rambus^{TM} DRAM interface is verified at-speed at 500 MHz using a Hewlett-Packard HP-8300, model F660 tester, which can clock at rates up to 660 MHz. The data pin rate must be 2 ns/bit, and this imposes a 100 ps accuracy requirement on the tester. This final test is low cost, because it only is applied to the interface logic, and therefore uses $< 100K$ vectors. Various timing and voltage corner tests require only a few seconds of tester time, so the added test cost is a few cents per device.

Gasbarro [239] believes that high-speed DRAM will become more prevalent, and that tester costs may actually decline, because memory testers can be built to take advantage of the high-speed port into the DRAM. Also, new testers may be designed to test both the core and the interface at speed. Also, a dramatic ATE cost reduction is possible by using high-speed DRAM chips within the ATE itself.

However, the analog aspects of high-speed DRAM testing are daunting [240]. These test issues were addressed:

1. The *device-under-test* (DUT) socket had to be redesigned for low inductance.

2. Long cables from device to tester degrade analog signals, and had to be shortened.

3. Noise causes jitter in the DRAM PLL, which requires testing.

4. Very high ATE accuracy is required, so the DUT fixture utilized a *dual transmission line* (DTL) with impedances matched with both the driver and the receiver. Rambus [240] designed a 50 Ω micro-strip probe card on which the DRAM device was mounted either by directly clamping its leads to the micro-strip or by using a 3 GHz bandwidth socket and a custom, open-cavity ceramic package. A *time domain reflectometer* (TDR) with ±80 ps accuracy in the ATE was calibrated to both drivers and receivers. However, the shorting device on the probe card had to be redesigned to reduce an unacceptable 300 ps error. Also, the receiving end of the transmission line had to be terminated at the logic V_{th} voltage, rather than V_{SS}, to further reduce error. The trace length had to be shortened by redesigning the probe board, and a new probe board

dielectric had to be used to make the high frequency loss acceptable. These corrections brought the error from an unacceptable ± 300 ps to an acceptable ± 120 ps.

5. The tester had to be adjusted to test various critical paths in the high-speed interface for propagation delay.

6. The TDR had to be used to characterize the DRAM pin capacitance.

This experience illustrates the complexity and high engineering expertise required for high-speed delay testing with an external ATE.

9.7.6 Functional ROM Chip Testing

ROM testing differs from RAM testing, in that the correct data that the ROM should contain is already known. The SAF model used for ROMs is sometimes a *restricted* SAF model in that only *undirectional SA faults* [604] can occur, meaning that any given chip will either have only SA0 faults or only SA1 faults. This is based on a ROM fault model where only "opens" occur, which are missing connections resulting in either all SA0 faults or all SA1 faults. The ROM functional testing model is that of Figure 9.5 with the write logic deleted from the read/write logic block. Please review Appendix A on *cyclic redundancy code* (CRC) theory before reading further.

The preferred ROM testing method is to cycle the ROM through all of its addresses and compress the output bit stream at the ROM outputs using a *linear feedback shift register* (LFSR) in the *automatic test equipment* (ATE.) This system is based on *cyclic redundancy codes* (CRCs). The LFSR treats the output data stream(s) from the circuit as a Boolean polynomial(s), where the first bit received over time is the 2^0 position coefficient, the next is the 2^1 position coefficient, and so on. In this system, shifting by one clock period is equivalent to multiplying a signal by x. The LFSR (or *multiple-input signature register* (MISR) if there are multiple bits in each ROM word) divides the polynomial representing the Boolean output stream from the circuit by a *primitive* polynomial, and produces the Boolean number corresponding to the remainder of this polynomial division. After reading out all of the ROM data, the ATE accesses the ROM location containing the CRC code for the ROM. If the accessed ROM CRC code does not agree with the one computed while reading out the data, then the ROM is faulty.

Table 9.24 [688] compares the ROM testing effectiveness of parity checking, checksumming over Modulo-2^{N+B}, checksumming over Modulo-2^B, and a CRC. Parity requires storage of parity bits in the ROM, checksumming requires storage of check sums in the ROM, and the CRC method requires storage of the CRC in the ROM. *Redundancy* measures the hardware overhead percentage to store these extra bits. Notice that $G(x)$, the *generator polynomial* for the CRC, must have more than one non-zero term, must contain the term 1, must include the factor $x+1$, and must satisfy the condition $((x^k + 1)/G(x)) \neq 0$. Clearly, the CRC method is far superior to the others, usually in hardware redundancy and always in fault coverage, so we do not even discuss any other methods. CRC coding should always be used for ROM testing.

Table 9.24: Comparison of ROM chip tests.

ROM chip test	Redundancy †		Fault coverage
	General	$N = 10; B = 8$	
Parity checking	$1/(B+1)$	0.1111	Single-bit errors, odd-bit errors
Modulo-2^{N+B} checksum	$(N+B)/(B \cdot 2^N)$	0.0022	All unidirectional errors
Modulo-2^B checksum	$1/(2^N)$	0.0010	Various types of unidirectional errors depending on columns
CRC with CCITT polynomial	$16/(B \cdot 2^N)$	0.0020	Single-bit errors, double-bit errors, odd-bit errors, multiple adjacent errors

† For a ROM of 2^N data words of B bits each.

9.7.7 Electrical Parametric Testing

Figure 9.34 [442] shows a typical RAM system organization. Electrical testing is done without an understanding/analysis of the underlying fabrication process and chip layout. The exceptions to this are the technology and layout-related parametric faults: *sleeping sickness* and *voltage imbalance* in the divided RAM bit-line architecture.

DC Parametric Tests.

Voltage Bump Test. Power supply voltage fluctuations (*voltage bumps*) can cause the RAM to read out erroneous data, and DRAMs are rigorously tested for this. A *positive voltage bump* happens when V_{CC} during a write exceeds V_{CC} during a read (a read error.) DRAM capacitors have 40 to 50 fF, and this test measures whether any of those capacitors are shorted to the supply rail. A positive voltage bump causes 0s to be stored at too high a voltage, so they read out as 1s. Storage of 1s is not affected.

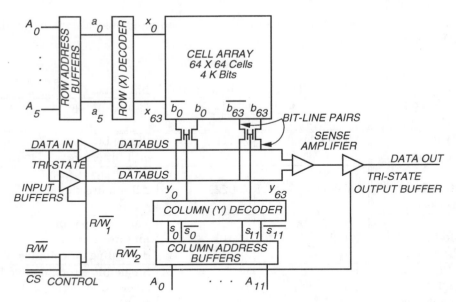

Figure 9.34: RAM system organization.

Method:
1. Zero out memory.
2. Increase the power supply above V_{CC} in steps of 0.01 V. For each voltage setting, read the memory. Stop as soon as a 1 is read from any location, and record this supply voltage as V_{high}.
3. Fill the memory with 1s.
4. Slowly decrease the supply below V_{CC} in steps of 0.01 V. For each setting, read the memory. Stop as soon as a 0 is read from any location, and record this supply voltage as V_{low}.
Possible test outcomes:
1. V_{high} and V_{low} are inconsistent with data book values (fails.)

DC parametric tests are simple and inexpensive, and measure worst case loading. They are inadequate as functional or timing tests, and only test a few devices connected to the terminals, leaving the rest of the chip untested.

Leakage Test. This finds the worst case input leakage current (*input leakage*) or tristated output leakage current (*tristated leakage*.)

9.7 Memory Testing

Method:
1. Apply a high to chip select, to deselect the chip.
2. Apply a high to output enable, and a low to write enable (set the chip pins to be tristated.)
3. Force a logic high on each data-out line and measure I_{OZ} (output leakage.)
4. Force a low voltage on each data-out line and measure I_{OZ}.
5. Select the chip (apply a low to chip select.)
6. Set the chip in read mode, force a high logic voltage on each address and data-in line, and measure I_I.
7. Set the chip in read mode, force a low logic voltage on each address and data-in line, and measure I_I.
Possible test outcomes:
1. $I_{OZ} < 10 \ \mu A$ and $I_I < 10 \ \mu A$ (passes.)
2. $I_{OZ} \geq 10 \ \mu A$ (fails.)
3. $I_I \geq 10 \ \mu A$ (fails.)

AC Parametric Tests. In AC parametric testing for SRAMs and DRAMs, we apply alternating voltages at some set of frequencies to the chip and measure the terminal impedance or dynamic resistance (reactance.) We select a DC bias level for these tests, which determine chip delays caused by input and output capacitances. However, the tests give no information on functional data or DC parameters.

Address Set-Up Time Sensitivity. This tests for excessive address decoder delay.

Method:
1. For every cell: a. Write a 1 in the first memory cell. b. Flip each address bit and write a 0 at the new address. c. Flip each bit back again and read the original cell. d. Repeat Steps a-c with complementary data.
Possible test outcomes:
1. The data read from the cell does not match the most recently written data to the cell (fails – excessive address set-up time.)

Access Time Tests. These tests characterize the RAM time delay between setting up a new address and retrieving its data. We can characterize the access time from an address with an external tester.

Method:
1. Split the memory into two halves.
2. Write 0s in one half and 1s in the other half.
3. Read the entire memory and compare it to the expected values.
4. Alternate between addresses in the two halves.
Possible test outcomes:
1. Speed up the read access time until reading fails, and then record the access time delay.

We can characterize the access time using a functional march test.

Characterization method:
1. Repeat the MATS++ test (or some other march test), with increasingly shorter access times, until memory fails. The true access time is between the times for the next-to-last and last tests.
Alternate characterization method:
1. Run a superset of the MARCH C− test. This measures access time, by successively reading two different addresses with opposite data (the worst case, since it captures address decoder + read logic delay.) Do this with increasingly shorter access times, until memory fails.
Production test method:
1. Run the MATS++ march test at the specified access time, and see if the memory fails.

Running Time Tests. These tests measure the fastest speeds for read/write operations.

Method:
1. Perform read operations of 0s and 1s from alternating addresses at a specified rapid speed.
Alternate characterization method:
1. Alternate read operations at increasingly rapid speeds until an operation fails.

Tests for Sense Amplifier Recovery Fault. Sense amplifiers can become saturated after reading/writing a long string of identical data values, at which point they are too slow to read the opposite data value.

Method:
1. Write the repeating pattern *dddddddd̄* to memory locations (*d* is 0 or 1.)
2. Read a long string of 0s (1s) starting at the first location up to the location with $\bar{d}$.
3. Read a single 1 (0) from the location with the $\bar{d}$.
4. Repeat Steps 2 and 3, but writing rather than reading in Step 2.

Test for Write Recovery Fault. Write recovery faults occur when a write is followed by a read/write at a different address. The two types are *read-after-write* and

9.7 Memory Testing

write-after-write.

Method:
1. Write a 0 at address a.
2. Complement all address bits and write a 1 at address $\bar{a}$.
3. Complement all address bits and read address a (reading a 1 indicates a fault.)
4. Repeat Steps 1 through 3 for each cell in one-half of the memory.
5. Repeat Steps 1 through 4 with complementary data.

Dual-Port SRAM Tests. Dual-ported SRAMs and DRAMs permit simultaneous access by two independent sets of address, control, and input/output lines. A video DRAM chip uses one port for the CPU and the other port for raster scan hardware for a PC or work station color display.

Standby Current Test.

Test method:
1. Check all 4 possibilities for the voltage combinations at the two ports. Four additional combinations come from each port having either TTL or CMOS level inputs.
Possible test outcomes:
1. The test fails if one port does not meet the current specification.

Circuit-Dependent Tests. Most memory testers cannot provide dual pattern generation and dual timing checking for dual-ported devices, so one of two *ad hoc* testing solutions is needed.

1. Tie one port address bus to the complement of the other. *Advantage*: There is no address contention. *Disadvantages*: This cannot test port arbitration circuits (no contention), and inverter delays reduce test timing accuracy.

2. X/Y Separation of Ports. Use the X address drivers to manipulate the left port, and the Y address drivers for the right one. This requires a tester with twice as many pins as the device, and with 6 clock drivers. *Advantage*: There is independent and simultaneous testing of both ports. *Disadvantages*: It is difficult to write topologically correct patterns into memory, and it is difficult to meet the address format control requirements. *Additional testing needs*: Software is needed to ensure writing of topologically correct patterns, and special test hardware is needed for format control for contention testing.

Interrupt test method:
1. Simultaneously write data into the interrupt location and monitor the $\overline{INT}$ output of the other port.

Arbitration Test. In multi-port RAMs, there is a contention circuit that continuously monitors addresses applied to the two ports, and activates a $\overline{BUSY}$ flag when they match. A set of semaphore latches controls which port gets *read* and which port gets *write* access when the two ports are in contention (access the same RAM address.) The contention circuit sets the semaphores, and arbitration depends only on relative signal timings.

1. Contention testing [535]. Each chip has a timing parameter called the *arbitration priority set-up time* (T_{aps}), which is the minimum time that the address applied to one port must be stable before the same address arrives at the other port, in order for the first to receive priority. The contention test must ensure that $T_{aps} = 5\ ns$. If the tester has separate X and Y timing generators, this can be guaranteed. Otherwise, we must use address formatting (see Figure 9.35 [442, 535].) We disable the Y-complementer exactly when the address has stopped arriving, so that the Y address leads the X address by T_{aps}.

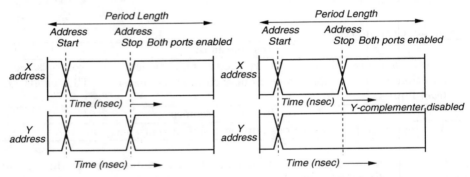

Figure 9.35: X/Y port separation for dual-port RAM testing.

2. Semaphore testing.

Method:
1. For each port, request, verify, and release each semaphore latch.

9.8 Summary

No single type of test (march, NPSF, DC parametric, AC parametric) is sufficient for current RAM testing needs, so a combination of various tests is used. Also, *inductive fault analysis* is now necessary, to ensure that the actual defects that are occurring are mapped into a fault model, and then appropriate tests can be selected for that fault model.

Problems

9.1 *State coupling faults.* Prove that the following MARCH C− test catches *state coupling faults.*

$$\{\ M0 :\updownarrow (w0);\ M1 :\Uparrow (r0, w1);\ M2 :\Uparrow (r1, w0); \tag{9.1}$$

$$M3 :\Downarrow (r0, w1); M4 :\Downarrow (r1, w0); M5 :\Updownarrow (r0) \}$$

9.2 *Address decoder faults.* Prove that a test for a *neighborhood pattern sensitive fault* cannot detect the *address decoder fault*, in which addresses a and b both access the contents C_b of location b. When location a is accessed, the memory produces a random combination of the contents of locations a and b (Hint: Try a proof by counter-example.)

9.3 *Transition faults.* What would happen if there were a $\uparrow$ *transition fault* ($<\uparrow /0>$) on the $\overline{CS}$ signal of a DRAM?

9.4 *Port arbitration faults.* Design a test for a Cache DRAM fault that causes the port arbitration hardware to be broken, such that both the SRAM and the DRAM access ports are allowed to write to the same location at the same time.

9.5 *ROM testing.* Analyze the time complexity of the CRC ROM testing algorithm.

9.6 *Graphs.* What is the difference between a Hamiltonian and an Eulerian graph traversal?

9.7 *Stuck-open faults.* Prove that the following IFA-13 march test algorithm detects all *stuck-open faults* in the memory.

$$\{ \Uparrow (w0); \Uparrow (r0, w1, r1); \Uparrow (r1, w0, r0); \Downarrow (r0, w1, r1); \quad (9.2)$$
$$\Downarrow (r1, w0, r0); Delay; \Uparrow (r0, w1); Delay; \Uparrow (r1) \}$$

9.8 *Test types.* Why are separate Probe and Contact Tests done during Electrical RAM chip Testing? The alternative would be to combine these tests with the Functional and Layout-Related Test.

9.9 *Idempotent coupling faults.* Rigorously prove that the MARCH C– test detects all idempotent coupling faults. Please indicate the testing time complexity for MARCH C– in terms of n, the number of bits in the memory.

9.10 *Fault modeling.* Please define the fault models for ONLY 8 of the following faults:

(a) State Coupling faults.
(b) Inversion Coupling faults.
(c) Idempotent Coupling faults.
(d) Dynamic Coupling faults.
(e) Transition faults.
(f) Active Neighborhood Pattern Sensitive faults.

(g) Passive Neighborhood Pattern Sensitive faults.

(h) Static Neighborhood Pattern Sensitive faults.

(i) Data Retention faults.

(j) Address Decoder faults.

9.11 *Memory test algorithms.* Rigorously prove that the MARCH C− test detects all inversion coupling faults $<\uparrow;\Updownarrow>$. Please indicate the testing time complexity for MARCH C− in terms of n, the number of bits in the memory.

9.12 *Stuck-at faults.* Rigorously prove that MATS++ catches all stuck-at faults.

9.13 *Dynamic coupling faults.* Rigorously prove that the MARCH C− test detects all dynamic coupling faults.

9.14 *Data retention faults.* Rigorously prove that the IFA-13 [193] test catches all data retention faults.

9.15 *SRAM physical faults.* What is the appropriate fault model(s) for a $\overline{bit}$ line shorted to a *word* line in an SRAM?

9.16 *DRAM physical faults.* What is the appropriate fault model(s) for two cell capacitors shorted together in a DRAM?

9.17 *Neighborhood PSFs.* Why cannot the two group method be used with Type-2 neighborhoods for a pattern sensitive fault test?

9.18 *Data retention faults.* What physical DRAM layout defect leads to a Data Retention Fault (*sleeping sickness*)?

9.19 *Write recovery faults.* What physical circuit problem leads to a sense amplifier Write Recovery Fault?

9.20 *Bridging faults.* Prove that a march test for a CFid will also detect the AND and OR bridging faults.

9.21 *State coupling faults.* Prove that a CFid march test will also detect state coupling faults.

9.22 *Neighborhood PSFs.* Write the steps (in pseudo-code) for a PNPSF test to detect the faults $<1,0,1,0;\uparrow/0>$ and $<0,1,0,1;\downarrow/1>$ using the two-group method and Type-1 neighborhoods. Your test need not be the optimal one.

9.23 *Neighborhood PSFs.* Write the steps (in pseudo-code) for a PNPSF test to detect the faults $<0,0,0,0;\uparrow/0>$ and $<1,1,1,1;\downarrow/1>$ using the two-group method and Type-1 neighborhoods. Your test need not be the optimal one.

Chapter 10

DSP-BASED ANALOG AND MIXED-SIGNAL TEST

> *"In the last few years, digital signal processing (DSP) has profoundly altered the design and use of automatic test equipment (ATE.) . . . the ATE computer, instead of simply controlling and monitoring hardware instruments, can now emulate and replace them."*
> — Matthew Mahoney [412].

The goals of testing are to discard defective devices and learn information to improve the fabrication process yield and reduce costs [699]. A *perfect* test fails all unacceptable units and passes all acceptable units. Perfect analog tests are overly expensive, so a *practical* test minimizes the number of bad units passed and the number of good units failed. Test processes are also used to sort parts into bins, according to their performance. Analog testing is more difficult than digital testing, because of the continuous range of analog circuit parameters, and the lack of well-accepted fault models.

10.1 Analog and Mixed-Signal Circuit Trends

Why do we need increased emphasis on analog testing? ICs with analog, digital, and mixed-signal circuits on the same substrate are now common [321]. Designers want to integrate analog and digital devices on the same chip to reduce circuit packaging and assembly costs. Applications include wireless communication, networking, multi-media information processing (sometimes in a personal computer), process control, and real-time control systems. The growth in these applications areas is explosive. Mixed-signal hardware systems have digital cores, frequently for digital signal processing, surrounded by analog filters, *A/D converters* (ADCs), and *D/A converters* (DACs.) Figure 10.1 shows the mixed-signal circuit testing problem. The analog portions interface the digital chip portions to the real world. Analog transistors and components are vastly larger than digital transistors, but the analog circuit contains fewer than 100 devices, whereas the digital part now contains

Chapter 10. DSP-BASED ANALOG AND MIXED-SIGNAL TEST

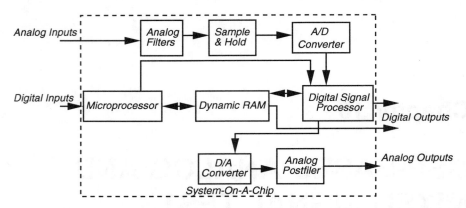

Figure 10.1: Mixed-signal testing problem and system-on-a-chip.

millions of devices. Mixed-signal circuits make the testing cost even more of a problem. Analog circuit signal observability is reduced in a mixed-signal system. Also, analog test has become a larger part of the system cost [584]. For *micro electromechanical systems* (MEMS), both mechanical and electromechanical components are integrated onto a single substrate with the analog and digital electronic signal processing system. An example is the air-bag controller chip now found in new automobiles, where various accelerometers exist on the same chip as the electronics that processes their signals. The advantage of this, from a system point of view, is reduced distance between the transducer and the point at which the measurement is taken. This leads to a more reliable measurement, since there is less opportunity for noise to interfere with the parameter reading. Other systems include chemical and optical sensors. At present, testing costs for analog circuits exceed 30% of the manufacturing cost of these circuits. Even more disconcerting is the certainty that these costs will rise in the near future. Analog *automatic test equipment* (ATE) is inherently much more costly than digital circuit ATE [465].

Designers are also moving from non-linear to linear analog circuits in their designs, since linear analog circuits are much easier to design and test than non-linear ones. Instead, designers move the non-linear signal processing function into a DSP processor in the digital part of the system. The benefits of this move are:

1. More accuracy, because a DSP signal processor has more accuracy and avoids various errors and information loss due to unwanted analog components or to analog components that are operating outside of their specifications.

2. Easier design and testing, because the remaining analog circuit portion is linear. However, incoming analog signals need linear filtering and A/D conversion, so there is a limit to how much analog circuitry can be eliminated. Also, there are significant tradeoffs between analog and digital implementations. An analog implementation of a filter can use significantly less power than a DSP processor, and it may be much faster. This could be critical for portable, wireless devices.

10.1 Analog and Mixed-Signal Circuit Trends

In addition, analog designers now may operate their analog MOSFET circuits in the *saturation* transistor region, rather than the *linear* transistor region. This is because channel length modulation in the saturation region provides a linear variation of channel current with respect to the drain-to-source voltage (V_{ds}), and the device behavior in saturation is often more linear than its behavior in the so-called linear transistor region. The saturation mode can be modeled as a current source in parallel with a resistor. However, this change introduces an additional testing requirement: We need to test for DC biasing faults in analog circuits, to ensure that the transistor operates in the intended transistor mode.

Contrast with Digital Testing. We repeat Vinnakota's contrast of analog testing with digital testing to illustrate the various problems [699].

1. *Size* is not a limitation. Analog circuits have relatively few devices, at most 50 to 100, unlike digital circuits, which have reached 50 million transistors. The number of analog inputs and outputs is small.

2. *Modeling* is far more difficult than in digital circuits:
   - There is no widely-accepted analog fault model like the digital stuck-at and path-delay fault models.
   - There is infinite analog signal range, and a range of good signal values.
   - Acceptable signal tolerances depend on process variations and measurement inaccuracies.
   - Modeling accuracy during analog fault simulation is crucial, unlike during digital fault simulation.
   - Analog circuits include noise, which must be modeled and also tested for.
   - Measurement error occurs, due to the load on the ATE pin, impedance of the pin, random noise, etc.
   - Capacitive coupling between digital and analog circuit substrates in a *mixed-signal circuit* (MSC) is another source of noise.
   - Absolute analog component tolerances can vary by ± 20%, but relative component matching can be as good as ± 0.1%. Circuit functionality is designed to depend on component ratios [699].
   - A multiple fault model is mandatory, but faults in multiple components can cancel each other's effect, so not every multiple fault is a real fault. The multiple fault set is too large to enumerate.
   - There is no unique direction of information flow in an analog circuit, whereas there usually is in digital CMOS circuits, with the exception of CMOS C-switch busses. The C-switch realizes a simple connect/disconnect switch between two sub-circuits [699].

3. *Decomposability*: Sub-components cannot be tested individually in an analog IC in the way that they can be in a digital IC.

4. *Test busses* are much harder to realize in analog than in digital circuits:

    - Transporting an analog signal to an output pin may alter the analog signal and the circuit functionality.
    - Reconfiguring an analog circuit during test is often unacceptable, unlike the case of digital circuit testing. The reconfiguration hardware can unacceptably change the analog circuit transfer function, even when it is turned off.

5. *Testing Methods*:

    - Both analog and digital circuits have *structural* ATPG methods, but because of a lack of well-accepted analog fault models and the lack of a mapping between structural analog faults and analog specifications, structural analog ATPG is not widely used.
    - *Specification-based* (*functional*) test methods exist for both analog and digital circuits. However, functional testing is rarely used for digital circuits, because the number of tests is intractable. Conversely, in analog testing, specification-based tests are most often used, because they are tractable and need no fault model.
    - Digital circuits can be tested separately for logic functionality (stuck-faults) and timing performance (path-delay faults.) However, these two types of tests cannot be separated in analog circuits, and are combined.

Present-Day Analog Testing Methods. Present analog circuit testing relies mostly on DSP-based analog testers. These instruments overcame the limits of purely analog instruments, and gave these additional benefits:

- *Accuracy*: The DSP-based ATE is nearly always more accurate than a pure analog test instrument, because the error in a set of waveform samples is much less after digital signal processing than the error in an individual sample. Crosstalk, noise, and signal drift are greatly reduced in the DSP-based tester, because analog waveforms are digitized at the earliest opportunity. Non-linearity in analog components is significantly reduced in the DSP-based tester, because the DSP processor is not subject to various analog device linearity errors, except in the front-end circuit. This contains a sample-and-hold device and a high-speed *A/D converter* (ADC), which have only linear analog subcircuits, so accuracy is improved. Aging of components is far less of a problem, because digital components are more likely to retain their specified performance as they age than analog components are. Thermal effects are less troublesome in the DSP-based tester than in the pure analog ATE. Analog components can exhibit a non-linear fluctuation in their transfer function with temperature.

- *Speed*: The DSP-based tester can acquire one set of samples for a *device-under-test* (DUT) using one relay switching period, one sampling period, and

10.1 Analog and Mixed-Signal Circuit Trends

one settling time. Then, analysis using a *discrete Fourier transform* (DFT) or *fast Fourier transform* (FFT) can produce many different emulated analog test instrument measurements. For multiple measurements, the DSP-based tester is more time efficient than using multiple analog instruments, *provided* that one needs to make multiple measurements. For example, the DSP-based ATE can do one FFT calculation, and then derive many emulated instrument measurements from that. The DSP-based tester eliminates some filter settling time that is inherent in the analog ATE.

- *Ease of Operation*: Testing is more repeatable between DSP-based ATE, because more of it is digital, rather than analog. Calibration of DSP-based ATE is simpler, again because more of the instrument is digital, rather than analog. Software can store offsets and gain adjustment factors and apply these automatically when measurements begin. Calibration is much more repeatable than with a pure analog tester. The DSP-based ATE has reduced maintenance, again because more of it is digital, rather than analog, so there are fewer continuous variable electrical adjustments to be made by a technician.

- *Modeling Convenience*: The DSP-based ATE can more readily model the ideal and flawed device than the pure analog ATE. The DSP-based ATE is extremely flexible. The operator can change test conditions from a computer console, rather than by adjusting many analog controls on the instrument.

- *More Measurement Information*: The DSP-based ATE provides additional information along with the desired parameter, e.g., the DSP peak detector reports not only the peak value, but also the peak location in time [412].

- *Size and Power*: A general-purpose DSP-based ATE is smaller, cheaper, and uses less power than a conventional analog ATE.

However, the DSP-based ATE also has the following liabilities:

1. It is expensive, although the cost is coming down because of VLSI technology. The number of required bits of precision and the bandwidth requirement lead to a costly digital signal processor. The large number of required FFT computations for functional measurements and the precision components required also compound the expense. For mixed-signal ATE, however, the main cost is becoming the digital pins.

2. When only making one measurement, a conventional purely analog tester is significantly cheaper. The DSP-based ATE is advantageous when multiple measurements must be made, which is almost always the case.

3. The flexibility of the instrument is a problem for unskilled operators. They need to know instrument theory.

4. The test engineer is required to know physical and mathematical principles underlying each test, the test error sources, and the DUT error sources.

In the future, these trends will become pronounced:

1. *Virtual test* is a method that simulates the effect of the combination of a DSP-based analog tester, measurement error, and a proposed test waveform for a DUT to see if the waveform really does test the DUT effectively. The benefit is that we avoid tying up a very expensive ATE for waveform evaluation. However, it is still mostly in the research stage.

2. Present-day, chip-area-intensive ADC self-testing mechanisms will evolve to require far less chip-area.

3. We will continue to reduce the cost of DSP-based ATEs by using VLSI technology in the ATE. Possibly *structural* analog circuit testing, rather than *functional* testing with the DSP-based ATE, can reduce cost. However, for this to happen, we must first establish a mapping between the proposed structural analog circuit fault models and the specification tests currently used.

4. New methods will be invented to reduce the time that an analog circuit must sit in the DSP-based ATE. Reduced time means less testing cost.

5. Industry will move to automatic analog test waveform generation methods, in order to reduce the non-recurring engineering cost of designing the testing mechanism. At present, analog test engineers must manually craft test waveforms.

10.2 Definitions

- *ADC* – An *analog-to-digital converter*.

- *ATE* – An *automatic test equipment*.

- *Catastrophic Fault* – An analog component either shorted or completely open.

- *DAC* – A *digital-to-analog converter*.

- *DFT* – Discrete Fourier transform.

- *Dither* – A useful oscillation of small amplitude, introduced to overcome the effects of friction, hysteresis, or clogging [339].

- *DUT* – An analog *device-under-test*.

- *FFT* – Fast Fourier transform.

- *Frequency Division Multiplexing* – Assigning different frequency zones to signals originally having the same frequency zone (usually for wireless transmission.)

- *Glitch Area* – The area in the DAC output represented by glitching pulses.

10.2 Definitions

- *Heterodyning* – The process of receiving a high-frequency wave, where it is combined in a non-linear device (often an analog multiplier or a band-pass filter) so that the output contains frequencies equal to the sum and difference of the combining frequencies.

- *Intermodulation* (IM) – A situation where the non-linear response of an analog device gives rise to a spectral line that is at a frequency that is the sum or difference of two tone frequencies (sinusoids at different frequencies) in the analog stimulus waveform to the device. This often indicates unwanted signal multiplication in the device.

- *Intrinsic Parameter* – A parameter defining a specification of the DUT.

- *Jitter* – Time-related, abrupt spurious variations in the duration of any specified interval of a repetitive wave. This is frequently applied to the position of clock edges during time measurement, but jitter can also occur in amplitude, frequency, or phase measurements. The jitter can be measured as average, root-mean-square, or peak-to-peak measurements [339].

- *ks/s* – Kilo-samples per second.

- *Measurement* – The result of measuring an output parameter of an analog circuit and quantifying it with a number. We can deduce properties of the device for testing purposes from measurements.

- *Measurement Error* – The error introduced into a measured signal by the process of measuring it.

- *Multi-Tone Testing* – A method of analog testing where the DUT is stimulated by a composite analog test waveform composed of multiple tones, each of which is a pure sinusoid with individual frequency, phase, and amplitude.

- *Noise* – Low-level electrical disturbances that usually corrupt the *least significant bits* of any digitized analog measurement.

- *Non-Deterministic Device* – An analog device whose behavior can be measured in the continuous variable domain, but with the problems that measurements are not exactly repeatable due to electrical noise in the device or in the measurement system.

- *Parametric Fault* – An analog component whose value deviates from its specified value by at least the element tolerance.

- *Pass/Fail Test* – A simple test that indicates whether a device should be accepted (or not), by comparing its behavior against another device. This test has the advantage of not requiring a measurement.

- *Phase-Locked Loop* (PLL) – A clocking circuit with feedback so that it automatically adjusts the output clock to keep it at the desired phase relationship to the input clock.

- *Pink Noise* – Random noise whose spectrum has a negative slope of 10 dB per decade [339].

- *Primitive Band* – The frequency band $0 \leq f \leq N\Delta/2$ that contains all information about the sampled waveform.

- *Primitive Frequency* (Δ) – $1/unit\ test\ period$. All tone frequencies during multi-tone testing must be a multiple of Δ.

- *Quantization Error* – The error introduced into a measured signal by the process of discretely sampling it.

- *Quantum Voltage* – The voltage corresponding to a flip of the least significant bit of the converter.

- *s/s* – Samples per second.

- *Settling Time* – The time needed for the reconstruction filter on the DAC output to settle to its correct value.

- *Single-Tone Testing* – A method of analog testing where the DUT is stimulated by a single, pure sinusoid analog test waveform.

- *Test* – A combination of a *stimulus waveform* for testing, along with a *measurement* of an accessible device voltage or current, with an error *tolerance* for the measurement. The DUT fails the test if the measurement falls outside the tolerance. The test is used to grade and sort devices, to determine their acceptability for a given application.

- *Tone* – A pure sinusoid of frequency f, amplitude A, and phase ϕ that is a component of an analog test waveform. Each component of the test can have a different frequency, amplitude, and phase.

- *Transmission Parameter* – (also called *performance parameter*) A test parameter that indicates how the channel where the analog circuit is embedded affects a multi-tone test signal.

- *UTP* – The *unit test period*, a common joint period for all test signals and sampling during DSP-based testing of an analog device.

- *White Noise* – Random or impulsive noise that has a flat frequency spectrum in the frequency range of interest [339].

10.3 Functional DSP-Based Testing

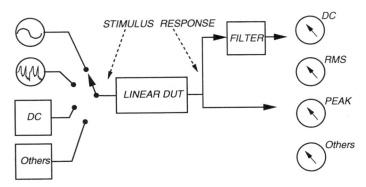

Figure 10.2: Traditional analog *transmission testing*.

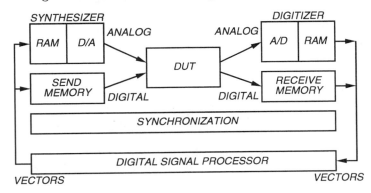

Figure 10.3: DSP-based analog tester.

10.3 Functional DSP-Based Testing

Nearly all analog circuits are presently tested using *functional* DSP-based tests applied by DSP-based automatic test equipment. These ATEs use emulated instruments, derived from FFT or DFT calculations, rather than actual analog instruments, in order to reduce *measurement error* and eliminate undesirable filter settling times during testing.

10.3.1 Concept

The basic concept of DSP-based testing is to replace purely analog instruments with a high-speed A/D converter, a memory for storing readings, and a DSP-processor controlled by software routines. Figure 10.2 [412] shows the conventional analog ATE from the 1930s and 1940s. Note that there was no frequency-time synchronization between the stimulus generation and response analysis parts of the conventional instrument. Conversely, Figure 10.3 [412] shows the present-day DSP-based test system, in which the waveform synthesizer and the response analyzer are synchronized. DSP has replaced the conventional analog instruments, and can emulate a wide variety of existing instruments. We view this instrument as processing *numerical vectors*, rather than as actually measuring voltages or currents. The vectors can represent sampled waveforms, frequency spectra, filter responses, etc. The test engineer creates a series of vectors as the test waveform. The *waveform synthesizer* feeds the vectors to a DAC, usually in a continuous loop. The DAC output is de-glitched (hazards are removed) and passed through a *reconstruction* filter to obtain continuous, band-limited waveforms. The filter output is applied to the DUT, which generates response output analog signals. The *waveform digitizer* digitizes the analog response coming from the DUT using a high-speed A/D converter, and stores the samples in a RAM. When the DUT is a mixed-signal circuit, digital device outputs are collected digitally in the *receive memory* and digital test waveforms are sent to the DUT from *send memory*. The DSP-based ATE has a *burst mode* vector transmission capability, in which a large number of vectors can be transmitted, along with a starting address and a vector length. This is also called *vector* bus architecture, and typically it takes 1 ms to transfer 1000 16-bit integers. For comparison, it takes about 5 ms to connect or disconnect a test circuit by switching relays.

Example 10.1 *Consider an example [412] of an audio amplifier needing testing for gain and distortion at 1000 Hz. The test involves connecting an audio synthesizer to the DUT, connecting an audio digitizer to the DUT output, and waiting for all circuits to settle to steady-state output. With a DSP-based ATE, the waveform is then digitized and analyzed by the DSP-processor. With an analog-based ATE, the signal is analyzed by an analog detector circuit that produces a DC voltage proportional to the magnitude of the parameter. The voltage from the detector is then digitized. So, digitization occurs* before *analysis in the DSP-based ATE, but* after *analysis in the conventional ATE. The analysis time is usually large compared to the intrinsic test time, during which the actual measurement is made. Table 10.1 [412] compares these times. The conventional ATE must repeat this entire sequence for every frequency component to be measured, whereas the DSP-based ATE will use the FFT for simultaneously analyzing multiple spectral lines in roughly 4 to 20 ms. Some measurements require capture of many cycles of the parameter, but for simple analog tests, only a few cycles are needed, and this can be achieved in 30 ms. The DSP-based ATE will have a speed advantage when the DUT must be tested for many parameters, because relay switching, loading and starting the synthesizer, and digitization of the DUT response are done only once, unlike in the conventional ATE. Also, the DSP-based ATE can overlap DSP array processing of a measurement with DUT handling, relay switching, digitizing, etc. The DSP-based ATE has higher throughput. If 10 parameters were evaluated, the DSP-based tester would take 3.5 ms per test, on average, whereas the conventional tester would take 50 ms per test. Also, it is easier to filter out power line and ripple components from measurements in the DSP-based ATE before processing the samples.*

10.3 Functional DSP-Based Testing

Table 10.1: Time comparison for analog vs. DSP-based ATE.

Component of 1 kHz amplitude measurement	Conventional analog ATE	DSP-based ATE
Relay switching	5 ms	5 ms
Load & start synthesizer	N/A	5 ms
Synthesizer + DUT settling	N/A	1 ms
Filter + detector + DUT settling	35 ms	N/A
Digitization interval	N/A	1 ms
Transfer time	N/A	1 ms
Computer overhead	10 ms	N/A
DSP processing/overhead	N/A	15 ms
Total	50 ms	28 ms

10.3.2 Mechanism of DSP-Based Testers

The DSP-based ATE is nearly always more accurate than analog ATE. In particular, it is more accurate for non-linear analog circuit measurements, because vector processing increases the accuracy of a set of vectors compared to the accuracy of an individual sample.

We need a very fast array processor with a pipeline in the DSP processor. We also need much higher precision in the DSP processor arithmetic than the precision in the vector samples, because the DSP processor multiplies and divides the various samples. Vectors have many samples, N, so signal-to-quantization noise of the entire vector can be better than that of a single sample by as much as $\sqrt{N}$ [412]. Given 1024 uniformly distributed samples over a prime number of signal cycles, quantization noise in any one spectral location is reduced 32 times, or nearly 30 dB. This gives five additional bits of resolution. Even more mathematical precision is needed because many DSP algorithms produce cumulative error. A rule of thumb [412] is that the mathematical processor should have at least three decimal orders of precision beyond what is desired in the end result. This translates into 31 bits of precision, so internal computational processes in the DSP-based analog tester need more precision than 32-bit floating point – they must have 40 or 64 bits, which requires double-precision floating point arithmetic.

Phase-Lock Synchronization Requirement. Vector samples must fall in exactly the right places in the right time interval, or testing will be very inaccurate. The digitizing window must be coordinated precisely with each clock, signal, and distortion component. This is implemented with a common reference frequency, controlled by a *phase-locked loop* (PLL). This reference frequency is divided into master clocks, which are further conceptually divided by programmable ratios to ensure uniformly distributed clocks and precisely timed windows. This also allows rates and times to be programmed in integer ratios, usually involving prime numbers. This is called *M/N synchronization* [412], *integer-ratio synchronization*, or *prime-ratio locking*.

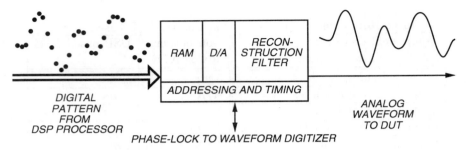

Figure 10.4: Waveform synthesizer concept.

Synchronization gives the DSP system a *coherence* property, in which all frequency and time functions are programmably related in exact whole-number ratios. We must restrict waveform analysis to a whole number of cycles, which provides highly repeatable results, and makes phase and delay measurements practical in production. Rosenfeld [549] lucidly covers ATE clock programming, and also proposes an alternative scheme using a low-frequency master clock and programmable frequency multipliers to achieve M/N synchronization.

10.3.3 Waveform Synthesis

Figure 10.4 [412] shows the waveform synthesizer part of the DSP-based ATE. It has a programmable reconstruction filter for a 16-bit DAC to produce continuous analog waveforms, by eliminating staircasing in the analog output signal. We also need $\sin(x)/x$ (sinc) correction applied to the signal spectrum during digital pattern synthesis (for the reason why, see Section 10.5.2.) Clocking of the pattern will be from one of the clock dividers following the phase-locked loop that generates the 10 MHz signal. Figure 10.5 shows the *Pacemaker* (PM) phase-locked loop clock generator with 8 clock dividers *PM1-8* (only two are shown here) for the LTX FUSION ATE. The *waveform source* (WS) unit (shown as DAC) and the *waveform measure* (WM) unit (shown as ADC) use this clock generator to synchronize themselves. Note that the *test head* is the actual test fixture with the DUT in it. Switching the clock from one divider to the other provides a phase-continuous *frequency-shift keyed* (FSK) signal. In addition, a practical synthesizer needs a subdivided local memory in order to switch on-the-fly from one wave shape to another.

We send the waveform from the DAC through programmable course and fine attenuators to set the proper level. These attenuators keep the DAC working at its full-scale value to minimize quantization noise. This also allows us to add a programmable DC offset to the synthesized signal.

Frequency synchronization requires that clocking for the digital pins, analog waveform synthesizer, and analog waveform sampler be derived from a single crystal reference. This allows different modules in the ATE to use different frequencies, which have known fixed integer ratio relationship between them. All of the clocking for the WS and WM comes from the Pacemaker, and is chosen for the waveform

10.3 Functional DSP-Based Testing

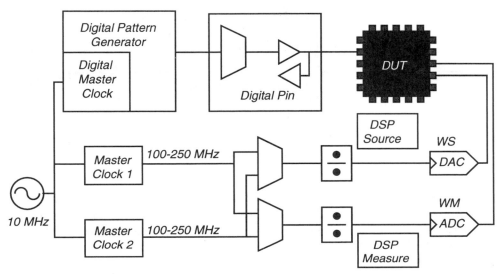

Integrated clock generator in digital pattern generator
Dual analog clock to each DSP (SMS) source and measurement unit
and selected pins

Figure 10.5: LTX FUSION PM practical phase-locked loop.

source unit in the LTX enVision++ operating system language. This language includes the Cadence *test programming language* (TPL) [407], and has the following syntax for selecting the master clock:

$$\text{set master clock} \ \left| \begin{array}{c} 1 \\ 2 \end{array} \right| \ \left| \begin{array}{c} \text{frequency} \\ \text{period} \end{array} \right| \ \text{to <double>} \ \left| \begin{array}{c} \text{times} \\ \text{over} \end{array} \right| \ \text{<integer>}$$

The vertical bars denote choices, one of which must be selected. The Fusion unit has multiple clock generators, all operating from the same 10 MHz reference frequency. The **times** option allows the user to program the master clocks in terms of primitive periods (discussed later), with an integer scalar. It has this syntax for setting up the clock connections to pins and the clock division rate:

clock ws main mem with pm clock <word1> divide by <word2>

or for the waveform measure unit with:

set wm to pm clk <word1> divide by <word2>

where < word1 > refers to PM line 1 to 8 and < word2 > ranges from 2 to 256. The *digital pin master clock* (DP MASTER CLOCK) in the Pacemaker can be connected to the PM lines 1-8 with this command:

$$\textbf{connect dp master clock} \ \left\{ \begin{array}{c} \text{internal reference} \\ \text{doubled reference} \\ \text{source1} \\ \text{source2} \end{array} \right\} \ \textbf{to pm line <word1>}$$

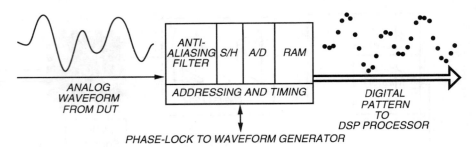

Figure 10.6: Waveform digitizer.

The enVision++ operating system provides a rich set of graphical debugging tools for stepping the tester through the TPL, setting breakpoints, etc. The Cadence TPL has a rich set of data types (integer, boolean, float, double, pin), storage classes (constant, local, static, global) and aggregate types (sets, lists, strings, arrays, multisite.) The language syntax most resembles that of VHDL.

10.3.4 Waveform Sampling and Digitization

Figure 10.6 [412] shows a conceptual version of the waveform digitizer. A typical unit achieves over 100 ks/s ($kilosamples/s$.) It is complex, and contains a differential buffer, a *programmable-gain amplifier* (PGA), and a programmable anti-aliasing filter. It needs a built-in track-hold circuit in the *analog-to-digital converter* (ADC.) A typical instrument has three independent PLLs, which are needed in telecommunications testing to generate and coordinate several different clock rates at once. It needs independent output dividers on each PLL to allow coherent clocking of devices at different rates from this single PLL.

10.4 Static ADC and DAC Testing Methods

At present, we use specialized testing methods for ADC and *digital/analog converter* (DAC) testing. ADCs are not inverses of DACs, so each requires different definitions of parameters and different tests. The ideal ADC discards data, while the ideal DAC does not lose information. This can be seen in Figure 10.7(a) where the ideal ADC transfer function has an uncertain domain of input voltages that map into a specific digitized range value. Any voltage within that input range maps into the same output code. The ideal DAC of Figure 10.7(b) has no such domain uncertainty. The ADC and DAC, therefore, are not inverse functions of each other. The ADC cannot be tested by a DC test and examination of the digitized output for correct codes, because noise, statistical converter behavior, and converter slew rate errors may cause such a test to pass defective ADCs, or to fail good ADCs. ADC testing is non-deterministic and must involve more statistical analysis than DAC testing. Even good ADCs have noise and occasional bursts of extraneous (*sparkle*) codes. Figure 10.7 [288] shows the ideal transfer functions of good ADCs and DACs,

10.4 Static ADC and DAC Testing Methods

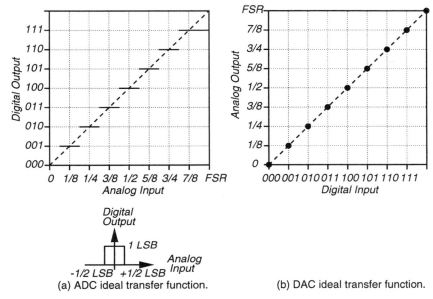

Figure 10.7: Ideal ADC (a) and DAC (b) transfer functions.

while Figure 10.8 [288] shows a DC offset error, Figure 10.9 [288] shows a gain error, and Figure 10.10 [412] shows DAC non-linearity errors. Another term for the DAC transfer function is the *transfer map*, since the DAC transfer function consists of discrete points.

10.4.1 Transmission vs. Intrinsic Parameters

Mahoney [412] defines *transmission parameters* (or *performance parameters*), which indicate how the channel where the converter is embedded affects a multi-tone test signal. The *gain, signal-to-distortion ratio, intermodulation* (IM) *distortion, noise power ratio* (NPR), *differential phase shift*, and *envelope delay distortion* are transmission parameters.

Intrinsic parameters define the specifications of the DUT. The *full scale range* (FSR) of the converter input voltage, the *gain*, the *number of bits*, the *static linearity* (differential and integral), the *maximum clock rate*, and the *code format* are common to ADCs and DACs. The *settling time* and the *glitch area* are relevant only to DACs. Settling time indicates how long the reconstruction filter on the DAC output takes to settle to its correct value. Glitch area is the area in the DAC output represented by glitching pulses. As frequencies and conversion rates increase, transmission parameters become more useful than intrinsic parameters for testing.

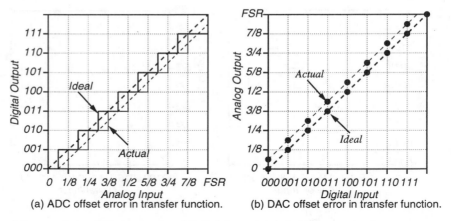

Figure 10.8: Offset error in ADC (a) and DAC (b) transfer functions.

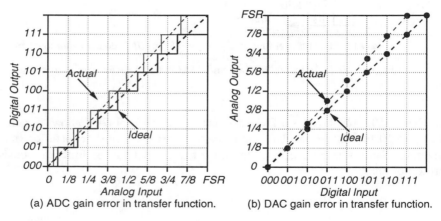

Figure 10.9: Gain error in ADC (a) and DAC (b) transfer functions.

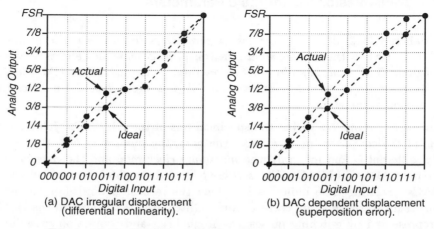

Figure 10.10: DAC transfer function non-linearity errors.

10.4 Static ADC and DAC Testing Methods

10.4.2 Uncertainty and Distortion in Ideal ADCs

Mahoney [412] shows how even an ideal ADC has uncertainty and distortion. Rotate Figure 10.7(a) so that the transfer slope is horizontal, and then the error function is discernible as a sawtooth waveform. Recall that the step widths of the ADC are statistical, and that there is uncertainty as to where the actual voltages cause a change to the next code. The RMS *uncertainty* of quantization is the RMS amplitude of this sawtooth, and is also the RMS distortion value introduced into analog signals transmitted through the converter. The sawtooth has a peak-to-peak height of 1 LSB, and the waveform power is 1/12 LSB squared. When Q is the *quantum* (LSB) *voltage*, then the RMS quantization distortion voltage is:

$$D = \frac{Q}{\sqrt{12}} V, RMS \tag{10.1}$$

There are 2^n code levels in the FSR of an n-bit linear, binary ADC, but the two end steps have no outer bounds. AT&T uses the convention of assigning *virtual edges*, or decision levels where edges occur if the transfer function continued beyond the FSR points. These edges represent *clipping levels* of the ADC.

When one applies a sinusoid to the ideal ADC such that the peaks just touch the virtual edges, then the RMS amplitude becomes:

$$\text{FS Sine Amplitude} = \frac{Q \times 2^n}{\sqrt{8}} V, RMS \tag{10.2}$$

By dividing the RMS amplitude into the RMS quantization distortion, we get:

$$\left[2^n \times \sqrt{1.5}\right]^{-1} \tag{10.3}$$

Converting this to decibels, one obtains:

$$\text{Relative Distortion Level in } dB = -(6.02\,n + 1.761) \tag{10.4}$$

To reiterate, this is the ADC distortion relative to a full-scale sinusoid level (only in linear conversion.) The distortion power is independent of signal level as long as the signal is at least several LSBs in amplitude. Distortion power can approximate the signal-to-noise ratio, after an adjustment to reflect the test signal actual power.

10.4.3 DAC Transfer Function Error

It is inaccurate to estimate offset and gain errors using only the two end points of the converter transfer characteristic. Instead, the vertical intercept of the transfer characteristic at the y-axis and the slope of the best-fit straight line are better indicators. *Differential non-linearity* (DNL) measures the difference between the actual and linearized increments for each step in the converter transfer curve. The *differential linearity error* (DLE) function lists step-by-step the differences between actual converter increments and linearized increments (actual vertical projections

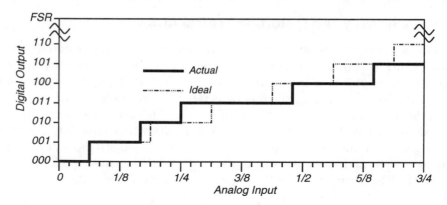

Figure 10.11: Example sampled converter transfer function.

onto the best-fit reference line.) DNL is defined as the root-mean-square value of the DLE function. We calculate increment step values, and then subtract the slope × step width of the best-fit line from these steps. *Integral non-linearity* (INL) measures the difference between the reference line and the actual converter transfer function. The *integral linearity error* (ILE) function is computed by integrating the different DLE error vectors from the lowest code through the code of interest, in order of increasing codes. INL is the root-mean-square value of the ILE function. *Superposition error* results from non-constant bit weights in DACs. Time-related errors occur in DACs because they are subject to random noise, so the real DAC has fuzzy vertical lines in its transfer function. Some designs have *hysteresis*, and some have glitching or even ringing at analog outputs due to variations in timings between currents switching off and those turning on. Finally, RC rolloff of analog circuits brings about a settling output delay before steady-state is achieved.

10.4.4 ADC Transfer Function Error

The ADC DLE is a plot of how each code step width differs from the ideal step width. Differential linearity is defined differently from DACs, because the ADC has comparators, and the DAC does not. For known decision levels, we compute the statistical step (the quantum size for each code) and subtract the average step size. This yields a plot of DLE values. If the DLE is less than -1, the code is instead declared as *missing*. Typically we compute the DLE without actually finding code edges by tallying the code in a *histogram*, which is the number of times each digital output code appears. The ADC ILE is the integration of these code step width differences from the ideal converter characteristic, measured using the best-fit straight line. Figure 10.11 shows the ADC ideal and measured transfer characteristics, and Table 10.2 shows the DLE and ILE functions. The transfer characteristic shows how each code step width differs from the ideal step width. Assume that each small X axis division represents a sample in Figure 10.11. Table 10.2 gives the statistics for the first five codes of this converter. Each matrix (T, D, C, E) is indexed by the output ADC code. The ideal step width is 6, and the average actual step width is

10.4 Static ADC and DAC Testing Methods

Table 10.2: Statistics for converter in Figure 10.11.

Code tally (counts)	T (0) 3 + 3 = 6	T (1) 5	T (2) 4	T (3) 11	T (4) 8
DLE (LSB fraction)	D (0) -0.1176	D (1) -0.2647	D (2) -0.4118	D (3) 0.6176	D (4) 0.1765
DNL (RMS LSB)	0.3650					
Transfer characteristic (counts)	C (0) 0	C (1) 5.5	C (2) 10	C (3) 17.5	C (4) 27
ILE (LSB fraction)	E (0) 0	E (1) -0.19115	E (2) -0.5294	E (3) -0.4265	E (4) -0.02945
INL (RMS LSB)	0.3161					

6.8.

10.4.5 Flash ADC Testing Methods

Flash A/D converters operate at 20 to 100 *Mega samples/s* (*Ms/s*), and even up to 250 *Ms/s*. They are code-dominated converters, so each step can have an error different from any other step. There are no major carries in the converter, so each code must be tested. The input signal frequency and wave shape affect dynamic linearity. Figure 10.12 shows the flash ADC. Normally, the flash converter decoding logic consists of a bank of flip-flops that latch the comparator *thermometer code*, followed by a digital priority encoder (implemented here as NOR gates.) This converter type can produce binary output codes unrelated to the best-fit horizon, usually a few *least significant bits* (LSBs) off, and called a *glitch code*. However, many false codes may be all zeros or all ones and are then known as *sparkle codes* because they produce bright pixels in video displays. The ADC functions as its own digitizer during test.

Static Linear Histogram Technique with a Triangle Wave

For DLE and ADC transfer function measurement we use a very slow triangle wave that ramps from just below the minimum full-scale voltage to a voltage just above the maximum full-scale voltage and back down again. The analog triangle wave must be slow to increase the tally resolution, or there will be quantization error in the DLE function of the measured converter transfer function. The ramp should be slow enough so that for each code, it will be digitized roughly 10 times before the ramp moves to the next code. We want to ensure that each code of the ADC is fully tested. Also, the ramp may be repeated for up to 150 periods, since the converter behaves statistically and we want its average behavior. We sample the codes coherently using the methods described earlier. Then, the step width is directly proportional to the tally of each code [412]. Let L be the 8-bit ADC output vector, so K is the tally (count) of all 256 unique codes, while T is the tally of the

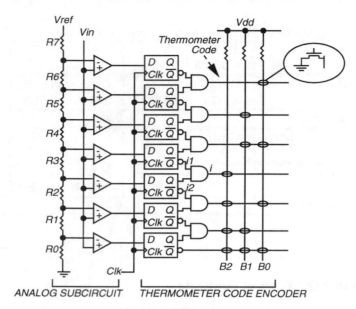

Figure 10.12: Typical flash ADC.

254 valid (double-bounded) steps (discarding codes 0 and 255.) The following code fragments are written in the LTX Cadence language, a *test programming language* for the LTX FUSION ATE family.

```
local
        double:   K [256]
        double:   L [2048]   -- Array of size sample_size
        double:   T [256]
end_local
for   i = 1   to sample_size
              K [ L [i]] = K [ L [i]] + 1  -- Increment the count
                 -- of the array element
                 -- corresponding to the captured code
end_for
T = K [1:254]
```

T is the code histogram (see Figure 10.13 [412].) The histogram has no notion of the code ordering – we could permute the occurrence of the codes produced by the ADC in time, and this histogram remains the same. This is a known weakness of the histogram technique. To find DLE, we subtract the average count from each tally, and express that in units of LSBs.

```
            A = avg (T)
            A = A [1]         -- (the average)
            D = - A + T
            D = (1 / A) * D   -- (to convert to units of LSBs)
```

10.4 Static ADC and DAC Testing Methods

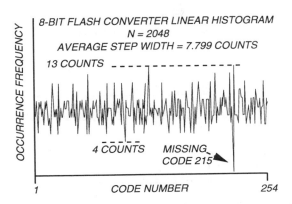

Figure 10.13: Linear histogram of 8-bit flash ADC.

We repeat the test waveform 100 to 150 times. One problem is that the histogram ignores non-monotonic ADC codes – the sparkle and glitch codes. Figure 10.14 [412] shows the DLE (D) of an 8-bit ADC [412]. Step inequality is caused by random displacements of decision levels due to comparator noise. However, this test will detect missing converter codes.

If the transfer function is monotonic, which will be true when the ramp is slow and comparator noise is minimal, then T represents the *derivative function* of the ADC transfer curve, since it gives the code step widths in their order of occurrence. We integrate T to obtain the ADC transfer function, C, which is the list of code centers relative to the center of the first step.

```
C [1]    = <CONSTANT>
for      i = 2   to 254 do
              C [i] = C [i - 1] + (T [i] + T [i - 1]) / 2
end_for
```

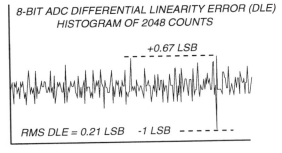

Figure 10.14: DLE of 8-bit flash ADC.

Derivation of *Integral Linearity Error* (ILE.) Matrix E is the ILE function of the ADC. We compute it from the DLE vector (D), by integration using the center of the first step as a reference.

330 Chapter 10. DSP-BASED ANALOG AND MIXED-SIGNAL TEST

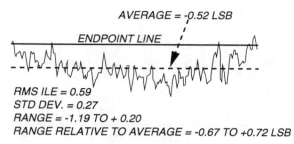

Figure 10.15: ILE of 8-bit flash ADC.

```
E [1]   = 0
for     i = 2   to 254 do
            E [i] = E [i - 1] + (D [i] + D [i - 1]) / 2
end_for
```

Figure 10.15 [412] shows the ILE. One problem with this ILE computation is that it loses evidence of missing codes. The weighted code center method, in which codes are weighted by the position range of their occurrence in the histogram, solves this problem (see Mahoney for details [412].)

High-Frequency Histogram Technique with a Sine Wave

We need a high-frequency test for ADCs to catch the sparkle and glitch codes. At high f, triangle waves are hard to generate, so we use a sine wave, and coherently test the converter with the previously-described DSP technique. The dynamic linearity of the ADC worsens with increasing input slew rate, and this test detects such problems, and also lets us examine the spectral response of the converter. Note that steps around the center of the input range receive less than average resolution, so N must be 2 to 4 times that needed for a good linear (triangle wave) histogram. $N = 4096$ may be adequate, but $8\,K$ to $16\,K$ is better for 8-bit ADCs. Vector $T1$ is again the tally of sinusoidal code counts. Since this histogram is non-linear, the computation must also use a different scalar reference for each of the 254 steps in the histogram (vector $T2$.) $T2$ is the reference step width for each code for an ideal converter with the sinusoidal input applied. $T2$ is computed by finding the offset and amplitude of the best-fit sinusoid.

10.4 Static ADC and DAC Testing Methods

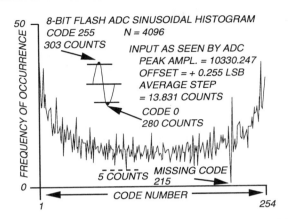

Figure 10.16: Sinusoidal histogram of 8-bit flash ADC.

```
N                 =   Number of counts, including end codes
N1                =   Number of maximum code counts
N2                =   Number of zero code counts
C1                =   cos (180 * N1 / N)
C2                =   cos (180 * N2 / N)
V0                =   127 * (C2 - C1) / (C2 + C1)
Signal_Offset     =   V0 LSBs
V                 =   (127 - V0) / C1
Signal_Amplitude  =   V LSBs
```

We can now compute D, the dynamic DLE, as follows:

$$D = T1 - T2$$
$$D = D / T2$$

Figure 10.16 [412] shows the sinusoidal histogram for an 8-bit flash ADC. The histogram is sampled coherently using M prime relative to N. M must not be low, or elements of matrices $T1$ and $T2$ will be nearly equal in value, leading to roundoff errors. This may cause the ILE error curve to tilt upward, which it must not do. This error can be eliminated by averaging D and subtracting the average from D, to force integration to produce a proper ILE curve.

We avoid quantizing the reference tally, by making it a list of floating-point reference widths for each step. In the following example, asn is the arc sine mathematical function. We use vector $T3$, computed below, in place of $T2$:

```
for  i = 1  to 254 do
     T3 [i] = (N / 180) * (asn ((i - 127 - V0) / V)
            - asn ((i - 128 - V0) / V) )
end_for
```

This complicated calculation merely ensures that $T3$ contains, for each code, the floating point step width. Figure 10.17 [412] shows the correct ILE function computed this way.

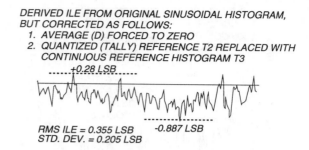

Figure 10.17: ILE derived from sinusoidal histogram.

10.4.6 DAC Testing Methods

For testing the simpler and slower D/A converters, it suffices to digitize the output steps of the DAC, average them, and compute the DNL and INL. For high-performance DACs, however, approaches such as the following must be used to ensure a high-quality product.

Need for Indirect Voltage Measurement. Mahoney [411] describes a procedure for computing INL for a DAC. A 16-bit D/A converter has a step size of 15 *parts per million* (*ppm*), which requires a measurement error below 2 *ppm*. It is necessary to use indirect measurement to achieve this. DAC transfer function errors (see Figure 10.10) appear as vertical displacements of the points. A direct measurement of the DAC transfer map is inappropriate for these reasons:

- A 16-bit DAC has 65,536 codes, while a 24-bit DAC has 16,777,216 codes. It takes too long to look at all of these codes – we need only look at a subset.

- A direct measurement demands excessive accuracy, since we restrict measurement error to ±1/10 LSB.

We can measure each interval in the transfer map, but we cannot measure each step directly, because the voltages defining it happen at different time instants.

V_x: smaller adjacent voltage	V_y: larger adjacent voltage
C: repeatable measurement error	e: nonrepeatable peak measurement error (voltmeter tracking error, drift, noise, etc.)

We want to find $(V_y + C + e) - (V_x + C + e) = V_y - V_x \pm 2e$. For 14 bits, the step size becomes 61 *ppm* (610 μV in a 10 V range DAC.) We limit $2e$ to 61 μV to keep measurement error to 6.1 *ppm*, so e is only 30.5 μV – this is too hard and expensive to do. The solution is to use indirect measurement, and shun measurements with large absolute values with respect to the differences being computed. Random noise aggravates this problem – e.g., 10 μV RMS Gaussian noise would reach the 30.5 μV error limit above. We decrease bandwidth to avoid this problem by using a controllable bandwidth test fixture, although this lengthens testing time.

10.4 Static ADC and DAC Testing Methods

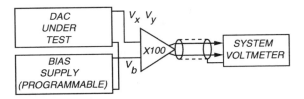

Figure 10.18: Differential fixture concept.

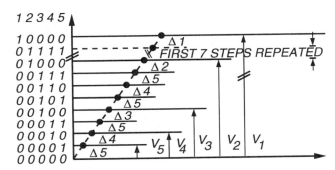

Figure 10.19: Superposition of unit steps in a 5-bit DAC.

Approach. We only measure the $V_y - V_x$ difference, not the absolute V_y and not V_x, by comparing the difference to a biased, programmable supply V_b (see Figures 10.18 and 10.19 [411].) V_b need not be accurate – just stable, quiet, and resolvable within a few millivolts. We measure $V_x - V_b$ and $V_y - V_b$, which is of the same order as the difference $V_y - V_x$. V_b should be a 0.1% accuracy programmable DC source, with even better resolution. We amplify the $V - V_b$ difference measurement 100 times with a low noise instrumentation amplifier, and then send it through a balanced, shielded cable to the system voltmeter. The benefits of this are:

- The voltmeter and amplifier offset cancel at the instrumentation amplifier.

- Gain error only happens in the difference (of order 1 LSB.)

- The range setting and location of the two readings determine the voltmeter tracking error (non-linearity), and we have greatly reduced the range setting. As an example, we measure a 610 μV step, which with 100× magnification leads to a voltmeter range of 0.5 V. With a 0.02% (200 ppm) voltmeter tracking accuracy for full scale range, we get only ±100 μV worst case error, which is ±1 μV reflected back at the DUT, or 1/600 LSB.

- We reduce tracking error further by keeping the two readings close together.

- Since the voltmeter and cabling noise after preamplification is 100 μV RMS, this is only 1 μV RMS reflected back at the DUT.

- The major accuracy limitation is the DUT output random noise.

Table 10.3: Necessary equations.

B_n = contribution of LSB = $V_{max}/(2^n - 1)$	
B_i = contribution of bit i = $FSR/2^i = 2^{n-i} B_n$	
FSR – full scale range	*electrical span* – $FSR - 1\ LSB$ (since
e_i = bit error = $V_i - B_i$	FSR is the one bit beyond the MSB)
$(+)$ – positive error set	$(-)$ – negative error set

To reduce noise, we average a number of readings at the voltmeter (provided that the noise effects on samples are independent.) We use an 8000 s/s sampling rate, and average 100 samples. Digital crosstalk is eliminated by silencing the digital bus for 1 ms before operating the voltmeter.

Derivation of INL Measure. The method measures $\Delta_i = \Uparrow \left\{ \frac{0001000000}{0000111111} \right\}$, the analog voltage change caused by one code increment in bit i, where 1 (MSB) $<= i <= n$ (LSB). From all measured Δs, we reconstruct the DAC point map. Error is the difference between the actual bit contribution and the nominal value. $V_{max} = \sum V_i = \sum B_i$ so $\sum e_i = 0$ meaning that errors have both positive and negative signs and sum to 0. When positive and negative errors are grouped separately, $\sum(+) + \sum(-) = 0$. The two sums define the limits of *integral non-linearity* (INL), and are averaged to limit the effect of measurement error:

$$INL = \frac{\sum(+) - \sum(-)}{2B_n} \qquad (10.5)$$

So, INL comes from the difference of sets V_i and B_i, which come from the increment voltage set Δ_i measured at the test fixture.

The algorithm removes non-noise errors:

- The large denominator $2^n - 1$ is twice the size of the positive coefficients and twice the negative coefficient sum. An error of 1% in a Δ_i value causes only 1/2% error in DNL.

- The sum of coefficients in the numerator is zero, so if all Δ_is have the same constant error, it vanishes in e_i.

- In Equation 10.5, if all Δ_i change by the same proportional error, the numerator and denominator change by the same proportion, so the INL is still correct (this eliminates voltmeter and amplifier gain error during measurement.) The remaining errors are random noise and amplifier/voltmeter tracking error.

V_i is obtained from summing the immediately preceding step, Δi, and binary multiples of all lower-order Δs [411]. The first n elements of the series:

$$1 + 1 + 2 + 4 + 8 + 16 + \ldots$$

10.5 Realizing Emulated Instruments Using Fourier Transforms

Table 10.4: Delta values measured during 5-bit DAC testing.

Coefficient bit	Magnitudes					Denominator
	$\Delta 1$	$\Delta 2$	$\Delta 3$	$\Delta 4$	$\Delta 5$	
1	+15	−1	−2	−4	−8	
2	−8	+15	−1	−2	−4	
3	−4	−8	+15	−1	−2	31
4	−2	−4	−8	+15	−1	
5	−1	−2	−4	−8	+15	

are the coefficients. Figure 10.19 shows how this technique, with a 5-bit converter, leads to these equations:

$$V_1 = \Delta_1 + \Delta_2 + 2\Delta_3 + 4\Delta_4 + 8\Delta_5 \quad (10.6)$$
$$V_2 = \Delta_2 + 1\Delta_3 + 2\Delta_4 + 4\Delta_5$$
$$V_3 = 1\Delta_3 + 1\Delta_4 + 2\Delta_5$$
$$V_4 = 1\Delta_4 + 1\Delta_5$$
$$V_5 = 1\Delta_5$$

First, we find $V_{max} = \sum V_i = 1\Delta_1 + 2\Delta_2 + 4\Delta_3 + ...$ We next find the B_i using equations in Table 10.3. We next obtain the MSB error:

$$e_1 = \frac{(2^{n-1} - 1)\Delta_1 - \Delta_2 - 2\Delta_3 - 4\Delta_4 - (2^{n-2})\Delta_n}{2^n - 1} \quad (10.7)$$

So, for the $n = 5$ example:

$$e_1 = \frac{15\Delta_1 - \Delta_2 - 2\Delta_3 - 4\Delta_4 - 8\Delta_5}{31}$$

The remaining e_i coefficients [411] are given by a circular shift of the numerator coefficients, as seen in Table 10.4. After each e_i is calculated, the positive and negative values are separately grouped, and the two sums are generated to compute INL. Mahoney's method can measure non-linearity to 3 *ppm* of full scale, but the absolute accuracy requirement is only 1%.

Souders and Flach describe the National Institute of Standards service for high-precision DAC and ADC calibration [626] for testing purposes.

10.5 Realizing Emulated Instruments Using Fourier Transforms

Conventional analog measurement instruments have the weakness of using a central ADC as the master DC voltmeter, and for measuring AC parameters, this ADC is preceded by a detector as shown in Figure 10.20 [412]. The detector diminishes AC measurement accuracy, because analog detectors usually employ at least one non-linear function (square law, square root, logarithm, exponential, rectification, clipping, multiplication/division by another waveform.) Unfortunately, analog circuits cannot reproduce non-linear characteristics nearly as well as linear ones, so the typical accuracy is 0.1 to 1%. Calibration is also a problem. In addition, the detector has a long time constant, low-pass filter to produce a smooth DC output, which slows down testing. Analog detectors may take 25 to 100 periods at the lowest-rated frequency to settle to 0.1%.

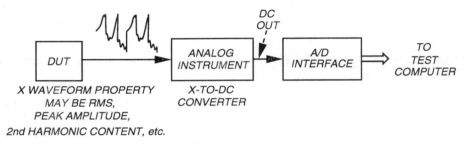

Figure 10.20: Digitization in the conventional analog ATE.

In the DSP-based ATE of Figure 10.21 [412], instead the detector is placed after the ADC, rather than before the ADC, and is implemented as a software program in a DSP processor. This is called an *emulated instrument*. Since there are no longer any non-linear analog functions in the analog portion of the ATE, accuracy is improved and both AC and DC measurements will now have similar accuracy. Also, statistical techniques can boost dynamic accuracy beyond the accuracy of individual samples. In addition, the DSP-based ATE will synchronize the detector with the signal source. This greatly increases detection speed and eliminates ripple, because the filter in the detector is replaced with a *timed integrator* whose integration interval must exactly span a whole number of cycles. In the continuous domain, DC, absolute average AC, and true *root mean square* (RMS) measurements are given by the following equations, respectively:

$$V(DC) = \frac{1}{P} \int_P V_{in} dt \tag{10.8}$$

$$V\begin{pmatrix} abs. \\ avg. \end{pmatrix} = \frac{1}{P} \int_P |V_{in}| dt \tag{10.9}$$

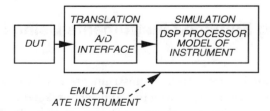

Figure 10.21: Digitization in the DSP-based analog ATE.

10.5 Realizing Emulated Instruments Using Fourier Transforms

$$V(RMS) = \sqrt{\frac{1}{P} \int_P V_{in}^2 dt} \qquad (10.10)$$

Here, P is the *test period* or integration interval. The integrator eliminates any filter settling time, and speeds up measurement.

Coherent Measurement. Coherent measurement means that we must process *exactly* an integer number of periods of the sampled waveform. Since emulated instruments in the DSP-based ATE operate on digitized waveform samples and use timed integration, rather than filtering, we must use the following discrete waveform equations (analogous to Equations 10.8 through 10.10):

$$V(DC) = \frac{1}{N} \sum_{I=1}^{N} V(I) \qquad (10.11)$$

$$V\left(\frac{abs.}{avg.}\right) = \frac{1}{N} \sum_{I=1}^{N} |V(I)| \qquad (10.12)$$

$$V(RMS) = \sqrt{\frac{1}{N} \sum_{I}^{N} V(I)^2} \qquad (10.13)$$

where N is the number of samples.

Unit Test Period. The integration interval P must contain an integer number of *signal cycles* (M) and an integer number of *sampling intervals* (N.) DSP algorithms often restrict M and N to certain ratios, and certain DUTs use sampling (e.g., a switched-capacitor filter), so this imposes a clock coordination requirement between the DUT and the ATE, which further restricts M/N. Digitization often requires sampling over multiple signal cycles to improve accuracy. In Figure 10.22, we define the *unit test period* (UTP) as the interval over which unique signal information is spread. The UTP contains M signal cycles and N sampling intervals, where M and N must be relatively prime. In coherent sampling, the sequence of recorded sample values (vectors) ultimately repeats itself, and an infinite coherent vector is numerically periodic [412]. Let F_t be the tone frequency, and F_s be the sampling rate.

Example 10.2 *Consider a circuit test [412] where the unit test period (UTP) is 257 signal cycles, or 256 samples. Relatively prime rate ratios produce uniform, high-resolution sampling without using high sampling rates or incremental time-delay circuits. We clock the digitizer at 50 ks/s, but must lock the sine wave generator at 257/256 times this rate. We use two clock dividers, PM_1 and PM_2, in the ATE. The voltage-controlled oscillator (VCO) is set to 257×50 kHz (12850 kHz), and register PM_2 is set to divide by 257. The result is delayed a programmable amount by the phase delay register in PM_2 and used as an ADC strobe. The external sine wave generator is then locked by setting divider PM_1 to 256, producing an output*

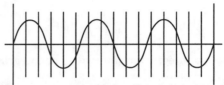

The sample sequence repeats itself regularly in coherent testing.
The Unit Test Period contains N samples, and M signal cycles.
Here, there are 3 signal cycles, so M = 3.
There are 16 sampling intervals, so N = 16.
Here, the primitive period is 16, since that is the shortest number of
Δ periods that gives an integral number of waveform cycles.
M and N must be relatively prime.

Figure 10.22: The unit test period.

of $50\ kHz \times 257/256 = 50.1953125\ ks/s$. We use one PLL as a common clocking source for dividers to minimize clock jitter.

Example 10.3 *Derivation of UTP for CODEC testing. A CODEC encoder is the serial ADC used in digital telephony (see Section 10.6.) AT&T standards dictate a sampling rate of 8000 s/s and a working audio frequency range of 300 to 3400 Hz. We will measure gain and signal-to-distortion ratio with a single tone, or sinusoidal waveform, at a test frequency of 1000 Hz. The CODEC output is converted to parallel and collected until N words are received. A DSP algorithm in the ATE computes the power of the fundamental tone and the total power of the remaining spectrum. The CODEC has an 8-bit floating point (FP) μ law A/D converter with 255 unique codes (see Section 10.6.) There must be at least as many samples, but with a sinusoidal test waveform, there must be more samples, because the μ law step distribution does not match the amplitude distribution. AT&T requires $N \geq 400$. Then:*

$$F_t = 1000\ Hz$$
$$F_s = 8000\ s/s$$
$$P = 50\ ms$$
$$M = 50\ cycles$$
$$N = 400\ samples$$

With these test parameters, we immediately encounter the problem that samples fall on the test waveform at only certain phases so that after eight samples, the pattern repeats. This means that only 8 of the 255 CODEC steps will be evaluated, so testing will not discern much about the CODEC transfer characteristic. If the first sample falls at 22.5°, only 4 steps are sampled. Therefore, we see that M and N must be relatively prime, which we achieve by factoring out 50 to get:

$$M = 1\ cycle$$
$$N = 8\ samples$$
$$UTP(P) = 1\ ms$$

10.5 Realizing Emulated Instruments Using Fourier Transforms

Here, with $F_t = 1000\ Hz$, the only information that we really get is obtained in the first eight samples. With $P = 50\ ms$, the information of the eight samples is simply repeated 50 times.

Since the 400 samples must fall uniformly over the entire CODEC transfer curve, we could set $F_s = 400\ ks/s$. This is impossibly fast for the CODEC, so another solution is to adjust F_t slightly, so that the signal is sampled at different points. Without coherent sampling (signal synchronization), this becomes similar to random sampling [412]. Mahoney states that for a CODEC, 2000 random samples give the same precision level and measurement repeatability that could be obtained with only 400 correctly distributed samples. We would rather sample 400 times, but guarantee that we obtain the same information content as the 2000 random samples hold.

We can maximize the average information per sample for unknown wave shapes and for arbitrary vector length using *coherent testing* with relatively prime M and N. We choose an offset F_t to make samples fall at different points in later waveform cycles, but the coherence property makes the sample locations controllable and repeatable. When all sampled signal periods are superimposed, the result is N samples equi-probably distributed in time over one waveform cycle [412], which is called the *primitive cycle*. The *primitive spacing* is $360/N$ degrees.

These equations define the necessary relationships:

$$F_t = M \times \Delta \qquad (10.14)$$
$$F_s = N \times \Delta \qquad (10.15)$$
$$\Delta = 1/UTP \qquad (10.16)$$

M and N must be relatively prime, and Δ is called the *primitive frequency*. These equations lead to the fundamental coherence requirement:

$$\frac{F_t}{F_s} = \frac{M}{N} \qquad (10.17)$$

Example 10.4 *Corrected CODEC Testing.* With $N = 400$ and $F_s = 8000\ Hz$, Equation 10.15 now gives $\Delta = 20\ Hz$, so all F_t waveforms must be multiples of Δ. Equation 10.17 then eliminates $F_t = 1000\ Hz$, since that would give $M = 50$, which is not relatively prime with $N = 400$. We instead choose $M = 51$, so $F_t = 1020\ Hz$, and we get these parameters:

$$
\begin{array}{llll}
F_t & = 1020\ Hz & UTP & = 50\ ms \\
F_s & = 8000\ s/s & \Delta & = 20\ Hz \\
M & = 51 & N & = 400
\end{array}
$$

With M and N relatively prime, the information content of the samples is maximized.

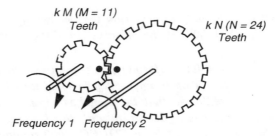

Figure 10.23: Gear train analogy to DSP sampling.

Coherence Methods. Figure 10.23 [412] provides a mechanical system analogy with a rotating gear system where every element can run at a different rate, as needed, but is coordinated in time so that over a time interval all gears make an integer number of rotations. The programmer can choose any number of gears, arrange them on different shafts, set arbitrary gear sizes, and rotate them to initial phase offests. However, every gear must make an integer number of rotations during the total testing time period.

Coherence with a Phase-Locked Loop. We use two frequencies and one unit test period. $F1$ is the primary frequency, while $F2$, the secondary frequency, is generated from $F1$ by a PLL. The requirement of sampling an integer number of cycles means that:

$$F2 = F1 \times M/N \qquad (10.18)$$

Example 10.5 *Let $F1 = 20\ Ms/s$ be the sampling rate of an ADC-under-test. We want to test the ADC near the Nyquist frequency, such that $N = 1024$ (the number of samples.) We then select $M = 511$ so that M and N are relatively prime. Then, the test frequency has to be $F_t = 20 \times 511/1024 = 9.98046875\ MHz$. Let the signal source be a sine wave generator operating at $20\ MHz$ to generate $F_s = 20\ MHz$. Then, we configure the DSP-based ATE so that the PLL produces the $9.98^+\ MHz$ frequency relative to F_s. This is done by setting the ATE feedback divider (forward multiplier in the PLL) to M and the combined forward dividers to $N = 1024$.*

Coherence with Parallel Division. Another way to produce coherent frequencies is to use a common high-frequency clock, and two different dividers set in the ratio of M/N. Parallel division is better than PLL synchronization, because it is simpler, has lower jitter, and can more rapidly respond to new M/N settings. However, in the above Example 10.5, parallel division will only work if the master clock is $511 \times 20\ MHz = 10.22\ GHz$. This is now feasible, and is often done. We see that the PLL has the advantage of providing the effect of a $10.22\ GHz$ clock without the need to generate it. We define the *implicit clock rate* as $F1 \times M$.

Vector Periodicity. In Figure 10.23, notice that there are two marks on the two gears. Imagine that we use a strobe light to observe the relative positions of

10.5 Realizing Emulated Instruments Using Fourier Transforms

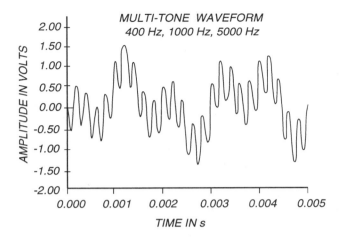

Figure 10.24: A multi-tone waveform.

the marks at a time instant t. As we observe the system, the gears appear in the same state at $(t+P), (t+2P)$, etc. P is the *primitive period* corresponding to the *unit test period* (UTP.)

$$\text{effective sampling rate} = M(F_s) = N(F_t) \tag{10.19}$$

For high-speed undersampling, this is the theoretical rate necessary to distribute N relatively prime samples of 1 cycle of frequency F_t.

Information in a Vector. The information content of a sampled waveform, when N and M are relatively prime, is proportional to N and independent of M [412]. This means that there is no lower limit to F_s in coherent testing. This is not possible with conventional, non-coherent sampling (or out-of-band sampling), where F_s cannot be less than the Nyquist limit, $2W$. With completely periodic signals, however, one collects $1/M$th of the information on each cycle, until all M cycles are sampled. The prime ratio of M/N ensures that each cycle contains unique, independent information.

Example 10.6 *In coherent testing, the UTP contains N unique samples and M essentially identical signal cycles. Figure 10.22 [412] shows an example waveform over its UTP. Figure 10.24 [412] shows a multi-tone test waveform. Figure 10.25 [412] shows the frequency spectrum of the sample set, which is not necessarily the same as the frequency spectrum of the actual waveform. In these spectra, there are $N/2+1$ discrete locations, or bins, where each bin frequency is a multiple of the primitive frequency, $\Delta = 1/UTP$. All information about the sampled waveform is in the band $0 \leq f \leq N\Delta/2$, called the primitive band. Each bin has amplitude and phase information, except for the two end bins. $N\Delta/2$ is the Nyquist frequency bin, which contains misleading information, so it must not be used as a test frequency.*

Chapter 10. DSP-BASED ANALOG AND MIXED-SIGNAL TEST

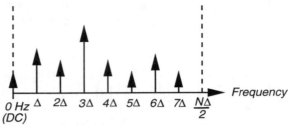

The frequency spectrum of the sample set is not necessarily the same as that of the original waveform. There are N/2 + 1 discrete bins in the sample spectrum. Each frequency bin is a multiple of Δ, the primitive frequency.

$$\Delta = 1 / UTP$$

The primitive band, from 0 to N Δ/2, contains all information. Except for the two end bins, all bins have amplitude and phase information. NΔ/2 is the Nyquist frequency. The Nyquist bin must not be used as a test frequency, as it contains misleading information.

Figure 10.25: The primitive frequency.

Example 10.7 *Figure 10.26 [412] shows another example of a 6-bit spectral test of an 16-bit ADC in which only the 6 most significant bits are sampled. Figure 10.27 [412] shows improper DSP-based testing, in which M and N are NOT relatively prime. Figure 10.28 [412] shows proper DSP-based testing, in which $M = 103$ and $N = 1024$ are relatively prime, so this is a valid testing scheme. Finally, Figure 10.29 shows the major spectral components of the ADC sample set.*

Coherent Filtering. Note that for the DSP-based ATE approach to be useful, we must eliminate filter settling times, as well as non-linear analog circuits, between the output of the DUT to be sampled and the ADC in the test instrument. Otherwise, we lose most of the DSP-based ATE speed advantage over conventional analog ATE. It is frequently tempting to put a band-pass filter in between the DUT and the digitizer in the ATE. However, this ruins the effectiveness of the ATE, because the filter introduces a signal settling time much longer than one signal period. For simple ATE filters, Mahoney's rule-of-thumb [412] is a settling time of 5 to 10 times the reciprocal of the 3 dB bandwidth, to settle to 0.1%. Coherent filtering is based on waveform correlation, where we digitize the unfiltered DUT output, and then use emulated instruments in the DSP software to derive measurements.

Correlation. Coherent correlation R of functions A and B is defined as [412]:

$$R(\tau) = G \cdot \int^P A(t) \cdot B(t - \tau) dt \qquad (10.20)$$

A and B may be functions of any variable, τ is a programmable delay, and G is the gain or scale factor. If the delay τ is positive, signal B is made to lag A, and vice

10.5 Realizing Emulated Instruments Using Fourier Transforms

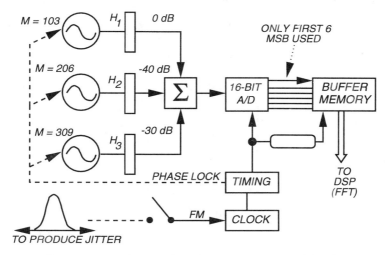

Figure 10.26: Example 6-bit spectral test of ADC.

versa when τ is negative (B leads A.) For any given delay, the correlation indicates the degree to which two waveforms are alike in shape. For *normalized correlation*, $-1 \leq R \leq +1$, regardless of the actual amplitudes of A and B. $R = +1$ indicates identical waveforms, $R = -1$ indicates inverted waveforms, and $R = 0$ indicates that the waveforms are statistically unrelated. Correlation is normalized with this scale factor:

$$G = \frac{1}{RMS(A) \times RMS(B) \times UTP} \qquad (10.21)$$

Cross-correlation indicates a comparison of two different signals, whereas *autocorrelation* is the comparison of the same waveform at both inputs of the *correlator* (the correlation hardware.)

It is possible, by combining two correlators with various other functions, to construct a rapid filter/measurement unit called the *Fourier voltmeter*. This construction depends on two principles that we now describe.

Fourier's first principle: In the correlator of Figure 10.30 [412] consider two steady-state analog signals, A and B in Figure 10.31 [412], of different frequencies. Let P denote the test period, which is not necessarily the UTP. If P is infinite, the correlation R between A and B is 0. If P is finite, and if P contains an integer number of A cycles (M_A) and an integer number of B cycles (M_B), then the cross-correlation R is 0, regardless of phase or amplitude. This principle allows us to force zero cross-correlation between two sinusoids. It applies to every component of a complex wave formed by the linear addition of coherent sinusoidal waveforms.

Example 10.8 *Mahoney [412] illustrates Fourier's first principle. Apply sinusoid A to both inputs of the correlator, and $R = +1$. Linearly add another coherent frequency sinusoid to the lower channel without changing G. The correlation is unaffected, and the two A components still correlate completely, while cross-correlation*

344   **Chapter 10. DSP-BASED ANALOG AND MIXED-SIGNAL TEST**

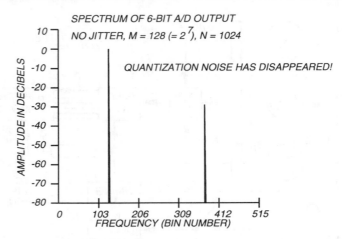

Figure 10.27: Spectrum of 8-bit ADC output.

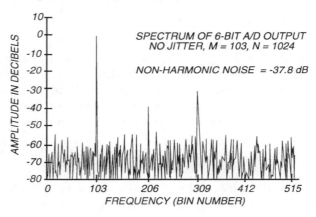

Figure 10.28: Spectrum of 6-bit ADC output.

with the new A component is 0. The total correlation is the combined correlation of + 1. The correlator in a coherent system can function as an infinitely selective band-pass filter, so we can ignore all frequency components in a signal, other than the one matching the frequency of a reference input.

Figure 10.32 [412] illustrates *Fourier's second principle*. If two sinusoids A and B of the same frequency are displaced 90° in phase, and there is a whole number of signal cycles J in the test interval P, then the cross-correlation is zero, regardless of amplitude or the starting point of the test interval P. By holding G and the reference amplitude constant, we can use the single correlator to determine the relative strength of an in-phase matching component. This principle allows a pair of correlators to measure the relative amplitude of a matching component, regardless of its phase. This is because any steady-state sinusoid locked to a reference frequency can be resolved into two parts: one in phase with the reference signal and one

10.5 Realizing Emulated Instruments Using Fourier Transforms

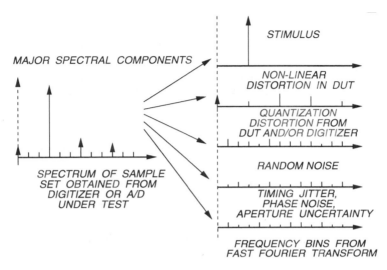

Figure 10.29: Major spectral components in DSP-based testing.

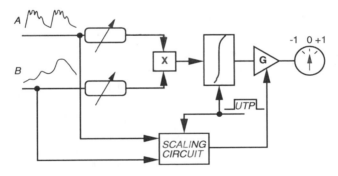

Figure 10.30: Analog model of correlation.

that lags the reference by 90°. If the reference is a cosine wave, a locked matching component can be resolved into a cosine part and a sine part. When two correlators are driven by quadrature reference signals, one output indicates the cosine part strength while the other indicates the sine part strength. The two correlator outputs vary with the phase of the test component. However, their power sum is independent of test component phase and proportional to the power of the matching component. The phase relative to the cosine channel is given by:

$$\text{phase angle} = \arctan\left(\frac{\sin x}{\cos x}\right) \tag{10.22}$$

10.5.1 Fourier Voltmeter

Mahoney [412] provides the following excellent tutorial discussion. The pair of quadrature correlators discussed above forms the *Fourier voltmeter* (FVM), and

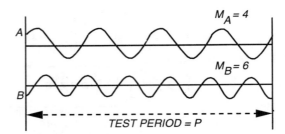

Figure 10.31: Waveforms showing Fourier's first principle.

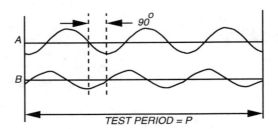

Figure 10.32: Waveforms showing Fourier's second principle.

provides an infinitely sharp band-pass filter, an AC voltmeter, and a phase meter. A single FVM can measure the magnitude and phase of any arbitrary spectral component in a periodic waveform. Thousands of parallel FVM units, each tuned to a different frequency component (in units of J), can perform detailed spectral analysis of a signal in a split second. Figure 10.33(a) [412] shows the rectangular output (cosine and sine) version of the FVM, and Part (b) shows the polar output (magnitude and phase) version. The polar version requires additional software for trigonometric operations, and a square root function to convert power into magnitude. In Figure 10.33(a) the rectangular form is made from a pair of quadrature correlators with a constant scale factor given by Equation 10.21. We must calibrate this unit to denormalize R such that each of its two parts shows the *actual amplitude in volts*, so we must multiply G by E, the actual signal component amplitude. In the DSP system, the reference cosine and sine functions A have unit peak amplitude, so the RMS amplitude value is $\sqrt{2}/2$. B represents the signal being compared with reference frequency A. So, $RMS(A) = \sqrt{2}/2$, $RMS(B) = E\sqrt{2}/2$, and:

$$G(FVM) = \frac{E}{(\frac{\sqrt{2}}{2})(\frac{E\sqrt{2}}{2}) \times UTP} = \frac{2}{UTP} \qquad (10.23)$$

Figure 10.34 [412] shows the analog equivalent of the FVM.

Software Version of Fourier Voltmeter. In an actual DSP system, the FVM is actually a software algorithm operating in the DSP processor. The signal vector is $\mathbf{X}$ with individual components $\mathbf{X}(I)$. I is the index or position, $1 \leq I \leq N$. The quadrature computation is done by multiplying $\mathbf{X}$ by a cosine vector $\mathbf{C}$ and by a

10.5 Realizing Emulated Instruments Using Fourier Transforms

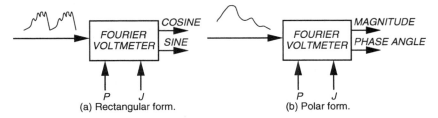

Figure 10.33: Two forms of Fourier voltmeter.

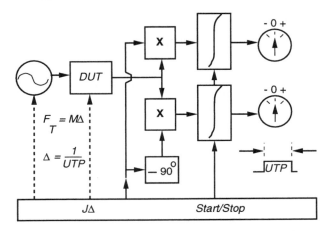

Figure 10.34: Analog Fourier voltmeter equivalent.

sine vector **S** using a • vector dot product operation:

$$\text{cosine part} = \frac{2}{N} \sum_{I+1}^{N} \mathbf{X}(I) \bullet \mathbf{C}(I) \tag{10.24}$$

$$\text{sine part} = \frac{2}{N} \sum_{I+1}^{N} \mathbf{X}(I) \bullet \mathbf{S}(I) \tag{10.25}$$

Index position I defines time, and the analog scale factor $2/UTP$ is replaced by $2/N$.

There are exactly J cycles in vectors **C** and **S**, which are precomputed and stored in memory the first time that a new value of J is encountered in a DSP program. The rectangular form of the FVM is faster and more accurate than the polar form, due to less computation. Test engineers can use the rectangular form by using signal *power* wherever possible, rather than voltage amplitudes. So, Equation 10.26 is preferred to Equation 10.27 for expressing performance as a dB ratio:

$$\text{Number of } dB = 10 \log \left(\frac{P2}{P1}\right) \tag{10.26}$$

$$\text{Number of } dB = 20 \log \left(\frac{V2}{V1}\right) \tag{10.27}$$

Chapter 10. DSP-BASED ANALOG AND MIXED-SIGNAL TEST

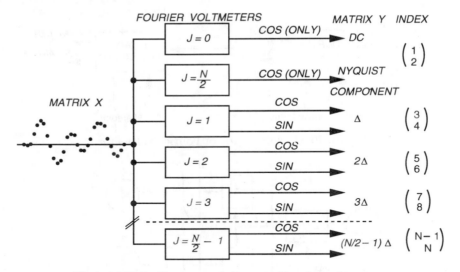

Figure 10.35: Conceptual discrete Fourier voltmeter.

Since the FVM reports peak voltage *amplitude*, power computed from this is twice the *true power*, so the following adjustment must be made:

$$\text{Average sine wave power} = \frac{\text{peak power}}{2} \qquad (10.28)$$

DFT, FFT, and Vector Synthesis. We often need to know amplitude and phase of many components. If N is a power of 2, the *Fast Fourier Transform* (FFT) effectively applies a full set of Fourier voltmeters at once for each spectral position. The FFT is slower than a single *Discrete Fourier Transform* (DFT), but much faster than using a separate DFT for each spectral line. The FFT provides $N/2$ FVMs and a DC voltmeter. The FFT turns time waveforms into frequency spectra, and this process is reversed to create a vector for the ATE waveform synthesizer by applying the *inverse* FFT. This operation works correctly only for select values of N (usually powers of 2.) If the UTP is not a power of 2, one cannot truncate the vector or fill out a short one by adding duplicate or zero samples. These techniques appear in textbooks, but they destroy coherence and remove orthogonality. Instead, we must alter F_s, F_t, and/or M to obtain an N that is a power of 2.

Orthogonal Signals. The coherence requirement carries an additional benefit in DSP-based testing: When two or more sinusoidal components are in the response, they can be *statistically orthogonal*. Two orthogonal complex functions or waveforms have zero cross-correlation, the same as two sinusoids at right angles. In the DSP domain, two vectors are orthogonal if the sum of their index-by-index products is zero [412]. Two orthogonal waveforms or vectors have these additional properties:

1. They are statistically independent.

10.5 Realizing Emulated Instruments Using Fourier Transforms

2. Each conveys separate, unique information.

3. If linearly added, they may be unambiguously separated.

4. When linearly added, their resulting signal power is the arithmetic sum of the individual component powers.

The first two properties permit application of many different test signals at once and allow simultaneous examination of different DUT properties. The third property lets us separate and measure each individual signal and distortion component accurately, even when there are thousands. Since random noise correlates poorly with coherent signals, true noise can be distinguished and separately measured from DUT quantization and distortion components. The final property is most important. Total power cannot be computed by addition unless the components are orthogonal. This allows us to directly compute spectral line power from the rectangular FVM, so total power is the *sum of the squares* (SSQ) of all of the individual FVM outputs for each FVM in the spectrum, if and only if measurement is coherent.

More than two hundred years ago, Fourier noted that sinusoidal orthogonality is guaranteed by these two conditions:

1. Different frequency sinusoids are orthogonal over a given interval if each produces a *whole number* of cycles, regardless of their relationships.

2. Sinusoids of identical frequency are orthogonal over a given interval if they produce a *whole number* of cycles and the two waveforms are out of phase by one-quarter cycle.

Figure 10.35 [412] shows a conceptual discrete Fourier voltmeter.

Frequency Leakage. The FVMs are accurate only if all components to be analyzed or synthesized are whole multiples of Δ. If the stimulus waveform satisfies this, then all output harmonics, all *intermodulation* (IM) products, and all quantization distortion components are also whole multiples of Δ. Random noise does not satisfy this requirement, but it is usually not problematic because its correlation with the FVM reference frequency is very small with large N. Instead, problems occur when the test engineer overlooks a periodic component, so it has no FVM location.

Example 10.9 *Mahoney [412] gives the example of a telecommunications notch filter program written for North America, with primitive frequency = 20 Hz and a UTP = 50 ms. This tests the filter rejection at 60 Hz. However, if the program is used in Europe, the notch is tested at 50 Hz, midway between two FVM bins. This causes the measurement to be wrong, and the 50 Hz component leaks into every other bin in the FFT spectrum. This leakage happened because sinusoids without coherence over the integration interval are no longer orthogonal to the FVM reference frequencies. Figure 10.36 [412] shows that leakage is greatest in bins closest to the offending non-coherent tone. In this figure, all components that are integer multiples of Δ fall into nulls, except the one at J (the one we are measuring.) The bin*

Chapter 10. DSP-BASED ANALOG AND MIXED-SIGNAL TEST

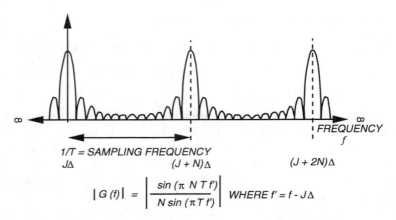

Figure 10.36: Voltage swept response of FVM.

frequences are Δ (FVM 1), and $2 \times \Delta$ (FVM 2.) Bins are numbered from 0 to $N/2$. Bin 0 is the DC voltage, and bin $N/2$ contains the magnitude of the cosine component at the Nyquist frequency. Bins 1 through $N/2 - 1$ have number pairs, either cosine-sine rectangular parts, or magnitude-phase pairs. Bin 1 describes the primitive frequency. The fundamental component of F_t is in bin M.

For example, if the primitive frequency is $20~Hz$ and there are no signal components at 40 or $60~Hz$, bins 2 and 3 are empty. If a $50~Hz$ component is present ($3\Delta/2$), bins 2 and 3 have leakage, as do all bins. For testing, the real problem is that these other bins contain components that we are trying to measure, so the leakage component power and that of the valid bin component are no longer additive because they are not orthogonal. There is no easy way to correct for leakage.

Example 10.10 *A magnitude FFT is applied to a signal sampled from an Analogic MP2735 ADC. Figure 10.37 [412] shows the spectrum of this signal. The input was a coherent sine wave generated by driving a 16-bit digital-to-analog converter (DAC) at $1076 +$ Hz. A very low distortion bandpass filter smoothed the DAC output, and had no harmonic above -115 dB. $N = 1024$ and the ADC sampling rate was $F_s = 44100~s/s$, the rate used for compact disc recording.*

10.5.2 Testing of Analog Devices Using Non-Coherent Sampling

DSP testing involves two kinds of sampling: coherent and non-coherent. The latter arises from CODEC, switched-capacitor filter, ADC, or DAC devices that are designed to sample non-coherent waveforms (e.g., speech, music, or video signals.) Therefore, we must discuss non-coherent sampling (or out-of-band sampling) for DSP-based testing.

Example 10.11 *Consider a band-pass filtered speech waveform that is being sent over telephone wires. The spectrum sharply rolls off below $300~Hz$ and above*

10.5 Realizing Emulated Instruments Using Fourier Transforms

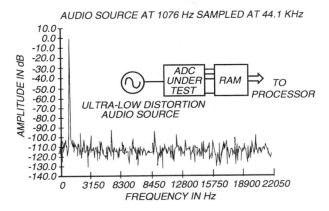

Figure 10.37: Spectrum of ADC converter.

3400 Hz, and power is effectively 0 at DC and above 4000 Hz. Figure 10.38(a) [412] shows the time-domain waveform, Figure 10.38(b) shows the frequency spectrum, and Figure 10.38(c) shows that if $F_s = 8000$ s/s, no information is lost. In Figure 10.38(d), we see that sampling replicates the spectrum indefinitely, as mirror images, so that the spectrum is periodic with F_s as its period.

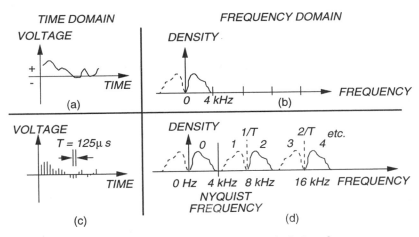

Figure 10.38: Spectrum of sampled signal.

Reconstruction. Although it seems that signal variations between sampling points are lost, the full sample set contains all necessary information to *reconstruct* the continuous, time-varying function. This is achieved by suppressing all portions of the periodic frequency spectrum except the original one using a *reconstruction filter*. This causes the time function to be continuous and assume its original shape. Digital reconstruction filters can only approximate the original curve by adding extra, interpolated points, but a classical analog filter can completely reconstruct the waveform. For digital reconstruction, the output clock must be greater than the primary sampling rate, which leads to the *oversampling filter*.

Time and Spectral Vectors. The sampled, time-varying signal is expressed as a numerical vector, and then we compute its spectrum. Since the time vector is discrete, the spectrum is *periodic* but continuous, so we must also sample the spectrum as well as the time-varying signal, since we cannot store an infinite number of spectral points. However, by sampling the spectrum, we make the original waveform periodic, whereas, in reality, it is not.

Imaging and Noncoherent Undersampling. All sampled signals, aperiodic or periodic, have periodic frequency spectra of the sampling frequency, F_s. For reconstruction, only half of the period is needed, from DC to $F_s/2$, and this is called the *Nyquist interval*. The second half of the period is merely the in-phase reversal of the first half. The DSP processor chooses the Nyquist interval by default for FFT and DFT routines. This use of the Nyquist interval is called *undersampling*.

Heterodyning and Reconstruction. Any one of the replicated half period signals in the frequency spectrum can be used to reconstruct the original signal, by playing it back through a band-pass filter. The filter merely shifts the frequencies downward. In RF applications, this technique is called *heterodyning*, where the local oscillator acts as the sampler. Heterodyning is also used in *frequency division multiplexing* (FDM) to assign different frequency zones to signals originally having the same frequency zone. Heterodyning requires low sampling jitter, low aperture uncertainty, and low DAC glitch area. These requirements are set by the actual test frequency F_t. The sampled signal must not spill over into an adjacent zone from the Nyquist interval, or images will overlap and become ambiguous. Interfering images are called *aliases*. The test engineer must keep track of the image locations to avoid aliasing.

Rules of Imaging. Here, f^0 represents any specific frequency between $0\,Hz$ and $F_s/2$. $f^i|_{i>0}$ represents the image of the specific frequency f^0 in the periodic spectrum in one of the images, as labeled in Figure 10.38(d). $F_s/2$ is the *Nyquist frequency*, and the zone from 0 to $F_s/2$ is the *Nyquist region*. For coherent testing, the spectrum in this region is the *primitive spectrum*. Table 10.5 shows some example zones. Then, $f^1 + f^0 = F_s$, $f^2 - f^0 = F_s$, $f^3 + f^0 = 2F_s$, and $f^4 - f^0 = 2F_s$, etc.

Table 10.5: Frequency zones for spectral images.

Image name	Frequency zone
f^0	Signal component in 0 to $F_s/2$
f^1	Corresponding image component in $F_s/2$ to F_s
f^2	Corresponding image component in F_s to $3F_s/2$
f^3	Corresponding image component in $3F_s/2$ to $2F_s$

10.5 Realizing Emulated Instruments Using Fourier Transforms

Sampling Rates – Shannon's Theorem. In order to handle imaging, we must use Shannon's Sampling Theorem [595].

Theorem 10.1 *Shannon's sampling theorem. If a function of time $f(t)$ contains no frequencies higher than W Hz, it is* completely determined *by giving the function value at a series of points spaced $1/(2W)$ seconds apart.*

This gives a sufficient condition for complete determination of the original signal, so that *all* information is preserved. This condition is sufficient for any time-varying function. There is also no restriction on the phase or time origin of the samples, so during testing if the DUT has unknown phase shift, we can still measure magnitude and/or phase. The absolute phase of the reconstructed signal is not affected by the phase of the sample set relative to the signal [412]. The theorem establishes the condition of regular sample spacing, with no gaps, jitter, frequency modulation, or pulse modulation. We cannot use irregular sampling, with *average* rate of $1/2W$, unless we sample at a much higher rate, and know exactly where the samples occurred in time. Uniform sampling requires the fewest samples for waves from stochastic sources. Shannon's theorem applies to the limiting case of classical functions, $f(t)$, where t ranges from $-\infty$ to ∞. With finite length vectors, the regularly-spaced points of the theorem must be closer than $1/2W$ apart. For sampling periodic waveforms in analog testing, one extra sample is sufficient.

Nyquist's Limit. The following *Nyquist limit* is a *necessary* condition on the signal bandwidth, rather than its highest frequency.

$$F_s > 2W \tag{10.29}$$

This limit guarantees the preservation of information in the original signal.

Universal Rule for Non-Coherent Sampling. If all signal spectral energy is in a spectrum of width $W = f_H - f_L$, then F_s must be chosen so that the interval $[f_L, f_H]$ falls within two adjacent harmonics of $F_s/2$, as in Figure 10.39 [412].

$$\text{If } f_L > \frac{nF_s}{2}, \text{ then } \frac{(n+1)F_s}{2} > f_H \tag{10.30}$$

These two inequalities lead to the *universal rule for non-coherent sampling*:

$$\frac{2f_L}{n} > F_s > \frac{2f_H}{n+1} \tag{10.31}$$

where n, the image or zone number, must be an integer ≥ 0. For $n = 0$, we have the *low-pass* case, while for $n \geq 1$, we have the *band-pass* case. The band-pass case is not always possible, and will be limited to a few zones for F_s. We compute n^*, the highest usable value for n:

$$n^* = \text{integer} \left[\frac{f_L}{W}\right] \tag{10.32}$$

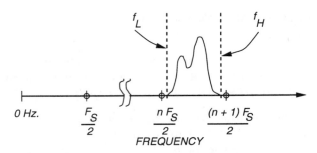

Figure 10.39: Universal rule for non-coherent sampling.

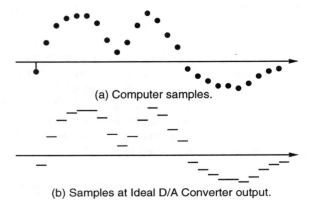

Figure 10.40: Samples with finite pulse width.

sin x/x (sinc) Distortion and Correction. The zero-width samples (see Figure 10.40(a)) that we have assumed are more or less valid for vectors coming from DSP system digitizers and high-speed ADCs. However, when these samples are played back through DACs, they have a finite width (see Figures 10.40(b) and (c).) We reprint Mahoney's discussion of this issue [412]. When a time-varying signal is represented by uniformly spaced pulses of width W_p that is tiny compared to the spacing T, the spectral images from sampling have equal strength (see Figure 10.41(a).) The dotted line is more-or-less flat, indicating that the spectral images extend indefinitely to $+\infty$, with no attenuation. When W_p widens, more power goes into the low end of the spectrum, and less into the high end, so the spectrum rolls off with higher frequencies. It drops to zero at $1/W_p$, $2/W_p$, $3/W_p$, etc. This distorts each of the individual images in amplitude and phase, and changes the relative amplitude of different images. This resulting envelope follows the $\sin(x)/x$ curve, where x is a function of W_p and frequency. For any frequency f, it is attenuated by:

$$\left| \frac{\sin(\pi f W_p)}{(\pi f W_p)} \right| \tag{10.33}$$

and phase shifted by:

$$-\pi f W_p \ radians \tag{10.34}$$

10.5 Realizing Emulated Instruments Using Fourier Transforms

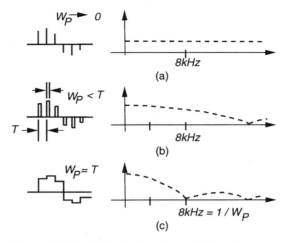

Figure 10.41: Effect of pulse width on images.

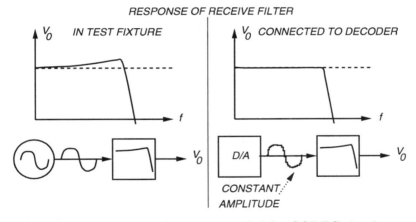

Figure 10.42: Filter adjustment needed for CODEC circuit.

Figure 10.41 [412] shows three pulsed waveforms, all with flat tops. The level is the zero-order derivative, the actual function, at the time instant of the *leading edge* of the sampling pulse. This is called *zero-order-hold sampling*. The information content of all three sampled waveforms is identical.

Example 10.12 *In order to correct $\sin(x)/x$ rolloff, the hardware reconstruction filer, or receive filter, in a CODEC must have a filter curve that rises in the system passband to offset the $\sin(x)/x$ droop. The filter curve must drop sharply through the Nyquist frequency to remove the adjacent image (see Figure 10.42 [412].)*

Figure 10.43 [412] shows a test setup for a telecommunications circuit, where the analog DUT is mostly digital. The DUT uses an internal ADC to sample at $F_s = 8000$ s/s, does digital signal processing, and then converts the data back into analog form with a DAC. The testing digitizer examines the DUT output from 0 to

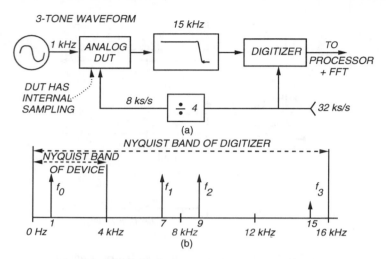

Figure 10.43: Testing setup for mixed-signal circuit.

16 kHz by clocking the digitizer at 32 ks/s (see Figures 10.43(a) and (b) [412].)

The digital anti-aliasing filter will cut off components above 15 kHz. Because the DUT itself samples at one quarter the rate of the digitizer, any input f has three significant images f^1, f^2, and f^3. Here, three equal amplitude frequencies (1000, 200, and 3500 Hz) were applied simultaneously and coherently with the DUT clock.

10.5.3 Coherent Multi-Tone Testing

In a coherent testing environment, the test stimulus is composed of many mutually orthogonal sinusoidal tones. These may be simultaneously sent through the DUT to produce a response, from which many different parameters can be measured.

Figure 10.44 shows a test where the stimulus is the linear sum of two tones of frequencies $3f$ and $7f$, respectively. The period P is any integer multiple k of the UTP. Here, $k = 1$. In this example, the DUT input and output are digitized for 1 UTP each, under phase lock to maintain coherence. Each sampled vector is then converted into a spectrum by the polar FFT. This gives four measured parameters: gain and phase shift at both $3f$ and $7f$.

Distortion Measurement. The method of Figure 10.44 [412] is often used as a fast screening test for low, mid, and high-frequency amplifier response, but with the magnitude FFT and three to five tones. Much more information, however, can be provided if prime multiples of tones are chosen. With prime tones $3f$ and $7f$, second harmonic distortion appears in bins 6 ($2 \times 3f$) and 14 ($2 \times 7f$), and third harmonic distortion appears in bins 9 and 21. The actual frequency requirement is that the set of all tone frequencies, which includes F_s, be irreducible (unfactorable.)

Intermodulation distortion (IM) can be measured with two or more tones. Second-order (sum-and difference) components occur in bins 4 ($7f - 3f$) and 10

10.5 Realizing Emulated Instruments Using Fourier Transforms

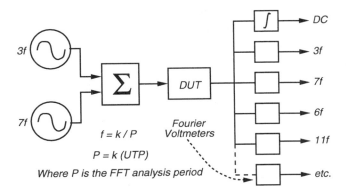

Figure 10.44: Coherent multi-tone testing.

$(7f + 3f)$, and third-order components fall in bins 1 $(4f - 3f)$, 13 $(3f + 10f)$, 11, and 17. The dynamic offset is in bin 0 (the DC component with a signal applied.) If input tone frequencies are prime multiples, we minimize the likelihood that more than one component will fall in the same bin.

The FFT of a vector of N samples has a spectral vector with $(N/2) + 1$ bins. With two or three tones as stimulus, the number of bins is far larger than the number of interesting harmonic and IM components. Many bins will be unoccupied, and are called *non-harmonic* bins. Nonetheless, these bins are helpful, because quantization distortion and random noise spread over all bins, so we can measure their effect independently of non-linear distortion. We estimate combined quantization and random power by measuring the average power per bin in the non-harmonic bins and scaling by $(N/2) + 1$. Since most non-harmonic power is removed from the harmonic bins, we obtain a more accurate non-linear distortion measurement.

One limitation in ADC testing is that with a test period $P = 1\ UTP$, it is not possible to separately measure quantization distortion and random noise. This is because the distributions of the two are similar. However, when P is a multiple of the UTP, the quantization error pattern repeats from 1 UTP to the next, whereas random errors do not. With $P = 2\ UTP$, we can correlate the first half with the second, and calculate the ratio of quantization distortion to random noise.

Separation of quantization error and random noise is done in the frequency domain. Let $P = k \times UTP$. Then we have kN time samples, and the magnitude FFT has $(kN/2) + 1$ bins and $\Delta = f/k$, where f is the primitive frequency. Refer to Figure 10.44. With no random errors, quantization distortion falls only in bins that are multiples of k. Symmetrical quantization produces only odd k components, while non-symmetrical quantization produces components at all integral values of k. This would be caused by DC offset or non-linearity in the ADC. With no random errors, bins that are not integral multiples of k are empty. So, if $k = 4$, there are three empty bins for each one that contains quantization distortion, and the empty bins are used to measure random noise. We add the power in these empty bins, scale it by $k/(k-1)$, and obtain the estimated total random noise of the converter under

Chapter 10. DSP-BASED ANALOG AND MIXED-SIGNAL TEST

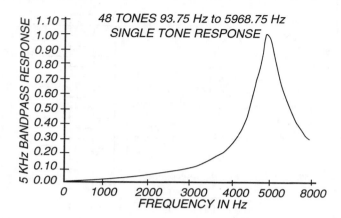

Figure 10.45: MF10 Magnitude results of 48 single-tone tests.

test. This includes random voltage (Boltzmann) noise and jitter-induced sampling errors. We estimate quantization distortion by adding the power of every kth line, but k must be big enough so that the average power in the random bins is much smaller than the average power in the quantization bins.

Another benefit of having $k > 1$ is the minimization of interference of external frequencies on a test, when those frequencies cannot be changed. For example, a test using tones at 1020 and 1104 Hz, with $F_s = 8000\ s/s$, gets FFT leakage errors if there is 50 Hz power interference. However, with $k = 2$, FFT bins occur at 10 Hz multiples, thus creating a valid bin for 50 Hz, so the leakage goes away.

Total Harmonic Distortion. A frequent number obtained from testing is *total harmonic distortion* (THD), which measures the amount of power appearing in the harmonics (H_2, H_3, ... in dB) of the fundamental tone (H_1 in dB) as a percentage of the power that is in the fundamental frequency in the response spectrum.

$$THD = \frac{\sqrt{10^{\frac{H_2}{10}} + 10^{\frac{H_3}{10}} + ... + 10^{\frac{H_{10}}{10}}}}{10^{\frac{H_1}{20}}} \qquad (10.35)$$

Frequency Measurement. Normally, multi-tone testing uses, at most, four tones. Large numbers of tones, however, are used to replace conventional *swept* frequency measurements. Figure 10.45 [412] shows the single-tone response measurement for a National Semiconductor MF10 switched-capacitor band-pass filter. The data was gathered by measuring the gain at 48 frequencies, one at a time. Figure 10.46 [412] shows a multi-tone measurement, in which all 48 tones were superimposed and applied at the same time. Obviously, the result is the same, except that the multi-tone measurement would take 1/48th as much time on the tester. In Figure 10.47 [412] the phase versus frequency-response is presented, for the same experiment. For in-band tones, the relative phase accuracy was better than 0.1°, which is the limit of an analog bench phase meter used for comparison.

10.5 Realizing Emulated Instruments Using Fourier Transforms

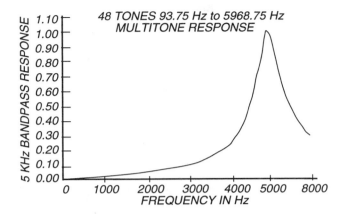

Figure 10.46: MF10 Magnitude results of one multi-tone test.

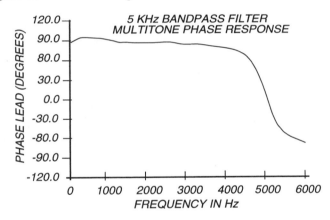

Figure 10.47: MF10 Phase lead result of one multi-tone test.

Contrasts with Single-Tone Testing. Multi-tone testing is sometimes the only practical technique for a given test. For example, it is essential for IM distortion testing, we need two tones for CCITT* tests, and we need four tones for AT&T standard telecom tests. AT&T standard tests for *peak-to-average ratio* (P/AR) require a 16-tone set with precise phases and amplitudes. CCITT CODEC tests for noise use multi-tone stimuli with 25 or more randomly-placed spectral lines. Also, multi-tone testing is required for DSP-based CCITT envelope or group delay tests.

Multi-tone testing is faster for DSP phase measurement than single-tone tests. In addition, multi-tone testing avoids the possibility of phase drift as the DUT warms up in the test fixture as single-tone tests are applied serially. Multi-tone tests must use the DFT when N is not a power of two. For example, telecom tests for trunk signaling require N to be a multiple of 6 or 12. Also, the DFT is a quick alternative to the FFT in ATE that lack an array processor.

---

*International Telegraph and Telephone Consultative Committee (CCITT).

For amplitude or phase measurement, single-tone tests may be superior to multi-tone tests. For example, a coherent sinusoidal test can rapidly and accurately measure amplitude using an RMS operation in the time domain. This is much faster than the DFT or the FFT, and works with any N. Multi-tone tests are faster for measuring frequency response, because they can apply all of the tones concurrently, rather than serially, over time, to the DUT.

Error Sources. Multi-tone measurements are less accurate than single-tone tests, when all other things are equal, because the relative individual tone amplitudes must be small to prevent the peak-to-peak composite waveform swing from exceeding the input range of the DUT. This unfortunately means that each tone has a poorer *signal-to-noise* (S/N) ratio.

When the DUT has no quantization or digital filtering, however, multi-tone tests are just as accurate as single-tone analog instruments, because the benefits of the DSP unit (describer earlier) cancel the poorer S/N ratio. To compare accuracies, consider the MF10 filter test over 20 runs, as described in Figure 10.46. The 48-tone test had 0.02 dB uncertainty, a single-tone analog analyzer had 0.03 dB uncertainty, and a single-tone DSP test had only ± 0.003 dB uncertainty.

Device Uncertainty. Scatter is defined using the *mean* and the standard deviation σ (RMS value) or peak deviation value. The variance is σ^2. Scatter should only happen from variations among tested devices, but the mean also shifts by calibration error, and σ increases due to quantization and random noise in the ATE and the test fixture. Quantization in the DUT also increases σ. Excess scatter is caused by dynamic uncertainties in the DUT and the test system, above and beyond the production spread of the test parameter. We call this *measurement uncertainty*.

Modern VLSI mixed-signal circuits use discontinuous time sampling functions and discontinuous amplitude quantization functions (e.g., ADC, DAC, comb filter, CODEC, digital filter, delta modulator, DSP, etc.) These functions interact with test signals and measurement procedures in the DSP-based ATE. Measurement is less certain than for earlier, all-linear analog circuits, so good devices may appear to fail. Measurement uncertainty sources include:

1. Synchronous interference

2. Discontinuous or non-linear functions.

This uncertainty does not behave like random noise, so it does not cause non-repeatable testing, and escapes notice.

In a coherent test, quantization errors repeat every run, leading to a constant error, which is not removed by averaging. This may not alter the mean, but just σ, so this error cannot be calibrated out. Variability in the DUT can be seen as we alter the test conditions, so this is *measurement uncertainty*. We discover this by slightly varying test conditions and recalculating the tested parameter. Figure 10.48 [412] shows variability in gain error during a test for a single DUT as the

10.5 Realizing Emulated Instruments Using Fourier Transforms

input test waveform amplitude varies. Since the average is zero, the ATE is correctly calibrated. A detailed study of a batch of devices would allow us to compute the *range of uncertainty* in measurement.

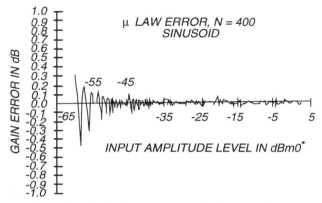

*Power level in decibels measured relative to the zero transmission level point of a telephone central office exchange, which is the power at the center of the switch.

Figure 10.48: Illustration of measurement uncertainty.

Devices are designed to have small measurement uncertainty during single-tone tests. During multi-tone testing, this uncertainty is magnified. Telephone μ law and A law encoders with integral filters have a tightly-specified linear filter, followed by a low-resolution ADC. This is one of the worst cascaded circuits to test. Multi-tone testing is acceptable when the filters are separately testable. When the filter is followed by an integral μ law or A law ADC, DUT uncertainty greatly exceeds digitizer uncertainty. Figure 10.48 [412] shows that μ law encoding causes $[0.01, 0.03]$ dB peak single-tone uncertainty at upper levels. A 12-tone test will magnify this.

Accuracy Factors. The following tone set adjustments improve test accuracy:

1. *Tone Pruning.* With $3f$ and $7f$ tone inputs in a multi-tone test, a $6f$ tone is undesirable, since it hides the weaker second harmonic of the $3f$ input. *Rules:*

    (a) Remove all even multiples in a broad-band tone set of two or more octaves, to avoid conflicts with second-order harmonics and IM products.

    (b) Remove all multiples divisible by three in a broad-band tone set of two or more octaves.

    (c) Choose multiples that are relatively prime, unless this eliminates necessary test tones.

    (d) Use odd multiples for narrow-band multi-tone sets under two octaves in frequency. (Enough bins are still present for measurement.)

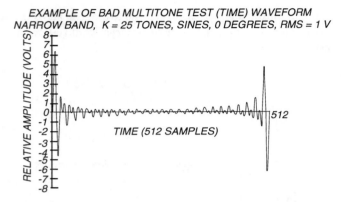

Figure 10.49: Multi-tone waveform with unacceptable S/N ratio.

(e) In bands under one octave in the CCITT noise-based test, all spectral lines are allowed.

2. *Narrow-Band Tone Sets.* These have better S/N ratios for each tone, but then a broad-band measurement cannot be made in one test. *Rule: In testing band-pass filters, use tone sets with only in-band or out-of-band tones. The digitized signal at the ATE can then be maximized through DUT gain or digitizer gain adjustments.*

3. *Peak-to-RMS Ratio.* Use a test waveform with a low peak-to-RMS ratio. Figure 10.49 [412] shows a test waveform synthesized from all sinusoids with a high peak-to-RMS ratio. When we adjust the zero time amplitude to avoid clipping by the DUT, the amplitude in between peaks falls into the noise range. An all-cosine test waveform has the same problem. *Rules:*

    (a) Using equally-spaced tones, make differential phase vary linearly from 0 to 360° (see Figure 10.50 [412].)

    (b) Do not use tone sets with minimum peak-to-RMS ratios, because they are extraordinarily sensitive to non-linear phase in the DUT. As they pass through the DUT, the peak grows relatively to the RMS value, and may burn out the DUT.

    (c) Use waveforms with near-Gaussian amplitude distribution and pseudo-random phase distribution, with peak-to RMS ratio of 10 to 11 dB.

4. *Dynamic Overload.* Multi-tone waves often have greater maximum dv/dt than ordinary single-tone test signals with equal amplitude peaks. The multi-tone signal may introduce excessive non-linear distortion, and increase measurement uncertainty, because of slew limiting, oscillation, etc. in the DUT. *Rules:*

    (a) Try a somewhat smaller test signal, and see if the measurement accuracy and repeatability drastically improves.

10.5 Realizing Emulated Instruments Using Fourier Transforms

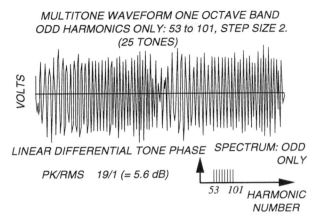

Figure 10.50: Multi-tone waveform with acceptable S/N ratio.

(b) Use narrow-band multi-tone waveforms, since wide-band waves have greater maximum slope values.

5. *Out-of-Band Measurement Uncertainty.*

$$\text{individual tone measurement uncertainty } \alpha \ \frac{1}{\text{tone amplitude}} \quad (10.36)$$

In the multi-tone test of Figure 10.46 [412], tones above/below the pass-band are greatly attenuated, and therefore inaccurate. *Rule: If out-of-band testing must be very accurate, use single-tone tests or use separate in-band and out-of-band multi-tone tests.*

Table 10.6: Relationships between various uncertainties.

RMS (standard deviation)	1	Peak error (95% of samples)	±2
Probable error:	0.06745	Peak error (99.7% of samples)	±3
Absolute average error:	0.7979	Peak-to-peak error (common)	6

Accuracy Estimation. For mostly random noise errors, measure the uncertainty $U\%$ for a single tone and extrapolate it to a multi-tone test set. This is $G \times U$ for a tone component of relative amplitude $1/G$. This is acceptable also if U has units of dB, if it is under 1 dB. Always express uncertainty as RMS, peak, peak-to-peak, probable error, or average error. For Gaussian noise uncertainty, the measurements relate as in Table 10.6. Table 10.7 [412] shows variation in amplitude measurements of a steady tone with expected RMS uncertainty of 0.0208 dB, whereas the actual observed uncertainty was 0.0212 dB.

For quantization distortion measurement error in FFT analysis using a DSP-based ATE, estimate multi-tone measurement uncertainty as follows [412]:

Table 10.7: Types of uncertainty in multi-tone tests.

100 Successive measurements of amplitude of one tone out of 48. Each measurement is done with a 1024-point DFT. Error is due to random noise 40 dB below power of multi-tone test signal. Error range is $[-0.0572\ dB, +0.0453\ dB]$	
Mean of 100 measurements	$0.0028\ dB$
Mean absolute error	$0.0164\ dB$
Standard deviation	$0.0212\ dB$
Peak-to-peak error	$0.1025\ dB$

1. Find whether the DUT or the digitizer is the main source of quantization distortion.

2. Express *integral non-linearity error* (INL) (see Section 10.4.3) with the number of linear-equivalent bits. For a good 15 or 16-bit ADC, 14 bits of integral linearity is typical.

3. For a DUT with non-linear coding, or dominated by noise, estimate the number of *equivalent* bits by expressing the S/N ratio in dB, and then dividing by 6 dB. E. g., a μ law CODEC has 40 dB S/N, and therefore 6.5 bits of equivalent linear quantization error.

4. Calculate the effective number of quanta in the full-scale range of the quantization source from the number of *equivalent* bits.

5. Express the peak-to-peak *individual tone* amplitude by the number of equivalent quanta, Nq. If the tone spans $F_s/10$ of a 14-bit converter, then $Nq = 16384/10 = 1638$.

6. Express this as a standard deviation as follows:

Measured amplitude	σ
Single sinusoid	$\frac{1}{3Nq}\ dB$
1 Tone in a multi-tone complex	$\frac{1}{4Nq}\ dB$
Measured phase, 1 tone in a multi-tone complex	$\frac{4}{Nq}\ °$

Quantization is less severe in a tone in a multi-tone complex, than in the same amplitude, single-tone complex, because the other tones in the tone set dither the signals, which moves the samples pseudo-randomly with respect to ADC decision levels. Figure 10.51 [412] shows, for the same amplitude tone, the uncertainty when it is embedded in a 48-tone test waveform.

10.5.4 ATE Vector Operations

Table 10.8 briefly gives the flavor of programming emulated instruments in the LTX Cadence test programming language for the LTX FUSION DSP-based ATE.

10.5 Realizing Emulated Instruments Using Fourier Transforms

Table 10.8: Significant DSP-based analog ATE matrix operations.

Description	LTX Cadence code
Matrix add/subtract	R = D ± U
Root mean square	V = **rms** (X)
Integer-floating conversion	K = **integer** (S)
Boolean logic operations	L = J **xor** K
Set all matrix elements to a constant 1.026	X = 1.026
Adding a constant 1.026 to all matrix elements	W = 1.026 + W
Indexing operations – Sum of 31st to 220th positions	S = **sum** (X [31:220])
Fourier voltmeter	Y = **fvm** (data, no_samples, harmonic)
Returns a 2-element array containing the cosine and sine of the *harmonic* of the *data* containing *no_samples*	
Discrete Fourier transform	Y = **dft** (result, samples, test_tone_freq, sampling_freq, no_harmonics_desired)
result [1]: Total signal RMS $\times \sqrt{2}$	
result [2]: Non-harmonic RMS $\times \sqrt{2}$	
result [3]: DC Voltage	
result [4]: Peak Amplitude 1st harmonic of test tone F_t	
result [5]: Peak Amplitude sine of 1st harmonic	
result [6]: Peak Amplitude cosine of 1st harmonic	
result [7]: Peak Amplitude 2nd harmonic of test tone F_t	
result [8]: Peak Amplitude sine of 2nd harmonic	
result [9]: Peak Amplitude cosine of 2nd harmonic ...	
Fast Fourier transform	freq_domain = **fft** (time_domain)
freq_domain [1]: DC component of *time_domain*	
freq_domain [2]: cosine component at frequency $F_s/2$ of *time_domain*	
freq_domain [3]: cosine component of multiples of Δ of *time_domain*	
freq_domain [4]: sine component of multiples of Δ of *time_domain* ...	
Inverse FFT	T = **inverse_fft** (F)
Magnitude	Y = **mag_fft** (X)
Power spectrum	power_results = **power_fft** (time_domain)
Phase	polar_coord = **polar** (fft (samples))
Converts an array of cosine-sine pairs into amplitude-angle data	
μ-law CODEC encoding	array1 = **mucode** (array2)
μ-law CODEC decoding	array1 = **mudec** (array2)
A-law CODEC encoding	array1 = **acode** (array2)
A-law CODEC encoding	array1 = **adec** (array2)
Normalized correlation	I = **correlation** (samples_1, samples_2)

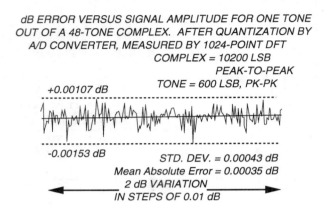

Figure 10.51: Multi-tone amplitude measurement quantization uncertainty.

An even richer set of additional mathematical subroutine and function calls are available for use with the high-speed vector processor. The language also contains primitives for device movement, handling, and binning. The letters i, j, k, l, m, and n denote integer subscripts. The bracket notation, [ ], indicates an array subscript. Variable j is the multiple of Δ that we wish to measure, X labels the time series to be analyzed, Y is the result vector, and n is the number of samples in the UTP. Although vectors in the DSP processor can represent anything one wants, they usually represent waveforms, spectra, and filters. Matrix and vector operations using the built-in functions in Table 10.8 are a replacement for test instrument hardware.

10.6 CODEC Testing

Definition of a CODEC. Figure 10.52 [412] shows a CODEC pair of coding and decoding functions to implement the *pulse code modulation* (PCM) telephone channel over the *voice frequencies*. The CODEC allows multiplexing of analog voice and touch tone signals, and switching and transmission of this information digitally. Note that while each subscriber has a CODEC pair, he will not be talking to himself, but instead to someone else at the other end of the telephone line. It is, therefore, critical to test each half of the CODEC separately, since by testing the pair together a fault in the CODEC encoder may be masked by a fault in the CODEC decoder. The full channel test randomly pairs send and receive CODEC units.

The analog telephone signal is digitized after passing through a *subscriber loop* to a *central office* (CO) or a *private branch exchange* (PBX). The communications path is separated into separate *transmit* and *receive* paths by the *subscriber loop interface circuit* (SLIC). Voice signals are transmitted by limiting the pass band to 300 to 3400 Hz, removing 60 Hz power and phone ringing frequencies, and removing the signal energy at and above 4000 Hz. The filtered signal is sampled at 8000 s/s, and converted into an 8-bit word, which is transmitted serially and multiplexed with other telephone conversation signals.

10.6 CODEC Testing

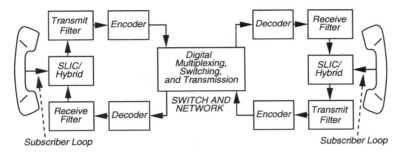

Figure 10.52: Telephone *pulse code modulation* (PCM) channel.

The 8000 s/s requirement leads to a data rate of 64 $kbits/s$, but samples are instead transmitted at 1544 or 2048 $kbits/s$ to permit time multiplexing of a *primary group* consisting of 24 to 32 channels. The receiving end must reverse this process, and then reconstruct the digitized signal as shown in Figure 10.53 [412]. It also corrects amplitude and phase to account for the $\sin(x)/x$ frequency rolloff due to sampling time quantization (see Section 10.5.2.)

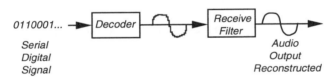

Figure 10.53: Digitized signal reconstruction.

μ Law Encoding (Companding.) The μ law logarithmic encoding function is shown in Figure 10.54 [412]. Each half approximates a semi-logarithmic function and has eight *chords*, each having 16 uniform steps. Advancing from the origin, each successive chord doubles in width, so every step is double the width of the 16th previous step[†]. Note that the points on the y-axis in Figure 10.54 are equally spaced. The benefit of μ law encoding is that quantization is proportional to the signal size, which leads to a rather uniform signal-to-quantizing noise ratio, 40 dB over wide amplitude ranges. Since resolution increases with small signals, there is nearly an 80 dB dynamic range. The result is the dynamic range of a 13-bit ADC using only an 8-bit ADC.

At the receiving end, the CODEC decoder performs the inversion function, signal *expansion*. Any mismatch between the encoder and decoder in the transmission pair appears as non-linear distortion or variable gain with changing input levels. Telephone *signaling* transmits subscriber loop status by adjusting every sixth frame of the code format to replace the LSB with a signal bit. This unfortunately changes audio behavior of the subscriber loop, and requires additional testing.

[†]The A law encoding used only in Europe keeps the first two chords uniform, but subsequent chords double in width as with the μ law.

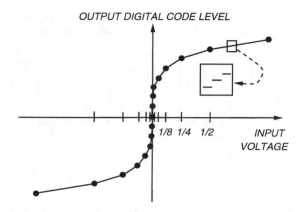

Figure 10.54: Transfer function of μ law encoder.

The number of units of μ law encoding is given by:

$$2^C \times (2S + 34) - 33 \tag{10.37}$$

where S is the step number (0 to 15) and C is the chord number (0 to 7.)

Decibel Units. The *decibel* (dB) is a ratio unit N for power signals $P1$ and $P2$:

$$N = 10 \log \left(\frac{P_2}{P_1}\right) \; dB \tag{10.38}$$

Voltage signals V_1 and V_2 from CODECs lead to:

$$N = 20 \log \left(\frac{P_2}{P_1}\right) \; dB \tag{10.39}$$

since most CODEC measurements are voltages. For telephony applications, signal level is measured with respect to a signal power of 1 $mWatt$ (*dBm*.) However, the power at the measurement point is not meaningful; instead, we use the power at the center of the central office switch. This is referred to as the *zero transmission level point* (0TLP). Measurements at CODECs must be translated to the 0TLP. We use *dBm0* units, which measure the signal level with respect to a signal level that will cause the 0TLP at the central office. These special *voice frequency* (VF) performance parameters are given in decibel ratios in units of *dBm0*.

Example 10.13 *A signal of* 1.5 *V RMS at a CODEC causes* 0 *dBm at the 0TLP [412]. We measure* 2.0 *V RMS at the CODEC, so the relative level in* dBm0 *is* $20 \log(2.0/1.5) = 2.5$ *dBm0. The advantage of this unit is that we need not know any impedance or power at the CODEC. There is a voltage standard defined relative to full scale. The* virtual edge *is the voltage at which the CODEC begins to clip, measured in* normalized voltage units such that all quantization levels are integers. *The μ law CODEC has edges at $\pm$ 8159 units. A sinusoid touching the μ law edges has a relative power level of* +3.172 *dBm0. However, the difference between two levels in the same units is given in dB, not in* dBm0.

10.6 CODEC Testing

10.6.1 Considerations for CODEC Performance Tests

Gain/Loss Tests. A connection in the telephone network should ideally have no gain or loss, so it must be tested to within ± 0.1 dB. Since the ATE has many sources of measurement error, such a test requires identification of any measurement error sources contributing 0.001 dB or more of error. Since CODEC quantization distortion does not allow input and output waveform shapes to be the same, gain must be redefined as the ratio of the fundamental channel output component amplitude to the pure input signal amplitude, expressed in dB. *Gain tracking* is used for absolute level measurements (in $dBm0$) and indicates whether the encoder and decoder are complementarily symmetric. The full-channel gain test is shown in Figure 10.55, where the bandpass filter passes only the signal fundamental. For a half-channel *decoder* gain test, we drive the decoder with a synthesized output of an ideal encoder, and then digitize and analyze the analog decoder output. For measuring half-channel encoder gain, the ATE parallelizes the encoder output bit stream into 8-bit words, and collects N of them. The ATE decodes the vector into 13-bit words using lookup hardware or by running a software algorithm in the DSP ATE. The function *mudec* decodes μ law vectors.

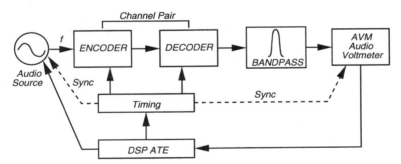

Figure 10.55: Full channel gain test fixture.

Frequency Choice and Distribution. Analog telephone systems are tested with standardized 800 and 1000 Hz frequencies, which are inappropriate for CODEC testing because they are not relatively prime with the CODEC sampling rate, 8000 Hz (see Section 10.5.) To solve this problem, the tone frequencies are offset and the unit test period is extended to provide many samples over many different CODEC levels. A further complication is that a single CODEC quantization step is between 3 and 6 per cent of the sampled voltage, so quantization uncertainly exceeds permissible gain error. Figure 10.22 shows a proper test frequency selection of 1500 Hz. The unit test period is 3 cycles (16 samples), which is $16/800 = 2$ ms. A realistic CODEC test will extend the UTP to 101 or more cycles to cover all of the most significant codes. AT&T standards require gain tests at 1004 Hz, not a power of 2, so the DFT is required.

Example 10.14 *AT&T gain test [412].* Let $F_t = 1004\ Hz$ and $F_s = 8000\ s/s$.

$$\frac{1004}{8000} \Rightarrow \frac{251}{2000} \Rightarrow \frac{M}{N}$$

There are 251 cycles in the UTP of the test waveform, 2000 samples, and its length is $N/F_s = 250\ ms$.

A problem arises with quantization distortion in CODEC testing, which inserts undesired energy into the reconstructed channel output spectrum. When this energy adds to the spectral components important to a gain test, the measurement is in error. The noise spectrum distribution and phase vary greatly with encoder gain and the phase of sampling. Therefore, even the amount of error in a given measurement is uncertain. To solve this, we shift the noise frequency components away from the important spectral components for the test. This means that we must allow many spectral lines where noise can appear, so that it is spread out over many FFT bins, and not concentrated in the important test frequency bins. This cannot be done when F_t divides $F_s = 800\ s/s$ because then $M = 1$. The noise becomes periodic with F_t and falls into the same bins as F_t and its harmonics.

Figure 10.56 [412] shows the solution where $M > 1$, so we have created an enormous number of noise line locations in the pass band, spaced by $\Delta = F_t/M$. Figure 10.56(a) shows the case with $M = 1$ while Figure 10.56(b) shows the case for $M = 51$. The latter case leads to $N/2$ possible noise lines (excluding bin 0) in the 0 to 4000 Hz band, each spaced by Δ. Note that when F_t was 1000 Hz, there were 4 possible lines, but with $N = 8$ in that case, the quantization was symmetric and even lines were excluded. This forced all quantization energy to fall into the spectral lines F_t and $3F_t$. In Figure 10.56(b) $F_t = 1020\ Hz$, so there are 200 bins as well as DC. Perfect CODECs fill alternate lines, but real ones fill all lines to some extent. Since the noise has so many bins to fall into, the errors at F_t and $3F_t$ are grossly reduced. A classical analog ATE measurement method would use a bandpass filter tuned to F_t, but unfortunately *bandwidth α 1/ filter settling time*. A fast filter would have leaked components on both sides of F_t, so F_t becomes a carrier frequency with amplitude modulation. The coherent DSP ATE eliminates this problem by replacing the bandpass filter with Fourier filtering to isolate the F_t component, while avoiding the excessive filter settling time.

Intrinsic vs. Extrinsic Error. Gain measurement (*intrinsic*) error at the standard CODEC testing frequencies (μ law or A law) masks some of the manufacturing defects (*extrinsic* errors.) The intrinsic error comes from quantization of signals during digitizing. This measurement error decreases as M and N are increased during coherent measurement. The peak error is $\pm/3N$ in dB with inputs above $-30\ dBm0$, and the measurement error is constant for larger N. Non-coherent measurements have a wider measurement error, and the gain error does not flatten until $N = 2000$. $N = 200$ is sufficient for coherent measurements, but $N = 2000$ is necessary for non-coherent measurements. The error stops dropping with increasing N because μ law quantization error only falls in certain spectral bins, and does not

10.6 CODEC Testing

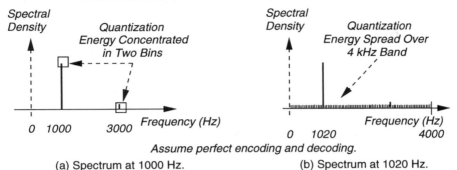

Figure 10.56: Importance of CODEC test frequency selection.

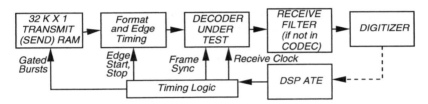

Figure 10.57: Decoder test fixture for DSP ATE.

continue to spread. Figure 10.48 shows the theoretical gain error of a perfect μ law channel as the input level varies from $-60\ dBm0$ to $+3\ dBm0$ at $F_t = 1020\ Hz$. The advantage of $N > 400$ with coherent testing is that the error becomes almost independent of signal phase delay from the encoder clock.

Half-Channel Tests. A half-channel test fixture is like a full-channel test fixture, but with an error-free reference unit replacing the half not under test. The CODEC encoder is its own digitizer, and outputs a serial PCM bit stream. More modern CODECs may output all eight digitized bits in parallel. An ADC test fixture is used, with a hardware serial-to-parallel converter added. The serial-to-parallel converter would be eliminated if CODEC bits appear in parallel. For a decoder test, the ideal encoder bit stream is computed for a given audio input. This is stored in a *transmit RAM* (see Figure 10.57 [412].) The following code creates and encodes the vector stream for a 1020 Hz sinusoid of amplitude E and phase T as vector M:

```
        for  i = 1  to 400 do
                  A [i] = E * sin (45.9 * (i - 1) + T)
                  M = mucode (A)
        end_for
```

The frequency increment, $45.9°$, comes from the expression $360 \times 1020/8000$, since $F_t = 1020\ Hz$ and $F_s = 8000\ s/s$. One of the problems with estimating full-channel behavior by combining two half-channel tests is that the quantization error will appear twice, yet it is a property only of the encoder ADC, and not of the decoder DAC. *Excess quantization distortion* comes from testing each half separately, and then combining the tests, because the CODEC decoder is tested with a synthesized perfect encoder digital waveform, and therefore has excess quantization error. This is corrected by subtracting the intrinsic μ law gain tracking error (Figure 10.48) from the actual gain tracking curve (Figure 10.58 [412]) to get the result in Figure 10.59.

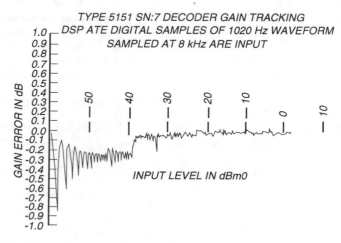

Figure 10.58: Excessive quantization distortion in decoder gain tracking test.

10.6.2 CODEC Tests

PCM Performance (Transmission) Tests

These test the DUT for transmission of voice or data through a complete channel, and must conform to various standards[‡]. The most important tests are the transmit half-channel (analog-to-digital) and receive half-channel (digital-to-analog) tests. There is also a full-channel (analog-to-analog) test.

Loss Variability. CODEC pairs may have any given gain, provided that every pairing of that type of CODEC is consistent. *Loss variability* measures the gain consistency between units, and measures how much a specific connection deviates from the average gain for that type. AT&T requires ± 0.5 dB for 99% of all pairing.

Gain Tracking. As the input level varies, *gain tracking* measures the individual channel gain deviation as the input level varies. AT&T obtains the reference gain value at 0 $dBm0$, and roughly six levels are tested in production.

[‡]International Telegraph and Telephone Consultative Committee (CCITT) standards, AT&T standards (USA), or Bell Northern Telecom (NORTEL) standards (Canada).

10.6 CODEC Testing

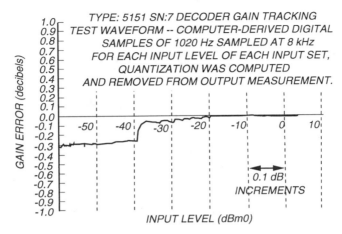

Figure 10.59: Gain tracking characterization test.

Attenuation Distortion. *Attenuation distortion* measures gain variation as a function of frequency. This parameter indicates problems with filters. When the filter is integrated into the CODEC, there is additional quantization. Also, CODEC filters are *sampling* filters, and therefore produce image frequencies beyond 200 kHz. Consult Section 10.5.2 for appropriate undersampling methods.

Signal-to-Distortion. This test evaluates the magnitude of distortion introduced into a signal by a channel. The test measures the total noise of a test signal, relative to the test signal amplitude, in dB, and is called the *signal-to-total distortion* (S/TD.) Figure 10.60 shows the test fixture. The weighting filter is a *C-message* filter (see Mahoney [412], Figure 12.16) for μ law encoders, and approximates the effect of noise on the human ear. S/TD is computed by measuring the fundamental components of F_t using the same bandpass filter as for the gain measurement. A second measurement is done with a complementary *notch* filter. A DSP ATE performs the filtering operations mathematically. This test happens are various different input levels.

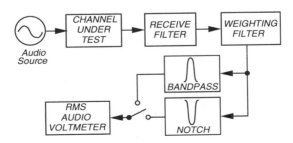

Figure 10.60: Measurement of signal-to-total distortion.

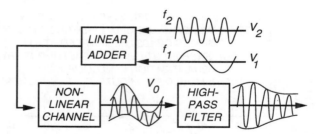

Figure 10.61: Intermodulation distortion test waveforms.

Idle Channel Noise. When there is no signal to the CODEC, the channel is in *idle* state, and the CODEC will generate a random idle pattern, due to its internal noise. The CODEC decoder turns this into analog noise, adding its own internal noise. The final signal is the *idle channel noise*. Various telephone companies either use a low-amplitude sinusoid, or random noise, to dither the signal to keep the CODEC signal alive.

Non-linear Distortion. *Signal-to-total distortion* cannot differentiate between intrinsic non-linearity of the μ law quantizer and distortion caused by manufacturing defects. *Harmonic* and *intermodulation distortion* tests differentiate between these non-linearities, because quantization noise appears widely in the spectrum, whereas non-linear distortion appears in just a few positions. This can be examined by looking at only a few Fourier bins.

Harmonic Distortion. Increased energy at $2F_t$ and $3F_t$ indicate low-order non-linearity in the channel. Asymmetric gain around the origin appears as second harmonic distortion. Symmetric non-linearity appears as third harmonic distortion. The receive filter nulls higher harmonics when $F_t > 1000\ Hz$. Harmonic distortion is part of the AT&T *single frequency distortion* tests.

Intermodulation Distortion. In this test, at least two sinusoids are applied simultaneously, to generate a great variety of distortion products. Figure 10.61 [412] shows an *intermodulation distortion test* (IMD). Signal V_1 carries and amplitude modulates signal V_2, which appears in different gain regions. The modulation is found by separating the modulated high-frequency component V_0 from V_1 using a high-pass filter, and then examining the resultant envelope. The modulation depth and shape show the magnitude and type of non-linear distortion. For CODEC testing, it is imperative to examine specific spectral lines in the V_0 spectrum. Lines at $f_2 + f_1$ and $f_2 - f_1$ indicate *second-order* IMD. Lines at $2f_2 + f_1$ and $(2f_1 \pm f_2)$ indicate third-order intermodulation distortion. With three-tone testing, lines at $(f_1 \pm f_2 \pm f_3)$ also indicate third-order IMD. AT&T requires a *four-tone* test that distinguishes second and third-order intermodulation distortion [58, 59, 320].

10.6 CODEC Testing

Digital Functional Tests

These test whether a component's digital logic functions are working properly (signaling, power down, time slot assignment, encoding law selection, test mode selection, etc.) Digital stuck-fault tests would be appropriate (see Chapter 7.)

Parametric DC and AC Tests

These tests are done only one one port of the circuit, and involve testing for shorts, opens, leakage, logic levels, threshold, ripple rejection, impedance, supply sensitivity, etc. Typical tests of this sort appear at the ends of Chapters 2 and 9.

Margin and High-Speed Parametric Tests

High-speed parametric tests determine specification parameters (e.g., rise time, propagation delay.) Margin testing determines the effect of drifting clocks and on-chip coupling on the CODEC, and determines the exact phase shift needed to cause the CODEC to fail. This test requires test equipment with programmable edge delays, clock frequencies, clock phases, strobe widths, and variable timing.

Device Characterization Tests

These tests are not used in daily manufacturing; instead, they provide information for design analysis and improvement, fabrication process improvement, and part evaluation. They are more detailed versions of the transmission tests.

Figure 10.59 [412] shows a gain tracking characterization curve of a CODEC decoder, with input levels from $-60\ dBm0$ to $+3\ dBm0$ at increments of $0.1\ dB$, at test frequency $1020\ Hz$. Figure 10.62 [412] shows a S/TD characterization test of the ideal μ law channel at $1020\ Hz$, using the C-message weighting curve.

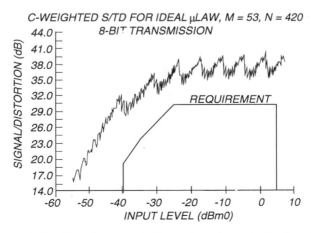

Figure 10.62: Signal to total distortion characterization test.

Chapter 10. DSP-BASED ANALOG AND MIXED-SIGNAL TEST

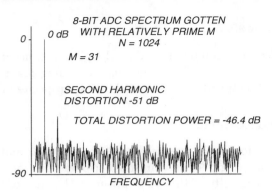

Figure 10.63: Spectrum of 8-bit ADC with relatively-prime M.

10.7 Dynamic Flash ADC Testing FFT Technique

We can use the previously-discussed multi-tone harmonic coherent testing and apply an FFT to the ADC spectrum. We can also separate quantization noise from random noise. To achieve this, we perform two tests – the first with prime M (see Figure 10.63 [412]), and the second with an adjacent submultiple of N (see Figure 10.64 [412].) The *quantization noise* bunches when $N_{(FFT)} = K \times N_{(UTP)}$. The quantization noise appears in bins $K, 2K, 3K, \ldots$ This allows us to extend the dynamic range of the DSP signal source, by using an M/N ratio that is not mutually prime. We take all quantization error, organize it as harmonics, and filter it out. The remaining bins contain *random noise power*, usually ADC comparator noise, timing uncertainty, and jitter noise. Excess quantization noise divided by the ideal quantization noise is the *distortion power due to dynamic non-linearity*. Harmonic distortion, IM distortion, *signal-to-noise ratio* (SNR), and the Noise Power Ratio can be computed by manipulating the FFT spectrum from coherent testing. For information on envelope delay distortion testing and differential gain testing, see Mahoney [412].

Example 10.15 *Extended dynamic range testing. We extend Example 10.4. We first test the CODEC as in Example 10.4. In a second test session, we take 400 samples using these parameters for the digitizer:*

$$M = 51 \quad F_s = 8 \ KHz$$

The waveform source uses 408 samples with:

$$M = 51 \quad F_s = 8 \ KHz \times 51/50 = 8160 \ Hz$$

The primitive periods remain the same, but the quantization noise bunches, so we can now examine the bins that do not receive quantization noise to compute the random noise power.

Computed distortion is different from *measured distortion*, because some of the distortion power falls into the same FFT bins as the test tones. Only the part in non-multi-tone bins can be recognized as noise in a typical DSP test.

10.8 Advanced Topics

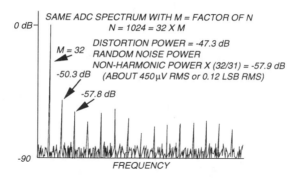

Figure 10.64: Spectrum of 8-bit ADC with $M = $ factor of N.

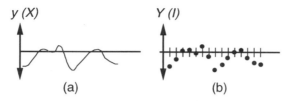

Figure 10.65: Continuous function (a) and its sample set (b).

10.8 Advanced Topics

We now discuss event digitization and noise measurement techniques.

10.8.1 Event Digitization

Information in a Sample Set. A sampled numerical vector is equivalent to a continuous function, which can represent voltage, current time, etc. We index the sample set with the index I. Figure 10.65(a) [412] shows a continuous function and Figure 10.65(b) shows a sample set that represents the function. While the vertical axis is quantized, it is effectively continuous, because the quantization of the vertical axis is much smaller than the gaps in the horizontal axis due to sampling. The continuous function $y(x)$ is fully defined everywhere by a properly chosen sample set. This remains true regardless of which axis is discretely sampled.

Implicit, Instead of Explicit, Digitization. Instead of sampling the quantity on the vertical axis at discrete horizontal axis time points, we will set horizontal line values for the vertical axis. When the actual signal crosses the horizontal line, instead of sampling the signal, we sample and record the time instant, instead. This is *implicit*, rather than *explicit*, digitization.

In Figure 10.66(a) [412] we show the conventional way of sampling the continuous gain at specific time instants. Figure 10.66(b) [412] instead shows a process of determining the gain values that we are interested in, and then sampling the exact time instant where these values are measured. In Figure 10.66(a) [412] we see that

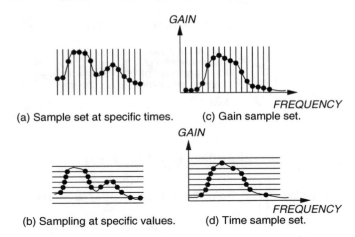

Figure 10.66: Continuous function sampling at specific times or values.

explicit sampling has no ambiguity, because for each time instant, there is one and only one gain value. Figure 10.66(b) [412] shows, on the other hand, that with implicit sampling, when we specify the gain value of interest, there may be multiple time values where that gain is observed.

In long DSP-based tests, such as occur when testing CD players and audio equipment, we encounter the problem that explicit digitization collects huge amounts of data, often far more that the ATE is equipped to process. In these situations, implicit sampling is far more effective, because we instead sample the time instants where a parameter crosses the desired value. We obtain a far shorter vector of samples that is just as useful as the voluminous data that would be produced by explicit digitization. This makes long time interval tests of mechanical recording devices affordable.

Example 10.16 *We want to find the center frequency, the peak gain, and the 3 dB frequencies of the filter in Figure 10.66(c). With explicit sampling, we step the test frequency F_t through the 1 KHz region, and repeatedly measure output amplitude. A multi-tone test permits us to apply all of the frequencies simultaneously. However, most likely none of the preselected frequency values would fall exactly at the 3 dB gain point, so we would need accurate interpolation and curve fitting to calculate these parameters.*

However, if instead we use implicit sampling, the gain axis is discrete and the frequency axis is continuous. We place horizontal lines near the expected peak output. Each line either has one value of x, in which case it exactly hits the center frequency, or two values of x, so the center frequency lies between the two x values. We analyze the highest line intersected. Having found the center frequency, we return to explicit sampling by setting the frequency to that value and recording the gain. Next, we return to implicit sampling to find the 3 dB frequency. We place a horizontal line 3 dB below the peak gain, and then sample implicitly to find the frequency that gives that 3 dB value.

10.8 Advanced Topics

Example 10.17 *We wish to measure amplifier distortion at 1 Watt output explicitly. An explicit digitization test would measure the DUT gain by applying a small signal. The DSP-based ATE could then calculate, from that gain, the exact input for 1 W output, and the measure the distortion. We also wish to find the amplifier power at 1% distortion implicitly. An implicit digitization test would apply an input and measure distortion. Were the distortion to be less than 1%, we would increase the input and repeat the measurement until the distortion becomes 1% (within a tolerance.) Should the distortion go above this level, then we would reduce the input. When distortion becomes 1%, then we measure the output power.*

Conventional ATE are very poor at measuring rise time, fall time, propagation delay, overshoot, settling time, glitch size, wow and flutter, and timing jitter, because they use an explicit measurement. These measurements require high-accuracy measurements of event times that are defined by waveform intersections with a few horizontal lines. Next, we show how the DSP-based ATE can be augmented for implicit measurement.

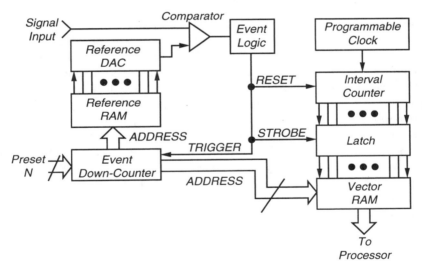

Figure 10.67: Block diagram of typical event digitizer in ATE.

Event Digitizer Hardware. Figure 10.67 [412] shows event digitization hardware in the DSP-based ATE. In the LTX Fusion ATE, this is known as the analog *time measurement unit* (TMU.) This samples continuous values of time at which predetermined events occur. Time is not regularly sampled, but only recorded with high resolution when an event happens. Events are defined as crossings of vertical axis parameters (usually a DUT output voltage) with a horizontal threshold. For measuring distortion, y instead could be a Fourier voltmeter output, and the input could instead be an input with successively increasing amplitude, x. Event digitizers do not need an ADC because no voltage is measured. Instead, high-speed comparators determine waveform intersections, and the time of the intersection is digitized. The sample vector consists of thousands of time measurements. We do not need many comparators, because no analog waveform can cross multiple voltage levels at the same time instant. The *reference RAM* in Figure 10.67 provides this reference level (horizontal line), which is converted to an analog voltage by the *reference DAC* and compared with the signal input by the *comparator*. However, since the comparator takes time to settle to a new state, we need a pair of comparator channels, programmed to trigger in some sequence. In practical systems, one pair may trigger on up-going crossings, while another pair triggers on down-going crossings. The *interval counter* resets to zero at each event, and is latched when the next event occurs. This *time difference* is stored in the *vector RAM*. The event down-counter is preset to the desired number of intervals, and decrements every time an interval count is latched. When it reaches zero, it interrupts the DSP computer. Thus, vector length is defined by the desired number of events, and not by time. A timeout circuit intervenes if no event occurs during a programmed time limit.

10.8.2 Measuring Random Noise

The internal sources of noise in circuits are every amplifier component. Transistors generate *shot noise*, which is associated with the probability of a carrier crossing a *pn* junction, and *flicker noise*, which is caused by defects in the emitter-base region of bipolar transistors and imperfections in the channels of MOSFETs. Resistors generate *thermal noise* from the thermal motion of electrons. Since all analog components generally have some resistive component, it is present in all discrete analog components and transistors. *Substrate coupling noise* is caused by crosstalk through the semiconductor substrate between the digital and analog portions of a design residing on a common chip. *Voltage noise* is the combined effect of internal components and noise due to dynamic short circuiting of inputs. We model this as an internal noise voltage generator (with no impedance) in series with upper input terminal of the DUT. *Current noise* is produced by junctions and components in the input path. We place a noise current generator with infinite internal impedance across two input terminals to model this noise. Another important source of noise is the actual analog test equipment itself [330].

Measurement Method

We measure noise at the DUT output terminals through a bandpass filter restricting the frequency range to an interesting one. We digitize the output noise, and mathematically analyze vectors. We calculate *equivalent input noise* or *noise referred to input* (RTI.) The output noise voltage depends on noise sources, circuit amplification, and the noise bandwidth. We normalize the output noise voltage – we divide the measured output noise by the gain of the DUT. We usually express the noise in terms of the noise that would be measured by using a $1Hz$ bandpass filter. This indicates the *noise density* at the center frequency, in units of $V/\sqrt{Hz}$. We com-

10.8 Advanced Topics

pute the *spectral density of noise* – a derived measurement. During measurement, use the widest practical measurement band, because measurement uncertainty is inversely proportional to the measurement bandwidth. The DSP approach actually measures relative *noise power* in the measurement band – we derive the power-to-bandwidth ratio in V^2/Hz or $Watts/Hz$ (assuming a 1 Ω load resistance.) This is important for a *white* (ideal) noise measurement, because it is the noise power density and is independent of frequency or bandwidth. Usually, the measurement frequency f_n is the mean of the upper f_1 and lower f_2 frequency limits. f_n is the logarithmic mean of f_1 and f_2 for *pink* noise, and for *brown* noise it is the geometric mean (see Mahoney [412].)

The DSP noise measurement procedure digitizes broadband noise, and then processes it to set the average to 0. We apply a *root mean square* (RMS) function to determine the standard deviation of the noise, which is also the RMS noise voltage. We use the FFT to obtain the noise power spectrum. We compute the noise power in local bands by a *sum of squares* (SSQ) operation. We normalize the vector power density in each region to 1 Ω by dividing the power by the bandwidth in Hertz, and then we normalize it to the DUT input.

It is important to select the correct gain $G2$ of the noise measurement fixture for the DUT in order to accurately sample the noise. We must set the digitizer input filter to the full band of interest. When $G1$ is the DUT gain and $G2$ is the noise measurement fixture circuit gain, the combined gain $G1 \times G2$ must be great enough so that broadband noise spans most of the digitizer's input range, but not so great that noise is clipped. See Mahoney [412] for details.

Noise signals are continuous but aperiodic, so multiple local measurements will differ from each other. The noise uncertainty is always inversely proportional to the square root of the local measurement interval. DSP measurement also introduces quantization uncertainty, because we only see discrete samples of noise. The noise bandwidth B_n is the cutoff frequency of an imaginary *brick wall* (infinitely sharp) filter, with the frequency integral of the noise power of the DUT equaling its power-frequency area. So, one must make the DSP noise sampling rate $F_s > 2B_n$, which is hard to achieve because a DSP-based tester has limited vector memory size. Rosenfeld describes the effect of noise on amplitude and phase measurements using multi-tone test sets [548].

Statistical Sampling

Statistical sampling for noise measurement has the advantage of reducing the amount of data to be processed by the DSP-based tester. Sparse sampling is called *statistical sampling*, which normally requires statistically-independent samples. We achieve statistical noise sampling in the DSP-based tester by lowering F_s, the sampling frequency. However, DSP-tester samples of noise are not statistically independent. This loss of data damages the spectral distribution, but not the amplitude distribution, of the noise. As long as N, the number of samples, is still large, time-domain parameters and total power can be determined from sparse samples [412].

10.9 Summary

Analog testing is increasing greatly in its importance, due not only to the nearly universal adoption of electronic switching for the telephone system, but also due to the tremendous proliferation of A/D and D/A converters, particularly in wireless cellular telephone devices. Other important applications are in consumer high-fidelity electronics, automotive electronics, and personal computer multi-media units for sound effects, internet telephony, digital compact disc playing, etc. Therefore, the system test engineer needs to give more emphasis to the DSP analog test methods, which are used for most of the analog tests. It is important to understand that analog circuit testing is non-deterministic, and therefore the testing process is statistical and must also deal with electrical noise. As a proportion of total testing costs, the percentage due to analog testing is generally increasing.

Problems

10.1 *Unit test period.* For an analog circuit, the test waveform frequency $F_t = 2010\ Hz$ and the sampling frequency in the DSP ATE is 8000 s/s. Please compute the minimum *unit test period* and the corresponding *primitive frequency*.

10.2 *Unit test period.* For an analog circuit, the *primitive frequency* for testing is $\Delta = 20\ Hz$ and the test waveform frequency $F_t = 2020\ Hz$. What DSP ATE sampling frequency will guarantee 600 unique samples during coherent testing?

10.3 *Unit test period.* How many unique waveform samples are obtained for the following test waveform using a DSP ATE with these parameters:
  $F_t = 2000\ Hz$ $\qquad F_s = 16000\ s/s$ $\qquad UTP = 50\ ms$
Is this an adequate sample set, given that the waveform is an amplitude modulated sine wave?

10.4 *Unit test period.* A DSP ATE can sample at 20000 s/s. Given a UTP of 40 ms and the desire to have a test waveform tone in the vicinity of 400 Hz, select a test waveform tone frequency that will maximize the number of unique samples.

10.5 *Unit test period.* A CODEC is to be tested on a DSP ATE with $F_s = 8000\ s/s$. Originally, $P = 40\ ms$, and $N \geq 400$. Select a test waveform frequency as close to 2000 Hz as possible that still generates $N \geq 400$ unique samples. How many test waveform cycles (M) will there be in the *primitive period*?

10.6 *Correlation.* Use MATLAB to calculate the normalized correlation of these two signals:
  $A(t) = 4\sin(2\pi t)$ $\qquad B(t) = 8\cos(2\pi t)$

Problems

10.7 *Correlation.* Use MATLAB to calculate the normalized correlation of these two signals:

$A(t) = 16\sin(6\pi t)$ $B(t) = 14\sin(6\pi t + 10)$

10.8 *Multi-tone testing.* A filter is tested with tones at frequencies $6f$, $19f$, and $27f$. Compute these frequencies in the FFT spectrum:

(a) All first harmonics.

(b) All second harmonics.

(c) All third harmonics.

(d) All fourth harmonics.

(e) All second-order intermodulation products.

(f) All third-order intermodulation products.

10.9 *CODEC testing.* A CODEC was tested with a single tone sinusoid at frequency $500\ Hz$. FFT spectral bins at $500\ Hz$ showed $2\ mW$ energy, $1000\ Hz$ had $0.5\ mW$ energy, and $1500\ Hz$ had $0.2\ mW$ energy. All other bins only had a small noise energy. Please calculate the *Total Harmonic Distortion* in decibels.

10.10 *ADC quantization error.* For the ideal ADC transfer function of Figure 10.7(a), please draw the quantization error on a graph. Both x and y axes should be labeled in *Volts* from 0 to *full scale range* (FSR.)

10.11 *ADC DLE and ILE.* For the ADC of Figure 10.8(a):

(a) Calculate and graph the DLE function.

(b) What is the DNL?

(c) Calculate and graph the ILE function.

(d) What is the INL?

10.12 *ADC DLE and ILE.* For the ADC of Figure 10.9(a):

(a) Calculate and graph the DLE function.

(b) What is the DNL?

(c) Calculate and graph the ILE function.

(d) What is the INL?

10.13 *DAC INL.* For the DAC with Δ_i given in Table 10.4, calculate the INL.

10.14 *Multi-tone testing.* Figure 10.44 shows a typical set-up for DSP-based multi-tone testing of a *device-under-test* (DUT.)

(a) The *primitive frequency* for testing is $\Delta = 20\ Hz$ and the test waveform frequency is $F_t = 2020\ Hz$. What DSP ATE sampling frequency F_s will guarantee 600 unique samples during coherent testing?

(b) Unfortunately, this sampling frequency leads to a very expensive analog tester. Please explain how to nearly halve the required sampling frequency, while still guaranteeing 600 unique samples, and recompute Δ, M, N, and F_s.

(c) During analog testing, the circuit is stimulated with a multi-tone waveform with frequencies $5f$, $9f$, and $17f$. What FFT frequency bins should be examined to look at 2nd-order *Harmonic Distortion*?

(d) Again using the test frequencies $5f$, $9f$, and $17f$, what FFT frequency bins should be examined to look at 2nd-order *Intermodulation Distortion*?

(e) Please provide a word definition of *Total Harmonic Distortion*, and explain how it is measured.

Chapter 11

MODEL-BASED ANALOG AND MIXED-SIGNAL TEST

> *"Research in analog and mixed-signal (AMS) test has been strongly influenced by the advances in digital IC test techniques and algorithms. . . . However, digital test is still far more advanced than is AMS IC test. The primary reason is the absence of a widely accepted paradigm for analog and mixed-signal circuit test."*
> — Bapiraju Vinnakota and Ramesh Harjani [699].

At present, there are relatively few CAD tools to assist in analog test design, so this is usually done by hand. A frequent problem is that the functional analog tests produced manually are not effective for manufacturing test of the IC on the available tester, so the tests must be redesigned, again by hand. The cost of analog testers is determined by the number of digital pins required for the tester, and by the number of analog instruments added to the tester for analog testing. Functional *digital signal processing* (DSP) based test sets are quite large in prototype testing. Also, with 22-bit *analog-to-digital converters* (ADCs) appearing, a standard histogram test for this converter would require a huge number of tests and a very long test time to collect sufficient samples for statistical analysis [621]. So, advances in analog and mixed-signal test are needed to lower costs [699]. This chapter on structural analog circuit testing requires an understanding of matrix algebra and the methods for representing systems of partial differential equations as Jacobian matrices [61, 203].

History. Early analog test efforts began in the 1960s. Analog circuits usually are tested functionally against their specifications, since they have few inputs and outputs and relatively few devices. Test inputs are generated from the specifications. Test application is expensive, however, because the number of specifications is large. Early research focused on discrete analog circuits [699]. Since these components were often unreliable, many failed during operation, and analog diagnosis was essential to be able to repair defective circuits.

With the advent of mixed-signal ICs, the observability of the analog circuit portions has been significantly reduced, when they feed on-chip ADCs. Conversely, the controllability of analog circuit portions has also been significantly reduced, when they are driven from on-chip digital circuits. The conventional DSP-based approach to analog circuits is still favored by industry. Recently, there is evidence that manufacturers only switch from separate digital and linear ICs to a mixed-signal chip when forced to by cost competition. However, one can expect competition to continually increase in microelectronics. *System-on-a-chip* (SOC) devices provide a level of integration well beyond traditional mixed-signal devices. Putting entire systems on a chip should eventually reduce I/O ports even further, thus reducing the observability of increasingly complex circuits. For example, there are recent efforts to put entire cell phones and personal computers on a single IC. This will lead to very low pin count devices, which would have I/O for antenna, keypad, display, microphone, speaker, and battery. At present, we do not know how to test the increased number of internal components with just those inputs and outputs!

11.1 Analog Testing Difficulties

Fault Modeling Problems. The major difference between analog *structural* test and analog *functional* test is the fault derivation and modeling procedure [621]. Functional test often assumes that the components are faulty and generates the fault list using component deviations and catastrophic faults (see Section 10.1.) Structural test uses manufacturing defect statistics, and the fault list may be either catastrophic or parametric.

Analog circuits have complex relations between input and output signals. Many analog circuits are non-linear systems (e.g., the MOSFET transistor, used as an amplifier.) The circuit parameter values vary widely, even in good circuits. Deterministic models are inefficient for analog circuits. Therefore, signals are specified by a nominal value, along with an acceptable *range* of values around the nominal value. Simulation and measurement inaccuracies and IC manufacturing process variations determine the acceptable signal value *tolerances* [699]. Finally, statistical distributions of analog faults generally are not known with enough precision to accurately predict fault coverages of a test set. Soma [620] reports evidence of analog circuits with catastrophic faults passing a conventional manufacturing test.

Simulation Error. Expected analog circuit signal values are computed by simulation, whose accuracy is limited by the numerical accuracy of the simulation algorithm, the simulation assumptions, and by the accuracy of the models of the parasitic analog devices. Also, process variations cause even good circuits to exhibit a range of different behaviors.

Tester Measurement Error. Measurement errors at the analog circuit tester come from analog offsets, the effect of the load of the measurement probe on the analog circuit behavior, and the impedance of the analog probe. Also, random noise is a problem, so analog testers are limited in bandwidth and measurement accuracy [699].

For mixed-signal chips, transporting internal analog signals to output pins may alter the signal and the circuit functionality. Capacitive coupling between high-frequency digital signals and analog signals causes additional analog circuit noise. Analog tests must create a difference in an analog output between the good and bad machines that lies outside the measurement error of the test fixture and the ATE [621]. Otherwise, the fault effect is masked by measurement error.

Test Accessibility Problems. Circuit complexity and the inaccessibility of internal components restrict the use of conventional analog ATE.

Manufacturing Process Variations. The various device parameters in large volume manufactured integrated circuits follow statistical distributions. These process variations can significantly impact component parameter values [699]. Analog design and circuit layout techniques exist to minimize the effect of temperature and diffusion gradients in circuit layouts. A *parametric fault* refers to component value variations. However, the analog testing literature also refers to *output* parameter variations, for example the variation in an amplifier gain. Both uses are correct, and the meaning will be clear from the context. For analog devices, multiple parametric faults (involving several minor component variations) are just as or more significant than large single parameter variations or catastrophic faults.

Information Flow. It is difficult to test circuits by individually testing subcircuits. Consider the case of two cascaded single-input, single-output analog circuits, C_1 and C_2, with analog voltage transfer functions $H1$ and $H2$. C_1 and C_2 may behave unacceptably when tested individually, due to manufacturing imperfections that distort their transfer functions. However, when cascaded, it could happen that the distortion in $H1$ is cancelled by the distortion in $H2$, which might be, in some sense, the inverse of the distortion in $H1$. Therefore, the cascaded combination of C_1 and C_2 may actually be acceptable. Conversely, individually acceptable analog circuits, when cascaded, may produce an unacceptable circuit.

11.2 Analog Fault Models

The conventional fault models for analog circuits are *catastrophic* or *hard* faults, where an analog component becomes open or shorted, and the *parametric* or *soft* faults, where an analog R, L, C, or transistor trans-conductance ($K = K_P W/L$) value changes sufficiently that it moves outside its tolerance box and causes unacceptable performance degradation of the analog circuit. Sometimes the additional faults stuck-at-V_{SS} and stuck-at-V_{DD} are also included in the catastrophic faults [699]. Catastrophic faults are easy to test, and parametric faults are difficult to test.

Single parametric faults are interesting in multi-chip module interconnects, as they will be termination resistances or important components such as precision off-chip inductors used in RF circuits. Linear analog ICs are designed so that the analog

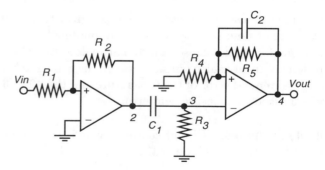

Figure 11.1: Amplification circuit.

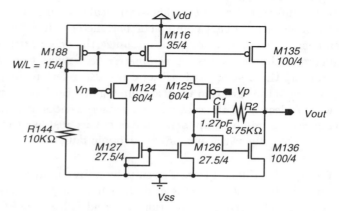

Figure 11.2: Compensated OPAMP circuit.

performance depends on ratios of components, so *multiple parametric* faults are most interesting in such chips. Many analog circuits are designed using Block's negative feedback principle, where an OPAMP is configured with an input impedance and a feedback impedance. Generally, it is the *ratio* of these two impedances that determine whether the OPAMP circuit is an integrator, a differentiator, or a buffer. Therefore, a multiple parametric fault model is the most useful fault model in this situation. Figure 11.1 [275, 276] shows an amplification circuit where the first amplification stage comprises R_1, R_2, and the first OPAMP, the second stage is a high-pass filter comprising C_1 and R_3, and the final stage is a low-pass filter comprising R_4, R_5, C_2, and the second OPAMP. The functional parameters of interest during test are shown in Table 11.1. Only two of these are single parametric faults, and the remaining ones are multiple parametric faults. However, we are not interested in all multiple parametric faults. The total number of double parametric faults is $\binom{7}{2} = 21$ and the total number of triple parametric faults in this circuit is $\binom{7}{3} = 35$. Testing all of these possibilities is expensive and unnecessary. Figure 11.2 shows a fully compensated OPAMP, and the relevant transistor-level faults are listed in Table 11.1.

In DSP-based analog testing methods (see Chapter 10), no analog fault models

Table 11.1: Structural analog faults in amplifier and OPAMP.

Functional parameter	Determining components
Amplification circuit	
First stage gain	R_2 / R_1
High-pass filter gain	R_3 and C_1
High-pass filter cutoff frequency	C_1
AC voltage gain of the low-pass filter	R_4, R_5, and C_2
DC voltage gain of the low-pass filter	R_4 and R_5
Low-pass filter cutoff frequency	C_2
OPAMP circuit	
Biasing current	R_{144}, K_{M188}, K_{M116}
Differential linearity	K_{M124}, K_{M125}, K_{M127}, K_{M126}
Output voltage gain	K_{M135}, K_{M136}, K_{M188}
Compensation	C_1, R_2

are used. In this chapter, the above fault models are used, and tests are generated for specific multiple faults. At present, the structural testing methods are gaining acceptance as a supplement to the DSP-based methods, but DSP-based methods remain the most important.

11.3 Levels of Abstraction

It is useful in analog circuit testing, as in digital testing, to look at the circuit from various different views. For analog testing, the *transistor level* of abstraction provides detailed models and structural interconnections for analog devices. At this level, the SPICE netlist, complete with transistor models, provides a *structural view*. The system of non-linear partial differential equations describing the netlist provides a *behavioral view*.

However, analog circuit testing can be done at a higher level of abstraction, the *functional level*, in which we model resistors, capacitors, inductors, and *ideal OPAMPs*, which have infinite gain and are considered to be fault-free. The benefits of this higher level of abstraction are modeling convenience and computational efficiency, and the liability is that OPAMPs may have faults, which should be tested for. At the functional level, the *structural view* of the circuit is provided by a *signal flow graph*, which graphically represents the idealized system of equations. Alternatively, a *behavioral view* is provided by the network transfer functions.

11.4 Types of Analog Testing

Specifications. Each class of analog circuits (A/D converters, D/A converters, filters, phase-locked loops, etc.) has its own separate set of specifications [699]. For each circuit class, there already exist accepted and specific functional tests for prototype test, and smaller test sets for production test [699]. There is no universal set of performance specifications. Also, there are no general design techniques for all analog circuits.

Analog circuit tests can be classified into these three categories:

- *Design characterization*, to determine whether the design meets specifications.

- *Diagnostics*, which determine the cause of a device failure when it fails a test.

- *Production tests* used for large volumes of linear or mixed-signal circuits.

We will focus on *fault-model* based analog circuit tests. At present, functional analog circuit testing without a particular fault model still reigns supreme, but it is now being augmented by fault-model based analog circuit tests.

There is a further taxonomy of analog tests. *Specification-based* tests are generated directly from the circuit specifications, without reference to an analog fault model. This approach is easily adapted to wide varieties of circuits. However, with large numbers of specifications, test application has become most expensive, and its cost must be reduced. The test set can be reduced by locating dependencies between specifications and eliminating unnecessary testing. *Structural fault-model based* tests will target a specific set of modeled faults. This allows quantification of a set of analog tests in terms of their *fault coverage*, so test sets can be *graded* [699]. The models also reduce the test set size, since test waveforms that detect faults already covered by other waveforms can be deleted. However, advocates of structural tests have been unable to establish a link between the fault coverage and satisfaction of the design specifications [699]. This makes designers reluctant to accept structural analog testing.

We will focus first on analog fault simulation, followed by automatic analog test-pattern generation. For further information, consult Vinnakota's book [699].

11.5 Analog Fault Simulation

Analog fault simulation is needed to evaluate the fault coverage, and effectiveness, of a set of analog test waveforms, which may be manually or automatically generated. Here we discuss three different kinds of analog fault simulation:

- *DC fault simulation of non-linear circuits*

- *AC fault simulation of linear circuits*

- *Transient* or *time-domain fault simulation*.

We first apply DC tests to analog circuits, and only if the circuit passes these do we apply AC tests, which are more difficult to generate and more costly to apply. Finally, only if the circuit passes AC testing do we apply *transient* or time-domain tests.

11.5 Analog Fault Simulation

11.5.1 Motivation

Analog circuit fault simulation is used for analyzing the analog fault coverage of a particular testing method, for fault grading, for analog fault collapsing, and for analyzing the effectiveness of analog BIST.

At present, no viable analog circuit synthesis tools exist, so analog design is accomplished manually by experienced analog designers, who design circuits using rules of thumb. For these designers, analog fault simulation is extremely useful for *what-if* analysis, where the designer asks the question, "What would happen if the resistance of resistor R_{17} were out of specification by 3.5%?" Answering this question requires analog fault simulation, which is far more computationally intensive than ordinary analog circuit simulation. It is important to take advantage of the structure of the circuit equations to concurrently simulate many analog faults.

11.5.2 DC Fault Simulation of Nonlinear Circuits

DC testing of analog circuits is attractive, because it requires less expensive testers and less testing time. DC fault simulation is useful to analyze how well a DC test can detect a given fault list. It is done by solving a set of non-linear equations using PSPICE or similar analog simulators, which converge only after many iterations. Simulation of catastrophic faults often fails to converge or causes the system matrix to be singular. Several techniques reduce fault simulation CPU time while improving convergence.

Complementary Pivot Method

Lin and Elcherif developed the *complementary pivot* method [398, 640] for DC fault simulation. They model the circuit linearity by ideal diodes. They also model transistor faults using switches, and these models are solved by an operations research method call *complementarity pivoting*. This method does not have the problem of Newton-Raphson iteration, which suffers from the extreme non-linearity coming from analog circuit and fault modeling.

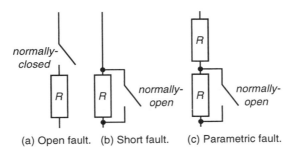

(a) Open fault. (b) Short fault. (c) Parametric fault.

Figure 11.3: Analog fault modeling using switches.

Complementarity pivoting consists of: (1) Modeling all non-linear devices with piecewise-linear $I - V$ characteristics (ideal diodes.) (2) Representing open, short, and parametric faults using switches. (3) Formulating fault simulation as the complementarity problem using n-port network theory. (4) Solving the resulting complementarity problem with Lemke's complementarity pivoting algorithm. In theory, non-linear analog circuits can be built so that all non-linearities come from two-terminal non-linear resistors [166]. These, in turn, are represented by a *piecewise linear* (PWL) $I - V$ curve, usually a 2 or 3-segment approximation to the diode $I - V$ curve. Figure 11.3 shows how switches can model open, short, and parametric faults (where the value of a resistance is halved.) The circuit-under-test is modeled with linear resistors, controlled sources, DC independent sources, switches, and ideal diodes. With d diodes and m switches, assume that there are k test nodes where voltage measurements are taken, and l branches where currents are measured. The voltages v_k and currents i_l are associated with k zero-valued independent current sources and l zero-valued independent voltage sources. These $k + l$ measurement ports, along with the d diodes and m switches, lead to $k + l + d + m$ pairs of *complementarity* variables (port currents and voltages.)

Vinnakota [699] illustrates the details of how this system is solved using Lemke's pivoting algorithm [390, 398, 640]. Others have provided additional solution methods [692]. Figure 11.4 illustrates how the ports are abstracted from the CUT.

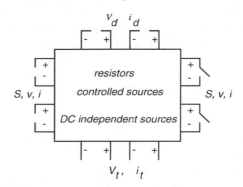

Figure 11.4: Formulation for a resistive n-port.

One-Step Relaxation Fault Simulation

The success of simple stuck-at fault models for digital testing inspired *one-step relaxation* [671] for DC fault simulation. Here, the exact modeling of the analog fault is avoided, as long as the fault simulator still can relatively correctly predict the faulty circuit behavior, even in the presence of measurement error. This method also avoids some of the numerical problems of analog circuit simulation.

For DC fault simulation, one solves the following equation system:

$$\mathbf{f}(\mathbf{x}) = 0 \qquad (11.1)$$

where $\mathbf{x}$ is the circuit variable vector (node voltages and branch currents) and $\mathbf{f}$ is a non-linear system function. In SPICE, Newton-Raphson iteration solves this equa-

11.5 Analog Fault Simulation

tion iteratively, starting from an initial point $\mathbf{x}^{(0)}$ and iterating until the difference between $\mathbf{x}^{(k)}$ and $\mathbf{x}^{(k+1)}$ converges. The exact algorithm is:

- Guess: $\mathbf{x}^{(0)}$

- Solve: $\mathbf{J}(\mathbf{x}^{(i)})(\mathbf{x}^{(k+1)} - \mathbf{x}^{(k)}) = -\mathbf{f}(\mathbf{x}^{(k)})$ for $k = 0, 1, 2, ...$, until it converges. $\mathbf{J}(\mathbf{x}^{(k)})$ is the *Jacobian* matrix of $\mathbf{f}(\mathbf{x}^{(k)})$ and is calculated as:

$$\mathbf{J}(\mathbf{x}^{(k)}) = \frac{\partial \mathbf{f}(\mathbf{x}^{(k)})}{\partial \mathbf{x}^{(k)}} \quad (11.2)$$

In *one-step relaxation*, the Newton-Raphson algorithm is operated for only one step using the good circuit solution as the starting point. Thus, the method solves the equation:

$$\mathbf{J}_f(\mathbf{x}_g)(\mathbf{x}_f^{(1)} - \mathbf{x}_g) = -\mathbf{f}_f(\mathbf{x}_g) \quad (11.3)$$

where $\mathbf{J}_f(\mathbf{x}_g)$ and $\mathbf{f}_f(\mathbf{x}_g)$ are the Jacobian matrix and function vector $\mathbf{f}_f(\mathbf{x})$ calculated at the point $\mathbf{x}_g$. $\mathbf{x}_f^{(1)}$ represented the vector to be solved. This method takes the linearized circuit at the good circuit solution point and the fault model. The fault models are linearized models of the actual faulty circuits at the good circuit solution point. Also, Householder's matrix updating formula [308] can be used (see Section 11.5.3), since the faulty circuit Jacobian matrix differs only slightly from the good circuit Jacobian matrix. Experiments on 29 MCNC benchmark circuits confirmed the validity of approximate fault simulation by one-step relaxation. For a majority of the circuits, one-step relaxation yielded the same fault coverage as exact fault simulation. Where there were differences, they amounted only to a few per cent [699]. This applied to both voltage measurements and supply current monitoring.

Simulation by Fault Ordering

Tian and Shi [670] proposed simulation by fault ordering, in order to improve simulation convergence and reduce the number of iterations required by Newton-Raphson iteration for exact DC fault simulation. The simulation algorithm convergence and convergence speed depends critically on the initial starting point. They *order* faults so that the results of one simulation become a "good" starting point for the next simulation. Their greedy fault-ordering strategy is very effective for parametric fault simulation and DC power supply testing where DC suply voltages are varied [624]. The results on a set of MCNC circuits show that the method could reduce the total number of iterations by 5 to 10 times. It is particularly useful when many difficult-to-detect faults have similar behaviors.

For selecting an initial condition, SPICE sets $\mathbf{x}^0$ to be $\mathbf{0}$ [484]. With *homotopy/simulation continuation* [374, 552, 737], one constructs a slightly different circuit whose solution is a "good" initial point for the original circuit simulation. They also modify the circuit so that circuit simulation converges easily using *Gmin* and *source stepping*.

Chapter 11. MODEL-BASED ANALOG AND MIXED-SIGNAL TEST

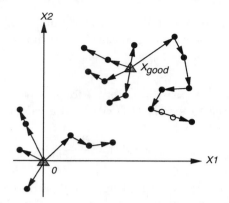

Figure 11.5: Fault ordering via one-step relaxation.

Tian and Shi order the fault list so that the result from one fault simulation reduces the number of iterations for simulating the next fault. Figure 11.5 shows the approach where **0** or the good circuit solution may be the best initial point for fault simulation. The solid circles represent faulty circuit solutions that must be simulated, and the two shadowed triangle points represent the good circuit solution and **0**, respectively. The unfilled circles represent *pseudo* faulty circuit solutions added to help SPICE simulation converge.

They use a simple greedy heuristic the order the faults.

- The faults are simulated using one-step Newton-Raphson iteration for all faults with the good circuit solution point. These results are estimates of the actual faulty circuit solutions, and can be obtained with Householder's formula.

- A simple formula quantifies heuristically how close any two faulty responses are or the distances between the good circuit and a faulty circuit. Given a non-zero solution $\mathbf{x}_i$ and a solution vector $\mathbf{x}_j$, define the *normalized absolute distance* as:

$$t_{ij} = \frac{1}{n} \sum_{k=1}^{n} \left| \frac{(\mathbf{x}_i)_k - (\mathbf{x}_j)_k}{(\mathbf{x}_i)_k} \right| \qquad (11.4)$$

$(\mathbf{x}_i)_k$ and $(\mathbf{x}_j)_k$ denote the kth components of the vectors $\mathbf{x}_i$ and $\mathbf{x}_j$. The vector dimensions are n. This is an effective metric of the closeness of responses of two faults, and is computed using one-step faulty response prediction using Householder's inverse matrix formula.

- Select the fault with the minimum distance and simulate it. If the distance is < 1, simulate this fault using the good circuit solution. Otherwise, use **0** as a starting point.

- After the new simulation result is obtained, calculate the distances of remaining faults in the list, and find the fault with the minimum distance.

11.5 Analog Fault Simulation

- If this distance is less than the distance between the next fault in the list and the good circuit solution, and is also < 1, exchange the next fault with the fault with minimum distance and simulate it. Otherwise, simulate the next fault in the list.

Vinnakota provides the details of this method [699].

11.5.3 Linear Analog Circuit AC Fault Simulation

Linear analog circuit fault simulation is used on analog circuits designed for video and image processing, DSP control, communications, etc.

Householder's Formula Method

This method uses *modified nodal analysis* (MNA) to analyze the analog circuit. We write the system equations as:

$$\boldsymbol{Tx} = \boldsymbol{w} \tag{11.5}$$

$\boldsymbol{T}$ is the circuit matrix, $\boldsymbol{x}$ is the nodal voltages and currents in voltage sources, and $\boldsymbol{w}$ is the contribution of voltage and current sources. The equivalent faulty circuit equation is:

$$\boldsymbol{T_f x_f} = \boldsymbol{w_f} \tag{11.6}$$

Observation: Only one or a few components simultaneously fail, so $\boldsymbol{T_f}$ differs in only a few elements from $\boldsymbol{T}$. We exploit this with Householder's formula [308]:

$$(\boldsymbol{A} + \boldsymbol{USW})^{-1} = \boldsymbol{A}^{-1} - \boldsymbol{A}^{-1}\boldsymbol{U}(\boldsymbol{S}^{-1} + \boldsymbol{WA}^{-1}\boldsymbol{U})^{-1}\boldsymbol{WA}^{-1} \tag{11.7}$$

where $\boldsymbol{A}$ is an $n \times n$ matrix, $\boldsymbol{U}$ is an $n \times m$ matrix, $\boldsymbol{S}$ is an $m \times m$ matrix, and $\boldsymbol{W}$ is an $m \times n$ real matrix. An additional advantage of Householder's formula is that it avoids the numerical problems of direct fault circuit simulation. For fault simulation, m components are faulty, and changes enter the MNA equations as $t_i \boldsymbol{p}_i \boldsymbol{q}_i^\top$ where

$$\boldsymbol{p}_i = \boldsymbol{e}_{i_j} - \boldsymbol{e}_{i_{j'}} \tag{11.8}$$

$$\boldsymbol{q}_i = \boldsymbol{e}_{i_k} - \boldsymbol{e}_{i_{k'}}$$

and $\boldsymbol{e}_i$ is an error vector of 0s, except for a 1 in the ith entry showing the presence of a fault.

We assume that m components denoted as $h_1, h_2, ..., h_m$ are subject to large changes $\delta_1, \delta_2, ..., \delta_m$. Then $\boldsymbol{T_f} = \boldsymbol{T} + \sum_{i=1}^{m} \delta_i \boldsymbol{p}_i \boldsymbol{q}_i^\top$ or $\boldsymbol{T_f} = \boldsymbol{T} + \boldsymbol{P\Delta Q}^\top$. $\boldsymbol{P} = [\boldsymbol{p}_1, \boldsymbol{p}_2, ..., \boldsymbol{p}_m]$ and $\boldsymbol{Q} = [\boldsymbol{q}_1, \boldsymbol{q}_2, ..., \boldsymbol{q}_m]$ are $n \times m$ matrices with 0 and ± 1 entries, respectively. $\boldsymbol{\Delta} = diag[\delta_i]$ is a diagonal $m \times m$ matrix.

Shi [600] applies Householder's formula with the substitutions $\boldsymbol{A} = \boldsymbol{T}$, $\boldsymbol{U} = \boldsymbol{P}$, $\boldsymbol{S} = \boldsymbol{\Delta}$, and $\boldsymbol{W} = \boldsymbol{Q}^\top$, and gets:

$$\boldsymbol{T_f}^{-1} = \boldsymbol{T}^{-1} - \boldsymbol{T}^{-1}\boldsymbol{P}(\boldsymbol{\Delta}^{-1} + \boldsymbol{Q}^\top \boldsymbol{T}^{-1}\boldsymbol{P})^{-1}\boldsymbol{Q}^\top \boldsymbol{T}^{-1} \tag{11.9}$$

$$\boldsymbol{x_f} = (\boldsymbol{T}^{-1} - \boldsymbol{T}^{-1}\boldsymbol{P}(\boldsymbol{\Delta}^{-1} + \boldsymbol{Q}^\top \boldsymbol{T}^{-1}\boldsymbol{P})^{-1}\boldsymbol{Q}^\top \boldsymbol{T}^{-1})\boldsymbol{w_f} \tag{11.10}$$

$$= \boldsymbol{T}^{-1}\boldsymbol{w_f} - \boldsymbol{T}^{-1}\boldsymbol{P}(\boldsymbol{\Delta}^{-1} + \boldsymbol{Q}^\top \boldsymbol{T}^{-1}\boldsymbol{P})^{-1}\boldsymbol{Q}^\top \boldsymbol{T}^{-1}\boldsymbol{w_f}$$

In any SPICE-like simulator, T^{-1} is not available, but its LU factors are known. $y = T^{-1}w_f$ is obtained by one forward substitution and one back substitution. $Q^\top y$ is a sparse vector of order m obtained by inspection. $T^{-1}P$ is obtained from the LU factors of T with m forward and back substitutions, with the columns of P serving, in turn, as right-hand side vectors. We obtain $Q^\top(T^{-1}P)$ by selecting matrix entries. This method obtains a solution z to the equation $(\Delta^{-1} + Q^\top T^{-1}P)^{-1}Q^\top y$, which is equivalent to solving $(\Delta^{-1} + Q^\top T^{-1}P)z = Q^\top y$. T^{-1} is explicitly represented when many (say η) fault simulation runs are required. Then, the number of operations for simulating each faulty circuit is:

$$\frac{n^3}{\eta} + n^2 + \frac{m^3}{3} + m^2 + mn + \frac{2}{3}m \tag{11.11}$$

where $m \ll n$. Temes [653] explains that for a 30-component analog circuit, Householder's formula accelerates simulation 1000 times over repeated equation solving and 10 times over sparse-matrix techniques.

If only a single faulty component must be simulated, then $w_f = w$. Notice that $x_f = T_f^{-1}w$ and $x = T^{-1}w$ so the above equations simplify to:

$$T_f^{-1} = T^{-1} - \frac{T^{-1}e_i e_j^\top}{(\Delta t_{ij})^{-1} + (T^{-1})_{ji}}T^{-1} \tag{11.12}$$

$$x_{fk} = x_k - \alpha_k x_j \text{ and } k = 1, 2, ..., n \tag{11.13}$$

$$\alpha_k = \frac{[(T)^{-1}]_{ki}}{(\Delta t_{ij})^{-1} + (T_{ij}^{-1})} \tag{11.14}$$

These additional simplifications occur for catastrophic analog faults:

- For *short* faults, matrix entry T_{ij} becomes 0, $\Delta t_{ij} = -T_{ij}$, and:

$$\alpha_k = \frac{[(T)^{-1}]_{ki}}{-T_{ij}^{-1} + (T_{ij}^{-1})} \tag{11.15}$$

- For *open* faults, matrix entry T_{ij} becomes ∞, $\Delta t_{ij}^{-1} = 0$, and:

$$\alpha_k = \frac{[(T)^{-1}]_{ki}}{(T_{ij}^{-1})} \tag{11.16}$$

Many fault simulators use Householder's formula [398, 148, 518, 602, 601, 669, 700].

Discrete Z-Domain Mapping Method

Nagi, Chatterjee, and Abraham [487] were the first to use analog signal flow graphs to represent the circuit and fault simulate it in the discrete Z-domain. They represented the s-domain complex frequency state equations as $s\mathbf{X}(s) =$

11.6 Analog Automatic Test-Pattern Generation

$\mathbf{AX}(s) + \mathbf{BU}(s)$, using signal flow graphs and dummy variables [600]. They approximated integration using recursive differences and used the bilinear transform of Equation 11.17 to map the s-domain equations into the Z-domain.

$$s = \frac{2z - 1}{t_s z + 1} \tag{11.17}$$

Using t_s as the sampling time, the original equation set is transformed into:

$$\mathbf{X}(z) = \mathbf{Z_A} z^{-1} \mathbf{X}(z) + \mathbf{Z_B}[z^{-1}\mathbf{u}(z) + \mathbf{u}(z)] \tag{11.18}$$

where z^{-1} represents a single delay. In the time domain, these become:

$$\mathbf{x}(t_k) = \mathbf{Z_A} \mathbf{X}(t_{k-1}) + \mathbf{Z_B}(\mathbf{u_{k-1}} + \mathbf{u_k}) \tag{11.19}$$

and the input signal $\mathbf{u}(t)$ is sampled at the rate $1/t_s$. They accelerated analog fault simulation 10 times by using behavioral models for OPAMPs. Fault mapping into the Z-domain is complex: a single fault generates multiple faults in the Z-domain, the fault changes the number of states, state-variable analysis has problems, the transformation works only for certain circuits, and it can be difficult to choose an effective sampling frequency.

11.5.4 Monte-Carlo Simulation

Monte-Carlo simulation has been extensively used in analog circuit and fault simulation. We perform the simulation for randomly-generated small variations in circuit component values. This is done because the actual IC manufacturing process will cause good circuits to deviate by such values. However, Monte-Carlo simulation is computationally expensive, unless the circuit is small with few statistical parameters [699]. Spinks and Bell used Monte-Carlo simulation only to compute the good circuit worst-case boundaries [629]. Then, they used the band fault simulation method [518] to compute the worst-case faulty-circuit boundaries. Although the method worked for some test circuits, in the general case good and faulty circuits have different worst-case corners. Also, Monte-Carlo simulation always underestimates the response bounds and produces non-robust test sets [699]. The simulator may say that a fault is detectable, when in fact it is not at that test point.

11.6 Analog Automatic Test-Pattern Generation

Analog automatic test-pattern generation is needed to relieve the analog designer of the tedium and extra test design iterations caused by manual design of analog test waveforms. Also, an automatic method can avoid generating unnecessary tests for analog faults that are already covered. We focus here on the automatic waveform generation methods involving sensitivities and signal flow graphs.

11.6.1 ATPG Using Sensitivities

Hamida and Kaminska developed a functional analog test generation method that is based on first-order sensitivity calculations [275, 276]. Others have also explored sensitivity-based ATPG [175, 488]. First-order *sensitivity* represents the relation between circuit *elements* and output *parameters*. Hamida and Kaminska deduce component deviations by measuring various *output parameters*, and through sensitivity analysis and tolerance computation. They identify tests for catastrophic and soft (parametric) faults, for both single and multiple fault models. This ATPG method is intended for production testing applications. In order to minimize the testing time, they find the circuit parameters to be tested, using the simplex method, that give the maximum analog fault coverage. They also determine the accuracy to which the output parameters should be measured. They combine time domain, harmonic, and static measurements to construct an analog test set.

Sensitivity

Sensitivity is a measure of the circuit's performance change for a change in the circuit element values [276]. The deviation in element values from their nominal values depends on the circuit manufacturing process and the element temperature. Sensitivity is also useful in choosing circuit element tolerances, and involves computing the first-order partial derivatives of network functions [547].

Sensitivity is the effect of a change in circuit element x_i on the change in the circuit transmission (performance) parameter T_j [113]. *Differential sensitivity* shows the effect of small variations in elements, and is defined as:

$$S_{x_i}^{T_j} = \frac{x_i}{T_j} \frac{\partial T_j}{\partial x_i} = \left. \frac{\Delta T_j / T_j}{\Delta x_i / x_i} \right|_{\Delta x_i \to 0} \tag{11.20}$$

Incremental sensitivity shows the effect of large element variations, and is defined as:

$$\rho_{x_i}^{T_j} = \frac{x_i}{T_j} \times \frac{\Delta T_j}{\Delta x_j} \tag{11.21}$$

Incremental sensitivity can be computed directly from the network transfer function of parameter T_j, or experimentally. The transfer function of T_j can be written as a fraction $T_j(w,x) = N(w,x)/D(w,x)$. For direct sensitivity computation, they use:

$$\rho_{x_i}^{T_j} = \frac{S_{x_i}^{T_j}}{1 + S_{x_i}^{D}(\Delta x_i / x_i)} \tag{11.22}$$

where $S_{x_i}^{D}$ is the differential sensitivity of the denominator D. $S_{x_i}^{T_j}$ is the differential sensitivity of a rational-form parameter. In complex circuits, sensitivity computation is difficult, so they use simulation and numerical computation of sensitivity [610]. In order to estimate sensitivities S_x^T and S_x^D, they measure incremental sensitivity ρ_x^T, and force component x into two deviations, Δx_a and Δx_b. Finally,

11.6 Analog Automatic Test-Pattern Generation

they measure the corresponding deviation values ΔT_a and ΔT_b of parameter T at the circuit output. Equation 11.22 then leads to:

$$\frac{\Delta T_a}{T_a} = \frac{S_x^T}{1 + S_x^D(\Delta x_a/x_a)} \times \frac{\Delta x_a}{x_a} \qquad (11.23)$$

$$\frac{\Delta T_b}{T_b} = \frac{S_x^T}{1 + S_x^D(\Delta x_b/x_b)} \times \frac{\Delta x_b}{x_b} \qquad (11.24)$$

These equations lead to:

$$S_x^T = \frac{x}{T}\left[\frac{\Delta T_a \Delta T_b \times (\Delta x_a - \Delta x_b)}{\Delta x_a \Delta x_b (\Delta T_a - \Delta T_b)}\right] \qquad (11.25)$$

$$S_x^D = x \times \frac{\Delta T_b/\Delta x_b - \Delta T_a/\Delta x_a}{\Delta T_a - \Delta T_b} \qquad (11.26)$$

Incremental sensitivity allows testing of both catastrophic and parametric faults, whereas differential sensitivity can only handle parametric faults.

Circuit Modeling

Analog testing must cover the relation between circuit elements and output parameters. The analog circuit has various elements x_i and various transmission (performance) parameters (parameters) T_j that characterize the circuit. The T_j may or may not be measurable from circuit outputs. Since the number of parameters is huge, they instead test the elements x_i by measurement of the output parameters. They build an analog circuit graph with two kinds of nodes: *parameter* and *element* nodes. The circuit connectivity matrix is the relation between parameters and elements, and is the sensitivity matrix. They use hierarchical sensitivity analysis [716]. A graph arc between a parameter and an element means that the parameter is sensitive to the element variation (the sensitivity $S_{x_i}^{T_j}$ is non-zero and is the graph edge weight.)

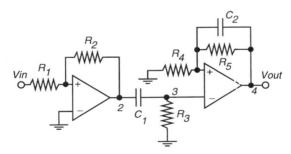

Figure 11.6: Amplification circuit.

Example 11.1 *Consider the amplifier in Figure 11.6. Its parameters are:*

A_1: First stage gain.
A_2: High-pass filter gain at known frequency.
A_3: AC voltage gain of third part.
fc_1: High-pass filter cutoff frequency.
A_4: DC voltage gain of third part.
fc_2: Low-pass filter cutoff frequency.

SPICE [461] computed the amplifier's incremental sensitivity matrix by simulation:

$$\begin{bmatrix} -0.91 & 1 & 0 & 0 & 0 & 0 & 0 & A_1 \\ 0 & 0 & 0.58 & 0.38 & 0 & 0 & 0 & A_2 \\ 0 & 0 & -0.91 & -0.89 & 0 & 0 & 0 & fc_1 \\ 0 & 0 & 0 & 0 & -0.96 & 0.48 & -0.48 & A_3 \\ 0 & 0 & 0 & 0 & -0.97 & -0.97 & 0 & A_4 \\ 0 & 0 & 0 & 0 & 0 & -0.88 & -0.91 & fc_2 \\ R_1 & R_2 & C_1 & R_3 & R_4 & R_5 & C_2 & \backslash \end{bmatrix} \quad (11.27)$$

Figure 11.7 is the equivalent bipartite graph of this incremental sensitivity matrix.

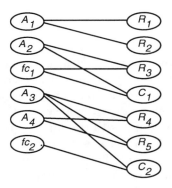

Figure 11.7: Bipartite graph of amplification circuit.

Sensitivity and Analog Circuit Testing

Graph modeling for sensitivity-based analog test generation has the advantages of reducing the complexity of relations between inputs and outputs, and overcoming system non-linearity. In this method, they transform the analog circuit testing problem into a graph theory flow problem.

Problem Formulation. They choose which accessible output parameters to measure, in order to cover all circuit elements. Since an element can be covered by multiple parameters, they choose parameters that guarantee maximal coverage of every element. *Element coverage* is the minimum circuit element deviation that can be observed at a primary output parameter. An element is considered faulty if its value is out of its tolerance box. They compute the relative deviation of an element

11.6 Analog Automatic Test-Pattern Generation

x_i using Equation 11.28. The maximum tolerance of the parameter T is:

$$\frac{\Delta T}{T} = \sum_{i=1}^{M} \left|S_{x_i}^T\right| \frac{\Delta x_i}{x_i} \qquad (11.28)$$

where M is the number of circuit elements. For a single parametric fault model, an element fault is related to multiple output parameters. Some transmission parameters (performances) are more sensitive to some element variations than others, so when the sensitivity of a transmission parameter (performance) T to the element variation x is small, the transmission parameter (performance) can tolerate more error than the element tolerance, so another transmission parameter (performance) T' should be used to observe x during testing. They assign the upper-bound value of element tolerance to fault-free elements to consider the worst case for the faulty element. They find the element tolerance for a number of parameters, and choose the output parameter most sensitive to the element for testing the element tolerance.

Test Method. The test method is as follows:

1. Perform sensitivity analysis.

2. Build the circuit bipartite graph.

3. Compute tolerances (relative deviations) of every circuit element. The relative deviation of fault-free elements comes from circuit data sheets. The tolerance of a faulty element is computed as a (min, max) pair.

4. Construct the bipartite optimization graph.

5. Select parameters (or performances) to be measured during testing using the *simplex* optimization method.

6. Perform testability analysis of the circuit by computing the analog fault coverage. For given tolerance values, analog *fault coverage* is the ratio of detected faults over all possible faults. They compute the real fault coverage, measure tolerances of output parameters, and then compute the maximal and minimal element tolerances. This gives the tolerance region where observation of the faulty element is guaranteed. An element deviation may not be observable if it falls between these maximal and minimal tolerance values.

7. Improve the circuit testability with *design for testability* (DFT) hardware.

Single-Fault Model. For a single parametric fault, the maximum tolerance value for circuit element x_i is:

$$\left(\frac{\Delta x_i}{x_i}\right)_{Max} = \frac{\left|\frac{\Delta T_k}{T_k}\right| + \sum_{j=1}^{i-1} \left|S_{x_j}^{T_k}\right| \left|\frac{\Delta x_j}{x_j}\right| + \sum_{j=i+1}^{M} \left|S_{x_j}^{T_k}\right| \left|\frac{\Delta x_j}{x_j}\right|}{S_{x_i}^{T_k}} \qquad (11.29)$$

The minimum tolerance value is:

$$\left(\frac{\Delta x_i}{x_i}\right)_{Min} = \frac{\left|\frac{\Delta T_k}{T_k}\right| - \sum_{j=1}^{i-1}\left|S_{x_j}^{T_k}\right|\left|\frac{\Delta x_j}{x_j}\right| - \sum_{j=i+1}^{M}\left|S_{x_j}^{T_k}\right|\left|\frac{\Delta x_j}{x_j}\right|}{S_{x_i}^{T_k}} \quad (11.30)$$

The minimum represents the x_i variation seen by observing T_k in the best case, while the maximum represents the variation seen in the worst case.

Once tolerances are computed, they build another weighted bipartite optimization graph relating *primary output* (PO) parameters to elements. Edge weights are the maximum tolerances computed from Equation 11.29. They use standard optimization software called LINDO [572] to select output parameters for observation during testing, by formulating this as a minimum cost flow problem using the simplex method.

The cost function to be minimized is the cost of the arcs between the elements and the parameters selected to test the elements:

$$f(X_{i,j}) = \sum_{i=1}^{N}\left[\sum_{j=1}^{M} C_{i,j} X_{i,j}\right] \quad (11.31)$$

$X_{i,j} = 1$ if there is an arc between parameter T_i and element x_j, otherwise it is 0. $C_{i,j}$ is the arc weight. N is the number of parameters and M is the number of elements.

The three optimization constraints for cost function minimization are:

1. An arc should be chosen only once. For each element x_j they compute the $X_{i,j}$ constraint. There are x_j of these constraints.

$$\sum_{i=1}^{N} X_{i,j} = 1 \quad (11.32)$$

2. The number of chosen arcs, for a parameter T_i, must be $\leq B_i$, the total number of arcs emanating from T_i. There are T_i of these constraints.

$$\sum_{j=1}^{M} X_{i,j} \leq B_i \quad (11.33)$$

3. $x_{i,j} \geq 0$.

LINDO selects arcs with minimum cost, which indicates what parameters to measure in order to have a complete analog test.

Multiple Fault Model. Hamida and Kaminska generalize the single-fault model to a multiple fault model. The maximum number of faults that can be detected depends on the number of independent parameters sensitive to the faulty elements.

11.6 Analog Automatic Test-Pattern Generation

For detecting m simultaneously faulty elements, one needs at least m parameters sensitive to the faulty elements.

Let elements x_i to x_{i+m-1} be faulty, and the remaining ones be fault-free. Maximum tolerance values are computed with these equations:

$$\left|\frac{\Delta T_k}{T_k}\right| = \left(-\sum_{j=1}^{i-1}|S_{x_j}^{T_k}|\left|\frac{\Delta x_j}{x_j}\right| - \sum_{j=i+m+1}^{M}|S_{x_j}^{T_k}|\left|\frac{\Delta x_j}{x_j}\right|\right) + \sum_{j=i}^{i+m}|S_{x_j}^{T_k}|\left|\frac{\Delta x_j}{x_j}\right|\bigg|_{Max}$$

$$\vdots$$

$$\left|\frac{\Delta T_{k+m}}{T_k}\right| = \left(-\sum_{j=1}^{i-1}|S_{x_j}^{T_k}|\left|\frac{\Delta x_j}{x_j}\right| - \sum_{j=i+m+1}^{M}|S_{x_j}^{T_k}|\left|\frac{\Delta x_j}{x_j}\right|\right)$$
$$+ \sum_{j=i}^{i+m}|S_{x_j}^{T_k}|\left|\frac{\Delta x_j}{x_j}\right|\bigg|_{Max} \quad (11.34)$$

Minimum tolerance values of faulty elements are computed as follows:

$$\left|\frac{\Delta T_k}{T_k}\right| = \left(\sum_{j=1}^{i-1}|S_{x_j}^{T_k}|\left|\frac{\Delta x_j}{x_j}\right| + \sum_{j=i+m+1}^{M}|S_{x_j}^{T_k}|\left|\frac{\Delta x_j}{x_j}\right|\right) + \sum_{j=i}^{i+m-1}|S_{x_j}^{T_k}|\left|\frac{\Delta x_j}{x_j}\right|\bigg|_{Min}$$

$$\vdots$$

$$\left|\frac{\Delta T_{k+m-1}}{T_{k+m-1}}\right| = \left(-\sum_{j=1}^{i-1}|S_{x_j}^{T_k}|\left|\frac{\Delta x_j}{x_j}\right| + \sum_{j=i+m}^{M}|S_{x_j}^{T_k}|\left|\frac{\Delta x_j}{x_j}\right|\right)$$
$$+ \sum_{j=i}^{i+m-1}|S_{x_j}^{T_k}|\left|\frac{\Delta x_j}{x_j}\right|\bigg|_{Min} \quad (11.35)$$

M is the total number of elements, and m is both the number of faulty elements and the size of the subset of parameters sensitive to the faulty elements. The *minimum relative variation* of elements x_i to x_{i+m-1} represents the variation in x_i to x_{i+m-1} that can be seen by observing T_k to T_{k+m-1} in the best case. The *maximum relative variation* of elements x_i to x_{i+m-1} represents the variation in x_i to x_{i+m-1} that can be seen by observing T_k to T_{k+m-1} in the worst case [276]. The tolerances of elements and parameters except x_i to x_{i+m-1} are supposed to be known, and their values are the maximum values allowed (in the tolerance box.)

Once all tolerances are computed, they construct the bipartite graph described above. For a double-fault model, the size of parameter subsets is two. C_2^L is the number of subsets to be considered, and l is the number of parameters sensitive to one or both defective elements. L, in the worst case, may be the total number of PO parameters, when all output parameters are sensitive to at least one defective element. The edge weights are the maximum tolerances from Equation 11.34. The minimum cost flow problem formation is exactly the same as for a single-fault model. The cost function for minimization is the cost of the arcs between the parameter subset selected for testing and the faulty elements (see Equation 11.31.) The number of parameter subsets is C_m^L.

After selection of parameters to be measured, they analyze the circuit testability by computing its fault coverage. They will measure these parameters, and if the circuit passes, then the estimated fault coverage was reached. For computing the *real* fault coverage, they measure the output parameter tolerances, and then compute the maximal element tolerance (using Equation 11.34) and the minimal element tolerance (using Equation 11.35.) The maximum tolerance gives the tolerance regions where faulty element observation is guaranteed. Element deviations below the minimal tolerance cannot be observed at any PO. In between these tolerances, the element observability is unknown. Relative deviations of parameters are measured and used with Equations 11.34 and 11.35 to estimate the real fault coverage.

DFT in Analog Circuits

Once the real fault coverage of the test set is known, they apply these techniques to improve analog circuit testability.

1. They add new POs to increase observability of untestable elements.

2. They select which element tolerances to increase. They start with elements unobservable from any parameter and insert an observation point (test point) to observe the element in at least one parameter.

3. They also may increase the number of parameters to be measured.

It is always necessary that the number of independent parameters sensitive to faulty elements be greater than or equal to the number of faulty equations. With too few equations, one must increase the number of internal test points.

Results

Hamida and Kaminska generated tests for the amplification circuit of Figure 11.6 to detect single and multiple faults.

Single-Faults. For testing catastrophic faults, the minimum set of parameters to be measured was $Test_1 = \{A_1, fc_1, A_3\}$, which made all circuit elements observable. For soft fault testing, A_3 should not be used, because its sensitivity to variations in R_5 and C_2 was small. The element tolerance was assumed to be 5%. The maximum tolerance for parameter A_3 was computed from Equation 11.28 to be $\Delta T/T = 9.6$. The relative deviations of C_2 and R_5 were 5%, and the tolerance box for R_4 was computed from Equations 11.29 and 11.30. The maximal tolerance of R_4 was $5 \leq \frac{\Delta R_4}{R_4} \leq 15$, so this meant that a fault less than 5% could not be tested and a fault greater than 15% could easily be tested, because it caused parameter A_3 to go out of range. For R_5 and C_2, $5 \leq \Delta R_5/R_5 \leq 35$, and $5 \leq \Delta C_2/C_2 \leq 35$. The test was bad for C_2 and R_5, because in the worst case R_5 increased by 35%, and the effect of this variation on A_3 was unobservable. The best and worst-case element deviations were:

11.6 Analog Automatic Test-Pattern Generation

$$A_1 \begin{cases} 5 \leq \frac{\Delta R_1}{R_1} \leq 15.98 \end{cases}$$

$$A_2 \begin{cases} 5 \leq \frac{\Delta R_2}{R_2} \leq 14.1 \\ 5 \leq \frac{\Delta R_3}{R_3} \leq 20.27 \\ 5 \leq \frac{\Delta C_1}{C_1} \leq 11.6 \end{cases}$$

$$A_4 \begin{cases} 5 \leq \frac{\Delta R_4}{R_4} \leq 15 \\ 5 \leq \frac{\Delta R_5}{R_5} \leq 15 \end{cases}$$

$$fc_1 \begin{cases} 5 \leq \frac{\Delta R_3}{R_3} \leq 14.81 \\ 5 \leq \frac{\Delta C_1}{C_1} \leq 15.2 \\ 5 \leq \frac{\Delta R_5}{R_5} \leq 14.65 \end{cases}$$

$$fc_2 \begin{cases} 5 \leq \frac{\Delta C_2}{C_2} \leq 13.96 \end{cases}$$

$$A_3 \begin{cases} 5 \leq \frac{\Delta R_4}{R_4} \leq 15 \\ 5 \leq \frac{\Delta R_5}{R_5} \leq 35 \\ 5 \leq \frac{\Delta C_2}{C_2} \leq 35 \end{cases}$$

Fig: 11.8 shows the bipartite optimization graph reflecting these equations.

The LINDO program computed the best test set for the amplifier to be $TEST = \{A_1, A_2, fc_1, A_3, fc_2\}$, which meant that all PO parameters should be measured, except A_4. Faults less than 15% in R_4 and faults less than 15.98% in R_1 could not be detected at any parameter. For $11.6\% \leq deviation \leq 14.1\%$, C_1, C_2, and R_2 were fully testable. Other element faults may not be testable, so the guaranteed fault coverage was 42.9%. There was no assurance that any fault less than 11.6% would be detected.

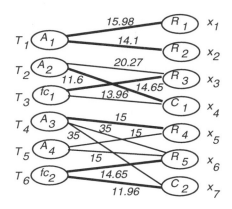

Figure 11.8: Weighted bipartite graph.

Multiple Faults. Only three element faults could be simultaneously defective in the amplifier. In the first stage, there are two components (R_1 and R_2) and only A_1 is sensitive to them, requiring a single-fault model. For the second stage, there are two components (R_3 and C_1) and both A_2 and fc_1 are sensitive to both, requiring both a single and a double fault model. In the third stage, there are three components (R_4, R_5, and C_2) and the A_3, A_4, and fc_2 parameters are sensitive to them, requiring single, double, and triple-fault models.

Double-Faults. LINDO determined that the test set that guarantees the maximum possible fault coverage of double-faults contains the subsets $\{A_2, fc_1\}$, $\{A_3, A_4\}$, and $\{A_4, fc_2\}$. So, the parameters $\{A_2, fc_1, A_3, A_4, fc_2\}$ should be measured, and four systems of equations should be solved using Equation 11.34. For example, when R_3 and C_1 are simultaneously faulty, A_2 and fc_1 are measured. This means that a variation in R_3, C_1, R_4, R_5, and C_2 greater than 15% can be easily detected using a double-fault model. Also, a simultaneous deviation in R_1 and R_2 cannot be measured, since they mask each other.

Triple-Faults. The only meaningful triple-fault is R_4, R_5, and C_2 simultaneously faulty. The measured transmission parameters (performances) are A_3, A_4, and fc_2. The maximum element deviation that can be seen by observing these parameters (performances) is 5% for the three elements, so deviations greater than 5% in the three components can be tested by a three-fault model.

Result Analysis. Analog catastrophic faults are easily detected with this method. Parametric (soft) faults are more difficult to detect with a single-fault model. There are usually three tolerance regions using this method: between 0% and the minimum tolerance value from Equation 11.35, where we cannot detect any fault. The second is between the minimum tolerance of Equation 11.35 and maximum tolerance of Equation 11.34, where we are unsure whether a fault can or cannot be detected. Finally, the third is where the fault has deviation greater than the maximal tolerance of Equation 11.34, where we are sure that all element faults can be detected.

The single-fault model has the disadvantages of fault masking and difficulty in testing some single-faults. Consider the amplifier gain, which is $-R_2/R_1$. Fault masking occurs when R_1 and R_2 are out of their tolerance range and increase or decrease by the same value. Faults in the ratio, however, can be detected by a multiple fault model, which is what we really care about in the case of the buffer.

11.6.2 ATPG Using Signal Flow Graphs

Ramadoss and Bushnell [533] proposed *structural* analog circuit testing, where they generate test waveforms that verify which component values or ratios of component values are within specifications. This method shortens tester time per circuit, by reducing the number of measurements.

They targeted analog/mixed-signal circuits [533] with these characteristics:

- They are linear, first-order blocks designed using *signal flow graphs* (SFGs). These are cascaded to realize second-order transfer functions.

- The analog part lies on the chip inputs and feeds the digital block after A/D conversion. The digital part may/may not have independent *primary inputs* (PIs.)

- The analog part is observable only at digital or analog outputs.

11.6 Analog Automatic Test-Pattern Generation

This important circuit class has applications in filtering and amplification. Circuit blocks can be cascaded to achieve a required transfer function, and are easily designed using SFGs. Unfortunately, integration of analog and digital circuits on the same chip reduces controllability and observability. One cannot sensitize the fault in the digital sense as there are an infinite number of faulty values; however, the range of good values is known.

Ramadoss and Bushnell provide this test generation solution: [534]

1. A user-specified output measurement tolerance leads to good and bad output values defining an output waveform envelope. They work backwards from the analog/digital interface with these values to generate structural analog component tolerances (or component ratio tolerances) that guarantee meeting the output specification This process is the *analog backtrace*.

2. The key element in analog backtracing is working backwards through the analog circuit, i.e., *reverse simulation*, to calculate the input test given the output(s.) They use SFG inversion and backwards traversal to compute (from digital testing requirements) corresponding analog test input waveforms, which excite fault(s) and propagate faulty analog signals through the analog and digital parts to digital outputs (the *analog backtrace*.) This is done for each path from input to output in the circuit SFG.

Signal flow graphs represent the circuit input-output relations in graph form, and our analog backtrace method over the inverted SFG is analogous to digital test generation. When analog blocks drive a digital circuit through an *A/D converter* (ADC), one can backtrace from the digital circuit inputs to calculate the analog inputs needed to justify digital tests.

Reverse Simulation

A *signal flow graph* (SFG) graphically represents the set of equations relating circuit state variables, inputs, and outputs of an active network circuit. Nodes can be *source* nodes, with outgoing edges only; *sink* nodes, with incoming edges only; or both. The value of any graph node (state variable) i is:

$$\text{value}(i) = \sum (\text{parent\_node\_value}) \cdot (\text{incoming\_edge\_weight}) \quad (11.36)$$

Consider the following equation with its corresponding SFG in Figure 11.9(a):

$$x_2 = a \cdot x_1 + b \cdot x_3 + c \cdot x_4 \quad (11.37)$$
$$x_5 = -d \cdot x_2$$

Only incoming edges effect the value of a node, so x_5 has no effect on Node x_2. Consider a rewritten version of Equation 11.37 and its SFG in Figure 11.9(b):

$$x_1 = \frac{1}{a} x_2 - \frac{b}{a} x_3 - \frac{c}{a} x_4 \quad (11.38)$$

They extended Balabanian's path inversion algorithm [61] to invert the path between an input and output of a SFG.

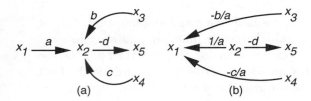

Figure 11.9: Signal flow graphs of (a) Equation 11.37 and (b) Equation 11.38.

Algorithm 11.1 *Signal Flow Graph Inversion Algorithm.*

1. Select a path in the SFG from PI x_1, through intermediate nodes x_2 through x_{n-1}, to PO x_n.

2. Start at the primary input x_1, a source node with only outgoing edges.

3. Reverse the direction of the outgoing edge from x_1 to x_2, and the new weight $1/a$ is the reciprocal of the old weight a. x_1 becomes a sink node (with incoming edges only.)

4. Redirect all edges incident on x_2 to x_1, multiply the original weights by the new weight $1/a$ on the reversed edge from x_2 to x_1, and change the sign.

5. Repeat Steps 3 and 4 for all source nodes x_i on the path from PI x_1 to PO x_n, until x_n becomes a source node. At this point, since all of the graph edges point towards the input, the graph is inverted.

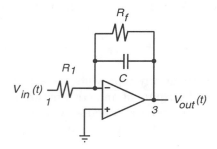

Figure 11.10: Real integrator circuit.

Example 11.2 *Consider the integrator circuit in Figure 11.10, its signal flow graph in Figure 11.11, and the equivalent Equation 11.39:*

$$Z_f = \frac{R_f \frac{1}{sC}}{R_f + \frac{1}{sC}} = \frac{R_f}{sR_f C + 1} \qquad (11.39)$$

$$\frac{V_{in}(s)}{R_1} = \frac{-V_{out}(s)}{Z_f}$$

$$V_{out}(s) = -V_{in}(s)\frac{Z_f}{R_1} = -V_{in}(s)\frac{1}{sR_1 C + \frac{R_1}{R_f}}$$

11.6 Analog Automatic Test-Pattern Generation

Nodes 1 and 3 of the SFG correspond to the circuit voltage input and output, respectively. Figure 11.11(a) shows the original SFG of the integrator, with a self-loop on the output. Node 2 in Figure 11.11(b) is a dummy node introduced to avoid a self-loop on the output node – it has a weight of $1/s$ that was common to both edges in the original graph. Differentiation (represented by the s operator) is a linear operation. Hence, multiple edges with non-zero s weights in the SFG of a single first-order block can be combined to produce multiple edges with ordinary numerical weights and one edge with non-zero s weight. This helps in reverse simulation, as numerical differentiation, an expensive operation, is done only once.

After Steps 3 and 4 of the algorithm, we have the intermediate SFG in Figure 11.11(c). The edge with weight $-1/R_1C$ has been reversed and its new weight is now $-R_1C$. The edge from the output, with weight $-1/R_fC$, has been redirected to the input node, and its new weight is $-(-R_1C \times -1/R_fC) = -R_1/R_f$. Only one more edge need be inverted: the intermediate edge to the output with weight $1/s$. It is inverted with the new weight becoming s; at this stage we stop, the output having become a source node and the input a sink node. Figure 11.11(d) shows final inverted graph and is equivalent to the equation:

$$V_{in}(s) = -R_1C \times s(V_{out}(s)) - \frac{R_1}{R_f}V_{out}(s) \qquad (11.40)$$

All cycles in the original SFG are broken in the inverted graph, which is now a feedforward network. All of node i's parents in the inverted graph will be closer to the circuit output than i. In SFGs, input and output are merely labels, and are interchangeable.

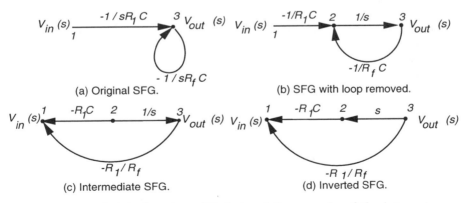

Figure 11.11: Original and modified signal flow graphs of the integrator.

One can simulate the inverted SFG with a chosen output to find the original network input. The s operator in the Laplace domain corresponds to differentiation in the time domain. The process starts with a set of output samples. The inverted graph weights are all either numerical scalars or s, the differential operator that can be numerically approximated. One can find all inverted graph node values using

Equation 11.36. This process finds the input corresponding to a given output, but non-linear networks must have an non-linear block input constrained to a constant before a SFG can be built for reverse simulation. One may have to search among many possible constant inputs to find an acceptable test. This provides a new ability to work backwards in an analog circuit during test generation. The method takes a set of actual SPICE output samples $f(x_i)$ and uses them to reverse simulate the original circuits in the time domain. Most samples were equally spaced points Δt apart in time.

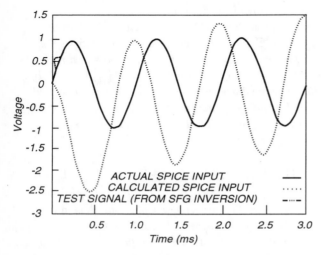

Figure 11.12: Comparison: SPICE vs. reverse simulator for integrator.

Figure 11.12 shows the reverse simulation results for the integrator circuit, with nominal values $R_f = 100\ K\Omega$, $R_1 = 10\ K\Omega$, and $C = 0.01\ \mu F$. The Actual Output is the SPICE simulation of the Actual Input, and the Calculated Input is the test waveform calculated using backtracing from the Actual Output. The Calculated Input test waveform follows the Actual Input waveform very closely, after some minor initial error due to a startup transient. Table 11.2 shows the component tolerances calculated for the integrator.

Table 11.2: Computed deviations for integrator circuit.

Component	Allowed value	Allowed deviation
R_1	$9.09 K\Omega$	$\pm 9.1\%$
R_f	$80.99 K\Omega$	$\pm 19.01\%$
C	$0.0093 \mu F$	$\pm 7\%$

Analog Backtrace

The reverse simulation algorithm can be used to define faults and generate analog circuit test waveforms. One starts with an output waveform tolerance, and works

11.6 Analog Automatic Test-Pattern Generation

backward with good and bad outputs. The good input found is the test signal, and the bad output and node values calculated along the way iet one calculate component deviations.

Test Generation Algorithm:

1. The user sets a circuit output voltage magnitude tolerance.

2. Parse the netlist (in SPICE format), build the SFG, and store edge weights symbolically as well as numerically.

3. Invert the SFG using Algorithm 11.1 and store edge weights numerically and symbolically.

4. Obtain a set of samples of the good machine output.

5. Perform good output reverse simulation and find values at all nodes, including the primary input (for all samples.)

    (a) Apply the good output waveform sample to the PO.

    (b) Traverse the SFG backwards from the PO in breadth-first order, setting all intermediate node values numerically using Equations 11.36 and a second-order approximation to the differential s operator [80]. If the inverted SFG has loops, iterate until signals converge.

    (c) The PI now has the correct good machine test waveform.

6. Perform bad output reverse simulation as in Step 5 and find all bad machine node values.

7. Calculate all internal structural component tolerances, either for single components or ratios of multiple components, using the method below.

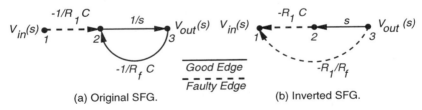

Figure 11.13: Original and inverted SFGs for integrator.

Analog Fault Modeling and Component Tolerance Calculation. SFG inversion can calculate the parameter tolerances that the circuit components must meet (during manufacturing), in order to ensure that the analog output waveforms remain within specifications. This new method avoids specifying analog components to tighter parametric tolerances than is necessary, and this reduces cost. Consider again the integrator circuit in Figure 11.10. Figure 11.13(a) shows the original integrator SFG, and Figure 11.13(b) shows the inverted graph. Consider a parametric tolerance on R_1, so the highlighted edges in both graphs show values that will change if R_1 has a fault. One calculates the bad value of resistance R_1 that will cause the analog output to go out of spec. ¿From the prior test generation algorithm, both good and bad machine analog magnitudes (voltages or currents) exist at all circuit nodes. This information, and the symbolic equation relationships in the inverted SFG, let one calculate the necessary component tolerance on R_1. One solves a linear equation symbolically to find the tolerance on R_1. In the inverted SFG, we use the *good* machine node values for nodes between the analog PIs and the faulty (dashed) edges in the SFG. For nodes in between a faulty edge and an analog PO, we use *bad* machine node values. For example, we use the good machine value for Node 1 and bad machine values for Nodes 2 and 3. Here is the equation for the bad machine edge in the inverted SFG in Figure 11.13(b):

$$\text{goodval}(1) = -R_1 C \times \text{badval}(2) - \frac{R_1}{R_f} \times \text{badval}(3)$$

$$\frac{\text{goodval}(1)}{-R_1 C} + \frac{\text{badval}(3)}{-R_f C} = \text{badval}(2)$$

After rearrangement, we obtain the parametric tolerance on resistor R_1:

$$\frac{\text{goodval}(1)}{-R_1 C} + \frac{\text{badval}(3)}{-R_f C} = \text{badval}(2)$$

$$\text{badval}(R_1) = \frac{-\text{goodval}(1)}{C \left(\text{badval}(2) + \frac{\text{badval}(3)}{R_f C}\right)}$$

$$\text{goodval}(R_1) = \frac{-\text{goodval}(1)}{C \left(\text{goodval}(2) + \frac{\text{goodval}(3)}{R_f C}\right)}$$

$$R_1 \text{ Tolerance} = \text{goodval}(R_1) - \text{badval}(R_1) \tag{11.41}$$

Thus, we have computed the maximum excursion permissible on R_1 that will still keep the output analog waveform within its envelope. They [534] proved that the test waveform for parametric single-faults also detects catastrophic faults.

Multiple-Input Multiple-Output Circuits. Multiple-input, multiple-output circuits can be tested by inverting a path between any one PI (PI_i) and PO (PO_j.) The method assumes a signal at the unconstrained analog PIs and uses the good and bad signals at PO_j to calculate the test signal PI_i. Ramadoss and Bushnell provide an example [534].

Multiple Parametric Faults. They also developed a multiple parametric fault model for analog circuits by assuming that all resistances were off by a certain percentage, instead of a single parametric fault model. To find this deviation, they used a variant on Newton's iterative method, which uses the iterative equation:

$$x_{n+1} = x_n - \frac{f(x_n)}{f\prime(x_n)} \tag{11.42}$$

to solve the equation $f(x) = 0$. They adapted this equation to their problem as:

$$R_{n+1} = R_n \pm \frac{V_{final} - V_n}{V_n - V_{n-1}} \times (R_n - R_{n-1}) \tag{11.43}$$

They expressed all resistances in the SFG as fractions of the smallest circuit resistance R. The $\pm$ sign is used to control iterations: the positive sign is used in systems where V_n increases as R_n decreases, and vice versa. This method also works for L and C components. Multiple fault models are more useful than single-fault models for process faults where all component values deviate in the same manner, as happens in analog ICs, when there is a doping variation in the process that affects all components of a particular type.

11.6.3 Additional Methods

Analog test generation has been extensively studied [5, 143, 162, 178, 200, 207, 222, 276, 400, 401, 431, 457, 466, 469, 487, 488, 489, 547, 627, 637, 679]. Analog fault diagnosis [609] and *design for testability* (DFT) [218] have also been investigated. There also exist DFT-based analog ATPG methods [87, 312, 313, 314, 619, 622, 693, 694, 695, 696]. Besides these methods, many additional algorithms exist [62, 70, 200, 201, 251, 431, 471, 470, 473, 505, 542, 623, 679]. Other methods are based on automatic test selection [220, 316, 401, 464, 465, 466, 628].

11.7 Summary

Analog circuit fault simulation and automatic test-pattern generation methods based on fault models are beginning to gain limited acceptance in industry. A recent panel concluded that these models will now begin to augment the traditional functional, DSP-based analog tests that are derived without reference to a fault model [346]. The advantage of the fault models is the opportunity to shorten the number of tests, which is now important for reducing testing costs. Also, analog fault models allow computation of a fault coverage for analog circuit testing, which is also attractive. The factors that are increasing the costs of analog testing, and forcing the investigation of new methods, are the widespread advent of mixed analog/digital circuits, which have significantly raised testing costs, and the advent of higher precision, 22-bit converters, for which conventional testing methods are too expensive.

Problems

In the following problems, all single and multiple parametric faults are to be expressed in terms of the analog component values. Assume that all OPAMPs are ideal, and fault-free. If transistors are given, then faults may change K_p and K_n of the p and nFETs. The functional parameters to be tested in the analog circuits are the AC gain, DC gain, *and* filter cutoff frequency *for each stage of the analog circuit. You must provide the appropriate multiple fault model for these tests.*

11.1 *Parametric faults.* For the integrator circuit in Figure 11.10, please enumerate the meaningful single and multiple parametric faults. Only consider those faults that *really* affect the circuit behavior, and try to obtain the minimal fault set.

11.2 *Parametric faults.* For the biquadratic filter circuit in Figure 11.14, please enumerate the meaningful single and multiple parametric faults. Only consider those faults that *really* affect the circuit behavior, and try to obtain the minimal fault set. It may be useful to simulate the circuit with SPICE using a variety of minor adjustments to the component parameters to determine their effect on the circuit's function.

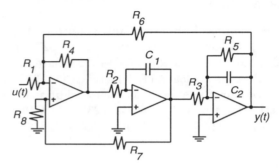

Figure 11.14: Biquadratic filter circuit for Problem 11.2.

11.3 *Parametric faults.* For the leapfrog filter circuit in Figure 11.15, please enumerate the meaningful single and multiple parametric faults. Only consider those faults that *really* affect the circuit behavior, and try to obtain the minimal fault set. It may be useful to simulate the circuit with SPICE using a variety of minor adjustments to the component parameters to determine their effect on the circuit's function.

11.4 *Parametric faults.* For the multiple input-output circuit in Figure 11.16, please enumerate the meaningful single and multiple parametric faults. Only consider those faults that *really* affect the circuit behavior, and try to obtain the minimal fault set.

11.5 *Transistor parametric faults.* For the OPAMP circuit in Figure 11.17, please enumerate the meaningful single and multiple parametric faults, including transistor gain faults. Only consider those faults that *really* affect the circuit behavior, and try to obtain the minimal set.

Problems

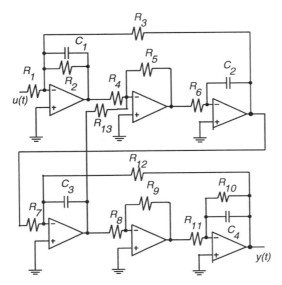

Figure 11.15: Leapfrog filter circuit for Problem 11.3.

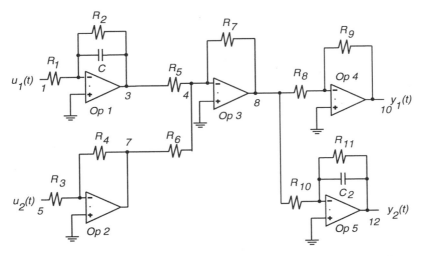

Figure 11.16: Multiple input/output circuit for Problem 11.4.

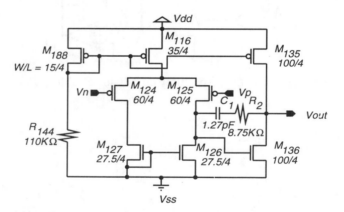

Figure 11.17: OPAMP circuit for Problem 11.5.

Chapter 12

DELAY TEST

> " . . . a new type of fault, called a delay fault, is introduced, and a model developed so that a test to detect this class of fault can be generated via conventional test generation techniques. . . . so that test generation is more of a science rather than a hit or miss process, and so that the correctness of results need not always be verified via simulation or physical fault injection." — Melvin Breuer, in a 1974 paper [91].

A stuck-at-0 fault on a signal means that the signal can be set to 0, but then cannot be changed to 1. Alternatively, this situation can be described by saying that the signal will take an "infinite" amount of time to rise from 0 to 1. Thus, a stuck-at fault is an infinite delay fault and, indeed, a circuit that passes stuck-at fault tests is not likely to have any infinite delay fault. For digital systems that work at any appreciable speed, this is not sufficient. The operation of such systems is usually synchronized by clock signals and it is necessary that all combinational logic elements attain steady state within some specified clock period. Application of stuck-at fault tests at higher speed can uncover some delay defects. However, at least four recent studies [108, 226, 438, 503] show that even that may not be sufficient and tests specifically generated to detect delay defects may be necessary.

This chapter is introductory and the reader may consult recent books on delay testing [371, 608] and timing analysis [452].

12.1 Delay Test Problem

Figure 12.1 shows a schematic of a digital system. Some inputs and outputs can be state variables connected to *flip-flops* (FF) (not shown) and others are *primary inputs* (PI) and *primary outputs* (PO.) All input changes are synchronized with a clock signal and all outputs are expected to attain their final steady state values within one clock period after the inputs change. Thus, for a correct operation the delay of the combinational logic should not exceed the clock period. Typical

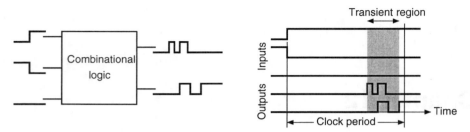

Figure 12.1: Delay fault problem defined.

outputs of logic circuits contain transients as shown in Figure 12.1. We make several observations:

- In order to examine the timing operation of a circuit we should examine signal transitions. The input signal in Figure 12.1 consists of two vectors: 010 → 100. *Delay tests consist of vector-pairs.*

- All input transitions occur at the same time in Figure 12.1. Thus, the duration of the transient region at the input is zero. This, of course, is an idealized illustration though it closely represents the real situation. The transient region at the output contains multiple transitions that are separated in time. As we will see in the next example, the position of each output transition depends upon the delay of some input to output combinational path.

- The right edge of the output transition region (grey shaded area in Figure 12.1) is determined by the last transition, or the delay of the longest combinational path activated by the current input vector-pair. Considering all possible input vector-pairs, "the longest delay combinational path" of the circuit is known as the *critical path*. There can be more critical paths than one if several paths meet the maximum delay criterion. The delay of critical paths determines the smallest clock period at which the circuit can function correctly.

- For a manufactured circuit to function correctly, the output transition region for any input vector-pair must not expand beyond the clock period. Otherwise, the circuit is said to have a *delay fault*. A delay fault means that the delay of one or more paths (not necessarily the critical path) exceeds the clock period.

Example 12.1 *Propagation of transitions.* Consider the circuit in Figure 12.2. Gates are modeled with equal rise and fall lumped delays that are integer multiples of some small time unit (nanosecond or picosecond.) We will also assume that these are purely transport delays and that the gates have negligible inertial delays. Thus, any waveforms are simply translated in time without distortion. More detailed delay models are discussed in Subsection 5.3.5. However, this simple model will suffice for the present discussion. Signal waveforms at various nodes are sketched in the figure, with the time of each transition shown under it. The time of input transitions is 0,

12.1 Delay Test Problem

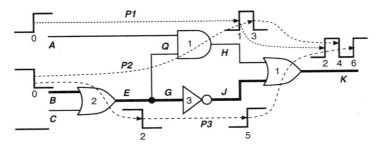

Figure 12.2: An example of transition propagation through paths.

making them the reference for all other transitions. The output has three transitions, brought via three paths. The reader can examine the progress of each transition by following the dashed lines with arrows. This circuit has five paths, which can potentially produce that many transitions at the output. The actual number depends on specific delays and the input stimuli.

Let us examine the three activated paths: Path P1: $A - H - K$, Path P2: $B - E - Q - H - K$, and Path P3: $B - E - G - J - K$. In the operation of this circuit, the input and output signals (irrespective of whether or not they are latched) are synchronized with a clock of period T. Given that these delays have been derived from the analysis of the design data (device parameters, routing capacitances, etc.), the critical path has a delay of 6 units in the fault-free circuit. Path P3 is one of the two critical paths. Suppose we choose $T = 7$. Any path will be faulty if its delay exceeds 7 units. Consider two cases:

1. *Single faulty path:* We examine the output at 7 units of time. As long as the delay of path P3 is 6 units or less, the output will have risen to logic 1 value irrespective of the delay of path P1 or P2. Thus, the delay faults of P1 and P2 will not be detected by this input vector pair. If the delay of path P3 exceeds 7 units, say, due to some manufacturing defect, then the last edge in the output will be shifted to the right and we will observe a 0 instead of 1. Thus, the delay fault of path P3 is detectable by this vector-pair.

2. *Multiple faulty paths:* Suppose all three paths have more than 7 units of delay. Then the entire waveform at the output will be translated to the right by more than 7 units and we will observe a failure. If P1 is not faulty but P2 and P3 are faulty, then the output will rise at 2 units and will remain high beyond 7 units. It may fall depending on the relative delays of P2 and P3. However, observing at 7 units, we will see no failure. In this case the fault of P2 interferes with the detection of the fault of P3. As we shall see later, this is because the present vector-pair is a "non-robust" test for the delay fault of P3.

We have considered only three paths that are activated by the given input vector-pair. Other paths, when activated, can be analyzed similarly.

12.2 Path-Delay Test

The path-delay fault is an important fault model used in delay testing. The following definitions characterize it.

Definition 12.1 *Path-delay fault. The delay defect in the circuit is assumed to cause the cumulative delay of a combinational path to exceed some specified duration. The combinational path begins at a primary input or a clocked flip-flop, contains a connected chain of gates, and ends at a primary output or a clocked flip-flop. The specified time duration can be the duration of the clock period (or phase), or the vector period. The propagation delay is the time that a signal event (transition) takes to traverse the path. Both switching delays of devices and transport delays of interconnects on the path contribute to the propagation delay.*

For each combinational path in a circuit, there are two path-delay faults corresponding to rising and falling transitions, respectively. These faults for a path consisting of gates a, b, and c are specified as $\uparrow a - b - c$ and $\downarrow a - b - c$, where the arrow gives the direction of the transition at the input of the path. The total number of path-delay faults is twice the number of physical paths in the circuit. In general, any combination of paths can be faulty. However, similar to the "single stuck-at" fault model (see Section 4.5) we consider delay faults of single paths. In practice, though, multiple paths can be faulty.

Definition 12.2 *Non-robust path-delay test. A test that guarantees to detect a path-delay fault, when no other path-delay fault is present, is called a non-robust test for that path. A path-delay fault for which a non-robust test exists is called a "singly-testable path-delay fault [245]."*

A non-robust path delay test applies a transition (two-vectors) at the input of the path and measures the output value after a specified interval (clock period.) For the test to be an effective measure of the path delay, the "expected or correct" output value must be uniquely controlled by the transition propagating through the path. Consider the path-delay fault $\downarrow P3$ shown with bold lines in Figure 12.2. Signals B, E, G, J, and K are called the *on-path signals*. Signals that are not in the path $P3$ but feed the gates on the path are called *off-path signals*. Thus, C and H are off-path signals for $P3$. A non-robust test consists of a vector-pair $V1, V2$, such that:

1. The change $V1 \rightarrow V2$ initiates the appropriate transition at the beginning of the path under test. For example, in Figure 12.2 the vector-pair $(V1, V2) = (010, 100)$ produces a falling transition at B to test the fault $\downarrow P3$.

2. All off-path input signals for the path under test assume non-controlling values (0 when the signal feeds into an OR or NOR gate, and 1 otherwise) in the steady-state following the application of the second vector $V2$. This condition is known as *static sensitization* of a path. We may point out that the static sensitization of paths should not be confused with the "static timing analysis,"

12.2 Path-Delay Test

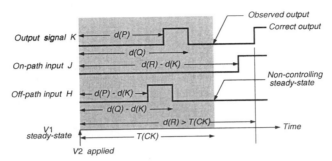

Figure 12.3: Non-robust path-delay test output for ↓ P3 being tested in Figure 12.2.

which simply refers to a topological analysis of physical paths without the application of any signals.

In Figure 12.2, transitions are applied to faults ↑ $P1$ and ↓ $P3$ but static sensitization is achieved only for the latter. Therefore, only the fault ↓ $P3$ is non-robustly tested.

The fact that the two conditions listed above indeed produce a non-robust test can be easily verified. First, by the definition of non-robust test, only a single path is faulty. Hence, all transitions arriving through other paths ending at the same destination must arrive prior to completion of the clock period (shown as the grey region in Figure 12.3.) This implies that by the end of the clock period, all signals other than the on-path signals of the path under test must be in their steady-state. Since the off-path steady-state signals sensitize the entire path under test, the path destination signal is uniquely controlled by the transition propagating through the path. If the path delay exceeds the clock period, then the observed value at the path destination at the end of the clock period will differ from the steady-state output due to $V2$, which is the correct expected value. This is illustrated in Figure 12.3.

Example 12.2 *Non-robust test. Figure 12.4 shows a non-robust delay test for the path delay fault* ↑ $A-B-C$. *The AND gate has rise and fall delays of one unit each, shown as 1/1. The rise and fall delays of the inverter are 2 units. A vector-pair* $(0,1)$ *is derived to satisfy the conditions of a non-robust test and is derived without the consideration of the specific gate delays. The first three waveforms are sketched for the fault-free circuit. The last two waveforms show the signals for a delay fault caused by the inverter delay increasing to 4 units. We notice that the test does not produce a steady-state signal change in the output, which is 0 for all inputs. This logically trivial circuit is a pulse generator whose pulse width is controlled by the inverter delay. If the position and width of the pulse have timing requirements with respect to the clock period, then the delay fault in the inverter path may be important and such a test would be useful. Since this is a non-robust test, it is not guaranteed to work when other paths are faulty. For example, if an additional delay fault* ↑ $A-C$ *is present (either due to increased routing delay or due to increase in the delay of the AND gate), then the signal C may remain as constant 0. In some delay distributions the output pulse will be produced but will be pushed to the right*

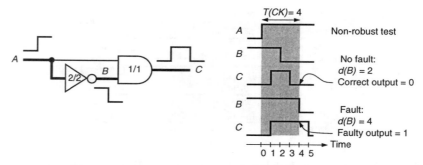

Figure 12.4: An example of a non-robust test.

and out of the clock period (grey region.) In either case, the correct logic value 0 will be observed at the end of the clock period. The presence of the fault ↑ $A - C$, therefore, "invalidates" the non-robust test for fault ↑ $A - B - C$. We also observe that a non-robust test is not possible for the fault ↑ $A-C$, because when we apply the rising transition at A, the off-path input B of the AND gate assumes the controlling value 0.

The notion of robust delay test, though implicit in Smith's 1985 paper [611], was formally defined by Lin and Reddy [397]. The following is an important concept in delay testing:

Definition 12.3 *Robust path-delay test. A robust path-delay test guarantees to produce an incorrect value at the destination if the delay of the path under test exceeds a specified time interval (or clock period), irrespective of the delay distribution in the circuit.*

Figure 12.5 shows a hypothetical (though typical) output waveform produced by combinational logic when a vector-pair $(V1, V2)$ is applied at the input. If this logic is a part of a clocked sequential circuit, the output value at the end of the clock period $T(CK)$ is of interest. The initial value (0) is the steady-state output of $V1$ and the final value is the steady-state output of $V2$. Each transition produced by the vector-pair can potentially propagate through some path and produce a transition at the output at a time determined by the delay of that path. The transitions propagating through paths whose delays are smaller than $T(CK)$ are shown as "fast transitions" and those propagating through paths with delays greater than $T(CK)$ are shown as "slow transitions." If the delay of a path increases, the corresponding transition at the output will move to the right. If the delay reduces, the transition will move to the left. When two neighboring transitions form a pulse, the pulse width equals the difference between the delays of the corresponding paths. If the pulse width is zero or negative (i.e., falling edge arrives earlier for a positive pulse), both transitions will disappear. In other words, the position of an output event is determined by the delay of the path the event travels through, while the existence of the event at the output depends upon the delays of other paths. A robust test that measures the delay of a path should produce an event at the output with following properties:

12.2 Path-Delay Test

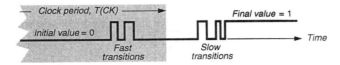

Figure 12.5: Output events produced by combinational logic.

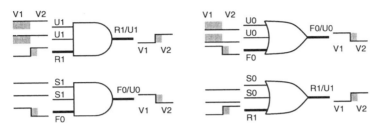

Figure 12.6: Robust path delay sensitization for rising and falling transitions.

1. It should be a "real event" defined as a transition from the initial value to the final value. This is because a real event can exist without the help of any other event. For a falling transition in Figure 12.5, to appear it must be preceded by another event (a rising transition.) Notice that the falling event at the output in Figure 12.3 is not a real event.

2. It should be a "controlling event." A controlling event permits no other events to appear prior to its own appearance. Thus, the output will remain at the initial value until the controlling event occurs at the output.

Having set the requirements for the event the test must produce at the output, we construct the test by recursively moving backward along the path under test. The on-path input of the gate contains the source of the output transition. It is a real transition of the same or the opposite type depending on whether or not the gate has an inversion. If the on-path event is a transition from the controlling value to non-controlling value, then it will prevent any output events prior to its own occurrence. So, there is no specific requirement for off-path inputs in $V1$. To ascertain that the output has a real event, all off-path inputs of the gate should have non-controlling value in $V2$. When the on-path event is a transition from non-controlling value to controlling value, all off-path inputs must have a steady non-controlling value in both $V1$ and $V2$. This is because any transition (even a glitch) can be propagated to the output from the off-path input. These conditions are illustrated in Figure 12.6 for AND and OR gates. The reader can easily work them out for other types of gates. The grey regions in waveforms are the times when "don't care" values or transients (glitches) can occur. We notice that glitches are permitted in on-path signals (shown in bold lines.) This is because these are fault detection tests and not "diagnostic tests." That means the output will not change from the initial value (due to $V1$) during an interval that equals the delay of the path under test. However, an incorrect output at the end of the clock period can also be due to some delayed

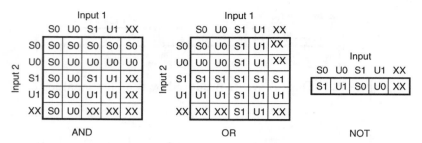

Figure 12.7: Five-valued algebra for path-delay tests.

transition or glitch propagating through off-path inputs.

The signal values shown in Figure 12.6 are due to Lin and Reddy [397]. S0 and S1 are steady (without glitch) 0 and 1 values for both vectors $V1$ and $V2$. U0 and U1 specify the final value as 0 and 1, respectively, and leave the initial value as don't care or X. F0 and R1 are falling and rising transitions on the on-path signals. For an off-path signal, F0 and R1 are treated same as U0 and U1, respectively. In addition, XX is used to denote both vectors in the don't care state. The value set (S0, U0, S1, U1, XX) is a five-valued algebra. With a careful examination, the reader can easily obtain the the truth-tables for AND, OR, NOT, NAND and NOR gates, the first three of which are shown in Figure 12.7 [397].

Multi-valued algebras have useful applications in delay testing. Bose *et al.* [84] give a theoretical treatment leading to optimal algebras for specific cases of test generation and fault simulation in combinational and sequential circuits.

12.2.1 Test Generation for Combinational Circuits

Generation of a test for a path-delay fault requires placing the appropriate transition at the origin of the path and justifying the required off-path inputs of all gates on the path. This is easily accomplished using the five-valued algebra.

Example 12.3 *Robust test generation. Consider the path-delay fault ↓ P3 in Figure 12.2. We proceed as follows:*

1. *Place a transition at the path origin, $B = F0$.*

2. *Propagate value F0 to line E, from Figure 12.6, $C = U0 \Rightarrow E = F0$.*

3. *$G = F0 \Rightarrow J = R1$.*

4. *F0 is interpreted as U0 for off-path logic, $Q = U0$.*

5. *Propagate value R1 from J to K, using Figure 12.6 set $H = S0 \Rightarrow K = R1$.*

6. *Justify $H = S0$, from Figure 12.7 set $A = S0$.*

7. *Test is $A = S0$, $B = F0$, $C = U0$; or $V1 = 01X$, $V2 = 000$.*

12.2 Path-Delay Test

We should remember that the value S0 implies that input A should hold its value steady for two vectors. We observe that this test is different from the one considered in Example 12.1. This test is robust and the reader can verify that it will not be invalidated irrespective of the delay of P2.

The procedure of the above example can be implemented in many ways. The path can be sensitized starting at the output, or all off-path signals can be set at once and then justified. We chose to sensitize the path from input because if sensitization becomes impossible at some gate, then we can immediately conclude that no robust test is possible. This simple example has only limited choices. With increasing number of inputs of a gate, justification choices also increase. In general, when the circuit has reconvergent fanouts, the test generation procedure frequently has to use backtracks in the same way as described in Chapter 7.

For some paths, robust tests are not possible and we must generate non-robust tests. As discussed before, non-robust tests only require static sensitization. That means all signals except the origin of the path under test can have arbitrary values in the first vector ($V1$). This condition is easily incorporated in the multi-valued algebra by simple substitutions, S0 ← U0 and S1 ← U1.

Example 12.4 *Non-robust test generation. Let us try to generate a robust test for path-delay fault ↑ P2 in Figure 12.2. We proceed as follows:*

1. *Place a transition at path origin, $B = R1$.*
2. *Propagate $R1$ to E, from Figure 12.6 set $C = S0$.*
3. *$R1$ is interpreted as $U1$ for off-path logic, $G = U1 \Rightarrow J = U0$.*
4. *$Q = R1$, Propagate $R1$ to H, from Figure 12.6 set $A = U1$.*
5. *$H = R1$, Propagate $R1$ to K, from Figure 12.6, must set $J = S0 \Rightarrow$ conflict since $J = U0$ in step 3.*
6. *Since no step has any alternatives, a robust test is not possible.*

For a non-robust test we change S0 and S1 to U0 and U1, respectively (static sensitization.) Now The Step 5 requirement becomes $J = U0$, which is consistent with Step 3. The non-robust test is $A = U1$, $B = R1$, $C = U0$ (changed from S0); or $V1 = X0X$, $V2 = 110$.

An alternative and simpler method for generating non-robust tests is to derive *single input change* (SIC) tests. For a SIC test, the two vectors $V1$ and $V2$ in the test differ in exactly one bit. We first find $V2$ to statically sensitize the entire path using any combinational ATPG procedure (see Chapter 7.) $V1$ is then obtained by just changing one bit in $V2$ that corresponds to the origin of the path. It can be easily shown that every non-robustly testable path must have a SIC test [245].

The procedure of Example 12.4 attempts to find a non-robust test only when a robust test is impossible. In view of the fact that the reliability of non-robust tests is questionable (see Example 12.2), there is merit in finding as many robust tests as possible. The presence of robust tests for some paths can improve the reliability of non-robust tests for other paths. For example, in Figure 12.2 six path-delay faults, $\uparrow P1$, $\downarrow P1$, $\uparrow P3$, $\downarrow P3$, $\uparrow C-E-G-J-K$ and $\downarrow C-E-G-J-K$, are robustly testable. Example 12.4 shows that $\uparrow P2$ only has a non-robust test. By including the six robust tests we can ensure that if the circuit passes those, there will be no delayed signal at off-path inputs of the path $P2$. We can conclude that in the presence of the other four tests, the non-robust test for $\uparrow P2$ is as good as a robust test. Such a test is called a *validatable non-robust* (VNR) test [371, 539, 632].

Example 12.5 *Untestable path delay fault. Consider the path-delay fault $\downarrow P2$ in Figure 12.2. A falling transition (F0) is placed at B and is easily propagated to H by setting appropriate values on A and C. However, a forward implication sets the off-path input of the output OR gate to U1 (i.e., controlling value in V2.) This path-delay fault has no test.*

A path for which both (rising and falling) path-delay faults (PDFs) are singly (i.e., non-robustly) testable is called a *testable path*. A path having one signly testable PDF and one singly untestable PDF is called a *partially testable path* [244]. When no non-robust test exists for both PDFs of a path, that path is called a *singly-untestable path*. Such a path can be eliminated by circuit transformations that preserve the logic function.

An untestable path is (and a partially testable path may be) associated with one or more redundant single stuck-at faults [415]. The function-preserving transformation such as redundancy removal, as discussed in Chapter 7, eliminates such paths. The fault considered in Example 12.5 is on a partially testable path. We observe that the fault Q stuck-at-1 in this path is redundant. Removal of this fault, removes the AND gate H feeding input A directly to the OR gate K. This eliminates the path completely.

In general, a partially testable path may not have a redundant stuck-at fault. However, there are procedures for modifying the circuit [244, 352] to expose redundant faults that can be removed. The resulting circuit always has fewer paths, a greater percentage of testable paths, and lower overall delay, but can be larger in size.

A combinational circuit may have paths whose delays cannot affect the time of signal change at the output. These paths are called *false paths* [244]. The paths of singly-untestable PDFs are not always false paths. For example, a singly-untestable PDF may be co-sensitized (sensitized simultaneously) with other singly-untestable PDFs and the timing of the circuit would be affected if all co-sensitized paths have excess delays. These paths belong to the classes of *multiply-testable PDFs* [245] and *functionally sensitizable PDFs* [155]. That is the reason why the delays of paths whose PDFs may be untestable are still taken into account while determining the clock period of the circuit (a point in favor of the static timing analysis).

12.2 Path-Delay Test

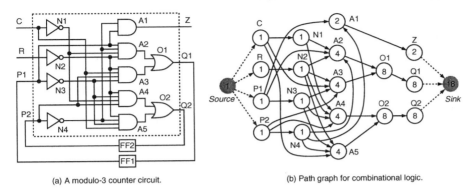

(a) A modulo-3 counter circuit.   (b) Path graph for combinational logic.

Figure 12.8: An illustration of path counting.

12.2.2 Number of Paths in a Circuit

An often-cited hurdle for path-delay testing is the large number of paths that a combinational circuit can have. Pomeranz and Reddy [526] give an example of a circuit to show that the number of paths can be an exponential function of gates (see Problem 12.10.) The benchmark circuit c6288 is known to have 1.98×10^{20} paths. It is a parallel multiplier with 32 inputs, 32 outputs, and 2,406 gates. Many, though not all, practical circuits have such behavior. Pomeranz and Reddy point out that *path counting* algorithms work efficiently in spite of the exponential number of paths.

Example 12.6 *Path counting. The combinational circuit of a modulo-3 counter is enclosed in a dotted-line box in Figure 12.8(a). To count paths, we construct a directed acyclic graph (DAG) as shown in Figure 12.8(b). Every primary input (PI), pseudo-primary input (PPI), gate, primary output (PO), and pseudo-primary output (PPO) is a vertex in this graph, with the name written alongside. Directed edges in this DAG represent signal flow paths. Two extra vertices, source and sink, are added. Source node has 0 indegree and it has directed edges to all PI vertices. All PO vertices have directed edges to the sink vertex, which has a 0 outdegree. Each vertex is given a label (written inside) that equals the number of paths arriving from the source. Source label is initialized to 1 and all others to 0. Vertices in an "active list" are processed in arbitrary order. The processing of a vertex involves two operations: (1) updating its label to be the sum of labels of fanin vertices, and (2) adding its fanout vertices to the active list. The active list is initialized to contain the source vertex and the process begins. Thus, all PI labels are changed to 1. Consider vertex C whose processing places fanout vertices N1, A3 and A5 in the active list. It is possible that they will be processed before nodes such as N2, N3, and N4, that are yet to be added to the active list. So, labels of N1, A3, and A5 will change to 1. This is the final count for N1, but the counts of A3 and A5 will further change as these vertices will be revisited. The maximum number of times a vertex can be visited equals its indegree. The procedure ends when the active list becomes empty. The label of sink is then the number of paths in the circuit. In this case it is 18.*

Since a vertex can potentially be visited as many time as its indegree, the worst-case complexity of path counting is $O(N^2)$ for a DAG with N vertices. This is because the maximum indegree of a node is $O(N)$. For practical circuits the gate fanin has some upper bound that does not grow with the number of gates. Therefore, the complexity of path counting for a circuit with N gates is usually $O(N)$.

Graph theorists [743] often consider the path counting algorithm as a folklore and thus explain its absence from books on the subject. Since Pomeranz and Reddy [526] proposed its application to delay testing, several new techniques for PDF coverage evaluation have been developed [246, 292, 344, 514].

Path counting does not help in test generation for PDFs. To contain the so-called "path explosion," several criteria for selecting paths have been proposed. Li et al. [396] give an algorithm that identifies a set of paths including the longest delay path through every line. Realizing that some selected paths may be untestable, Park and Mercer [507] generate tests for longest delay "testable" paths through all lines. Majhi et al. [414] have defined such tests as *line delay tests*. The upper bound on the number of line delay tests is twice the number of lines in the circuit. However, as the authors of these papers point out, such tests offer no guarantee to cover all types of delay defects. Besides, the problem of circuits such as c6288 is still unsolved. That circuit has a very large number of "longest delay" paths very few of which are testable.

Several classification schemes for path delay faults have been proposed [371], each identifying a subset of paths that must be tested to guarantee the coverage of all delay defects. However, often the large size of the target path set and the absence of robust tests for many paths voids any guarantee.

12.3 Transition Faults

A simpler delay fault model, known as the *transition fault*, was defined in Chapter 4. A transition fault on a line makes the signal change on that line slow. The two possible faults are *slow-to-rise* and *slow-to-fall* types. For detecting a slow-to-rise fault on a line, we take a test for a stuck-at-0 fault on that line. This test will set the line to 1 in the fault-free circuit and propagate the state of the line to a primary output. Let us call this vector $V2$ and precede it with any vector $V1$ that sets the line to 0. Now the vector-pair $(V1, V2)$ is a test for the slow-to-rise transition fault on the line. Note that $V1$ sets the line to 0 and $V2$ sets it to 1. $V2$ also creates an observation path to a primary output. If the line is slow to rise then that effect will be observed as a 0 at the output instead of the expected value of 1. The basic assumption in this test is that the faulty delay of the signal rise has to be large, since the observation path may be, and often is, a short path. Besides, the effects of hazards and glitches can interfere with the observation of the output value. As a result, the tests for transition faults can detect localized (spot) delay defects of large (gross) delay amounts. Because of sensitization of short paths these tests may fail to detect distributed defects, where small delay increases in a large number of

12.4 Delay Test Methodologies

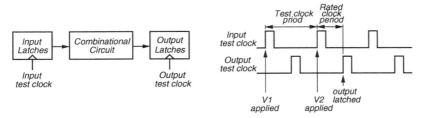

Figure 12.9: Skewed slow-clock test application to a combinational circuit.

gates cause a long path to fail. The advantages of the transition fault model are:

- The number of faults has an upper bound of twice the number of lines.

- Tests are easy to generate. A stuck-at fault test generator can be easily modified to produce tests for transition faults [371, 392].

- Circuits that either have, or are modified to have, a high stuck-at fault coverage usually also have high transition fault testability [707].

Transition fault tests have been used in the industry. It is generally recommended that they be augmented by some path delay tests, at least by tests for critical paths.

12.4 Delay Test Methodologies

Practical application of delay tests depends on the type of circuit under test and the DFT hardware used. We will describe five different test methodologies.

12.4.1 Slow-Clock Combinational Test

This procedure is applicable to combinational circuits or to those sequential circuits that are internally combinational with flip-flops only at PI and PO. The test architecture is shown in Figure 12.9. Input and output latches can be either part of the circuit or provided by the automatic test equipment (ATE.) Input and output test clocks control the application of vectors and latching of combinational outputs, respectively. These clocks should be independently controllable to allow a phase delay or skew. A two-vector delay test assumes that all signals due to the first vector $V1$ will have reached their steady state when $V2$ is applied. If this assumption is not valid, then the actual circuit may still have some transient signals when $V2$ is applied. These transients can interfere with the testing of the targeted path. To avoid this problem, vectors are applied at a slower than the rated clock frequency. In the timing diagram of Figure 12.9, the output clock is skewed by an amount that equals the rated-clock period, which is the time allowed for the $V1 \rightarrow V2$ transitions to flow through the combinational logic. If the delay of the activated path is longer than the rated-clock period, then the output produced by $V1$ will be captured in the output latch and an observation of its state will detect the fault.

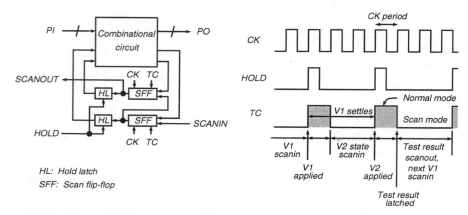

Figure 12.10: Enhanced-scan delay test application to sequential circuit.

In practice the test clock frequency can be slowed down by any amount at the cost of increased test time. This method is useful when the ATE cannot apply the vectors at the rated speed. Tests can be generated without restriction since any arbitrary vector-pair can be applied. However, only the circuits that conform to the architecture of Figure 12.9 can by tested.

12.4.2 Enhanced-Scan Test

This method is applicable to scan types of sequential circuits. Its main advantage is that any arbitrary vector-pair can be applied. So, delay tests can be generated by considering the combinational logic alone, making the test generation easier. However, a normal scan circuit should be enhanced by inserting *hold latches* and an additional $HOLD$ signal. The design and operation of the enhanced flip-flop, known as *scan-hold flip-flop* (SHFF), is discussed in Section 14.4.

Figure 12.10 shows an enhanced scan circuit and the timing diagram of test application. Each vector contains two parts, namely, bits corresponding to PI and bits corresponding to state variables. The state portion of $V1$ is serially shifted in the scan register via the $SCANIN$ terminal by setting test control $TC = 0$ and applying the clock CK. Often, scan is done using a slow-clock to reduce the power dissipation. However, it is also necessary that any delay faults in the scan path do not interfere with the vector. Scanned $V1$ bits are then transferred to hold latches (HL) by activating the $HOLD$ signal while the PI bits of $V1$ are applied at PI. As signals due to $V1$ stabilize, the state bits of $V2$ are scanned in. Next, simultaneous activation of $HOLD$ and application of $V2$ bits to PI provides a $V1 \rightarrow V2$ transition at the input of combinational logic. Test control $TC = 1$ sets the circuit in normal mode for exactly one rated-clock period, at the end of which the clock CK latches the combinational outputs in flip-flops. This one cycle of clock must have the rated period. PO signals are directly observed and flip-flop states are scanned out. As usual, scanout can be overlapped with the scanin of the next test. As we pointed out, this architecture has the full flexibility of combinational circuit test. Test time

12.4 Delay Test Methodologies

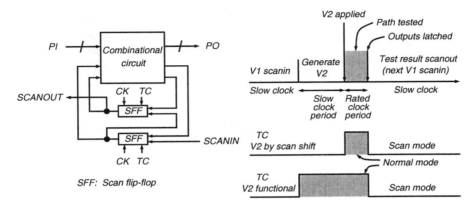

Figure 12.11: Normal-scan delay test application to sequential circuit.

is similar to that of full scan design (see Chapter 14.) However, scan area overhead is increased due to the hold latch, which also adds some delay in the signal path.

12.4.3 Normal-Scan Sequential Test

Normal full-scan circuits (with no hold latches) can be tested for delay faults, but the vector-pairs must be especially generated [156]. Here, the first vector $V1$ is scanned in (usually with a slow scan clock) and is then replaced in the scan register either by (a) applying a one-bit shift to the scan register, or (b) propagating $V1$ through the combinational logic in the normal mode. Figure 12.11 illustrates the two methods.

In the first method, known as *scan-shift delay test* or *skewed-load delay test* [564, 565], scan in of $V1$ is followed by one extra cycle of slow-clock while the circuit is still in the scan mode $TC = 0$. The test is so designed that $V2$ is obtained by a one-bit translation of $V1$. So, only the bit in the flip-flop closest to the $SCANIN$ input is supplied externally. PI bits of two vectors, which are not restricted, are directly applied at the proper time. As soon as $V2$ is applied, the mode is changed from scan to normal ($TC = 1$) and one application of clock CK with the rated period latches the outputs. The observation of PO is synchronized with the rated-clock and flip-flop states are scanned out for observation.

In the second method, the state portion of $V2$ must be justified by $V1$ applied to the combinational logic. When the vector-pair has this characteristic, scanning in of $V1$ and application of its PI bits automatically produce the state portion of $V2$ at the output of the combinational logic. An application of the clock in the normal mode and application of the PI part of $V2$ creates the $V1 \rightarrow V2$ transition. Keeping the circuit in the normal mode, one rated-clock period later, the outputs are observed at PO or scanned out. This method is also called *broad-side delay test* [566].

Both methods are used for testing path-delay and transition faults. However, a high fault coverage is dependent on the circuit and cannot be guaranteed due to the

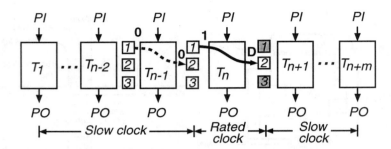

Figure 12.12: Delay test application to non-scan sequential circuit.

correlation between the two vectors.

12.4.4 Variable-Clock Non-Scan Sequential Test

Testing of a delay fault in a non-scan sequential circuit requires more than two vectors [416]. First, the vector-pair should be like the one used in the second method of the last subsection. That is, $V2$ should be justified by $V1$ through the combinational function. Second, $V1$ should be generated, by a set of vectors starting at some initial state. We will call this set a justification sequence. Third, if the path destination is a flip-flop then the state should be propagated to some PO. We will call this part of the test as propagation sequence. This test scenario is depicted in Figure 12.12 by the time-frame expansion as was used in Chapter 8 for sequential circuit ATPG.

In Figure 12.12, vector $V1$ is applied in time-frame T_{n-1} and the target path is partially sensitized. Recall that the path has to be fully sensitized only in $V2$. That is why we have shown the path with a broken line. In T_n, $V1$ produces $V2$ states in the three flip-flops. However, to ensure that $V2$ is correctly produced and is not affected by some delays in T_{n-1}, we must allow extra time for $V1$ to propagate through the circuit. This is done by using a slow-clock, which is also used for the same reason for justification and propagation sequences. Thus, only one vector, i.e., $V2$, in the entire test sequence uses the rated-clock. This procedure is known as the *slow-clock* or *variable-clock delay testing*.

The slow-clock prevents the delays in the circuit from interfering with the detection of the target fault. In practice, the clock can be slowed down by a factor of two or greater. One problem, however, remains. In the time-frame T_n, the path destination in Figure 12.12 is flip-flop 2. Since the rated-clock is used in this time-frame, other path delays can also affect the signals and the state of flip-flops 1 and 3 (shown in grey) can also have faulty values. It is very difficult to generate a test to ensure that paths leading to those flip-flops are not activated. Chakraborty et al. [119] provide several alternatives.

Definition 12.4 Combinational robust test. *A test that guarantees to detect a target delay fault of a combinational circuit in the presence of arbitrary delays in the circuit is a combinational robust delay test. Such a test consists of a vector-pair*

12.4 Delay Test Methodologies

$(V1, V2)$ that must satisfy the conditions we discussed in previous sections. Test application must ensure that sufficient time is allowed for changes due to $V1$ to stabilize before $V2$ is applied.

When a combinational circuit is embedded among flip-flops, a combinational robust test is not guaranteed to work in a non-scan circuit.

Definition 12.5 Sequential robust test. *A test that guarantees to detect a target delay fault of a sequential circuit in the presence of arbitrary delays in the circuit is a sequential robust test.* For variable-clock test application, which applies the rated-clock only to a single time-frame, we consider three fault models:

- *Fault model A. All flip-flops other that the path destination are assumed to be unaffected by any delay fault. The propagation sequence can be invalidated if the assumption is not true. The tests are not sequential robust.*

- *Fault model B. All flip-flops other that the path destination are assumed to be in the unknown state (X) following the rated-clock time-frame. Tests are sequential robust, though the model is pessimistic.*

- *Fault model C. At the end of the rated-clock time-frame all non-destination flip-flops that have a steady (hazard-free) input during the period of both vectors $(V1, V2)$ take that value as their state at the end of the rated-clock time-frame. All other flip-flops assume unknown (X) state. The tests for this model are robust.*

Chakraborty et al. [119] propose a 13-valued algebra for analyzing these fault models. Algorithms for deriving path-delay tests for sequential circuits using the variable-clock mode have been described [17, 199]. According to the reported data [17], the sequential benchmark circuit s1494 has 976 paths. Of the 1,952 PDFs tests could be obtained only for 37% faults in the variable-clock non-scan mode. In the enhanced-scan mode, 98.7% PDFs were tested. Path delay test generation for non-scan sequential circuits continues to be a difficult problem. Fortunately, in delay testing, the issue of PDF coverage is not as important as the testing of long paths.

Path-delay fault simulators can analyze the variable-clock mode. Such results give data on the length of paths that are tested. Reported data [514] gives the following result for the benchmark circuit s35932: Number of vectors = 2,124; Total PDFs = 394,282; Tested PDFs = 26,228 (6.65% coverage); Longest PDF = 29 gates; Longest tested PDF = 27 gates. These vectors were randomly generated by letting the fault simulator select useful vectors. In spite of a low coverage, the longest path tested is within two gates from the longest physical path. One problem with such tests is the long test time. In actual test application, the vector sequence should be repeated 2,124 times. In each application a different vector will have the rated-clock while all the rest will have slow-clocks. Thus, a total of $2,124 \times 2,124 = 4,511,376$ vectors will be applied.

12.4.5 Rated-Clock Non-Scan Sequential Test

This is the most natural form of test. All vectors, either functional or those generated to cover any types of faults, are applied at the rated speed. A target delay fault can be activated in several time frames. If robust detection is desired, one must consider all delay combinations to be potentially possible. Even fault simulation requires massive computation. Nevertheless, it shows a much reduced PDF coverage for vectors generated for variable-clock test. This is basically due to the pessimism one assumes in simulating the rated-clock test. Bose *et al.* [83] give a 41-valued algebra for generating rated-clock tests. Because of high complexity, it will be fruitless to attempt rated-clock test generation for all PDFs. However, inclusion of such tests for critical paths may be extremely useful.

12.5 Practical Considerations in Delay Testing

Timing correctness of VLSI circuits is as important as their logical correctness. The design of a VLSI chip is verified via timing simulation (see Section 5.3) and timing analysis. The timing analysis examines combinational paths in the circuit topology. Delays of gates and interconnects are obtained from the simulator database. All paths are examined without consideration to their sensitizability. Because no signal values are used in this analysis, it is also known as "static timing analysis [22, 301, 647]." The results of timing analysis are used in several ways to improve the design and test:

- *Timing simulation.* Critical paths are simulated using circuit-level or timing simulators. Stimulus for activation of those paths are normally provided by the designer using the functional knowledge of the circuit (See Chapter 5.) The results of simulation are used to modify the design, if necessary, to meet the timing specification.

- *Critical path tests.* The critical path of a circuit is the longest delay combinational path between clocked flip-flops. Critical path delay determines the clock period. Tests of a chip normally include test vectors that propagate signal transitions through critical paths.

- *Layout optimization.* Critical path data is used for placement of standard cells and custom blocks (memories, etc.) and for prioritizing interconnects for routing [208]. Critical path data are also used for transistor-sizing [223]. Chip area can be reduced by shrinking the devices in the gates on non-critical paths, letting their delays increase, while larger devices are selectively used only on critical paths.

A careful timing design increases the manufacturing yield but cannot guarantee that every chip will function correctly. Tests developed for detecting stuck-at faults, when applied at a slower than the rated-clock frequency, uncover many manufacturing defects. However, such tests only ascertain the logical correctness of the circuit.

12.5 Practical Considerations in Delay Testing

Slow clock testing can be due to the limitation of the ATE and test fixturing (cables, probes, etc.) Since stuck-at faults do not model delay faults, this form of testing is often termed "DC testing" or "static testing."

12.5.1 At-Speed Testing

At-speed testing in which vectors are applied and responses observed at the rated-clock speed is essential unless the timing design is too pessimistic and process tolerances are extremely tight. Both of these attributes are not possible for today's VLSI chips that drive extremely high speed systems and are manufactured through leading edge processes. Application of the stuck-at fault test vectors at the rated-clock speed, though used frequently, is not the best strategy. This is because those vectors may not have a high delay fault coverage. Path-delay tests for critical paths should be included in the at-speed testing. Considering the possibility of a very large number of paths, critical path testing is a good approach. These tests are very good at uncovering "correlated defects." These defects are caused by variations in the manufacturing process and affect all components on the chip in a similar way. For example, the resistivity of all interconnects may increase, or all transistors may slow down. Obviously, longest delay paths will be the first to fail.

Another class of defects is referred to as *spot defects* or *gross defects*. These affect localized regions of the chip. Though physically small they can cause just a few devices or interconnects to have grossly excessive delays. Transition fault tests provide the capability of detecting such defects. Besides, the number of transition faults is a linear function of the circuit size (similar to that of stuck-at faults) and so a high percentage can be covered. Therefore, a combination of critical path-delay tests and transition tests provides adequate at-speed testing.

Built-in self-test (BIST) for Delay-faults. Considering the fact that a high-speed ATE is expensive and sometimes unavailable, BIST is an alternative method of at-speed testing. This requires on-chip circuitry for vector generation and response analysis. Only a high-speed clock, which determines the speed of testing, is supplied from outside. While pseudo-random vectors provide good coverage (sometimes with test points) of stuck-at faults, they can also cover a large percentage of transition faults if applied at high speed. Coverage of path-delay faults frequently requires additional modifications in the combinational logic, especially if robust tests are desired (see Section 15.4.) There are several problems that require careful considerations in the implementation of at-speed BIST. First, if there are unscanned flip-flops in the logic being tested then their initialization must be examined. The response signature register should be initialized only after all flip-flops have been initialized. Second, when pseudo-random patterns are applied to combinational logic, some long combinational paths that are non-functional in the sequential mode can be activated. If that happens, BIST can produce timing failures even in a circuit that meets the functional timing requirements. In such cases, the clock rate of BIST should be lowered below the specification. A suitable clock rate can be found by timing simulation. A recent paper reports the results on a *boundary-scan master* (BSM2) chip [299], designed for a 65 MHz clock rate. As determined by timing simulation, BIST was operated at a 40 MHz clock rate to avoid failures due to the activation of false-paths. Third, the power consumption of at-speed BIST can exceed the power rating of the chip. This is because of the high signal activity that random vectors cause in some circuits. Both peak and average power for BIST should be analyzed and corrected, if necessary. Increased average power can cause heating of the device under test and increased peak power can produce noise-related failures.

12.6 Summary

In the 1974 paper, Breuer [91] was mainly concerned with the delay faults in asynchronous circuits. That is a very difficult problem. In the following years, the industry largely adopted the synchronous (clocked) design style. That did not simplify the delay test problem because sizes of VLSI chips grew beyond anything imaginable in the early days. In 1980, Lesser and Shedletsky [391] studied the problem of path delay testing for scan circuits. They used *single-input-change* (SIC) vector-pairs to measure delays of a subset of combinational paths. The measured data were analyzed to characterize the delays of all paths of interest. Besides finding a suitable set of tests, difficulties in this method also arise in the application of tests. This is because the arrival times of events at combinational outputs should be exactly measured rather than just determining that the events occur prior to some clock edge. Still, in view of the fact that path delay testing must deal with a large number of paths, there has been renewed interest in this type of technique [651].

All technology indicators point to the increasing use of delay testing in the future. The field of delay testing has expanded recently. Due to the limited space, we had to focus on fundamentals, leaving out many important techniques. We end by listing a set of topics on which an interested reader may continue the study: *delay fault models, classification of delay faults, multi-valued algebras, path delay fault simulators, path counting techniques,* and *design for delay fault testability.*

Problems

12.1 *Non-robust path-delay test.* Does the exclusive-OR circuit in Figure 12.13 have any redundant stuck-at fault? How many paths does the circuit have? Derive non-robust tests for all *path-delay faults* (PDFs). Are there any singly-untestable PDFs? Note: Any untestable PDFs can be eliminated by cir-

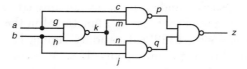

Figure 12.13: Exclusive-OR circuit for Problem 12.1.

Problems

cuit transformation knows as the KMS (Keutzer, Malik, and Saldanha) algorithm [352].

12.2 *Robust path-delay tests.* Remove the redundant fault Q stuck-at-1 from the circuit of Figure 12.2 and verify that all path-delay faults in the irredundant circuit are robustly testable.

12.3 *Robust path-delay tests.* Show that a robust path-delay test must produce a "real" transition (different initial and final values) at the output. Is any path in the circuit of Figure 12.4 robustly testable?

12.4 *Single-input change (SIC) tests.* Prove that every singly-testable (i.e., non-robustly testable) fault has a single-input change test.

12.5 *Path-delay tests.* Consider the path $C-F-G$ in the circuit of Figure 12.14(a):

  (a) Derive a test for a rising transition at C.

  (b) Will the above test work if a falling transition is applied at B?

  (c) Sketch all signal waveforms for the situation in (b) when all gates have one unit of delay. Assuming that the permitted delay for the circuit is 2.5 units, interpret the result of the test. Can you locate the faulty path?

  (d) How will you diagnose the faulty path?

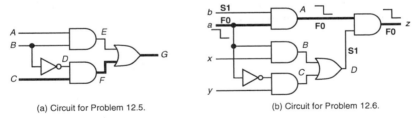

(a) Circuit for Problem 12.5.   (b) Circuit for Problem 12.6.

Figure 12.14: Path-delay fault testing circuits for Problems 12.5 and 12.6.

12.6 *Path-delay test robustness.* Consider the path shown in bold lines in the circuit of Figure 12.14(b). Suppose that we choose a test: $x = S1$, $y = S1$, and $b = S1$.

  (a) The permitted circuit delay is 3.5 units. Assuming that the gate A has a delay of 5 units and that all other gates have one unit of delay, sketch the relevant waveforms to show that the test is not robust.

  (b) Can you derive a robust test?

12.7 *Off-path signals.* Specify the off-path signal states for delay testing of a two-input XOR gate.

12.8 *Logical and timing conditions.* Boolean gates have inputs A, B, C, ..., and output Z. To test the propagation delay from A to Z, a transition is propagated through the gate. Consider both rising and falling transitions at A. Find the least restrictive off-path signal states to satisfy these conditions:

(a) Logical Condition – The pre-transition and post-transition steady states of Z must be the same as the respective states of A, irrespective of the values at other inputs of the gate.

(b) Timing Condition – The signal at Z must not change from its pre-transition steady state as long as A is steady in the pre-transition state.

12.9 *Path counting.* Write an algorithm to count the number of paths in a combinational circuit. What is the complexity of your algorithm?

12.10 *Pomeranz-Reddy example* [526]. Show that the number of paths in the circuit of Figure 12.15 is $3 \times 2^k - 2$.

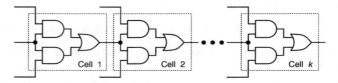

Figure 12.15: Circuit for path counting in Problem 12.10.

12.11 *Sequential path-delay fault testing.* Prove that in the circuit of Figure 12.16(a), there is no robust path-delay fault test for the path d-e-f-g.

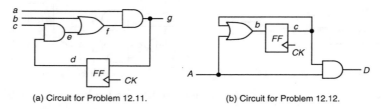

(a) Circuit for Problem 12.11.   (b) Circuit for Problem 12.12.

Figure 12.16: Circuits for Problems 12.11 and 12.12.

12.12 *Sequential path-delay fault ATPG.* In the circuit of Figure 12.16(b), determine whether there are robust path-delay tests for faults $\uparrow A - D$ and $\downarrow A - D$.

Chapter 13

IDDQ TEST

> " . . . the superior testability of the CMOS class of integrated circuits relative to other classes . . . arises from the nominally zero supply current of CMOS, which is exploited in a new set of test methods which are capable of determining existence and functionality for every part of a VLSI circuit. The set of test vectors . . . is expected to be significantly smaller than the usual functional set . . ."
> — Mark W. Levi, in 1981 paper, "CMOS Is Most Testable" [393].

IC quality and reliability dramatically improved in the 1980s, due to pressure from the auto industry and strong foreign competition [450, 451]. In the early 1990s, failure rates ranged from 1000 *dpm* (defects per million) to 50 *dpm*. IBM's stated goal is to achieve a failure rate of 3.4 *dpm* [537]. This approach is called *zero defects* or *six sigma*, and is considered mandatory for semiconductor companies to be competitive in the U.S. and international markets. Conventional approaches to meeting this goal involve increasing test fault coverage, increasing burn-in coverage, increasing ESD (*electrostatic discharge*) damage awareness, and other methods. I_{DDQ} current testing has been effective in achieving low defect levels. The stuck-fault model fails to identify certain analog circuit defects that are not severe enough to cause a logical fault in digital circuits, but instead may be resistive bridging faults. I_{DDQ} testing exposes some of these faults. Sandia Labs. [305] and Ford Microelectronics [382] found I_{DDQ} testing to be more sensitive than stuck-fault testing. Segura and Rubio [579, 580] found that gate oxide shorts elevated I_{DDQ} current to 300 to 450 μA.

13.1 Motivation

CMOS ICs are intrinsically designed for current testability [617]. I_{DDQ} testing can significantly improve quality, decrease chip production cost, and is very useful for *failure effect analysis* (FEA.) Wanlass invented *complementary MOS* (CMOS) logic gates and observed that the gates draw very little power in standby mode [543, 714]. Subsequent research demonstrated that in static CMOS circuits, standby power higher than a nanoWatt, and dependent on the signal states, indicates a defect.

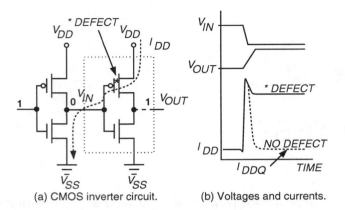

(a) CMOS inverter circuit. (b) Voltages and currents.

Figure 13.1: Basic principle of I_{DDQ} testing.

Figure 13.1(a) shows a CMOS inverter, with a * indicating a defect in the pFET that causes its input impedance to drop from infinity to a finite value. Now, the DC current flows in steady state along the path indicated by the arrow, and this elevates the steady state current, since current can still flow through the defective pFET [617]. Figure 13.1(b) shows the input and output voltages, and the drain current I_{DD} that flows through the transistors. After switching completes, this current is referred to as the *quiescent* current and called I_{DDQ}. In the good circuit I_{DDQ} falls to a negligible value, whereas in the defective circuit, I_{DDQ} remains elevated long after switching is over. We detect such faults by measuring I_{DDQ} at the time instant shown by the arrow. The I_{DDQ} current is measured through the V_{SS} bus of the circuit, although it could also be measured through the V_{DD} bus of the circuit. The automatic test equipment (ATE) can measure the I_{DDQ} current at the V_{SS} pin of the circuit, or, alternatively, we can build in a current measurement device (see Section 13.7) into the V_{SS} bus on the chip. However, a current measurement takes milliseconds, far longer than a voltage measurement, because all current transients must subside before the steady-state current measurement can be taken. I_{DDQ} tests were used for functional tests [181, 225], for testing delay faults [493], and to detect pattern sensitive memory failures [617]. I_{DDQ} testing measures a current, which is inherently much slower than measuring a voltage, as in stuck-fault testing. However, I_{DDQ} testing has problems as chip minimum feature sizes go to 0.18 μm. The sub-threshold conduction current of MOSFETs increases, and the increased density of chips with 50 to 100 million transistors makes it increasingly difficult to distinguish the defective I_{DDQ} current from normal devices with somewhat elevated leakage currents. Many have used I_{DDQ} testing to improve IC reliability, reduce manufacturing costs by 50%, improve field quality, and cut burn-in failures [450, 451, 519, 520]. The defects and mechanisms elevating I_{DDQ} current often greatly reduce reliability [268, 282, 450, 451, 615]. High I_{DDQ} current can also cause premature battery failure in portable devices.

13.2 Faults Detected by I_{DDQ} Tests

I_{DDQ} testing can sometimes detect transistors stuck-open, transistors stuck-closed, transistor gate oxide shorts (exhibiting a diode behavior), interconnect bridging shorts, and unpowered interconnect opens. A gate oxide short has resistive behavior if both shorted terminals are doped the same type (n or p), but has diode behavior otherwise. A *defect* is a physical occurrence in a semiconductor device, and a *fault* is a defect manifestation. On average, one defect causes more than one fault, often of different types [437]. A pinhole short through the gate oxide of a pFET in a digital NOR gate can cause an input-to-output bridging fault, a slow-to-rise delay fault, and an excessive I_{DDQ} current fault [437]. The probability of zero defects is the probability of zero faults (which is the yield, Y.) The probability of n defects differs from the probability of n faults, because the probability of one defect is greater than the probability of one fault. Maxwell and Aitken [437] give a formula for the probability of n defects.

Stuck-at Faults.

Bridging Defects with Stuck-at Fault Behavior. Levi noticed that an I_{DDQ} test detected the bridging of a logic node to V_{DD} or V_{SS} [393], which is a true stuck-at fault. These are a small percentage of the real defects. I_{DDQ} test vectors must drive all logic gate input/output nodes to logic 0 and 1 [227] for detecting all such faults. A transistor gate oxide short of 1 $K\Omega$ to 5 $K\Omega$ to source also causes a stuck-at fault. This is detected by an I_{DDQ} test [281]. Defects causing both n and pFET transistors in a gate to be on are almost always detected by I_{DDQ} tests.

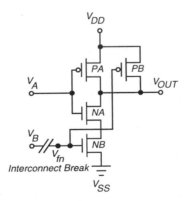

Figure 13.2: NAND open circuit defect.

Floating Gate Defects. Many defects causing open circuits elevate I_{DDQ} current [617]. These defects in transistor gates usually do not fully turn off the transistor [422, 541]. A small break (100 to 200 $\overset{\circ}{A}$) in logic gate inputs (see Figure 13.2 [617]) still allows signal coupling between the two wire fragments by electron

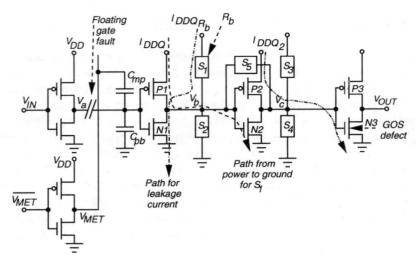

Figure 13.3: Circuit illustrating floating gate, GOS, leakage, and bridging faults.

tunneling [290]. This causes a delay fault, and elevation of I_{DDQ} current, which can be tested with an I_{DDQ} test. A large open results in a stuck-at fault, which sometimes can be tested by an I_{DDQ} test. In Figure 13.2, the open leaves V_{fn} at a voltage that is a function of the circuit parasitics. The output then may behave as a stuck-at fault, and may have either a weak or strong logic voltage because of the logic gate analog voltage gain. A weak output voltage happens when $V_{tn} < V_{fn} < V_{DD} - |V_{tp}|$, and is detected by an I_{DDQ} measurement. If $V_{fn} < V_{tn}$ or $V_{fn} > V_{DD} - |V_{tp}|$, then a stuck-at fault occurs, which is probably not detected by an I_{DDQ} current test, and must be detected by a voltage test, instead.

When large line breaks occur, tunneling effects are negligible. The floating gate voltage is determined by capacitive coupling of the broken polysilicon path to the metal lines crossing it. For certain floating gate defects, the transistor conducts, and for others it remains stuck-open [135, 138, 422]. Figure 13.3 [580, 579] shows the circuit schematic modeling the severed transistor gate with C_{pb}, the capacitance from poly to bulk, and C_{mp}, the overlapped metal wire to floating poly capacitance. The floating gate voltage depends on these capacitances and node voltages. If the nFET gets enough voltage at its gate to turn it on, then a path from V_{DD} to ground exists if the pFET is on, so the abnormal I_{DDQ} current can be sensed. The floating gate voltage is sufficient to activate the faulty transistor over many conditions [580, 579]. Segura et al. fabricated five defective inverter chains with deliberate defects causing a floating gate on a 2 μm n-well CMOS process. Here, 1.95 $fF <= C_{pb} <= 5.3 \ fF$ and 1.1 $fF <= C_{mp} <= 5.4 \ fF$. Figure 13.4 [580] shows the static transfer characteristics of the five defective circuits. When the pFET gate nears 0 V, and the defective nFET is influenced by an overlapping metal track at 5 V, I_{DDQ} current testing detects the fault. If the metal were at 0 V, I_{DDQ} testing would not work.

13.2 Faults Detected by I_{DDQ} Tests

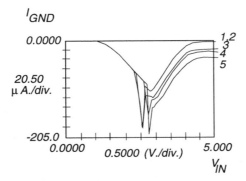

Figure 13.4: Transfer characteristics of I_{DDQ} currents.

Bridging Faults. A logic gate bridging fault may be an absolute short (< 50 Ω resistance) or a higher resistance short between two logic gates. For voltage testing, the fault is activated with a test vector that places opposite logic values across the fault (at the two different logic gates), and then sensitizes two paths from the logic gate outputs to two *primary outputs* (POs.) A deviant voltage at one of the POs reveals the fault. With I_{DDQ} testing, propagation is not required.

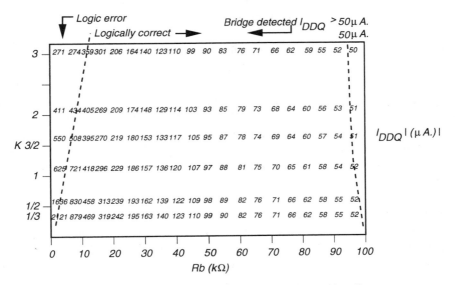

Figure 13.5: I_{DDQ} dependence on K, R_b for $S1$.

Segura *et al.* evaluated I_{DDQ} testing using a three CMOS inverter chain, shown in Figure 13.3 [580]. The sizes of the second and third inverters were held constant, while the size of the first was varied using a parameter K.

$$K = \frac{W_{p2}}{W_{p1}} = \frac{W_{n2}}{W_{n1}} \quad \text{where} \quad W_{n2} = W_{n3} = W_n \quad \text{and} \quad W_{p2} = W_{p3} = W_p \quad (13.1)$$

Figure 13.5 shows the I_{DDQ} dependence on K and R_b (bridging resistance in S_1) values. For the small R_b range, the I_{DDQR_b} dependence on the relative size is high because I_{DDQR_b} values directly depend on the resistance offered by the transistor in the conducting path. There is also an I_{DDQR_b} dependence when $R_b > 50~K\Omega$. For $0 <= R_b <= 100~K\Omega$ the I_{DDQR_b} current changes from $2~\mu A$ to $50~\mu A$, so bridges exhibit good current testability. Below the critical resistance, the circuit is voltage and current testable. Above the critical resistance, the circuit is only current testable or delay fault testable. The largest I_{DDQR_b} deviation is when V_{in} is $5~V$, because the bridged nodes have opposite logic values. The maximum I_{DDQ} current occurred when $R_b = 3.5~K\Omega$, because of the current contribution of the non-defective inverter stages [580]. A maximum deviation of 8.5% of the total I_{DDQ} current happened for low resistive shorts. For high resistive shorts, the agreement was excellent for all bridging faults in the inverter chain with deviations below 0.1%.

CMOS Stuck-Open Faults. CMOS transistors stuck-open cause high impedance states at a logic gate output, and under certain situations, I_{DDQ} is elevated and the fault can be detected [617]. I_{DDQ} testing does not guarantee detection, but works in practice [618] because the floating output node is capacitively coupled into the substrate, as well. The coupling often results in an intermediate voltage on the node [332, 616]. Mao and Gulati select I_{DDQ} vectors for testing the case of transmission gates driving multi-input logic gates [428].

Delay Faults. Most random CMOS defects cause a timing delay fault, rather than a catastrophic failure [332, 614, 616]. Many of these defects elevate I_{DDQ} current, which also changes the signal rise and fall times [617]. Many delay faults are detected with few I_{DDQ} test vectors. However, there is a defect subset that causes a delay fault, but does not elevate I_{DDQ} current, by increasing interconnect resistance at vias, or by increasing the transistor threshold voltage [614]. Motorola used I_{DDQ} tests to detect delay faults [271]. Metal briding shorts in 256 K SRAMs caused a $3~ns$ increase in data path delay, and increased the I_{DDQ} current 2 to 3 times. I_{DDQ} testing may be an inexpensive way to detect some delay faults.

Leakage Faults. Mao and Gulati [428] developed the *leakage fault* model, which accounts for leaking current between the *gate*, *source*, *drain*, or *bulk* terminals of a MOSFET. There should be no leakage in a healthy MOSFET, except between source and bulk and also between drain and bulk, and this must be less than a specified value. Gate oxide shorts can cause leakage between gate and source or between gate and drain. They proposed these six leakage faults for each MOSFET:

f_{GS} – between gate and source f_{BS} – between bulk and source
f_{GD} – between gate and drain f_{BD} – between bulk and drain
f_{SD} – between source and drain f_{BG} – between bulk and gate

They assume that the leakage faults do not change circuit logic values, which is true for gate oxide shorts during production test. However, these shorts later develop

13.2 Faults Detected by I_{DDQ} Tests

hard faults, which cause field failures. Mao and Gulati selected test vectors for single leakage faults [428] so that they also cover multiple leakage faults.

Weak Faults. An n MOSFET cannot pass logic 1 (traditionally 5 V) without degrading it to $5\ V - V_{tn}$. Similarly, a p MOSFET cannot pass logic 0 (0 V) without degrading it to $|V_{tp}|$. In the CMOS C-switch device, an $nFET$ and a $pFET$ are connected in parallel. A *weak fault* causes one of the devices to fail to turn on, so the signal passed from the source to the drain of the C-switch is degraded (has a *weak* voltage), which increases propagation delays and increases noise [428]. A path between the n_i and n_j nodes is a *transmission path* if a chain of channel connected transistors exists between n_i and n_j. Channel connected means that all of the channels of the transistors could be shorted together if all transistors were turned on. In a *conducting path* all transistors in a transmission path are turned on. A *normal-1 (0) transmission path* has only pMOS (nMOS) transistors. A path with at least one pMOS (nMOS) transistor is a *weak-0 (1) transmission path*. Similarly, we define *normal-1 (0) conducting paths* and *weak-0 (1) conducting paths*.

Figure 13.6 shows transistors *N4* and *N5* forming a normal-0 transmission path from V_{SS} to node A [428]. *P2* and *P3* form a normal-1 transmission path from V_{DD} to *O1*. When $(I1, I2) = (1, 0)$, *P2* and *P3* form a normal-1 conducting path, and *P2* and *N3* form a weak-1 conducting path, from V_{DD} to *O1*.

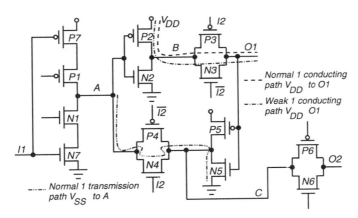

Figure 13.6: Circuit with transmission, normal, and weak conducting paths.

There are two *necessary* and *sufficient* conditions to have a weak-0 (1) fault [428]. There must be at least one normal-0 (1) conducting path between n_i and V_{SS} (V_{DD}) in the good circuit, and all normal-0 (1) conducting paths should be blocked in the bad circuit. There must be at least one weak-0 (1) conducting path between node n_i and V_{SS} (V_{DD}) in the bad circuit [272]. A node may have multiple weak-0 (1) faults, which must be considered separately because detection of one does not imply detection of the others. We assume that a weak-0 (1) fault at a node is caused by blocking of all normal-0 (1) conducting paths due to a single defective transistor.

Transistor Stuck-Closed Faults.

Gate Oxide Short Failures. In CMOS technology, *gate oxide short* (GOS) failures happen frequently, resulting in an undesirable current path through the MOSFET gate oxide [580]. The gate is no longer isolated from the channel. Figure 13.7 shows an $nFET$ with a gate oxide short. The original transistor is split into two transistors, separated by a rectifying barrier between the channel and the gate. Parameter k is the location of the defect in the channel, and R_s is the short resistance.

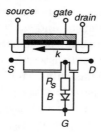

Figure 13.7: Circuit level gate oxide short defect model.

Figure 13.3 shows a three inverter chain, with a gate oxide short in the channel of the last inverter $nFET$ (*N3*.) Segura *et al.* simulated defects, as a function of R_s and k, to obtain the results in Figure 13.8 [580]. The I_{DDQ2} current results from various positions of the GOS (k parameter) and from various short resistances. In the good circuit, I_{DDQ2} current is 10 to 30 nA. The I_{DDQ2} current from the defect increases as the defect moves nearer to the source and is elevated 3 or 4 orders of magnitude by the defect, so I_{DDQ} testing works. As R_s increases, the I_{DDQ2} current decreases, so I_{DDQ} testing fails only for very high R_s values ($M\Omega$ range.) Only a small part of the failing machine graph is covered by logical voltage testing. Voltage testing will find the GOS defect when $R_s = 50$ Ω, but not when R_s is significantly higher. For these defects, I_{DDQ} testing is superior to logical voltage testing.

13.3 I_{DDQ} Testing Methods

13.3.1 I_{DDQ} Fault Coverage Metrics

Malaiya and Su were the first to propose a new *conductance fault model* [418], in which current monitoring is used to detect all of the *leakage faults* in transistors during application of conventional single stuck-at fault test vectors. Their model handled open wires, open/shorted transistors, and leakage faults. An on transistor was modeled with a high conductance, and an off transistor was modeled with a leakage conductance. A leaky transistor has a minimum conductance greater than the maximum allowable conductance, and a normal transistor has a maximum conductance less than the allowable maximum conductance. The great significance of

13.3 I_{DDQ} Testing Methods

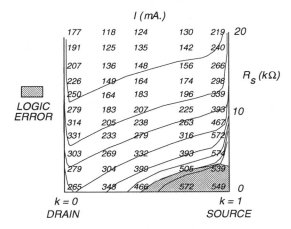

Figure 13.8: Logic and I_{DDQ} testing zones vs. k and R_s.

their work was a proof that a stuck-fault test set can be used to generate the minimum complete leakage fault test set. A conventional logic fault simulator is modified to simulate these effects, by requiring each stuck-at fault to propagate through a single logic gate, rather than to a primary output. The *pseudo-stuck-at fault coverage* definition appeared later and is a voltage stuck-at coverage that represents the internal transistor shorts coverage [227, 437] and hard stuck-at faults. Complex CMOS gates are modeled using AND, NAND, OR, NOR, and NOT primitives, and the stuck-at fault effect must propagate to the output of the actual gate, not just through a primitive.

Stuck-fault test coverage of an I_{DDQ} test vector set is obtained when a node is driven to both 0 and 1 logic values during the test (the *toggle test*) [227]. A better metric is based on shorts or bridges, since current exists when two connected nodes have different values. Maxwell and Aitken propose a fault coverage based on shorts between transistor poles that are always connected, such as the source and the inverter substrate, and which cannot be detected by any voltage test [437]. We compute these coverages with leakage and weak fault tables [428].

The short fault coverage handles intra-gate bridges, but inter-gate bridges can also happen. Maxwell and Aitken modeled bridges between adjacent metal lines. The bridges were modeled in lines that were separated by the minimum metal pitch in the layout. When I_{DDQ} measurements are used, a bridge is detected if the two nets have opposite values in the good circuit. This I_{DDQ} current testing method can detect resistive bridges that may not cause a logic error at low frequency [545].

Maxwell and Aitken generated an I_{DDQ} test set using an *automatic test-pattern generator* (ATPG) modified to use the pseudo-stuck-at fault model [437]. They targeted all single stuck-at faults on gate inputs. Figure 13.9 [437] shows fault coverages for these three I_{DDQ} fault models: pseudo-stuck-at, transistor short, and bridges [437]. Pseudo-stuck-at faults do not track transistor shorts well, and the pseudo-stuck-at and short fault models never converge. However, the coverages for transistor shorts and bridges closely match.

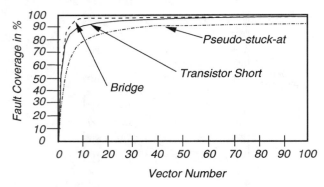

Figure 13.9: Fault coverage profiles for I_{DDQ} fault models.

13.3.2 I_{DDQ} Test Vector Selection from Stuck-Fault Vector Sets

Vector Selection with Full Scan. Perry used voltage testing vectors and full-scan for I_{DDQ} testing [519]. He measured I_{DDQ} current whenever the voltage test vector set hit an internal scan chain boundary (see Chapter 14), so all internal nodes, chip inputs, and outputs were in a known state and made the I_{DDQ} test repeatable. He used only the first 10 scan ring boundaries for making I_{DDQ} measurements, by applying a vector that minimized I_{DDQ} current. This vector sets all chip inputs with pull-up resistors to high, disables all clocks, and sets all tristate outputs into high-impedance mode. Stopping the clock holds the internal chip state constant. The minimum I_{DDQ} test vector also set all bidirectional inputs to a known state, if they were in an output mode when the I_{DDQ} measurement was to be made. The remaining inputs were held in their present state for the I_{DDQ} measurement. He found that 10% of his I_{DDQ} testing rejects came from faults in input/output circuits. Perry recommends eliminating pull-ups on clock inputs or tristate control pins and adding outside controls to set internal RAM blocks into standby mode during I_{DDQ} testing. He avoids pull-ups wherever possible on chip inputs to ensure that the largest possible number of input pin states can be tested. He waited 30 ms while making an I_{DDQ} current measurement to allow proper settling of the circuit. The minimum I_{DDQ} test vector allows measurement of the lowest possible input leakage current against a fixed I_{DDQ} limit. Perry succeeded in measuring I_{DDQ} current against 75 μA limits with 1 μA. accuracy [519].

Vector Selection from Complete Stuck-Fault Tests. Mao and Gulati [428] developed leakage and weak fault models to test for gate oxide shorts and opens. They select a subset of a complete logic-level test vector set for I_{DDQ} fault tests.

Fault Detection. Leakage fault detection depends on the circuit's logic state. In Figure 13.3 [428] transistor $N2$ has a gate to source leakage fault f_{GS}, which is

13.3 I_{DDQ} Testing Methods

detected only if node V_b is at logic 1 and input V_{IN} is at logic 1. This creates the I_{DDQ} leakage current through the fault (see Figure 13.3 [428].) Detection requires that two transistor terminals with leakage must have opposite logic values, and be at driving strengths. Non-driving or high-impedance states are insufficient since a current path cannot be made through them.

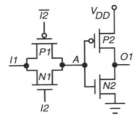

Figure 13.10: Example circuit.

The weak fault in Figure 13.10 [428] is detected if logic 1 (0) is applied when transistor *P1* (*N1*) is open. This degrades the input voltage at the load inverter, so both inverter transistors remain partially turned on, elevating I_{DDQ} current from 0 μA to 56 μA. In Figure 13.11 [428], where the transmission gate feeds one NAND gate input, the I_{DDQ} current will not be elevated unless the other NAND gate input is also set to logic 1. No I_{DDQ} path can exist if the second NAND input is 0.

A weak-0 (1) fault at a node becomes sensitized by a vector if a defective transistor disconnects all normal-0 (1) conducting paths to the node, but at least one weak-0 (1) path is maintained. A sensitized weak fault can be detected by an I_{DDQ} measurement only if the fault effect is propagated by a conducting path from V_{DD} to V_{SS} through complementary devices partly turned on by the weak voltage level. The states of other nodes determine whether such a path is established. Some weak faults cannot be propagated and are undetectable.

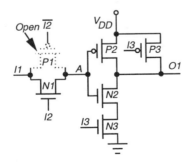

Figure 13.11: Sensitization and propagation of weak-1 fault at node *A*.

Hierarchical Test Vector Selection. Mao and Gulati avoid full circuit switch-level logic simulation with a hierarchical methodology, using logic-level simulation [428]. They characterize each logic component to relate input/output logic values and internal states to detection of leakage faults or the sensitization/propagation of weak faults. This step involves switch-level simulation, but is done only once for each component type to characterize it. This information is stored in *leakage* and *weak fault* tables. After that, a logic simulator captures input/output and internal state values of each component instance due to the current test vector. The previously-generated fault tables indicate which leakage/weak faults are detected by each vector, without the need for more switch-level simulation.

Leakage Fault Table. This table is an $m \times n$ matrix M, where $m = 2^k$ (k is the number of component I/O pins) and n is the number of component transistors. Each component logic state is a unique matrix row. Each entry m_{ij} of the matrix is an octal value containing leakage fault information. Each bit represents a leakage fault (see Table 13.1 [428]) and is 1 if the corresponding leakage fault will be detected by measuring I_{DDQ} in the component I/O state for the row.

For the NOR gate of Figure 13.12(a) [428] and its simulation in Figure 13.12(b), consider Figure 13.12(c) [428]. Entry m_{53} in row 5 ($i = 4$) means that when $(I_1, I_2, O_1) = (1, 0, 0)$, transistor *P1* (column 3) has entry $(26)_8$ (in octal), meaning that the leakage faults f_{BD}, f_{SD}, and f_{GD} in *P1* are detected by an I_{DDQ} test.

Table 13.1: Representation of transistor leakage faults.

f_{BG}	f_{BD}	f_{BS}	f_{SD}	f_{GD}	f_{GS}

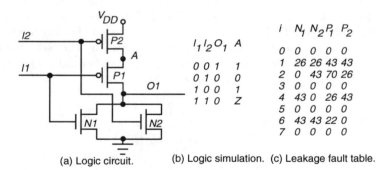

(a) Logic circuit. (b) Logic simulation. (c) Leakage fault table.

Figure 13.12: NOR gate (a), logic simulation (b), and leakage fault table (c).

Weak Fault Tables. The sensitization of a weak fault with an internal fault node depends only on I/O states of the faulty component. Weak faults on component boundaries are sensitized by the I/O states of the component, but their propagation is determined by either the I/O states of the component with the weak fault or the I/O states of components driven by the node with the weak fault.

After characterization, a *weak fault detection table* is generated containing all detected weak faults associated with internal nodes or nodes with weak faults. The *weak fault sensitization table* contains all sensitized weak faults, associated boundary

13.3 I_{DDQ} Testing Methods

nodes, and sensitized but undetected weak faults associated with internal boundary nodes for each component I/O state. The *weak fault propagation table* shows the inputs that propagate a sensitized weak fault at the component input.

I_{DDQ} Test Vector Selection. The circuit undergoes gate-level simulation using the production stuck-fault test patterns. Logic levels at inputs/outputs of all components are compared with the entries of the leakage and weak fault tables to determine which leakage and weak faults are tested. If a vector tests even one new leakage/weak fault, it is selected for I_{DDQ} measurement. In Figure 13.13 [428], only the italicized production test vectors in Table 13.2 [428] needed I_{DDQ} measurements to cover all leakage and weak faults.

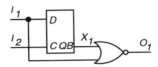

Figure 13.13: Example for I_{DDQ} test vector selection.

Table 13.2: Logic and I_{DDQ} test vectors (shaded) for circuit of Figure 13.13.

Time	I_1	I_2	X_1	O_1	Time	I_1	I_2	X_1	O_1
99	0	1	1	0	799	0	0	0	1
199	0	0	1	0	899	0	1	1	0
299	1	0	1	0	999	1	1	0	0
399	1	1	0	0	1099	1	0	0	0
499	0	1	1	0	1199	0	0	0	1
599	1	1	0	0	1299	1	0	0	0
699	1	0	0	0	1399	1	1	0	0

Component boundaries influence I_{DDQ} test vector selection. A node can have a weak fault due to a normal conducting path in one component, and a weak conducting path in another. The two components must be analyzed together in order to obtain correct results. All normal and weak conducting paths associated with a weak fault must exist in a single component. This is guaranteed when all inputs of a component type only drive gates of transistors in the component, and outputs of different components do not simultaneously drive the same node [272]. A CMOS transmission gate cannot be separate, but must be combined with other components that it influences. Table 13.3 [428] presents QUIETEST results.

13.3.3 Instrumentation Problems

Measurement of I_{DDQ} current requires analog circuits that can measure current below 1 μA at clock rates above tens of kHz [617]. Many I_{DDQ} test setups use circuits on the tester load board or inside the tester. The circuits use the Keating and

Table 13.3: QUIETEST results.

Ckt.	# of transistors	# of leakage faults	% selected vectors	Leakage fault coverage	# of weak faults	% selected vectors	Weak fault coverage
1	7584	39295	0.50 %	94.84 %	1923	0.35 %	85.30 %
2	42373	220571	0.99 %	90.50 %	1497	0.21 %	87.64 %

Meyer floating pin technique [350, 712] or circuits in the tester power supply [176]. Off-chip I_{DDQ} measurements are degraded by the pulse width of the CMOS IC transient current, impedance loading of the tester probe (from 20 to 200 pF), and current leakages into/out of the tester. Also, the high noise of the tester load board [176, 283] is caused by impedances on probes that vary with the probe voltage. This can be eliminated if output pins are disconnected or put in high impedance mode during the I_{DDQ} test. Many testers let one multiplex one probe between two chip pins, but this increases tester noise, and requires a slower I_{DDQ} testing rate.

13.3.4 Current Limit Setting

Production I_{DDQ} current testing needs a pass/fail value for the current limit, and it is difficult to pick a correct value. One should evaluate test data from representative circuits and characterize I_{DDQ} current using every vector from a functional vector set and a slow, *precision measurement unit* on a tester [283].

Figure 13.14 [617] shows the relative I_{DDQ} scatter for various vectors. Most devices have low I_{DDQ} current (they are good), but there is a distribution of high-current devices due to gate oxide shorts. There may be a multi-modal distribution with enough separation between the peaks to indicate an appropriate limit for I_{DDQ} current. One should drive the I_{DDQ} limit to < 1 μA. The I_{DDQ} data expose many undetected defects [282, 450, 451, 519, 520]. It is common in production testing to have 1 μA $<=$ I_{DDQ} $limit$ $<=$ 20 μA. A 0.5 to 1 mA limit for I_{DDQ} current on a few vectors can find defects not caught by scan-based stuck-fault voltage testing with 99.6% stuck-at fault coverage [520].

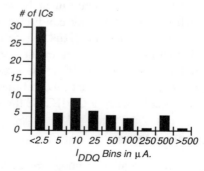

Figure 13.14: I_{DDQ} current histogram for a 32 bit μprocessor.

13.4 Surveys of I_{DDQ} Testing Effectiveness

Industrial Chips. Table 13.4 [437] presents reject rate data for a Hewlett-Packard (HP) static CMOS standard cell design with 8577 logic gates and 436 flip-flops. The chip was tested with 2 MHz functional vectors, and vector subsets were run at 20 MHz and 32 MHz. The table also presents data on 5000 static RAM tests from Sandia National Laboratory [451]. Figure 13.15 [437] shows the distributions of the failures in each category of testing for the HP chip. We must use a mixture of testing methods to achieve high reliability. There is less correlation between I_{DDQ} failures and timing failures than between I_{DDQ} failures and voltage failures in general [437]. Many timing failures are not caught by I_{DDQ} tests.

Table 13.4: Reject rate for various tests.

Company	Reject rates (%)	Scan and functional tests			
		Neither	No scan/ functional	Scan/ no functional	Both
Hewlett-Packard	Without I_{DDQ}	16.46	6.36	6.04	5.80
	With I_{DDQ}	0.80	0.09	0.11	0.00
		Functional tests			
Sandia Laboratories	Without I_{DDQ}	5.562			
	With I_{DDQ}	0			

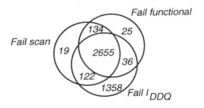

Figure 13.15: Distribution of failures in each test class.

In Figure 13.16 McEuen [451] shows a correlation between high I_{DDQ} currents during testing and functional chip failures after 1000 hours of life testing at Ford Microelectronics. I_{DDQ} testing time increases in Figure 13.17 not only with circuit density but with the resolution demanded of the I_{DDQ} current test [451].

SEMATECH Experiment. The SEMATECH* experiment compared various test methods used in IC manufacturing [501, 502]. The chip for the experiment was an IBM graphics controller ASIC of fully complementary static CMOS having 166,000 standard cells. The line width was 0.45 μm for effective channel length. V_{DD} was 3.3 V, and the frequency varied between 40 MHz and 50 MHz. The chip had 304 pins and 249 signal input/outputs. It used full-scan LSSD [211] design (see

---

*SEMATECH *Test Methods Evaluation* Project Number S-121, 1994.

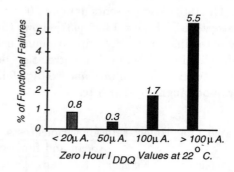

Figure 13.16: Percentage functional failures after 100 hrs. of life testing.

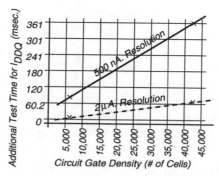

Figure 13.17: Lower and upper I_{DDQ} test time limits vs. I_{DDQ} current resolution.

13.5 Limitations of I_{DDQ} Testing

Chapter 14) with 5280 scan latches and eight scan chains. All pins had boundary scan (see Chapter 16.) The chip had two clocks.

The tests compared were [501]:

- Scan-based stuck-at fault tests with 99.7% *stuck-at fault* (SAF) coverage,
- Functional tests (design verification patterns) with 52% SAF coverage,
- Scan-based delay fault tests with 90% transition delay fault coverage, and
- I_{DDQ} tests (at least 125) with 96% I_{DDQ} pseudo-stuck-at fault coverage.

They also did burn-in testing, characterization testing, and physical failure analysis. All test methods uniquely detected some defect class, so none can be dropped. Figure 13.18 shows chips passing one testing method but failing others [502]. Many passed functional test, but failed all others. Many passed all tests, but failed I_{DDQ} test. Many passed stuck-at and functional test, but failed delay and I_{DDQ} tests. Many failed stuck-at and delay tests, but passed I_{DDQ} and functional tests.

		I_{DDQ} (5 µA limit)					
		pass	pass	fail	fail		
	pass		6	1463	7	pass	
Scan-based	pass	14	0	34	1	fail	scan-based
stuck-at	fail	6	1	13	8	pass	delay
	fail	52	36	1251		fail	
		pass	fail	pass	fail		
			Functional				

Data for devices failing some, but not all, tests.

Figure 13.18: Diagram of pass/fail data for first package test.

13.5 Limitations of I_{DDQ} Testing

I_{DDQ} testing has been highly useful in detecting bridging faults, which are hard to detect using voltage testing [42, 109, 132, 394, 500, 664]. Emerging submicron technologies have increased leakage currents, mainly due to sub-threshold conduction of MOSFETs. It is increasingly hard to determine an appropriate I_{DDQ} current threshold separating good and bad ICs [57, 221]. Many object to using a single DC value for I_{DDQ} testing [241, 242, 439, 440, 502]. Others predict that I_{DDQ} testing will soon disappear [731]. I_{DDQ} testing detects passive defects (involving non-switching circuit nodes) and active defects (involving switching nodes) simultaneously [664]. This is possible when the average defect-induced current is significantly higher than the average good IC I_{DDQ}, and there is small variation in I_{DDQ} over a test sequence and between chips. IC technology scaling makes these two conditions less likely, and the two averages are converging [731].

13.6 Delta I_{DDQ} Testing

Thibeault recently proposed using Δ (or differential) I_{DDQ} testing [663, 664, 665, 666]. He uses the derivative of I_{DDQ} at test vector i as a current signature:

$$\Delta I_{DDQ}(i) = I_{DDQ}(i) - I_{DDQ}(i-1) \qquad (13.2)$$

This method was derived from I_{DDF} testing [661]. ΔI_{DDQ} testing is excellent for diagnosis [662, 666], and can greatly improve the resolution of I_{DDQ} testing [663].

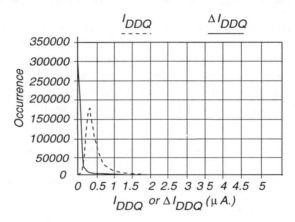

Figure 13.19: Histograms of $I_{DDQ}(i)$ and $\Delta I_{DDQ}(i)$.

Figure 13.19 [664] shows the why ΔI_{DDQ} testing is better than I_{DDQ} testing, and why it may significantly extend the usefulness of current testing. The ΔI_{DDQ} histogram is narrower than the I_{DDQ} histogram built from the SEMATECH experimental data using a 5 μA threshold current. The I_{DDQ} histogram is affected by chip-to-chip and wafer-to-wafer variations in current measurements, which may be larger than variations in vector-to-vector measurements. ΔI_{DDQ} testing eliminates this confusing variation between chips. Declaring a good IC to be bad results in a *yield loss* and declaring a bad IC to be good results in a *test escape*. Thibeault compared the probability of false test decisions P for I_{DDQ} and ΔI_{DDQ} testing in Figure 13.20 [664]. The solid line is the current distribution of good ICs, and the dashed one is for bad ICs. Area A represents the test escapes and area B represents the yield loss. The vertical line separating A and B is the decision threshold (x). Table 13.5 [664] shows parameters used to compare I_{DDQ} with ΔI_{DDQ} tests.

Thibeault's estimation results appear in Table 13.5 [664], and account for the distribution of peak values among ICs. The results also depend on yield, but a 25% change in yield causes a variation of only 12% in the P_{iddq}/P_{delta} ratio. The assumption that the distributions in Figure 13.20 [664] were Gaussian increased P_{iddq} (about 10%) with respect to the regular I_{DDQ} histogram, so Table 13.5 [664] is slightly optimistic. Thibeault found that the variance (σ_i^2 or σ_d^2) is the same with or without defects by analyzing the SEMATECH data.

13.6 Delta I_{DDQ} Testing

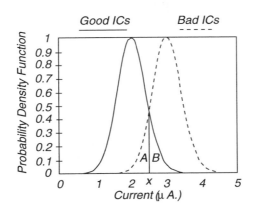

Figure 13.20: Generic current distribution example.

Table 13.5: Parameters for estimating P and estimates of P.

Parameter (μA or μA^2) of distribution	I_{DDQ}		ΔI_{DDQ}	
	Symbol	Value	Symbol	Value
P	P_{iddq}	See below	P_{delta}	See below
μ_g (good mean)	μ_{gi}	0.696 [665]	$\mu_{gd}(\approx 0)$	$-2e^{-4}$ [665]
Δ_{def} (min. $|\Delta I_{DDQ}|$ peak due to active defect)	Δ_{def}	0.4 [665]	Δ_{def}	0.4 [665]
μ_b (bad mean)	$\mu_{gi} + \Delta_{def}$	1.096 [665]	$\mu_{gd} + \Delta_{def}$	0.4 [665]
σ^2 (variance for both)	σ_i^2	0.039 [665]	σ_d^2	0.004 [665]
	Values of P for different Δ_{def} Values			
	Δ_{def}	P_{iddq}	P_{delta}	P_{iddq}/P_{delta}
	0.3	0.059	$7.3e^{-4}$	81
	0.4	0.032	$4.4e^{-5}$	721
	0.5	0.017	$1.7e^{-6}$	10000

Figure 13.21 [664] shows a ΔI_{DDQ} histogram. Thibeault uses $|\Delta I_{DDQ}(i)|$ instead of $\Delta I_{DDQ}(i)$ values, which provides better peak resolution by doubling the point count. This method reduces the effective variance caused by vector-to-vector variation during I_{DDQ} tests. The abscissa location of a peak is usually very close or equal to the average current value caused by a given defect mechanism. The average absolute error between these two values was $< 3\%$. This reduced the variance around the peak by a factor of $2 \times n$ (n is the number of measurement points included in the peak), which reduces the probability of a bad test decision.

In *pre-analysis* Thibeault first builds a differential I_{DDQ} histogram for a given set of ICs, then he builds a second histogram with gradual elimination of ICs with high peaks, and finally he analyzes the correlation between peaks and vectors. The second histogram and a failure analysis step are used to set the ΔI_{DDQ} limit between good and bad ICs. Thibeault's paper gives another method to mathematically set the ΔI_{DDQ} threshold decision value when experimental data is not available [664].

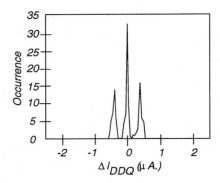

Figure 13.21: Differential I_{DDQ} histogram example.

Table 13.6 [664] compares I_{DDQ} testing and the ΔI_{DDQ} test procedure. An advantage of ΔI_{DDQ} testing is that any constant (vector-insensitive) current increase due to process drift is eliminated by the differential operation, but process drift may still increase the measurement variation between vectors.

Table 13.6: I_{DDQ} testing vs. new ΔI_{DDQ} histogram-based procedure (HBTP.)

Item	I_{DDQ}	HBTP
R_{YL} (yield loss ratio)	$4.4e^{-4}$	$3.5e^{-3}$
R_{TE} (test escape ratio)	$1.8e^{-1}$	$2.1e^{-3}$
$P(=R_{YL}+R_{TE})$	$P_{iddq} = 1.8e^{-1}$	$P_{delta} = 5.6e^{-3}$
Gain in test quality	$P_{iddq}/P_{delta} = 31$	

13.7 I_{DDQ} Built-In Current Testing

Built-in current testing (BIC) alleviates the need for special purpose I_{DDQ} testing hardware and increases the testing rate [423, 424, 425, 426, 427]. BIC uses current sensors to monitor the quiescent current in the power lines of the *device-under-test* (DUT.) In Figure 13.22 [426] the BIC sensor has a *voltage drop device* (V_{dro}) and a voltage comparator. At the end of each clock, the device compares the

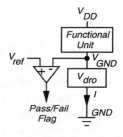

Figure 13.22: BIC sensor.

13.7 I_{DDQ} Built-In Current Testing

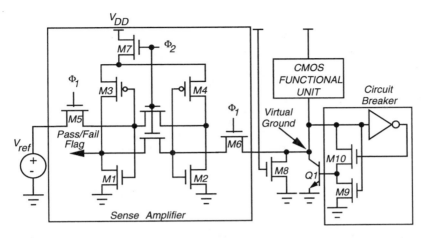

Figure 13.23: CMOS technology BIC sensor.

virtual ground voltage V_{GND} with the voltage V_{ref}, chosen so that $V_{GND} < V_{ref}$ for good circuits and $V_{GND} > V_{ref}$ for bad ones. The BIC device can also sit between the functional unit and V_{DD} by changing V_{ref}. Voltage drops across the voltage drop device must be small, even in defective circuits with shorts. However, the voltage drop caused by a GOS must be large enough to be measured by the comparator. A non-linear I-V characteristic [426] is required, so a bipolar device is used to obtain high resolution for current detection. Figure 13.23 [426] shows a CMOS BIC sensor using a substrate *npn* transistor as the voltage drop device ($Q1$) [421]. The unit has a sense amplifier and a circuit breaker, and disconnects the defective functional unit from power when abnormal currents occur due to V_{DD}-GND shorts or radiation generated latch-up [499]. The circuit breaker activates when the virtual ground voltage becomes sufficiently high to turn on the inverter in the circuit breaker. This turns on $M9$ and turns off $M10$, which shorts the base of $Q1$ to ground to turn it off. A MOSFET could also be the voltage drop device.

During probe or package testing, BIC can detect permanent defects, regardless of whether vectors are generated on or off-chip [425, 498, 499]. The IC must be partitioned into functional units, each with a single BIC sensor. If functional units are too large, the combined leakage currents may erroneously trigger the BIC sensor. We minimize the number of BIC sensors in the IC, to minimize test circuit area. The choice of N_{max}, the number of transistors controlled by one BIC sensor, is key. I_{defmin} is the smallest current indicating a defect, and $I_{noisemax}$ is the maximum noise-related peak current in the supply. The total quiescent current in the good DUT must be less than I_{defmin}. The minimum area sensor design is located at the intersection of I_{defmin} and I_{DDQ} (see Figure 13.24 [426].) The good machine current components are the drain area leakage currents, all well leakage currents, and the sub-threshold conduction leakage currents. In the final BIC sensor layout, we minimize the parasitic capacitance in the sensor, to minimize delay and decay times. Proportional BIC sensors also exist [544].

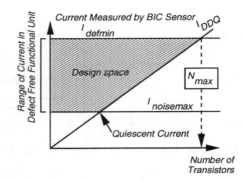

Figure 13.24: Optimal transistor count in a functional unit.

13.8 I_{DDQ} Design for Testability

We very briefly describe circuit design precautions that make it easier to test for I_{DDQ} faults. A fully static CMOS design style should be used, and floating nodes should be eliminated from the design. There should be no n-channel only pass gates, and there should be no floating or tristated busses. All analog circuits and memories should be provided with an I_{DDQ} testing mode [557].

13.9 Summary

I_{DDQ} testing improves reliability, and identifies certain defects that cause delay faults, bridging faults, weak faults, and chips damaged by electro-static discharge. It may be used with built-in self-testing. The SEMATECH experiment concluded [502] that there was no natural breakpoint for setting an I_{DDQ} current threshold. High fault coverages in all fault categories were necessary for the maximum testing benefit, which requires high testability. It was hard to find a point differentiating good and bad devices for delay test. Traditional at-speed functional testing required full pin contact and complex timing support. It was difficult, expensive, and reduced yield. Many failures that were found (45) did not degrade device functionality, only circuit timing performance. For all test methods, the predicted fault coverage curves did not match the actual device fallout measured on the tester. One conclusion is that we now need both I_{DDQ} and delay fault testing. I_{DDQ} measurements have been used for leakage fault [42] and bridging fault [133] diagnosis. It is still uncertain whether I_{DDQ} testing will remain of future use in VLSI processes as minimum feature sizes shrink further, because of increased sub-threshold MOSFET conduction, which confuses I_{DDQ} testing.

Problems

13.1 *Leakage fault tests.* For the example in Figure 13.12, consider the complete set of stuck-at fault tests. Use the leakage fault table given in Figure 13.12 to generate a complete set of leakage fault tests for the circuit, and identify the minimum number of stuck-fault test vectors needed to test all leakage faults.

13.2 *Tester time.* A digital *automatic test equipment* (ATE) supports chips with 512 pins, operates at 750 MHz frequency, and costs $ 9,000 per probe. Calculate the total cost of the ATE. The ATE is run at maximum throughput, 24 hours a day, 7 days a week. You wish to perform stuck-at fault tests and I_{DDQ} tests on a 512 pin microprocessor chip with 200,000 logic gates. The stuck-at fault test vector set has 100,000 vectors. I_{DDQ} vector selection chooses 2% of these vectors for I_{DDQ} measurements. ¿From Figure 13.17 determine the minimum time needed per vector for a coarse I_{DDQ} measurement. Calculate the total test application time for the test vector set, assuming at-speed testing at 750 MHz, except when I_{DDQ} current is measured. Compute the testing cost per chip, assuming that the ATE has a lifetime of 10 years, and that ATE maintenance costs $ 50,000 per year.

13.3 I_{DDQ} *threshold.* An ASIC chip can have a 1% field failure rate. ¿From the data in Figure 13.16, determine the highest I_{DDQ} threshold that you can test with, and still achieve this failure rate (assuming that all field-related faults cause excessive I_{DDQ} current.) Testing costs are $ 40.00 per microprocessor, but a field failure costs $ 300.00 in warranty costs, handling, shipping, and replacement of the microprocessor. At a volume of 20 million microprocessors per year, more stringent I_{DDQ} testing to screen out all I_{DDQ} defects would add $ 15.00 per microprocessor test. Are you financially better off shipping the defective product, and absorbing the costs of field returns, or would it be less costly to institute more stringent I_{DDQ} testing?

13.4 *Built-in current testing.* We use BIC sensors for on-chip I_{DDQ} testing. Each transistor on the chip draws a leakage current of 0.01 nA around its drain, each transistor also has a sub-threshold conduction current of 0.005 nA, and each $n$well has a 0.5 nA leakage current. The chip has 50 million transistors. Assume that each clump of 50 pFETs exists together in one nwell, and that there are equal numbers of nFETs and pFETs on the chip. Compute the number of nwells in this chip, and then compute the total chip leakage current. If your BIC sensor pops its circuit breaker when the I_{DDQ} current exceeds 5 μA, calculate the maximum number of nFET and pFET pairs that can be serviced by one BIC sensor in your design. Find the total number of BIC sensors needed to provide I_{DDQ} testing for this entire chip.

13.5 *Built-in current testing.* We use *built-in current testing* (BIC) sensors for on-chip I_{DDQ} testing. Each transistor on the chip draws a leakage current of 0.011 nA around its drain, each transistor also has a sub-threshold conduction current of 0.006 nA, and each $n$well has a 0.55 nA leakage current. The chip has 50 million transistors. Assume that each clump of 30 pFETs exists together in one nwell, and that there are equal numbers of nFETs and pFETs on the chip. Compute the number of nwells in this chip, and then compute the total chip leakage current. If your BIC sensor pops its circuit breaker when the I_{DDQ} current exceeds 3 μA, calculate the maximum number of nFET and pFET pairs that can be serviced by one BIC sensor in your design. Find the total number of BIC sensors needed to provide I_{DDQ} testing for this entire chip.

Part III

DESIGN FOR TESTABILITY

Chapter 14

DIGITAL DFT AND SCAN DESIGN

> ". . . We . . . summarize the techniques applied to the System/360 Model 50 processor, as of late 1965. . . . Special hardware for testing purposes only was included . . . Some of this hardware is used to provide an extremely complete facility, both to reset the processor to a specified state, and to record its state. This facility is invoked by executing either of two machine functions, which we may call Scan-In and Scan-Out. Execution of Scan-Out causes the state of every memory element (independent feedback loop) in the processor to be recorded in a special, fixed area of core memory. Scan-In is the inverse function; . . . an execution of Scan-Out when a malfunction is first detected will save the entire processor state, for later analysis by program or by the maintenance man. . . . The hard-core, or fraction of the processor which must be operable for testing to proceed, is approximately 10 percent of the processor. . . . it seems fair to call this a self-diagnosis procedure." – H. Y. Chang, E. G. Manning and G. Metze, in a 1970 publication (perhaps the first book!) on digital testing [139].

A VLSI chip is manufactured through a series of steps that involve chemical, metallurgical, and optical processes. With careful control on process variations, economically viable yields can be obtained. If the yield of good chips is 75%, then on an average 25% of manufactured chips will be faulty. Thus, at the end of the VLSI manufacturing process we always have "testing," which isolates the good chips from bad ones. Inadequate testing will have some faulty chips shipped to the customer. At the same time, the cost of testing directly increases the overall cost of the chip.

For a VLSI chip to be manufactured, we must have a verified design and a set of tests. The following questions characterize testing of complex systems [302]:

- Can tests that detect all faults be assured?

- Can test development time be kept low enough to be economical?

- Can test execution time be kept low enough to be economical?

Design for testability (DFT) refers to those design practices that allow us to answer the above questions in the affirmative.

Electronic systems contain three types of components: (a) digital logic, (b) memory blocks, and (c) analog or mixed-signal circuits. There are specific DFT methods of each type of component. In this chapter, we discuss DFT techniques for digital logic. *Built-in self-test* (BIST), which is also used for digital logic as well as for memory blocks, is the subject of Chapter 15. A system (a printed circuit board or a multi-chip module) consists of an interconnect of the three types of components. Thus, the component-level DFT methods, as discussed here and in the next chapter, are not sufficient for producing a testable system. Special techniques, known as *boundary-scan* and *analog test bus*, provide test access to components embedded in a system. These are discussed in Chapters 16 and 17, respectively. Finally, it all comes together in Chapter 18 where we discuss systems test.

14.1 Ad-Hoc DFT Methods

Keeping in line with current practices, we will focus on DFT techniques that aim at improving the testability of stuck-at faults. An interested reader may review recent literature on DFT methods for other fault models such as delay faults [371] and physical faults [557].

Logic DFT takes one of two possible routes: *ad-hoc* and structured. The ad-hoc DFT relies on "good" design practices learned from experience. Some of these are [7, 721]:

- *Avoid asynchronous logic feedbacks.* A feedback in the combinational logic can give rise to oscillation for certain inputs. This makes the circuit difficult to verify and impossible to generate tests for by automatic programs. This is because test generation algorithms are only known for acyclic combinational circuits.

- *Make flip-flops initializable.* This is easily done by supplying *clear* or *reset* signals that are controllable from primary inputs.

- *Avoid gates with a large number of fan-in signals.* Large fan-in makes the inputs of the gate difficult to observe and makes the gate output difficult to control.

- *Provide test control for difficult-to-control signals.* Signals such as those produced by long counters require many clock cycles to control and hence increase the length of the test sequence. Long test sequences are harder to generate.

These and many other similar situations cause poor controllability and observability of signals and usually result in long test sequences and low fault fault coverage. Experienced designers could easily spot the problem areas on a logic schematic. For very large circuits, approximate testability measures have been used (see Chapter 6.)

Once testability problems are found, either the circuit is modified or test points are inserted.

There are difficulties with the use of ad-hoc DFT methods. First, circuits are too large for manual inspection. Second, human testability experts are often hard to find, while the algorithmically generated testability measures are approximate and do not always point to the source of the testability problem [38]. Finally, even after DFT modifications are made, tests must be produced to enhance the fault coverage. Manual test generation is far too labor-intensive and the ad-hoc techniques do not guarantee good results from *automatic test pattern generators* (ATPG.) Due to these reasons, the use of ad-hoc DFT is usually discouraged for large circuits.

As the size and complexity of digital systems grew, an alternative form of DFT, known as *structured DFT* [210], gained popularity. In structured DFT, extra logic and signals are added to the circuit so as to allow the test according to some pre-defined procedure. Apart from the normal functional mode, such a design will have one or more test modes. Commonly used structured methods are scan (discussed in this chapter) and built-in self-test (discussed in Chapter 15.)

14.2 Scan Design

The main idea in *scan design* is to obtain control and observability for flip-flops. This is done by adding a test mode to the circuit such that when the circuit is in this mode, all flip-flops functionally form one or more shift registers. The inputs and outputs of these shift registers (also known as scan registers) are made into primary inputs and primary outputs. Thus, using the test mode, all flip-flops can be set to any desired states by shifting those logic states into the shift register. Similarly, the states of flip-flops are observed by shifting the contents of the scan register out. All flip-flops can be set or observed in a time (in terms of clock periods) that equals the number of flip-flops in the longest scan register. In the following discussion we will assume a single scan register. In practice, however, a design can have any number of them.

Although the idea of scan had been used for system test (see the quotation at the beginning of this chapter), its application to hardware test was detailed in the 1973 paper by Williams and Angell of Stanford University [722]. Various implementations of the concept were used in companies like IBM, NEC and others. Perhaps Eichelberger and Williams [210] of IBM have contributed most to advance it within and outside their company.

An alternative way of accomplishing the scan function is called *random-access scan* (RAS) [733]. In that design, flip-flops work as addressable memory elements in the test mode in a similar fashion as a *random access memory* (RAM.) This approach reduces the time of setting and observing the flip-flop states, but the design requires a large overhead both in gates and test pins.

In this section, we will assume that all flip-flops of the circuit have the scan capability. Sometimes specified as *full-scan*, that is the way the traditional scan design was understood. A variation, known as *partial-scan*, will be discussed in

468   **Chapter 14. DIGITAL DFT AND SCAN DESIGN**

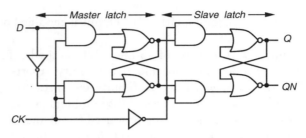

Figure 14.1: A D flip-flop.

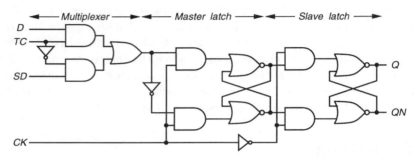

Figure 14.2: A single-clock scan flip-flop.

Section 14.3.

For a circuit to have the scan capability, first the designer uses only *D type flip-flops* (DFF) with one or more clock signals, all of which are controlled from primary inputs. A typical DFF is shown in Figure 14.1. Once the circuit is functionally verified, the DFFs are replaced by *scan flip-flops* (SFF). One typical SFF is shown in Figure 14.2. Here a multiplexer and two new signals, scan-data SD and test control TC, are added to the *D flip-flop* (DFF.) The original data input D is stored in the flip-flop when TC is 1 and SD is stored when TC is 0.

Another popular design style uses two non-overlapping clock signals. Figure 14.3 shows a scan flip-flop with two function clocks, MCK and SCK. When MCK is high, data D is latched in the master latch. When SCK is high, the state of master latch is copied in the slave latch. For a proper operation of a general sequential circuit, MCK and SCK are never turned high, simultaneously. In the scan mode, MCK is held low and scan data SD is latched in by using clocks TCK and SCK as master and slave clocks, respectively [210]. The MC inputs of all scan flip-flops are supplied by a new primary input. The SD input of one SFF is supplied by another new primary input $SCANIN$. All SFFs are chained by connecting the Q output of one SFF to the SD input of the next SFF. The Q output of the last SFF in the chain is a new primary output $SCANOUT$. The complete design is given in Figure 14.4, with the wiring added for scan design shown in broken lines. This design has the advantage of reducing the effort of test generation. Especially for the case of *full-scan*, where all flip-flops are scanned, a combinational ATPG program (much simpler than sequential ATPG) can produce tests for all stuck-at faults in

14.2 Scan Design

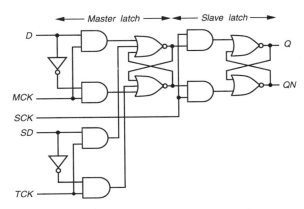

Figure 14.3: A two-clock scan flip-flop.

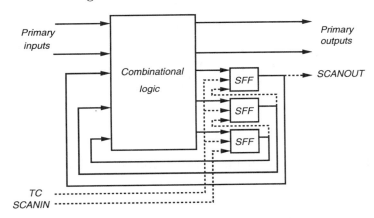

Figure 14.4: A scan design schematic.

the circuit.

14.2.1 Scan Design Rules

A circuit is designed to meet its functional requirements. After the functional correctness of the design is verified, it is modified to include the scan function. In order to be able to make it scan-testable, the designer must adhere to certain rules during the functional design. In general, these rules depend upon the specific design environment, which may dictate choices such as single versus multiple clocks, etc. The following four rules, however, are found to be useful:

R-1: *Only D-type master-slave flip-flops should be used.* This rule prohibits the use of other types of flip-flops (JK, toggle, etc.) or other forms of asynchronous logic (unclocked RS latches, combinational feedback elements.)

R-2: *At least one primary input pin must be available for test.* In general, flip-flops can be connected as multiple scan registers (see Section 14.2.3), each of which

470 Chapter 14. DIGITAL DFT AND SCAN DESIGN

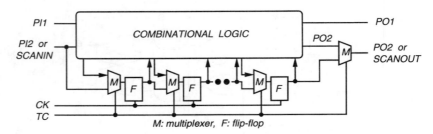

Figure 14.5: Scan design with a single extra pin for *test control* (TC).

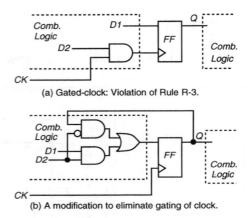

Figure 14.6: An example of a work-around to satisfy rule R-3.

will require a scan-in and a scan-out terminal. If extra pins are not available, then any normal primary input can be used as scan-in and any primary output pin can be multiplexed as scan-out. This is illustrated in Figure 14.5 where TC (test control) is the only pin added. Primary input $PI2$ serves as $SCANIN$ and primary output pin $PO2$ also outputs $SCANOUT$. Note that the cost of saving these pins is just one extra multiplexer used for $SCANOUT$.

R-3: *All flip-flop clocks must be controllable from primary inputs.* This rule is necessary for filp-flops to function as a scan register. Some violations of this rule, if they exist, can be removed by a simple *work-around*. Figure 14.6 shows an example. Here the clock signal (CK) is gated by a combinational signal, $D2$. Thus, when $D2 = 0$, the clock is inhibited and the flip-flop FF retains its state Q. When $D2 = 1$, the clock CK stores $D1$ as the new state. In Figure 14.6(b), the clock is applied directly to FF and a multiplexer is added to the combinational logic to regenerate data for FF. The two circuits are functionally identical and the modified circuit satisfies the rule R-3. It can be converted into scan design as described in the previous section.

R-4: *Clocks must not feed data inputs of flip-flops.* A violation of this rule can potentially lead to a race condition in the normal mode. Thus the value

14.2 Scan Design

captured in the flip-flop cannot be guaranteed to be the state of the signal produced by the combinational logic. In scan design, flip-flops play a dual role. They capture combinational data in the normal mode and then carry the data out for observation in the scan mode. The test procedure relies on the flip-flop correctly capturing data in the normal mode and hence no race condition is permitted.

14.2.2 Tests for Scan Circuits

Testing of scan circuits is done in two phases. The first phase tests the scan register by a *shift test*. The circuit is set in scan mode by setting $TC = 0$. All flip-flops now form a shift register between $SCANIN$ and $SCANOUT$. A *toggle sequence*, 00110011 ..., of length $n_{sff} + 4$, where n_{sff} is the total number of flip-flops, is applied at $SCANIN$. The toggle sequence is clocked through the shift register using the normal clock signal. This sequence produces all four transitions, $0 \to 0$, $0 \to 1$, $1 \to 1$, and $1 \to 0$, in each flip-flop and shifts the outputs to the observable output $SCANOUT$. It covers most, if not all, single stuck-at faults in the flip-flops, and verifies the correctness of the shift operation of the scan register.

The shift test is used in both single-clock and two-clock designs. For the two-clock scan flip-flop of Figure 14.3, an additional *flush test* is possible [86]. In this test, master clock MCK is held low and the other two clocks, TCK and SCK, are simultaneously set high. Thus the scan register provides a continuous path between $SCANIN$ and $SCANOUT$. The test then consists of application of 0 and 1 signals at $SCANIN$ and their observation at $SCANOUT$ after a delay that equals the total gate delays in the scan register path. The flush test is not possible for the single-clock design because the continuous path cannot be created for any state of the clock.

In the second phase of testing, single stuck-at faults in the combinational logic are targeted. A combinational ATPG program is used to generate test vectors assuming that all flip-flop outputs are completely controllable and all flip-flop inputs are observable. Figure 14.7(a) shows this combinational ATPG scenario. Three input vectors and the corresponding outputs are shown in the figure. Each input vector contains two parts, primary input parts, $i1$, $i2$, and $i3$, and state variable parts, $s1$, $s2$, and $s3$. Similarly, the corresponding outputs of the combinational logic contain two parts, primary outputs, $o1$, $o2$, and $o3$, and next state outputs, $n1$, $n2$, and $n3$. The combinational vectors are converted into scan sequences shown in Figure 14.7(b) before they can be applied to the actual circuit. We assume the single-clock design of Figure 14.4, where $TC = 1$ sets the circuit in the normal mode and $TC = 0$ sets it in the scan mode. The test sequences contain primary inputs (including TC) that are applied one vector per clock, using the normal system clock. Also the expected output response at primary outputs (including $SCANOUT$) is specified at each clock. The grey area in these scan sequences corresponds to *don't care* signals. In the input sequence, don't cares can be filled with all 0, all 1, or random bits. In the output sequence, they can be either left as don't cares in which case they will not be compared with the chip outputs by the ATE, or filled with

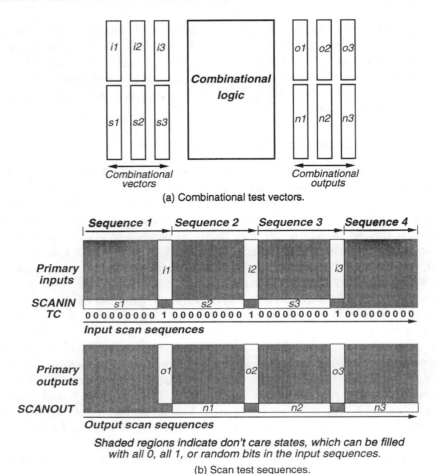

Figure 14.7: Scan test sequences for single-clock design.

their expected values obtained by simulation.

The example illustrated in Figure 14.7 has nine flip-flops. Each combinational input vector produces an input sequence of ten vectors. Sequence 1, produced from the vector ($i1$, $s1$), has nine scan-in vectors and one normal-mode vector. In the scan-in vectors, $TC = 0$ and the bits of $s1$ are serially applied to the $SCANIN$ input. All other primary inputs remain in the *don't care* state. After the ninth vector, all flip-flops are in the states specified by $s1$. On the tenth vector, $TC = 1$ and $i1$ is applied to primary inputs. Thus, the circuit is set in the normal mode and all inputs of the combinational logic are as specified by the vector ($i1$, $s1$). At the end of the tenth vector, the expected values of $o1$ appear at primary outputs and the state $n1$ is clocked into flip-flops. Any faults affecting the primary outputs would have been detected at this time. The faults affecting the states of flip-flops will be detected via Sequence 2.

14.2 Scan Design

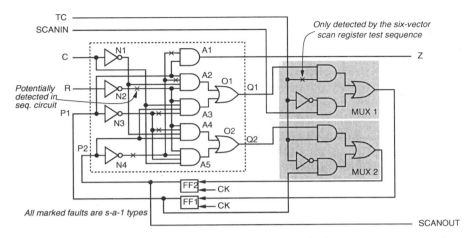

Figure 14.8: A modulo-3 circuit illustrating scan design and test generation.

Following a procedure similar to the one above, Sequence 2 loads $s2$ in the scan register. However, as the bits of $s2$ are loaded in the register, the bits of $n1$ are pushed out through the $SCANOUT$ terminal and are observed for any fault effects produced by the first vector $(i1, s1)$ at the next state outputs of the combinational logic. Thus, the scan out of the first vector overlaps with the scan in of the second vector. This process continues until all combinational input vectors have been applied (via Sequences 1, 2, and 3 in this example.) Since Sequence 3 leaves the last state $n3$ still unobserved, a final scan-out sequence (Sequence 4) of length 9 is added.

For three combinational vectors, the total scan test length is 39 clock periods. Considering also the shift test discussed before, the length of the scan test can be generalized as:

$$\text{Scan test length} = n_{sff} + 4 + (n_{comb} + 1) \times n_{sff} + n_{sff}$$
$$= (n_{comb} + 3)n_{sff} + 4 \quad \text{clock periods} \qquad (14.1)$$

where n_{comb} is the number of combinational test vectors and n_{sff} is the number of flip-flops in the scan register. Here we have assumed that all flip-flops form a single scan register. Such a design can lead to a long test time. For example, consider a circuit with 2,000 flip-flops. If all faults in the combinational logic are tested by $n_{comb} = 500$ vectors, then the complete scan test will run for 1,006,004 clock periods.

Example 14.1 *A modulo-3 counter with scan. Figure 14.8 shows a modulo-3 counter circuit with scan added. The circuit has two normal inputs C and R, and one output Z. When $R = 1$, the counter is set to its initial state 00. For $R = 0$ and $C = 1$, the state advances as $00 \to 01 \to 10 \to 00 \ldots$ with each clock. The output Z becomes 1 only for the state 10 and remains 0 for all other states. When $R = C = 0$, the counter retains its state. The combinational logic (in dotted line box) is designed to be combinationally irredundant. The scan logic consists of two* multiplexers (MUX 1 and MUX 2) enclosed in grey boxes. Two inputs, TC (test control) and SCANIN, and one output, SCANOUT, have been added. When $TC = 0$, a two-bit shift register is formed between SCANIN and SCANOUT. For $TC = 1$ (normal mode), Q1 feeds into FF1 and Q2, into FF2. If we remove the scan circuitry and directly connect Q1 and Q2 to FF1 and FF2, the sequential circuit has six untestable s-a-1 faults (marked on Figure 14.8.) One of these, associated with the reset signal R, is potentially detectable. Untestable and potentially detectable faults are quite typical of sequential circuits. A sequential ATPG program [72] generated 35 vectors to cover 36 of 42 faults in the non-scan circuit. The ATPG program produces 12 vectors for the combinational circuit in the dotted-line box, having four inputs, C, R, P1, and P2, and three outputs, Z, Q1, and Q2. Converted to scan sequences including a six-vector shift register test, these give a scan test sequence of 34 vectors. Fault simulation shows that this sequence detects all faults, including those in multiplexers and the six untestable faults in the original sequential circuit. The scan sequence without the six-vector shift register test left one fault in MUX1 undetected. Obviously, faults in multiplexers were not targeted by the ATPG program. Even if they were, some of those tests can be invalidated if the fault interferes with the scan-in or scan-out operations. That is why the shift register sequence is important. Example 14.2 illustrates the scan design of a larger circuit.

14.2.3 Multiple Scan Registers

To reduce the time of scan test, sometimes flip-flops are arranged in multiple scan registers. Each scan register requires separate SCANIN and SCANOUT pins. If extra pins are not available, added fanouts from normal primary input pins can provide SCANIN signals to scan chains. This is possible because the normal primary inputs and SCANIN are never simultaneously used (see Figure 14.5.) Similarly, the SCANOUT signals can be multiplexed with the normal primary output pins under the control of the test control (TC) signal. In general, multiple scan registers can have varying lengths. The length of scanin and scanout sequences depends on the longest register. The scan test length is computed from Equation 14.1 by substituting the length of the longest scan register for n_{sff}.

14.2.4 Overheads of Scan Design

The use of scan design has two types of penalties. The scan hardware increases the chip size (area overhead) and slows the signals down (performance overhead.) In the early stages of a design, it is useful to estimate the gate overhead.

Gate overhead. Suppose a circuit has n_g gates and n_{ff} flip-flops. Assume that the original flip-flops were as shown in Figure 14.1 and all are replaced by the one shown in Figure 14.2. Then each flip-flop adds an overhead of four gates (used in the multiplexer.) The total gate overhead is computed as follows:

$$\text{Gate overhead of scan} = \frac{4 \times n_{sff}}{n_g} \times 100\% \qquad (14.2)$$

14.2 Scan Design

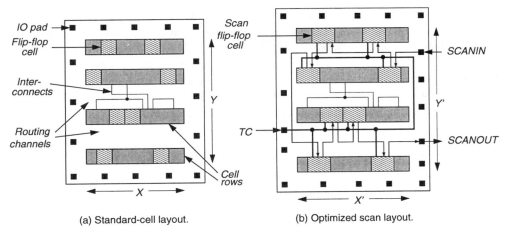

Figure 14.9: Scan implementation in standard-cell chip layout.

where n_{sff} is the number of scan flip-flops, which in this case (full-scan) equals n_{ff}. As an example, if prior to scan insertion, the circuit has 100,000 gates and 2,000 flip-flops, then the gate overhead will be 8%. The actual area of a gate depends upon the layout detail. Besides, scan routing also occupies chip area. Despite gross approximation, Equation 14.2 is a reasonable estimate for the early design stage.

Area overhead. Scan design requires a significant amount of routing that can impact the chip area. The test control signal (TC) is routed to all flip-flops and the output of each flip-flop is routed to the scan data (SD) input of the next flip-flop in the scan register chain. The impact of scan routing on the chip area increase can be reduced by: (a) flip-flop placement on the layout for optimum routing and (b) selecting the flip-flop order in the scan chain.

We will calculate the area overhead for the standard-cell layout shown in Figure 14.9(a). In this layout style, building blocks, called *standard cells*, are pre-designed and saved in a cell library. Combinational and sequential cells of up to about 30-40 gate complexity allow implementation of most chips. These cells are designed with a fixed height, so that they can be placed in *cell rows*. Spaces between rows, known as *routing channels*, are used for interconnects between cells [598]. A channel contains a layer of metal wires or *tracks* and another layer of low-resistance polysilicon connections that run perpendicular to tracks. This allows a horizontal-vertical (two-layer) routing. The physical design of a chip is a two-step process: cell-placement and routing. These steps may be repeated for iterative improvements.

In an optimized scan design, the cells are first placed without the scan wiring. This placement ensures that the scan wiring will not adversely affect the functional interconnects. Figure 14.9(b) shows the scan additions. First, flip-flop cells (uniformly distributed among cell rows and shown cross-hatched) are replaced by corresponding scan flip-flop cells. These cells are wider due to the added multiplexer (see Figure 14.2.) Next, two types of interconnects are added. A test control (TC)

signal feeds all scan flip-flop cells. As shown in Figure 14.9(b) by bold lines, this takes at most one track in every alternate routing channel. The second set of interconnects forms a chain between the $SCANIN$ and $SCANOUT$ pins. When flip-flops are suitably ordered, as the wires with small arrows show, this is also accomplished by one track per alternate channel. Thus, two tracks per alternate channel, or an average of one track per channel, can accommodate the scan wiring. The track overhead can go higher if the scan layout is not optimized [34].

We will only consider the *active area*, which is the area of cells and routing. The pad area has no significant overhead. So, our estimate will give an upper bound on the area overhead. The active area in Figure 14.9 is a rectangle of linear dimensions X and Y, which expand to X' and Y', respectively. The increase in X is due to wider scan cells and that in Y is due to the extra tracks used for the scan wiring. The cell width is measured in a normalized unit called a *grid*. Similarly, the channel height is measured in the number of *tracks*. We will use the following notation:

y: Track width in length units.

C: Combined width of combinational cells.

S: Combined width of sequential (flip-flop) cells in the non-scan circuit.

s: Fraction of cell area under flip-flops, $s = S/(C+S)$.

α: Fractional width increase of a scan flip-flop over a non-scan cell.

r: Number of cell rows or routing channels (assumed to be equal.)

β: Fraction of active area occupied by routing channels (routing fraction.)

T: Cell height in number of tracks.

We can easily obtain:

$$X = \frac{C+S}{r} \quad \text{and} \quad X' = \frac{C+S+\alpha S}{r} \quad (14.3)$$

We have assumed that there are as many routing channels as there are cell rows. The combined height of a cell row and one adjoining routing channel is $T+T\beta/(1-\beta) = T/(1-\beta)$ tracks. Therefore, the number of routing channels, $r = Y(1-\beta)/(yT)$. As explained earlier, the scan routing on average adds one track per routing channel. Therefore:

$$Y' = Y + ry = Y + Y(1-\beta)/T \quad (14.4)$$

Upon substitution and simplification, we get:

$$\begin{aligned}
\text{Area overhead of scan} &= \frac{X'Y' - XY}{XY} \times 100\% \\
&= \left[(1+\alpha s)\left(1 + \frac{1-\beta}{T}\right) - 1\right] \times 100\% \\
&\approx \left(\alpha s + \frac{1-\beta}{T}\right) \times 100\% \quad (14.5)
\end{aligned}$$

14.2 Scan Design

The first term is the cell area overhead, which grows with the fraction (s) of flip-flop area. The second term is routing overhead. Depending on the cell height (T) that is around 8-10 tracks, this part can contribute up to about 10% in a cell-dominant chip. As the routing fraction (β) increases, routing overhead drops.

Eichelberger et al. [210] discuss the overhead for the *level-sensitive scan design* (LSSD) style of IBM. In modern VLSI technologies that use several metal layers, as well as over-the-cell routing [598, 599], the routing overhead of scan is significantly reduced over the single-layer metal technologies of the past. The area overhead is typically found to be in the range of 5 to 10%.

Performance overhead. Scan design also has a performance overhead. The multiplexer of the scan flip-flop adds delay equivalent to two gate-delays in all clocked paths. In addition, flip-flop outputs have one extra fanout, which increases the capacitive loading of the signal. In general, scan design can reduce the clock speed by 5 to 10%. Notice that the two-clock design of Figure 14.3 does not insert any gate delay in the data path. This results in reduced performance penalty. However, the routing of the two system clocks (MCK and SCK) and the test clock (TCK) requires extreme care. Relative skews between these clock signals should be carefully controlled for the correct operation in both normal and scan modes.

14.2.5 Design Automation

The full-scan design is considered the *best* DFT discipline. It can be completely automated using commercially available design tools. Over the years, it has gained wide-spread acceptability in system design environments.

Figure 14.10 shows a typical scenario of scan design. DFT practically impacts every aspect of design. First, scan design audits are applied during the design phase. Audit programs analyze the topology of the circuit. For example, by simply tracing the clock signal through the netlist, one can easily find any violations of rules R-3 and R-4 of Section 14.2.1. Algorithms for design rule checking have been described by Bhavsar [78]. In practice, the rules of Section 14.2.1 may be augmented to suit the needs of specific design environments. For example, Godoy et al. [255] describe audit tools developed at IBM and Agrawal et al. [34] discuss those at Bell Labs. Some of today's automatic synthesis systems can be restricted to make only the scan-compatible design choices so that no post-synthesis auditing is necessary. The *Test Compiler* system developed by Synopsys, Inc., is an example.

In manual and semi-automatic design environments, the importance of audits is far greater. In fact, post-layout audits that check for the integrity of scan chains, or provide the information about inversions in the scan chains, have been used [34].

Once the design is completed in the form of a verified *netlist*, design and test activities can proceed in parallel. Automatic scan implementation consists of two phases. First, all flip-flops are replaced by corresponding scan versions, which are generally available in standard-cell libraries. If the SFF module is not available in the library, then multiplexers can be added to all flip-flops in the netlist.

The second phase in the scan implementation is to connect flip-flops into shift register chains. Chip layout programs optimally place flip-flops to minimize the

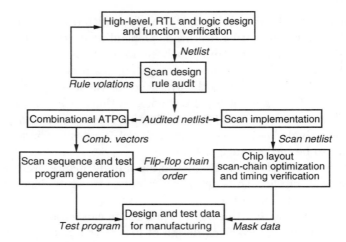

Figure 14.10: A flow-chart of automated scan design.

routing area and delay. As discussed in the previous section, a good layout strategy for scan chips is to first let the program optimally place the modules without any scan wiring. This placement relies on the critical net information generated from the functional timing requirement [208]. Once placed, the flip-flops are chained using some suitable, e.g., closest neighbor, criteria. In routing optimization, again, functionally critical nets are given precedence over scan connections. These strategies tend to reduce the area and timing penalties of scan [34].

Independent of the physical design, a combinational circuit netlist is generated by removing flip-flops and clocks from the *audited netlist*. Flip-flop output signals appear as *pseudo-primary inputs* (PPIs) and flip-flop input signals appear as *pseudo-primary outputs* (PPOs) in the combinational circuit. A combinational ATPG program is used to generate test vectors for all single stuck-at faults in this circuit.

Next, the combinational vectors and the flip-flop chain order information obtained from the layout are used to generate scan-in and scan-out sequences as described in Section 14.2.2. Certain models of *automatic test equipment* (ATE) provide special support for scan testing. In those cases, combinational vectors and the scan register topology may be directly incorporated in the test program.

A careful examination of the methodology of Figure 14.10 shows a minimal impact on the design flow. This is because all test related activities run parallel to design activities. Combinational ATPG often provides 100% or close to 100% fault coverage. Besides the CAD support, there are other reasons for the acceptance of scan. First, a clearly-defined structured approach decouples the functions of design verification and testing. This is also the reason for the acceptance of scan in automatic synthesis systems. Second, scan tests have been found to be useful in diagnosing and repairing electronic assemblies. However, there is a certain segment of digital chips where full-scan is found to be too expensive, either in hardware overhead or in delay penalty or in test time. In addition, there are cases where functional requirements force the designer to violate the rules of scan design. For such chips, which usually fall in the *application specific integrated circuit* (ASIC) category, partial-scan design has gained popularity.

14.2.6 Physical Design and Timing Verification of Scan

The physical design of a scan chain requires timing verification because its integrity is crucial to the application of scan tests. Several problems should be checked:

1. *Very small delay in scan path.* Because there is no logic gate in the scan path, signal propagation between two consecutive flip-flops of the scan registers may be very quick. A comparatively larger delay (skew) of the clock signal at the second flip-flop can produce a race condition.

2. *Large delay in scan path.* Because the functional routing gets precedence over scan routing in the physical design, some parts of the scan path can be slow. The timing of the scan path should be analyzed to determine a proper clock frequency used for the scan operation. Sometimes, the scan clock is run slower than the rated clock of the chip for similar reasons.

3. *Dynamic multiplexers.* Scan multiplexers can be economically implemented with dynamic logic (transmission gates.) However, a potential skew between the TC (test control) and $\overline{TC}$ signals can create a temporaty short circuit between the two data inputs of the multiplexer. If the the two input signals are generated by flip-flops that are in different states, such a short can change the state of one of them. A static design of a multiplexer requires more transistors, but does not create a short due to the skew. It should be preferred. Alternatively, the multiplexer should be integrated within the flip-flop cell and carefully analyzed for any signal delay problems.

4. *Power dissipation during scan.* Scan operation produces many changes at the inputs of the combinational logic. These can cause significantly higher power dissipation than the power rating of the device. Such activity can be analyzed by a power analysis tool like PowerMill [198]. Both average and peak power consumption should be controlled. Average power that is responsible for heating of the chip may be reduced by slowing down the scan clock. Increased peak power can cause a drop in the supply voltage and create noise problems in the chip. Its reduction may require redesign of test vectors.

14.3 Partial-Scan Design

Cheng and Agrawal [154] invoked an observation by the early researchers, which says that feedback in sequential logic may cause test problems. In a synchronous circuit, all feedback paths contain clocked flip-flops. While such a design is functionally acceptable, setting flip-flops to specific states can take long sequences of vectors. It is the generation of such sequences that makes ATPG difficult.

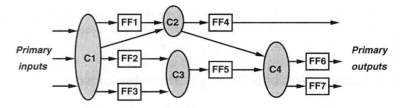

(a) A feedback-free sequential circuit.

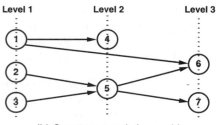

(b) Structure graph (s-graph).

Figure 14.11: A feedback-free sequential circuit and its s-graph.

The *structure graph* (s-graph) of a sequential circuit clearly shows the difficulty caused by feedback. Each flip-flop is represented by a vertex in the s-graph. A directed edge from vertex i to vertex j means that a combinational path from flip-flop i to flip-flop j exists. Figure 14.11 illustrates the construction of the s-graph for a feedback-free circuit. Shaded regions marked as C1, C2, C3, and C4 are combinational logic blocks. Flip-flops FF1 through FF7 are represented by vertices 1 through 7 in Figure 14.11(b). Primary inputs and primary outputs are not represented in the s-graph. Since the circuit has no feedback, its s-graph is a *directed acyclic graph* (DAG [41, 363].)

The DAG of Figure 14.11(b) can be levelized. Vertices without any incoming edge are placed in level 1. For any other vertex, the level is one higher than the level of the highest level vertex that feeds into it. For example, vertex 6 is fed by 1 and 5, of which vertex 5 belongs to the highest level of 2. Thus, vertex 6 has a level 3. The maximum level in the s-graph (3 in this example) is called the *sequential depth*. Notice that all flip-flops corresponding to level 1 vertices can be directly controlled by a single vector applied at primary inputs. The control of a level 2 flip-flop requires that some flip-flops in level 1 be controlled, and it will therefore require the application of two vectors at primary inputs. Thus, a sequence of vectors that controls all flip-flops must be as long as the sequential depth. In general, as the sequential depth increases the length of the test sequence to detect a fault increases proportionately. Also, a sequential ATPG program takes more time to produce a test sequence for a target fault. The sequential depth of a purely combinational circuit is 0. The reader should refer to Chapter 8 for sequential ATPG.

Next, consider the circuit of Figure 14.12(a), which is similar to that of Figure 14.11(a) with added feedback paths from FF5 and FF7. Notice that the circuit

14.3 Partial-Scan Design

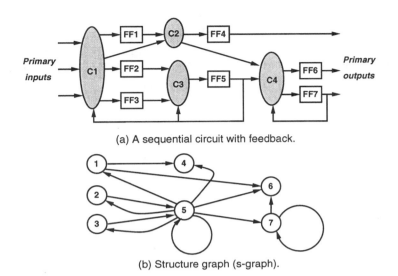

(a) A sequential circuit with feedback.

(b) Structure graph (s-graph).

Figure 14.12: A sequential circuit with feedback and its s-graph.

does not have any combinational feedback and hence is a proper synchronous sequential circuit. The corresponding s-graph is shown in Figure 14.12(b). This is no longer a DAG as it contains several cycles. Because of cycles, it is not possible to levelize the graph. Hence, we cannot talk about a sequential depth or bound the length of sequences to control flip-flops. In general, the test length and ATPG run time for such circuits can be quite large (see Chapter 8.)

The complexity of a circuit with feedback can be reduced by scanning a selected set of flip-flops. Since a scan flip-flop is controllable and observable via a test mode, its output is treated like a primary input and its input is treated like a primary output. Thus, vertices corresponding to scan flip-flops are deleted from the s-graph. For the circuit of Figure 14.12, if we scan FF5 and FF7, then vertices 5 and 7 will be deleted and the s-graph becomes acyclic with a sequential depth of 2. Finding the smallest set of vertices is known as the *minimum feedback vertex set* (MFVS) problem [41]. Since the problem has an exponential complexity, an exact solution is not advisable. Many practical heuristics are available in the literature [24].

Cheng and Agrawal [154] recommend that at least all large cycles be broken. They suggest that if the number of self-loops (cycles with a single vertex) is large, those may not be broken to keep the scan overhead low. They also point out that the clock signal of scan flip-flops must be separate from the clock of non-scan flip-flops. This is necessary since the circuit remains sequential in the scan mode and the states of all non-scan flip-flops must be held unchanged while scan sequences are being shifted in and out of the scan flip-flops. They report that a sequential ATPG program can achieve a fault coverage in excess of 95% when about 25 to 50% of the flip-flops are scanned.

Example 14.2 *Full and partial-scan designs of s5378 circuit. This benchmark*

Table 14.1: Full and partial-scan designs of s5378 circuit.

s5378	Original	Full-scan	Partial-scan
Number of gates	2,781	3,497	2,901
Number of non-scan flip-flops	179	0	149
Number of scan flip-flops (n_{sff})	0	179	30
Gate overhead (Equation 14.2)	0.0%	25.75%	4.31%
Number of faults	4,603	4,603	4,603
PI/PO for ATPG	35/49	214/228	65/79
Fault coverage (Equation 7.12)	70.0%	99.1%	93.7%
Fault efficiency (Equation 7.13)	70.9%	100.0%	99.5%
Test generation time (SUN Ultra II, 200MHz)	5,533 s	5 s	727 s
Number of ATPG vectors	414	585	1,117
Test sequence length (Equation 14.1)	414	104,719	33,604

circuit [99] has 2,781 logic gates and 179 flip-flops. Table 14.1 gives the results of test generation for three versions of the circuit. For partial scan, a set of 30 flip-flops was selected using a minimum feedback vertex set (MFVS) program [129]. We clearly notice that the gate overhead is significantly lower for partial-scan. Both types of scan designs improve the fault coverage and fault efficiency to adequately high levels. However, two problems remain. First, the scan test sequences are very long due to the scan-in and scan-out operations. The sequence length is significantly higher for full-scan than for partial-scan. Second, while the test generation time is very small for the full-scan design, it is still two orders of magnitude higher for partial-scan. This is because we must use sequential ATPG for partial-scan as opposed to combinational ATPG used for full-scan. All test generation results in Table 14.1 were obtained by Bell Labs' GENTEST sequential ATPG program [72].

As the above example shows, one disadvantage of the partial scan method is that it requires the use of a sequential ATPG program. It is possible to design partial scan using combinational ATPG. We can derive combinational tests assuming that all flip-flop outputs are in *unknown* or *don't care* states and that all flip-flop inputs are unobservable. In general, this technique will provide tests for very few faults in a sequential circuit. It is, however, possible to make simple modifications in a combinational ATPG program such that it will produce a single test vector for a target fault and use a minimum number of flip-flops. To apply such tests, the selected flip-flops will then be scanned. Such methods [32] rely on a good set of functional tests, which can achieve about 75% fault coverage. For the remaining faults, a modified combinational ATPG program generates tests and selects the scan flip-flops. Almost 100% fault coverage is attainable, though as many as 70 to 80% of flip-flops are scanned in some reported cases [32]. The use of sequential ATPG is also reduced in a technique in which all non-scan flip-flops are shorted [468]. This method also requires a higher scan percentage because no self-loops are permitted.

14.4 Variations of Scan

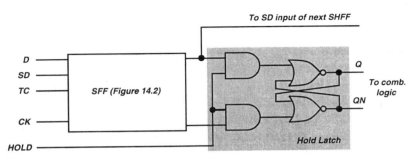

Figure 14.13: Scan-hold flip-flop (SHFF.)

Other ingenious partial-scan methodologies based on combinational ATPG have been developed by Gupta et al. [274], and Kunzmann and Wunderlich [379].

14.4 Variations of Scan

There have been many variations of scan and we will discuss two of those. The first, known as the *scan-hold flip-flop*, is useful in delay testing and the second, called *random-access scan*, reduces the scan test time.

Scan-Hold Flip-Flop (SHFF.) A scan flip-flop with an additional hold latch is shown in Figure 14.13. The idea of adding the hold capability to SFF was proposed by Das Gupta et al. [184]. In Figure 14.13, a hold latch is cascaded with the SFF of Figure 14.2. The hold latch retains its state when the control signal $HOLD$ is held low. For $HOLD = 1$, the hold latch becomes transparent. When a circuit is designed with SHFFs, an extra input pin supplies the $HOLD$ signal to all flip-flops. The scan register is formed by chaining only the SFFs.

In the normal mode, $TC = HOLD = 1$. In the scan mode, $TC = 1$ and $HOLD = 0$. This isolates the combinational logic from the scan register activity. The state inputs of combinational logic driven by the hold latch remain *frozen* at their pre-scan values. Once the desired values are scanned in, a 0 to 1 change on $HOLD$ applies the new state variables to the combinational logic.

The SHFF was originally designed to isolate the scan and non-scan portions of a circuit. This is accomplished by placing SHFFs on the boundary between the two portions. Thus, while the scan logic is tested, the inputs to the non-scan logic (which may be asynchronous) remain fixed. Wagner and Willams [703] propose a similar application in mixed-signal circuits. Another application of scan, which has become popular recently, is in delay testing (see Subsection 12.4.2.) Delay testing requires the application of vector-pairs to a combinational logic problem. The normal scan structure (with SFFs) places severe restrictions on the vector-pairs that can be produced.

The use of the hold latch converts the delay testing problem completely into a combinational logic problem. Efficient algorithms are available [371] for deriving vector-pairs to detect delay faults in combinational logic. The procedure for the application of a vector-pair (V_1, V_2) is as follows:

1. Set $HOLD = 0$ and $TC = 0$, and scan the state variable bits of V_1 into the scan register using the clock CK.

2. Set $TC = 1$.

3. Set $HOLD = 1$ and apply the primary input portion of V_1. Thus the entire vector V_1 appears at the inputs of the combinational logic.

4. Change $HOLD$ to 0.

5. Repeat Steps 1, 2, and 3 for V_2. This produces a $V_1 \rightarrow V_2$ transition at the inputs of the combinational logic.

6. Change $HOLD$ to 0, and capture the output of the combinational logic in SFF by applying the clock CK.

7. Set $TC = 0$ and apply clocks to scan out the contents of flip-flops. This completes the application of one vector-pair delay test.

In general, delay tests will contain several vector-pairs. The scanout following the second vector of one pair can be overlapped with the scanin of the first vector of the next pair.

The hardware overhead of a SHFF consists of the hold latch, which increases the size of the scan flip-flops by about 30%. One extra input pin is needed for the $HOLD$ signal, which must be routed to all flip-flops. The technique also has a performance overhead. Although hold latches remain transparent in the normal operation ($HOLD = 1$) of the circuit, they add 1 to 2 gate delays in signal paths. Despite these penalties, the SHFF offers a feasible DFT discipline for delay testing.

Random-Access Scan (RAS.) In this technique, the scan function is implemented like a random-access memory (RAM) [7, 721, 724]. A general architecture is given in Figure 14.14(a). Here all flip-flops form a RAM in the scan mode. In general, a subset of flip-flops can be included in the RAM if partial-scan is desired. In the normal mode ($TC = 1$), all flip-flops receive data from the combinational logic under the control of the clock CK. Flip-flop outputs directly feed into the combinational logic. A typical design of a cell in this RAM is given in Figure 14.14(b). In the scan mode, this scheme allows reading or writing of any selected flip-flop. The flip-flop address, which may contain $log_2 n_{ff}$ bits when there are n_{ff} flip-flops in the RAM, is serially loaded into an *address shift register* (ASR) using an address clock ACK. The address decoder now produces the select signal $SEL = 1$ for the addressed flip-flop. The SEL signals to all other flip-flops remain 0. The $SCANOUT$ signals of all flip-flops are tied together to the $SCANOUT$ pin. Thus, the content of the addressed flip-flop appears at this output. An advantage of this method is that any flip-flop can be observed even when the circuit is in the normal mode ($TC = 1$.)

For scanning data into a flip-flop, the scan mode ($TC = 0$) is used. Assuming that the select signal ($SEL = 1$) has been generated for the addressed flip-flop,

14.5 Summary

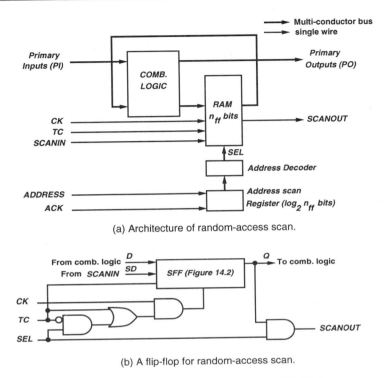

Figure 14.14: Random-access scan (RAS.)

the state of SD is latched in that flip-flop by the clock CK. The SD inputs of all flip-flops in RAM are tied together to the $SCANIN$ pin. However, the $SCANIN$ signal is not latched in the unaddressed flip-flops because $SEL = 0$ inhibits the clock CK.

The ability to control and observe individual flip-flops has the advantage of reducing the length of the test sequence. RAS has been used in practice [53], but has not gained popularity perhaps due to a large overhead. First the *scan flip-flop* (SFF) requires additional logic shown in Figure 14.14(b). Then, an address decoder and *address shift register* (ASR) are added. The routing of SEL signals, if flip-flops are distributed over a VLSI chip, can also take a significant amount of area.

Despite its drawbacks, the RAS technique may have benefits in partial-scan where only a few flip-flops need scanning, or in delay testing where a change of single flip-flop states can provide good tests [371].

14.5 Summary

Scan design has been the backbone of design for testability in the industry for about three decades. It continues to be the most popular technique. It is often supported by design automation tools that can insert scan logic into a predesigned circuit and then generate tests. Recent variations of the technique are partial-scan and boundary scan. In general, partial-scan requires the use of sequential ATPG. However, there are partial-scan techniques that use combinational ATPG. Such techniques either assume unknown states for unscanned flip-flops [32], or make the circuit "completely" acyclic by removing even the self-loops [274, 379]. In either case, the scan percentage can be 50% or higher.

A scan overhead of about 10 to 15% is generally considered acceptable in ASIC chips. In addition, there may be a speed penalty, which can be less than 5%. These costs are justified by the high quality of devices obtained in a short design time.

Problems

14.1 *Importance of initialization.* The circuit in Figure 14.15 produces a toggling output when the input is 1. Input 0 holds the output at 0. Assuming the state of the flip-flop (FF) to be unknown, derive a test for the s-a-1 fault shown. Explain why this fault is potentially detectable. What design change is needed to make the test deterministic?

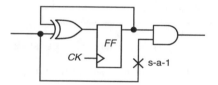

Figure 14.15: A toggle circuit.

14.2 *MUX design.* Design a two-to-one multiplexer using three Boolean gates.

14.3 *Multiple scan chains.* Rederive Equations 14.1 and 14.2 when flip-flops are distributed in n_{chain} scan chains of equal length. Assume that only one extra pin is available for test, but the number of primary input and output data pins is not limited.

14.4 *Scan tests.* Suppose that your chip has 100,000 gates and 2,000 flip-flops. A combinational ATPG program produced 500 vectors to fully test the logic. According to Section 14.2.2, a single scan-chain design will require about 10^6 clock cycles for testing. Find the scan test length if 20 scan chains are implemented. Given that the circuit has 20 primary input and 20 primary output data pins, and only one extra pin can be added for test, how much more gate overhead will be needed for the new design?

14.5 *Scan-testable counter circuit.* Repeat Example 14.1 for a five-bit counter circuit that counts from 0 to 4 and then resets to 0. A single output becomes 1 only when the count is 4.

Problems

14.6 *Full-scan design.* Convert the modulo-3 counter of the previous problem into a full-scan circuit by inserting multiplexers at the inputs of flip-flops. Derive tests for the combinational logic and derive scan sequences. Use a fault simulator to simulate the scan sequences in the sequential mode. Are all faults detected? If not, what can be done to detect the remaining faults?

14.7 *Scan overhead.* Rederive Equation 14.5 assuming that scan routing requires k tracks per routing channel.

14.8 *Partial-scan.* Draw the s-graph for the circuit in Figure 14.16. Identify the flip-flop to be scanned such that all cycles are broken.

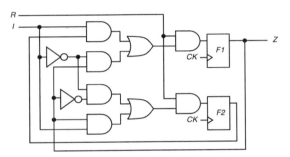

Figure 14.16: A circuit for partial-scan.

14.9 *Partial-scan.* Redraw the circuit of Figure 14.16 after removing the flip-flop selected for scan in the previous problem. Make the input signal of the removed flip-flop a new PO, and its output signal a new PI. Use a sequential ATPG to generate tests for this circuit. Convert those tests into partial-scan sequences and simulate using a fault simulator. Are all faults detected?

14.10 *Partial-scan.* Consider a circuit with no self-loops and an s-graph that is a *complete graph*. In a complete graph, a directed edge from vertex v_i to vertex v_j exists for all i and j. Show that to eliminate all cycles, we must scan all but one flip-flops.

14.11 *Partial-scan.* Consider a circuit with no self-loops and an s-graph that is a *strongly connected graph* (SCC.) In an SCC, a directed path exists from any vertex to any other vertex. Find a simple heuristic to select a small set of vertices whose deletion will make the graph free from cycles.

14.12 *Partial-scan overhead.* Rederive Equation 14.5 for a partial-scan chip in which $p\%$ of flip-flops are scanned.

14.13 *Scan-hold flip-flops.* Rederive Equation 14.2 for a scan design where scan-hold flip-flops (SHFF) are used. For a circuit with 100,000 gates and 2,000 flip-flops, connected in a single scan chain, how much more is the overhead compared to the 8% for the SFF design of Section 14.2.4?

14.14 *Random-access scan.* Design random-access scan for a circuit with 1,000 flip-flops. The circuit has 20 primary input and 10 primary output data pins. The main design consideration is to have fast access to memory elements. *Hint: Use word-addressable memory.*

Chapter 15

BUILT-IN SELF-TEST

> "In a digital instrument designed for troubleshooting by signature analysis, this method can find the components responsible for well over 99% of all failures, even intermittent ones, without removing circuit boards from the instrument."
> — Robert A. Frohwerk,
> in "Signature Analysis: A New Digital Field Service Method" [228].

Frohwerk ushered in a new era of determining the correctness of a circuit by examining a *signature*, which is some statistical property of the circuit. He applied the work of Peterson and Weldon [521] and Golomb [264] on error-correcting codes and shift register sequences to *built-in self-testing* (BIST) of ICs. Frohwerk also provided the first analysis for transition count testing and BIST to determine the probability of a signature indicating the circuit was good, when it fact it was faulty.

A digital system is tested and diagnosed during its lifetime on numerous occasions. Test and diagnosis must be quick and have very high fault coverage. One way to ensure this is to specify test as one of the system functions, so it becomes self-test. At the highest level of systems test, the testing function is frequently implemented in software. Many digital systems designed at AT&T circa 1987 had self-test, usually implemented in software [33]. Its most common use was in maintenance and repair diagnostics. Although this approach provided flexibility, it also had disadvantages. The fault coverage and the diagnostic resolution of those software-implemented tests were not as high as desired. The diagnostic resolution may be poor because the software must test parts that are difficult to test, and therefore it may not effectively determine which part is at fault. Also, software tests can be long, slow, and expensive to develop. Therefore, it becomes increasingly attractive to build the self-test function into the hardware [36, 37]. It is also most effective to consider testing as early in the design cycle as possible. Otherwise, costly prototyping turns (cycles of redesign and refabrication of the prototype) result, and these lead to schedule slippages for product introduction.

Systems designed without an integrated test strategy (covering all levels from the entire system to components) can best be described as *chip-wise and system-foolish*. With properly designed BIST, the cost of added test hardware will be more than balanced by the benefits in terms of reliability and the reduced maintenance cost [33].

Thus, these benefits, rather than the cost, can be passed on to the customers. The savings from BIST include reduced test generation effort at all levels, reduced test effort at chip through system levels, improved system-level maintenance and repair, and improved component repair. Gordon and Nadig [266] described the economic impact of signature analysis and BIST on the first systems that used BIST: two Hewlett-Packard digital voltmeters, one of them called the HP 3455A. Development time and costs rose roughly 1%. There was a 1% increase in parts cost due to added jumpers and extra ROM space required in the electronics for signature analysis, but total factory costs dropped, because of a 5% decrease in other materials costs. For example, it was no longer necessary to divide the product into small, replaceable modules. Also, BIST reduced service-module inventory at the factory, and decreased administrative and handling costs for failing units returned to the factory.

15.1 The Economic Case for BIST

These are some chip-level testability problems of the late 1990s [532]:

1. There is an extremely high and still increasing logic-to-pin ratio on the chip. This increasingly makes it harder to accurately observe signals on the device, which is essential for testing.

2. VLSI devices are increasingly dense and faster with sub-micron feature sizes.

3. There are increasingly long test-pattern generation and test application times.

4. Prohibitive amounts of test data must be stored in the *automatic test equipment* (ATE).

5. There is increasing difficulty in performing at-speed (rated clock) testing using external ATE. For clock rates approaching $1\ GHz$, at-speed testing with an ATE is very expensive due to pin inductance and high tester pin costs.

6. Designers are unfamiliar with the gate-level structure of their designs, since logic is now automatically synthesized from the *VHDL* or *Verilog* hardware description languages. This compounds the problem of testability insertion.

7. There is a lack of skilled test engineers.

Complexity. One unfortunate property of large VLSI circuits is that testing cannot be easily partitioned. Consider two cascaded devices. There is frequently no simple way to obtain tests for the complete system from tests for the individual parts. In fact, even though each part is fully testable and has a test set that gives 100% stuck-fault coverage, the cascaded connection of the two parts will often have untestable and redundant hardware and much lower stuck-fault coverage. In other words, testing is a global problem. It is well known that there is no simple way to create tests for an entire *printed circuit board* (PCB) from tests for the chips on the board. For

15.1 The Economic Case for BIST 491

design and test development effort, BIST provides a way to hierarchically decompose the electronic system-under-test, so this allows sub-assemblies to be first run through a BIST cycle, and if there are no faults, then boards in the system are run through a BIST cycle. Finally, if there are no board faults, then the entire system can be run through a BIST cycle. As an example, consider a system containing boards, which in turn contain chips. For a chip test, the system sends a control signal to the PCB, which then activates self-test on the desired chip, and sends the test result back to the system. BIST efficiently tests embedded components and interconnect, thus reducing the burden on system-level test, which now only needs to verify the synergy among the functional components [36]. When faults occur, the BIST hardware should be designed to indicate via an error signal or bus which sub-assembly is faulty. This greatly reduces repair costs.

Quality. Typical quality requirements are 98% single stuck-fault coverage or 100% interconnect fault coverage. The *reject ratio* is the percentage of faulty parts in the number of parts passing a test. The goal of testing at many companies is a low reject ratio, e.g. 1 in 10,000, at reasonable cost. The Motorola *six sigma* project has a goal of lowering this number to 1 in 100,000 [532]. In huge systems, this is attainable only through *design for testability* (DFT), and BIST is the preferred form of DFT.

Test Generation Problems. It is difficult to carry a test stimulus involving hundreds of chip inputs through many layers of circuitry to the chip-under-test, and then convey the test result back through the many circuit layers to an observable point. BIST localizes testing, which eliminates these problems.

Test Application Problems. In the past, *in-circuit testing* (ICT) [69] used a *bed-of-nails* fixture customized for the PCB-under-test. The bed-of-nails tester applied stimuli to the solder balls on the back of the PCB where the component leads were soldered to the PCB. Power was applied only to the component under test – all others in the PCB were left unpowered. It was effective for chip diagnosis and board wiring tests. However, ICT is not effective unless the PCB is removed from the system, so it is not helpful in system-level diagnosis. Also, *surface-mount technology* (SMT) components are often mounted densely on both sides of the board, and the PCB wire pitch is also too small for accurate probing of the back of the board by the bed-of-nails tester. Therefore, ICT is no longer a solution. BIST, however, solves these problems by eliminating expensive ATE, and BIST also lets us use the same tests and test circuits that are used at the system level [36, 37]. With BIST, there can be virtually unlimited circuit access via test points designed into the circuit through scan chains, resulting in an *electronic bed-of-nails* [33]. Another advantage of BIST is that the testing capability grows with the VLSI technology, whereas with external testing, the test capability always lags behind the VLSI technology capability. Logic gates and transistors are relatively cheap compared to the labor needed to develop test programs, the cost of automatic test equipment, and the cost of real time for the tests to be run on production chips with ATE [33].

An additional benefit of BIST is lower test development cost, because BIST can be automatically added to a circuit with a CAD tool. Also, BIST generally provides a 90 to 95% fault coverage, and even 99% in exceptional cases [33]. The test engineer need no longer worry about backdriving problems of in-circuit test (where electrical stimuli provided to the middle of the circuitry damage outputs of logic gates), or how much memory is available in the ATE.

Table 15.1: Built-in self-testing costs.

Level	Design & test	Fabri-cation	Prod. test	Mainte-nance test	Diagnosis & repair	Service interruption
CHIPS	+/−	+	−			
BOARDS	+/−	+	−		−	
SYSTEMS	+/−	+	−	−	−	−

+ cost increase; − cost reduction; +/− cost increase ≈ cost saving (neutral or slight saving)

Table 15.1 [36, 37] shows the relative BIST costs at the chip, board, and system levels of packaging. BIST always requires added circuit hardware for a *test controller* to operate the testing process, *design for testability* hardware in the circuit to improve fault coverages during BIST, a hardware *pattern generator* to generate test-patterns algorithmically during testing, and some form of hardware *response compacter* to compact the circuit response during testing. We see an increase in fabrication costs at all three levels of circuit packaging. The BIST cost is frequently measured in terms of the added chip/board area required for the BIST hardware. The relative costs of added logic gates are declining, because hardware continues to become cheaper, but the relative costs of added long wires for test mode control are not really decreasing. This cost can also include added circuit delay, due to the extra device loads and delays from the test hardware. This may require a slight increase in the clock rate, and additional electrical adjustments to the design. Also, the test hardware can consume extra power, which is an additional cost. Since the BIST circuitry uses chip area, a final BIST cost is a decrease in the chip yield and chip reliability, due to the increased chip area [33]. BIST feasibility for a system must be evaluated using benefit-cost analysis, in the context of assessing total life cycle costs. Table 15.2 [33] gives more detailed metrics for evaluating BIST costs.

15.1.1 Chip/Board Area Cost vs. Tester Cost

At all three levels, BIST reduces testing costs. In order to understand why, consider the example of a 1 GHz microprocessor on a chip with 800 pins. For reliable stuck-fault and limited transition-delay fault testing, we should conduct the test at the rated clock speed. This forces us to use the Advantest Model T6682 1 GHz ATE, which can sample circuit outputs at this rate. The tester costs 800 *pins* ×

15.1 The Economic Case for BIST

Table 15.2: Metrics for evaluating BIST.

Fault characteristics	Fault classes tested –
	   Single stuck-at faults in functional circuitry
	   Sequential faults in functional circuitry
	   Delay faults
	   Single stuck-at faults in BIST circuitry
	 Fault Coverage –
	   % of faults detected in functional circuitry
	   % of faults detected in the BIST circuitry

Associated costs	Area overhead – Additional active area and interconnect Pin overhead – Additional pins. At least 1 pin is required to control whether BIST operates or not. One can design circuits that put the chip into BIST mode without using an extra pin, by applying a voltage level not normally used (e.g., 12 V instead of 0 or 5 V) to a pin. Additional input (output) pins for BIST are obtained by pin multiplexing, where input (output) pins are multiplexed into the BIST circuit when in BIST mode. The MUX adds a slight performance penalty. Performance overhead – added path delays due to BIST Yield loss – due to increased area or more chips in the system Reliability reduction – due to increased area Increased design effort and time Testability of the BIST hardware. The BIST hardware complexity increases when the BIST hardware is made testable.
Associated benefits	Reduced cost of testing and maintenance Lower test generation cost Reduced storage and maintenance of test patterns Simpler and less costly ATE Ability to test many units cost-effectively in parallel Shorter test application times Ability to test at system speed
Other characteristics	Degree to which BIST structure is function independent Diagnostic resolution Effect of engineering changes on BIST structure

$6,000$ *per pin* $= \$ 4,800,000$, but there is no chip area cost due to testing, because we do not use on-chip BIST hardware. Therefore, there is a huge initial capital cost for the ATE, but there is no recurring chip area cost on each chip for test hardware. If, instead, we provide BIST hardware, then the need for a very high-speed ATE is eliminated, except to test the wires from the circuit pins to the *Input MUX*, and from the circuit outputs to the output pins. The number of tests for that is very short, say perhaps 7 or 8 patterns and measurements per pin, and the cost of this can be safely ignored in this analysis. Therefore, with BIST doing all stuck-fault and transition-delay fault testing, we need a 1 GHz signal oscillator to clock the chip, and we need the ATE only to provide DC command signals to tell the microprocessor to perform BIST. Finally, we need an ATE to read out the success or failure DC signal for BIST from a circuit pin. In this case, we can use an inexpensive, 20 MHz ATE that costs roughly \$391 per pin, so our cost is 800 *pins* $\times$ \$391 *per pin* $= \$312,800$, a savings of \$6,887,200. This example is hardly far fetched. On-chip clock rates are expected to rise above 1 GHz, and at present, no ATE exists to test a circuit above 1 GHz.

For design and test development, BIST significantly reduces the costs of *automatic test-pattern generation* (ATPG), and reduces the likelihood of disastrous product introduction delays because a fully-designed system cannot be tested. Such a delay has occurred in the Intel Merced^{TM} project due to unexpected delays in inserting testability hardware into the chip, and fabrication line problems. There is a slight cost increase due to BIST in design and test development, because of the added time required to design and add pattern generators, response compacters, and testability hardware. However, our experience is that this is less costly than test development with ATPG.

15.1.2 Chip/Board Area Cost vs. System Downtime Cost

The true economic benefits of BIST show up in the last three columns of the Table 15.1. Without BIST, maintenance test requires the presence of an expensive ATE at the site of the failing system, and this is a significant cost. With BIST, there is no need for an ATE, so this reduces system test cost. For boards and systems, BIST drastically reduces the diagnosis and repair cost, by quickly determining and indicating which sub-assembly or component is faulty, without the extensive labor and equipment normally required. This great reduction in diagnosis and repair time naturally leads to a major shortening in service interruption, particularly at the system level. An example of an application where service interruptions must be minimal is the # 5 ESS telephone exchange designed by Lucent Technologies. The exchange is designed to have, at most, a fraction of a second down time per year, because the lost revenue to an operating phone company when long distance calls cannot be made is quite severe. Another example is the credit card operations of American Express, which must be completed in a timely way so that the company knows which of its customers should be allowed additional credit and which should not. A final example is the Quotron system of the New York Stock Exchange for electronic trading of stocks. Down time of this system can cost billions of dollars per hour in lost trading opportunities. When viewed in this light, the added costs of BIST hardware may be very minor compared with the severe cost of service interruption. A further benefit of BIST is its cost reduction for diagnosis and repair. Whenever any consumer home appliance breaks, the minimal cost just to have a repairman come out and look at the appliance is usually $100, and the total repair cost may exceed $175. It may be possible to justify BIST on a benefit-cost basis in personal computers and other home electronic appliances costing more than $1,500.

The reader may recall the rule of ten in testing costs from Chapter 3. If it costs $0.50 to detect a fault at the chip level, it will cost $5 to detect the same fault at the board level, $50 at the system test level, and $500 for a field repair for the same fault (because of the expense of sending out a serviceman to a customer site.) Therefore, test cost reductions due to BIST in Table 15.1 in the board and system columns are particularly important. It is most economical to detect problems early.

15.2 Random Logic BIST

15.2.1 Definitions

- BILBO – *Built-In Logic Block Observer.* This is a bank of circuit flip-flops with added testing hardware, which can be configured to make the flip-flops behave like a scan chain, an *linear feedback shift register* (LFSR) pattern generator, an LFSR-based response compacter, or merely as D flip-flops.

- Concurrent Testing – A testing process that detects faults during normal system operation.

- CUT – *Circuit-Under-Test*

- Exhaustive Testing – A BIST approach in which all 2^n possible patterns are applied to n circuit inputs.

- Irreducible Polynomial – A Boolean polynomial that cannot be factored.

- LFSR – *Linear Feedback Shift Register.* This is hardware that generates an exhaustive or pseudo-random pattern sequence of test patterns, and can also be used as a response compacter.

- Non-Concurrent Testing – A testing process that requires suspension of normal system operation to test for faults.

- Primitive Polynomial – A primitive Boolean polynomial $p(x)$ has the property that we can compute increasing powers of x *modulo* $p(x)$, and obtain all possible non-zero polynomials of degree less than $p(x)$. We compute the remainders of $x/p(x)$, $x^2/p(x)$, and so on. A primitive polynomial defines a mathematical number system that has the properties of a mathematical *field*.

- Pseudo-Exhaustive Testing – A BIST approach in which a circuit having n *primary inputs* (PIs) is broken into smaller, overlapping blocks, each with $< n$ inputs. Each of the smaller blocks is tested exhaustively.

- Pseudo-Random Testing – A BIST pattern generator that produces, via an algorithm, a subset of all possible tests that has most of the properties of randomly-generated tests. The random patterns must have a statistically high enough fault coverage to ensure a good test.

- TPG – Hardware *Test-Pattern Generator*

15.2.2 BIST Process

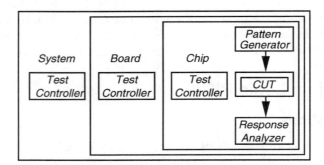

Figure 15.1: BIST hierarchy.

Figure 15.1 shows the BIST system hierarchy and all three levels of packaging mentioned earlier. The system has several PCBs, each of which, in turn, has multiple chips. The system *Test Controller* can activate self-test simultaneously on all PCBs. Each Test Controller on each PCB can activate self-test on all chips on the PCB. The Test Controller on a chip executes self-test for that chip, and then transmits the result to the PCB Test Controller, which accumulates test results from all chips on the board and sends the results to the system Test Controller. The system Test Controller uses all of these results to isolate faulty chips and boards [36, 37].

System diagnosis is effective only if the self-test procedures are thorough. For BIST, fault coverage is a major issue. Other issues are chip area overhead, its impact on chip yield, the cost of the additional chip pins required for test, the performance penalty in terms of added circuit delay, and extra power requirements. For BIST, the test engineer frequently, but not always, modifies the chip logic to make all latches and flip-flops controllable, perhaps by using the scan technique [36, 37].

BIST Implementations

Figure 15.2 shows typical BIST hardware in more detail. Note that the wires from PIs to the *Input MUX* and the wires from circuit outputs P to *primary outputs* (POs) cannot be tested by BIST. These wires, instead, require another testing method, such as an external ATE or JTAG Boundary Scan hardware. Figure 15.2 also shows how a comparator compares the signature produced by the data compacter with a reference signature stored in a ROM during BIST. This comparator and ROM hardware can frequently be implemented with a single logic gate with 32

15.2 Random Logic BIST

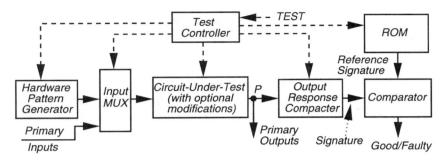

Figure 15.2: BIST process.

or fewer inputs. This is acceptable only when the comparison can occur at extremely low rates of circuit operation, since this logic gate is exceedingly slow.

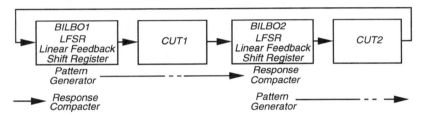

Figure 15.3: Multi-purpose registers in a BIST implementation.

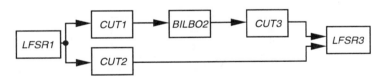

Figure 15.4: Complex BIST implementation.

Figure 15.3 shows a BIST implementation using *built-in logic block observers* (BILBOs) [364]. A BILBO is a bank of D flip-flops in the CUT that has test hardware added to make it behave in one of four modes:

- As ordinary D flip-flops.
- As a *linear feedback shift register* (LFSR) hardware pattern generator.
- As an LFSR configured to compact a circuit response.
- As a scan chain.

$BILBO1$ is configured as an LFSR pattern generator to test $CUT1$ in the circuit, while $BILBO2$ is configured as a response compacter to compact the responses of $CUT1$. During this process, the behavior of $CUT2$ is ignored. $BILBO2$ is configured as an LFSR pattern generator to test $CUT2$ in the circuit, while $BILBO1$ is configured as a response compacter to compact the responses of *CUT2*. During this second process, the behavior of *CUT1* is ignored. For the normal system function, both *BILBO1* and *BILBO2* are configured to behave as simple D flip-flops.

Figure 15.4 shows a more complicated BIST system. Here, *LFSR1* is used to simultaneously generate patterns to test *CUT1* and *CUT2*. *BILBO2* is configured as a response compacter for *CUT1*, while *LFSR3* is configured as a response compacter for *CUT2*. In this mode, inputs to *CUT3* must be held steady so that the outputs of *CUT3* remain stable. In the second test mode, *BILBO2* is configured as a pattern generator for *CUT3*, while *LFSR3* is configured as a response compacter for *CUT3*. The outputs of *CUT1* are ignored, and LFSR1 must be set so that the outputs of *CUT2* are held steady during this second mode. Finally, Figure 15.5 shows a bus-oriented BIST implementation. The *self-test control* broadcasts test-patterns to each of the CUTs over their common bus. The self-test control then awaits bus transactions from each CUT that indicate the CUT's response to the stimulus pattern broadcast over the bus. This allows for a certain amount of concurrency during self-test. In this mechanism, pattern generation for all of the CUTs can happen in parallel, but response compaction is serialized over the bus.

The strategy for overall test control is the most difficult part of BIST design [33]. Care must be taken so that the BIST circuitry, as well as the circuit-under-test, can be tested for stuck-at faults. The BIST circuitry that must operate correctly for BIST to work is referred to as the *hard-core*. Either the hard-core must be minimized, or it must be made testable.

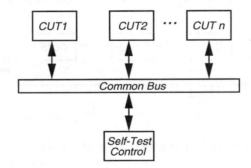

Figure 15.5: Bus-based BIST implementation.

15.2.3 BIST Pattern Generation

The following hardware pattern generation approaches have been used.

1. *ROM*. One method is to store a good test-pattern set (from an ATPG program) in a ROM on the chip, but this is prohibitively expensive in chip area, and will not be discussed further.

2. *LFSR*. Another method is to use a *linear feedback shift register* (LFSR) to generate pseudo-random tests. This frequently requires a sequence of 1 million

15.2 Random Logic BIST

or more tests to obtain high fault coverages, but the method uses very little hardware and is currently the preferred BIST pattern generation method.

3. *Binary Counters.* A binary counter can generate an exhaustive test sequence, but this can use too much test time if the number of inputs is huge. For example, with 64 inputs and the test-pattern generator clocked at 100 MHz, this takes 51,240,955.8 hours of test time to generate all 2^{64} patterns, which is impractical. Therefore, this type of pattern generator must be partitioned. Also, the binary counter requires more hardware than the typical LFSR pattern generator.

4. *Modified Counters.* Modified counters have also been successful as test-pattern generators, but they also require long test sequences.

5. *LFSR and ROM.* One of the most effective approaches is to use an LFSR as the primary test mode, and then generate test-patterns with an ATPG program for the faults that are missed by the LFSR sequence. These few additional test-patterns can either be stored in a small ROM on the chip for a second test epoch, they can be embedded in the output of the LFSR, or they can be embedded in a scan chain in order to augment the stuck-fault coverage to 100%.

6. *Cellular Automaton.* In this approach, each pattern generator cell has a few logic gates, a flip-flop, and connections only to neighboring gates. The cell is replicated to produce the cellular automaton.

Exhaustive Pattern Generation

Exhaustive testing methods are intended to show that:

1. Every intended circuit state exists, and

2. Every state transition works.

For an n-input circuit, one must test the circuit with all 2^n input vectors, which causes exhaustive testing to be impractical when $n > 20$. Figure 15.6 shows an exhaustive pattern generator implemented with a binary counter.

Example 15.1 Exhaustive Test. *In the Intel 80386, exhaustive test was used for three control PLAs and for one control ROM using multiple-input signature registers (MISRs) for response compacters.*

Pseudo-Exhaustive Pattern Generation

We discuss four methods of pseudo-exhaustive testing:

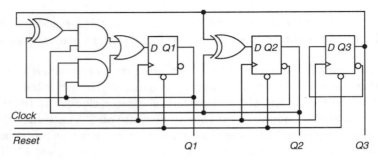

Figure 15.6: Exhaustive pattern generator.

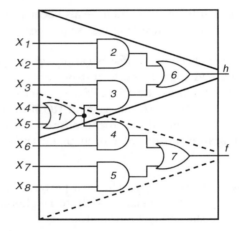

Figure 15.7: Backtracing for pseudo-exhaustive testing.

1. In Figure 15.7, exhaustive testing is made practical by *verification testing* (cone segmentation) [447]. We partition a large circuit into fanin cones by backtracing from each PO through the circuit to the input gates that influence that output. In fact, the fanin cones can often be tested in parallel to reduce test time. In this case PO h is influenced by inputs X_1, X_2, X_3, X_4, and X_5 while PO f is influenced by inputs X_4, X_5, X_6, X_7, and X_8. Therefore, we can design a pseudo-exhaustive test-pattern generator that in one mode stimulates all combinations of X_1, X_2, X_3, X_4, and X_5 while in a second mode it stimulates all combinations of X_4, X_5, X_6, X_7, and X_8. The benefit of pseudo-exhaustive testing is that the number of patterns is reduced from 256 to $2 \times 32 = 64$, which represents a testing cost reduction. However, observe that testing of the stuck-faults on the fanout branch from gate *1* to gate *4* will be missed in the first mode, but will instead be done in the second mode, so we will still obtain 100% stuck-fault coverage. Now, if we delete inputs X_4 and X_5 from the second testing epoch, in the interest of shortening the test pattern length by 24 patterns, then observe that this fanout branch fault will no longer be tested. This is because it still is not tested in the first test

15.2 Random Logic BIST

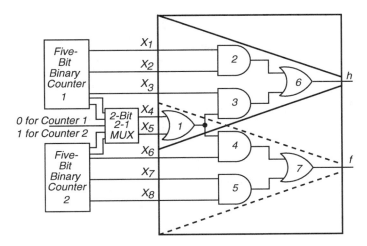

Figure 15.8: Pseudo-exhaustive pattern generator.

mode, and since the second test mode no longer controls inputs X_4 and X_5, it cannot be tested there, either. Finally, note that nearly any digital ATE can be configured to generate pseudo-exhaustive tests, apply them to a CUT, and observe the circuit responses. This is *ex situ* testing, in which the patterns are applied externally. BIST is a form of *in situ* testing, in which the patterns are applied internally by hardware in the circuit. However, ex situ testing lacks many of the economic benefits of BIST, such as a field-test capability and reduced tester cost.

2. A second method is *hardware partitioning* (physical segmentation) [448] in which we add extra circuit logic in order to divide the CUT into smaller subcircuits, each directly controllable and observable. Each of these is tested exhaustively.

3. A third method is *sensitized path segmentation* [136, 147, 448, 681], in which the circuit is partitioned so that sensitizing paths are set up from PIs to the partition inputs, and then from the partition outputs to the POs. Each partition is tested individually while the remaining partitions are simulated, so that non-controlling signals are set in the CUT to sensitize and propagate signals in the partition-under-test.

4. A final method is *partial hardware partitioning* [682], which combines *hardware partitioning* for observing signals in the partition-under-test, and *sensitized path segmentation* to control inputs to the partition-under-test.

Example 15.2 Pseudo-Exhaustive Pattern Generator. *Figure 15.8 shows a pseudo-exhaustive pattern generator using binary counters for the circuit of Figure 15.7. Pseudo-exhaustive testing is practical when each partitioned CUT element*

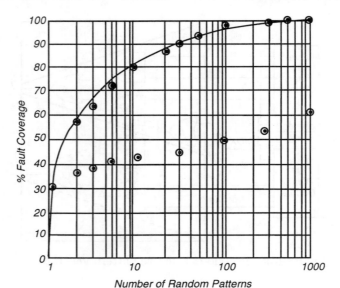

(a) Top curve -- random pattern testing with acceptable fault coverage.
(b) Bottom curve -- unacceptable random pattern testing.

Figure 15.9: Random-pattern testing and fault coverages.

can be made controllable from no more than k elements, where $k < n$, the number of PIs.

Testability vs. Random Pattern Count

Random pattern testing was investigated by Agrawal and Agrawal [15, 16, 20], Parker and McCluskey [512], and Eichelberger and Lindbloom [209]. Figure 15.9(a) shows how the stuck-fault coverage rises in a logarithmic fashion towards 100%, but at the cost of enormous numbers of random patterns. However, certain circuits, notably *programmable logic arrays* (PLAs), instead exhibit the behavior shown in Figure 15.9(b), and these are called *random-pattern resistant* circuits. Such circuits either require extensive insertion of testability hardware, or a modification of *random pattern generation* (RPG) with *weighted pseudo-random pattern generation* in order to obtain an acceptable fault coverage. At present, industry appears to require at least 98% stuck-fault coverage for reliability.

One can estimate the number of pseudo-random patterns needed from the desired fault coverage and either the set of hard-to-detect faults [67] or the circuit testability [591]. A million or more patterns is common for BIST, so fast fault simulation is essential. An appropriate technique for combinational circuits is the *parallel-pattern, single-fault propagation* (PPSFP) technique [210]. If the test sequence is too long for fault simulation, one can instead use deterministic ATPG to create test patterns for the hard-to-test faults, and embed these patterns in a ROM on the chip for a second episode of BIST. Alternatively, one can modify the circuit

15.2 Random Logic BIST

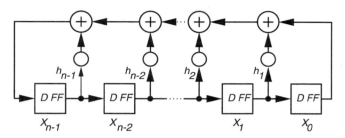

Figure 15.10: Standard linear feedback shift register.

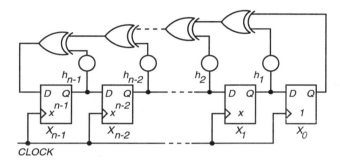

Figure 15.11: Actual digital circuit for standard LFSR.

to improve testability [210].

Pseudo-Random Pattern Generation

The *linear feedback shift register* (LFSR) pattern generator is most commonly used for pseudo-random pattern generation [67]. These patterns have all of the desirable properties of random numbers, but are algorithmically generated by the hardware pattern generator and are therefore repeatable, which is essential for BIST. We no longer cover all 2^n input combinations, but long test-pattern sequences may still be necessary to attain sufficient fault coverage. In general, pseudo-random pattern generation requires more patterns than deterministic ATPG, but fewer than exhaustive test [33].

Standard LFSR and Equations. Figure 15.10 shows a *standard, external exclusive-OR*, or *Type 1 linear feedback shift register* (LFSR) [206, 264]. It consists of D flip-flops and linear *exclusive-OR* (XOR) gates. It is an external exclusive-OR LFSR because the feedback network of XOR gates feeds externally from X_0 to X_{n-1}. The theory of these devices is described in Appendix A, which should be read before attempting this section. There are n flip-flops ($X_{n-1}, ..., X_0$), so this is called an n-stage LFSR. A properly-designed LFSR can function as a near-exhaustive test-pattern generator, as it can cycle through $2^n - 1$ distinct states, the only omitted state being the all 0 state in the flip-flops. This is known as a *maximal length* LFSR.

Maximal-length sequences from such LFSRs have these additional properties:

- There is one pattern of n consecutive ones and one pattern of $n-1$ consecutive zeroes.

- There is an *autocorrelation property*. Any two sequences, the original and the circularly shifted sequence (in the same LFSR), will be identical in $2^{n-1} - 1$ bit positions and will differ in 2^{n-1} positions [532].

In Figure 15.10, each tap coefficient h_i indicates the presence or absence of feedback from that particular flip-flop position into flip-flop position X_{n-1}. This is indicated by setting h_i ($0 \leq i \leq n - 1$) to 1 if the feedback exists, and to 0 if there is no feedback in that particular position. Figure 15.11 shows the actual hardware implementing this circuit. In the actual hardware, if an h_i is 0, then there is no XOR gate in the feedback network for that bit position; otherwise, the XOR gate is included. Remember that in this Galois field number system, multiplication by x is equivalent to a right shift in the LFSR register by one bit, and the addition operation is the XOR ($\oplus$) operator. Therefore, addition is equivalent to XOR subtraction, so $0 - 0 = 0$, $0 - 1 = 1$, $1 - 0 = 1$, and $1 - 1 = 0$. This is because there are no carries or borrows in XORing arithmetic. The following matrix system of equations describes the system:

$$\begin{bmatrix} X_0(t+1) \\ X_1(t+1) \\ \vdots \\ X_{n-3}(t+1) \\ X_{n-2}(t+1) \\ X_{n-1}(t+1) \end{bmatrix} = \begin{bmatrix} 0 & 1 & 0 & \cdots & 0 & 0 \\ 0 & 0 & 1 & \cdots & 0 & 0 \\ \vdots & \vdots & \vdots & & \vdots & \vdots \\ 0 & 0 & 0 & \cdots & 1 & 0 \\ 0 & 0 & 0 & \cdots & 0 & 1 \\ 1 & h_1 & h_2 & \cdots & h_{h-2} & h_{n-1} \end{bmatrix} \begin{bmatrix} X_0(t) \\ X_1(t) \\ \vdots \\ X_{n-3}(t) \\ X_{n-2}(t) \\ X_{n-1}(t) \end{bmatrix} \quad (15.1)$$

This system is written as:

$$\boldsymbol{X}(t+1) = \boldsymbol{T_S X}(t) \quad (15.2)$$

The first column of $\boldsymbol{T_S}$ is 0, except for the last row, to indicate that the flip-flops shift right. The 2nd through nth columns and 1st through $n-$1st rows are the identity matrix, to indicate that X_0 receives input from X_1, etc. Finally, the nth element in the first column is 1 to indicate that X_0 always feeds back into X_{n-1} through the XOR feedback network. The remaining elements in the nth row are the feedback coefficients h_i, which indicate whether the remaining flip-flops feed back into X_{n-1} or not. We also see why this LFSR cannot be initialized to all zeroes. If that were done, the feedback network and the right shifts of the flip-flops would always produce all zeros, and the LFSR would hang in the all-zero state. Note that the + operator implied in this matrix system is actually the XOR ($\oplus$) operator.

If $\boldsymbol{X}$ is the LFSR initial state, the LFSR will progress through the states: $\boldsymbol{X}$, $\boldsymbol{T_S X}$, $\boldsymbol{T_S^2 X}$, $\boldsymbol{T_S^3 X}$, etc. The *matrix period* is the smallest integer k such that:

$$\boldsymbol{T_S^k} = \boldsymbol{I} \quad (15.3)$$

15.2 Random Logic BIST

where I is the identity matrix, k is the LFSR cycle length (note that $k = 0$ for $X = 0$), and T_S is known as the *companion matrix*. Recall that multiplication by x is equivalent to shifting a bit through the D flip-flop register of this LFSR. Therefore, we view X_0 as the constant 1, $X_1 = x \times X_0 = x$, $X_2 = x \times X_1 = x^2$, ... $X_n = x \times X_{n-1} = x^n$, etc. This hardware system can be described by the *characteristic polynomial*:

$$f(x) = |T_S - I X| \qquad (15.4)$$
$$= 1 + h_1 x + h_2 x^2 + \cdots + h_{n-1} x^{n-1} + x^n$$

Thus, polynomial algebra can be used to predict the LFSR behavior.

Pattern Length and Detection Probability for LFSRs. Seth et al. [591] developed a method that can estimate both the fault coverage of a testing process, as well as the test length. The fault *detection probability* is the probability of detecting the fault by a random vector, and is represented by the distribution $p(x)$ of detectable faults:

$$p(x)\, dx = \text{Fraction of detectable faults with probability}$$
$$\text{of detection between } x \text{ and } x + dx. \qquad (15.5)$$

$p(x)$ is non-zero and positive only when $0 \leq x \leq 1$. Also, $\int_0^1 p(x)\, dx = 1$. This assumes that all faults have non-zero detection probabilities. If redundant faults are present, then the formulation can be modified by adding a *Dirac delta function*, $\delta(x)$, to $p(x)$ such that the integral still evaluates to 1. The magnitude of the delta function will then be the fraction of redundant faults. For BIST pseudo-random vectors, there are $p(x)\, dx$ faults with detection probability x. The mean coverage among those faults by a pseudo-random vector is $x\, p(x)\, dx$. The mean fault coverage of the first pseudo-random vector is:

$$y_1 = \int_0^1 x\, p(x)\, dx$$

After removing the fault detected by the first vector, the detection probability distribution becomes $(1-x)\, p(x)$. This leads to the mean fault coverage of n vectors:

$$y_n = 1 - \int_0^1 (1-x)^n\, p(x)\, dx = 1 - I(n) \qquad (15.6)$$

where $I(n)$ represents the above integral. If vectors are generated deterministically, and Y is the total number of faults, then the fault coverage after the first vector is generated is:

$$y_1 = \frac{1}{Y} + \left(1 - \frac{1}{Y}\right) \int_0^1 x\, p(x)\, dx$$

where the first term is the coverage due to the targeted fault, and the second term is the random coverage from the remaining faults. Proceeding recursively, one obtains an approximation for the coverage after n vectors are generated:

$$y_n \approx 1 - I(n) + \frac{n}{Y} \qquad (15.7)$$

where $1 \ll n < Y$. This is valid for a large number of vectors, provided that the vector set is significantly smaller than the fault set.

Fault coverage and vector length can now be estimated by a fault simulation of a sampled set of faults, with fault dropping. The *random-first-detection* (RFD) variable of a fault is the vector number at which that fault is first detected. We obtain w_i as the number of faults whose RFD value is i. If N random vectors are generated, then the complete detection probability distribution is:

$$p(x) = \frac{1}{n_s} \sum_{i=1}^{N} w_i \, p_i(x)$$

where n_s is the number of faults in the sample that is simulated, and:

$$w_0 \triangleq n_s - \sum_{i=1}^{N} w_i.$$

This expression is valid when N is large. The integral can now be evaluated as:

$$I(n) = \frac{w_0(N+1)}{n_s(n+N+1)} + \frac{1}{n_s} \sum_{i=1}^{N} \frac{i(i+1)w_i}{(n+i)(n+i+1)}. \qquad (15.8)$$

Accuracy improves as N increases, so one obtains the w_is from fault simulation using fault sampling, and then $I(n)$ is computed. The *fault coverage* is estimated using Equation 15.6 for random vectors and Equation 15.7 for deterministically-generated vectors. The *test length* to obtain a prespecified fault coverage can be predicted by inverting Equation 15.6 or 15.7, and solving it numerically. The functions $p(x)$ and $I(n)$ represent the circuit testability. This can be determined topologically (see Chapter 6) where it represents the testability by random vectors, but it can also be determined from the fault coverage data from fault sampling given a particular BIST vector set, using the above method. This may be more useful to the designer than a topological estimate.

Wagner et al. [702] provide a method of estimating the test length needed to achieve a given test confidence, which is the probability of detecting a fault with L pseudo-random vectors. However, that requires an exhaustive fault simulation, with no fault dropping, to determine the confidence accurately. Rajski and Tyszer [532] derive a test length for purely random testing, using assumed fault detection probabilities, which are generally not known. This would again require a lengthy fault simulation. Williams [723] and Savir and Bardell [567] also provide methods to estimate the test length.

15.2 Random Logic BIST

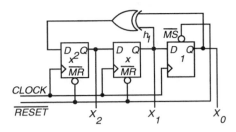

Figure 15.12: Example external-XOR standard LFSR.

Table 15.3: Pattern sequence for LFSR of Figure 15.12.

X_0	1	0	0	1	0	1	1	1	0
X_1	0	0	1	0	1	1	1	0	0	...
X_2	0	1	0	1	1	1	0	0	1

Example 15.3 Standard LFSR. *Figure 15.12 [732] shows an example external XOR standard LFSR with characteristic polynomial $f(x) = 1 + x + x^3$. Table 15.3 shows the pattern sequence generated by this LFSR when $X_2 X_1 X_0$ are set to "001." The characteristic polynomial $f(x)$ of the external-XOR LFSR is read from right to left, so since the rightmost flip-flop is always tapped, this polynomial has a $1(x^0)$ and since the middle flip-flop is tapped, it also has an x term, since $h_1 = 1$. However, there is no x^2 term, because the leftmost flip-flop is not tapped. There is always an x^n term in the characteristic polynomial for an n-bit standard LFSR, so we include the x^3 term. So, $f(x) = 1 + x + x^3$. We see the pattern repeating after the first seven patterns are generated. The student should convince him/herself of the correctness of the pattern sequence by simulating the circuit logic. This LFSR implements the equation system:*

$$\begin{bmatrix} X_0(t+1) \\ X_1(t+1) \\ X_2(t+1) \end{bmatrix} = \begin{bmatrix} 0 & 1 & 0 \\ 0 & 0 & 1 \\ 1 & 1 & 0 \end{bmatrix} \begin{bmatrix} X_0(t) \\ X_1(t) \\ X_2(t) \end{bmatrix} \qquad (15.9)$$

During BIST, as will be explained below, it is essential that the circuit be excited once and only once with a particular pattern. This is because a given pattern causes an *error vector* to appear at the faulty circuit outputs, which are read by the BIST response compacter, and repeating the pattern later causes the same error vector to appear again. Since the response compacter is an XORing system as well, the two erroneous responses from that error vector will cancel and leave the BIST system with only the good-machine response. This will cause the testing hardware to accept a faulty circuit as a good circuit. Therefore, we must avoid repeating any of the LFSR patterns more than once, and we must not initialize the LFSR to all zeros, or it will hang indefinitely in the all zero state.

Modular LFSR and Equations. The *modular, internal exclusive-OR*, or *Type 2* LFSR is described by a companion matrix $T_M = T_S^T$, which is the transpose of

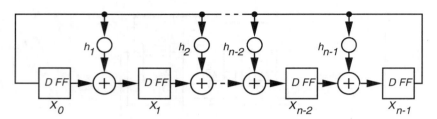

Figure 15.13: Modular LFSR example.

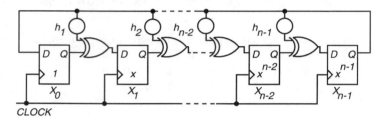

Figure 15.14: Digital circuit for modular LFSR example.

T_S. It is called an internal XOR LFSR because the feedback XOR gates are located between adjacent flip-flops. The modular LFSR can run somewhat faster than the standard LFSR, because it has at most one XOR gate delay between adjacent flip-flops. However, this is usually not a serious consideration in testing, because actual circuits always have more logic gates between flip-flops than there are XOR gates in the feedback network of the external XOR LFSR.

Figure 15.13 shows the modular LFSR, and Figure 15.14 shows the actual LFSR circuit implementation. This hardware implements these equations:

$$\begin{bmatrix} X_0(t+1) \\ X_1(t+1) \\ X_2(t+1) \\ \vdots \\ X_{n-3}(t+1) \\ X_{n-2}(t+1) \\ X_{n-1}(t+1) \end{bmatrix} = \begin{bmatrix} 0 & 0 & 0 & \cdots & 0 & 0 & 1 \\ 1 & 0 & 0 & \cdots & 0 & 0 & h_1 \\ 0 & 1 & 0 & \cdots & 0 & 0 & h_2 \\ \vdots & \vdots & \vdots & & \vdots & \vdots & \vdots \\ 0 & 0 & 0 & \cdots & 0 & 0 & h_{n-3} \\ 0 & 0 & 0 & \cdots & 1 & 0 & h_{n-2} \\ 0 & 0 & 0 & \cdots & 0 & 1 & h_{n-1} \end{bmatrix} \begin{bmatrix} X_0(t) \\ X_1(t) \\ X_2(t) \\ \vdots \\ X_{n-3}(t) \\ X_{n-2}(t) \\ X_{n-1}(t) \end{bmatrix} \quad (15.10)$$

This system is written as:

$$X(t+1) = T_M X(t) \quad (15.11)$$

This hardware system can be described by the *characteristic polynomial*:

$$\begin{align} f(x) &= |T_M - IX| \\ &= 1 + h_1 x + h_2 x^2 + \cdots + h_{n-1} x^{n-1} + x^n \end{align} \quad (15.12)$$

In this LFSR of Figure 15.13, a right shift is equivalent to multiplying the register contents by x, and then dividing its value by the characteristic polynomial and

15.2 Random Logic BIST

storing the remainder. Every LFSR can be realized either in standard or modular form. Both use m XOR gates, where m is the number of non-zero h_i feedback coefficients in the LFSR.

Example 15.4 *Modular LFSR. Figure 15.15 shows an example modular LFSR and its characteristic polynomial, which implement this equation system:*

$$\begin{bmatrix} X_0(t+1) \\ X_1(t+1) \\ X_2(t+1) \\ X_3(t+1) \\ X_4(t+1) \\ X_5(t+1) \\ X_6(t+1) \\ X_7(t+1) \end{bmatrix} = \begin{bmatrix} 0 & 0 & 0 & 0 & 0 & 0 & 0 & 1 \\ 1 & 0 & 0 & 0 & 0 & 0 & 0 & 0 \\ 0 & 1 & 0 & 0 & 0 & 0 & 0 & 1 \\ 0 & 0 & 1 & 0 & 0 & 0 & 0 & 0 \\ 0 & 0 & 0 & 1 & 0 & 0 & 0 & 0 \\ 0 & 0 & 0 & 0 & 1 & 0 & 0 & 0 \\ 0 & 0 & 0 & 0 & 0 & 1 & 0 & 0 \\ 0 & 0 & 0 & 0 & 0 & 0 & 1 & 1 \end{bmatrix} \begin{bmatrix} X_0(t) \\ X_1(t) \\ X_2(t) \\ X_3(t) \\ X_4(t) \\ X_5(t) \\ X_6(t) \\ X_7(t) \end{bmatrix} \quad (15.13)$$

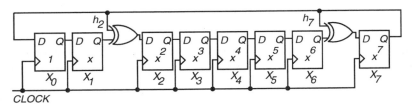

Figure 15.15: Modular LFSR for $f(x) = 1 + x^2 + x^7 + x^8$.

Primitive Polynomials. It is highly desirable that the LFSR generate all possible $2^n - 1$ patterns. Peterson and Weldon [521] have established the conditions necessary to satisfy this requirement, or to have a *primitive* polynomial for the LFSR:

1. The polynomial must be *monic*, which means that the coefficient of the highest-order x term of the characteristic polynomial must be 1. For the modular LFSR, this means that all D flip-flops must right shift through XOR gates from X_0 through X_1, ..., through X_{n-1}, which must then feed back directly into X_0. For the standard LFSR, this means that all D flip-flops must right shift directly from X_{n-1} through X_{n-2}, ..., through X_0, which must then feed back into X_{n-1} through the XORing feedback network.

2. The characteristic polynomial must divide the polynomial $1 + x^k$ for $k = 2^n - 1$, but not for any smaller value of k.

Bardell et al. [67] have published tables of characteristic polynomials that are primitive, and therefore suitable for LFSRs (see Appendix B.)

To conclude, pseudo-random is the most frequently used method for BIST pattern generation [33]. However, it may require detailed fault simulation to evaluate the fault coverage.

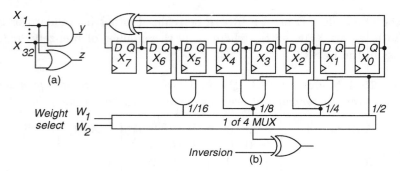

Figure 15.16: Weighted pseudo-random pattern generator.

Weighted Pseudo-Random Pattern Generation

The desire to achieve higher fault coverages with shorter test lengths, and therefore shorter test times, led to the invention of the weighted pseudo-random pattern generator. With this sort of pattern generator, one can adjust the probabilities of generating 0s or 1s at each input, in order to improve the fault coverage while shortening test length. Figure 15.16(b) [532, 706] shows a weighted pseudo-random pattern generator implemented with programmable probabilities of generating zeroes or ones at the PIs. This pattern generator effectively deals with hard-to-detect faults. In pseudo-random test, each input bit has probability 0.5 of being either a 0 or a 1. In weighted pseudo-random BIST, the probabilities are adjusted to make it more likely that tests will detect hard-to-detect faults. One method uses software to determine a single or multiple-weight set by probabilistically analyzing the hard-to-detect faults [210]. Another method uses a heuristic to generate an initial weight set, and relies on ATPG systems to produce additional weight sets. Weights are realized either by logic or are stored in a ROM. Over 98% fault coverage was obtained [210] for 10 designs, the same coverage as deterministically-generated vectors.

For the 32-bit arithmetic generate-propagate circuit of Figure 15.16(a), in order to detect the fault y stuck-at-0, all inputs must be 1. If uniformly-distributed pseudo-random patterns are used, the detection probability is 2^{-32}, because there are 32 inputs. This will lead to unacceptably long testing times. Pseudo-random patterns rarely achieve 100% fault coverage. If, instead, we could set each input to 1 with probability of 31/32, then the fault y stuck-at-0 could be detected with probability $(31/32)^{32} = 0.362$ [532]. This means that, on average, only three vectors would have to be generated to detect the fault. However, then each AND gate input stuck-at-1 fault would be detected with probability $(1/32)(31/32)^{31} = 0.01168$, so on average 86 vectors would be needed to detect such faults [532]. Figure 15.16(b) shows an adjustable weight pseudo-random pattern generator [532]. If each LFSR stage has probability of 0.5 of being either 0 or 1, and is statistically independent of the values on the other LFSR bits, then ANDing k LFSR signals results in a 1 value at the AND gate output with probability 0.5^k. OR gates and INVERTERS can be used to obtain other probabilities. In this example, the rightmost bit entering

15.2 Random Logic BIST

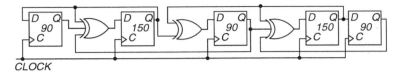

Figure 15.17: Five-stage hybrid cellular automaton pattern generator.

the MUX has a probability of 1/2 of being 1, and the second bit from the right has a probability of 1/4 of being 1, since it is the AND of two bits each with probability 1/2 of being 1. The third bit from the right has probability of 1/8 of being 1, since it is the AND of three bits each with probability 1/2 of being 1, and so on. If we program the *Weight select* lines to select the 1/16 probability of generating a 1, but we set the *Inversion* signal to 1, then we instead get a signal with 15/16 probability of being 1. This is because the XOR gate functions here as a programmable inverter. For the fault y stuck-at-0, each input should receive a 1 with probability of $1 - (31/32) = 1/32$ [532]. There is no common weight set for both faults, so two different weights must be stored for each circuit input. The main problem with weighted pseudo-random BIST is that several different weight sets are required in order to obtain 100% stuck-fault coverage, and for each set one must generate a number of random patterns [74, 347, 517, 706, 739, 740]. The multiple weight sets lead to excessive pattern generator hardware overhead.

Cellular Automaton Pattern Generation

Cellular automata (CA) are excellent for pattern generation, because they have a better randomness distribution than LFSRs [307]. A cellular automaton is a cell collection with regular connections [306, 353, 532, 691]. Each cell can only connect to its local neighbors. The connections are expressed as rules, which determine the next state based on the state of the cell's neighbors. If cell c can talk only with its neighbors, $c-1$ and $c+1$, then the following rule, called *rule 90*, can be established based on the following state transition table:

$x_{c-1}(t)x_c(t)x_{c+1}(t)$	111	110	101	100	011	010	001	000
$x_c(t+1)$	0	1	0	1	1	0	1	0

$$2^6 + 2^4 + 2^3 + 2^1 = 90$$

The term *rule 90* comes from the decimal equivalent of the binary code for the next state of cell c [738]. In this case, $x_c(t+1) = x_{c-1}(t) \oplus x_{c+1}(t)$. Another relation, *rule 150*, is implemented as $x_c(t+1) = x_{c-1}(t) \oplus x_c(t) \oplus x_{c+1}(t)$. Figure 15.17 [532] shows a hybrid cellular automaton alternately using rules 90 and 150 in its cells. Serra et al. [585] demonstrated an isomorphism between a one-dimensional linear hybrid cellular automaton and the LFSR having the same irreducible characteristic polynomial. However, state sequencing may still differ between the CA and the LFSR. CAs have no shift-induced bit value correlation, whereas LFSRs do. A LFSR pattern generator can be made more random, however, by using linear phase shifters [182].

Test Pattern Augmentation

In certain applications, notably telephone exchange testing, 100% stuck-fault coverage is required. This, in turn, requires one to augment the set of patterns generated by an LFSR in a BIST testing method, in order to include test-patterns for the faults missed by the LFSR pattern sequence. We now describe the more frequently used methods for augmenting test-patterns.

Adding a Secondary ROM. The simplest way to increase stuck-fault coverage to 100% for a BIST pattern generator is to generate the patterns with an ATPG tool for the random-pattern resistant faults in the CUT that the BIST pattern generator misses, store the generated patterns in a ROM, and set up a second on-chip test pattern generation epoch. The Input MUX needs an extra control mode, and can connect the input pins, the BIST pattern generator, or the ROM output to the PIs. For ROM pattern generation, a simple hardware counter cycles the ROM through all of its addresses. The disadvantage of this method is the hardware overhead. With this method, it is extremely important to use a test-pattern compaction program to compact the test-patterns from the ATPG program into the minimum number needed to detect the random-pattern resistant faults.

Additional Methods. One method, *test pattern diffraction* [532] involves computing a good test pattern and then using a *hardware diffractor* to generate a cluster of vectors in the neighborhood of the pattern, which is stored in a ROM. Other approaches embed the deterministic patterns to detect the random-pattern resistant faults in the LFSR pattern sequence [676, 741]. Additional approaches [144, 675] transform the patterns produced by a pseudo-random pattern generator into a new test vector set providing the desired fault coverage. The STAR-BIST scheme embeds a small number of patterns to test random-pattern-resistant faults in the scan chain [678].

15.2.4 BIST Response Compaction

During BIST, it is necessary to reduce the enormous number of circuit responses to a manageable size that can be stored on the chip. For example, consider a circuit with a hardware pattern generator that computes 5 million test patterns during testing, and where there are 200 POs. The total number of responses will be $5,000,000 \times 200 = 1,000,000,000$ *bits*! This amount of information cannot be economically stored on the CUT, so the circuit responses must be compacted.

Definitions

- Aliasing – During circuit response compaction, because of the information loss, it is possible that a signature of a bad machine may match the good machine

15.2 Random Logic BIST

signature, which is called aliasing. In such cases, a failing circuit will pass the testing process.

- Compaction – A method of drastically reducing the number of bits in the original circuit response during testing in which some information is lost.

- Compression – A method of reducing the number of bits in the original circuit response during testing in which no information is lost, so the original output sequence can be fully regenerated from the compressed sequence.

- Signature – A statistical property of a circuit, usually a number computed for a circuit from its responses during testing, with the property that faults in the circuit usually cause the signature to deviate from that of the good machine.

- Signature Analysis – A method of circuit response compaction during testing, whereby the entire good circuit response is compacted into a *good machine signature*. The actual circuit signature is generated during the testing process on the CUT, and then compared with the good machine signature to determine whether the CUT is faulty.

- Transition Count Response Compaction – A method of response compaction in which the number of transitions from 0 to 1 and 1 to 0 at circuit POs are counted to create a testing signature.

In this matter, we must distinguish between *compression* and *compaction*. Circuit response compression is lossless, because the original output sequence (10^9 bits in the above example) can be completely regenerated from the compressed sequence. Compaction, however, results in information loss, so regenerating the original circuit response information is not possible. Compression schemes, at present, are impractical for BIST response analysis, because they inadequately reduce the huge volume of data, so we use only compaction schemes. In mathematical words, compression functions are *invertible*, but compaction functions are not. *Signature analysis* is the process of compacting the circuit responses into a very small bit length number, representing a statistical circuit property, for economical on-chip comparison of the behavior of a possibly defective chip with a good machine chip. Frohwerk [228] invented *signature analysis* in 1977 at Hewlett-Packard. Also, the signature must preserve as much of the fault information contained in the circuit output response before compaction as possible, and the circuitry used to implement the compacter should be small [33]. All compaction techniques require that the fault-free circuit signature be known.

Some trivial schemes for response compaction are:

1. Parity checking, where we form parity across all circuit responses, and

2. Ones counting, where we count the number of ones in the output responses from the circuit. Savir [562] pioneered *syndrome testing*, in which pattern generation must be exhaustive, and ones counting is used for response compaction.

Aliasing occurs when the compacted response of the bad machine matches the compacted response of the good machine, and is always a problem with compaction because information is lost. In parity checking, aliasing frequently happens. Also, with ones counting, it is possible to permute the placement of ones in the circuit's Karnaugh map, and still obtain a correct ones count, so it is also very prone to aliasing and also requires significant arithmetic hardware.

Transition Count Response Compaction

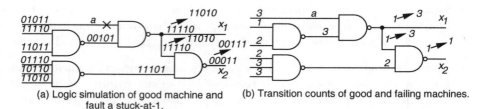

(a) Logic simulation of good machine and fault a stuck-at-1.

(b) Transition counts of good and failing machines.

Figure 15.18: Example of transition counting response compaction.

Hayes [285] described *transition count testing*. The *transition count* $C(R)$ is the number of times signals in the circuit response R change during BIST. Figure 15.18 shows an example circuit with the fault a stuck-at-1. In Figure 15.18(a), we see the circuit responses to five test-patterns, where the faulty machine response is shown above the good machine response. Figure 15.18(b) shows the sum of $0 \to 1$ and $1 \to 0$ transitions, with those of the faulty machine shown above those of the good machine. In Figure 15.18(b) at PO x_1, the good machine has a transition count of 1, but the faulty machine count is 3. One advantage of transition count compaction is that $|C(R)|$, the number of bits to represent $C(R)$, is $|C(R)| \leq \lceil log_2|R| \rceil$. $|R| = \#$ *bits in* R. r_i is the circuit output response at time i. Then:

$$C(R) = \sum_{i=1}^{m}(r_i \oplus r_{i-1}) \; for \; R = r_1 r_2 \ldots r_m \qquad (15.14)$$

In order to maximize the test set fault coverage, in transition count testing we must make $C(R_0)$, the transition count of the good machine, as large or as small as possible. Transition count testing aliases less than ones counting, because it not only checks for the correct number of ones and zeroes in the circuit output response, but also partially tests for the correct *ordering* of the ones and zeroes in the response.

LFSR for Response Compaction

Frohwerk [228] introduced the LFSR for response compaction by signature analysis. The *signature* is any statistical property of the circuit that is used for checking its correct operation. He used the data compaction method of the *cyclic redundancy check* (CRC) code generator, which requires an LFSR hardware device (see Appendix A.) In this method, the circuit output data stream is treated as a descending order coefficient polynomial. The output response compacter LFSR performs

15.2 Random Logic BIST

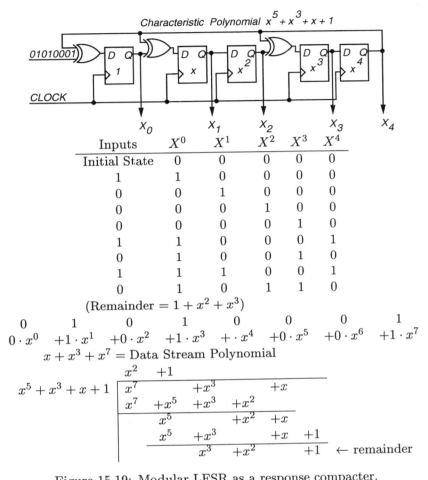

Figure 15.19: Modular LFSR as a response compacter.

polynomial division of this data stream polynomial by the characteristic polynomial of the LFSR. The final state of the modular LFSR is the polynomial remainder of this division. The final state of the standard LFSR is not always the polynomial remainder of this division, but is related to the true remainder through a different state assignment. The error detection hypothesis is that a faulty data stream changes the output data stream, and hence the remainder of this polynomial division, which is used as the signature in this compaction method. The LFSR must be initialized to the *seed* value, and after data compaction, the signature must be observed and compared with the known good-machine signature [33]. The signature analyzer circuit is easily testable.

Modular LFSR Response Compaction. Figure 15.19 shows a modular LFSR that has an extra XOR gate at the input to the flip-flop driving the least significant

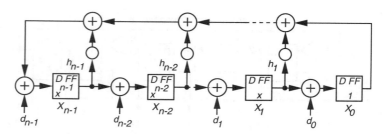

Figure 15.20: Example multiple input signature register.

bit X_0. This XOR gate XORs the circuit output response stream, "01010001" in this case, into the least significant bit of the modular LFSR. Before response compaction occurs, the LFSR flip-flops must be initialized to all zeroes. Here, "01010001" is interpreted as $0 \times x^0 + 1 \times x^1 + 0 \times x^2 + 1 \times x^3 + 0 \times x^4 + 0 \times x^5 + 0 \times x^6 + 1 \times x^7 = x + x^3 + x^7$. Reading the LFSR tap coefficients from left to right in Figure 15.19, we see that the characteristic polynomial of this modular LFSR is $1 + x + x^3 + x^5$. The figure shows how eight clock periods are simulated after the LFSR is initialized to "00000." It also shows the long division of the reversed data stream polynomial by the reversed characteristic polynomial of the LFSR. The remainder of the division, $1 + x^2 + x^3$, also matches the remainder left after eight clock periods in the LFSR, because only X_0, X_2, and X_3 are ones. Thus, we have agreement between the signature predicted by polynomial division and the signature produced by logic simulation.

Multiple Input Signature Register. In the example of Figure 15.19 [186] we see that one primary circuit output required an LFSR for signature analysis with 5 flip-flops and 3 XOR gates. However, consider the case where the above circuit has 200 outputs. Then, we would need $200 \times 5 = 1000$ flip-flops and more than $200 \times 3 = 600$ XOR gates. This is a serious hardware overhead. Fortunately, we can exploit the fact that the hardware pattern generation and response compaction system using LFSRs is a *linear* system, obeying the equation $X(t+1) = T_S X(t)$. Therefore, because of its linearity, this system also obeys the *superposition principle*. If we superimpose all of the responses of the 200 circuit outputs in the *same* LFSR for response compaction, then the final remainder will be the sum (under XOR logic arithmetic) of the remainders due to all of the circuit outputs. This is highly advantageous, as it reduces the flip-flop count from 1000 to 200 and the XOR gate count from more than 600 to approximately $3 + 200$. The 200 added XOR gates are needed to XOR all of the circuit outputs into different bits of the LFSR, where there must be one bit for each circuit PO, called d_i. This new response compacter is known as a *multiple-input signature register* (MISR), and an example is shown in Figure 15.20. The alternative to using the MISR structure is to provide only one simple LFSR for one circuit output, but multiplex it among the 200 different outputs. This then requires 200 different testing epochs, where for each epoch the LFSR compacts the response from a different circuit output. It is much more attractive to use the MISR, because it eliminates a 200 to 1 MUX, and also because

15.2 Random Logic BIST

the response compaction time with the MISR is 200 times less than the time with a multiplexed LFSR. The MISR can be represented by a system of equations:

$$\begin{bmatrix} X_0(t+1) \\ X_1(t+1) \\ \vdots \\ X_{n-3}(t+1) \\ X_{n-2}(t+1) \\ X_{n-1}(t+1) \end{bmatrix} = \begin{bmatrix} 0 & 1 & \cdots & 0 & 0 \\ 0 & 0 & \cdots & 0 & 0 \\ \vdots & \vdots & & \vdots & \vdots \\ 0 & 0 & \cdots & 1 & 0 \\ 0 & 0 & \cdots & 0 & 1 \\ 1 & h_1 & \cdots & h_{h-2} & h_{n-1} \end{bmatrix} \begin{bmatrix} X_0(t) \\ X_1(t) \\ \vdots \\ X_{n-3}(t) \\ X_{n-2}(t) \\ X_{n-1}(t) \end{bmatrix} + \begin{bmatrix} d_0(t) \\ d_1(t) \\ \vdots \\ d_{n-3}(t) \\ d_{n-2}(t) \\ d_{n-1}(t) \end{bmatrix} \quad (15.15)$$

The vector of $d_i(t)$ values represents the circuit outputs at time t on PO i.

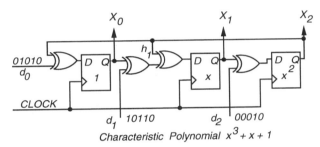

Figure 15.21: Multiple input signature register.

Example 15.5 MISR Example. *Figure 15.21 shows a modular LFSR converted into a MISR, by XORing a different circuit primary output into each flip-flop position. The resulting signature, since this system is linear, is the XORing of the three different signatures due to the polynomial division from each of the three POs. It implements the following equation system:*

$$\begin{bmatrix} X_0(t+1) \\ X_1(t+1) \\ X_2(t+1) \end{bmatrix} = \begin{bmatrix} 0 & 0 & 1 \\ 1 & 0 & 1 \\ 0 & 1 & 0 \end{bmatrix} \begin{bmatrix} X_0(t) \\ X_1(t) \\ X_2(t) \end{bmatrix} + \begin{bmatrix} d_0(t) \\ d_1(t) \\ d_2(t) \end{bmatrix} \quad (15.16)$$

Multiple Signature Checking. Hassan and McCluskey [280] have proposed using multiple signatures to reduce the likelihood of aliasing. Two separate testing epochs are run, one with a MISR with one particular primitive polynomial, and the second using the same MISR but with a different primitive polynomial implemented in the feedback network. The hardware cost of this is low, usually a few XOR gates for a second MISR feedback network and a 2-to-1 MUX to select which feedback network will feed back into the External-XOR MISR. In some cases, it may be necessary to increase the bit width of the MISR in order to obtain a different primitive polynomial. This method is highly effective, since it is unlikely that the eigenvectors of the two different feedback networks would coincide, so aliasing is unlikely.

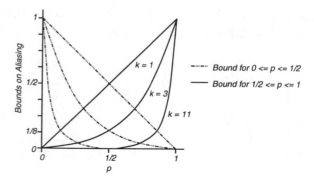

Figure 15.22: Aliasing error probability.

Aliasing Analysis

Williams *et al.* [726, 727, 728, 730] have extensively analyzed the aliasing probabilities of BIST LFSR response compacters. Earlier analyses also exist [114, 612, 630]. Williams *et al.* consider only the error sequence $e(n)$ at a circuit PO, which is obtained by *exclusive-XORing* (XORing) the output sequences of the good and faulty CUTs. A 1 in the error sequence indicates a manifestation of a fault. Let p be the probability of a 1 in $e(n)$. Let P_{al} be the probability of aliasing. If $0 < p \leq 1/2$ then $p^k \leq P_{al} \leq (1-p)^k$, but if $1/2 \leq p \leq 1$, then $(1-p)^k \leq P_{al} \leq p^k$ [729]. If all error sequences are equally likely, which is usually not true, then $P_{al} = 2^{-k}$. Figure 15.22 shows the weak aliasing bounds as a function of p, the probability of an error, and k, the number of bits in the response compacter. Williams *et al.* analyzed the dynamic properties of aliasing to confirm that primitive polynomials in response compacters alias less than non-primitive polynomials.

Theorem 15.1 *Assuming that each circuit PO d_{ij} has probability p of having an error, and that all outputs d_{ij} are independent, in a k-bit MISR, the aliasing probability is $\frac{1}{2^k}$, regardless of the initial condition of the MISR.*

The assumption of independent outputs is not true in the general case since circuit outputs are correlated, but the correlated outputs have little impact on this analysis. In the situation where each circuit PO d_{ij}, $j = 1, 2, ..., k$ had a unique probability of error p_j, the result still holds [730].

Theorem 15.2 *Assuming that each circuit PO d_{ij} has probability p_j of having an error, where the p_j probabilities are independent, and that all outputs d_{ij} are independent, in a k-bit MISR, the aliasing probability is $\frac{1}{2^k}$, regardless of the initial condition of the MISR.*

The aliasing probability is $1/2^k$, and therefore the error detection probability is:

$$1 - \frac{1}{2^k} \qquad (15.17)$$

15.2 Random Logic BIST

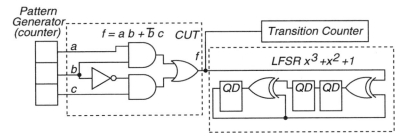

Figure 15.23: Example circuit with transition counter and LFSR.

These theorems say that it is just as good for testing purposes to just look at the effect of the last pattern, which is an unfortunate conclusion for testing.

Note that a MISR has an additional aliasing source, compared with a single-output signature analyzer. An error in CUT output d_j at time t_i followed by an error in output d_{j+h} at time t_{i+h} has no effect on the signature, if there is no feedback tap in the MISR between the outputs Q_j and Q_{j+h}. The probability of this aliasing can be reduced, by using a different primitive polynomial so that there is a feedback tap between outputs Q_j and Q_{j+h} [33]. Others have also analyzed aliasing probabilities [179, 180, 325, 349].

Example 15.6 *In the circuit of Figure 15.19 [186], we see that one primary circuit output required an LFSR for signature analysis with 5 flip-flops. The aliasing probability here would be $1/2^5 = 3.125$ %. If we add 3 more bits to the LFSR, the probability reduces to $1/2^8 = 0.39$ %, which is much more acceptable.*

Example 15.7 *Comparison of Transition Count and LFSR Compaction. Figure 15.23 shows a BIST system where both a transition counter and an LFSR with characteristic polynomial $f(x) = x^3 + x^2 + 1$ are used to compact the testing responses. The circuit function is $f = a \cdot b + \overline{b} \cdot c$, and the hardware pattern generator is an exhaustive three-bit binary counter. Table 15.4 shows the results of testing. The column labeled abc gives the test-pattern, and the other columns give the fault-free response and the responses for faults a stuck-at-one, f stuck-at-one, and b stuck-at-one. The table clearly shows the LFSR aliasing for the fault f stuck-at-one, and the transition count compacter aliasing for the fault a stuck-at-one.*

15.2.5 Built-In Logic Block Observers

As mentioned earlier, the BILBO combines the functionality of the D flip-flop, a testing hardware pattern generator (for the circuit portion driven by the BILBO Q outputs), a testing response compacter (for the circuit portion driving the BILBO D inputs), and a scan chain function. The scan chain BILBO can be reset to zero by shifting in an all-zero pattern into the BILBO in serial scan chain mode. Figure 15.24 shows a BILBO with characteristic polynomial $f(x) = 1 + x + \ldots + x^n$. Note the use

Table 15.4: Comparison of transition count and LFSR response compacters.

| Pattern | Responses | | | |
abc	Fault-free	a stuck-at-one	f stuck-at-1	b stuck-at-1
000	0	0	1	0
001	1	1	1	0
010	0	1	1	0
011	0	1	1	0
100	0	0	1	1
101	1	1	1	1
110	1	1	1	1
111	1	1	1	1
Signatures				
Transition count	3	3	0	1
LFSR	001	101	001	010

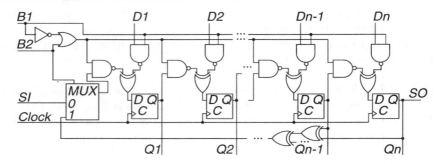

Figure 15.24: BILBO example.

of NAND gates in this BILBO in order to accelerate its speed over implementations with AND and OR gates. Table 15.5 shows the control line modes for this BILBO.

Table 15.5: Control modes for the BILBO of Figure 15.24.

$B1$	$B2$	Mode	$B1$	$B2$	Mode
0	0	Serial scan chain	1	0	Normal D flip-flop
0	1	LFSR pattern generator	1	1	MISR response compacter

Figure 15.25(a) shows a testing configuration with three subcircuits to be tested, $CUTA$, $CUTB$, and $CUTC$; two BILBOs, $BILBO1$ and $BILBO2$; an input $LFSR$; and an output $MISR$ for testing hardware. Figure 15.25(b) shows how the testing hardware is used to test the three parts of the circuit.

Figure 15.26 shows the effective BILBO hardware in serial scan mode $(B1B2) = (00)$, Figure 15.27 shows the hardware in LFSR mode $(B1B2) = (01)$, Figure 15.28 shows the hardware in D flip-flop mode $(B1B2) = (10)$, and Figure 15.29 shows the hardware in MISR mode $(B1B2) = (11)$. Bold lines show the enabled data path.

15.2 Random Logic BIST

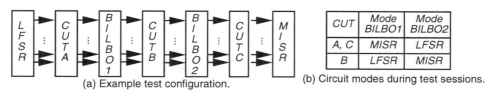

(a) Example test configuration. (b) Circuit modes during test sessions.

Figure 15.25: Circuit configured with BILBOs.

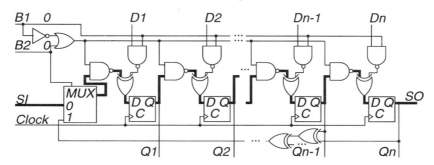

Figure 15.26: Example BILBO in serial scan mode.

15.2.6 Test-Per-Clock BIST Systems

In a *test-per-clock* BIST system, some new set of faults is tested during every clock period. The advantage of this BIST system is that it has the shortest possible pattern length, which may or may not be important. Consider a BIST pattern sequence of 10 million vectors, applied at the system operating speed of 200 MHz. Testing takes only $10,000,000/200 \times 10^6 = 0.05$ s. However, a major issue for BIST pattern length is fault simulation time. We can only know the actual fault coverage for a BISTed system from a stuck-fault simulation. The lengthy computation time worsens as the BIST pattern sequence increases, because the entire sequence must be simulated for the good machine and all failing machines. Figure 15.30(a) shows a test-per-clock system. In Figure 15.30(b) there are large numbers of PIs, so it is feasible to apply an LFSR to a subset of these inputs and then serially shift the most significant bit coming out of the LFSR into a shift register to provide pattern stimulation for the remaining inputs. The saving in test hardware is extremely minor. The shift register portion of the circuit has the same flip-flop hardware as the LFSR, but it lacks the few XOR gates used to form a LFSR feedback network.

15.2.7 Test-Per-Scan BIST Systems

In a *test-per-scan* BIST system, each new set of faults that is tested requires one clock to conduct the test and a series of shifts of the scan chain to complete that test and read out all of the test results. Test-per-scan, therefore, takes significantly more time than a test-per-clock method to detect the same number of faults in a given circuit. The advantage of test-per-scan systems over test-per-clock systems is that a judicious combination of scan chains and a MISR can lead to a significantly smaller

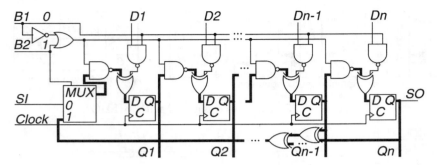

Figure 15.27: Example BILBO in LFSR mode.

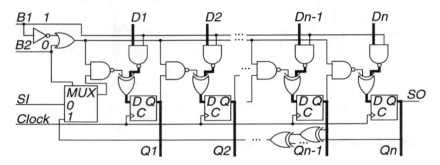

Figure 15.28: Example BILBO in normal D flip-flop mode.

MISR than in a test-per-clock system. However, this saving occurs at the expense of greatly increased BIST test-pattern length, which can cause a major computational bottleneck in fault simulation. Also, the advantages of hardware savings in the MISR are far less important now than they were a decade ago.

One problem in test-per-scan systems is that usually a new set of circuit input patterns is generated using a pseudo-random or exhaustive technique. However, because the input patterns are time shifted and repeated to the circuit through the scan chain, the patterns become correlated. This, in turn, reduces the pattern effectiveness for fault detection, so very frequently it is necessary to provide an input network of XOR gates to phase shift the inputs and de-correlate them.

Figure 15.31 shows a test-per-scan system called STUMPS* [66], in which the LFSR generates pseudo-random patterns, which are then fed through full-scan chains ($SR1$, $SR2$, ..., SRn) on various chips in the system-under-test. The scan chains drive the inputs of these chips. The chip outputs are collected in another scan chain, which is serially driven by all of the chip outputs. The chip output scan chains then drive a MISR. The advantage of this system is that if there are collectively 5,000 chip outputs, but they are sampled by 25 scan chains, each of length 20, then the output MISR needs to have only 25 bit positions in it, one for each scan chain, rather than 5,000. A scan chain for the chip outputs requires only 1

---

*Self-Test Using a MISR and Parallel Shift register sequence generator.

15.2 Random Logic BIST

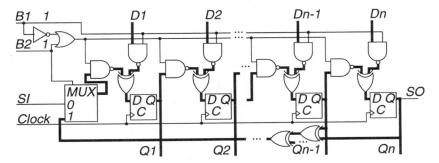

Figure 15.29: Example BILBO in MISR mode.

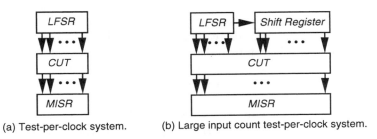

(a) Test-per-clock system.   (b) Large input count test-per-clock system.

Figure 15.30: Test-per-clock scheme.

flip-flop and one MUX per output. However, a MISR for the chip outputs requires 1 flip-flop per output, 1 XOR gate per output, reset hardware for the flip-flops, and a number of XOR gates for the MISR feedback network. Again, because of the recent drastic decrease in hardware cost, the hardware savings of STUMPS over a MISR may be less important than they were a decade ago. A single test requires that we scan in patterns from the LFSR into all of the scan chains, and then we switch the system into normal functional mode and clock it once with the *system clock*. Finally, we scan out the contents of the scan chains into the MISR, where the scan chain contents are compacted. Scan out of the prior test results can be

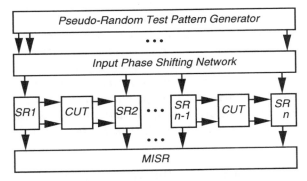

Figure 15.31: STUMPS test-per-scan testing system.

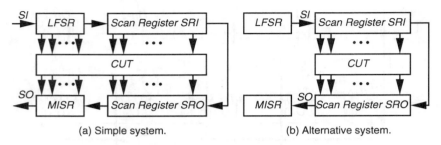

(a) Simple system. (b) Alternative system.

Figure 15.32: Test-per-scan systems.

overlapped with scan in of the next test pattern into the scan chains. This test method requires that every system input be driven by a scan chain, and requires that every system output either be caught in a scan chain or drive another chip in the system, where the other chip is also participating in the STUMPS testing system. Figure15.32(a) shows an alternative test-per-scan system and Figure15.32(b) shows a final alternative test-per-scan system.

Example 15.8 BILBO Versus STUMPS Versus External ATE at 325 MHz. *We now compare the performance of a BILBO system, a STUMPS system, and a conventional testing system using an external ATE. Here,* level-sensitive scan design (LSSD) *is assumed, where each flip-flop in the system is modified so that it cannot generate hazards at its outputs, and all clock lines are rigidly controlled during design so that there are no gated clocks and so that there are distinct clocks for normal mode operation and scan chain operation. We define the following variables:*

P = *number of patterns* L = *maximum scan chain length*
CP = *Clock period* k *ratio* = *self-test speed / LSSD tester speed*

We assume a system clock rate of 325 MHz, and an LSSD tester speed also of 325 MHz, so $k = 1.0$. We assume that there are 2,000,000 BIST patterns and a maximum scan chain length of 100 bits. Table 15.6 shows the results of this gestalt experiment. Here, we see that the External ATE and STUMPS take the same amount of test time, but the BILBO system only takes 1/100 as much time, because it obtains a test-per-clock, rather than a test-per scan. This is a test cost reduction. Although fixturing setup times for chip testing are long, most production ATE has multiple chip fixtures and is capable of applying tests to one chip while several other chips are being loaded into or unloaded from other fixtures. Therefore, this reduced test time could possibly lead to an increase in chip testing throughput at the ATE, provided that there are enough different fixtures in the ATE to fully exploit this test time reduction. Also, fault simulation time will be 100 times longer for STUMPS and the External ATE system than for the BILBO system.

Example 15.9 BILBO Versus STUMPS Versus External ATE at 1 GHz. *Now let us consider another example involving IBM's experimental 50 million transistor*

15.2 Random Logic BIST

Table 15.6: BILBO, STUMPS, and external ATE comparisons.

	Testing method		
	BILBO (test-per-clock)	STUMPS (test-per-scan)	External ATE (test-per-scan)
Test time at 325 MHz	$P \times CP$ 0.00615 s	$P \times L \times CP$ 0.61538 s	$P \times L \times CP \times k$ 0.61538 s
Test time at 1 GHz	$P \times CP$ 0.002 s	$P \times L \times CP$ 0.2 s	$P \times L \times CP \times k$ 0.615384 s

chip, which operates a 1 GHz clock rate. No existing ATE can operate at this speed, so we will assume that the ATE still operates at 325 MHz. Now, k = 3.07692 and $CP = 10^{-9}$ s for the BILBO and STUMPS methods, and $CP \times k = 3.076923 \times 10^{-9}$ s for the external ATE approach. We continue to assume that P = 2,000,000 and L = 100. Table 15.6 shows the results of this gestalt experiment. We see here that external testing with the ATE takes 307 times longer than with the BILBO, in addition to requiring a far more elaborate and expensive ATE than the BILBO system requires. We also see that STUMPS takes 100 times longer in testing time than the BILBO, because of the need to conduct a test-per-scan, rather than being able to conduct a test-per-clock.

Example 15.10 *IBM RISC/6000 Pseudo-Random STUMPS Test. This testing system was used for embedded RAM test and transition-delay fault testing. The STUMPS system was hierarchical. Delay testing was performed using two clock phases, and operated at-speed, between scans of the scan chain.*

15.2.8 Circular Self-Test Path System

Figure 15.33 [367] shows the *circular self-test path* (CSTP) BIST configuration. In this testing system, the hardware pattern generator and response compacter are combined into a single hardware device, which is the entire circular flip-flop path. Therefore, this is a non-linear mathematical BIST system, so superposition no longer holds. Some of the flip-flops are converted into self-test cells (see Figure 15.33(a)), where in *TEST* mode, the cell XORs its D input with the state from the immediately prior flip-flop in the CSTP chain. After initialization of the registers, in the *TEST* mode, the circuit runs for a number of clock cycles and then the signature is read out of the circular register path. The entire path can be regarded as a MISR with characteristic polynomial $f(x) = x^n + 1$. However, the non-linear nature of this system makes it difficult to compute the fault coverage.

Example 15.11 Circular BIST. *At Lucent Technologies circular BIST has been used to test application-specific integrated circuits (ASICs) [37]. This example has four ASICs. On three, everything was tested using BIST except for the input/output buffers and the Input MUXes. For all four devices, the average logic gate count was*

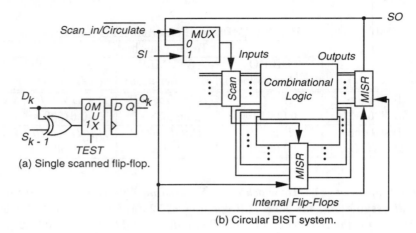

Figure 15.33: Full-circular BIST configuration.

14,150, the BIST logic overhead (per cent testing logic gates) was 20%, and the chip area overhead was 13%. The BIST stuck-fault coverage was 92%.

15.2.9 Circuit Initialization

It is very important in random logic BIST to initialize all flip-flops in the circuit when BIST is used with partial scan. Otherwise, X logic values in a 3-valued logic system will be clocked into the MISR. In the real hardware, different chips will randomly initialize their flip-flops to different values. Initialization problems can be discovered by setting all flip-flops initially to the X state, running the BIST cycle, and simulating the system in a 3-valued logic simulator. If the MISR or other response compacter finishes the test session with bits in the X state, then initialization is not correct. All such uninitializable flip-flops must then be initialized by adding master set or reset lines to them. Another approach is to break all cycles (loops of flip-flops) in the circuit, and then apply a partial BIST pattern sequence that synchronizes all flip-flops to a known state. Then, the response compacter can be turned on to compact the circuit's response. If the BISTed circuit uses a full-scan chain, then it is important to initialize the flip-flops in scan mode before initiating BIST. The initialization hardware significantly increases the chip area overhead of BIST.

15.2.10 Device Level BIST

It is necessary to isolate the BIST circuitry and the circuit-under-test from normal system data during testing. Inputs to the CUT can be isolated by multiplexers (known as the *Input MUX*) or by blocking gates. The Input MUX switches input to the CUT from the normal system inputs (port 0) to the BIST pattern generator (port 1.) When blocking gates, such as AND gates, are used, a constant '0' would

15.2 Random Logic BIST

be applied to the second AND gate input, to block the normal system input arriving from the first AND gate input. Most importantly, note that neither the Input MUX nor the blocking gate will be thoroughly tested by the BIST hardware. The untested Input MUX or blocking gate hardware can be tested by additional, short external testing sequences, or by use of the Boundary Scan Standard (see Chapter 16.)

One useful technique during design simulation and verification of the BIST circuitry is to initiate BIST in an uninitialized circuit and apply unknown values to the CUT inputs during the simulation. If there is a logical "leak" of the system data into the BIST circuitry, this usually results in a signature full of unknown values [33]. If this is observed, then the isolation mechanism must be redesigned.

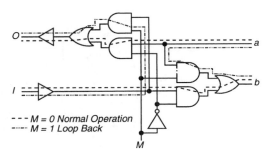

Figure 15.34: Circuit pack input/output loop back.

Another possibility for isolation is to loop back circuit outputs into circuit inputs [33]. A signal from outside the circuit indicates that the circuit should perform a self-test. The test controller sets a series of loop backs on the chip to allow data to circulate through the VLSI chip and not interfere with the rest of the system, as in Figure 15.34. The BIST procedure would exercise each VLSI chip using the loopback circuit. Then, the boundary scan chain (see Chapter 16) can be used to test the interconnections between the VLSI chips. Alternatively, the interconnections can also be tested by sending data from the test controller circuit through the subcircuits, looped I/O, and back to the test controller, as shown in Figure 15.35.

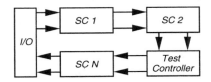

Figure 15.35: Functional test path.

It may be necessary to distribute BIST capabilities throughout the chip due to mixtures of general sequential logic, RAMs, ROMs, PLAs, etc., on a single chip. The BIST capabilities will differ significantly, and incorporating the results of the various BIST schemes into a single pass/fail result status is complex but highly desirable. Additional circuitry and test development effort is needed. One can instead provide the capability for routing the signatures of the individual BIST functions to the

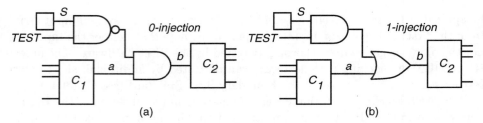

Figure 15.36: Control points to force 0 and 1.

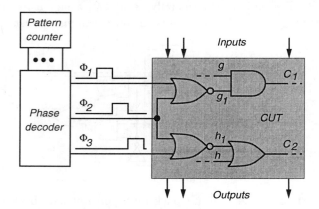

Figure 15.37: Test point activation.

device-under-test outputs [33]. If multiple VLSI devices on the same circuit pack are using BIST, then the hardware to incorporate the signatures from the various devices into a single BIST result can be included on one of the VLSI chips [33].

BIST can come close to, but not quite meet, the benefits of *level-sensitive scan design* (LSSD) in terms of fault coverage and area overhead [33]. One method uses an LFSR pseudo-random pattern generator, an LFSR for signature analysis, and LSSD. Fault coverage of BIST using this scheme can be greater than 98% [735]. Totten describes BIST approaches for general sequential logic [673].

15.2.11 Test Point Insertion

When random-logic BIST is used in a circuit, often not all of the faults are detected, either because of aliasing or because the test-pattern set from the hardware pattern generator is not rich enough to excite all faults. Touba and McCluskey [674] proposed modifying the circuit during logic synthesis to make it fully testable, using a two-level circuit representation. However, practical circuits are multi-level, and have already been optimized for function, timing, and power consumption. Re-synthesis would ruin the prior optimizations. Post-synthesis methods, instead, introduce observation points into the BISTed circuits to achieve 100% fault coverage. These methods use either exact fault simulation [101, 326, 676] or approximate testability measures [157, 561, 581]. Fault simulation finds signal correlations that

15.3 Memory BIST

block fault propagation, but is extremely expensive in computation time. Rajski and Tyszer [532] show how to improve fault detection probabilities by inserting logic gates to improve signal controllabilities in Figure 15.36 [532]. The Part (a) circuit controls signal b to 0 when both $TEST$ and S are 1, while the Part (b) circuit controls b to 1 when both $TEST$ and S are 1. Observation points are inserted by tapping a signal with poor observability, and routing the wire from the signal to an extra flip-flop in the scan chain BILBO. Tamarapalli and Rajski [650] partition the entire random-logic BIST test set into multiple phases, each contributing to the goal of 100% stuck-fault coverage. Figure 15.37 shows how each phase activates a group of control and observation points [532]. They use probabilistic fault simulation to select the optimal set of control and observation points that gives the maximum improvement in fault coverage using the fewest test points.

Example 15.12 *Test point activation. In Figure 15.37, there are four test epochs, ϕ_0, ϕ_1, ϕ_2, and ϕ_3. A phase decoder enables different test points during each phase, during which a specific number of test-patterns are applied. Here, signal g_t is 0 in phases ϕ_1 and ϕ_2, forcing c_1 to be 0, independently of g. During phases ϕ_0 and ϕ_3, g_t is 1, so g controls c_1 during those phases. Similarly, h_t is 1 during phases ϕ_2 and ϕ_3, forcing c_2 to 1, independently of h. During ϕ_0 and ϕ_1, h_t is 0, allowing h to control c_2.*

The probabilistic fault simulator determines which signals receive control and observation points. It finds, for each circuit node, the fault list that propagated to it, along with the fault detection probabilities (called the *propagation profile*.) The detection probability is computed from analytical equations, and represents for all cases whether the fault effect is D or $\overline{D}$ at that node. The probabilistic fault simulation is set up to meet a user-specified threshold for the detection probability of a maximum number of faults. A greedy heuristic selects observation points first, and then an estimation technique computes two estimates, E_0 and E_1, of the number of faults that could potentially be detected by placing a 0-controllability or 1-controllability point at each circuit node. Nodes with E_0 or E_1 greater than a threshold are considered further, by injecting, for each candidate node, a 0 or 1 determining the improvement in the detection profile. Candidates are ranked by the number of additional faults that propagate to POs or observation points, as a result of control point insertion. Results of experiments show that near-complete fault coverage can be achieved by inserting very few test points and using a two-phase testing scheme. Even better results occur when a multi-phase testing scheme is used.

15.3 Memory BIST

We have entered an era of integration of various layouts or *cores*, from different companies, onto a single chip. For example, one custom VLSI chip may now contain an embedded RAM, a microprocessor, a DSP processor, and various analog circuit layouts. Embedded RAM memories are perhaps the hardest type of digital circuit to test, because memory testing requires delivery of a huge number of pattern stimuli to the memory and the readout of an enormous amount of cell information. The difficulty and time required to propagate all of that information through the various glue logic and busses in an embedded core chip almost forces the use of memory BIST. The reader may refer to Chapter 9 for a complete discussion of memory march tests and the various memory fault models, as we will not repeat that information. With memory *design for testability* (DFT), the most time-consuming part of a memory test algorithm is implemented on-chip, and reduces the memory test time by an order of magnitude [688]. Kraus *et al.* [368] give a 1% area overhead for memory DFT for a 4 *Mb* DRAM. With memory BIST, the entire memory testing algorithm is implemented on-chip, and operates at the speed of the circuit, which is 2 to 3 orders of magnitude faster than a conventional memory test [688]. A 2% chip area overhead for memory BIST can be expected. Most memory BIST schemes exploit the parallelism within the memory device to achieve a massive reduction in test time (and therefore cost.) This is done by a test mode where more than one memory cell is accessed with each address, usually by accessing the entire row of cells on a word line for a single read or write operation. For n cells in the memory, with $\sqrt{n}$ rows and $\sqrt{n}$ columns, this reduces test time by a $\sqrt{n}$ factor. However, the parallel mechanism makes it difficult to test for memory coupling faults between cells in the same row, so it may not be appropriate. The regularity of the march tests makes them most suitable for memory BIST. We will not discuss random or pseudo-random memory BIST here, because the march tests achieve higher fault coverages with shorter pattern sequences than random or pseudo-random memory tests. However, sometimes the complexity of march test BIST implementation is too great, so longer memory BIST test algorithms are used, because they have less chip area overhead. ROM testing, as described in Chapter 9, can easily be extended to ROM BIST.

15.3.1 Definitions

- Concurrent BIST – A memory test mechanism where the memory can be tested concurrently with normal system operation.

- Non-Concurrent BIST – A memory test mechanism that requires interruption of the normal system function in order to perform the testing. The original memory contents are lost.

- Transparent Testing – A memory test mechanism that requires interruption of the normal system function for testing. The original memory contents are preserved in the memory after testing is finished.

Memory BIST requires an *address generator* or stepper (often an LFSR) and a *data generator*. As pointed out by van de Goor [688], an LFSR is better for march test BIST than a binary counter, because it uses substantially less area, and can easily be made self-testable [494]. Furthermore, the LFSR can be adjusted to provide the all-zero pattern and the forward and exact reverse LFSR

15.3 Memory BIST

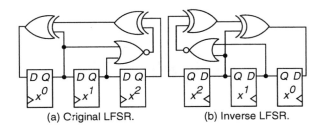

Figure 15.38: LFSRs that count up/down in inverse order.

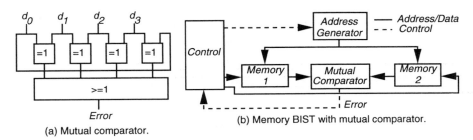

Figure 15.39: Mutual comparator for memory BIST.

sequences [67, 688]. Therefore, it satisfies all of the address ordering conditions for detecting address decoder faults with march tests. A reverse sequence LFSR generator has to have a characteristic polynomial that is the *reciprocal characteristic polynomial* of the LFSR, and it must shift in the opposite direction from the original LFSR. This is achieved by numbering the cells of the LFSR in the reverse order. Figure 15.38(a) shows the LFSR with the characteristic polynomial $G(x) = x^3 + x + 1$, while Figure 15.38(b) shows the inverse LFSR with inverse characteristic polynomial $G(x) = x^3 + x^2 + 1$, which also can generate the all-0 pattern because of the extra NOR gate and XOR gate. The Part (a) LFSR generates the sequence $1 \to 0 \to 4 \to 6 \to 7 \to 3 \to 5 \to 2$ when initialized to 1, while the second generates the sequence $1 \to 2 \to 5 \to 3 \to 7 \to 6 \to 4 \to 0$ when initialized to 1. The NOR gate forces the LFSR into the all-zero state. These two LFSRs can be combined into a single LFSR, by adding a few additional logic gates. Another advantage of the LFSR over a counter is that the probability of an address bit changing is equal for all address bits. This enables detection of *write recovery faults* (see Chapter 9.) The test data can either be produced by a finite state machine or from the address. Response data evaluation is often carried out by deterministic comparison.

The *mutual comparator* [688] (see Figure 15.39(a)) is useful in memory BIST when the memory system has multiple arrays. We test two or more arrays (in this case 4) simultaneously, by applying the same test commands and addresses to all 4 arrays. The mutual comparator asserts the *Error* signal when one of d_0 through d_3 disagrees with the other data coming out of the memory arrays. The comparator eliminates the need to generate the good machine response, and implicitly assumes that only a minority of the memory array outputs are incorrect at any given time.

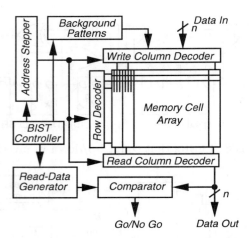

Figure 15.40: Parallel memory BIST.

The march tests are appropriate for SRAM testing. However, for DRAM testing, a *neighborhood pattern sensitive fault* (NPSF) testing model is more appropriate, since it provides better DRAM fault coverage [688]. Since the operation count is much longer for NPSF tests than for march tests, the benefit of BIST is greater when the NPSF tests are implemented on-chip. However, no NPSF test can detect address decoder faults, whereas all march tests can. Therefore, an appropriate scheme would be to put both test algorithms in the BIST hardware. We discuss march test BIST here, and the reader can refer to the literature for NPSF BIST descriptions.

15.3.2 March Test SRAM BIST

Nadeau-Dostie *et al.* [483] proposed a method to handle BIST of n-bit word RAMs. They provide a serial access design by adding multiplexers to the inputs of the write drivers. Each MUX selects between normal data input and a latch driven by the sense amplifier output of its left neighbor bit. This creates a shift register in memory BIST mode where a write causes each cell to copy data from its left neighbor, so only the leftmost bit in the memory word is directly controlled. A read causes only the rightmost bit in the word to be observed, while all other cells in the row are copied to the right.

Any march test can now be generalized to test an n-bit word memory row (called *line-mode* testing.) In the SMARCH test below, which is derived from MARCH C (see Tables 9.14 and 9.15), $(x)^n$ refers to operation x being repeated n times.

$$\{\updownarrow (w0)^n (r0, w0)^n; \Uparrow (r0, w1)^n (r1, w1)^n; \Uparrow (r1, w0)^n (r0, w0)^n;$$
$$\Downarrow (r0, w1)^n (r1, w1)^n; \Downarrow (r1, w0)^n (r0, w0)^n; \Downarrow (r0, w0)^n (r0, w0)^n\}$$

Memories can be daisy-chained together, so that the serial output of one memory drives the serial input of the next, such that the memories appear to be one very

15.3 Memory BIST

Table 15.7: Signals needed for march test BIST.

Signal	Type	Meaning
$\overline{WRITE/READ}$	OUTPUT	If 1, write $Data\_Out$ to memory, else read memory contents into $Data\_In$
$Count\_Up$	OUTPUT	If 1, count up in the $Address\ Stepper$; otherwise, count down
$COUNT$	OUTPUT	If 1, connect $Memory\ Decoders$ to the $Address\ Stepper$, else connect them to the normal memory address pins
$First\_Address$	INPUT	If 1, the memory $Address\ Stepper$ is currently at the first memory address
$Last\_Address$	INPUT	If 1, the memory $Address\ Stepper$ is currently at the last memory location
$CLEAR$	OUTPUT	If 1, clear the $Address\ Stepper$ to the first address; otherwise, do nothing

large block. We present parallel memory BIST in Figure 15.40 [532]. The memory must be equipped with this test hardware:

1. A memory *BIST Controller*.

2. An *Address Stepper*.

3. A MUX circuit feeding the memory during self-test from the controller.

4. A *Comparator* for response checking.

5. A *Background Pattern* inserter or *Data Generator* for inserting test patterns into memory columns.

Example 15.13 MATS+ March Test RAM BIST. *We wish to design BIST hardware for the memory system of Figure 15.40 to implement the MATS+ march test { M0 : $\Updownarrow$ (w0); M1 : $\Uparrow$ (r0, w1); M2 : $\Downarrow$ (r1, w0) } in hardware. Table 15.7 shows the necessary signals that the hardware of Figure 15.40 must provide. Note that the Up/Down LFSR Address Stepper of Figure 15.40 must also include an address MUX, to switch the memory address control lines from their normal external pin inputs to the outputs of the Address Stepper. Generally, the number of states required for a hardware implementation of a march test in a finite state machine is equal to 2 × the number of march elements + 3. The extra three states are a START state, where the controller sits until the TEST command is asserted to start a test; an ERROR state, where the controller sits when a memory error is detected (and the Address Stepper shows the location in error); and a CORRECT state, where the controller ends up if the memory passes the march test. Figure 15.41 shows the state transition diagram for the march test BIST controller. Because memories are*

Chapter 15. BUILT-IN SELF-TEST

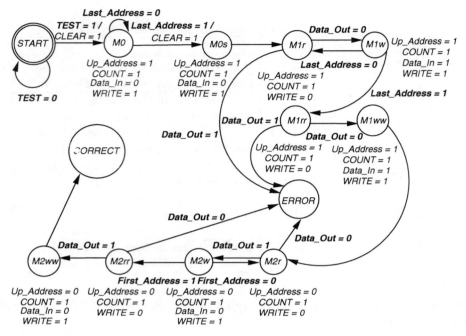

Figure 15.41: State transition diagram for the MATS+ memory BIST test.

huge, and the march test controller is very simple, the area overhead of memory BIST with march tests is one to two per cent, so memory BIST is widely used.

Example 15.14 *March Test RAM BIST with Mutual Comparator.* Nicolaidis [494] partitioned a memory into two arrays, and tested it with the BIST system in Figure 15.39(b). The Address Generator is an Up/Down LFSR. Memories 1 and 2 each contain either half of the original number of rows (columns.) The control block steps through the states for the selected march test (see Figure 15.41.) The address generator can be tested with a parity generator [688], and the mutual comparator is tested with generated vectors (see van de Goor [688] Section 12.2.2), which are often embedded in the memory BIST RAM output sequence.

15.3.3 SRAM BIST with MISR

A memory BIST scheme due to Jain and Stroud [337] accounts for the memory layout, and the memory address scrambling from the logical address space to the physical memory address space. The memory location is written with the value of its address or the bitwise complement of its address, so a data generator is not needed. Each location contains unique data. A series of up and down marches are performed using the Up/Down LFSR to detect stuck-at faults and transition faults. Another scheme uses a binary counter to provide addresses, input test data, and testing control signals.

15.3 Memory BIST

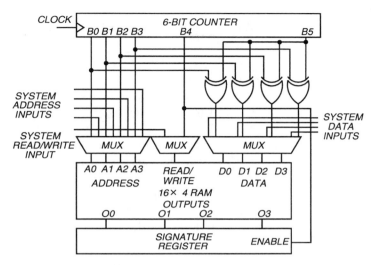

Figure 15.42: Counter test technique for RAM BIST.

Example 15.15 BIST of embedded RAM *[33]*. *Figure 15.42 shows the test hardware and RAM together. Here, rather than directly comparing the memory output to the expected output during testing, instead a MISR is used to compress the output from the memory. The MISR must be initialized before BIST, and must be disabled when shifting or loading new data into the RAM, since the output of the RAM is unknown during writes to the RAM. The same counter bit that enables writing also disables the LFSR in the MISR. Aliasing is controlled either by: (1) Providing a second set of feedback taps on the MISR, resulting in division by two different polynomials, or (2) Repeating the RAM test patterns in reverse order. This is done by using a counter bit to invert the RAM address during one of the multiple read sequences. This reduces aliasing, because the MISR is most apt to alias when output errors lie along a diagonal in the output pattern array. By reversing the RAM test patterns, the diagonal error pattern will lie along an orthogonal diagonal during testing, thus ensuring detection of the fault.*

Table 15.8 shows characteristics of six memory chips tested with memory BIST. They used a 6-bit binary pattern generation counter, in which the 4 LSBs were used for data and address generation during test, the 5th bit controlled reading or writing of the RAM, and the 6th bit inverted the RAM data during test. We extend march test notation to describe the actual test performed. In the following, (w $Address$) means write the address lines into the addressed location as data, and (r $Address$) means read the address lines from the addressed location without checking. However, all read-out data checking is done by a MISR, and not by the march test. The system implements the following march test: $\{\Uparrow (w\ Address); \Uparrow (r\ Address); \Uparrow (w\ \overline{Address}); \Uparrow (r\ \overline{Address}); \Downarrow (r\ \overline{Address}); \Downarrow (w\ Address); \Uparrow (r\ Address); \Downarrow (r\ Address)\}$. Note that this method is not proven to detect coupling faults and address decoder faults, but that march test BIST is

Table 15.8: Memory chips designed with BIST.

Aspect	Explanation
BIST application:	Embedded RAM, ROM, PLA
Test speed:	System operational speed
Benefits:	Vertical testing capability, less diagnostic run time, less run time, less test code development, less test complexity
Method:	Use extended counter as pattern generator, LFSR as response compacter, write memory address as data, read it back, then write & read back complemented address.
Example RAM size:	16 4-bit locations
Faults tested:	Stuck-faults and transition faults
Technology:	CMOS 2.5 μm
Aliasing probability:	0.000015 in a 16-bit LFSR

Device	Size	Type	# Flip-flops	Area overhead (%)		
				Logic area	Active area	Total area
A	8K	DRAM	427	9.8	5.5	3.4
B	8K	DRAM	171	18.0	7.6	5.0
C	16K	DRAM	284	20.0	5.5	3.8
D	4K	SRAM	210	11.0	4.2	2.9
E	–	–	–	11.8	6.8	4.2
F	4K	DRAM	239	6.8	4.6	3.2

able to do that, because march tests are proven to cover both types of faults.

15.3.4 Neighborhood Pattern Sensitive Fault Test DRAM BIST

Kinoshita and Saluja [358, 359] present an algorithm to test DRAMs for *static neighborhood pattern sensitive faults* (SNPSFs) (see Chapter 9.) The test has two parts: a MATS+ test (see Section 15.3.2) to test the address decoder, and a test for SNPSFs in a Type-1 neighborhood using the two-group method. The algorithm does not use parallelism, and can be also implemented on an ATE. The response compressor uses three simple count functions, described below The operation count is $58 \cdot n$, and the chip area overhead is 0.09% for a 1 Mb DRAM. This test is preferred over a pure march test for DRAMs, because its fault model matches the actual faults encountered in production.

Fault Model. The algorithm uses a static *weight-sensitive fault* (WSF) model. This fault changes the content of the base cell, depending on how many 1s exist in the deleted neighborhood. A t-*WSF* is a WSF which occurs when any deleted neighborhood pattern has t cells at logic one and the remaining $k - t - 1$ cells at logic 0, assuming neighborhood size k. A *positive* (*negative*) WSF only allows the base

15.3 Memory BIST

Step 0: {Assume all cells are initialized to 0};
Step 1: {Deleted neighborhood p2}
write 1 to all cells-A and all cells-B of group-1;
read all base cells 'b' of group-1;
write 0 to all cells-B of group-1;
Step 2: {Deleted neighborhood p3}
write 1 to all cells-D of group-1;
read all base cells 'B' of group-1;
write 0 to all cells-A of group-1;
Step 3: {Deleted neighborhood p5}
write 1 to all cells-C of group-1;
read all base cells 'b' of group-1;
write 0 to all cells-C of group-1;
Step 4: {Deleted neighborhood p6}
write 1 to all cells-B of group-1;
read all base cells 'b' of group-1;
write 0 to all cells-D of group-1;
Step 5: {Deleted neighborhood p4}
write 1 to all cells-C of group-1;
read all base cells 'b' of group-1;
write 0 to all cells-B of group-1;
Step 6: {Deleted neighborhood p1}
write 1 to all cells-A of group-1;
read all base cells 'b' of group-1;
write 0 to all cells-A and all cells-C of group-1;
Step 7-12: **Repeat** Steps 1-6 for group-2;

Figure 15.43: Algorithm to locate all positive, static 2-WSFs.

cell to change in the direction $0 \to 1$ $(1 \to 0)$ due to a fault. A test that detects all positive and negative static t-WSFs (for $0 \leq t \leq 4$) also detects all SNPSFs [688].

SNPSF Algorithm. This algorithm detects and locates SNPSFs, and has a separate part for each value of t to locate the static t-WSFs (see Figure 15.43.) The part for $t = 0$ has been deleted because it is trivial, and we only show the algorithm for locating positive, static WSFs, as the algorithm for negative ones is very similar. It uses the 2-group method (see Figure 9.28.) See van de Goor [688] for the detailed operation analysis of this algorithm [358, 359].

Test Response Compression. Kinoshita and Saluja used three *count functions* to compress the responses of the DRAM from read operations. Here, r_i represents the value returned by read operation i, and c represents the number of times a read

Table 15.9: Count function values for good/failing machines.

Entry number	Permutation of response string	Count function		
		$C1(\mathbf{R})$	$C2(\mathbf{R})$	$C3(\mathbf{R})$
1 (good machine)	0011	2	1	1
2 (bad)	1100	2	0	1
3 (bad)	1010	2	1	3
4 (bad)	0101	2	2	3

has been performed by a specific algorithm.

$$C1(\mathbf{R}) = \sum_{i=1}^{c} r_i \qquad (15.18)$$

$$C2(\mathbf{R}) = \sum_{i=1}^{c-1} \overline{r_i} \cdot r_{i+1} \qquad (15.19)$$

$$C3(\mathbf{R}) = \sum_{i=1}^{c-1} r_i \oplus r_{i+1} \qquad (15.20)$$

A count function detects a memory fault if, for the given test, the value of the count function differs from the reference, fault-free value. For $C1(\mathbf{R})$, the reference count values are 0 for 0-WSFs through 4-WSFs, and value n for *address decoder faults* (AFs.) The $C2(\mathbf{R})$ and $C3(\mathbf{R})$ reference counts are 1 for AFs. AFs are detected by $C1(\mathbf{R})$, $C2(\mathbf{R})$, and $C3(\mathbf{R})$. $C1(\mathbf{R})$ counts the number of $0 \rightarrow 1$ transitions. $C2(\mathbf{R})$ counts both $0 \rightarrow 1$ and $1 \rightarrow 0$ transitions. Table 15.9 shows all 3 count function values when a MATS+ test is applied to a memory with $n = 2$.

Implementation. This BIST method requires no changes to the memory cell array.

1. It has been implemented using a microcoded ROM control having 665 6-bit words, a row and column address counter, a $\log_2(n) + 1$ bit counter for $C1(\mathbf{R})$, a 2-bit counter for $C2(\mathbf{R})$, and a 2-bit counter for $C3(\mathbf{R})$ [359, 559]. The area overhead for a 64 kb RAM is 1.85% [559].

2. It was also implemented with a custom logic control, instead of the control ROM [559]. The area overhead decreases rapidly with the memory size n, because only the address counter size grows with increasing memory size. The area overhaed for a 6 kb RAM was 1.21%, for a 256 kb RAM it was 0.32%, and for a 1 Mb RAM it was 0.09%.

Several additional NPSF BIST algorithms exist. Majumder contributed *restricted* SNPSF algorithms for a Type-2 neighborhood [443, 444]. Static and dynamic *neighborhood pattern sensitive faults* (NPSF) in the Type-2 nine-cell neighborhood are detected in a parallel BIST environment. Multiple bit lines are selected in the BIST mode by a modified column decoder, and the same data is simultaneously written to several cells of the same word line. During reading, an additional

15.3 Memory BIST

multi-bit comparator checks the contents of the bit lines. Mazumder also gave an algorithm for ANPSFs and SNPSFs in a Type-1 neighborhood [444], and another BIST method based on pseudo-random techniques [441].

15.3.5 Transparent Memory BIST Tests

Transparent BIST eliminates the trouble of restoring the RAM contents after the system function has been interrupted for a periodic memory testing episode. The basic principle is that during testing, the memory stored data is complemented an even number of times. Assume that cell c contains bit v. Nicolaidis [495] then modifies the memory testing algorithm as follows:

1. Add initial memory read operations to the original algorithm.

2. Replace every $write - x$ operation on cell c with a $write - (x \oplus v)$ operation.

3. If the last write on cell c stored $\overline{v}$, add an extra read and an extra write operation to complement the cell data.

The sequence of values produced by the read operations is compacted into a signature. First, the test is run without any write operations, to calculate the signature. Then, the test is rerun with both the read and write operations. After all test patterns are applied, the actual signature is compared with the reference one.

The BIST controller must be augmented to generate not only the test sequence, but also the signature predicting sequence before the actual test. The controller must switch between the test application and signature production modes, to avoid write cycles during signature production. An extra register is needed to store the contents of addressed cells, which are used for test data generation. The transparent BIST area overhead for a 32 $k-$byte RAM with the MARCH C test was 1.2% [495], only 0.2% more than conventional memory BIST with MARCH C.

15.3.6 Complex Examples

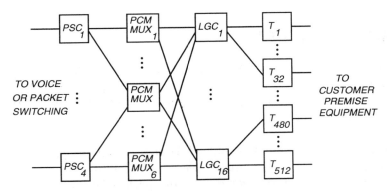

Figure 15.44: BIST of telephone exchange switch.

Example 15.16 Subsystem testing of a switching system at Lucent Tech. *This subsystem provides the Integrated Services Data Network (ISDN) capability in a telephone exchange [33]. Figure 15.44 shows a block diagram of this hardware. The inputs on the left side come from the digital voice or packet switching units. BIST was incorporated into the four PSCi chips, along with a hardware loopback capability so that any output data bus on the right side of the diagram could be looped back into any data input on the left side. In addition, intermediate points in the switch could be selected for loop back. More hardware is tested when the loop back includes the PSCi and Ti units. The BIST increased the PSC chip active area by 19.7%, but since the chip size had been increased to the minimum size for manufacturing, BIST did not increase its overall size. In terms of the total circuitry in this system, the overhead was only 4%. However, the extra hardware caused a minor decrease in yield, though in actuality, the cost was minor. The BIST area overhead at the circuit pack level was 1%. The system provided a 60% stuck-fault coverage, which was significantly better than what could be achieved by sourcing patterns at its inputs. This improvement in fault coverage simplified the writing of diagnostics, and reduced diagnostic run time by a factor of 8.*

Example 15.17 Combined sequential circuit and embedded RAM BIST [33]. *Table 15.10 shows an example of BIST used in a telephone exchange. The control RAM (CRAM) provides a control function. Figure 15.45 shows the block diagram of the CRAM memory system, and Figure 15.46 shows the automated circular BIST approach used. The* Pseudo LFSR *is a modified BILBO used simultaneously for both pattern and response compaction, built out of system flip-flops with added logic to XOR in the shifted bits from the prior BILBO stage during test. The method achieved nearly 98% fault coverage of stuck-at, transition, and neighborhood pattern sensitive faults for the memory and stuck-at faults for the random logic. Although this is outstanding, there is no guarantee here that the more exotic memory coupling faults and data retention faults will be detected, since the pattern sequence has not been proven to cover these.*

15.4 Delay Fault BIST

It is also possible to test circuits for timing delays using BIST. A delay fault BIST testing system has the standard BIST architecture, but with a hybrid pattern generator optimized to test both stuck-faults and delay faults replacing the standard LFSR pattern generator.

Motivation. For delay fault built-in self-testing, one of the problems that must be addressed is hazards in the circuit. Figure 15.47 illustrates the nature of this problem. Here, we are testing for distributed path delay for a falling transition (from logic 1 to 0.) The delay is caused by wire delays from A to the AND gate, delay in the AND gate, and additional wire delay from the AND gate to F. The

15.4 Delay Fault BIST

Table 15.10: BIST of embedded RAMs.

Aspect	Explanation
BIST application:	Full chip random sequential logic & embedded RAM
Levels of testing:	VLSI device, circuit pack, & system test
Product:	Electronic Switching System
Specific function:	Control RAM
Detail:	Controls 8K bit RAM read sequentially by a counter RAM written into/read from a processor interface
Technology:	Old: 2.5 μm CMOS, New: 1.25 μm CMOS
Redesign reason:	Technology update
BIST control:	Only 1 pin
Faults tested:	Stuck-at, RAM transition, RAM NPSF
BIST approach:	LFSR in a BILBO, used simultaneously for pattern generation & response compaction – non-linear circular BIST system, with loopback circuit
Approach reason:	Lower overhead than with linear BIST
Problem:	Signal correlation in pattern generation
Input isolation:	MUXes at input buffers driving combinational logic Blocking gates at input buffers driving flip-flops
Test sequencing	Two BIST sequences, combined into a single one
BIST patterns:	RAM initialized using RAM address as data, Signals from each LFSR propagate through circular chain – combined in Output Data Register. The same LFSRs compact RAM BIST & random logic BIST results
Signature:	Read from Output Data Register by processor
Signature Requirement:	*Good machine* signature must match old product. Done by seeding Output Data Register before BIST
Logic overhead:	29%

Area overhead:	CRAM function	RAM BIST	Logic BIST	Input isolation	Inte-gration	Total
(% Active area)	90.3 %	0.6 %	6.5 %	0.8 %	1.7 %	9.7 %
Fault coverage:	BIST 94 %	I/O Buffer external test 3.7 %			Total 97.7 %	Prior 96 %

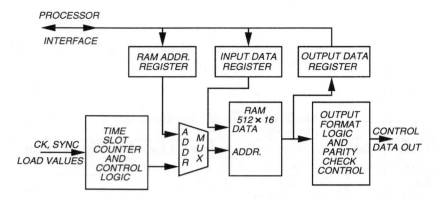

Figure 15.45: CRAM block diagram.

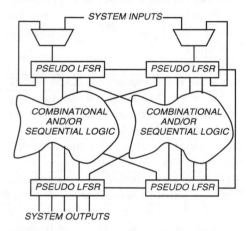

Figure 15.46: Automated circular BIST approach.

testing method is to apply two patterns, separated by the δt shown between the dotted lines. The path-delay fault testing requirement is that the off-path input (the OR gate output) must be a 1 in both time-frames. In the good machine timing diagram, the output F is sampled at a time given by the sum of δt and the *Path Delay Specification*. However, in the failing machine, where the signal is late propagating from A, a hazard appears at the logical sampling time, caused by the switch in the logic 1 value from signal B to signal C. Since the hazard arrives at the sampling time, testing is fooled into accepting a circuit with a timing defect as good. This *test invalidation problem* can be alleviated by using a delay fault BIST pattern generator that reduces the circuit hazard activity.

Delay Fault Testing Pattern Generation. One way to avoid hazards is to use a hardware pattern generator that creates *single-input changing* (SIC) patterns, so only one input changes during each clock period. This reduces the likelihood of generating hazards during testing, although some will still occur [593, 594]. Breuer and

15.5 Summary

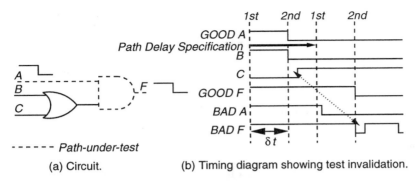

(a) Circuit.   (b) Timing diagram showing test invalidation.

Figure 15.47: Test invalidation by hazards during delay fault testing.

Nanda [97] used a Gray code pattern generator, which must be partitioned, because otherwise for n PIs, this pattern generator goes through a cycle of 2^n patterns, which is intractable for large n. Figure 15.48 shows a hybrid pattern generator patented by Shaik and Bushnell [593, 594]. When the MUX is set to the top feedback path, this is an ordinary External-XOR LFSR pattern generator. When the MUX is set to the lower feedback path, this uses a Johnson or moebius counter of length n, which leads to $2 \times n$ patterns.

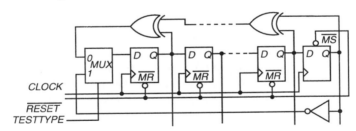

Figure 15.48: Hybrid delay fault testing hardware pattern generator.

15.5 Summary

BIST is now becoming more accepted as the preferred method of VLSI circuit testability insertion. This is because BIST hardware overheads have come down, particularly for memory BIST (1-3%), but also because BIST enables partitioning of the testing problem for large hardware systems. At present, memory BIST is widely used. Linear feedback shift registers, multiple-input shift registers, and built-in logic block observers are the most commonly used schemes to provide pattern generation and response compaction for BIST. Also, random logic BIST, while not yet fully accepted, has overheads of 13 to 20%, and experimental random logic BIST [111, 515, 516] has a chip area overhead of 6.5%. Random logic BIST has been used by IBM and Lucent Technologies. Finally, experimental systems for path-delay fault BIST also exist [515].

Problems

15.1 *Test length.* If $N = 15$ patterns are produced by an LFSR, and 2 of those patterns detect a given fault, say e stuck-at 0, what is the average test length T to detect e stuck-at-0?

15.2 *Standard LFSR.* Implement a standard LFSR for the characteristic polynomial $f(x) = x^8 + x^7 + x^2 + 1$. Write the system of equations with the *companion matrix* for this LFSR.

15.3 *Modular LFSR.* Implement a modular LFSR for the characteristic polynomial $f(x) = x^3 + x + 1$. Write the system of equations with the *companion matrix* for this LFSR.

15.4 *Standard LFSR.* Compute the first eight patterns generated by the standard LFSR with characteristic polynomial $f(x) = x^8 + x^7 + x^2 + 1$ and an initialization of "00000001," with the one in the least significant bit.

15.5 *Modular LFSR.* Compute the first eight patterns generated by the modular LFSR with characteristic polynomial $f(x) = x^3 + x + 1$ assuming that the LFSR was initialized to "001" with the one in the least significant bit.

15.6 *MISRs.* Figure 15.49 shows a multiple-input signature register of the STANDARD (external XOR) type. This MISR takes outputs from the circuit, labeled as A and B, and compacts their responses. Please convert this signature register into the equivalent MODULAR (internal XOR) type and draw this equivalent signature register. Although the equations representing these two response compacters are the same, the signatures will be different, so explain the relationship between the two signatures.

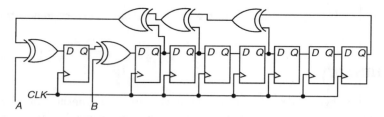

Figure 15.49: MISR for Problem 15.6.

15.7 *Weighted random patterns.* Apply four bits of the weighted pseudo-random pattern generator of Figure 15.16(b) to the four-input circuit: $f = (a \oplus b) \cdot (c \oplus d)$. For each of the four inputs, you can either chose one of X_7, X_6, X_5, X_4, X_3, X_2, X_1, X_0 (1/2 probability), $X_2 \wedge X_0$ (1/4 probability), $X_4 \wedge X_2 \wedge X_0$ (1/8 probability), or $X_6 \wedge X_4 \wedge X_2 \wedge X_0$ (1/16 probability.) If necessary, you may activate the *Inversion* signal to invert the probability. How many weight sets are needed to obtain 100% stuck-fault coverage for the faults in this circuit?

Problems

15.8 *Weighted random pattern generator.* Design a weighted pseudo-random pattern generator with programmable weights 1/2, 1/4, 11/32, and 1/16.

15.9 *Cellular automaton.* Build a four flip-flop rule 150 *cellular automaton* (CA), and compute its pattern sequence. Seed the CA pattern generator with "0001." What is the period of the cellular automaton? Compute the pattern sequence of the four flip-flop LFSR with characteristic polynomial $f(x) = 1 + x^4$. What is the LFSR's period? Is the CA better than the LFSR, and if so, then why?

15.10 *Maximal LFSR.* Design a 3-bit maximal LFSR, but please add hardware to map the test-pattern "010," which is not useful, into the pattern "000," which detects several circuit faults.

15.11 *Aliasing probability.* Using Figure 15.22, please compute the probability of aliasing for an error vector e with error probability $p = 0.3$, where a 15-bit LFSR is used for response compaction.

15.12 *Fault detection.* In Figure 15.23, can the transition counter detect the multiple stuck-at fault both b and c stuck-at-0? Can the LFSR detect this fault?

15.13 *LFSR enhancement.* Design test pattern embedding hardware to control an LFSR with the characteristic polynomial $f(x) = 1 + x + x^3$ to produce an all-zero test pattern. This problem also requires you to design the actual LFSR. Is this less hardware than just implementing a 3-bit binary counter?

15.14 *Aliasing analysis.* Consider the BIST system in Figure 15.50. Circuit inputs are A, B, and C, which are generated by the LFSR, and the outputs are Y and Z, which are compacted by the output MISR. The LFSR is initialized to 001 (i.e., the bottom LFSR flip-flop is set to 1 and the other two are cleared) and the MISR is initialized to 000. The circuit is clocked for eight periods to produce this test sequence:

$$\begin{array}{c|c} L_1 & 0\ 1\ 0\ 1\ 1\ 1\ 0\ 0 \\ L_2 & 0\ 0\ 1\ 0\ 1\ 1\ 1\ 0 \\ L_3 & 1\ 0\ 0\ 1\ 0\ 1\ 1\ 1 \end{array}$$

The LFSR and the MISR are wired to the same clock line, and are fault-free. Explain why aliasing does not occur for the fault e stuck-at 0, even though it is expected since the test 001 for this fault is repeated twice. What are the final good machine and bad machine signatures for the fault e stuck-at-0, eight clock periods after the LFSR and the MISR were initialized?

15.15 *Fault detection.* For Problem 15.14, find which of these faults are detected:

A sa0	A sa1	B-e sa0	B-e sa1	C-e sa0	C-e sa1

15.16 *Fault detection.* For Problem 15.14, find which of these faults are detected:

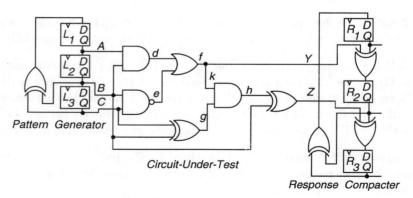

Figure 15.50: BIST system for Problem 15.14.

| B sa0 | B sa1 | B-g sa0 | B-g sa1 | f sa0 | f sa1 |

15.17 *Fault detection.* For Problem 15.14, find which of these faults are detected:

| C sa0 | C sa1 | C-g sa0 | C-g sa1 | f-Y sa0 | f-Y sa1 |

15.18 *Fault detection.* For Problem 15.14, find which of these faults are detected:

| B-d sa0 | B-d sa1 | B-Z sa0 | B-Z sa1 | f-k sa0 | f-k sa1 |

15.19 *Signature computation.*

(a) For the circuit in Figure 15.51, please design an external-XOR LFSR pattern generator implementing the characteristic polynomial $1 + x^2 + x^3$ and an Input MUX for testing.

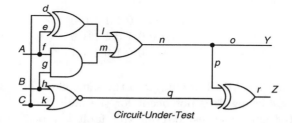

Figure 15.51: Circuit for BIST Problem 15.19.

(b) Express the linear system of matrix equations describing this pattern generator.

(c) Now, assume an exhaustive counter-based pattern generator and the given response compacter for the circuit in Figure 15.52. Compute the good machine signature for the circuit and design a single NAND gate

Problems

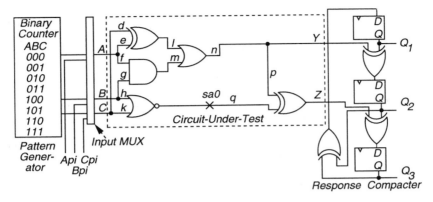

Figure 15.52: Circuit for Problem 15.19 with BIST hardware.

signature comparator that outputs a logic 0 on the $\overline{GOOD}$ signal when the circuit is good, and a logic 1 when it is faulty. The output MISR is initialized to "000" before testing.

(d) For the same circuit, compute the bad machine signature for the fault q stuck-at-0. Does the test hardware alias for this fault?

15.20 *STUMPS.* For the circuit of Figure 15.53, devise a STUMPS testing system. Drive the three scan chains with a maximal 3-bit LFSR, initialized to "001," and compress the output of the scan chains with a 3-bit MISR, initialized to "000." Simulate the system for 12 clock periods and provide the final signature. Discuss the pros and cons of STUMPS.

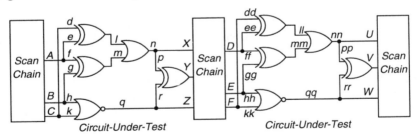

Figure 15.53: Circuit for BIST Problem 15.20.

15.21 *MATS+ memory BIST.* Implement the logic level hardware realizing the test controller documented in Figure 15.41 for the MATS+ memory BIST system.

15.22 *MARCH X memory BIST.* Implement the state transition diagram for the memory BIST controller for the MARCH X test described in Table 9.15.

15.23 *BIST system.*

(a) For the circuit in Figure 15.54, please design a 4-bit internal-XOR (modular) LFSR pattern generator implementing the characteristic polynomial

$1 + x + x^4$ and an Input MUX for testing. Note that one of the bits of the pattern generator will be wasted, as there are only 3 circuit inputs. Please include appropriate reset hardware.

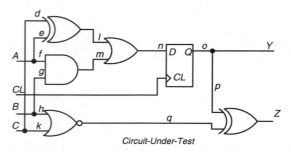

Figure 15.54: Circuit for built-in self-testing.

(b) Express the linear system of matrix equations describing this pattern generator.

(c) Now, design an external-XOR (standard) MISR for this circuit, using the same characteristic polynomial $1+x+x^4$. Note that only two of the MISR bits take circuit outputs (since the circuit has only two outputs.) Add appropriate initialization hardware for the MISR. What is the advantage of using a 4-bit MISR, rather than a 2-bit one?

(d) Express the linear system of matrix equations describing this MISR.

(e) In production testing, it was found that a significant number of these circuits failed due to the incorrect signature. Yet, *failure effect analysis* indicated that there were no defects (nothing was wrong) in these rejected circuits. Please explain what is wrong with this BIST testing process, and how to fix it.

15.24 *Up/Down LFSR.* Design a 4-bit LFSR using the tables in Appendix B and the up/down LFSR counting method of Section 15.3. The LFSR should take an $Up/\overline{Down}$ input signal to indicate whether it counts up or down. Also, design the LFSR to initialize to 0001.

Chapter 16

BOUNDARY SCAN STANDARD

"*Circuitry that may be built into an integrated circuit to assist in the test, maintenance, and support of assembled printed circuit boards . . . includes a standard interface through which instructions and test data are communicated. A set of test features is defined, including a boundary-scan register, such that the component is able to respond to a minimum set of instructions designed to assist with testing of assembled printed circuit boards.*" — Test Technology Standards Committee of the IEEE Computer Society, in "IEEE Standard Test Access Port and Boundary-Scan Architecture" [318].

In the 1970s, the *in-circuit testing* (ICT) method appeared, in which *printed circuit boards* (PCBs) are tested by probing the backs of the boards with nails. The component technology of that era was *dual in-line* (DIP) packages, which were attached to a PCB by drilling two rows of holes, inserting the DIP into the holes, and wave soldering the leads to the hole (or via), which was plated on the inside with metal. This testing mechanism relies on *nails* in a *bed-of-nails* tester. The nails are positioned to hit various solder bumps on the back of the PCB and they force various signal values on the component inputs, and measure the component output signals at various other solder bumps. This very significant testing development, however, was only partially automated, and still relied extensively on ad hoc testability insertion. Scan design was another main testing development of the 1970s that initially succeeded in large, vertically integrated electronics companies. Its main contributions were a great improvement in testability of state machines, and also an automated mechanism for inserting testability.

Parker [511] points out that the IEEE/ANSI* JTAG† 1149.1-1990 [318, 436] *boundary scan* standard gives us a standard mechanism for very different segments of the electronics industry to support testing. The main virtue of the 1149.1 standard and the more recent 1149.4 *analog test bus* standard is that they can be used by board designers, IC designers, and systems designers, without the need for members

---
*Acronym for *American National Standards Institute.*
†Acronym for *Joint Test Action Group.*

of each design community to fully understand the testing problems of the other communities. In addition, the *standard test port* features of these standards provide a standard way to deliver test vectors to electronics sub-assemblies. However, the port also greatly facilitates *built-in self-testing* (BIST), particularly when it must be implemented hierarchically over several levels of packaging, such as chips, boards, and systems. Boundary scan is actually a collection of *design rules*, applied at the IC level, but benefiting the testing at all of the above-mentioned system packaging levels. The port can even be used in functional operation modes of electronics to read out internal status information that has nothing to do with testing. Parker states that at present, a majority of programmable devices and custom ICs use the 1149.1 standard [511]. Keep in mind that miniaturization continues to be the main driver of the electronics industry, because the cost reductions that it produces open up vast new consumer markets. Consider the effect of notebook PCs, cellular telephones, and personal data assistants. The miniaturization, in turn, also reduces costs for segments of the industry that do not need miniaturization, so all segments benefit. Another benefit of boundary scan is that the standard enables testing software to effectively and quickly deal with board testing problems that previously required ad-hoc testing methods [733, 734]. According to Kajitani [345], a complex board test was created at least ten times more rapidly than expected, without any debugging required, due to the use of the boundary scan standard.

16.1 Motivation

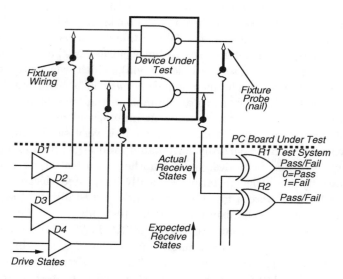

Figure 16.1: Bed of nails tester concept.

In-circuit testing with bed-of-nails tester has become increasingly untenable. Figure 16.1 [511] shows the concept of the tester, and Figure 16.2 [511] shows an

16.1 Motivation

actual tester. The divide-and-conquer strategy of the bed-of-nails tester lets us test components in the circuit as if they were actually standing alone. However, this has the weakness that an open in an input line to an IC will cause the IC to appear to fail, when in fact it is good. Nonetheless, the technique has been enormously successful [172, 509]. Throughout the 1970s and 1980s, DIP packages dominated

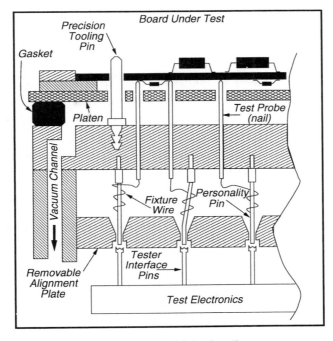

Figure 16.2: Actual bed of nails tester.

IC packaging, so every board signal was visible on the bottom of the board and the pins were spaced on 100 mil (tenth-inch) centers. Therefore, in-circuit testing nails[‡] were arranged on 100 mil centers. *Surface-mount technology* (SMT) with flat-pack chip packages replaced dual in-line packages, because of a need to reduce inductance in the PCB by reducing the package height above the ground plane. Board inductance is particularly an electrical problem at higher digital clock rates. Another benefit of SMT was further miniaturization and, therefore, a cost reduction. Components are now soldered onto a single side of the PCB without drilling holes in the PCB, so we can now mount components on both sides of PCBs to further reduce cost. SMT technology means that there are no through-hole pin targets with solder bumps for nails to hit. Also, some signals may never appear on the bottom side of a PCB (particularly if they are routed in intermediate layers of multi-layer boards), so these signals are simply unavailable for in-circuit testing. Mounting components

---

[‡]A nail actually was circular and had a waffle pattern of sharpened points, which eventually collected solder flux and formed a resistive barrier. More recent nails are designed for specific test pads on a PCB and have a single, very sharp point to pierce any resistive coatings [511].

on both sides of the PCB is the *coup de grace* (mortal stroke) for in-circuit testing. The reduced spacing between PCB wires with SMT technology is not necessarily a problem for in-circuit tester nails if special test pins are provided. All of this resulted in the need to replace the PCB test delivery system to the component. It also has led to a resurgence in interest in *functional* automatic test-pattern generation methods, as in-circuit testing becomes more difficult. The JTAG 1149.1 Boundary Scan Design Standard fulfills the former need for a PCB test delivery system to the component. Parker [511] claims that boundary scan may actually prolong the life of in-circuit testing methods, since it greatly reduces the number of test pads needed on the PCB.

The additional benefits of a standard System Test Port and Bus are that the system designer can integrate components from different vendors, and still have a standard testing mechanism. Also, one chip can supply the test hardware to support several other chips.

16.1.1 Purpose of Standard

Boundary scan provides the following major modes of operation:

- In *non-invasive* mode, the standard provides resources guaranteed to be independent of the rest of the logic (called the *system logic*) in the IC [511]. These resources enable asynchronous communication with the outside world to serially read in test data and instructions or serially read out test results. The activities are invisible to the normal IC behavior.

- The *pin-permission* modes of the standard take control of the IC input/output pins, thus disconnecting the system logic from the outside world. These modes allow testing of the system interconnect separately from component testing, and also allow testing of components separately from system interconnect testing. The testing activities totally disrupt the normal IC behavior.

All ICs adhering to the 1149.1 standard must be designed to power-up in non-invasive mode. After an IC is switched to one of the pin-permission modes, great care must be exercised in returning the IC to non-invasive mode. Bus driver conflicts (simultaneously driving a bus to both logic 0 and 1) can easily occur, unless the IC is properly reset.

The 1149.1 standard also allows delivery of built-in self-test mode commands (e.g., *RUNBIST*) through the JTAG hardware to the component-under-test. This eliminates excessive shifting that occurs when pure external testing and boundary scan are used to deliver the test-vectors to a component. Instead, an on-chip pattern generator can be activated by the JTAG *RUNBIST* command to eliminate all of that vector shifting. The 1149.1 standard operates at the chip, PCB, and system test levels. It also lets the test engineer control tri-state signals during testing to avoid system burnout. The JTAG standard also lets the test engineer use other chips on the PCB to collect responses from chips-under-test.

The 1149.1 standard is highly extensible, so designers can add non-invasive or pin-permission modes of operation to the JTAG hardware. This not only gives greater benefit from the hardware invested in supporting the JTAG standard, but also allows the creative invention of additional testing and non-testing system modes of operation.

16.2 System Configuration with Boundary Scan

Figure 16.3 [511] shows an integrated circuit that is compliant with the 1149.1 boundary scan standard. Note that on each pin of the chip, there is internal hardware that provides a register at that pin position. The serial connection of these registers around the periphery of the chip at the pins is known as the *boundary register*. Input pins can drive the internal system circuit through this internal pin hardware, or the *boundary register cell* for the particular input pin can be loaded by serially shifting a pattern into the boundary register, and the value at that pin can be used to drive the system circuitry. Similarly, the output of the system circuitry can directly drive an output pin, or the output of the system circuitry can be caught in the boundary register cell for that pin, and then serially shifted out of the chip. The *TDI* pin is the serial input to the boundary register, and the *TDO* pin is the serial output from the boundary register. Between *TDI* and *TDO*, a number of registers provided by the boundary scan hardware can be connected, depending on the current mode of the test hardware.

16.2.1 TAP Controller and Port

We now explain the remaining boundary scan control signals and registers. Several *boundary scan data registers*, including the *boundary register*, the *Device ID register*, and the *bypass register*, can be connected serially between *TDI* and *TDO*. Also, the *instruction* register can be connected serially between *TDI* and *TDO*. The *Device ID* register provides the device identification. The *bypass* register bypasses the boundary register for this component. This is useful when all boundary registers of all components on the PCB are chained together into one long shift register, and it is desired to reduce the length of the register by ignoring hardware on components that are not involved in the current test. The *instruction* register can be loaded with an instruction, which enables various different operation modes of the test hardware. Several instruction modes are mandatory, others are optional, and user-defined instructions can be added, subject to the constraints of the JTAG standard. The *TCK* pin provides the test clock for the boundary scan hardware, and must be capable of operating at an independent clock rate from the system clock rate, asynchronously from the system circuitry. The *TMS* pin provides the *test mode select* signal, which causes the testing hardware to enter various testing modes. Finally, the **optional** $TRST^*$ signal provides an asynchronous reset capability for the boundary scan hardware. The TDI, TDO, TCK, TMS, and $TRST^*$ pins form the *Test Access Port* (TAP), and may not be shared with any other system function.

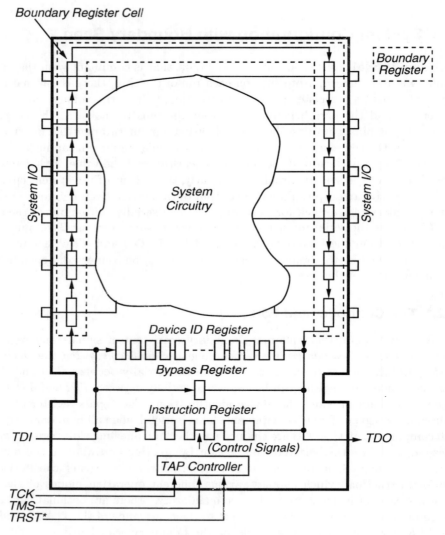

Figure 16.3: Schematic of system test logic.

16.2 System Configuration with Boundary Scan

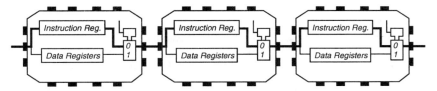

Figure 16.4: Loading of instruction registers via JTAG.

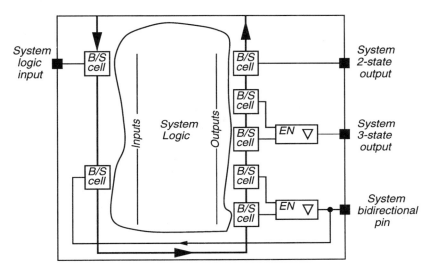

Figure 16.5: System view of interconnect.

Active-high signals have no asterisk, where as active-low signals do. These pins are used with a simple protocol to communicate with on-chip boundary scan logic [511]. The JTAG standard requires that *TMS*, *TDI*, and *TRST** float to logic 1, perhaps by the use of an internal pull-up resistor, if they are unconnected in a PCB or *multi-chip module* (MCM.) This property enhances fail-safe behavior of the system. Finally, the *TAP Controller* is a simple finite state machine that recognizes the boundary scan communication protocol and controls the operation of the boundary scan test hardware through internal signals. Only the *TCK*, *TMS*, and *TRST** signals are allowed to influence the TAP Controller. Figure 16.4 shows how the instruction registers in multiple chips can be loaded through the TAP, by shifting the instructions for multiple chips serially through the *TDI* and *TDO* pins connecting the chips [504].

The boundary scan standard allows considerable flexibility in configuring the test data paths in a PCB/MCM. Figure 16.5 [318] shows the conceptual view of the system interconnect when using the boundary scan standard. Figure 16.6 [318] shows a view of the entire system from the boundary scan chain point of view. Figure 16.7 [318] shows the hardware in the elementary boundary scan cell. Boundary scan chains can be configured in many different ways in a PCB/MCM. Fig-

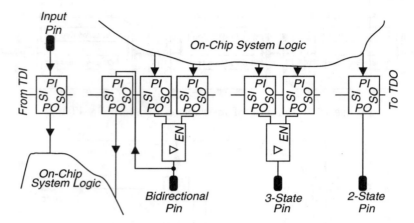

Figure 16.6: Boundary scan chain view of system.

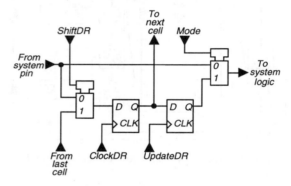

Figure 16.7: Elementary boundary scan test cell.

ure 16.8 [318] shows how a PCB/MCM is configured so that there is only one boundary scan chain in the entire PCB/MCM, and all cell boundary scan data registers are serially loaded and read from this single chain. The advantage of this configuration is that only two pins on the PCB/MCM are needed for boundary scan data register support. This disadvantage is very long shifting sequences to deliver test patterns to each component, and to shift out test responses. This leads to excessive time on the external tester, and expensive tests. Figure 16.9 [318] shows how the single long boundary scan chain was broken into two boundary scan chains, which share a common test clock (*TCK*.) The extra PCB/MCM pin overhead is one more pin (a second *test mode select* (*TMS2*) pin.) The advantage is that there are now two boundary scan chains, so the test patterns are half as long, and test time is roughly halved. Note that both chains share common *TDI* and *TDO* pins, so when the two top chips are being shifted using *TMS1*, the bottom two chips must be disabled so that they do not drive their *TDO* lines. The opposite must hold true when the bottom two chips are being tested. Finally, Figure 16.10 [318] shows a configuration where every chip has an independent boundary scan chain,

16.2 System Configuration with Boundary Scan

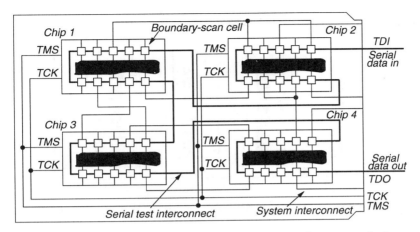

Figure 16.8: Board/MCM with one serial boundary scan chain.

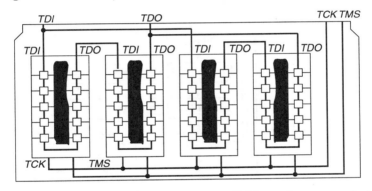

Figure 16.9: Board/MCM with two parallel boundary scan chains.

but all chips share common *TCK* and *TMS* signals. The extra overhead for this is $2 \times no\_chips$, since each chip has its own $TDIi$ and $TDOi$ pins. However, test pattern and response reading shift sequences are as short as they can possibly be. Note in these configurations that all components controlled by a common *TMS* pin will always be in the exact same TAP Controller state at the same time.

The *TAP Controller* must adhere to the standard finite state machine state diagram shown in Figure 16.11 [318]. Also, the TAP Controller must follow the timing of its various signals shown in Figure 16.12 [318]. Figure 16.13 [318] shows the necessary power-up reset logic, which is needed to bring the TAP Controller into a known initial state at system power-up.

16.2.2 Boundary Scan Test Instructions

We now simply describe the various JTAG TAP Controller test instructions. Note that the *IDCODE* and *USERCODE* instructions, as well as some of the other instructions, can be useful in normal system mode operation, and not just in test

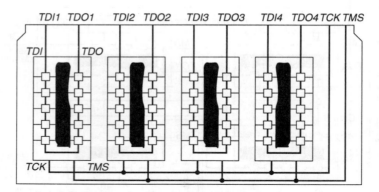

Figure 16.10: Board/MCM with independent path boundary scan chains.

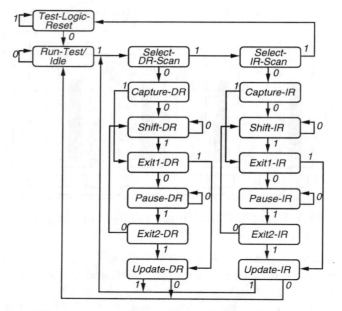

Figure 16.11: TAP controller state diagram.

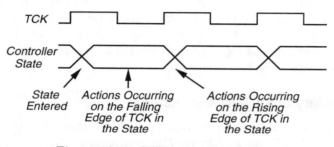

Figure 16.12: TAP controller timing.

16.2 System Configuration with Boundary Scan

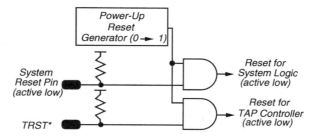

Figure 16.13: TAP controller power-up reset logic.

mode operation.

SAMPLE/PRELOAD Instruction

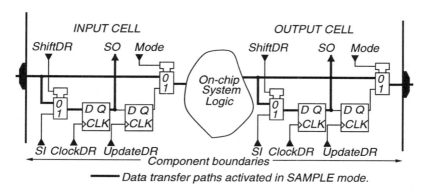

Figure 16.14: *SAMPLE* mode.

The purpose of the *SAMPLE/PRELOAD* instruction is to obtain a snapshot of the normal component input and output signals and store them in the first of the two master-slave flip-flops in the boundary scan ring. This is shown in Figure 16.14 [318], where the pin inputs are passed to the system logic, and the system logic outputs are also passed to the pin outputs. In addition, however, the input and output pin values are also captured in the first master-slave flip-flop controlled by the *ClockDR* signal, which sits on the boundary scan ring. This instruction puts data onto the boundary scan register before another instruction occurs. The boundary scan ring also prevents the shifting of signals on the boundary scan chain, and the glitching that they would cause, from being passed directly to the on-chip system logic. This is accomplished by the second master-slave hold flip-flop, clocked by the *UpdateDR* signal, which appears between the boundary scan ring flip-flop and the MUX just before the on-chip system logic on the input pin circuit. For the output pin circuit, this flip-flop lies between the boundary scan ring flip-flop and the MUX just before the output pin. It is extremely desirable that transient signals appearing on the boundary scan ring be isolated from the system logic by this flip-

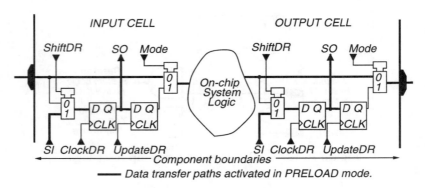

Figure 16.15: *PRELOAD* mode.

flop to avoid driving busses with conflicting 0 and 1 signals simultaneously. When the boundary scan ring is correctly loaded with the signal, the *UpdateDR* signal activates the filtering flip-flop and stores the signal to be passed to the system logic or the output signal pin in it, as shown in Figure 16.15 [318].

EXTEST Instruction

The purpose of the *EXTEST* instruction is to test off-chip circuits and board-level interconnections independently of the chip. This is achieved by capturing the signals coming into the chip in the boundary scan register, and also by driving the signals coming out of the chip from the boundary scan register. The hold latches in the boundary scan register are held at their prior values, to avoid disturbing the on-chip system logic with the various hazards and changing signals caused by the various shifts of the boundary scan register. In this testing mode, the test vector is shifted into the boundary scan register, and also applied at the PCB/MCM inputs. The response to the test vector is then captured in the boundary scan register and the PCB/MCM outputs. It will be necessary to shift the response out of the boundary scan registers of all chips on the board to see what the total response was to the vector. Figure 16.16 [318] shows the signal flows when the system is in this testing mode. *EXTEST* may leave the on-chip system logic in an indeterminate state, so it may be necessary to reset this logic to return to normal (system) operation.

INTEST Instruction

The purpose of the **optional** *INTEST* instruction is to conduct a test of the on-chip system logic when the chip is assembled onto the PCB/MCM, by the use of externally-applied test vectors shifted into the chips through the boundary scan register. This instruction also facilitates shifting of the response of the on-chip system logic to the vector out through the boundary scan register. In Figure 16.17 [318], we see that the first phase of the *INTEST* instruction causes the hold registers (those clocked with *UpdateDR*) to be updated with the present contents of the boundary scan register (those flip-flops clocked by *ClockDR*.) Then, in the second phase, the

16.2 System Configuration with Boundary Scan

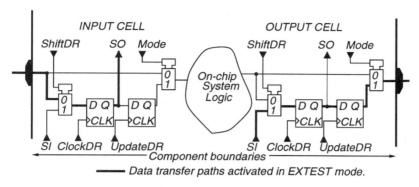

Figure 16.16: $EXTEST$ mode.

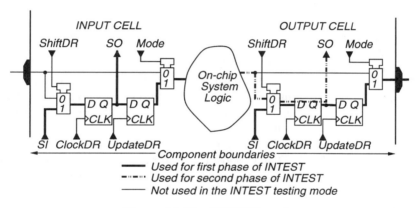

Figure 16.17: $INTEST$ mode.

boundary scan register (those flip-flops clocked by $ClockDR$) is updated with the values coming in from the SI signal. In this mode, arbitrary numbers of TCK pulses can be applied, as necessary, to shift the boundary scan register appropriately to shift in the test vector. Then, the on-chip system logic is single stepped by applying *only one* system clock pulse. The system response is captured in the boundary scan register. Then, another arbitrary number of TCK pulses are applied to shift the chip response out of the chip. This mode lets an external tester shift component responses out of the component-under-test. With the $INTEST$ instruction, the chip output pins can either be forced into the high-impedance state, or they can be driven by the boundary scan hold register. In Figure 16.17, the first phase corresponds to the thick black line path, while the second phase corresponds to the hatched data movement path. Figure 16.18 shows how even the system clock can be driven from the boundary scan data register.

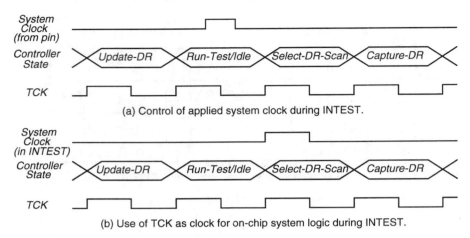

Figure 16.18: *INTEST* instruction clocking.

RUNBIST Instruction

The purpose of the **optional** *RUNBIST* instruction is to issue BIST commands to a component through the JTAG hardware. The test logic can control the state of the component output pins, which can be determined by the pin boundary scan cell, or the output pin can be forced into the high-impedance state. The BIST result indicating success or failure can be left in a boundary scan cell or in an internal cell in the on-chip system logic. The BIST result can be shifted out through the boundary scan register. Since *RUNBIST* may leave the chip pins in an indeterminate state, a system reset may be required before normal system operation can resume.

CLAMP Instruction

The purpose of the **optional** *CLAMP* instruction is to force component output pin signals to be driven by the boundary-scan register. This instruction bypasses the boundary scan chain between *TDI* and *TDO* by using the one-bit *bypass register* instead. One may have to reset the on-chip system hardware to prevent circuit damage caused by shorting zeroes and ones simultaneously onto internal busses after the *CLAMP* instruction has been used.

IDCODE Instruction

The purpose of the *IDCODE* instruction is to connect the component *device identification register* serially between the *TDI* and *TDO* pins in the *Shift-DR* TAP Controller state. This allows a board-level test controller or external tester to read out the JEDEC [§] component ID. This JTAG instruction is **required** whenever a JEDEC identification register is included in the chip design. Figure 16.19 shows the required layout of the device identification register.

[§] Acronym for *Joint Electron Device Engineering Council.*

16.2 System Configuration with Boundary Scan

MSB						LSB
31	28	27	12	11	1	0
Version		Part number		Manufacturer identity		'1'
(4 bits)		(16 bits)		(11 bits)		(1 bit)

Figure 16.19: JEDEC code for Device ID register.

USERCODE Instruction

The *USERCODE* instruction is intended for user-programmable components, such as *field-programmable gate arrays* (FPGAs) and *electrically erasable and programmable ROMs* (EEPROMs). The *USERCODE* instruction allows an external tester to determine the user programming of a programmable component. This is necessary because a user-programmable component can be personalized to act like an arbitrary large number of different devices, so the *USERCODE* instruction reads out the register indicating which programming is present in the programmable device. The *USERCODE* instruction selects the *device identification register* to be serially connected between *TDI* and *TDO* JTAG pins. The user-programmable ID code is loaded into the device identification register on the rising *TCK* edge. This instruction switches the component test hardware to its system function. The *USERCODE* instruction is required when the device ID register is included in the user-programmable component.

HIGHZ Instruction

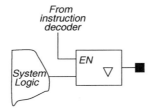

Figure 16.20: Conceptual pin control by *HIGHZ* instruction.

The **optional** *HIGHZ* instruction puts all component output pin signals into the *high-impedance* (Z) state. This prevents damage to logic on this particular chip and to other components in the PCB/MCM when the various JTAG test instructions are used. It may be necessary to reset the component after the *HIGHZ* instruction is issued, before returning to the system mode of operation. Figure 16.20 [318] shows the operation of the *HIGHZ* instruction.

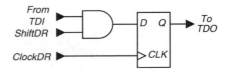

Figure 16.21: Bypassing of boundary scan chain for a single component.

BYPASS Instruction

The purpose of the *BYPASS* instruction is to bypass the boundary scan chain with a one-bit *bypass register*. This is useful in PCBs/MCMs where all components have their boundary scan chains connected serially, but only one component is being tested. The *BYPASS* instruction makes all other components appear to be have only one-bit long boundary scan registers. Figure 16.21 shows how the entire boundary scan chain on a chip looks conceptually in bypass mode. Figure 16.22 shows how testing is done for IC57 by bypassing the pins of IC56 and IC58 [504]. Table 16.1 lists all of the optional and required JTAG instructions.

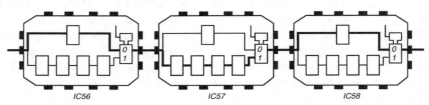

Figure 16.22: Bypassing ICs in BYPASS mode.

Table 16.1: Optional/required JTAG instructions.

Instruction	Status
BYPASS	Mandatory
CLAMP	Optional
EXTEST	Mandatory
HIGHZ	Optional
IDCODE	Optional
INTEST	Optional
RUNBIST	Optional
SAMPLE/PRELOAD	Mandatory
USERCODE	Optional

16.2.3 Pin Constraints of the Standard

Observe-Only Scan Cell

Figure 16.23 [318] shows how the JTAG standard allows *observe-only* scan cells in the boundary scan register. Figure 16.23 shows such a cell, in which the hold

16.2 System Configuration with Boundary Scan

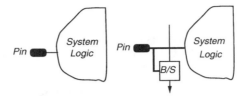

Figure 16.23: Observe-only scan cell.

register, the data path from the hold register through the MUX closest to the on-chip system logic, and the MUX itself have been deleted. This type of cell can only sample the signal coming into the chip pin, but it cannot control the signal going into the on-chip system logic. Contrast this with the normal *control-and-observe* scan cell in Figure 16.24 [318].

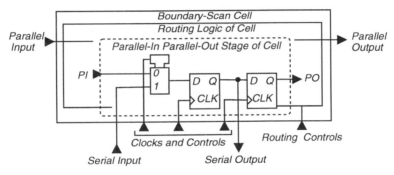

Figure 16.24: Control-and-observe scan cell.

Bidirectional Pins

Figure 16.25 [318] shows the required treatment of bi-directional pins under the JTAG standard. One bi-directional pin requires three boundary scan cells in the boundary scan register, one to read the pin in *input* mode, one to drive the pin in *output* mode, and a third to drive the pin *enable* signal, which determines whether the pin is driven by the output boundary scan cell or left in the high-impedance state.

One-Pin Control of Multiple Tri-State Pins

Since large numbers of bi-directional pins consume huge numbers of boundary scan cells, the JTAG standard allows one boundary scan enable cell to control multiple output drive enable signals for multiple pins. Figure 16.26 [318] shows how one drive *Enable* line controls four tri-state cells, so the total boundary scan cell requirement is only five cells. However, Figure 16.27 [318] shows an example of illegal boundary scan cell use. In this example, an input pin signal from a boundary

Chapter 16. BOUNDARY SCAN STANDARD

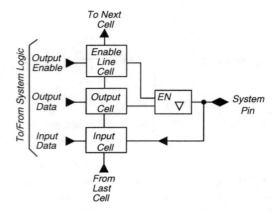

Figure 16.25: Bidirectional pin.

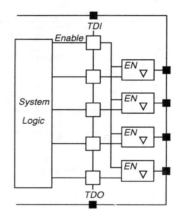

Figure 16.26: One-pin control of multiple tri-state pins.

scan register is used both to control an output pin, and as an input directly to the system logic, which is illegal under the standard.

Data Non-Inversion Requirement

Figure 16.28 [318] shows how data must not be inverted (using the boundary scan standard) as it flows from input pins into the on-chip system logic (Figure 16.28(a)), as it flows from input pins into the boundary scan register (Figure 16.28(b)), and as it flows from the boundary scan register into the system logic (Figure 16.28(c).)

Cell Delay Constraints and Measurements

The delay between the falling TCK edge and changes at the component output pins may be skewed. Since boundary scan cells contain flip-flops, they have setup times, hold times, and propagation delays from D to Q. Figure 16.29 [318] shows the

16.2 System Configuration with Boundary Scan

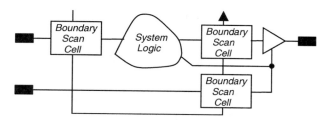

Figure 16.27: Illegal cell use.

method for measuring pin setup and hold times. Figure 16.30 [318] shows how to measure pin propagation delay. One may also need to avoid simultaneous component output switching to save power and avoid component burnout during testing.

Note that scan register cells with latched parallel outputs may be reset to either logic '0' or '1' when the *Test-Logic-Reset* TAP Controller state is entered or on the first falling *TCK* edge in the *Test-Logic-Reset* TAP Controller state.

Board-Level Bus Testing

Figure 16.31 [318] shows how a bus is tested in the PCB/MCM. There are four components with pins labeled A, B, C, and D, respectively, on the bus. Pins A, B, or C can drive the bus. In each test session, one and only one of these three drivers must be allowed to drive the bus. Pin D can read the bus. One possible test would be to enable A, disable B and C, and drive a 0 through A's data boundary scan cell, and read the other end of the bus at pin D. The *EXTEST* instruction, along with the *SAMPLE/PRELOAD* instruction, would be used for this test. First the 6-bit control patterns for A, B, and C are shifted into the boundary data register. Then, pin D is read into the boundary scan register, and the reading from D is shifted out through the *TDO* pin where an external tester can check if a 0 was correctly transferred from A to D. This test is repeated to test data transfer of a 1 from A to D. Data transfer from B to D, and from C to D, is checked in a similar fashion.

Figure 16.32 [318] shows how a memory system bus can be burnt out during testing by careless use of the boundary scan hardware to test a system. The boundary scan data register between *TDI* and *TDO* is used to shift in vectors for the system logic, clock the system logic once, capture the response of the system logic in the boundary scan data register output pins, and shift the output response serially through *TDO* where an external tester can check its correctness. Unfortunately, the test engineer did not take care to place the three system logic output pins in the logic '1' state by loading '1's in the boundary scan pin hold register. Notice that these three pins control active-low *chip-select* (*CS*) pins on three memory chips, whose outputs are bussed together. Therefore, during testing random signals are presented to the three *CS* pins, and frequently multiple *CS* pins are set to logic '0', which enables the corresponding chip. As a result, multiple memory chips are active, and frequently drive conflicting '0' and '1' signals onto the same pin of the *Bus* at the right of Figure 16.32. This ultimately burns out the transistors on the

Chapter 16. BOUNDARY SCAN STANDARD

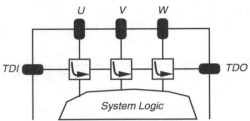

(a) Data must not invert as it moves from input pins into the boundary scan register.

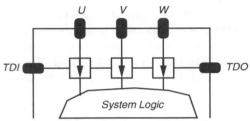

(b) Data must not invert as it moves from input pins into the system logic.

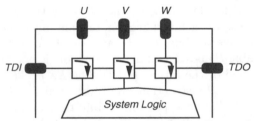

(c) Data must not invert as it moves from boundary scan register into the system logic.

Figure 16.28: Data non-inversion requirement.

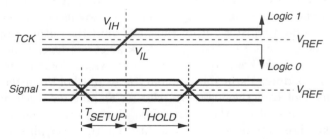

Figure 16.29: Setup and hold timing measurement method.

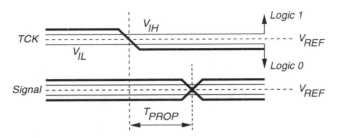

Figure 16.30: Propagation delay measurement method.

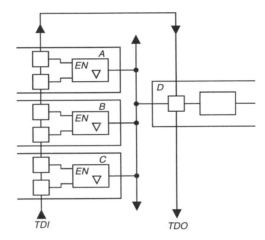

Figure 16.31: Board-level bus testing.

memory chips driving the *Bus*.

16.3 Boundary Scan Description Language

The *Boundary Scan Description Language* (BSDL) [318] was added to the JTAG Boundary Scan Standard to provide a standard means of communicating information about the boundary scan hardware on a chip to users of the chip and to CAD tools through the VHDL hardware description language. BSDL facilitates communication of information describing test logic in parts between companies and CAD tools. BSDL can be used by *automatic test-pattern generators* to generate chip test patterns, and by *high-level* and *logic synthesis* tools to synthesize test logic. BSDL is not usable as a simulation model, since it cannot describe voltages, currents, or timing. It is implemented as a subset of VHDL, but may need modification for certain VHDL tools.

In the *Boundary Scan Description Language* (BSDL), one can describe the length and structure of boundary scan registers and the availability of the optional $TRST^*$ (TAP Controller reset) pin. Also, the physical locations of TAP pins, the available

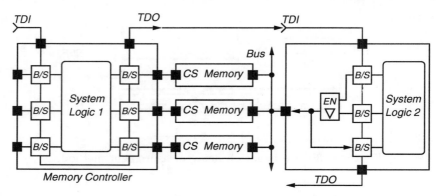

Figure 16.32: Example memory system bus burnout problem.

JTAG instruction codes on this chip, and the device identification code can be described. However, the TAP Controller state diagram cannot be described (that must be the standard one.) Also, the bypass register cannot be described, as it is always present and always one bit long. The length of the device identification register cannot be described, because that is set by the JEDEC standard. The *SAMPLE/PRELOAD*, *BYPASS*, and *EXTEST* instructions cannot be described, since these are always present. Finally, the operation of user-defined instructions cannot be described.

16.3.1 BSDL Description Components

The BSDL *entity description* describes the component-specific test logic parameters. It describes the standard VHDL package and package body and the BSDL subset of VHDL. It defines the commonly-used boundary scan cell types and the user-specified VHDL packages and package bodies. Figure 16.33 shows an example BSDL chip description.

```
entity    diff is
generic   (  Physical_Pin_Map: string := "Pack");
port      (  TDI, TMS, TCK: in bit;
             TDO: out bit; IN1, IN2: in bit;
             OUT1: out bit; OUT2: buffer bit;
             OUT3: out bit_vector (1 to 8);
             OUT4: out bit_vector (4 downto 1);
             BIDIR1, BIDIR2, BIDIR3: inout bit;
             GND, VCC: linkage bit);
use STD_1194_1_1994.all;
attribute BOUNDARY_REGISTER of diff:entity is
          ...
```

Figure 16.33: BSDL example.

16.3 Boundary Scan Description Language

16.3.2 Pin Descriptions

The following standard *USE* statement is always required:

use STD_1149_1_1994.all;

These boundary scan pin types are available:

- *in* (input-only)
- *out* (may be tri-state or open-collector)
- *buffer* (active, two-state, and always driven)
- *inout* (bi-directional)
- *linkage* (power, ground, analog, or non-connected pins)

The BSDL language relates logical signals to physical pins on the chip package. It allows one to group ports into differential voltage or current pairs, where one signal is always the complement of the other.

The TAP descriptions specify which logical signals comprise the TAP. They also specify which input port logic values enable JTAG compliance. That is, the part can either conform to the JTAG standard or refuse to conform to the standard. This situation arises when multiple test standards (e.g., both IBM LSSD and JTAG) are used in the same hardware system.

In BSDL, the instruction register is described by specifying its length and the allowable OPCODES (the user can add optional user-defined instruction OPCODES.) Also, one specifies the mapping from bit patterns to instruction OPCODES, the private instructions, and the bit pattern captured in the *Capture-IR* TAP Controller state (but the 2 least significant bits are always "01.") Finally, one specifies the *IDCODE* and *USERCODE* register contents.

The scan cell definitions define the existence and length of the boundary register cells. Boundary scan register cells have these types:

- INPUT – control and observe, or observe-only cell.
- CLOCK – cell at a clock input.
- OUTPUT2 – drives a two-state output.
- OUTPUT3 – drives a tri-state output.
- CONTROL – controls a tri-state output.
- CONTROLR – disabled in *Test-Logic-Reset* TAP Controller state.
- INTERNAL – not associated with a digital pin.
- BIDIR – reversible cell of a bidirectional pin.
- OBSERVE_ONLY – a single input observe-only cell.

The BSDL language also lets one define which registers get accessed by which instructions.

16.4 Summary

The boundary scan standard has become absolutely essential. It is no longer possible to test many of the newer PCBs exclusively with a bed-of-nails tester. It is not possible to test MCMs at all without the boundary scan standard. The standard supports external testing with an ATE, and boundary scan chain reconfiguration as a pattern generator and response compressor for *built-in-self-testing* (BIST.) The standard is beginning to get widespread usage.

Problems

16.1 *Boundary scan design for testability.* Given a VLSI circuit with:
$g = $ # logic gates $pi = $ # functional input pins
$f = $ # internal flip-flops $po = $ # functional output pins
Calculate the percent gate overhead for:

(a) Full-scan design over the original design. Sketch the logic-level schematic for the scan flip-flop with testability hardware.

(b) Boundary-scan design over the full-scan design. Sketch the logic-level schematic for the Boundary-scan pin circuit.

Ignore overhead from the TAP Controller, ID register, and bypass register.

16.2 *Boundary scan economics.* Estimate the cost of the boundary scan hardware for a chip with 256 pins. Transistors cost 525 μcents/transistor [584]. One logic gate uses 4 transistors, one master-slave flip-flop uses 24 transistors, and the JTAG TAP controller has 262 transistors. Now, given a test vector set of 512,000 vectors for the chip, and a test clocking rate of 200 MHz, estimate the cost of testing this chip using the JTAG hardware, given that the tester cost is 4.5 cents/s.

16.3 *Boundary scan test time.* The printed circuit board in Figure 16.8 has four chips with these characteristics:

Chip	# Test patterns	# Pins	Chip	# Test patterns	# Pins
Chip1	500,000	1024	Chip3	1,000,000	1024
Chip2	1,000,000	512	Chip 4	512	500,000

Compute the test time for four testing episodes at a test and board clock rate of 512 MHz, without using the JTAG BYPASS instructions, to test only Chips 1-4 (and not the interconnect) using an external tester applying the patterns through the JTAG boundary scan chain. Now, recompute the testing time for only Chips 1-4 using the JTAG BYPASS instruction.

16.4 *Interconnect test time.* Assume the same system of Problem 16.3, and again use Figure 16.8 for the PCB description. Compute the test time for interconnect test, assuming that half of the chip pins are outputs, and half are inputs.

Problems

Each output requires two test patterns for a sa0 and a sa1 test. Each interconnect is sampled during testing by all chips that receive the signal. During interconnect test, all output pins on the chip driving the interconnect will be enabled, and all input pins on all other chips must latch the signal. Describe the JTAG instruction sequence needed to test the interconnect, and calculate the test time at a 100 MHz clock rate.

16.5 *Interconnect delay test time.* Use the same system of Problem 16.4, but now compute the test modes and test time to perform interconnect delay fault test. This requires, for each interconnect being tested, a two-pattern test for a rising transition, and another for a falling transition delay test. Describe the JTAG instruction sequence needed to test the interconnect, and calculate the test time at a 100 MHz clock rate.

16.6 *Boundary scan delay fault test.* Figure 16.17 shows two boundary scan cells surrounding the on-chip system logic. We test the path from the INPUT boundary scan cell, through the on-chip system logic, and ending at the OUTPUT boundary scan cell. The JTAG commands are: *SAMPLE, PRELOAD, EXTEST, INTEST, RUNBIST, CLAMP, IDCODE, USERCODE, HIGHZ,* and *BYPASS*. Please explain the sequence of these commands used for delay fault testing of this particular path.

16.7 *Test controller.* Implement a logic-level finite state machine realizing the test controller state diagram in Figure 16.11. The controller has output active high signals indicating when it is in the state *Shift-DR, Shift-IR, Update-DR, Update-IR, Capture-DR, Capture-IR, Run-Test/Idle,* or *Test-Logic-Reset*. These signals control the boundary scan register. If a sequential automatic test-pattern generator is available, generate tests for your test controller and determine whether the controller itself is testable.

16.8 *Bus and controller testing.* In Figure 16.32, the three memory chips are pre-tested. However, 512 patterns must be applied by an external tester to the *System Logic 1* chip. Describe the test instructions, configurations, and procedures for the *System Logic 1* and *2* chips for stuck-fault testing of the *System Logic 1* chip, without burning out the *Bus* or Memory chips. Calculate the test time with a clock rate of 100 MHz and an external tester. Now, describe a test configuration and procedure to test the wires from the *System Logic 1* chip to each memory chip, and to test the Bus between the memory chips and the *System Logic 2* chip, under the same test conditions.

16.9 *Memory testing.* Test a 1 Gb DRAM using boundary scan, the MARCH C−test (see Chapter 9), and an external tester. The chip has these signals:

Signal	# Bits	Signal	# Bits
CS	1	Address	26
R/$\overline{W}$	1	Data	4

Calculate the number of test patterns required. The JTAG boundary scan chain length is 64 bits, so calculate the total test time if external testing (including boundary scan chain shifting), is done at 200 MHz.

16.10 *Full scan and boundary scan.* Redraw Figure 16.3 with updated hardware to show how the JTAG hardware can provide access to two full scan chains, each 256 bits long, that are embedded in the *System Circuitry*.

Chapter 17

ANALOG TEST BUS STANDARD

> "This ... proposal for an Analog Testability Bus ... could be used as the basis for a Standard such as IEEE P1149.4. The proposed testability structure is imposed on the I/O pin cells of the analog and mixed technology ICs. . . This allows for the testing of interconnect failures such as shorts and opens, the testing of discrete analog components and networks between ICs, and supports the testing of analog functions within the ICs."
> — Kenneth P. Parker, John E. McDermid, and Stig Oresjo [513].

We are entering an era where a single VLSI chip will be a complete electronic system. *Printed circuit boards* (PCBs) of today will soon give way to *systems-on-a-chip* (SOC). The chip will contain analog, digital logic, and memory subsystems, each with differing test needs. The chip will also contain interconnects between those modules. This shift means several changes in the test methodology. Traditionally, PCBs are assembled under assumptions that the printed wiring, as well as chips, were tested prior to mounting chips on the board. Thus, only a limited set of faults (mostly related to insertion of wrong chips, incorrect orientation, pin soldering, etc.) can occur. These faults are effectively tested by the *in-circuit test* (ICT) [69] technique, in which all pins of a chip mounted on the board are directly probed for test. When a SOC is manufactured, assumptions of pretested modules and interconnects are not valid. Besides, internal probing of modules is impractical. Thus, ICT cannot be applied to a SOC. Fortunately, there have been developments in design methods that facilitate testing of SOC.

A possible design for testability strategy would be to insert the boundary scan structure, based on the IEEE 1149.1 and 1149.4 Standards, to partition the three types of subsystems. This structure will allow an efficient test of the interconnects. Since carrying lengthy test sequences through the boundary scan structure will be inefficient, self-test of subsystems, supported by internal full or partial scan will be almost necessary to reduce the time and cost of testing. With shrinking circuitry, test logic and latches are becoming much less expensive and less intrusive than probe pads [504].

The basic process of chip testing involves sending test signals through the chip pins and then receiving the response again via pins. Modules that are surrounded by other modules cannot be accessed directly from the chip pins. Methods of providing access for digital and analog test signals to embedded modules are needed.

17.1 Analog Circuit Design for Testability

Analog circuit testing differs from both logic and memory testing. In short, analog circuits are tested for their *specifications*, which are expressed in terms of functional parameters, such as voltage levels, frequency response, inter-modulation distortion, etc. Here is an incomplete list of analog blocks, often found in mixed-signal devices [217]: analog switches, *analog-to-digital* (A/D) converters, comparators, *digital-to-analog* (D/A) converters, operational amplifiers, precision voltage references, phase-locked loops, resistors, capacitors, inductors, and transformers.

There are two types of measurements associated with analog circuit testing. In the first type, we measure component values and ascertain that they are within the specified range. In the other type of measurement, signal characteristics (levels, frequency spectra, harmonics, etc.) are measured. These require applying carefully generated signal waveforms to the inputs and analyzing the outputs. Both functions are generally accomplished by *digital signal processing* (DSP) techniques.

The main idea of *design for testability* (DFT) for analog circuits is to provide access to selected nodes for testing. Traditionally, signal generation and DSP were provided by the expensive *automatic test equipment* (ATE.) With the current trend of integrating analog and digital circuitry on the same chip, often the DSP function is available on-chip. On-chip digital functions have also been used for creating self-test for the analog portion.

We will discuss analog circuit test access in Section 17.2. The method of analog boundary structure, described there, works intimately with the digital boundary scan structure, to partially solve the DFT problem for mixed-signal devices.

17.2 Analog Test Bus (ATB)

The *analog test bus* (ATB) (IEEE Standard 1149.4) architecture is illustrated in Figure 17.1 for a mixed-signal device [214, 387, 405, 435, 510, 511, 554, 658, 719]. This bus replaces the in-circuit tester, because it provides electronic access and multiplexing hardware in order to probe various digital and analog components in a mixed-signal chip, or various external impedances connected to the pins of the chip.

We use the P1149.4 *analog test bus* [319] with the JTAG P1149 digital *boundary scan standard* [318] to add testability to test some analog circuits. The advantages are that:

1. We gain analog circuit observability, particularly at the interface between the analog and digital parts of the chip.

17.2 Analog Test Bus (ATB)

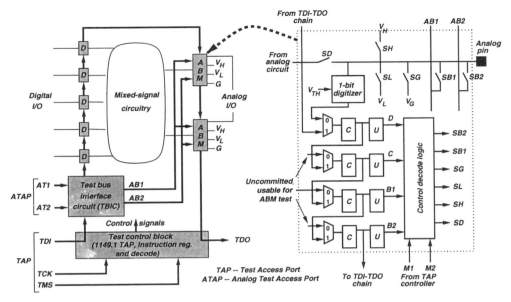

Figure 17.1: The analog test bus architecture.

2. We eliminate a large analog chip area needed for extra test points.

The disadvantages are that:

1. The bus may have a 5% measurement error.

2. The C-switch sampling devices couple all probe points capacitively, even when the test bus is not in use. To avoid this unwanted change in the analog circuit transfer function, we use more elaborate analog C-switches in the P1149.4 standard.

3. There is a stringent limit on how far one can move analog data through a system using the P1149.4 standard before having to digitize it to preserve accuracy.

17.2.1 Targeted Analog Faults

Figure 17.2 [511] shows the types of analog circuit faults that must be tested for, including shorts and opens in the interconnect (usually due to solder problems.) Other problems include misloaded discrete analog components or components that are out of their tolerance range. Be aware that shorts can occur between analog and digital pins. The 1149.4 standard also handles interconnect testing, and classifies interconnect as *simple interconnect* (wires), *differential interconnect* (where a pair of wires transmits a signal), and *extended interconnect*, where there are non-wire connections between two IC pins, e.g., through a capacitor. This is illustrated in Figure 17.2. Discrete analog components will continue to be used in extended interconnect because:

578 Chapter 17. ANALOG TEST BUS STANDARD

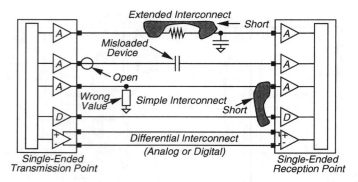

Figure 17.2: Defects in a mixed-signal circuit.

- Impedance matching of transmission lines is necessary, and merchant ICs will not have on-chip impedance matching resistances. This is because on-chip resistances limit the applications for the IC, so resistors will remain off-chip.

- Discrete resistors dissipate significant power, which may prevent them from being integrated into a chip.

- It is presently impossible to make accurate, high-valued inductors on chips, even with modern spiral inductors. That is why cellular telephones still use off-chip inductors. We still cannot make a good transformer on a chip.

- Integrated analog R, C, and L components never have as precise values as external, discrete components.

- Some ICs can be extended into additional functions if one can customize the IC by changing the value of an external R, C, or L connected to it.

We will slowly see more of these external analog components being integrated onto the mixed-signal IC, but there will always be a need for external, discrete analog components.

We saw in Chapter 10 how a DSP-based analog tester can perform complex measurements such as analog gain, harmonic distortion, inter-modulation distortion, etc. The goal of the 1149.4 standard is NOT to provide such a capability, but instead to provide a simpler capability to find shorts, opens, and misloaded analog components in mixed-signal circuits. The analog test bus will NOT eliminate the need for the analog tester, but will instead provide access to the mixed-signal chip and interconnect for the tester. The 1149.4 standard simply integrates part of the analog ATE's measurement bus and multiplexing system into the IC, eliminating the bed-of-nails tester (see Chapters 15 and 16 for discussions of the problems with the bed-of-nails, or in-circuit, tester.) The conventional in-circuit test allowed one to test the PCB by disconnecting it from the power source, and then probing the analog component to be tested and measuring voltages and currents. With the 1149.4 standard, it is now necessary to test the device with power on, and also multiplexing

17.2 Analog Test Bus (ATB)

is now done with silicon devices in the chip, rather than with external reed relays and wire. This introduces additional, unwanted impedances during testing, and additional current leakage paths to ground. Table 17.1 [511] shows how silicon switches are less desirable than the conventional reed-relay switches in the analog ATE. Notice how the CMOS switch of the 1149.4 standard has an on-resistance four or more decades higher than a reed relay, but is 4 million times smaller and hundreds of times faster. Also, the CMOS switch is somewhat non-linear over larger signal swings. These effects must be accounted for when using the 1149.4 bus, or inaccurate test measurements will be made. Figure 17.3 [511] shows how linear ICs can be chained along the 1149.4 test bus. Lofstrom [504] states that 1149.4 will not be used for directly measuring the quality of a signal, because switches are too slow and non-linear. The 1149.4 bus is slow, and usually has less than 1 MHz bandwidth. Components are usually tested with 10 $kHz <= f <= 100\ kHz$, but these frequencies are still adequate for measuring very small capacitances and inductances [504]. One needs to limit the number of analog measurements, possibly by using the pseudo-1149.1 test on analog pins to check for shorts and opens, because accurate current and voltage measurements are slow (many ms per reading) and therefore expensive [504]. The pseudo test switches the pin between V_H and V_L, and uses the digitizing receiver at the other end of the interconnect (driven by the pin) to check the interconnect for shorts and opens.

Table 17.1: Switches in various technologies.

Parameter	Mechanical relay (surface mount)	CMOS switch (0.35 μm)	Bipolar switch (0.35 μm)
On-resistance	$10^{-2}\ \Omega$	10^2 to $10^3\ \Omega$	Varies
Off-resistance	$10^{12}\ \Omega$	$10^{12}\ \Omega$	$10^{10}\ \Omega$
Bidirectional?	Yes	Yes	No
Switching time	$\geq 500\ \mu s$	$< 1\ \mu s$	$< 1\ \mu s$
Area	$96.7 \times 10^6\ \mu m^2$	$20\ \mu m^2$	100 to 5000 μm^2

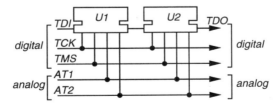

Figure 17.3: Chaining of 1149.4 ICs.

17.2.2 Analog Test Access Port (ATAP)

The *analog test access port* (ATAP) has four required and one optional signals from the 1149.1 digital boundary scan standard (see Chapter 16.) These signals are *TDI*, *TDO*, *TCK*, *TMS*, and *TRST* (optional.) The 1149.4 standard adds two required analog signals *AT1* and *AT2* (see Figure 17.1.) Usually, *AT1* is used for an analog stimulus of the 1149.4 IC, and *AT2* is used for transmitting the ICs response back to a measurement unit in the ATE. However, *AT1* and *AT2* can be partitioned, so that different *AT1* and *AT2* signals are used for different mixed-signal ICs. This is necessary, since one IC may generate substantial internal noise, which ruins test measurements of another IC, so partitioning is needed. The digital portion of the ATAP is identical to the 1149.1 digital boundary scan standard, except that the boundary register now includes the following:

- The *test bus interface circuit* (TBIC), which interfaces the mixed-signal chip to analog bus signals *AT1* and *AT2*.

- The collection of digital boundary scan cells for digital pins of the mixed-signal chip. Since several boundary scan cells are required to control a tri-state digital pin, the multiple cells associated with one pin are grouped together into the *digital boundary module* (DBM), which is labeled *D* in Figure 17.1.

- A set of digital boundary scan cells required to control a single analog pin, referred to as the *analog boundary module* (ABM) in Figure 17.1. Note that an ABM should not be used on power supply pins – they should be directly connected to supply [504].

17.2.3 Test Bus Interface Circuit (TBIC)

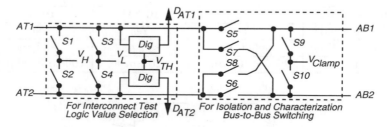

Figure 17.4: Test Bus Interface Circuit.

Figure 17.4 [511] shows the *test bus interface circuit* (TBIC), which can:

- Connect or isolate internal analog measurement buses *AB1* and *AB2* within the mixed-signal chip from the external analog bus *AT1* and *AT2*.

- Perform 1149.1 digital interconnect tests on the *AT1* and *AT2* pins, by treating them as digital pins.

- Support characterization processes needed to improve the accuracy of analog measurements.

17.2 Analog Test Bus (ATB)

In the TBIC, switches $S1$ through $S4$ and digitizers generating digital signals D_{AT1} and D_{AT2} support 1149.1 digital boundary scan style interconnect tests of signals on $AT1$, $AT2$, $AB1$, and $AB2$. Digitization occurs relative to a threshold V_{TH}, and is coarse. Generally, $V_L < V_{TH} < V_H$ and a short between $AT1$ and $AT2$ will not yield a voltage near V_{TH}. This capability allows interconnect testing of a PCB with 1149.1 and 1149.4 compliant devices. First, a digital integrity test checks the TAP signals, then one tests the wiring for interconnect defects, and then the 1149.4 hardware can check each ATn port for shorts and opens. Later, the ATAP can be used for analog tests, since the analog connectivity has been shown to be functional.

Table 17.2: TBIC switching patterns for Figure 17.4.

P#	\multicolumn{10}{c}{Switch state ($S1 - S10$)}	Function	
	1	2	3	4	5	6	7	8	9	10		
0	0	0	0	0	0	0	0	0	0	1	ATn disconnect (High Z), clamp ABn	
1	0	0	0	0	0	1	0	0	1	0	Connect $AT2$ to $AB2$	$P1 - P3$
2	0	0	0	0	1	0	0	0	0	1	Connect $AT1$ to $AB1$	for analog
3	0	0	0	0	1	1	0	0	0	0	Connect ATn to ABn	measurement
4	0	0	1	1	0	0	0	0	1	1	$AT1/2$ drive 00 out	$P0$ & $P4 -$
5	0	1	1	0	0	0	0	0	1	1	$AT1/2$ drive 01 out	$P7$ for 1149.1
6	1	0	0	1	0	0	0	0	1	1	$AT1/2$ drive 10 out	interconnect
7	1	1	0	0	0	0	0	0	1	1	$AT1/2$ drive 11 out	tests
8	0	0	0	0	1	1	0	1	0		For characterization	
9	0	0	0	0	1	0	0	1	0	1	For characterization	

Switches $S5$ and $S6$ connect or isolate the $AT1/AT2$ signals to/from internal $AB1/AB2$ signals. Note that isolated ABn signals float and can pick up noise. Optional switches $S9$ and $S10$ clamp the internal ABn signals to safe voltages to eliminate undesirable noise or parasitics when the ABn signals are out of use. Switches $S9$ and $S5$ and complementary, as are $S6$ and $S10$. One closes $S5$ and $S7$ (or $S6$ and $S8$) to characterize the $AT1/AT2$ pathway using a loopback test. This test finds the parasitic devices in the bus, so that we can correct our analog measurements to account for the parasitics. Table 17.2 [511] shows the switching patterns $P0$ through $P9$ for the TBIC of Figure 17.4.

Figure 17.5 [511] shows the minimal TBIC control structure for the switches of Figure 17.4. This hardware uses four boundary register cells and two control mode signals $M1$ and $M2$ (from the TAP controller), which are set by the instruction currently in the Instruction Register. The four boundary register cells are *Calibrate* (*Ca*), *Control* (*Co*), *Data1* (*D1*), and *Data2* (*D2*). The capture states of $D1$ and $D2$ capture signals D_{AT1} and D_{AT2}. Cells *Co*, *D1*, and *D2* are used for 1149.1 interconnect tests of the $AT1/2$ pins. In order to understand the TBIC Control Decode Logic of Figure 17.5, we must understand Table 17.2 [511].

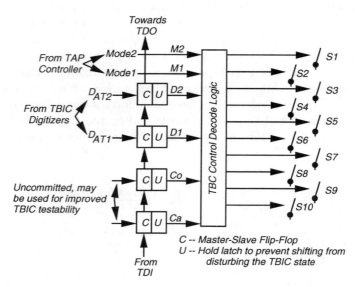

Figure 17.5: Control hardware for TBIC switches of Figure 17.4.

Table 17.3: TBIC switch patterns for boundary register cell values.

Cells $Ca/Co/D1/D2$	EXTEST, CLAMP, RUNBIST	PROBE, INTEST	HIGHZ	BYPASS, SAMPLE, PRELOAD, IDCODE, USERCODE
0000	$P0$	$P0$	$P0$	$P0$
0001	$P1$	$P1$	$P0$	$P0$
0010	$P2$	$P2$	$P0$	$P0$
0011	$P3$	$P3$	$P0$	$P0$
0100	$P4$	*	$P0$	$P0$
0101	$P5$	*	$P0$	$P0$
0110	$P6$	*	$P0$	$P0$
0111	$P7$	*	$P0$	$P0$
1000	$P0$	*	$P0$	$P0$
1001	$P8$	*	$P0$	$P0$
1010	$P9$	*	$P0$	$P0$
1011	*	*	$P0$	$P0$
1100	*	*	$P0$	$P0$
1101	*	*	$P0$	$P0$
1110	*	*	$P0$	$P0$
1111	*	*	$P0$	$P0$

17.2.4 Analog Boundary Module (ABM)

Figure 17.1 shows the detail of the *analog boundary module* (ABM), which supports 1149.1 interconnect tests and also allows measurement of analog component values that are off-chip. The 1149.4 standard requires every chip pin to have a DBM or an ABM. From the left of the figure, switch SD lets us disconnect the analog circuit from the analog pin. This *conceptual* switch may not physically exist as shown – instead, it may be incorporated into the analog circuit itself, or it may be unnecessary. The one-bit digitizer interprets the voltage on the analog I/O pin and passes it into the boundary register as signal D. Switches SL, SH, and SG allow us to connect voltages V_L, V_H, and V_G to drive the analog pin for 1149.1 interconnect tests. V_G is used for analog measurements, so it should be of *reference quality*, meaning that it is a stable voltage source able to source or sink current over a defined range without noticeable voltage changes. Switches SH and SL may also be conceptual. Further to the right, switches $SB1$ and $SB2$ can connect the analog pin to internal measurement bus wires $AB1$ and $AB2$. $AB1$ must be able to provide current to the pin, and $AB2$ must be able to monitor the pin voltage.

Table 17.3 [511] shows how the boundary register contents of Ca, Co, $D1$, and $D2$, along with the Instruction Register command, determine the switch pattern setting for the ABM. Table 17.4 [511] shows the 20 allowed switch settings. $P0$ isolates the pin, $P1$ through $P5$ test extended interconnect, and $P6$ and $P7$ characterize the quality of the V_G reference. $P0$, $P8$, and $P12$ are HI-Z, drive low, and drive high for 1149.1 interconnect tests. $P9$ through $P11$ and $P13$ through $P15$ allow characterization of V_L and V_H sources, or biasing of external devices. $P16$ (functional mode) connects the pin to the analog core and disconnects all test circuitry, while $P17$ through $P19$ support the $PROBE$ and $INTEST$ instructions (see below.)

The ABM is controlled by four boundary cells, C, D, $B1$, and $B2$, as shown in Table 17.5 [511], where the mode varies depending on the TAP Controller instruction. Cell D always captures the digitized state of the analog pin. The C and D cells form a two-cell bidirectional structure for 1149.1 interconnect tests. The capture stages of the other three cells are uncommitted, and can be used to capture hard-to-observe signals within the ABM to improve testability. Asterisks in Table 17.5 indicate that no switch pattern is yet assigned for that combination, but that it is reserved for future assignment, so these combinations should not be used. When $C = 0$, cell D determines whether V_G is connected to the pin. When $C = 1$, cell D controls the pin logic state for 1149.1 interconnect tests. C is a driver enable cell and D is a self-monitoring data cell for the pin. Cell $B1$ always controls the connection of $AB1$ to the pin, and $B2$ controls the connection of $AB2$ to the pin. Finally, Figure 17.6 (b). [511] shows how the electrostatic discharge protection circuit absolutely must be connected to avoid the measurement corruption by the series resistance R_P in the ESD protector when making analog measurements. Osseiran provides an outstanding description of how to validate the 1149.4 hardware [504] (pages 68-76), which must happen before any cores, interconnect, or external impedances can be tested with the 1149.4 hardware.

Table 17.4: ABM switching patterns for the six switches.

P #	Switch state (0 = open)						Pin state
	SD	SH	SL	SG	SB1	SB2	
0	0	0	0	0	0	0	Completely isolated (*core-disconnect state*)
1	0	0	0	0	0	1	Monitored (mon.) by $AB2$
2	0	0	0	0	1	0	Connected (conn.) to $AB1$
3	0	0	0	0	1	1	Conn. to $AB1$, mon. by $AB2$
4	0	0	0	1	0	0	Conn. to V_G
5	0	0	0	1	0	1	Conn. to V_G, mon. by $AB2$
6	0	0	0	1	1	0	Conn. to V_G & $AB1$
7	0	0	0	1	1	1	Conn. to V_G & $AB1$, mon. by $AB2$
8	0	0	1	0	0	0	Conn. to V_L
9	0	0	1	0	0	1	Conn. to V_L, mon. by $AB2$
10	0	0	1	0	1	0	Conn. to V_L & $AB1$
11	0	0	1	0	1	1	Conn. to V_L & $AB1$, mon. by $AB2$
12	0	1	0	0	0	0	Conn. to V_H
13	0	1	0	0	0	1	Conn. to V_H, mon. by $AB2$
14	0	1	0	0	1	0	Conn. to V_H & $AB1$
15	0	1	0	0	1	1	Conn. to V_H & $AB1$, mon. by $AB2$
16	1	0	0	0	0	0	Conn. to core, isolated from test
17	1	0	0	0	0	1	Conn. to core, mon. by $AB2$
18	1	0	0	0	1	0	Conn. to core & $AB1$
19	1	0	0	0	1	1	Conn. to core & $AB1$, mon. by $AB2$

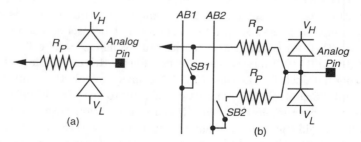

Figure 17.6: ESD protection circuit modifications for ABM.

17.2 Analog Test Bus (ATB)

Table 17.5: ABM switches set by boundary register contents.

Cells $C/D/B1/B2$	$EXTEST$, $CLAMP$, $RUNBIST$	$PROBE$, $INTEST$	$HIGHZ$	$BYPASS$, $SAMPLE$, $PRELOAD$, $IDCODE$, $USERCODE$
0000	$P0$	$P16$	$P0$	$P16$
0001	$P1$	$P17$	$P0$	$P16$
0010	$P2$	$P18$	$P0$	$P16$
0011	$P3$	$P19$	$P0$	$P16$
0100	$P4$	*	$P0$	$P16$
0101	$P5$	*	$P0$	$P16$
0110	$P6$	*	$P0$	$P16$
0111	$P7$	*	$P0$	$P16$
1000	$P8$	*	$P0$	$P16$
1001	$P9$	*	$P0$	$P16$
1010	$P10$	*	$P0$	$P16$
1011	$P11$	*	$P0$	$P16$
1100	$P12$	*	$P0$	$P16$
1101	$P13$	*	$P0$	$P16$
1110	$P14$	*	$P0$	$P16$
1111	$P15$	*	$P0$	$P16$

17.2.5 Instructions for 1149.4 Standard

The 1149.4 definitions of the $BYPASS$, $PRELOAD$, $IDCODE$, and $USERCODE$ instructions are identical to those for the 1149.1 standard in Chapter 16. These instructions all connect the analog pin to the analog core in functional mode. The $HIGHZ$ instruction is identical, but also includes the analog pins, which all float and are completely isolated from the core and the test circuits. The $SAMPLE$ instruction is also identical, but also digitizes the voltages appearing on the analog pins. The following instructions have expanded meanings for analog pins, but operate according to Chapter 16 for digital pins.

EXTEST Instruction. On an analog pin, $EXTEST$ can disable it, or connect it to V_L or V_H (see Table 17.4.) All analog and digital pins are disconnected from the core (*core-disconnect state*.) The *core-disconnect state* is individually programmable for each analog pin from settings in the boundary register. This is needed because bias voltage analog pins must never be disconnected, or all other analog measurements are meaningless. Also, low impedance pullup Rs or Ls often cannot be disconnected, because the necessary isolation transistor is either huge or inaccurate. Note that it is often not necessary, nor desirable, to implement the core-disconnect state with a switching transistor. Lofstrom [504] cites a validation chip for the 1149.4 standard, where explicit series core-disconnect switches reduced the operating frequency of a differential video amplifier from 140 MHz to 60 MHz. Instead, one may leave

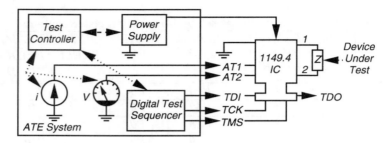

Figure 17.7: ATE system used to measure an externally connected impedance Z.

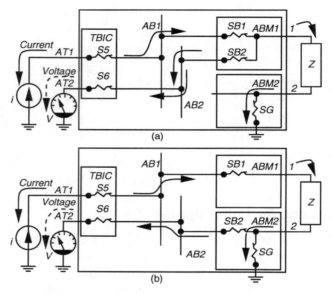

Figure 17.8: Measurement of external impedance using 1149.4.

much of the analog circuitry connected to the pin, but design it to be immune to test voltages applied on the pin and its interconnect. One may also use other functions in the analog circuitry to turn it off. The core-disconnect transistor often reduces analog driver performance [504].

External impedances may prevent the pin from behaving like a digital pin (e.g., low valued termination resistors.) Also, pins connected to capacitors behave like an open circuit under DC tests. Remember that the ABM can simultaneously control and observe a pin. For analog measurements using $EXTEST$, we employ switch patterns $P1 - P5$ in Table 17.4. The test engineer has the flexibility to give different switch patterns to different analog pins in $EXTEST$ mode, and this is usually quite necessary to stabilize the circuit. All ICs must interpret an all-zero instruction register code as an $EXTEST$ instruction.

Figure 17.7 [511] shows how an ATE is connected to a chip through the 1149.4 bus to measure an external impedance Z in $EXTEST$ mode. The ATE powers the

17.2 Analog Test Bus (ATB)

circuit, then it tests the TAP controller to verify that the boundary register works, and next it does 1149.1 interconnect tests on all of the board wiring. Finally, the ATE tests impedance Z. First, it produces a small current in its current source and injects it along $AT1$ to the IC TBIC, where the current is routed onto the internal $AB1$ bus. Then, the current travels to $ABM1$ where it goes out onto pin 1 and Z. The current travels through Z to pin 2 and back into the IC, where $ABM2$ routes the current to the reference supply V_G. We must measure the voltage across Z and use Ohm's law to calculate Z. Figure 17.8(a) [511] shows the ATE system voltmeter connected to $AT2$, and the TBIC connects $AT2$ to $AB2$. $ABM1$ connects $AB2$ to pin 1, completing a ground reference voltage measurement circuit. The ATE records the voltage at pin 1. In Figure 17.8(b), the ATE measures the voltage at pin 2. Subtracting these two voltages gives the voltage across Z due to a known current $Z = \frac{V_1 - V_2}{I}$. Using a 50 μA DC current, we get a 0.2525 V measurement at $AT1$ and only 2.5 mV across Z. If the system voltmeter has 10 μV resolution, there is 20 μV error in the two voltage measurements, which leads to $\pm 0.4\ \Omega$ of error, or 0.8% in the calculation of Z. It is necessary to select a current that causes all system voltages to be within operating range for probable values of Z. The current source must have a voltage compliance limit, because if there is an open circuit, or if Z has a much larger value than expected due to misloading, abnormally high voltages will result. If Z is a capacitor, then an AC voltage source is needed. Osseiran provides an excellent detailed $EXTEST$ interconnect test description [504] (pages 76-82.) Osseiran also describes analog parametric test in detail [504] (pages 82-86.) Lofstrom [504] notes that in $EXTEST$, all analog pins go into the core-disconnect state. This may break a control loop for a voltage regulator, eventually causing the regulator to fail, so this must be accounted for by the analog test designer.

CLAMP Instruction. The $CLAMP$ instruction disconnects all pins from cores and freezes analog pins and the TBIC circuit in their present state, which is useful for keeping parts of the circuit quiescent while measuring voltages and currents in other parts.

HIGHZ Instruction. The $HIGHZ$ disables all analog pins by opening the core-disconnect switch SB, disconnecting all test circuits, and disabling the TBIC.

PROBE Instruction. This new instruction provided by the 1149.4 standard is mandatory. It is analogous to the $SAMPLE$ instruction, but works for both digital and analog pins to allow continuous time sampling of digital or analog pins, while the core is in functional mode. Only one analog pin can be sampled at a time, because only one set of ABn wires exists for sampling. $PROBE$ sets all DBMs to connect the digital pins to the core circuits. All core-disconnect switches SD are closed so that all analog pins are connected to analog cores. The TBIC switches are set by the TBIC control register, so the test engineer can choose many different ATn to ABn connections. The $PROBE$ instruction uses ABM switch patterns $P16 - P19$ in Table 17.3 to connect none, one, or both ABn lines to an analog pin. If the analog

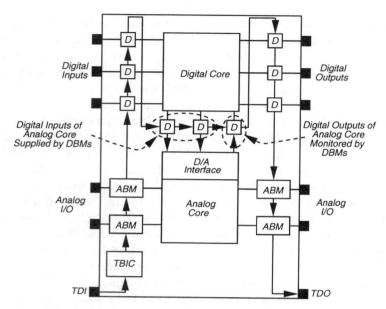

Figure 17.9: Analog core testing using *INTEST*.

pin is connected to $AB2$, and $AB2$ is connected to $AT2$, then the ATE can monitor the voltage on any analog pin in real time using $AT2$. In all other ways, the IC is fully operational, but the $AB2$ switch may add a small parasitic element into the circuit. If $AB1$ and $AT1$ are enabled, then one can inject a stimulus current of up to 100 μA into any analog pin while the IC is otherwise fully operational. This facility is most useful for noise measurements, and is usually capable of making frequency measurements only up to 10 kHz. Osseiran describes the use of *PROBE* for analog test [504] (pages 87-88) in detail.

RUNBIST Instruction. *RUNBIST* behaves according to the 1149.1 standard. Analog pins can either mimic the *HIGHZ* instruction, or the *CLAMP* instruction, during *RUNBIST*.

INTEST Instruction. The optional *INTEST* instruction tests IC circuitry while the IC is mounted on a board. Each I/O pin is replaced with one (or more) boundary register cells, which present inputs and collect outputs in parallel. If *INTEST* or *RUNBIST* is supported, then the boundary register MUST include interface cells between the digital and analog portions of the on-chip circuitry, as in Figure 17.9. This is necessary to control and observe signals passing between analog and digital portions during *INTEST*. With *INTEST*, digital pins can mimic either the *HIGHZ* or the *CLAMP* instruction. Digital pins (except clocks) are disconnected from the core. Analog pins remain connected to the analog core during *INTEST*, so switch SD is closed. Each analog pin may be connected to none, one or both ABn busses,

17.2 Analog Test Bus (ATB)

depending on the ABM control bits ($P16 - P19$ in Table 17.3.) Digital core testing proceeds exactly as with standard 1149.1 (see Chapter 16.) Figure 17.9 shows how the analog core is tested with $INTEST$. At any given time, only one analog pin can be stimulated, and only one observed. The difference between $PROBE$ and $INTEST$ is that $PROBE$ allows the internal digital circuits to interact with the analog core, while $INTEST$ disconnects the digital core and replaces it with set up patterns from the boundary register. Figure 17.9 shows an analog core being tested with the $INTEST$ instruction. The TBIC, ABM modules, and the three boundary register cells between the digital and analog cores are used to set up the test. If the digital core were to be tested instead, all of the cells marked D would be active, and the TBIC and ABM cells would be configured merely to pass through all signals from TDI to TDO. The analog input pins, and sometimes certain analog output pins, must have signals applied to them from off-chip for a meaningful 1149.4 analog test.

SAMPLE/PRELOAD Instruction. The required $SAMPLE$ instruction for analog pins digitizes the pin voltage and stores it as a bit in the boundary register [504]. If the pin voltage is $> V_{TH}$, it is digitized as 1, otherwise as 0.

17.2.6 Other 1149.4 Standard Features

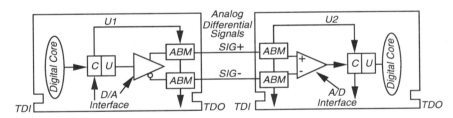

Figure 17.10: Differential interconnect with 1149.4.

Differential ATAP Port. Figure 17.10 [511] shows the differential I/O line permitted by the 1149.4 standard. A second TBIC must be added, to create a differential ATAP port, and a differential ABn bus. The second TBIC is connected to test pins $AT1N$ and $AT2N$, and internally to a second $ABnN$ bus labeled $AB1N$ and $AB2N$. The ABn bus is the positive side of the differential function pins and the $ABnN$ bus is the negative side. This allows a differential test measurement to be made, which is badly needed during system test in a noisy environment to remove common mode noise. In this differential setup, you cannot test common mode noise rejection over the differential bus using the $EXTEST$ instruction, because it causes the ABMs to disconnect the pins from the differential driver/receiver. Therefore, the $INTEST$ instruction is recommended for this noise rejection test between $U1$ and $U2$, although it was not intended for this purpose.

However, differential interconnect, while greatly increasing common-mode noise rejection, is not an unmitigated delight for test. Shorting or opening single lines or resisters still causes the differential interconnect to work, but at high speed the signal settling time is degraded, noise is increased, and there is pattern sensitivity [504]. Even worse problems happen when two different differential signals are shorted. These errors must be tested for by parametric measurements that look for small changes in the received voltage, or by examining the common-mode point at the receiver and looking for unexpected variations in the common-mode signal. A good technique is to measure the resistance of each of the two nets [504].

Partitioned Internal Test Buses. Figure 17.11 [511] shows a partitioned TBIC satisfying the 1149.4 standard, with a single set of ATn pins being distributed among multiple ABn bus sets. It may be undesirable to use a single ABn bus for all analog boundary modules, because the single bus becomes a pathway for noise to migrate from large-signal outputs back to small-signal outputs. If the IC has multiple power supplies, then various parts of the IC may have voltages incompatible with other parts, and so a single ABn bus covering the entire IC will mix incompatible voltages. The TBIC in the figure allows the ATn signals to connect either to the $ABna$ signals or $ABnb$ signals. This is done by expanding the $D1$ and $D2$ control signals in the boundary register for the TBIC into $D1a$, $D1b$, $D2a$, and $D2b$ signals. Note that you must also capture the differential voltage across the bus, which is done in Figure 17.10 by the rightmost CU digital boundary module. If it is necessary to capture a differential analog signal, then the digital boundary module would be replaced by an analog boundary module.

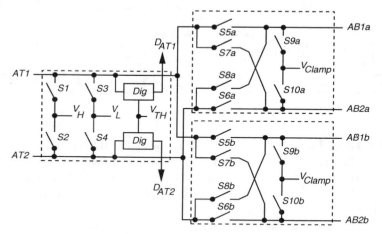

Figure 17.11: TBIC structure with one extension.

Bus Calibration. The 1149.4 test bus is calibrated by measuring a known quantity using the bus, and then when unknown quantities are measured, we subtract out the known quantity [504]. First, one measures the characteristics of the $AT1$ and $AT2$ buses themselves. We connect $AT1$ to $AT2$ using either the $AB1$ or $AB2$ on-chip bus. A known current is applied to $AT1$, a known load is connected via $AT2$, and the voltage at $AT2$ is measured (using $AB1$.) This provides the offset, gain, linearity, and bandwidth of the bus for current transmission. This path should produce a current into $AT2$ equaling the current driven by the tester into the virtual ground of $AT1$. Next, a known voltage is applied at $AT1$, and monitored at $AT2$ (using $AB2$) to calculate the same calibration parameters for voltage transmission. This path should produce a voltage on $AT2$ equaling the voltage driven by the tester into the high impedance $AT1$. However, it is hard to calibrate characteristics between a function pin and an ATAP pin on a board, because the exact value of the load device is not known, and the load device cannot be bypassed. The most important circuit elements needing calibration are the capacitance and leakage of the analog busses, the resistance of the 1149.4 switches, and the amplifier resistance for the driver of $AT2$. The 1149.4 standard has an accuracy objective of 1%.

Analog Boundary Module. It is permitted to use the ABM modules to test digital pins and interconnect with the 1149.4 analog standard.

Isolation of Analog and Digital Cores. The 1149.4 standard requires that a digital boundary module be present on each digital line between a digital and an analog core only when the *INTEST* or *RUNBIST* 1149.4 instructions are supported for the SOC. Otherwise, the DBMs may be eliminated between the cores.

17.3 Summary

Lofstrom [405, 406] and others [146] experimented with the 1149.4 standard. Systems with extremely high frequencies, small amplitudes, or high precision components may be hard to test with the 1149.4 standard. The 1149.1 and 1149.4 standards allow static tests, but non-static or feedback-dependent circuits may be difficult to test. If the digitizing threshold in the 1149.4 hardware cannot be reached by the circuit-under-test, the digitizer is useless. The 1149.4 interconnect test locates shorts [405], so the V_H and V_L switches in the ABM must be able to survive large voltage differences if they are not to be destroyed during test. A modified digital inverter [405] is completely unsuitable for the digitizer in the ABM. An analog input to a CMOS digital inverter would keep both the p and n FETs on, resulting in undesirable power consumption. If the inverter is biased near its threshold, the Miller capacitance from the inverter output to its input transfers charge into the input analog signal, causing noise and distortion. A custom digitizing receiver solves these problems [405]. The 1149.4 standard can be used to eliminate separate process monitor transistors and resistors included on wafers with separate probe pads purely for characterizing the semiconductor manufacturing process [405]. Test devices can now be placed inside the chip, and connected to $AB1$ and $AB2$. This eliminates the probe pads, giving better wafer tests with significant wafer area savings.

Proper sizing for switches in the 1149.4 standard is critical, if analog measurements are to be possible [405]. Large, low-resistance transistor switches are needed, particularly for V_L, because noise on the switch gate affects the switch resistance, causing common mode measurement errors. The switches heat up if currents are

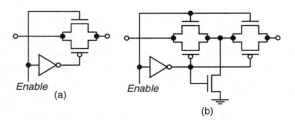

Figure 17.12: Improved analog transmission gate to reduce coupling.

large, causing the common-mode voltage across an impedance being measured to drift. Figure 17.12(a) [511] shows an analog switch used to sample an analog voltage. This switch still couples the two disconnected nodes, even when it is off, because of the significant coupling capacitance between sources and drains of the turned-off MOSFETs. Figure 17.12(b) shows a better switch with two analog C-switches. When the switch is off, the extra nFET grounds the middle node, significantly reducing signal coupling. The *AT1* and *AT2* signals must be guarded from each other and from system signals during analog measurements so that they do not couple capacitively. In Figure 17.13(a) [511], we ground a pin in between *AT1* and *AT2*, while in Figure 17.13(b) a long grounded wire separates them for their entirety. In Figure 17.13(c), the inverted *AT2* signal is a *driven guard wire* surrounding the entire *AT2* wire.

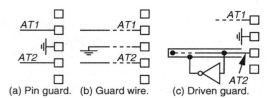

(a) Pin guard. (b) Guard wire. (c) Driven guard.

Figure 17.13: Guarding between multiple *ATn* signals.

DFT techniques will provide solutions to the test problems of SOCs. Because of the high system complexity, structured approaches are more apt to succeed. Most systems today have mixed-signals. A feasible approach is to partition analog, digital, and memory parts. For communicating test signals to partitions and for interconnect testing, boundary scan (IEEE 1149.1 and 1149.4) is the best available solution.

Problems

In the following problems, be acutely aware of the difference between DC and AC testing. In some cases, DC or AC testing may be impossible, and if that is the case, please explain why.

17.1 *Inductance measurement.* Devise a method of measuring the value of an external inductor attached to a linear IC as in Figure 17.8(a), where the inductor replaces the Z component. Please propose test waveforms and a series of measurements. Assume that switches $S5$, $S6$, $SB1$, $SB2$, and SG each introduce a 100 Ω parasitic resistor when they are on. If the inductor is supposed to be nominally 60 μH, needs to be accurate to 0.01%, and your system voltmeter has 20 μV error in each measurement, does measurement error exceed the inductor tolerance?

17.2 *Capacitance measurement.* Answer all questions in Problem 17.1, but instead measure the value of a 200 nF nominal capacitor for Z that must be held to a 0.01% tolerance.

17.3 *Resistance measurement.* Answer all questions in Problem 17.1, but instead measure the value of a 40 Ω nominal resistor for Z that must be held to a 1% tolerance.

17.4 *Analog core test.* Explain the difference between using the $INTEST$ instruction and using the $PROBE$ instruction to test an analog core.

17.5 *Bus impedance calibration.* Devise an analog test to measure the resistive impedance in the switches in the $AT1$ and $AT2$ busses.

17.6 *Bus impedance calibration.* Devise an analog test to measure the resistive impedance in the switches in the $AB1$ and $AB2$ busses.

Chapter 18

SYSTEM TEST AND CORE-BASED DESIGN

> *"No parts of the schedule are so thoroughly affected by sequential constraints as component debugging and system test. Furthermore, the time required depends on the number and subtlety of the errors encountered. Theoretically this number should be zero. Because of optimism, we usually expect the number of bugs to be smaller than it turns out to be. Therefore testing is usually the most mis-scheduled part of programming. . . . In examining conventionally scheduled projects, I have found that few allowed one-half of the projected schedule for testing, but that most did indeed spend half of the actual schedule for that purpose. Many of these were on schedule until and except in system testing."*
> — Frederick P. Brooks, Jr.,
> in a 1982 book, *The Mythical Man-Month* [102].

An organized group of components that performs some specified task is called a *system*. A car is a system made of components such as the engine, tires, battery, etc. A computer is a system whose components are chips, boards, disc drive, keyboard, and software. A system can be, and usually is, *hierarchical*. Thus, the disc drive may be a subsystem that is embodied within the computer system. The basic idea is that a system performs a much larger function that none of its components can do by themselves.

Systems are designed and built by *partitioning*. Thus, a computer system may be partitioned into a hardware box and a software program. Each of these may be further partitioned. For example, the hardware box would be partitioned into several boards and I/O components. Each board will be partitioned into several VLSI chips. While the system design follows a *top-down* hierarchy, that is, we start the process of partitioning from an overall function of the system, the system is actually built by a *bottom-up* process [123]. The lowest-level components in the system hierarchy, i.e., VLSI chips, are fabricated, and of course tested, first. These are then assembled on boards and other subsystems, which are also tested. Finally,

Type of test \ Application	Functional test	Diagnostic test LRU	Diagnostic test SRU
Manufacturing test	●	●	●
Maintainance test	●		
Field repair		●	
Shop repair			●

Figure 18.1: System tests and their applications.

the system is built. We will address the testing of such a system in this chapter.

18.1 System Test Problem Defined

System tests are used at various stages in the lifetime of a system. These tests can be classified into two broad categories (Figure 18.1):

1. *Functional test.* This test verifies the integrity of the system by testing for functional and performance-related specifications. This is basically a pass/fail type of test, with a limited, if any, diagnostic capability. In general, there can be several versions of functional tests, some more comprehensive than others. A typical example of a comprehensive functional test is the *acceptance test* of a computing system. This would be the test performed when the system is first delivered to its user. It may test for all or most system functions and check for performance parameters. A less comprehensive test may be run every time the system is powered up. This would test for the availability of all subsystems. Functional tests are used as the final step in manufacturing and as the first step in the system operation or maintenance (see Figure 18.1.) If a failure is indicated by these tests, then the next level of testing involves fault diagnosis.

2. *Diagnostic test.* A diagnostic test is designed for fault isolation or diagnosis. *Diagnosis* refers to the identification of a faulty part. A diagnostic test is characterized by its *diagnostic resolution*, defined as the ability to get closer to the fault. When a failure is indicated by a pass/fail type of test in a system that is operating in the field, a diagnostic test is applied. For an effective field repair, this test must have a diagnostic resolution of the *lowest replaceable unit* (LRU) [606]. For a computer system, the LRU may be a part such as a memory board, hard drive, or keyboard. The diagnostic test must identify the faulty part so that it can be replaced. The faulty part is then sent to the repair shop, where another diagnostic test having a higher diagnostic resolution will be used. This test will identify the faulty *shop replaceable unit* (SRU) [606], which may be a processor or a memory chip on the faulty board. As shown in Figure 18.1, diagnostic tests are also used during the manufacture. For example, if a the specification test shows a failure, then the field repair

procedure will be used. Further, if the system manufacturer has also assembled the faulty board, then the shop repair will follow.

18.2 Functional Test

A functional test performs the following tasks:

- Check existence and responsiveness of all subsystems.

- Check system specifications. For example, the amount of memory available in a computing system.

- Check critical functions of the system. This part can be more or less elaborate depending on the environment in which the test is applied. For example, the functional test at manufacturing may execute many of the tests that have been used during the simulation-based verification. A daily maintenance test, on the other hand, would include only a subset of those to limit the duration of the test.

Most electronic systems today are optimized via a hardware-software co-design process. Thus, hardware and software cannot be independently tested. The functional test is usually executed by the software of the system. Software test coverage criteria, such as *statement coverage*, *branch coverage*, and *path coverage* [338], can be used. However, in a system, the software is the "upper layer" and such coverages are not good indicators of the hardware fault detection capability.

The function of a system usually consists of an *operation* performed on some *data*. A system test is then a collection of operation-data pairs. Some heuristics used to generate tests are:

- *Instruction-set fault model* [660]. The instruction decoder is assumed to malfunction, causing a wrong instruction to execute. The data is selected so as to indicate an error in the result. These tests are found to thoroughly cover some portions of the hardware. However, they do not guarantee a good coverage of the complete system.

- *All instructions with random data* [187]. All instructions, usually a countable set, are exercised with randomly generated data. These tests can quickly cover some faults, but for other faults that have low probabilities of detection the coverage remains low. For adequate fault coverage, random tests tend to be too long to be practical.

These procedures are similar to those often used to generate design verification tests. Various heuristics are used to select a subset of verification tests, which otherwise tend to be lengthy. A good functional test should be as short as possible and yet comprehensive enough to cover all failures that are "likely to occur." The test is built through the design and development phases and its optimization continues through manufacturing, field trials, and system use. As actual faults are found and

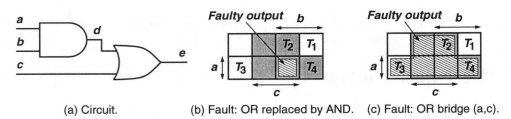

Figure 18.2: An example of gate-level diagnosis.

repaired, new test sequences are added. Also, some existing tests that are not found to be useful in detecting actual faults are dropped.

18.2.1 Microprocessor Test

Most systems have one or more microprocessors that are excellent resources for building test strategies. A microprocessor can be programmed to test itself as well as to control the test of the entire system. Microprocessor test methods are closely related to their verification [687]. The instruction set fault model is very effective for generating tests [660]. Portions of a microprocessor may be designed with scan or BIST. For the scan portions combinational ATPG is used. Partial scan and sequential ATPG have also been used for microprocessors [151]. BIST can be exhaustive or non-exhaustive. In the latter case, one needs to use fault simulation and test points if necessary to achieve high fault coverage. Tests for microprocessor chips evolve over time as the design passes through verification, characterization, and production phases. Other programmable chips such as digital signal processors (DSPs) may also follow a similar test development method. There are numerous techniques that are specific to microprocessor testing. Those are not included in this book for the lack of space. An interested reader should study the recent literature [2].

18.3 Diagnostic Test

The diagnostic test is applied after a system has failed. The aim of this test is to identify the faulty part that should be replaced. The environment of the system repair determines the level or the units to be identified. For example, in field maintenance, the *lowest replaceable unit* (LRU) would be a board, faulty disc drive, or an IO subsystem. The cardinality of the suspected set of LRUs that the test identifies is defined as its *diagnostic resolution*. An ideal test will have the diagnostic resolution of 1. Such a test will be able to exactly pinpoint the faulty unit and will allow the most efficient repair. In a repair shop, one may need to repair the faulty board and LRUs would be chips or discrete components on the board. The concepts of *fault dictionary* and *diagnostic tree* are relevant to any type of diagnosis.

18.3 Diagnostic Test

Table 18.1: Fault dictionary for the circuit of Figure 18.2.

Fault	Test syndrome			
	t_1	t_2	t_3	t_4
No fault	0	0	0	0
a_0, b_0, d_0	0	0	0	1
a_1	1	0	0	0
b_1	0	0	1	0
c_0	0	1	0	0
c_1, d_1, e_1	1	0	1	0
e_0	0	1	0	1

Table 18.2: Diagnosis using fault dictionary of Table 18.1.

Fault	Test syndrome				Diagnosis
	t_1	t_2	t_3	t_4	
$OR \rightarrow AND$	0	1	0	1	e_0
$OR\ bridge(a,c)$	0	0	1	0	b_1
$OR \rightarrow NOR$	1	1	1	1	c_1, d_1, e_1, e_0

18.3.1 Fault Dictionary

Application of a test simply tells us whether or not the system under test is faulty. We must further analyze the test result to determine the nature of the fault, so that it can be fixed. A *fault dictionary* contains the set of test symptoms associated with each modeled fault. The following example of gate-level diagnosis illustrates the concept.

Consider a test set that detects all single stuck-at faults in the circuit of Figure 18.2(a). It contains four vectors, (a, b, c): $T_1 = (010)$, $T_2 = (011)$, $T_3 = (100)$, $T_4 = (110)$. For diagnosis, we simulate faults without *fault dropping*. The simulation result is shown in the fault dictionary of Table 18.1, where a binary indicator t_i, associated with test T_i, is 1 only if that test detects the corresponding faults in the first column. We have used the subscript notation for faults. Thus, the fault a s-a-1 is denoted as a_1. For this fault, the 1 under t_1 indicates that test T_1 detects it. The 0s under t_2, t_3, and t_4 indicate that tests T_2, T_3, and T_4 do not detect the fault a_1. The vector 1000 is called the "test syndrome" of fault a_1. Note that some groups of faults share the same test syndrome. The reader will recognize all faults in such a group as "equivalent faults." By definition, equivalent faults cannot be distinguished from each other (see Chapter 4.) The first row of Table 18.1, 0000, is the test syndrome of the fault-free circuit.

We will consider three faults. The first is a fault caused from replacing the OR gate by an AND gate. Such a fault can occur in circuit assembly. This fault changes the output function to $e = abc$, as shown in Figure 18.2(b), which also gives the fault-free function, $e = ab + c$ (true minterms are shaded grey.) Two tests, T_2 and T_4, which produced a 1 output, will produce a 0 output for the faulty circuit. Thus, the test syndrome is 0101. This syndrome matches with the last entry in the dictionary of Table 18.1 giving a diagnosis of e_0 as summarized in Table 18.2. An exact diagnosis is impossible since the actual fault was not modeled and is not in the dictionary. The diagnosed fault, however, is associated with the faulty gate.

The second fault is an OR type of bridging between signals a and c. This means that both signals will behave as $a + c$. The output function, $e = a + c$, is shown in Figure 18.2(c). Only the test t_3 will show a failure and the test syndrome is 0010. This matches with the dictionary entry for b_1 (Table 18.1.) Thus, we have an incorrect diagnosis since signal b is not faulty. Such a situation is not uncommon when the dictionary is made for one kind of faults (stuck-at in this case) and the actual fault is of another kind.

The case of misdiagnosis for a bridging fault in this example is quite typical. For actual systems, the initial fault dictionary may still be based on stuck-at faults, because tools (fault simulators) are available for those. However, the dictionary is augmented with entries corresponding to actual faults found during manufacturing test. A system may have many users and may be installed at several sites. Maintenance and repair data from all sites may be passed on to the manufacturer for enhancing the fault dictionary.

On application of a diagnostic test it is also possible that the observed test syndrome does not match any entry of the dictionary. We then rely on heuristics for looking up the dictionary [597]. For example, suppose that due to a manufacturing error, the OR gate in the circuit of Figure 18.2 has been replaced by a NOR gate. The inverted output will fail all four tests, producing a syndrome 1111. An often used diagnostic procedure is to identify the dictionary entry at the smallest Hamming distance from the observed test syndrome as the most likely suspect. Thus, the last two entries in the dictionary of Table 18.1, which are both at Hamming distance 2 from the observed syndrome, give the suspects as c_1, d_1, e_1, and e_0 (Table 18.2.) This diagnosis does indicate a problem surrounding the OR gate but does not give a clear idea of the actual fault.

18.3.2 Diagnostic Tree

One disadvantage of the fault dictionary approach is that all tests must be applied before an inference can be drawn. Besides, for large circuits, the volume of data storage can be large and the task of matching the test syndrome can also take time. An alternative procedure, known as *diagnostic tree* or *fault tree*, is often found to be more efficient [356, 597]. In this procedure, tests are applied one at a time. After the application of a test a partial diagnosis is obtained. Also, the test to be applied next is chosen based on the outcome of the previous test.

A diagnostic tree for the circuit of Figure 18.2 is shown in Figure 18.3. The test T_4, shown at the root of the tree, is applied first. If it shows a failure, the diagnosis will follow the lower branch labeled $t_4 = 1$. At this point, the suspected fault set contains a_0, b_0, d_0, and e_0. Test T_2 is applied next. A failure ($t_2 = 1$) will diagnose the fault as e_0. In case of other faults, the diagnosis will follow other paths in the tree. Each leaf node provides the same type of diagnosis as given by the dictionary. However, the process can terminate before all tests are applied, as it did for the

18.3 Diagnostic Test

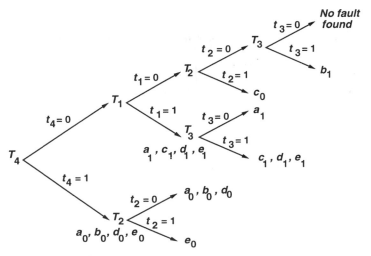

Figure 18.3: A diagnostic tree for the circuit of Figure 18.2.

fault e_0. The maximum depth of the diagnostic tree is bounded by the total number of tests, but it can be less than that also.

The depth of the diagnostic tree is the number of tests on the longest path from the root to any leaf node. It represents the length of the diagnostic process in the worst case. The depth of the tree in Figure 18.3 is four. That means all four tests may have to be applied. The presence of several shorter branches indicates that the process may terminate earlier in many cases.

The diagnostic tree can be arranged in several ways. One approach is to reduce the depth. We start with the set of all faults as "suspects." Tests are ordered such that the passing of each test will reduce the suspect set by the greatest amount. This may sometimes increase the depth of the tree on the side of failing tests. The overall depth can be reduced by dividing the fault set into equal halves by each test (although that may not be possible for given tests.) If that can be done, we get the theoretical lower bound on the length of the diagnosis as $\lceil \log_2(N_f + 1) \rceil$, where N_f is the number of distinguishable fault sets. In Table 18.1, $N_f = 6$ for the circuit of Figure 18.2, giving the lower bound on the depth of the diagnostic tree as three.

Misdiagnosis, as we found with the fault dictionary, is also possible with a diagnostic tree. For example, consider the fault, a NOR gate replacing the OR gate in the circuit of Figure 18.2. Table 18.2 (see last row) shows that all tests will indicate failure. When the diagnostic tree of Figure 18.3 is used, the procedure will terminate after two tests, giving e_0 as the fault.

An alternative procedure of arranging the diagnostic tree can result in substantially early termination when the *maintenance log*, which carries the history of system failures and repairs, is used. For example, if it is found that the fault c_0 is the most frequently occurring fault, then the tests are ordered to minimize the distance of the c_0 leaf node from the root. Thus, in the upper branch of the diagnostic tree

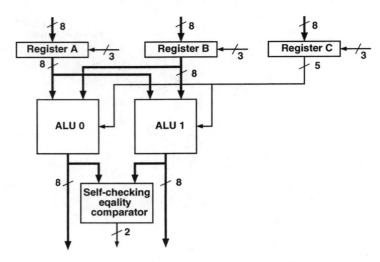

Figure 18.4: A system to illustrate the concepts of diagnostic tests [356].

of Figure 18.3, we can interchange T_1 and T_2, bringing c_0 one step closer to the root. As the maintenance history of a system builds up and occurrence probabilities of faults become known, the diagnostic tree can be rearranged to reduce the average termination time. Suppose that p_i is the occurrence probability of the fault (or fault set) represented by the ith leaf node. The probability p_0 of finding no fault is

$$p_0 = 1 - \sum_{i=1}^{N_f} p_i \qquad (18.1)$$

The average diagnostic length, i.e., the average number of test applications for termination of diagnosis, is given by

$$\text{Average diagnostic length} = \sum_{i=0}^{N_f} d_i \times p_i \qquad (18.2)$$

where d_i is the number of tests on the branch from the root to the ith leaf node.

One notable advantage of the diagnostic tree approach is that the process can be stopped any time with some meaningful result. If it is stopped before reaching a leaf node, a diagnosis with reduced resolution (a larger set of suspected faults) results. In the fault dictionary method, any early termination makes the dictionary look up practically impossible.

18.3.3 A System Test Example

We will use an example given by Kime [356] for illustrating the methods used in diagnostic tests. This system, shown in Figure 18.4, consists of two 8-bit ALUs whose outputs are fed to a self-checking equality comparator. Registers A and B

Table 18.3: Fault simulation of tests for system of Figure 18.4.

Faulty unit	Test syndrome				
	t_1	t_2	t_3	t_4	t_5
None	0	0	0	0	0
ALU0	1	1	0	x	0
ALU1	1	0	1	0	x
Register A	0	0	0	1	x
	0	1	1	1	x
Register B	0	0	0	1	x
	0	1	1	1	x
Register C	0	0	0	x	1
	0	1	1	x	1
Comparator	1	0	0	0	0

feed data and register C feeds control signals to ALUs. These registers can be reconfigured as counters to generate test data. The comparator, which provides error-checking during the normal operation, is also used in the test assuming a single ALU to be faulty. Five unit tests, T_1, T_2, T_3, T_4, and T_5, constitute a system test. Associated with each test T_i, there is a binary indicator t_i such that $t_i = 1$ if the application of T_i shows a failure, and $t_i = 0$ if the application of T_i shows no failure.

Test T_1 configures registers A, B, and C to apply an exhaustive set of vectors to the two ALUs and observes the output of the comparator. The other four tests, T_2, T_3, T_4, and T_5, are applied by the controlling software to test ALU0, ALU1, registers A and B, and register C, respectively. Each test consists of several vectors (clock cycles.) As should be obvious to the reader, these tests are not independent of each other. For instance, test T_5 for register C must pass data through the ALUs, whose inputs come from registers A and B. Thus, T_5 will also show failures caused by faults in registers A and B.

All single stuck-at faults in the entire system are simulated for five test sets. For a test T_i, the state of the indicator variable t_i is determined with respect to each unit. The result is shown in Table 18.3. Test T_1 detects all single faults in ALU0, ALU1, and the comparator. Hence, the column under t_1 shows a 1 for these units.

Test T_2 detects all faults in ALU0 and none in ALU1 or the comparator. It uses registers A, B, and C, and can detect some, but not all, faults in those units. Similarly, test T_3 detects all faults in ALU1 and none in ALU0 or the comparator. It also uses registers A, B, and C, and can detect some, but not all, faults in those units. Test T_4 covers all faults in registers A and B. This test uses ALU0 to pass the result to the output, hence, it can show failures for some faults in ALU0. It does not cover any fault in ALU1 or the comparator. It is sensitive to some faults in register C, which supplies control signals to ALU0. Test T_5 detects all faults in register C, none in ALU0 or the comparator, and detects some faults in the remaining units. Two types of entries in Table 18.3 require further explanation.

First, several entries are shown as x. These indicate an uncertain result of the

Table 18.4: Fault dictionary for system of Figure 18.4.

Test syndrome					Diagnosed set
t_1	t_2	t_3	t_4	t_5	of faulty units
0	0	0	0	0	None
0	0	0	0	1	Register C
0	0	0	1	0	Register A, Register B
0	0	0	1	1	Register A, Register B, Register C
0	1	1	0	1	Register C
0	1	1	1	0	Register A, Register B
0	1	1	1	1	Register A, Register B, Register C
1	0	0	0	0	Comparator
1	0	1	0	0	ALU1
1	0	1	0	1	ALU1
1	1	0	0	0	ALU0
1	1	0	1	0	ALU0

test. For example, $t_4 = x$ for ALU0 because, as we explained earlier, the test T_4 observes the faults of registers A and B through ALU0. Some faults of ALU0 are detectable by this test while others are not. If ALU0 was indeed faulty, this test may or may not indicate a failure, meaning an uncertain outcome of the test. Similarly, T_5 is a complete test for register C, but relies on the correctness of registers A and B, and ALU1. Our single faulty unit assumption means that when only register C is faulty, T_5 is a complete and reliable test for it. However, when register C is fault-free, any one of these three units could have a fault that is detectable by T_5. Yet, this test does not allow us to conclude that any unit other than register C is faulty, because some faults in those units escape detection.

Second, several faulty units have multiple test syndromes. Consider register A as the faulty unit. Both T_2 and T_3 produce an uncertain result and from the previous argument, we should collapse the two test syndromes into one by placing x's under t_2 and t_3. However, in this case, any fault of register A that is detected by T_2 is also detected by T_3. Also, any fault of register A that is not detected by T_2, is also not detected by T_3. Therefore, given that register A has a fault, any one of the two test syndromes shown in Table 18.3 can be produced, depending on the fault present.

Table 18.4 gives the fault dictionary produced from the fault detection data of Table 18.3. Each x entry in Table 18.3 splits the test syndrome into two, one with x replaced by 0 and the other with x replaced by 1. Thus, ALU0 has two test syndromes, 11000 and 11010.

18.4 Testable System Design

Traditionally, systems were designed using *printed circuit boards* (PCBs.) A board contains a number of VLSI chips and the wiring between them. During functional test, the board is placed in the actual system environment, e.g., by plugging the board in a cabinet with power supply, other boards, and I/O devices. A system-

18.4 Testable System Design 605

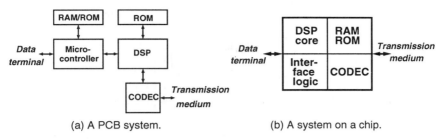

Figure 18.5: A mixed-signal CODEC system.

level diagnostic test will isolate the faulty board. Then the repair is performed by the *in-circuit test* (ICT) method [69]. In ICT, a test fixture makes direct contact with the pins of all chips on the board. This allows the application of tests to each chip and to interconnects between chips. The ICT fixture, known as the *bed-of-nails* fixture because of its numerous connector pins, has several disadvantages:

- *Cost.* The bed-of-nails fixture is expensive and must be personalized for each board. Changes in board design require changes in the fixture, as well.

- *Surface-mount technology* (SMT) [529] and *multi-chip modules* (MCMs) [744]. Many high-density boards have chips mounted on both sides and thus do not allow access to chip pins via the bed-of-nails fixtures. Similar test problems arise with multi-chip modules, which contain multi-layer silicon or ceramic routing substrates with directly mounted chips, all within the same package.

The above reasons have motivated the development of alternative methods such as *boundary scan* and the *analog test bus* that provide test access to chips and interconnects embedded on a board. These methods are discussed in Chapters 16 and 17, respectively.

In addition to the above problems, PCBs are bulky and have high assembly cost. The modern *system-on-a-chip* (SOC) provides fast and high reliability interconnects and offers a low cost system solution. Figure 18.5 shows an example of a mixed-signal modem as a PCB and as an SOC. The PCB version has three types of chips. The micro-controller and DSP (*digital signal processor*) are digital logic chips. RAMs and ROMs are memory chips, and the CODEC (*coder-decoder*) is a chip that contains mixed-signal (analog and digital) circuitry. As we have seen in previous chapters, these three types of chips are tested with different types of fault models and tests. Basically, the direct access to chips on a PCB in the ICT facilitates the application of the specific tests to each chip. Similarly, boundary scan and the analog test bus will allow the tests to be applied to each chip on the PCB via the edge connector.

The modern VLSI technology allows the entire mixed-signal system to be fabricated on a single chip as shown in Figure 18.5(b). Because of the varying test methodologies, the system is partitioned among digital logic, memory, and mixed-signal blocks. Test access can be provided by structures using the boundary scan (Chapter 16) and analog test bus (Chapter 17.) Notice, however, that the SOC of Figure 18.5(b) contains a DSP "core."

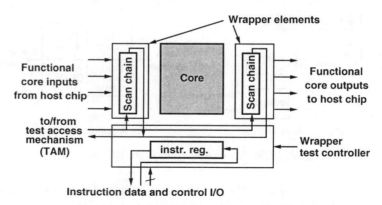

Figure 18.6: A test-wrapper for an externally tested core.

18.5 Core-Based Design and Test-Wrapper

A *core* is a predesigned and verified functional block that is included on a chip. Highly optimized (for high speed, small area, or low power consumption) core designs for functions such as DSP, microprocessor, and memory can be obtained from independent vendors. Basically, there are three types of cores:

1. *Soft cores* come in the form of a synthesizable RTL (*register-transfer level*) description.

2. *Firm cores* are supplied as gate-level netlists.

3. *Hard cores* or *legacy cores* are available as non-modifiable layouts.

A core is considered the *intellectual property* (IP) of the vendor, who may provide only limited details of the design to the user. Hard cores are also referred to as *IP cores*. In general, the vendor supplies tests for the core. However, the user (SOC designer) must provide the test access to a core embedded on the chip. If the core does not contain the test access (e.g., boundary scan), then testing of the core, as well as that of the surrounding logic, becomes difficult, especially when a hard core is used.

The core test problem, mentioned above, is solved by surrounding a hard core with test logic. This logic, known as a *test-wrapper* or *test-collar* [482], provides three modes of operation for each terminal of the core. For each input terminal of the core, these functions are:

- A normal mode where the core terminal is driven by the host chip.

- An external test mode where the wrapper element observes the core input terminal for interconnect test.

- An internal test mode where the wrapper element controls the state of the core input terminal for testing the logic inside the core.

18.6 A Test Architecture for System-on-a-Chip (SOC)

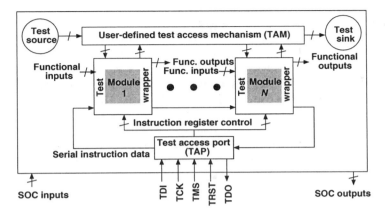

Figure 18.7: Test architecture for an SOC.

For each output terminal of the core, the wrapper provides:

- A normal mode where the host chip is driven by the core.

- An external test mode where the host chip is driven by the wrapper element for interconnect test.

- An internal test mode where the wrapper element observes the core output for the core test.

Figure 18.6 [482] shows the schematic of a test-wrapper for a core that is to be tested by external tests. Suitable modifications are made to customize the test-wrapper for cores that have internal scan chains or built-in self-test [482].

The test-wrapper of Figure 18.6 contains an element for each core I/O terminal. This element inserts a two-to-one multiplexer between the host chip signal and the core terminal. It also contains a flip-flop with a scan multiplexer. Thus, elements can be connected as a scan register. The test-wrapper includes a test controller with an internal scan chain that is connected to the *test access port* (TAP) of the chip. All scan registers work with the system test clock. This logic provides all functions outlined above. For details, the reader should refer to Nadeau-Dostie [482].

In the near future, the test specification of any IP core will be governed by the IEEE P1500 Standard, which will provide a *core-test language* (CTL) and a standardized core test-wrapper. CTL will offer similar functions as the current IEEE 1450 Standard, *the standard test interface language* (STIL). STIL supports a tester-independent description of timing, specifications, and vectors for full, partial, or boundary scan test application [88].

18.6 A Test Architecture for System-on-a-Chip (SOC)

The test architecture for testable systems borrows ideas from the research of previous decades [3, 354]. Figure 18.7 shows a generic architecture for a *system-on-a-*

chip (SOC) with N modules. Typical modules can be processor, ROM, RAM, DSP, combinational logic, *finite state machine* (FSM), ALU, multipliers, comparators, etc. Some modules can be cores as discussed in the previous section. A module is either designed with an internal test mode (such as a RAM with BIST) or enclosed in a test-wrapper. A set of test resources is shared among modules. This resource set consists of the following components [88, 430, 745, 746]:

- *Test source* – It provides test vectors. A typical test source, built on the chip can be a *linear feedback shift register* (LFSR), a counter, or a ROM with stored vectors. An alternative approach is to keep the test source off chip. For example, an external ATE can be used as a vector source.

- *Test sink* – Test sink is an on-chip signature analyzer or an output response comparison circuit. Alternatively, it can be an external sink such as an ATE providing output comparison with stored responses.

- *Test access mechanism* (TAM) – The TAM is a user-defined test data communication structure. A typical example is a bus. It carries test vectors from the test source to any module under test and communicates module outputs to the test sink. The TAM also allows testing of interconnects between modules via the test-wrapper elements.

- *Test controller* – This consists of the boundary scan *test access port* (TAP) (see Chapter 16) that permits the entire operation of the SOC test to be controlled from outside the chip. Instruction data are loaded serially into the instruction registers of all test-wrappers. These set the TAM drivers appropriately to conduct tests on one or more modules, or on interconnects.

An SOC chip would normally have its boundary scan tying up all I/O pins. This would allow the testing of the chip when embedded in a larger system and will also help in its own internal interconnect test. If the system contains analog modules, then the TAM would include analog test buses, *ATB1* and *ATB2*, and a *test bus interface circuit* (TBIC) is added on to the SOC (see Chapter 17.)

18.7 An Integrated Design and Test Approach

The test design of a system is a top-down process, carried out through the following steps [25]:

1. *System partitioning.* Partition the system for test into analog, logic, and memory blocks. In general, there can be several blocks of each type. In high-level design, the three types of function are separately described. Partitioning for test, therefore, does not cause any conflict with the functional design.

2. *Design blocks for testability.* Design each block either with self-test or assuming external test. If the block is a hard (legacy) core, then design a suitable

18.7 An Integrated Design and Test Approach

test-wrapper. Generate or obtain tests for those blocks that would be externally tested.

3. *Design test access.* Design boundary-scan and mixed-signal test bus structures for test access to all blocks. Design test controller circuits. The access to digital (logic and memory) blocks is provided through boundary scan sequences. An analog block test requires real-time access to its inputs and outputs. Boundary scan is used to set the control flip-flops to connect all active inputs and outputs of the block under test to separate ATB wires.

4. *Develop diagnosis procedures.* Develop test controller schedules for test application to blocks. Develop interconnect tests for partitioned blocks.

5. *Develop system test.* Develop a system test that will run in the normal mode and check critical functions and timing of the system.

The main idea in the above strategy is to separate the module (or macro) test from the test access. For digital logic and memory blocks the tests consist of a large number of vectors. Since the test access through boundary scan is serial, application of such tests will be slow. An acceptable solution is BIST, which incurs an area overhead of about 10% to 30% for those blocks.

Self-test for analog blocks is not as common. Since these blocks generally have smaller numbers of input and output terminals, tests can be efficiently applied by an external tester once the test access paths have been established. Alternatively, an on-chip *digital signal processor* (DSP), if present, can be used to test analog blocks (see Chapter 10.)

System-level test access requires the addition of one boundary-scan cell (a scan flip-flop) to each digital and memory block terminal and one analog access cell (two flip-flops and two transmission gates) to each analog interconnect. Thus the overhead of these additions is proportional to the number of terminals. The terminal statistic is empirically expressed by Rent's rule [680]:

$$T = K \times G^\alpha \qquad (18.3)$$

where T is the number of input and output terminals for a block containing G logic gates, K is a constant between 1 and 5, and the exponent α lies in the range 0.5 to 0.67. Example 1.3 gives a derivation of the Rent's rule for a special case. Memory blocks contain fewer terminals and can be described by α lower than 0.5. In general, K is higher for combinational circuits and lower for sequential circuits. Analog blocks tend to have fewer terminals. For our illustration, we will assume $\alpha = 0.5$ for a typical block of any type. Further assuming that the chip area, A, occupied by the block is proportional to G, we can express the number of terminals as $T = k\sqrt{A}$, with some other constant k. The area of test access logic is proportional to T and the overhead (usually defined as the fractional increase over the original area) is found upon dividing it by the area A. Thus:

$$\text{Test access logic overhead} \propto \frac{1}{\sqrt{A}} \qquad (18.4)$$

This clearly shows that as the block size increases, the overhead of the test access logic drops. The reason is that the access logic is non-intrusive and is tied to the input and output signals of blocks. Additional chip area is required for routing the boundary scan and analog test bus signals. These signals feed all blocks on the chip and can be accommodated in a fixed number of routing tracks. The overheads are similar to those for scan design (see Chapter 14) and can be kept low by special layout considerations. Further, the routing overhead fraction reduces as the total routing area becomes large, which will be the case for large SOCs.

Besides the area overhead, the test access logic also adds device delays on signal paths, which should be considered in the design. The interconnect delays, however, may be smaller than those in the alternative technologies where wires must be routed on a printed circuit board. The area and delay overhead penalties of the test access structure cannot be avoided if the system-on-a-chip has to be realized. Such a system cannot be tested by the in-circuit test method that is generally used for printed circuit boards. Thus, the economic benefits of higher level of integration must offset the disadvantages.

18.8 Summary

System test overlaps both hardware and software testing. In generating tests the designer uses the concept of "orthogonality" in a sense to cover all "degrees of freedom" in the system function [641]. Basically, this amounts to executing different types of functions that the system is capable of performing. However, what is "different" depends on designer's heuristic. Software test concepts such as *error seeding* [243] and *error rate statistics* [216, 480] have been applied to evaluate the coverage of functional tests. In this chapter, we have used an informal definition of *diagnostic resolution*. A more rigorous definition is provided by Kime [355]. Also, the concept of *t-diagnosability* [356], which means that a faulty unit will be exactly identified as long as the number of faulty units does not exceed t, is useful. Methods for generating diagnostic tests have recently been reported by Hartanto *et al.* [279]. The problem of diagnosing unmodeled faults, as we discussed in Section 18.3, is an important one. A new method, known as *X-list propagation* [81, 333], assumes that the fault produces ambiguous but unknown signals. Diagnosis based on the observation of these signals, though pessimistic, has proved to be more reliable.

In designing testable systems, the concept of *macro test* [71] is useful. Here, the system is partitioned hierarchically to the lowest level blocks, called *leaf-macros*. Each leaf-macro is designed with its own design for testability rules and is provided with a test procedure. The system hierarchy is specifically designed to execute tests for all macros in the system. The methods of boundary scan, analog test bus, and core test-wrapper are in fact quite consistent with the macro test approach.

Scheduling or management of system tests is another emerging area. Many tests produce intense logic activity in the circuit, causing a large consumption of power. Reducing the test time by testing several modules in parallel can damage the system if the power consumption exceeds the specified limits. Scheduling of system tests is a useful optimization problem [165].

Most board level testers cannot match the high clock rate afforded by the present-day *system-on-a-chip* (SOC) design. At-speed testing of these SOC devices is still a challenge, but methods such as built-in self-test and embedded-ATE hold promise [482]. In the latter technique, some or most *automatic test equipment* (ATE) functions are embedded on the SOC.

Problems

18.1 *Fault dictionary.* Diagnose the circuit of Figure 18.2 using the fault dictionary of Table 18.1 when a multiple fault (a_1, c_0) has occurred.

18.2 *Diagnosis.* Diagnose the circuit of Figure 18.2 using the diagnostic tree of Figure 18.3 when a multiple fault (a_1, b_1) has occurred.

18.3 *Diagnosis.* Construct a fault dictionary for single stuck-at faults in the exclusive-OR circuit of Figure 18.8.

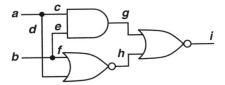

Figure 18.8: Exclusive-OR circuit for Problems 18.3 and 18.4.

18.4 *Diagnosis.* Construct a diagnostic tree for the exclusive-OR circuit of Figure 18.8.

18.5 *Diagnosis.* Repair shop data gives the following occurrence probabilities for faults in the circuit of Figure 18.2: $Prob(a_1) = 0.25$, $Prob(b_1) = 0.25$, $Prob(c_0) = 0.25$, $Prob(e_0) = 0.25$, and all other faults are never found to occur. Find the average number of tests to diagnose the circuit using the diagnostic tree of Figure 18.3.

18.6 *Diagnosis.* Redesign the diagnostic tree for the circuit of Figure 18.2 to minimize the average time of diagnosis for the fault occurrence probabilities given in Problem 18.5.

18.7 *Diagnosis.* Construct a diagnostic tree for the system shown in Figure 18.4. Is it possible to make all decisions with three or fewer tests?

18.8 Figure 18.9 shows an SOC in which a system test is conducted by the processor. Answer the following questions giving reasons: (a) Should the processor have a self-test? (b) The processor generates random vectors and output signatures in software for ASIC, FSM, and DSP. It stores correct signatures. If random vector fault coverages are low, what can be done to improve them? (c) If the SOC had an analog module, could that be tested?

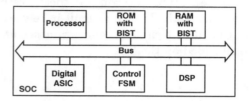

Figure 18.9: SOC for Problem 18.8.

18.9 *Rent's rule.* For Rent's rule, given by Equation 18.3, assume that $\alpha = 0.5$ and derive the value of the constant K to best fit the data on ten ISCAS'85 [100] combinational benchmark circuits given in Table 18.5.

Table 18.5: Size statistics for the ISCAS'85 combinational benchmark circuits.

Circuit name	Number of PIs	Number of POs	Number of gates
c432	36	7	160
c499	41	32	202
c880	60	26	383
c1355	41	32	546
c1908	33	25	880
c2670	233	140	1193
c3540	50	22	1669
c5315	178	123	2307
c6288	32	32	2406
c7552	207	108	3512

18.10 *Rent's rule.* For Rent's rule, given by Equation 18.3, assume that $\alpha = 0.5$ and derive the value of the constant K to best fit the data on ten largest ISCAS'89 [99] sequential benchmark circuits given in Table 18.6.

Table 18.6: Size statistics for the ISCAS'89 sequential benchmark circuits.

Circuit name	Number of PIs	Number of POs	Number of gates
s1423	17	5	657
s1488	8	19	653
s1494	8	19	647
s5378	35	49	2779
s9234	19	22	5597
s13207	31	121	7951
s15850	14	87	9772
s35932	35	320	16065
s38417	28	106	22179
s38584	12	278	19253

Chapter 19

THE FUTURE OF TESTING

"When faced with the dilemma, to test or not to test, ask not why, ask why not."

Over the last decade, rapid strides have been made in electronic systems design and integration. From the several million transistor chips of the 1990s, we now have industry prototypes of 50 million transistor chips, operating at 1 GHz frequency. Soon, we will have one billion transistor chips. This trend is expected to continue well into the next decade with the use of ultra high-performance, deep submicron devices operating at below 1 V supply voltages. In parallel, research in nano and quantum electronics will push digital circuit designs into new realms.

Future Designs. The ability of electronic circuit and system designers to design systems of massive complexity is rapidly outpacing their ability to verify the designs and test the assembled systems for manufacturing errors. As much as 50% of the manufacturing cost of high-end integrated parts today can be attributed to design verification and manufacturing test processes. This can dramatically worsen in the coming years. Very low supply voltages and 1 GHz clock speeds, coupled with aggressive device scaling, may necessitate the use of analog methods to analyze many aspects of digital circuit operation. Hitherto second-order effects such as crosstalk, ground bounce, and electromagnetic interference will significantly impact circuit performance. Packages with opto-electronic interconnect and devices and as many as 3000 pins are projected by 2004. In this context, reliable simulation of all such effects for verification, design debugging, and test development, particularly for worst-case behavior, appears extremely challenging. The required modeling accuracy has greatly increased, and the computational requirements have severely increased.

Over the next 20 years, we will integrate entire electronic systems consisting of digital, analog (including *radio frequency* (RF) circuits), optical, chemical, and *micro-electromechanical system* (MEMS) parts into a single chip. The problems of design verification and test of integrated *systems-on-a-chip* (SOC) will assume hitherto unseen proportions. This complicates manufacturing economics and the global competitiveness of the industry.

Implications for System Verification. Design verification will migrate from primarily logic verification (combinational and sequential) to the problem of coupled verification of digital, analog, MEMS, chemical, optical, thermal, and material properties of the assembled package.

Implications for Design and Test. The same issues will dominate test development and manufacturing test of SOCs. Logical test generation algorithms will gradually evolve from their present state to one that completely encompasses the physics of manufacturing such tightly integrated electronic systems. We need new test stimulus generation algorithms, currently unavailable, that generate digital and analog test stimuli and waveforms for low-cost and high-coverage test of SOC components. It will be impossible to map realistic failure modes into simple models. The presently popular stuck-at fault model must be augmented with delay fault models (both transition and path-delay) and I_{DDQ} elevated current fault models. This effect will continue, and a set of diverse models and associated tests will achieve high defect coverage through test diversity. Carefully optimized thermal integrated systems will also have to be concurrently designed to provide extensive *design for testability* (DFT), *built-in self-test* (BIST), performance verification, debugging, and diagnosis of early silicon prototypes through DFT and BIST.

We need new methods for concurrent fault simulation of integrated mixed-technology systems. Even with high-level behavioral models, simulation of fault-free mixed analog and digital systems can be prohibitively expensive.

We close with a listing of the key future unsolved challenges for VLSI design and test, as device feature sizes scale down and SOC advances:

- We must learn to project the effects of process variations and manufacturing-induced defects across technology boundaries (between digital and analog, between optical and RF circuits, etc.)

- We must model the system-on-a-chip at higher levels of abstraction to get acceptable simulation speed without losing accuracy. The analog electromagnetic effects must be captured at a very high digital level of modeling.

- Interconnect delay will continue to grow more important than logic gate delay.

- We need efficient behavioral models of digital, analog, MEM, and optical systems.

- We must reduce test costs, but the SIA National Technology Roadmap [584] projects testing costs holding steady, at best.

- We need to invent a diagnosis ability for optical, chemical, and MEMS faults.

- We need a new ability to predict thermal and mechanical stresses in the mixed-technology SOC across technology boundaries.

Appendix A

CYCLIC REDUNDANCY CODE THEORY

> *"It is hard to establish . . . who did what first. . . . E. N. Gilbert of the Bell Telephone Laboratories derived much of linear theory a year or so earlier than either Zierler, Welch, or myself, . . . the first investigation of linear recurrence relations modulo p goes back as far as Lagrange, in the eighteenth century, and an excellent modern treatment was given (. . . purely mathematical . . .) by Marshall Hall in 1937."* — Solomon W. Golomb [264].

Cyclic redundancy code theory is described in this appendix because it is important for *built-in self-testing* (BIST), and for Memory ROM testing. Cyclic redundancy codes have been extensively used for decades in the computer industry to guarantee correct recording of blocks on disk and magnetic tape drives. They also are extensively used for encoding and decoding communication channel *burst errors*, where a transient fault causes several adjacent data errors.

Preliminaries. We express the binary vector $R = r_m r_{m-1} \ldots r_0$ as the polynomial $r_m x^m + r_{m-1} x^{m-1} + \ldots + r_0$. As an example, 10111 is written as $1 + x^2 + x^3 + x^4$ [36, 37]. The *degree* of the polynomial is the superscript of the highest non-zero term. Polynomial addition and subtraction arithmetic is performed similarly to integer arithmetic, except that the arithmetic is modulo 2. In this case, addition and subtraction will have the same effect. As an example, let $q(x) = x^3 + x^2 + 1$ (degree 3) and $p(x) = x^2 + x + 1$ (degree 2.)

$$p(x) + q(x) = x^3 + 2x^2 + x + 2 = x^3 + x \tag{A.1}$$

$$p(x) \cdot q(x) = x^5 + 2x^4 + 2x^3 + 2x^2 + x + 1 = x^5 + x + 1 \tag{A.2}$$

We can describe the remainders of polynomials modulus a polynomial. Two polynomials $r(x)$ and $s(x)$ will be *congruent* modulus polynomial $n(x)$, written as $r(x) \equiv s(x) \bmod n(x)$ if $\exists$ a polynomial $q(x)$ such that $r(x) = n(x) \cdot q(x) + s(x)$. We find the residue (the remainder) by dividing $r(x)$ by $n(x)$. For example:

$$(x^2 + x + 1) \cdot (x^3 + 1) \equiv (x^5 + x^4 + x^3 + x^2 + x + 1) \bmod (x^4 + x) \qquad (A.3)$$

$$
\begin{array}{r}
x +1 \\
x^4 + x \overline{\smash{\big)}\, x^5 + x^4 + x^3 + x^2 + x + 1} \\
\underline{x^5 + x^2 } \\
x^4 + x^3 + x \\
\underline{x^4 + x } \\
x^3 + 1
\end{array}
$$

So, $(x^2 + x + 1) \cdot (x^3 + 1) \equiv x^3 + 1 \bmod (x^4 + x)$. *Irreducible polynomials* cannot be factored and are analogous to the prime integer numbers.

A.1 Polynomial Multiplier

Figure A.1(a) shows a polynomial multiplier, used as an encoding circuit to create a cyclic code. The code has the property that any end-around shift of a code word produces another code word. The code has these properties:

- It has a Boolean generator polynomial $G(x)$ of degree $n - k$ or greater, where n is the number of bits in the complete code word and k is the number of bits in the original information to be encoded. Coefficients of $G(x)$ are 0 or 1. This is called an (n, k) cyclic code, and it is able to detect all single errors and all multiple adjacent errors in transmission of the code word.

- The cyclic code represents data as a Boolean polynomial, as described above. The code word is the coefficients of a polynomial $V = (V_0, V_1, \ldots, V_{n-1})$ so $V(x) = V_0 + V_1 x + V_2 x^2 + \ldots + V_{n-1} x^{n-1}$ An n-bit code word is represented by a polynomial of degree $n - 1$ or less. $V(x)$ is called the *code polynomial* of the code word V.

We generate code polynomials by multiplying a polynomial representing the data to be encoded by the generator polynomial.

Example A.1 *We wish to encode the binary word* $(1101) = D(x) = 1 + x + x^3$. *In this case, the generator polynomial is* $G(x) = 1 + x + x^3$. *The code polynomial is:*

$$
\begin{aligned}
V(x) &= D(x) \times G(x) = (1 + x + x^3) \times (1 + x + x^3) \qquad (A.4) \\
&= 1 + x + x^3 + x + x^2 + x^4 + x^3 + x^4 + x^6 \\
&= 1 + x^2 + x^6
\end{aligned}
$$

The code is the coefficient set (1010001).

A.2 Polynomial Divider

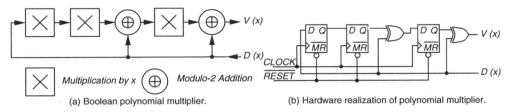

(a) Boolean polynomial multiplier.   (b) Hardware realization of polynomial multiplier.

Figure A.1: Boolean polynomial multiplier and realization.

Cyclic codes are *non-separable* codes, which means that it is not possible to decode the data by simply dropping extra code bits. The *Hamming distance* is the distance between adjacent code words, in terms of the number of bit changes. This is found for a cyclic code by comparing all possible code word pairs and determining the minimum Hamming distance between any two code words. The system of Figure A.1(a) performs polynomial multiplication for the generator $G(x) = 1 + x + x^3$.

In this number system, we implement multiplication by x with a time shift using a D flip-flop, and we implement the modulo-2 addition operator with an EXCLUSIVE-OR logic gate. The data to be encoded appears serially on $D(x)$, and the bits of the encoded word appear serially on $V(x)$. Initially, the registers in the circuit must be initialized to 0 before encoding can occur. Figure A.1(b) is an implementation of the polynomial multiplier of Figure A.1(a). Table A.1 shows the code words generated by this system.

Table A.1: Cyclic code for 4-bit information.

Information (D_0, D_1, D_2, D_3)				Code $(V_0, V_1, V_2, V_3, V_4, V_5, V_6)$			
Info.	Code	Info.	Code	Info.	Code	Info.	Code
0000	0000000	0100	0110100	1000	1101000	1100	1011100
0001	0001101	0101	0111001	1001	1100101	1101	1010001
0010	0011010	0110	0101110	1010	1110010	1110	1000110
0011	0010111	0111	0100011	1011	1111111	1111	1001011

A.2 Polynomial Divider

Figure A.2(a) shows a polynomial divider, used as a decoding circuit. It is the companion circuit to the encoder of Figure A.1(a). In the decoding procedure, we treat the code word as a Boolean polynomial, but now we divide it by the generator polynomial and determine the remainder, as well as the quotient, of the division. If the remainder is 0, then the code word was valid. A non-zero remainder indicates a transmission error. Here the code word $(V_0, V_1, V_2, \ldots, V_{n-1})$ again represents the polynomial $V_0 + V_1 + V_2 x^2 + \ldots + V_{n-1} x^{n-1}$. We obtained $V(x) = D(x)G(x) + S(x)$ when we encoded the data $D(x)$ by multiplying by the generator $G(x)$, using $S(x) = 0$. $S(x)$ is the *syndrome polynomial*, which should be 0 during encoding. Figure A.2(a) shows the conceptual division circuit, again using $G(x) = 1 + x + x^3$.

Here, $D(x) = V(x) + B(x)$ and $B(x) = x \times D(x) + x^3 \times D(x)$, so:

$$D(x) = V(x) + x \times D(x) + x^3 \times D(x) \quad (A.5)$$
$$D(x)(1 + x + x^3) = V(x)$$
$$D(x) = \frac{V(x)}{(1 + x + x^3)}$$

Since this is a modulo-2 addition system, adding the same quantity to both sides of the equation has the same effect as subtracting the same quantity from both sides of the equation. Figure A.2(b) shows the division circuit, and Table A.2 shows the results of decoding a code word $V(x)$.

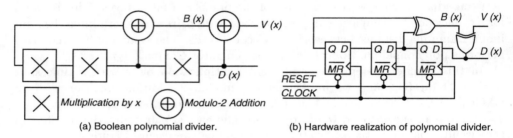

(a) Boolean polynomial divider.   (b) Hardware realization of polynomial divider.

Figure A.2: Boolean polynomial divider and realization.

Table A.2: Decoding process.

Clock period	Register values 1	2	3	$V(x)$	$B(x)$	$D(x)$
0	0	0	0			
				1	0	1
1	0	0	1			
				0	1	1
2	0	1	1			
				1	1	0
3	1	1	0			
				0	1	1
4	1	0	1			
				0	0	0
5	0	1	0			
				0	0	0
6	1	0	0			
				1	1	0
7	0	0	0			
	Syndrome			Code word		Original information

Appendix B

PRIMITIVE POLYNOMIALS OF DEGREE 1 TO 100

The primitive polynomials shown below in Table B.1 were taken from the book by Bardell *et al.* [67]. They, in turn, obtained the first 100 polynomials from Stahnke [633]. To decode the table, consider that the entry

`12: 7 4 3 0`

represents the polynomial $x^{12}+x^7+x^4+x^3+1$, as each table entry represents a term exponent. Since there is always a term for x^n, it is not represented. Notice that there is always an x^0 term, but it is shown for consistency reasons. In the unlikely event that a polynomial of degree greater than 100 is needed, consult Bardell *et al.* [67] for polynomials of degree up to 300.

Appendix B. PRIMITIVE POLYNOMIALS OF DEGREE 1 TO 100

Table B.1: Exponents of primitive binary polynomial terms.

n	a	b	c	d	n	a	b	c	d	n	a	b	c	d
1:	0				34:	15	14	1	0	67:	10	9	1	0
2:	1	0			35:	2	0			68:	9	0		
3:	1	0			36:	11	0			69:	29	27	2	0
4:	1	0			37:	12	10	2	0	70:	16	15	1	0
5:	2	0			38:	6	5	1	0	71:	6	0		
6:	1	0			39:	4	0			72:	53	47	6	0
7:	1	0			40:	21	19	2	0	73:	25	0		
8:	6	5	1	0	41:	3	0			74:	16	15	1	0
9:	4	0			42:	23	22	1	0	75:	11	10	1	0
10:	3	0			43:	6	5	1	0	76:	36	35	1	0
11:	2	0			44:	27	26	1	0	77:	31	30	1	0
12:	7	4	3	0	45:	4	3	1	0	78:	20	19	1	0
13:	4	3	1	0	46:	21	20	1	0	79:	9	0		
14:	12	11	1	0	47:	5	0			80:	38	37	1	0
15:	1	0			48:	28	27	1	0	81:	4	0		
16:	5	3	2	0	49:	9	0			82:	38	35	3	0
17:	3	0			50:	27	26	1	0	83:	46	45	1	0
18:	7	0			51:	16	15	1	0	84:	13	0		
19:	6	5	1	0	52:	3	0			85:	28	27	1	0
20:	3	0			53:	16	15	1	0	86:	13	12	1	0
21:	2	0			54:	37	36	1	0	87:	13	0		
22:	1	0			55:	24	0			88:	72	71	1	0
23:	5	0			56:	22	21	1	0	89:	38	0		
24:	4	3	1	0	57:	7	0			90:	19	18	1	0
25:	3	0			58:	19	0			91:	84	83	1	0
26:	8	7	1	0	59:	22	21	1	0	92:	13	12	1	0
27:	8	7	1	0	60:	1	0			93:	2	0		
28:	3	0			61:	16	15	1	0	94:	21	0		
29:	2	0			62:	57	56	1	0	95:	11	0		
30:	16	15	1	0	63:	1	0			96:	49	47	2	0
31:	3	0			64:	4	3	1	0	97:	6	0		
32:	28	27	1	0	65:	18	0			98:	11	0		
33:	13	0			66:	10	9	1	0	99:	47	45	2	0
										100:	37	0		

Appendix C

BOOKS ON TESTING

C.1 General and Tutorial

- M. Abramovici, M. A. Breuer, and A. D. Friedman, *Digital Systems Testing and Testable Design*. Piscataway, New Jersey: IEEE Press, 1994. Revised printing.
- V. D. Agrawal and S. C. Seth, *Tutorial: Test Generation for VLSI Chips*. Los Alamitos, California: IEEE Computer Society Press, 1988.
- J. DiGiacomo, editor, *VLSI Handbook*. New York: McGraw-Hill, 1989.
- N. G. Einspruch, editor, *VLSI Handbook*. Orlando, Florida: Academic Press, 1985.
- W. G. Fee, *Tutorial: LSI Testing*. Los Alamitos, California: IEEE Computer Society Press, 1978.
- A. D. Friedman and P. R. Menon, *Fault Detection in Digital Circuits*. Upper Saddle River, New Jersey: Prentice-Hall, 1971.
- H. Fujiwara, *Logic Testing and Design for Testability*. Cambridge, Massachusetts: MIT Press, 1985.
- Z. Kohavi, *Switching and Automata Theory*. New York: McGraw-Hill, 1978.
- L. Lavagno and A. Sangiovanni-Vincentelli, *Algorithms for Synthesis and Testing of Asynchronous Circuits*. Boston: Kluwer Academic Publishers, 1993.
- F. Lombardi and M. Sami, editors, *Testing and Diagnosis of VLSI and ULSI*. Boston: Kluwer Academic Publishers, 1988.
- R. E. Massara, editor, *Design & Test Techniques for VLSI & WSI Circuits*. London, United Kingdom: Peter Peregrinus, 1989.
- E. J. McCluskey, *Logic Design Principles With Emphasis on Testable Semicustom Circuits*. Upper Saddle River, New Jersey: Prentice-Hall, 1986.
- A. Miczo, *Digital Logic Testing and Simulation*. New York: Harper & Row, 1986.
- D. M. Miller, *Developments in Integrated Circuit Testing*. San Diego, California: Academic Press, 1987.
- H. K. Reghbati, *Tutorial: VLSI Testing & Validation Techniques*. Los Alamitos, California: IEEE Computer Society Press, 1985.
- G. Russel and I. L. Sayers, *Advanced Simulation and Test Methodologies for VLSI Design*. London, United Kingdom: Van Nostrand Reinhold, 1989.
- B. R. Wilkins, *Testing Digital Circuits, An Introduction*. Berkshire, United Kingdom: Van Nostrand Reinhold, 1986.

- T. W. Williams, editor, *VLSI Testing*. Amsterdam, The Netherlands: North-Holland, 1986.

C.2 Analog and Mixed-Signal Circuit Test

- A. Afshar, *Principles of Semiconductor Network Testing*. Boston: Butterworth-Heinemann, 1995.
- M. Burns and G. Roberts, *Introduction to Mixed-Signal IC Test and Measurement*. New York: Oxford University Press, 2000.
- B. Kaminska and B. Courtois, editors, *Special Issue on Analog and Mixed Signal Testing*, volume 9 of *Journal of Electronic Testing: Theory and Applications*. Boston: Kluwer Academic Publishers, August-October 1996.
- R. W. Liu, editor, *Selected Papers on Analog Fault Diagnosis*. Piscataway, New Jersey: IEEE Press, 1987.
- R. W. Liu, editor, *Testing and Diagnosis of Analog Circuits and Systems*. New York: Van Nostrand Reinhold, 1991.
- M. Mahoney, *DSP-Based Testing of Analog and Mixed-Signal Circuits*. Los Alamitos, California: IEEE Computer Society Press, 1987.
- A. Osseiran, *Analog and Mixed-Signal Boundary-Scan*. Boston: Kluwer Academic Publishers: 1999.
- T. Ozawa, editor, *Analog Methods for Computer-Aided Circuit Analysis and Diagnosis*. New York: Marcel Dekker, 1988.
- G. W. Roberts and A. K. Lu, *Analog Signal Generation for Built-In Self-Test of Mixed-Signal Integrated Circuits*. Boston: Kluwer Academic Publishers, 1995.
- M. Soma, editor, *Special Issue on Analog and Mixed-Signal Testing*, volume 4 of *Journal of Electronic Testing: Theory and Applications*. Boston: Kluwer Academic Publishers, November 1993.
- B. Vinnakota, editor, *Analog and Mixed-Signal Test*. Upper Saddle River, New Jersey: Prentice-Hall PTR, 1998.

C.3 ATE, Test Programming, and Production Test

- J. H. Arabian, *Computer Integrated Electronics Manufacturing and Testing*. New York: Marcel Dekker, 1989.
- J. Bateson, *In-Circuit Testing*. New York: Van Nostrand Reinhold, 1985.
- A. Buckroyd, editor, *Computer Integrated Testing*. Oxford, United Kingdom: Blackwell Scientific Publications, 1989.
- J. M. Cortner, *Digital Test Engineering*. Somerset, New Jersey: Wiley, 1987.
- R. J. Feugate and S. M. McIntyre, *Introduction to VLSI Testing*. Upper Saddle River, New Jersey: Prentice-Hall, 1988.
- J. P. Hanna, R. G. Hillman, H. L. Hirsch, T. H. Noh, and R. R. Vemuri, *Using WAVES and VHDL for Effective Design and Testing*. Boston: Kluwer Academic Publishers, 1997.
- J. T. Healy, *Automatic Testing and Evaluation of Digital Integrated Circuits*. Reston, Virginia: Reston Publishing Company, 1981.

- E. R. Hnatek, *Integrated Circuit Quality and Reliability*. New York: Marcel Dekker, 1987.
- F. Jensen and N. E. Petersen, *Burn-In*. Chichester, United Kingdom: John Wiley & Sons, 1982.
- R. Knowles, *Automatic Testing Systems and Applications*. London, United Kingdom: McGraw-Hill, 1976.
- W. Nelson, *Accelerated Testing: Statistical Models, Test Plans, and Data Analysis*. Somerset, New Jersey: John Wiley & Sons, 1990.
- K. P. Parker, *Integrating Design and Test: Using CAE Tools for ATE Programming*. Los Alamitos, California: IEEE Computer Society Press, 1987.
- C. Pynn, *Strategies for Electronics Test*. New York: McGraw-Hill, 1986.
- B. P. Richards and P. K. Footner, *The Role of Microscopy in Semiconductor Failure Analysis*. New York: Oxford University Press, 1992.
- A. K. Stevens, *Introduction to Component Testing*. Reading, Massachusetts: Addison-Wesley, 1986.
- A. C. Stover, *ATE: Automatic Test Equipment*. New York: McGraw-Hill, 1984.
- R. A. Witte, *Electronic Test Instruments: Theory and Applications*. Englewood Cliffs, New Jersey: PTR Prentice Hall, 1993.
- V. N. Yarmolik, *Fault Diagnosis of Digital Circuits*. Somerset, New Jersey: John Wiley & Sons, 1990.

C.4 Board and MCM Test and Boundary Scan

- R. G. Bennetts, *Introduction to Digital Board Testing*. New York: Crane Russak, 1982.
- R. G. Bennetts, editor, *Special Issue on Boundary Scan*, volume 2 of *Journal of Electronic Testing: Theory and Applications*. Boston: Kluwer Academic Publishers, February 1991.
- H. Bleeker, P. van den Eijnden, and F. de Jong, *Boundary-Scan Test, A Practical Approach*. Boston: Kluwer Academic Publishers, 1993.
- G. Ginsberg, *Surface Mount and Related Technologies*. New York: Marcel Dekker, 1989.
- S. W. Hinch, *Handbook of Surface Mount Technology*. New York: John Wiley & Sons, 1988.
- C. M. Maunder and R. E. Tulloss, *The Test Access Port and Boundary-Scan Architecture*. Los Alamitos, California: IEEE Computer Society Press, 1990.
- B. Nadeau-Dostie, editor, *Design for At-Speed Test, Diagnosis and Measurement*, Boston: Kluwer Academic Publishers, 2000.
- K. P. Parker, *The Boundary-Scan Handbook*. Boston: Kluwer Academic Publishers, 1998.
- R. P. Prasad, *Surface Mount Technology Principles and Practice*. New York: Van Nostrand Reinhold, 1989.
- S. F. Scheiber, *Building a Successful Board-Test Strategy*. Boston: Butterworth-Heinemann, 1995.
- Y. Zorian, editor, *Multi-Chip Module Test Strategies*. Boston: Kluwer Academic Publishers, 1997.

C.5 Built-In Self-Test

- P. H. Bardell, W. H. McAnney, and J. Savir, *Built-In Test for VLSI: Pseudorandom Techniques*. Somerset, New Jersey: Wiley, 1987.
- S. W. Golomb, *Shift Register Sequences*. Laguna Hills, California: Aegean Park Press, 1982.
- P. Pal Chaudhuri, D. Roy Chowdhury, S. Nandi, S. Chattopadhyay, *Additive Cellular Automata Theory and Applications*. Los Alamitos, California: IEEE Computer Society Press, 1997.
- S. Pilarski and T. Kameda, *A Probabilistic Analysis of Test-Response Compaction*. Boston: Kluwer Academic Publishers, 1995.
- J. Rajski and J. Tyszer, *Arithmetic Built-In Self-Test for Embedded Systems*. Englewood Cliffs, New Jersey: PTR Prentice Hall, 1998.
- C. Ronse, *Feedback Shift Registers*. Berlin: Springer-Verlag, 1984.
- V. N. Yarmolik and S. N. Demidenko, *Generation and Application of Pseudorandom Sequences for Random Testing*. Chichester, United Kingdom: John Wiley & Sons, 1988.
- V. N. Yarmolik and I. V. Kachan, *Self Testing VLSI Design*. Amsterdam, The Netherlands: Elsevier, 1993.

C.6 Delay Fault Test

- A. Krstić and K.-T. Cheng, *Delay Fault Testing for VLSI Circuits*. Boston: Kluwer Academic Publishers, 1998.
- M. Sivaraman and A. J. Strojwas, *A Unified Approach for Timing Verification and Delay Fault Testing*. Boston: Kluwer Academic Publishers, 1998.

C.7 Design for Testability

- V. D. Agrawal, editor, *Special Issue on Partial Scan Methods*, volume 7 of *Journal of Electronic Testing: Theory and Applications*. Boston: Kluwer Academic Publishers, August-October 1995.
- F. P. M. Beenker, R. G. Bennetts, and A. P. Thijssen, *Testability Concepts for Digital ICs: The Macro Test Approach*. Boston: Kluwer Academic Publishers, 1995.
- R. G. Bennets, *Design of Testable Logic Circuits*. Reading, Massachusetts: Addison-Wesley, 1984.
- A. L. Crouch, *Design-for-Test for Digital ICs and Embedded Core System*, Piscataway, New Jersey: IEEE Press, 2000. Co-published with Prentice Hall PTR.
- E. B. Eichelberger, E. Lindbloom, J. A. Waicukauski, and T. W. Williams, *Structured Logic Testing*. Upper Saddle River, New Jersey: Prentice-Hall, 1991.
- W. M. Needham, *Designer's Guide to Testable ASIC Devices*. New York: Van Nostrand Reinhold, 1991.
- C. C. Timoc, *Selected Reprints on Logic Design for Testability*. Los Alamitos, California: IEEE Computer Society Press, 1984.
- F. F. Tsui, *LSI-VLSI Testability Design*. New York: McGraw-Hill, 1986.
- J. Turino, *Design to Test*. Florence, Kentucky: Van Nostrand Reinhold, 1990.

C.8 Fault Modeling

- D. Bhattacharya and J. P. Hayes, *Hierarchical Modeling for VLSI Circuit Testing*. Boston: Kluwer Academic Publishers, 1990.
- N. K. Jha and S. Kundu, *Testing and Reliable Design of CMOS Circuits*. Boston: Kluwer Academic Publishers, 1990.
- R. Rajsuman, *Digital Hardware Testing: Transistor Level Fault Modeling and Testing*. Boston: Artech House, 1992.

C.9 Fault Tolerance and Diagnosis

- M. A. Breuer and A. D. Friedman, *Diagnosis & Reliable Design of Digital Systems*. Rockville, Maryland: Computer Science Press, 1976.
- H. Y. Chang, E. G. Manning, and G. Metze, *Fault Diagnosis of Digital Systems*. New York: Wiley-Interscience, 1970.
- M. Goessel and S. Graf, *Error Detection Circuits*. New York: McGraw-Hill, 1993.
- P. Jalote, *Fault Tolerance in Distributed Systems*. Englewood Cliffs, New Jersey: PTR Prentice Hall, 1994.
- B. W. Johnson, *Design and Analysis of Fault Tolerant Digital Systems*. Reading, Massachusetts: Addison-Wesley, 1988.
- I. Koren, editor, *Defect and Fault Tolerance in VLSI Systems*. New York: Plenum Press, 1989.
- P. K. Lala, *Fault Tolerant & Fault Testable Hardware Design*. London, United Kingdom: Prentice Hall International, 1985.
- M. Nicolaidis, Y. Zorian, and D. K. Pradhan, editors, *On-Line Testing for VLSI*. Boston: Kluwer Academic Publishers, 1998.
- D. K. Pradhan, editor, *Fault-Tolerant Computing Theory and Techniques*, volumes I and II. Upper Saddle River, New Jersey: Prentice-Hall, 1986.
- D. K. Pradhan, *Fault-Tolerant Computer System Design*. Upper Saddle River, New Jersey: Prentice Hall PTR, 1996.
- F. F. Sellers, Jr., M. Y. Hsiao, and L. W. Bearnson, *Error Detecting Logic for Digital Computers*. New York: McGraw-Hill, 1968.
- R. Spence and R. S. Soin, *Tolerance Design of Electronic Circuits*. Wokingham, England: Addison-Wesley, 1988.

C.10 Formal Verification

- L. Bening and H. Foster, *Principles of Verifiable RTL Design*. Boston: Kluwer Academic Publishers, 2000.
- E. M. Clarke, Jr., and O. Grumberg and D. A. Peled, *Model Checking*, Cambridge, Massachusetts: MIT Press, 1999.
- G. D. Hachtel and F. Somenzi, *Logic Synthesis and Verification Algorithms*. Boston: Kluwer Academic Publishers, 1996.
- G. J. Holzman, *Design and Validation of Computer Protocols*. Englewood Cliffs, New Jersey: Prentice Hall, 1991.

- S. Y. Huang and K.-T. Cheng, *Formal Equivalence Checking and Design Debugging*, Boston: Kluwer Academic Publishers, 1998.
- W. Kunz and D. Stoffel, *Reasoning in Boolean Networks: Logic Synthesis and Verification Using Testing Techniques*. Boston: Kluwer Academic Publishers, 1997.
- R. P. Kurshan, *Computer Aided Verification of Coordinating Processes*. Princeton, New Jersey: Princeton University Press, 1994.
- R. J. Linn and M. U. Uyar, editors, *Conformance Testing Methodologies and Architectures for OSI Protocols*. Los Alamitos, California: IEEE Computer Society Press, 1994.
- K. L. McMillan, *Symbolic Model Checking*, Boston: Kluwer Academic Publishers, 1993.
- J. P. Roth, *Computer Logic, Testing, and Verification*. Rockville, Maryland: Computer Science Press, 1980.

C.11 High-Level Test and Verification

- J. Bergeron, *Writing Testbenches*. Boston: Kluwer Academic Publishers, 2000.
- M. T. C. Lee, *High-Level Test Synthesis of Digital VLSI Circuits*. Boston: Artech House, 1997.

C.12 I_{DDQ} Test

- S. Chakravarty and P. J. Thadikaran, *Introduction to I_{DDQ} Testing*. Boston: Kluwer Academic Publishers, 1997.
- R. K. Gulati and C. F. Hawkins, editors, *I_{DDQ} Testing of VLSI Circuits*. Boston: Kluwer Academic Publishers, 1993.
- Y. K. Malaiya and R. Rajsuman, editors, *Bridging Faults and IDDQ Testing*. Los Alamitos, California: IEEE Computer Society Press, 1992.
- R. Rajsuman, *I_{DDQ} Testing for CMOS VLSI*. Boston: Artech House, 1995.

C.13 Memory Test

- P. Mazumder, editor, *Special Issue on Memory Testing*, volume 5 of *Journal of Electronic Testing: Theory and Applications*. Boston: Kluwer Academic Publishers, November 1994.
- P. Mazumder and K. Chakraborty, *Testing and Testable Design of High-Density Random-Access Memories*. Boston: Kluwer Academic Publishers, 1996.
- A. K. Sharma, *Semiconductor Memories: Technology, Testing, and Reliability*. Piscataway, New Jersey: IEEE Press, 1997.
- A. J. van de Goor, *Testing Semiconductor Memories Theory and Practice*. Chichester, United Kingdom: Wiley-Interscience, 1991.

C.14 Microprocessor Verification and Test

- M. S. Abadir, editor, *Special Issue on Microprocessor Test and Verification*, volume 16, number 1/2 of *Journal of Electronic Testing: Theory and Applications*. Boston: Kluwer Academic Publishers, February/April 2000.

C.15 Semiconductor Defect Mechanisms

- E. A. Amerasekera and D. S. Campbell, *Failure Mechanisms in Semiconductor Devices*. Chichester, United Kingdom: John Wiley & Sons, 1987.
- J. P. deGyvez and D. K. Pradhan, *Integrated Circuit Manufacturability*. Piscataway, New Jersey: IEEE Press, 1998.
- S. W. Director, W. Maly, and A. J. Strojwas, *VLSI Design for Manufacturing Yield Enhancement*. Boston: Kluwer Academic Publishers, 1989.
- A. P. Dorey, B. K. Jones, A. M. D. Richardson, and Y. Z. Xu, *Rapid Reliability Assessment of VLSI*, Boston: Plenum Publishers/Kluwer Academic Press, 1990.
- W. D. Greason, *Electrostatic Damage in Electronics*. Somerset, New Jersey: Wiley, 1987.
- M. J. Howes and D. V. Morgan, editors, *Reliability and Degradation – Semiconductor Devices and Circuits*. Chichester, United Kingdom: Wiley-Interscience, 1981.
- J. B. Khare and W. Maly, *From Contamination to Defects, Faults and Yield Loss: Simulation and Applications*. Boston: Kluwer Academic Publishers, 1996.
- D. J. Klinger, Y. Nakada, and M. A. Manendez, *AT&T Reliability Manual*. Florence, Kentucky: Van Nostrand Reinhold, 1990.
- J. M. Kolyer and D. E. Watson, *ESD A to Z Electrostatic Discharge Control for Electronics*. New York: Van Nostrand Reinhold, 1990.
- M. Sachdev, *Defect Oriented Testing for CMOS Analog and Digital Circuits*. Boston: Kluwer Academic Publishers, 1998.

C.16 System Test

- R. Rajsuman, *System-on-a-Chip: Design and Test*, Boston: Artech House, 2000.
- J. W. Sheppard and W. R. Simpson, *Research Perspectives and Case Studies in System Test and Diagnosis*. Boston: Kluwer Academic Publishers, 1998.
- W. R. Simpson and J. W. Sheppard, *System Test and Diagnosis*. Boston: Kluwer Academic Publishers, 1994.

C.17 Test Economics

- M. Abadir and A. P. Ambler, editors, *Economics of Electronic Design, Manufacture and Test*. Boston: Kluwer Academic Publishers, 1994.
- A. P. Ambler, M. Abadir, and S. Sastry, editors, *Economics of Design and Test for Electronics Circuits and Systems*. Chichester, United Kingdom: Ellis Horwood, 1992.
- B. Davis, *The Economics of Automatic Testing*. London, United Kingdom: McGraw-Hill, 1982.

- C. Dislis, J. H. Dick, I. D. Deer, and A. P. Ambler, *Test Economics and Design for Testability*. New York: Ellis Horwood, 1995.

C.18 Test Evaluation

- K. M. Butler and M. R. Mercer, *Assessing Fault Model and Test Quality*. Boston: Kluwer Academic Publishers, 1992.
- B. Ciciani, editor, *Manufacturing Yield Evaluation of VLSI/WSI Systems*. Los Alamitos, California: IEEE Computer Society Press, 1995.
- E. G. Ulrich, V. D. Agrawal, and J. H. Arabian, *Concurrent and Comparative Discrete Event Simulation*. Boston: Kluwer Academic Publishers, 1994.
- D. M. H. Walker, *Yield Simulation for Integrated Circuits*. Boston: Kluwer Academic Publishers, 1987.

C.19 Test Generation

- S. T. Chakradhar, V. D. Agrawal, and M. L. Bushnell, *Neural Models and Algorithms for Digital Testing*. Boston: Kluwer Academic Publishers, 1991.
- X. Chen and M. L. Bushnell, *Efficient Branch and Bound Search with Application to Computer-Aided Design*. Boston: Kluwer Academic Publishers, 1996.
- K.-T. Cheng and V. D. Agrawal, *Unified Methods for VLSI Simulation and Test Generation*. Boston: Kluwer Academic Publishers, 1989.
- R. David, *Random Testing of Digital Circuits Theory and Applications*. New York: Marcel Dekker, 1998.
- A. Ghosh, S. Devadas, and A. R. Newton, *Sequential Logic Testing and Verification*. Boston: Kluwer Academic Publishers, 1992.
- P. Mazumder and E. M. Rudnick, *Genetic Algorithms for VLSI Design, Layout, and Test Automation*. Upper Saddle River, New Jersey: Prentice-Hall, 1999.
- N. Singh, *An Artificial Intelligence Approach to Test Generation*. Boston: Kluwer Academic Publishers, 1987.

C.20 Periodicals

- *Design & Test of Computers*, IEEE Computer Society, practical interest articles since 1984.
- *Journal of Electronic Testing: Theory and Applications*, Kluwer Academic Publishers, theoretical and scholarly articles since 1990.
  http://www.wkap.nl/journalhome.htm/0923-8174
- *Journal of Solid State Circuits*, IEEE, mostly solid state circuit articles.
- *Test & Measurement World*, Cahners Publishing, articles on ATE products.
- *Transactions on Computers*, IEEE Computer Society, scholarly computer and testing articles.
- *Transactions on Computer-Aided Design*, IEEE Circuits and Systems Society, scholarly computer-aided design articles since 1982.

- *Transactions on Instrumentation and Measurement*, IEEE.
- *Transactions on VLSI Systems*, IEEE Computer Society and IEEE Circuits and Systems Society, scholarly VLSI design articles since 93.

C.21 Conferences and Workshops

- *Asian Test Symposium*, IEEE.
- *Autotestcon Conf.*, IEEE.
- *Custom Integrated Circuits Conf.*, IEEE.
- *Design and Test in Europe (DATE) Conf.*
- *Design Automation Conf.*, ACM/IEEE.
- *European Design Automation Conf.*
- *European Design and Test Conf.*
- *Int'l. Conf. on Computer Design*, IEEE.
- *Int'l. Conf. on Computer-Aided Design*, IEEE.
- *VLSI Test Symposium*, IEEE.
- *Int'l. Conf. on VLSI Design*, VLSI Society of India, IEEE.
- *Int'l. Fault Tolerant Computing Symposium*, IEEE.
- *Int'l. Mixed-Signal Test Workshop*, IEEE.
- *Int'l. Symposium on Circuits and Systems*, IEEE.
- *Int'l. Test Conf.*, IEEE.
- *North Atlantic Test Workshop*, IEEE.
- *Int'l. Test Synthesis Workshop*, IEEE.

Check the IEEE TTTC web site for other meetings.

C.22 Web Sites

- http://www.ieee.org
- http://www.computer.org
- http://www.computer.org/tttc
- http://www.itctestweek.org

BIBLIOGRAPHY

[1] M. Abadir and A. P. Ambler, *Economics of Electronic Design, Manufacture and Test.* Boston: Kluwer Academic Publishers, 1994.

[2] M. S. Abadir, editor, *Special Issue on Microprocessor Test and Verification. Journal of Electronic Testing: Theory and Applications*, vol. 16, Feb./Apr. 2000.

[3] M. S. Abadir and M. A. Breuer, "A Knowledge-Based System for Designing Testable VLSI Chips," *IEEE Design & Test of Computers*, vol. 2, no. 4, pp. 56–68, Aug. 1985.

[4] M. S. Abadir and J. K. Reghbati, "Functional Testing of Semiconductor Random Access Memories," *ACM Computing Surveys*, vol. 15, no. 3, pp. 175–198, Sept. 1983.

[5] A. Abderrahman, E. Cerny, and B. Kaminska, "CLP-based Multifrequency Test Generation for Analog Circuits," in *Proc. of the 15th VLSI Test Symp.*, Apr. 1997, pp. 158–165.

[6] E. M. Aboulhamid, Y. Karkouri, and E. Cerny, "On the Generation of Test Patterns for Multiple Faults," *Journal of Electronic Testing: Theory and Applications*, vol. 4, no. 3, pp. 237–253, Aug. 1993.

[7] M. Abramovici, M. A. Breuer, and A. D. Friedman, *Digital Systems Testing and Testable Design.* Piscataway, New Jersey: IEEE Press, 1994. Revised printing.

[8] M. Abramovici, J. J. Kulikowski, P. R. Menon, and D. T. Miller, "Test Generation in LAMP2: System Overview," in *Proc. of the International Test Conf.*, Nov. 1985, pp. 45–48.

[9] M. Abramovici, J. J. Kulikowski, P. R. Menon, and D. T. Miller, "SMART and FAST: Test Generation for VLSI Scan-Design Circuits," *IEEE Design & Test of Computers*, vol. 3, no. 4, pp. 43–54, Aug. 1986.

[10] M. Abramovici, P. R. Menon, and D. T. Miller, "Critical Path Tracing: An Alternative to Fault Simulation," *IEEE Design & Test of Computers*, vol. 1, no. 1, pp. 83–93, Feb. 1984.

[11] M. Abramovici, P. R. Menon, and D. T. Miller, "Checkpoint Faults Are Not Sufficient Target Faults for Test Generation," *IEEE Trans. on Computers*, vol. C-35, no. 8, pp. 769–771, Aug. 1986.

[12] Advantest Corp., Japan, *T6682 VLSI Test System Product Description*, 2000.

[13] V. K. Agarwal and A. F. S. Fung, "Multiple Fault Testing of Large Circuits by Single Fault Test Sets," *IEEE Trans. on Computers*, vol. C-30, no. 11, pp. 855–865, Nov. 1981.

[14] P. Agrawal, "Test Generation at Switch Level," in *Proc. of the International Conf. on Computer-Aided Design*, Nov. 1984, pp. 128–130.

[15] P. Agrawal and V. D. Agrawal, "Probabilistic Analysis of Random Test Generation Method for Irredundant Combinational Logic Networks," *IEEE Trans. on Computers*, vol. C-24, no. 7, pp. 691–695, July 1975.

[16] P. Agrawal and V. D. Agrawal, "On Monte Carlo Testing of Logic Tree Networks," *IEEE Trans. on Computers*, vol. C-25, no. 6, pp. 664–667, June 1976.

[17] P. Agrawal, V. D. Agrawal, and S. C. Seth, "Generating Tests for Delay Faults in Nonscan Circuits," *IEEE Design & Test of Computers*, vol. 9, no. 1, pp. 19–29, Jan. 1993.

[18] P. Agrawal, V. D. Agrawal, and J. Villoldo, "Test Pattern Generation for Sequential Circuits on a Network of Workstations," in *Proc. of the 30th Design Automation Conf.*, June 1993, pp. 107–111.

[19] P. Agrawal and W. J. Dally, "A Hardware Logic Simulation System," *IEEE Trans. on Computer-Aided Design*, vol. 9, no. 1, pp. 19–29, Jan. 1990.

[20] V. D. Agrawal, "When to Use Random Testing," *IEEE Trans. on Computers*, vol. C-27, no. 11, pp. 1054–1055, Nov. 1978.

[21] V. D. Agrawal, "Sampling Techniques for Determining Fault Coverage in LSI Circuits," *Journal of Digital Systems*, vol. V, no. 3, pp. 189–202, 1981.

[22] V. D. Agrawal, "Synchronous Path Analysis in MOS Circuit Simulator," in *Proc. of the 19th Design Automation Conf.*, June 1982, pp. 629–635.

[23] V. D. Agrawal, "A Tale of Two Designs: the Cheapest and the Most Economic," in M. Abadir and A. P. Ambler, editors, *Economics of Electronic Design, Manufacture and Test*, pp. 5–9, Boston: Kluwer Academic Publishers, 1994.

[24] V. D. Agrawal, editor, *Special Issue on Partial Scan Methods*, volume 7 of *Journal of Electronic Testing: Theory and Applications*. Boston: Kluwer Academic Publishers, Aug.-Oct. 1995. no. 1/2.

[25] V. D. Agrawal, "Design of Mixed-Signal Systems for Testability," *Integration, the VLSI Journal*, vol. 26, no. 1/2, pp. 141–150, Dec. 1998.

[26] V. D. Agrawal and P. Agrawal, "An Automatic Test Generation System for Illiac IV Logic Boards," *IEEE Trans. on Computers*, vol. C-21, no. 9, pp. 1015–1017, Sept. 1972.

[27] V. D. Agrawal, A. K. Bose, P. Kozak, H. N. Nham, and E. Pacas-Skewes, "Mixed-Mode Simulation in the Motis System," *Journal of Digital Systems*, vol. V, no. 4, pp. 383–400, 1981.

[28] V. D. Agrawal, M. L. Bushnell, and Q. Lin, "Redundancy Identification Using Transitive Closure," in *Proc. of the 5th Asian Test Symp.*, Nov. 1996, pp. 4–9.

[29] V. D. Agrawal and T. J. Chakraborty, "High-Performance Circuit Testing with Slow-Speed Testers," in *Proc. of the International Test Conf.*, Oct. 1995, pp. 302–310.

[30] V. D. Agrawal and S. T. Chakradhar, "Combinational ATPG Theorems for Identifying Untestable Faults in Sequential Circuits," *IEEE Trans. on Computer-Aided Design*, vol. 14, no. 9, pp. 1155–1160, Sept. 1995.

[31] V. D. Agrawal, K.-T. Cheng, and P. Agrawal, "A Directed Search Method for Test Generation using a Concurrent Simulator," *IEEE Trans. on Computer-Aided Design*, vol. 8, no. 2, pp. 131–138, Feb. 1989.

[32] V. D. Agrawal, K.-T. Cheng, D. D. Johnson, and T. Lin, "Designing Circuits with Partial Scan," *IEEE Design & Test of Computers*, vol. 5, pp. 8–15, Apr. 1988.

[33] V. D. Agrawal, R. Dauer, S. K. Jain, H. A. Kalvonjian, C. F. Lee, K. B. McGregor, M. A. Pashan, C. E. Stroud, and L.-C. Suen, *BIST at Your Fingertips Handbook*. AT&T, June 1987.

[34] V. D. Agrawal, S. K. Jain, and D. M. Singer, "Automation in Design for Testability," in *Proc. of the Custom Integrated Circuits Conf.*, May 1984, pp. 159–163.

BIBLIOGRAPHY

[35] V. D. Agrawal and H. Kato, "Fault Sampling Revisited," *IEEE Design & Test of Computers*, vol. 7, no. 4, pp. 32–35, Aug. 1990.

[36] V. D. Agrawal, C. R. Kime, and K. K. Saluja, "A Tutorial on Built-In Self-Test, Part 1: Principles," *IEEE Design & Test of Computers*, vol. 10, no. 1, pp. 73–82, Mar. 1993.

[37] V. D. Agrawal, C. R. Kime, and K. K. Saluja, "A Tutorial on Built-In Self-Test, Part 2: Applications," *IEEE Design & Test of Computers*, vol. 10, no. 2, pp. 69–77, June 1993.

[38] V. D. Agrawal and M. R. Mercer, "Testability Measures – What Do They Tell Us?," in *Proc. of the International Test Conf.*, Nov. 1982, pp. 391–396.

[39] V. D. Agrawal and S. C. Seth, *Test Generation for VLSI Chips*. IEEE Computer Society Press, 1988.

[40] V. D. Agrawal, S. C. Seth, and P. Agrawal, "Fault Coverage Requirements in Production Testing of LSI Circuits," *IEEE Journal of Solid-State Circuits*, vol. SC-17, pp. 57–61, Feb. 1982.

[41] A. V. Aho, J. E. Hopcroft, and J. D. Ullman, *The Design and Analysis of Computer Algorithms*. Reading, Massachusetts: Addison-Wesley, 1974.

[42] R. C. Aitken, "Diagnosis of Leakage Faults with I_{DDQ}," *Journal of Electronic Testing: Theory and Applications*, vol. 3, no. 4, pp. 367–375, Dec. 1992.

[43] S. B. Akers, "Binary Decision Diagrams," *IEEE Trans. on Computers*, vol. 27, no. 6, pp. 66–73, June 1978.

[44] S. B. Akers, "Functional Testing with Binary Decision Diagrams," in *Proc. of the International Fault-Tolerant Computing Symp.*, June 1978, pp. 82–92.

[45] S. B. Akers, C. Joseph, and B. Krishnamurthy, "On the Role of Independent Fault Sets in the Generation of Minimal Test Sets," in *Proc. of the International Test Conf.*, Sept. 1987, pp. 1100–1107.

[46] W. P. Albrecht Jr., *Microeconomic Principles*. Englewood Cliffs, New Jersey: Prentice-Hall, 1979.

[47] R. W. Allen, C. D. Chen, M. M. Ervin-Willis, K. R. Rahlfs, R. E. Tulloss, and S. L. Wu, "DORA: A System of CAD Post-processors Providing Test Programs and Automatic Diagnostics Data for Digital Device and Board Manufacture," in *Proc. of the International Test Conf.*, Oct. 1981, pp. 555–560.

[48] A. P. Ambler, M. Abadir, and S. Sastry, *Economics of Design and Test for Electronic Circuits and Systems*. Chichester, UK: Ellis Horwood Limited, 1992.

[49] A. P. Ambler, P. Agrawal, and W. R. Moore, editors, *Hardware Accelerators for Electrical CAD*. Bristol: Adam Hilger, 1988.

[50] A. P. Ambler, P. Agrawal, and W. R. Moore, editors, *CAD Accelerators*. Amsterdam: North-Holland, 1991.

[51] A. P. Ambler, M. Paraskeva, D. F. Burrows, W. L. Knight, and I. D. Dear, "Economically Viable Automatic Insertion of Self-Test Features for Custom VLSI," in *Proc. of the International Test Conf.*, Sept. 1986, pp. 232–243.

[52] E. A. Amerasekera and D. S. Campbell, *Failure Mechanisms in Semiconductor Devices*. Chichester, UK: John Wiley & Sons, Inc., 1987.

[53] H. Ando, "Testing VLSI with Random Access Scan," in *Proc. of the COMPCON*, Feb. 1980, pp. 50–52.

[54] J. H. Arabian, *Computer Integrated Electronics Manufacturing and Testing*. New York: Marcel Dekker, 1989.

[55] D. B. Armstrong, "On Finding a Nearly Minimal Set of Fault Detection Tests for Combinational Logic Nets," *IEEE Trans. on Electronic Computers*, vol. EC-15, no. 1, pp. 66–73, Feb. 1966.

[56] D. B. Armstrong, "A Deductive Method for Simulating Faults in Logic Circuits," *IEEE Trans. on Computers*, vol. C-21, no. 5, pp. 464–471, May 1972.

[57] S. P. Athan, D. L. Landis, and S. A. Al-Arian, "A Novel Built-In Current Sensor for I_{DDQ} Testing of Deep Submicron CMOS ICs," in *Proc. of the 14th VLSI Test Symp.*, May 1996, pp. 112–117.

[58] AT&T, Basking Ridge, New Jersey, *Bell System Publication 41009, Transmission Parameters Affecting Voiceband Data Transmission-Measuring Techniques*, May 1975.

[59] AT&T, Basking Ridge, New Jersey, *Bell System Publication 48501, Local Switching System General Requirements (LSSGR)*, Dec. 1980. Section 7.4.

[60] B. Ayari and B. Kaminska, "A New Dynamic Test Vector Compaction for Automatic Test Pattern Generation," *IEEE Trans. on Computer-Aided Design*, vol. 13, no. 3, pp. 353–358, Mar. 1994.

[61] N. Balabanian, T. A. Bickart, and S. Seshu, *Electrical Network Theory*, chapter 9, pp. 646–647. John Wiley & Sons, Inc., 1969.

[62] A. Balivada, J. Chen, and J. A. Abraham, "Analog Testing with Time Response Parameters," *IEEE Design & Test of Computers*, vol. 13, no. 2, pp. 18–25, summer 1996.

[63] P. Banerjee, *Parallel Algorithms for VLSI Computer-Aided Design*. Englewood Cliffs, New Jersey: PTR Prentice Hall, 1994.

[64] S. Banerjee, S. T. Chakradhar, and R. K. Roy, "Synchronous Test Generation Model for Asynchronous Circuits," in *Proc. of the 9th International Conf. on VLSI Design*, Jan. 1996, pp. 178–185.

[65] M. R. Barber, 2000. Personal communication.

[66] P. H. Bardell and W. H. McAnney, "Self-Testing of Multichip Logic Modules," in *Proc. of the International Test Conf.*, Nov. 1982, pp. 200–204.

[67] P. H. Bardell, W. H. McAnney, and J. Savir, *Built-In Test for VLSI: Pseudorandom Techniques*. New York: John Wiley & Sons, Inc., 1987.

[68] Z. Barzilai, D. K. Beece, L. M. Huisman, V. S. Iyengar, and G. M. Silberman, "SLS - A Fast Switch Level Simulator for Verification and Fault Coverage Analysis," in *Proc. of the 23rd Design Automation Conf.*, June-July 1986, pp. 164–170.

[69] J. Bateson, *In-Circuit Testing*. New York: Van Nostrand Reinhold Company, 1985.

[70] J. S. Beasley, H. Rammamurthy, J. Ramirez-Angulo, and M. DeYong, "I_{DD} Pulse Response Testing of Analog and Digital CMOS Circuits," in *Proc. of the International Test Conf.*, Oct. 1993, pp. 626–634.

[71] F. P. M. Beenker, R. G. Bennetts, and A. P. Thijssen, *Testability Concepts for Digital ICs: The Macro Test Approach*. Boston: Kluwer Academic Publishers, 1995.

[72] R. Bencivenga, T. J. Chakraborty, and S. Davidson, "The Architecture of the Gentest Sequential Test Generator," in *Proc. of the Custom Integrated Circuits Conf.*, May 1991, pp. 17.1.1–17.1.4.

[73] R. G. Bennetts, *Introduction to Digital Board Testing*. New York: Crane Russak, 1982.

[74] M. Bershteyn, "Calculation of Multiple Sets of Weights for Weighted Random Testing," in *Proc. of the International Test Conf.*, Oct. 1993, pp. 1031–1040.

[75] H. Bhatnagar, *Advanced ASIC Chip Synthesis*. Boston: Kluwer Academic Publishers, 1999.

[76] D. Bhattacharya and J. P. Hayes, "A Hierarchical Test Generation Methodology for Digital Circuits," *Journal of Electronic Testing: Theory and Applications*, vol. 1, no. 2, pp. 103–123, May 1990.

[77] D. Bhattacharya and J. P. Hayes, *Hierarchical Modeling for VLSI Circuit Testing*. Boston: Kluwer Academic Publishers, 1990.

[78] D. K. Bhavsar, "Design for Test Calculus: An Algorithm for DFT Rules Checking," in *Proc. of the 20th Design Automation Conf.*, June 1983, pp. 300–307.

[79] J. W. Bierbauer, J. A. Eiseman, F. A. Fazal, and J. J. Kulikowski, "System Simulation with MIDAS," *A T & T Technical Journal*, vol. 70, no. 1, pp. 36–51, Jan./Feb. 1991.

[80] E. K. Blum, *Numerical Analysis and Computation: Theory and Practice*, chapter 8. Addison-Wesley, 1972.

[81] V. Boppana and M. Fujita, "Modeling the Unknown! Towards Model-Independent Fault and Error Diagnosis," in *Proc. of the International Test Conf.*, Oct. 1998, pp. 1094–1101.

[82] S. Bose and P. Agrawal, "Concurrent Fault Simulation on Message Passing Multicomputers," *IEEE Trans. on VLSI Systems*, vol. 6, no. 2, pp. 332–341, June 1998.

[83] S. Bose, P. Agrawal, and V. D. Agrawal, "A Rated-Clock Test Method for Path Delay Faults," *IEEE Trans. on VLSI Systems*, vol. 6, no. 2, pp. 323–331, June 1998.

[84] S. Bose, P. Agrawal, and V. D. Agrawal, "Deriving Logic Systems for Path Delay Test Generation," *IEEE Trans. on Computers*, vol. 47, no. 8, pp. 829–846, Aug. 1998.

[85] D. C. Bossen and S. J. Hong, "Cause-Effect Analysis for Multiple Fault Detection in Combinational Networks," *IEEE Trans. on Computers*, vol. C-20, pp. 1252–1257, Nov. 1971.

[86] P. S. Bottorff, R. E. France, N. H. Garges, and E. J. Orosz, "Test Generation for Large Logic Networks," in *Proc. of the 14th Design Automation Conf.*, June 1977, pp. 479–485.

[87] A. H. Bratt, R. J. Harvey, A. P. Dorey, and A. M. D. Richardson, "Design-for-Test Structure to Facilitate Test Vector Application with Low Performance Loss in Non-Test Mode," *IEEE Electronics Letters*, vol. 29, no. 16, pp. 1438–1440, Aug. 1993.

[88] E. A. Bretz, "Technology 2000: Test & Measurement," *IEEE Spectrum*, vol. 37, no. 1, pp. 75–79, Jan. 2000.

[89] M. A. Breuer, "A Random and an Algorithmic Technique for Fault Detection Test Generation for Sequential Circuits," *IEEE Trans. on Computers*, vol. C-20, no. 11, pp. 1364–1370, Nov. 1971.

[90] M. A. Breuer, "Testing for Intermittent Faults in Digital Circuits," *IEEE Trans. on Computers*, vol. C-22, no. 3, pp. 241–246, Mar. 1973.

[91] M. A. Breuer, "The Effects of Races, Delays, and Delay Faults on Test Generation," *IEEE Trans. on Computers*, vol. C-23, no. 10, pp. 1078–1092, Oct. 1974.

[92] M. A. Breuer, "New Concepts in Automated Testing of Digital Circuits," in *Proc. of the EEC Symp. on CAD of Digital Electronic Circuits and Systems*, North-Holland Publishing Co., Nov. 1978, pp. 69–92.

[93] M. A. Breuer, "Test Generation Models for Busses and Tri-State Drivers," in *Proc. of the IEEE ATPG Workshop*, Mar. 1983, pp. 53–58.

[94] M. A. Breuer and A. D. Friedman, *Diagnosis and Reliable Design of Digital Systems*. Rockville, Maryland: Computer Science Press, 1976.

[95] M. A. Breuer and A. D. Friedman, "Functional Level Primitives in Test Generation," *IEEE Trans. on Computers*, vol. C-29, no. 3, pp. 223–235, Mar. 1980.

[96] M. A. Breuer and L. M. Harrison, "Procedures for Eliminating Static and Dynamic Hazards in Test Generation," *IEEE Trans. on Computers*, vol. C-23, no. 10, pp. 1069–1078, Oct. 1974.

[97] M. A. Breuer and N. K. Nanda, "Simplified Delay Testing for LSI Circuit Faults." U.S. Patent # 4,672,307, June 9, 1987.

[98] F. Brglez, "On Testability Analysis of Combinational Networks," in *Proc. of the International Symp. on Circuits and Systems*, May 1984, pp. 221–225.

[99] F. Brglez, D. Bryan, and K. Kozminski, "Combinational Profiles of Sequential Benchmark Circuits," in *Proc. of the International Symp. on Circuits and Systems*, May 1989, pp. 1929–1934.

[100] F. Brglez, P. Pownall, and R. Hum, "Accelerated ATPG and Fault Grading Via Testability Analysis," in *Proc. of the International Symp. on Circuits and Systems*, June 1985, pp. 695–698.

[101] A. J. Briers and K. A. E. Totton, "Random Pattern Testability by Fast Fault Simulation," in *Proc. of the International Test Conf.*, Sept. 1986, pp. 274–281.

[102] F. P. Brooks, Jr., *The Mythical Man-Month*. Reading, Massachusetts: Addison-Wesley, 1982.

[103] R. E. Bryant, "A Switch-Level Model and Simulator for MOS Digital Systems," *IEEE Trans. on Computers*, vol. C-33, no. 2, pp. 160–177, Feb. 1984.

[104] R. E. Bryant, "Graph-Based Algorithms for Boolean Function Manipulation," *IEEE Trans. on Computers*, vol. C-35, no. 8, pp. 677–691, Aug. 1986.

[105] R. E. Bryant, "A Survey of Switch-Level Algorithms," *IEEE Design & Test of Computers*, vol. 4, no. 4, pp. 26–40, Aug. 1987.

[106] J. A. Brzozowski and B. F. Cockburn, "Detection of Coupling Faults in RAMs," *Journal of Electronic Testing: Theory and Applications*, vol. 1, no. 2, pp. 151–162, May 1990.

[107] J. A. Brzozowski and H. Jurgensen, "A Model for Sequential Machine Testing," Research Report CS-88-12, U. of Waterloo, Dept. of Computer Science, Apr. 1988.

[108] O. Bula, J. Moser, J. Trinko, M. Weissman, and F. Woytowich, "Gross Delay Defect Evaluation for a CMOS Logic Design System Product," *IBM Journal of Research and Development*, vol. 34, no. 2/3, pp. 325–338, Mar./May 1990.

[109] D. J. Burns, "Locating High Resistance Shorts in CMOS Circuits by Analyzing Supply Current Measurement Vectors," in *Proc. of the International Symp. for Testing and Failure Analysis (ISTFA)*, Nov. 1989, pp. 231–237.

[110] M. L. Bushnell and J. Giraldi, "A Functional Decomposition Method for Redundancy Identification and Test Generation," *Journal of Electronic Testing: Theory and Applications*, vol. 10, no. 3, pp. 175–195, June 1997.

[111] M. L. Bushnell and G. Parthasarathy, "Random Logic Partial-Scan Delay-Fault Built-In Self-Test," in *IMAPS Multi-Chip Module Conf.*, Sept. 1998. Invited paper.

[112] K. M. Butler and M. R. Mercer, *Assessing Fault Model and Test Quality*. Boston: Kluwer Academic Publishers, 1992.

[113] W. J. Butler and S. S. Haykin, "Multiparameter Sensitivity Problems in Network Theory," *Proc. of the IEEE*, vol. 117, no. 12, pp. 2228–2236, Dec. 1970.

[114] W. C. Carter, "Signature Testing with Guaranteed Bounds for Fault Coverage," in *Proc. of the International Test Conf.*, Nov. 1982, pp. 75–82.

[115] G. R. Case, "Analysis of Actual Fault Mechanisms in CMOS Logic Gates," in *Proc. of the 13th Design Automation Conf.*, June 1976, pp. 265–270.

[116] C. W. Cha, W. E. Donath, and F. Ozguner, "9-V Algorithm for Test Pattern Generation of Combinational Digital Circuits," *IEEE Trans. on Computers*, vol. C-27, no. 3, pp. 193–200, Mar. 1978.

[117] T. J. Chakraborty. Personal communication.

[118] T. J. Chakraborty and V. D. Agrawal, "Robust Testing for Stuck-at Faults," in *Proc. of the 8th International Conf. on VLSI Design*, Jan. 1995, pp. 42–46.

[119] T. J. Chakraborty, V. D. Agrawal, and M. L. Bushnell, "On Variable Clock Methods for Path Delay Testing of Sequential Circuits," *IEEE Trans. on Computer-Aided Design*, vol. 16, no. 11, pp. 1237–1249, Nov. 1997.

[120] S. T. Chakradhar. Personal communication.

[121] S. T. Chakradhar, *Neural Network Models and Optimization Methods for Digital Testing*. PhD thesis, Computer Science Department, Rutgers University, New Brunswick, New Jersey, Oct. 1990.

[122] S. T. Chakradhar and V. D. Agrawal, "A Transitive Closure Based Algorithm for Test Generation," in *Proc. of the 28th Design Automation Conf.*, June 1991, pp. 353–358.

[123] S. T. Chakradhar and V. D. Agrawal, "VLSI Design," in A. Kent and J. G. Williams, editors, *Encyclopedia of Microcomputers, Volume 20*, pp. 97–111, New York: Marcel Dekker, 1997.

[124] S. T. Chakradhar, V. D. Agrawal, and M. L. Bushnell, "Automatic Test Pattern Generation Using Quadratic 0-1 Programming," in *Proc. of the 27th Design Automation Conf.*, June 1990, pp. 654–659.

[125] S. T. Chakradhar, V. D. Agrawal, and M. L. Bushnell, "Neural Net and Boolean Satisfiability Models of Logic Circuits," *IEEE Design & Test of Computers*, vol. 7, no. 5, pp. 54–57, Oct. 1990.

[126] S. T. Chakradhar, V. D. Agrawal, and M. L. Bushnell, "Toward Massively Parallel Automatic Test Generation," *IEEE Trans. on Computer-Aided Design*, vol. 9, no. 9, pp. 981–994, Sept. 1990.

[127] S. T. Chakradhar, V. D. Agrawal, and M. L. Bushnell, *Neural Models and Algorithms for Digital Testing*. Boston: Kluwer Academic Publishers, 1991.

[128] S. T. Chakradhar, V. D. Agrawal, and S. G. Rothweiler, "A Transitive Closure Algorithm for Test Generation," *IEEE Trans. on Computer-Aided Design*, vol. 12, no. 7, pp. 1015–1028, July 1993.

[129] S. T. Chakradhar, A. Balakrishnan, and V. D. Agrawal, "An Exact Algorithm for Selecting Partial Scan Flip-Flops," *Journal of Electronic Testing: Theory and Applications*, vol. 7, pp. 83–93, Aug. 1995.

[130] S. T. Chakradhar, M. L. Bushnell, and V. D. Agrawal, "Automatic Test Generation Using Neural Networks," in *Proc. of the International Conf. on Computer-Aided Design*, Nov. 1988, pp. 416–419.

[131] S. T. Chakradhar, S. G. Rothweiler, and V. D. Agrawal, "Redundancy Removal and Test Generation for Circuits with Non-Boolean Primitives," *IEEE Trans. on Computer-Aided Design*, vol. 16, no. 11, pp. 1370–1377, Nov. 1997.

[132] S. Chakravarty and M. Liu, "Algorithms for Current Monitor Based Diagnosis of Bridging and Leakage Faults," in *Proc. of the 29th Design Automation Conf.*, June 1992, pp. 353–356.

[133] S. Chakravarty and M. Liu, "Algorithms for I_{DDQ} Measurement Based Diagnosis of Bridging Faults," *Journal of Electronic Testing: Theory and Applications*, vol. 3, no. 4, pp. 377–385, Dec. 1992.

[134] S. Chakravarty and P. J. Thadikaran, *Introduction to I_{DDQ} Testing*. Boston: Kluwer Academic Publishers, 1997.

[135] V. H. Champac, R. Rodrigues-Montanes, J. A. Segura, J. Figueras, and J. A. Rubio, "Fault Modeling of Gate Oxide Short, Floating Gate and Bridging Failures in CMOS Circuits," in *Proc. of the European Test Conf.*, Apr. 1991, pp. 143–148.

[136] A. K. Chandra, L. T. Kou, G. Markowsky, and S. Zaks, "On Sets of Boolean n-Vectors with All k-Projections Surjective," *Acta Informatica*, vol. 20, no. 1, pp. 103–111, Oct. 1983.

[137] S. J. Chandra and J. H. Patel, "Experimental Evaluation of Testability Measures for Test Generation," *IEEE Trans. on Computer-Aided Design*, vol. 82, no. 17, pp. 93–97, Jan. 1989.

[138] R. Chandramouli, "On Testing Stuck-Open Faults," in *Proc. of the International Fault-Tolerant Computing Symp.*, June 1983, pp. 258–265.

[139] H. Y. Chang, E. G. Manning, and G. Metze, *Fault Diagnosis of Digital Systems*. New York: Wiley-Interscience, 1970.

[140] S.-J. Chang and M. A. Breuer, "A Fault-Collapsing Analysis in Sequential Logic Networks," *Bell System Technical Journal*, vol. 60, no. 9, pp. 2259–2271, Nov. 1981.

[141] S. G. Chappell, "Automatic Test Generation for Asynchronous Digital Circuits," *Bell System Technical Journal*, vol. 53, no. 8, pp. 1477–1503, Oct. 1974.

[142] S. G. Chappell, C. H. Elmendorf, and L. D. Schmidt, "LAMP: Logic-Circuit Simulators," *Bell System Technical Journal*, vol. 53, no. 8, pp. 1451–1476, Oct. 1974.

[143] A. Chatterjee, "Concurrent Error Detection in Linear Analog and Switched-Capacitor State Variable Systems Using Continuous Checksums," in *Proc. of the International Test Conf.*, Oct. 1991, pp. 582–591.

[144] M. Chatterjee and D. K. Pradhan, "A Novel Pattern Generator for Near-Perfect Fault Coverage," in *Proc. of the 13th VLSI Test Symp.*, April-May 1995, pp. 417–425.

[145] B. R. Chawla, H. K. Gummel, and P. Kozak, "MOTIS – An MOS Timing Simulator," *IEEE Trans. on Circuits and Systems*, vol. CAS-22, no. 12, pp. 901–910, Dec. 1975.

[146] D. Cheek and R. Dandapani, "Integration of IEEE Std. 1149.1 and Mixed-Signal Test Architectures," in *Proc. of the International Test Conf.*, Oct. 1995, pp. 569–576.

[147] C. L. Chen, "Exhaustive Test Pattern Generation Using Cyclic Codes," *IEEE Trans. on Computers*, vol. C-37, no. 2, pp. 225–228, Feb. 1988.

[148] H. S. M. Chen and R. Sacks, "A Search Algorithm for the Solution of the Multifrequency Fault Diagnosis Equations," *IEEE Trans. on Circuits and Systems*, vol. 26, no. 7, pp. 589–594, July 1979.

[149] J. E. Chen, C. L. Lee, and W. Z. Shen, "Single-Fault Fault-Collapsing Analysis in Sequential Logic Circuits," *IEEE Trans. on Computer-Aided Design*, vol. 10, no. 12, pp. 1559–1568, Dec. 1991.

[150] X. Chen and M. L. Bushnell, *Efficient Branch and Bound Search with Application to Computer-Aided Design*. Boston: Kluwer Academic Publishers, 1996.

[151] X. Chen, T. Snethen, J. Swnton, and R. Walther, "A Simplified Method for Testing the IBM Pipeline Partial-Scan Microprocessor," in *Proc. of the 8th Asian Test Symp.*, Nov. 1999, pp. 321–326.

[152] K.-T. Cheng, "On Removing Redundancy in Sequential Circuits," in *Proc. of the 28th Design Automation Conf.*, June 1991, pp. 164–169.

[153] K.-T. Cheng and V. D. Agrawal, *Unified Methods for VLSI Simulation and Test Generation*. Boston: Kluwer Academic Publishers, 1989.

[154] K.-T. Cheng and V. D. Agrawal, "A Partial Scan Method for Sequential Circuits with Feedback," *IEEE Trans. on Computers*, vol. 39, no. 4, pp. 544–548, Apr. 1990.

[155] K.-T. Cheng and H.-C. Chen, "Classification and Identification of Nonrobust Untestable Path Delay Faults," *IEEE Trans. on Computer-Aided Design*, vol. 15, no. 8, pp. 845–853, Aug. 1996.

[156] K.-T. Cheng, S. Devadas, and K. Keutzer, "Delay Fault Test Generation and Synthesis for Testability Under a Standard Scan Design Methodology," *IEEE Trans. on Computer-Aided Design*, vol. 12, no. 8, pp. 1217–1231, Aug. 1993.

[157] K.-T. Cheng and C.-J. Lin, "Timing-Driven Test Point Insertion for Full-Scan and Partial-Scan BIST," in *Proc. of the International Test Conf.*, Oct. 1995, pp. 506–514.

[158] W.-T. Cheng, "SPLIT Circuit Model for Test Generation," in *Proc. of the 25th Design Automation Conf.*, June 1988, pp. 96–101.

[159] W.-T. Cheng, "The BACK Algorithm for Sequential Test Generation," in *Proc. of the International Conf. on Computer Design*, Oct. 1988, pp. 66–69.

[160] W.-T. Cheng and T. J. Chakraborty, "Gentest: An Automatic Test Generation System for Sequential Circuits," *Computer*, vol. 22, no. 4, pp. 43–49, Apr. 1989.

[161] W.-T. Cheng and M.-L. Yu, "Differential Fault Simulation for Sequential Circuits," *Journal of Electronic Testing: Theory and Applications*, vol. 1, no. 1, pp. 7–13, Feb. 1990.

[162] K. R. Chin, "Functional Testing of Circuits and SMD Boards with Limited Nodal Access," in *Proc. of the International Test Conf.*, Aug. 1989, pp. 129–143.

[163] S. H. Chisholm and L. W. Nagel, "Efficient Computer Simulation of Distortion in Electronic Circuits," *IEEE Trans. on Circuit Theory*, vol. CT-20, no. 6, pp. 742–745, Nov. 1973.

[164] D. C. Choi, "Personal communication," July 2000. Samsung Electronics Company, Ltd.

[165] R. M. Chou, K. K. Saluja, and V. D. Agrawal, "Scheduling Tests for VLSI Systems Under Power Constraints," *IEEE Trans. on VLSI Systems*, vol. 5, pp. 175–185, June 1997.

[166] L. O. Chua, *Introduction to Nonlinear Network Theory*. New York: McGraw-Hill, 1969.

[167] B. Ciciani, *Manufacturing Yield Evaluation of VLSI/WSI Systems*. Los Alamitos, California: IEEE Computer Society Press, 1995.

[168] E. M. Clarke, Jr., O. Grumberg, and D. A. Peled, *Model Checking*. Cambridge, Massachusetts: MIT Press, 1999.

[169] B. F. Cockburn and J. A. Brzozowski, "Near-Optimal Tests for Classes of Write-Triggered Coupling Faults in RAMs," *Journal of Electronic Testing: Theory and Applications*, vol. 3, no. 3, pp. 251–264, Aug. 1992.

[170] J. Cocking, "RAM Test Patterns and Test Strategy," in *Proc. of the 1975 Semiconductor Test Symp.*, IEEE, Oct. 1975, pp. 1–8.

[171] D. R. Coelho, *The VHDL Handbook*. Boston: Kluwer Academic Publishers, 1989.

[172] F. Coombs, editor, *Printed Circuits Handbook*. New York: McGraw-Hill, fourth edition, 1995.

[173] C. Cooper and M. L. Bushnell, "Neural Models for Mixed-Level Test Generation," in *Proc. of the 12th VLSI Test Symp.*, Apr. 1994, pp. 208–213.

[174] F. Corno, P. Prinetto, M. Rebaudengo, and M. S. Reorda, "A Genetic Algorithm for Automatic Test Pattern Generation for Large Synchronous Sequential Circuits," *IEEE Trans. on Computer-Aided Design*, vol. 15, no. 8, pp. 991–1000, Aug. 1996.

[175] F. Corsi, M. Chiarantoni, R. Lorusso, and C. Marzocca, "A Fault Signature Approach to Analog Devices Testing," in *Proc. of the European Test Conf.*, Apr. 1993, pp. 116–121.

[176] C. Crapuchettes, "Testing CMOS I_{DD} on Large Devices," in *Proc. of the International Test Conf.*, Sept. 1987, pp. 310–315.

[177] W. Daehn, "Fault Simulation Using Small Fault Samples," *Journal of Electronic Testing: Theory and Applications*, vol. 2, no. 2, pp. 191–203, June 1991.

[178] H. Dai and T. M. Souders, "Time-Domain Testing Strategies and Fault Diagnosis for Analog Systems," *IEEE Trans. on Instrumentation and Measurement*, vol. 39, no. 1, pp. 157–162, Feb. 1990.

[179] M. Damiani, P. Olivio, M. Favalli, S. Ercolani, and B. Ricco, "Aliasing in Signature Analysis Testing with Multiple Input Shift Registers," *IEEE Trans. on Computer-Aided Design*, vol. CAD-9, no. 12, pp. 1344–1353, Dec. 1990.

[180] M. Damiani, P. Olivio, M. Favalli, and B. Ricco, "An Analytical Model for the Aliasing Probabitiliy of Signature Analysis Testing," *IEEE Trans. on Computer-Aided Design*, vol. CAD-8, no. 11, pp. 1133–1144, Nov. 1989.

[181] R. G. Daniels, "The Changing Demands of Microprocessor Testing," in *Proc. of the International Test Conf.*, Sept. 1990, p. 21. Keynote Address.

[182] A. K. Das, M. Pandey, A. Gupta, and P. P. Chaudhuri, "Built-In Self-Test Structures Around Cellular Automata and Counters," *IEE Proceedings, Computers and Digital Techniques*, vol. 37, no. 1, pp. 268–276, July 1990. Part E.

[183] D. Das, S. C. Seth, and V. D. Agrawal, "Accurate Computation of Field Reject Ratio Based on Fault Latency," *IEEE Trans. on VLSI Systems*, vol. 1, no. 4, pp. 537–545, Dec. 1993.

[184] S. DasGupta, R. G. Walther, and T. W. Williams, "An Enhancement to LSSD and Some Applications of LSSD in Reliability," in *Proc. of the International Fault-Tolerant Computing Symp.*, June 1981, pp. 32–34.

[185] B. Davari, R. H. Dennard, and G. G. Shahidi, "CMOS Scaling for High Performance and Low Power – The Next Ten Years," in A. Chandrakasan and R. Brodersen, editors, *Low-Power CMOS Design*, pp. 47–58, Piscataway, New Jersey: IEEE Press, 1998.

[186] R. David, "Signature Analysis of Multi-Output Circuits," in *Proc. of the International Fault-Tolerant Computing Symp.*, June 1984, pp. 366–371.

[187] R. David, *Random Testing of Digital Circuits Theory and Applications*. New York: Marcel Dekker, 1998.

[188] B. Davis, *The Economics of Automatic Testing*. London, United Kingdom: McGraw-Hill, 1982.

[189] P. de Jong and A. J. van de Goor, "Comments on 'Test Pattern Generation for API Faults in RAM'," *IEEE Trans. on Computers*, vol. C-37, no. 11, pp. 1426–1428, Nov. 1988.

[190] G. de Micheli, *Synthesis and Optimization of Digital Circuits*. New York: McGraw-Hill, 1994.

[191] J. T. de Sousa and V. D. Agrawal, "Reducing the Complexity of Defect Level Modeling using the Clustering Effect," in *Proc. of the Design, Automation and Test in Europe (DATE) Conf.*, Mar. 2000, pp. 640–644.

[192] I. D. Dear, C. Dislis, J. Dick, and A. P. Ambler, "Economic Effects in Design and Test," *IEEE Design & Test of Computers*, vol. 8, no. 4, pp. 64–77, Dec. 1991.

[193] R. Dekker, "Fault Modeling and Self-Test of Static Random Access Memories," TUD Research Report 1-68340-28(1987)25, Dept. of Electrical Eng., Delft University of Technology, Delft, The Netherlands, 1987.

[194] R. Dekker, F. Beenker, and L. Thijssen, "Fault Modeling and Test Algorithm Development for Static Random Access Memories," in *Proc. of the International Test Conf.*, Sept. 1988, pp. 343–352.

[195] R. Dekker, F. Beenker, and L. Thijssen, "A Realistic Fault Modeling and Test Algorithm for Static Random Access Memories," *IEEE Trans. on Computer-Aided Design*, vol. C-9, no. 6, pp. 567–572, June 1990.

[196] W. E. Deming, *Some Theory of Sampling*. New York: Dover Publications, Inc., 1950.

[197] W. E. Deming, *Out of the Crisis*. Cambridge, Massachusetts: MIT Center for Advanced Engineering Study, 1982.

[198] A. Deng, "Power Analysis for CMOS/BiCMOS Circuits," in *Proc. International Workshop on Low-Power Design*, Apr. 1994, pp. 3–8.

[199] S. Devadas, "Delay Test Generation for Synchronous Sequential Circuits," in *Proc. of the International Test Conf.*, Sept. 1989, pp. 144–152.

[200] G. Devarayanadurg and M. Soma, "Analytical Fault Modeling and Static Test Generation for Analog ICs," in *Proc. of the International Conf. on Computer-Aided Design*, Nov. 1994, pp. 44–47.

[201] G. Devarayanadurg and M. Soma, "Dynamic Test Signal Design for Analog ICs," in *Proc. of the International Conf. on Computer-Aided Design*, Nov. 1995, pp. 627–630.

[202] T. E. Dillinger, *VLSI Engineering*. Prentice-Hall, 1988.

[203] S. W. Director, *Circuit Theory: A Computational Approach*. New York: J. S. Wiley and Sons, 1975.

[204] C. Dislis, J. H. Dick, I. D. Dear, and A. P. Ambler, *Test Economics and Design for Testability*. New York: Ellis Horwood, 1995.

[205] K. Dosaka, Y. Konishi, K. Hayano, K. Himukashi, A. Yamazaki, H. Iwamoto, M. Kumanoya, H. Hamano, and T. Yoshihara, "A 100 MHz 4 Mb Cache DRAM with Fast Copy-Back Scheme," *IEEE Journal of Solid-State Circuits*, vol. 27, no. 11, pp. 1534–1539, Nov. 1992.

[206] C. Dufaza and G. Cambon, "LFSR-Based Deterministic and Pseudo-random Test Pattern Generator Structures," in *Proc. of the European Test Conf.*, Apr. 1991, pp. 27–34.

[207] P. Duhamel and J.-C. Rault, "Automatic Test Generation Techniques for Analog Circuits and Systems: A Review," *IEEE Trans. on Circuits and Systems*, vol. CAS-26, no. 7, pp. 411–440, July 1979.

[208] A. E. Dunlop, V. D. Agrawal, D. N. Deutsch, M. F. Jukl, P. Kozak, and M. Wiesel, "Chip Layout Optimization Using Critical Path Weighting," in *Proc. of the 21st Design Automation Conf.*, June 1984, pp. 133–136.

[209] E. B. Eichelberger and E. Lindbloom, "Random-Pattern Coverage Enhancement and Diagnosis for LSSD Logic Self-Test," *IBM Journal of Research and Development*, vol. 27, no. 3, pp. 265–272, May 1983.

[210] E. B. Eichelberger, E. Lindbloom, J. A. Waicukauski, and T. W. Williams, *Structured Logic Testing*. Englewood Cliffs, New Jersey: Prentice-Hall, 1991.

[211] E. B. Eichelberger and T. W. Williams, "A Logic Design Structure for LSI Testability," in *Proc. of the 14th Design Automation Conf.*, June 1977, pp. 462–468.

[212] K. L. Einspahr and S. C. Seth, "A Switch-Level Test Generation System for Synchronous and Asynchronous Circuits," *Journal of Electronic Testing: Theory and Applications*, vol. 6, no. 1, pp. 59–73, Feb. 1995.

[213] N. G. Einspruch, *VLSI Handbook*. Orlando, Florida: Academic Press, 1985.

[214] W. Eklow, "Optimizing Boundary Scan in a Proprietary Environment," in *Proc. of the International Test Conf.*, Oct. 1994, p. 1024.

[215] R. D. Eldred, "Test Routines Based on Symbolic Logical Statements," *Journal of the ACM*, vol. 6, no. 1, pp. 33–36, Jan. 1959.

[216] D. Farren and A. P. Ambler, "System Test Cost Modeling Based on Event Rate Analysis," in *Proc. of the International Test Conf.*, Oct. 1994, pp. 84–92.

[217] P. P. Fasang, "Analog/Digital ASIC Design for Testability," *IEEE Trans. on Industrial Electronics*, vol. 36, no. 2, pp. 219–226, May 1989.

[218] P. P. Fasang, D. Mullins, and T. Wong, "Design for Testability for Mixed Analog/Digital ASICs," in *Proc. of the Custom Integrated Circuits Conf.*, May 1988, pp. 16.5.1–16.5.4.

[219] W. Feller, *An Introduction to Probability Theory and Its Applications*. New York: John Wiley and Sons, Inc., 1950.

[220] E. Felt and A. L. Sangiovanni-Vincentelli, "Testing of Analog Systems Using Behavioral Models and Optimal Experimental Design Techniques," in *Proc. of the International Conf. on Computer-Aided Design*, Nov. 1994, pp. 672–678.

[221] A. Ferre and J. Figueras, "On Estimating Bounds of the Quiescent Current for I_{DDQ} Testing," in *Proc. of the 14th VLSI Test Symp.*, May 1996, pp. 106–111.

[222] J. K. Fidler, "Differential-Incremental Sensitivity Relationships," *IEEE Electronics Letters*, vol. 20, no. 10, pp. 626–627, May 1984.

[223] J. P. Fishburn and A. E. Dunlop, "TILOS: A Posynomial Programming Approach to Transistor Sizing," in *Proc. of the International Conf. on Computer-Aided Design*, Nov. 1985, pp. 326–328.

[224] P. L. Flake, G. Musgrave, and I. White, "A Digital System Simulator: HILO," *Digital Processes*, vol. 1, no. 1, pp. 39–53, 1975.

[225] J. Flood, *RCA Solid-State Integrated Circuits Databook, Application Note ICAN-6532*, chapter Fundamentals of Testing COS/MOS Integrated Circuits, pp. 622–630. RCA, 1978.

[226] P. Franco, S. C. Ma, Y.-C. Chang, R. Stokes, W. Farwell, and E. J. McCluskey, "Analysis and Detection of Timing Failures in an Experimental Test Chip," in *Proc. of the International Test Conf.*, Oct. 1996, pp. 691–700.

[227] R. R. Fritzemeier, J. M. Soden, K. R. Treece, and C. F. Hawkins, "Increased CMOS IC Stuck-at Fault Coverage with Reduced I_{DDQ} Test Sets," in *Proc. of the International Test Conf.*, Sept. 1990, pp. 427–435.

[228] R. A. Frohwerk, "Signature Analysis: A New Digital Field Service Method," *Hewlett-Packard Journal*, vol. 28, no. 9, pp. 2–8, May 1977.

[229] H. Fujiwara, "FAN: A Fanout-Oriented Test Pattern Generation Algorithm," in *Proc. of the International Symp. on Circuits and Systems*, July 1985, pp. 671–674.

[230] H. Fujiwara, *Logic Testing and Design for Testability*. Cambridge, Massachusetts: The MIT Press, 1985.

[231] H. Fujiwara, "Computational Complexity of Controllability / Observability Problems for Combinational Circuits," in *Proc. of the International Fault-Tolerant Computing Symp.*, June 1988, pp. 64–69.

[232] H. Fujiwara and T. Shimono, "On the Acceleration of Test Generation Algorithms," in *Proc. of the International Fault-Tolerant Computing Symp.*, June 1983, pp. 98–105.

[233] H. Fujiwara and T. Shimono, "On the Acceleration of Test Generation Algorithms," *IEEE Trans. on Computers*, vol. C-32, no. 12, pp. 1137–1144, Dec. 1983.

[234] R. K. Gaede, M. R. Mercer, K. M. Butler, and D. E. Ross, "CATAPULT: Concurrent Automatic Testing Allowing Parallelization and Using Limited Topology," in *Proc. of the 25th Design Automation Conf.*, June 1988, pp. 597–600.

[235] S. Gai and P. L. Montessoro, "Creator: New Advanced Concept in Concurrent Simulation," *IEEE Trans. on Computer-Aided Design*, vol. 13, no. 6, pp. 786–795, June 1994.

[236] S. Gai, P. L. Montessoro, and F. Somenzi, "MOZART: A Concurrent Multilevel Simulator," *IEEE Trans. on Computer-Aided Design*, vol. 7, no. 9, pp. 1005–1016, Sept. 1988.

[237] J. M. Galey, R. E. Norby, and J. P. Roth, "Techniques for the Diagnosis of Switching Circuit Failures," in R. S. Ledley, editor, *Proc. of the Second Annual Symp. on Switching Circuit Theory and Logical Design*, (Detroit), AIEE, Oct. 1961, pp. 152–160.

[238] M. K. Gandhi, *The Story of my Experiments with Truth*. Washington, DC: Public Affairs Press, 1948. Translated from Gujarati by Mahadev Desai.

[239] J. A. Gasbarro, "Testing High Speed DRAMS," in *Proc. of the International Test Conf.*, Oct. 1994, p. 361.

[240] J. A. Gasbarro and M. A. Horowitz, "Techniques for Characterizing DRAMs with a 500 MHz Interface," in *Proc. of the International Test Conf.*, Oct. 1994, pp. 516–525.

[241] A. E. Gattiker and W. Maly, "Current Signatures," in *Proc. of the 14th VLSI Test Symp.*, Apr. 1996, pp. 112–117.

[242] A. E. Gattiker, P. Nigh, D. Grosch, and W. Maly, "Current Signatures for Production Testing," in *Proc. of the IEEE International Workshop on I_{DDQ} Testing*, Oct. 1996, pp. 25–28.

[243] F. A. Gay, "Evaluation of Maintenance Software in Real-Time Systems," *IEEE Trans. on Computers*, vol. C-27, no. 6, pp. 576–582, June 1978.

[244] M. A. Gharaybeh, V. D. Agrawal, M. L. Bushnell, and C. G. Parodi, "False-Path Removal Using Delay Fault Simulation," *Journal of Electronic Testing: Theory and Applications*, vol. 16, no. 5, pp. 463–476, Oct. 2000.

[245] M. A. Gharaybeh, M. L. Bushnell, and V. D. Agrawal, "Classification and Test Generation for Path-Delay Faults Using Single Stuck-at Fault Tests," *Journal of Electronic Testing: Theory and Applications*, vol. 11, no. 1, pp. 55–67, Aug. 1997.

[246] M. A. Gharaybeh, M. L. Bushnell, and V. D. Agrawal, "The Path Status Graph with Application to Delay Fault Simulation," *IEEE Trans. on Computer-Aided Design*, vol. 17, no. 4, pp. 324–332, Apr. 1998.

[247] A. Ghosh, S. Devadas, and A. R. Newton, "Test Generation for Highly Sequential Circuits," in *Proc. of the International Conf. on Computer Design*, Nov. 1989, pp. 362–365.

[248] A. Ghosh, S. Devadas, and A. R. Newton, *Sequential Logic Testing and Verification*. Boston: Kluwer Academic Publishers, 1992.

[249] I. Ghosh, A. Raghunathan, and N. K. Jha, "A Design for Testability Technique for RTL Circuits Using Control/Data Flow Extraction," in *Proc. of the International Conf. on Computer-Aided Design*, Nov. 1996, pp. 329–336.

[250] P. Ghosh, "Transistor Stuck-Open Fault Test Generation for Switch-Level Circuits Using Energy Minimization Techniques," Master's thesis, ECE Dept., Rutgers University, May 1996.

[251] G. Gielen, Z. Wang, and W. Sansen, "Probabilistic Fault Detection and the Selection of Measurements for Practical Testing of Analog Integrated Circuits Using a Sensitivity-driven Enumeration of Common Spanning Trees," *IEEE Trans. on Circuits and Systems Volume II*, vol. 45, no. 10, pp. 1342–1350, Oct. 1998.

[252] J. Giraldi and M. L. Bushnell, "EST: The New Frontier in Automatic Test Pattern Generation," in *Proc. of the 27th Design Automation Conf.*, June 1990, pp. 667–672.

[253] J. Giraldi and M. L. Bushnell, "Search State Equivalence for Redundancy Identification and Test Generation," in *Proc. of the International Test Conf.*, Oct. 1991, pp. 184–193.

[254] A. Glaser and G. Subak-Sharpe, *Integrated Circuit Engineering*. Reading, Massachusetts: Addison-Wesley, 1977.

[255] H. C. Godoy, C. B. Franklin, and P. Bottorff, "Automatic Checking of Logic Design Structures for Compliance with Testability Ground Rules," in *Proc. of the 14th Design Automation Conf.*, June 1977, pp. 469–478.

[256] P. Goel, "An Implicit Enumeration Algorithm to Generate Tests for Combinational Logic Circuits," in *Proc. of the International Fault-Tolerant Computing Symp.*, Aug. 1980, pp. 145–151.

[257] P. Goel, "Test Generation Costs Analysis and Projections," in *Proc. of the 17th Design Automation Conf.*, June 1980, pp. 77–84.

[258] P. Goel, "An Implicit Enumeration Algorithm to Generate Tests for Combinational Logic Circuits," *IEEE Trans. on Computers*, vol. C-30, no. 3, pp. 215–222, Mar. 1981.

[259] P. Goel and B. C. Rosales, "Test Generation and Dynamic Compaction of Tests," in *Proc. of the International Test Conf.*, Oct. 1979, pp. 189–192.

[260] P. Goel and B. C. Rosales, "Dynamic Test Compaction with Fault Selection Using Sensitizable Path Tracing," *IBM Technical Disclosure Bulletin*, vol. 23, no. 5, pp. 1954–1957, Oct. 1980.

[261] D. E. Goldberg, *Genetic Algorithms in Search, Optimization, and Machine Learning*. Reading, Massachusetts: Addison-Wesley, 1989.

[262] L. H. Goldstein, "Controllability/Observability Analysis of Digital Circuits," *IEEE Trans. on Circuits and Systems*, vol. CAS-26, no. 9, pp. 685–693, Sept. 1979.

[263] L. H. Goldstein and E. L. Thigpen, "SCOAP: Sandia Controllability/Observability Analysis Program," in *Proc. of the 17th Design Automation Conf.*, June 1980, pp. 190–196.

[264] S. W. Golomb, *Shift Register Sequences*. Laguna Hills, California: Aegean Park Press, 1982.

[265] F. M. Gonçalves and J. P. Teixeira, "Sampling Techniques of Non-Equally Probable Faults in VLSI Systems," in *Proc. of the 16th VLSI Test Symp.*, Apr. 1998, pp. 283–288.

[266] G. Gordon and H. Nadig, "Hexadecimal Signatures Identify Troublespots in Microprocessor Systems," *Electronics*, vol. 50, no. 5, pp. 89–96, Mar. 1977.

[267] N. Gouders and R. Kaibel, "PARIS: A Parallel Pattern Fault Simulator for Synchronous Sequential Circuits," in *Proc. of the International Conf. on Computer-Aided Design*, Nov. 1991, pp. 542–545.

[268] C. W. Green, "PMOS Dynamic RAM Reliability – A Case Study," in *Proc. of the International Reliability Physics Symp.*, Apr. 1979, pp. 213–219.

[269] M. R. Grimaila, S. Lee, J. Dworak, K. M. Butler, B. Stewart, B. Houchins, V. Mathur, J. Park, L.-C. Wang, and M. R. Mercer, "REDO – Random Excitation and Deterministic Observation – First Commercial Experiment," in *Proc. of the 17th VLSI Test Symp.*, Apr. 1999, pp. 268–274.

[270] A. Grochowski, D. Bhattacharya, T. R. Viswanathan, and K. Laker, "Integrated Circuit Testing for Quality Assurance in Manufacturing: History, Current Status, and Future Trends," *IEEE Trans. on Circuits and Systems – II*, vol. 44, no. 8, pp. 610–633, Aug. 1997.

[271] T. Guckert, P. Schani, M. Phillips, M. Seeley, and N. Herr, "Design and Process Issues for Elimination of Device Failures Due to 'Drooping' Vias," in *Proc. of the International Symp. for Testing and Failure Analysis (ISTFA)*, Nov. 1991, pp. 97–101.

[272] R. K. Gulati, W. Mao, and D. K. Goel, "Detection of Undetectable Faults using I_{DDQ} Testing," in *Proc. of the International Test Conf.*, Sept. 1992, pp. 770–777.

[273] R. Guo, I. Pomeranz, and S. M. Reddy, "Procedures for Static Compaction of Test Sequences for Synchronous Sequential Circuits based on Vector Restoration," in *Proc. of the Design Automation and Test in Europe Conf.*, Feb. 1998, pp. 583–587.

[274] R. Gupta, R. Gupta, and M. A. Breuer, "A BALLAST Methodology for Structured Partial Scan Design," *IEEE Trans. on Computers*, vol. 39, no. 4, pp. 538–544, Apr. 1990.

[275] N. B. Hamida and B. Kaminska, "Analog Circuit Testing Based on Sensitivity Computation and New Circuit Modeling," in *Proc. of the International Test Conf.*, Oct. 1993, pp. 652–661.

[276] N. B. Hamida and B. Kaminska, "Multiple Fault Analog Circuit Testing by Sensitivity Analysis," *Journal of Electronic Testing: Theory and Applications*, vol. 4, no. 4, pp. 331–343, Nov. 1993.

[277] D. Harel and B. Krishnamurthy, "Is There Hope for Linear Time Fault Simulation?," in *Proc. of the 17th International Fault-Tolerant Computing Symp.*, 1987, pp. 28–33.

[278] L. R. Harriott, "A New Role for E-Beam: Electron Projection," *IEEE Spectrum*, vol. 36, no. 7, pp. 41–45, July 1999.

[279] I. Hartanto, V. Boppana, J. H. Patel, and W. K. Fuchs, "Diagnostic Test Pattern Generation for Sequential Circuits," in *Proc. of the 15th VLSI Test Symp.*, Apr. 1997, pp. 196–202.

[280] S. Z. Hassen and E. J. McCluskey, "Increased Fault Coverage Through Multiple Signatures," in *Proc. of the International Fault-Tolerant Computing Symp.*, June 1984, pp. 354–359.

[281] C. F. Hawkins and J. M. Soden, "Electrical Characteristics and Testing Considerations for Gate Oxide Shorts in CMOS ICs," in *Proc. of the International Test Conf.*, Nov. 1985, pp. 544–555.

[282] C. F. Hawkins and J. M. Soden, "Reliability and Electrical Properties of Gate Oxide Shorts in CMOS ICs," in *Proc. of the International Test Conf.*, Sept. 1986, pp. 443–451.

[283] C. F. Hawkins, J. M. Soden, R. R. Fritzemeier, and L. K. Horning, "Quiescent Power Supply Current Measurement for CMOS IC Defect Detection," *IEEE Trans. on Industrial Electronics*, vol. 36, no. 2, pp. 211–218, May 1989.

[284] J. P. Hayes, "Detection of Pattern-Sensitive Faults in Random-Access Memories," *IEEE Trans. on Computers*, vol. C-24, no. 2, pp. 150–157, Feb. 1975.

[285] J. P. Hayes, "Transition Count Testing of Combinational Logic Circuits," *IEEE Trans. on Computers*, vol. C-25, no. 6, pp. 613–620, June 1976.

[286] J. P. Hayes, "Testing Memories for Single-Cell Pattern-Sensitive Faults in Semiconductor Random-Access Memories," *IEEE Trans. on Computers*, vol. C-29, no. 3, pp. 249–254, Mar. 1980.

[287] J. P. Hayes, "An Introduction to Switch-Level Modeling," *IEEE Design & Test of Computers*, vol. 4, no. 4, pp. 18–25, Aug. 1987.

[288] H. Haznedar, *Digital Microelectronics*. Redwood City, California: Benjamin/Cummings Publishing Co., Inc., 1991.

[289] W. Heisenberg, *Gesammelte Werke (Collected Works)*. Berlin: Springer-Verlag, 1989.

[290] C. L. Henderson, J. M. Soden, and C. F. Hawkins, "The Behavior and Testing Implications of CMOS IC Logic Gate Open Circuits," in *Proc. of the International Test Conf.*, Oct. 1991, pp. 302–310.

[291] M. Henftling, H. Wittmann, and K. J. Antreich, "A Formal Non-Heuristic ATPG Approach," in *Proc. of the European Design Automation Conf.*, Sept. 1995, pp. 248–253.

[292] K. Heragu, *New Techniques to Verify Timing Correctness of Integrated Circuits*. PhD thesis, Dept. of EECS, University of Illinois, Urbana, Illinois, 1998.

[293] K. Heragu, V. D. Agrawal, and M. L. Bushnell, "An Efficient Path Delay Coverage Estimator," in *Proc. of the 31st Design Automation Conf.*, June 1994, pp. 516–521.

[294] K. Heragu, V. D. Agrawal, and M. L. Bushnell, "FACTS: Fault Coverage Estimation by Test Vector Sampling," in *Proc. of the 12th VLSI Test Symp.*, Apr. 1994, pp. 266–271.

[295] K. Heragu, V. D. Agrawal, and M. L. Bushnell, "Statistical Methods for Delay Fault Coverage Analysis," in *Proc. of the 8th International Conf. on VLSI Design*, Jan. 1995, pp. 166–170.

[296] K. Heragu, M. L. Bushnell, and V. D. Agrawal, "Fault Coverage Estimation by Test Vector Sampling," *IEEE Trans. on Computer-Aided Design*, vol. 14, no. 5, pp. 590–596, May 1995.

[297] D. Herrell, "Power to the Package," *IEEE Spectrum*, vol. 36, no. 7, pp. 46–53, July 1999.

[298] R. L. Hickling, "Tester Independent Problem Representation and Tester Dependent Program Generation," in *Proc. of the International Test Conf.*, Oct. 1983, pp. 476–482.

[299] F. P. Higgins and R. Srinivasan, "BSM2: Next Generation Boundary-Scan Master," in *Proc. of the 18th VLSI Test Symp.*, Apr. 2000, pp. 67–72.

[300] F. J. Hill and B. Huey, "SCIRTSS: A Search System for Sequential Circuit Test Sequences," *IEEE Trans. on Computers*, vol. C-26, no. 5, pp. 490–502, May 1977.

[301] R. B. Hitchcock, "Timing Verification and the Timing Analysis Program," in *Proc. of the 19th Design Automation Conf.*, June 1982, pp. 594–604.

[302] E. R. Hnatek, *Integrated Circuit Quality and Reliability*. New York: Marcel Dekker, 1987.

[303] J. H. Holland, *Adaptation in Natural and Artificial Systems*. Ann Arbor, Michigan: University of Michigan Press, 1975.

[304] J. J. Hopfield, "Artificial Neural Networks," *IEEE Circuits and Devices Magazine*, vol. 4, no. 5, pp. 3–10, Sept. 1988.

[305] L. K. Horning, J. M. Soden, R. R. Fritzemeier, and C. F. Hawkins, "Measurements of Quiescent Power Supply Current for CMOS ICs in Production Testing," in *Proc. of the International Test Conf.*, Sept. 1987, pp. 300–308.

[306] P. D. Hortensius, R. D. McLeod, and B. W. Podaima, "Cellular Automata Circuits for Built-In Self-Test," *IBM Journal of Research and Development*, vol. 34, no. 2/3, pp. 389–405, March/May 1990.

[307] P. D. Hortensius, R. D. McLeod, W. Pries, D. M. Miller, and H. C. Card, "Cellular Automata-Based Pseudorandom Number Generators for Built-In Self-Test," *IEEE Trans. on Computer-Aided Design*, vol. CAD-8, no. 8, pp. 842–859, Aug. 1989.

[308] A. S. Householder, "A Survey of Some Closed Methods for Inverting Matrices," *SIAM Journal on Applied Mathematics*, vol. 5, no. 3, pp. 155–169, 1957.

[309] M. J. Howes and D. V. Morgan, editors, *Reliability and Degradation - Semiconductor Devices and Circuits*. Chichester, UK: Wiley-Interscience, 1981.

[310] M. S. Hsiao, E. M. Rudnick, and J. H. Patel, "Sequential Circuit Test Generation using Dynamic State Traversal," in *Proc. of the European Design and Test Conf.*, 1997, pp. 22–28.

[311] M. S. Hsiao, E. M. Rudnick, and J. H. Patel, "Dynamic State Traversal for Sequential Circuit Test Generation," *ACM Trans. on Design Automation of Electronic Systems (TODAES)*, vol. 5, no. 3, July 2000.

[312] J.-L. Huertas, A. Rueda, and D. Vazquez, "Design for Testability Techniques Applicable to Analog Circuits," in *Proc. of the European Test Conf.*, Apr. 1993, pp. 522–523.

[313] J.-L. Huertas, A. Rueda, and D. Vazquez, "Improving the Testability of Switched-Capacitor Filters," *Analog Integrated Circuits & Signals Proceedings*, vol. 4, pp. 199–213, 1993.

[314] J.-L. Huertas, A. Rueda, and D. Vazquez, "Testable Switched-Capacitor Filters," *IEEE Journal of Solid-State Circuits*, vol. 28, no. 7, pp. 719–724, July 1993.

[315] J. L. A. Hughes and E. J. McCluskey, "Multiple Stuck-at Fault Coverage of Single Stuck-at Fault Test Sets," in *Proc. of the International Test Conf.*, Sept. 1986, pp. 368–374.

[316] S. D. Huss and R. S. Gyurcsik, "Optimal Ordering of Analog Integrated Circuit Tests to Minimize Test Time," in *Proc. of the 28th Design Automation Conf.*, June 1991, pp. 494–499.

[317] O. H. Ibarra and S. K. Sahni, "Polynomially Complete Fault Detection Problems," *IEEE Trans. on Computers*, vol. C-24, no. 3, pp. 242–249, Mar. 1975.

[318] IEEE Standards Board, 345 East 47th St. New York 10017, *IEEE Standard Test Access Port and Boundary-Scan Architecture*, 1994. IEEE/ANSI Standard 1149.1-1994 (revision b), includes supplements 1149.1a and 1149.1b.

[319] IEEE Standards Board, 345 East 47th St. New York 10017, *IEEE Standard for a Mixed-Signal Test Bus*, June 2000. IEEE Standard 1149.4.

[320] International Telecommunication Union, Geneva, *CCITT Yellow Book*, 1981. PCM parameter information is repeated in the later edition, the *Red Book*, 1984.

[321] M. Ismail and T. Fiez, editors, *Analog VLSI: Signal and Information Processing*. McGraw-Hill, 1994.

[322] N. Itazaki and K. Kinoshita, "Test Pattern Generation for Circuits with Three-state Modules by Improved Z-algorithm," in *Proc. of the International Test Conf.*, Sept. 1986, pp. 105–108.

[323] N. Itazaki and K. Kinoshita, "Test Pattern Generation for Circuits with Tri-state Modules by Z-algorithm," *IEEE Trans. on Computer-Aided Design*, vol. 8, no. 12, pp. 1327–1334, Dec. 1989.

[324] A. Ivanov and V. K. Agarwal, "Testability Measures – What Do They Do for ATPG?," in *Proc. of the International Test Conf.*, Sept. 1986, pp. 129–138.

[325] A. Ivanov and V. K. Agarwal, "An Analysis of the Probabilisitc Behavior of Linear Feedback Signature Registers," *IEEE Trans. on Computer-Aided Design*, vol. CAD-8, no. 10, pp. 1074–1088, Oct. 1989.

[326] V. S. Iyengar and D. Brand, "Synthesis of Pseudo-Random Pattern Testable Designs," in *Proc. of the International Test Conf.*, Aug. 1989, pp. 501–508.

[327] V. S. Iyengar, B. K. Rosen, and J. A. Waicukauski, "On Computing the Sizes of Detected Delay Faults," *IEEE Trans. on Computer-Aided Design*, vol. 9, no. 3, pp. 299–312, Mar. 1990.

[328] M. A. Iyer, *On Redundancy and Untestability in Sequential Circuits*. PhD thesis, ECE Department, Illinois Institute of Technology, Chicago, Illinois, July 1995.

[329] M. A. Iyer and M. Abramovici, "Low-Cost Redundancy Identification for Combinational Circuits," in *Proc. of the 7th International Conf. on VLSI Design*, Jan. 1994, pp. 315–318.

[330] M. K. Iyer and M. L. Bushnell, "Effect of Noise on Analog Circuit Testing," *Journal of Electronic Testing: Theory and Applications*, vol. 15, no. 1/2, pp. 11–22, Aug./Oct. 1999.

[331] J. Jacob and N. N. Biswas, "GTBD Faults and Lower Bounds on Multiple Fault Coverage of Single Fault Test Sets," in *Proc. of the International Test Conf.*, Sept. 1987, pp. 849–855.

[332] M. Jacomino, J. L. Rainard, and R. David, "Fault Detection by Consumption Measurement in CMOS Circuits," in F. Belli and W. Dorke, editors, *Proc. of the 3rd International Conf. on Fault Tolerant Computing Systems*, (Berlin), Springer Verlag, 1987, pp. 83–94.

[333] A. Jain, "Arbitrary Defects; Modeling and Applications," Master's thesis, Electrical and Comp. Eng. Dept., Rutgers University, New Brunswick, New Jersey, Oct. 1999.

[334] S. K. Jain and V. D. Agrawal, "STAFAN: An Alternative to Fault Simulation," in *Proc. of the 21st Design Automation Conf.*, June 1984, pp. 18–23.

[335] S. K. Jain and V. D. Agrawal, "Modeling and Test Generation Algorithms for MOS Circuits," *IEEE Trans. on Computers*, vol. C-34, no. 5, pp. 426–433, May 1985.

[336] S. K. Jain and V. D. Agrawal, "Statistical Fault Analysis," *IEEE Design & Test of Computers*, vol. 2, no. 1, pp. 38–44, Feb. 1985.

[337] S. K. Jain and C. E. Stroud, "Built-In Self Testing of Embedded Memories," *IEEE Design & Test of Computers*, vol. 3, no. 4, pp. 27–37, Aug. 1986.

[338] P. Jalote, *An Integrated Approach to Software Engineering*. New York: Springer-Verlag, 1991.

[339] F. Jay, editor, *IEEE Standard Dictionary of Electrical and Electonics Terms*, p. 470. J. Wiley and Sons, third edition, 1984.

[340] F. Jensen and N. E. Petersen, *Burn-In*. Chichester, UK: John Wiley & Sons, Inc., 1982.

BIBLIOGRAPHY

[341] N. K. Jha, "Detecting Multiple Faults in CMOS Circuits," in *Proc. of the International Test Conf.*, Sept. 1986, pp. 514–519.

[342] N. K. Jha and S. Kundu, *Testing and Reliable Design of CMOS Circuits*. Boston: Kluwer Academic Publishers, 1990.

[343] E. R. Jones and C. H. Mays, "Automatic Test Generation Methods for Large Scale Integrated Logic," *IEEE Journal of Solid-State Circuits*, vol. SC-2, pp. 221–226, Dec. 1967.

[344] D. Kagaris, S. Tragoudas, and D. Karayiannis, "Nonenumerative Path Delay Fault Coverage Estimation Based on Optimal Polynomial Time Algorithms," *IEEE Trans. on Computer-Aided Design*, vol. 16, no. 3, pp. 309–315, Mar. 1997.

[345] H. Kajitani, H. Sato, H. Saito, and S. Oresjo, "Practical Test Generation with IEEE 1149.1 Boundary Scan," in *Proc. of the Automatic Test Equipment & Interfacing Conf.*, (Anaheim, California), Jan. 1992, pp. 10–17.

[346] B. Kaminska, "Comments made during the Panel Session, *North Atlantic Test Workshop*," 1999.

[347] R. Kapur, S. Patil, T. J. Snethen, and T. W. Williams, "Design of an Efficient Weighted Random Pattern Generation System," in *Proc. of the International Test Conf.*, Oct. 1994, pp. 491–500.

[348] M. Karam and G. Saucier, "Functional versus Random Test Generation for Sequential Circuits," *Journal of Electronic Testing: Theory and Applications*, vol. 4, no. 1, pp. 33–41, Feb. 1993.

[349] M. G. Karpovsky, S. K. Gupta, and D. K. Pradhan, "Aliasing and Diagnosis Probability in MISR and STUMPS Using a General Error Model," in *Proc. of the International Test Conf.*, Oct. 1991, pp. 828–839.

[350] M. Keating and D. Meyer, "A New Approach to Dynamic IDD Testing," in *Proc. of the International Test Conf.*, Sept. 1987, pp. 316–321.

[351] T. P. Kelsey, K. K. Saluja, and S. Y. Lee, "An Efficient Algorithm for Sequential Circuit Test Generation," *IEEE Trans. on Computers*, vol. 42, no. 11, pp. 1361–1371, Nov. 1993.

[352] K. Keutzer, S. Malik, and A. Saldanha, "Is Redundancy Necessary to Reduce Delay?," *IEEE Trans. on Computer-Aided Design*, vol. 10, no. 4, pp. 427–439, Apr. 1991.

[353] M. Khare and A. Albicki, "Cellular Automata Used for Test Pattern Generation," in *Proc. of the International Conf. on Computer Design*, Oct. 1987, pp. 56–59.

[354] K. Kim, J. G. Tront, and D. S. Ha, "BIDES: A BIST Design Expert System," *Journal of Electronic Testing: Theory and Applications*, vol. 2, no. 2, pp. 165–179, June 1991.

[355] C. R. Kime, "An Analysis Model for Digital System Diagnosis," *IEEE Trans. on Computers*, vol. C-19, no. 11, pp. 1063–1073, Nov. 1970.

[356] C. R. Kime, "System Diagnosis," in D. K. Pradhan, editor, *Fault-Tolerant Computing Theory and Techniques*, volume II, chapter 8, pp. 577–632, Prentice-Hall, 1986.

[357] A. Kinoshita, S. Murakami, Y. Nishimura, and K. Anami, "A Study of Delay Time on Bit Lines in Megabit SRAMs," *IEICE Trans. on Electron Devices*, vol. E75-C, no. 11, pp. 1383–1386, Nov. 1992.

[358] K. Kinoshita and K. K. Saluja, "Built-In Testing of Memory Using On-Chip Compact Testing Scheme," in *Proc. of the International Test Conf.*, Oct. 1984, pp. 271–281.

[359] K. Kinoshita and K. K. Saluja, "Built-In Testing of Memory Using an On-Chip Compact Testing Scheme," *IEEE Trans. on Computers*, vol. C-35, no. 10, pp. 862–870, Oct. 1986.

[360] T. Kirkland and M. R. Mercer, "A Topological Search Algorithm for ATPG," in *Proc. of the 24th Design Automation Conf.*, June-July 1987, pp. 502–508.

[361] E. Kjelkerud and O. Thessen, "Generation of Hazard Free Tests using the D-Algorithm in a Timing Accurate System for Logic and Fault Simulation," in *Proc. of the 16th Design Automation Conf.*, June 1979, pp. 180–184.

[362] J. Knaizuk, Jr. and C. R. P. Hartmann, "An Optimal Algorithm for Testing Stuck-at Faults in Random Access Memories," *IEEE Trans. on Computers*, vol. C-26, no. 11, pp. 1141–1144, Nov. 1977.

[363] D. E. Knuth, *The Art of Computer Programming*, volume 1/Fundamental Algorithms. Reading, Massachusetts: Addison-Wesley, second edition, 1975.

[364] B. Köenemann, J. Mucha, and G. Zwiehoff, "Built-In Test for Complex Digital Integrated Circuits," *IEEE Journal of Solid-State Circuits*, vol. SC-15, no. 3, pp. 315–318, June 1980.

[365] Y. Konishi, T. Ogawa, and M. Kumanoya, "Testing 256k Word × 16 Bit Cache DRAM (CDRAM)," in *Proc. of the International Test Conf.*, Oct. 1994, p. 360.

[366] P. G. Kovijanic, "A New Look at Test Generation and Verification," in *Proc. of the 14th Design Automation Conf.*, June 1977, pp. 58–63.

[367] A. Krasniewski and S. Pilarski, "Circular Self-Test Path: A Low Cost BIST Technique for VLSI Circuits," *IEEE Trans. on Computer-Aided Design*, vol. CAD-8, no. 1, pp. 46–55, Jan. 1989.

[368] R. Kraus, O. Kowarik, K. Hoffmann, and D. Oberle, "Design for Test of Mbit DRAMs," in *Proc. of the International Test Conf.*, Aug. 1989, pp. 316–321.

[369] P. A. Krauss and K. J. Antreich, *Application of Fault Parallelism to the Automatic Test Pattern Generation for Sequential Circuits*, pp. 234–245. Lecture Notes in Computer Science 732, Springer-Verlag, 1993. Proc. 1993 Foundation of Software Technology and Theoretical Computer Science.

[370] P. A. Krauss, A. Ganz, and K. J. Antreich, "Distributed Test Pattern Generation for Stuck-At Faults in Sequential Circuits," *Journal of Electronic Testing: Theory and Applications*, vol. 11, pp. 227–245, Dec. 1997.

[371] A. Krstić and K.-T. Cheng, *Delay Fault Testing for VLSI Circuits*. Boston: Kluwer Academic Publishers, 1998.

[372] H. Kubo, "A Procedure for Generating Test Sequences to Detect Sequential Circuit Failures," *NEC Res. & Dev.*, vol. 12, no. 4, pp. 69–78, Oct. 1968.

[373] D. J. Kuck, *High Performance Computing Challenges for Future Systems*. New York: Oxford University Press, 1996.

[374] K. S. Kundert, *The Designer's Guide to SPICE & SPECTRE*. Boston: Kluwer Academic Publishers, 1995.

[375] C.-P. Kung and C.-S. Lin, "Parallel Sequence Fault Simulation for Synchronous Sequential Circuits," in *Proc. of the European Design Automation Conf.*, Mar. 1992, pp. 434–438.

[376] W. Kunz and D. K. Pradhan, "Recursive Learning: An Attractive Alternative to the Decision Tree for Test Generation in Digital Circuits," in *Proc. of the International Test Conf.*, Sept. 1992, pp. 816–825.

[377] W. Kunz and D. K. Pradhan, "Recursive Learning – A New Implication Technique for Efficient Solution to CAD Problems," *IEEE Trans. on Computer-Aided Design*, vol. 13, no. 9, pp. 1143–1158, Sept. 1994.

[378] W. Kunz and D. Stoffel, *Reasoning in Boolean Networks: Logic Synthesis and Verification Using Testing Techniques*. Boston: Kluwer Academic Publishers, 1997.

[379] A. Kunzmann and H.-J. Wunderlich, "An Analytical Approach to the Partial Scan Problem," *Journal of Electronic Testing: Theory and Applications*, vol. 1, no. 2, pp. 163–174, May 1990.

[380] R. P. Kurshan, *Computer Aided Verification of Coordinating Processes*. Princeton, New Jersey: Princeton University Press, 1994.

[381] P. Kurup and T. Abbasi, *Logic Synthesis Using Synopsys*. Boston: Kluwer Academic Publishers, 1995.

[382] M. Lane and S. McEuen, "I_{DDQ}: A Method for Detecting Potential Warranty Returns," Technical report, Ford Microelectronics Inc., Internal Report, Aug. 1990.

[383] T. Larrabee, "Efficient Generation of Test Patterns Using Boolean Difference," in *Proc. of the International Test Conf.*, Aug. 1989, pp. 795–801.

[384] T. Larrabee, "Test Pattern Generation Using Boolean Satisfiability," *IEEE Trans. on Computer-Aided Design*, vol. 11, no. 1, pp. 4–15, Jan. 1992.

[385] C. Y. Lee, "Representation of Switching Circuits by Binary Decision Diagrams," *Bell System Technical Journal*, vol. 38, pp. 985–999, July 1959.

[386] K. J. Lee, C. A. Njinda, and M. A. Breuer, "A Switch-Level Test Generation System for CMOS Combinational Circuits," *IEEE Trans. on Computer-Aided Design*, vol. 13, no. 5, pp. 625–637, May 1994.

[387] N.-C. Lee, "Practical Considerations for Mixed-Signal Test Bus," in *Proc. of the International Test Conf.*, Oct. 1993, pp. 591–592.

[388] T.-C. Lee, W. H. Wolf, N. K. Jha, and J. M. Acken, "Behavioral Synthesis for Easy Testability in Data Path Allocation," in *Proc. of the International Conf. on Computer Design*, Oct. 1992, pp. 29–32.

[389] D. Leet, P. Shearson, and R. France, "A CMOS LSSD Test Generation System," *IBM Journal of Research and Development*, vol. 28, no. 5, pp. 625–635, Sept. 1984.

[390] C. E. Lemke, "Bimatrix Equilibrium Points and Mathematical Programming," *Management Science*, vol. 11, pp. 681–689, May 1965.

[391] J. D. Lesser and J. J. Shedletsky, "An Experimental Delay Test Generator for LSI Logic," *IEEE Trans. on Computers*, vol. C-29, no. 3, pp. 235–248, Mar. 1980.

[392] Y. Levendel and P. R. Menon, "Transition Faults in Combinational Circuits: Input Transition Test Generation and Fault Simulation," in *Proc. of the International Fault-Tolerant Computing Symp.*, July 1986, pp. 278–283.

[393] M. W. Levi, "CMOS is Most Testable," in *Proc. of the International Test Conf.*, Oct. 1981, pp. 217–220.

[394] M. Lewitt, "ASIC Testing Updated," *IEEE Spectrum*, vol. 29, no. 5, pp. 26–29, May 1992.

[395] L. L. Lewyn and J. D. Meindl, "Physical Limits of VLSI DRAM's," *IEEE Journal of Solid-State Circuits*, vol. SC-20, no. 1, pp. 231–241, Feb. 1985.

[396] W. Li, S. M. Reddy, and S. K. Sahni, "On Path Selection in Combinational Logic Circuits," *IEEE Trans. on Computer-Aided Design*, vol. 8, no. 1, pp. 56–63, Jan. 1989.

[397] C. J. Lin and S. M. Reddy, "On Delay Fault Testing in Logic Circuits," *IEEE Trans. on Computer-Aided Design*, vol. CAD-6, no. 5, pp. 694–703, Sept. 1987.

[398] P. M. Lin and Y. S. Elcherif, "Analogue Circuits Fault Dictionary – New Approaches and Implementation," *International Journal of Circuit Theory and Applications*, vol. 13, no. 2, pp. 149–172, Apr. 1985.

[399] Q. Lin, "Efficient Techniques for a Transitive Closure-Based Test Generation Algorithm," Master's thesis, Rutgers University, ECE Dept., Jan. 1996.

[400] W. M. Lindermeir, "Design of Robust Test Criteria in Analog Testing," in *Proc. of the International Conf. on Computer-Aided Design*, Nov. 1996, pp. 604–611.

[401] W. M. Lindermeir, H. E. Graeb, and K. J. Antreich, "Design Based Analog Testing by Characteristic Observation Inference," in *Proc. of the International Conf. on Computer-Aided Design*, Nov. 1995, pp. 620–626.

[402] A. Lioy and M. Mezzalama, "On Parameters Affecting ATPG Performance," in *Proc. of the CompEuro '87*, May 1987, pp. 394–397.

[403] A. Lioy, P. L. Montessoro, and S. Gai, "A Complexity Analysis of Sequential ATPG," in *Proc. of the International Symp. on Circuits and Systems*, May 1989, pp. 1946–1949.

[404] C.-Y. Lo, H. N. Nham, and A. K. Bose, "Algorithms for an Advanced Fault Simulation System in MOTIS," *IEEE Trans. on Computer-Aided Design*, vol. CAD-6, no. 2, pp. 232–240, Mar. 1987.

[405] K. Lofstrom, "A Demonstration IC for the P1149.4 Mixed-Signal Test Standard," in *Proc. of the International Test Conf.*, Oct. 1996, pp. 92–98.

[406] K. Lofstrom, "Early Capture for Boundary Scan Timing Measurements," in *Proc. of the International Test Conf.*, Oct. 1996, pp. 417–422.

[407] LTX Corporation, *enVision++ Programming Manual*, 2000. Software Release 10.4.0.

[408] H.-K. T. Ma, S. Devadas, A. R. Newton, and A. Sangiovanni-Vincentelli, "Test Generation for Sequential Circuits," *IEEE Trans. on Computer-Aided Design*, vol. 7, no. 10, pp. 1081–1093, Oct. 1988.

[409] H.-K. T. Ma, S. Devadas, A. Sangiovanni-Vincentelli, and R. Wei, "Logic Verification Algorithms and their Parallel Implementation," in *Proc. of the 24th Design Automation Conf.*, June-July 1987, pp. 283–290.

[410] H.-K. T. Ma, S. Devadas, A. Sangiovanni-Vincentelli, and R. Wei, "Logic Verification Algorithms and their Parallel Implementation," *IEEE Trans. on Computer-Aided Design*, vol. 8, no. 2, pp. 181–189, Feb. 1989.

[411] M. Mahoney, "Automated Measurement of 12 to 16-Bit Converters," in *Proc. of the International Test Conf.*, Oct. 1981, pp. 319–327.

[412] M. Mahoney, *DSP-Based Testing of Analog and Mixed-Signal Circuits*. Los Alamitos, California: IEEE Computer Society Press, 1987.

[413] A. K. Majhi, *Algorithms for Test Generation and Fault Simulation of Path Delay Faults in Logic Circuits*. PhD thesis, Indian Institute of Science, Bangalore, India, Jan. 1996.

[414] A. K. Majhi, J. Jacob, L. M. Patnaik, and V. D. Agrawal, "On Test Coverage of Path Delay Faults," in *Proc. of the 9th International Conf. on VLSI Design*, Jan. 1996, pp. 418–421.

[415] S. Majumder, V. D. Agrawal, and M. L. Bushnell, "On Delay-Untestable Paths and Stuck-Fault Redundancy," in *Proc. of the 16th VLSI Test Symp.*, Apr. 1998, pp. 194–199.

[416] Y. K. Malaiya and R. Narayanaswamy, "Modeling and Testing for Timing Faults in Synchronous Sequential Circuits," *IEEE Design & Test of Computers*, vol. 1, no. 4, pp. 62–74, Nov. 1984.

[417] Y. K. Malaiya and R. Rajsuman, *Bridging Faults and IDDQ Testing*. Los Alamitos, California: IEEE Computer Society Press, 1992.

[418] Y. K. Malaiya and S. Y. H. Su, "A New Fault Model and Testing Technique for CMOS Devices," in *Proc. of the International Test Conf.*, Nov. 1982, pp. 25–34.

[419] S. Mallela and S. Wu, "A Sequential Circuit Test Generation System," in *Proc. of the International Test Conf.*, Nov. 1985, pp. 57–61.

[420] W. Maly, "Modeling of Lithography-Related Yield Losses for CAD of VLSI Circuits," *IEEE Trans. on Computer-Aided Design*, vol. CAD-4, no. 3, pp. 166–177, July 1985.

[421] W. Maly, "Design Methodology for Defect Tolerant Integrated Circuits," in *Proc. of the Custom Integrated Circuits Conf.*, May 1988, pp. 27.5.1–27.5.4.

[422] W. Maly, P. K. Nag, and P. Nigh, "Testing Oriented Analysis of CMOS ICs with Opens," in *Proc. of the International Conf. on Computer-Aided Design*, Nov. 1988, pp. 344–347.

[423] W. Maly and P. Nigh, "Built-In Current Testing of Integrated Circuits." U.S. Patent No. 5,025,344, June, 1991.

[424] W. Maly and P. Nigh, "Built-In Current Testing: Feasibility Study," in *Proc. of the International Conf. on Computer-Aided Design*, Nov. 1988, pp. 340–343.

[425] W. Maly and P. Nigh, "Built-In Current Testing for VLSI Circuits," in *Proc. of the SRC TECHCON '88 Conf.*, Semiconductor Research Corp., Oct. 1988, pp. 149–152.

[426] W. Maly and M. Patyra, "Design of ICs Applying Built-In Current Testing," *Journal of Electronic Testing: Theory and Applications*, vol. 3, no. 4, pp. 111–120, Dec. 1992.

[427] W. Maly and M. Patyra, "Design of ICs Applying Built-In Current Testing," *IEEE Journal of Solid-State Circuits*, vol. 27, no. 3, pp. 425–428, Mar. 1992.

[428] W. Mao and R. K. Gulati, "QUIETEST: A Methodology for Selecting I_{DDQ} Test Vectors," *Journal of Electronic Testing: Theory and Applications*, vol. 3, no. 4, pp. 349–357, Dec. 1992.

[429] M. Marinescu, "Simple and Efficient Algorithms for Functional RAM Testing," in *Proc. of the International Test Conf.*, Nov. 1982, pp. 236–239.

[430] E. J. Marinissen, R. Arendsen, G. Bos, H. Dingemanse, M. Lousberg, and C. Wouters, "A Structured and Scalable Mechanism for Test Access to Embedded Reusable Cores," in *Proc. of the International Test Conf.*, Oct. 1998, pp. 284–293.

[431] M. J. Marlett and J. A. Abraham, "DC_IATP – An Iterative Analog Circuit Test Generation Program for Generating DC Single Pattern Tests," in *Proc. of the International Test Conf.*, Sept. 1988, pp. 839–844.

[432] R. A. Marlett, "EBT: A Comprehensive Test Generation Technique for Highly Sequential Circuits," in *Proc. of the 15th Design Automation Conf.*, June 1978, pp. 335–339.

[433] R. A. Marlett, "An Effective Test Generation System for Sequential Circuits," in *Proc. of the 23rd Design Automation Conf.*, June-July 1986, pp. 250–256.

[434] MathWorks, Inc., Natick, Massachusetts, *MATLAB Computing Software Products*. See website: http://www.mathworks.com.

[435] J. S. Matos, A. C. Leao, and J. C. Ferreira, "Control and Observation of Analog Nodes in Mixed-Signal Boards," in *Proc. of the International Test Conf.*, Oct. 1993, pp. 323–331.

[436] C. M. Maunder and R. E. Tulloss, *The Test Access Port and Boundary Scan Architecture*. Los Alamitos, California: IEEE Computer Society Press, Sept. 1990.

[437] P. C. Maxwell and R. C. Aitken, "I_{DDQ} Testing as a Component of a Test Suite: The Need for Several Fault Coverage Metrics," *Journal of Electronic Testing: Theory and Applications*, vol. 3, no. 4, pp. 305–316, Dec. 1992.

[438] P. C. Maxwell, R. C. Aitken, R. Kollitz, and A. C. Brown, "IDDQ and AC Scan: The War Against Unmodelled Defects," in *Proc. of the International Test Conf.*, Oct. 1996, pp. 250–258.

[439] P. C. Maxwell and J. R. Rearick, "A Simulation-Based Method for Estimating Defect-Free I_{DDQ}," in *Proc. of the IEEE International Workshop on I_{DDQ} Testing*, Nov. 1997, pp. 80–84.

[440] P. C. Maxwell and J. R. Rearick, "Estimation of Defect-Free I_{DDQ} in Submicron Circuits Using Switch Level Simulation," in *Proc. of the International Test Conf.*, Oct. 1998, pp. 882–889.

[441] P. Mazumder, "An Efficient Design of Embedded Memories and their Testability Analysis using Markov Chains," in *Proc. of the Int'l. Conf. on Wafer Scale Integration*, Jan. 1989, pp. 389–400.

[442] P. Mazumder and K. Chakraborty, *Testing and Testable Design of High-Density Random-Access Memories*. Boston: Kluwer Academic Publishers, 1996.

[443] P. Mazumder and J. H. Patel, "An Efficient Built-In Self-Testing for Random Access Memory," in *Proc. of the International Test Conf.*, Sept. 1987, pp. 1072–1077.

[444] P. Mazumder and J. H. Patel, "Parallel Testing for Pattern-Sensitive Faults in Semiconductor Random-Access Memories," *IEEE Trans. on Computers*, vol. C-38, no. 3, pp. 394–407, Mar. 1989.

[445] P. Mazumder and E. M. Rudnick, *Genetic Algorithms for VLSI Design, Layout, and Test Automation*. Upper Saddle River, New Jersey: Prentice-Hall, 1999.

[446] W. J. McCalla, *Fundamentals of Computer-Aided Circuit Simulation*. Boston: Kluwer Academic Publishers, 1988.

[447] E. J. McCluskey, "Verification Testing – A Pseudo-Exhaustive Test Technique," *IEEE Trans. on Computers*, vol. C-33, no. 6, pp. 541–546, June 1984.

[448] E. J. McCluskey and S. Bozorgui-Nesbat, "Design for Autonomous Test," *IEEE Trans. on Circuits and Systems*, vol. CAS-28, no. 11, pp. 1070–1079, Nov. 1981.

[449] E. J. McCluskey and F. W. Clegg, "Fault Equivalence in Combinational Logic Networks," *IEEE Trans. on Computers*, vol. C-20, no. 11, pp. 1286–1293, Nov. 1971.

[450] S. D. McEuen, "I_{DDQ} Benefits," in *Proc. of the 17th VLSI Test Symp.*, Apr. 1991, pp. 258–290.

[451] S. D. McEuen, "Reliability Benefits of I_{DDQ}," *Journal of Electronic Testing: Theory and Applications*, vol. 3, no. 4, pp. 328–335, Dec. 1992.

[452] P. C. McGeer and R. K. Brayton, *Integrating Functional and Temporal Domains in Logic Design*. Boston: Kluwer Academic Publisher, 1991.

[453] K. L. McMillan, *Symbolic Model Checking*. Boston: Kluwer Academic Publishers, 1993.

[454] M. G. McNamer, S. C. Roy, and H. T. Nagle, "Statistical Fault Sampling," *IEEE Trans. on Industrial Electronics*, vol. 36, no. 2, pp. 141–150, May 1989.

[455] R. F. M. Meershoek, "Functional and I_{DDQ} Testing on a Static RAM," TUD Report 1-68340-28(1990)04, Dept. of Electrical Eng., Delft University of Technology, Delft, The Netherlands, 1990.

[456] R. F. M. Meershoek, B. Verhelst, R. McInerney, and L. Thijssen, "Functional and IDDQ Testing on a Static RAM," in *Proc. of the International Test Conf.*, Sept. 1990, pp. 929–937.

[457] A. Meixner and W. Maly, "Fault Modeling for the Testing of Mixed Integrated Circuits," in *Proc. of the International Test Conf.*, Oct. 1991, pp. 564–572.

BIBLIOGRAPHY

[458] P. R. Menon and S. G. Chappell, "Deductive Fault Simulation with Functional Blocks," *IEEE Trans. on Computers*, vol. C-27, no. 8, pp. 689–695, Aug. 1978.

[459] P. R. Menon, Y. H. Levendel, and M. Abramovici, "Critical Path Tracing in Sequential Circuits," in *Proc. of the International Conf. on Computer-Aided Design*, Nov. 1988, pp. 162–165.

[460] P. R. Menon, Y. H. Levendel, and M. Abramovici, "SCRIPT: A Critical Path Tracing Algorithm for Synchronous Sequential Circuits," *IEEE Trans. on Computer-Aided Design*, vol. 10, no. 6, pp. 738–747, June 1991.

[461] Meta-Software, Inc., *HSPICE User's Manual: H8801*, 1988.

[462] W. Meyer and R. Camposano, "Active Timing Multilevel Fault-Simulation with Switch-Level Accuracy," *IEEE Trans. on Computer-Aided Design*, vol. 14, no. 10, pp. 1241–1256, Oct. 1995.

[463] A. Miczo, *Digital Logic Testing and Simulation*. New York: Harper & Row, 1986.

[464] L. Milor and A. L. Sangiovanni-Vincentelli, "Optimal Test Set Design for Analog Circuits," in *Proc. of the International Conf. on Computer-Aided Design*, Nov. 1990, pp. 294–297.

[465] L. Milor and A. L. Sangiovanni-Vincentelli, "Minimizing Production Test Time to Detect Faults in Analog Circuits," *IEEE Trans. on Computer-Aided Design*, vol. 13, no. 6, pp. 796–813, June 1994.

[466] L. Milor and V. Visvanathan, "Detection of Catastrophic Faults in Analog Integrated Circuits," *IEEE Trans. on Computer-Aided Design*, vol. CAD-8, no. 2, pp. 114–130, Feb. 1989.

[467] H. B. Min, H. A. Luh, and W. A. Rogers, "Hierarchical Test Pattern Generation: A Cost Model and Implementation," *IEEE Trans. on Computer-Aided Design*, vol. 12, no. 7, pp. 1029–1039, July 1993.

[468] H. B. Min and W. A. Rogers, "A Test Methodology for Finite State Machines using Partial Scan Design," *Journal of Electronic Testing: Theory and Applications*, vol. 3, no. 2, pp. 127–137, May 1992.

[469] S. Mir, M. Lubaszewski, and B. Courtois, "Fault-Based ATPG for Linear Analog Circuits with Minimal Size Multifrequency Test Sets," *Journal of Electronic Testing: Theory and Applications*, vol. 9, no. 1/2, pp. 43–57, Aug./Oct. 1996.

[470] S. Mir, M. Lubaszewski, and B. Courtois, "Fault-Based ATPG for Linear Analogue Circuits with Minimal Size Multifrequency Test Sets," *Journal of Electronic Testing: Theory and Applications*, vol. 9, no. 1/2, pp. 43–57, Aug./Oct. 1996.

[471] S. Mir, M. Lubaszewski, V. Kolarik, and B. Courtois, "Optimal ATPG for Analogue Built-In Self-Test and Fault Diagnosis," in *Proc. of the IEEE International Mixed-Signal Test Workshop*, June 1995, pp. 80–85.

[472] R. L. Mitchell, *Engineering Economics*. Chichester, UK: John Wiley & Sons, Inc., 1980.

[473] Y. Miura, "Real-Time Current Testing for A/D Converters," *IEEE Design & Test of Computers*, vol. 13, no. 2, pp. 34–41, summer 1996.

[474] P. R. Moorby, "Fault Simulation Using Parallel Value Lists," in *Proc. of the International Conf. on Computer-Aided Design*, Sept. 1983, pp. 101–102.

[475] G. E. Moore, "MOS Transistors as Individual Devices and in Integrated Arrays," in *Proc. of the National Electronics Conf.*, Oct. 1965, pp. 25–30.

[476] G. E. Moore, "Trends in Silicon Device Technology," in *Int'l. Electron Devices Meeting*, 1969. *Keynote Address*.

[477] G. E. Moore, "Progress in Digital Integrated Electronics," in *Proc. of the Int'l. Electron Device Meeting*, Dec. 1975, pp. 11–13. Moore's Law.

[478] G. E. Moore, "Lithography and the Future of Moore's Law," in *Proc. of the Society of Photo-Optical Instrumentation Engineers (SPIE)*, (Santa Clara, CA), Feb. 1995, pp. 2–17.

[479] B. T. Murphy, "Cost-Size Optima of Monolithic Integrated Circuits," *Proc. of the IEEE*, vol. 52, no. 12, pp. 1537–1545, Dec. 1964.

[480] J. D. Musa, A. Iannino, and K. Okumoto, *Software Reliability Measurement, Prediction, Application*. New York: McGraw-Hill, 1987.

[481] P. Muth, "A Nine-Valued Circuit Model for Test Generation," *IEEE Trans. on Computers*, vol. C-25, no. 6, pp. 630–636, June 1976.

[482] B. Nadeau-Dostie, editor, *Design for At-Speed Test, Diagnosis and Measurement*. Boston: Kluwer Academic Publishers, 2000.

[483] B. Nadeau-Dostie, A. Silburt, and V. K. Agarwal, "Serial Interfacing for Embedded-Memory Testing," *IEEE Design & Test of Computers*, vol. 7, no. 2, pp. 52–63, Apr. 1990.

[484] L. W. Nagel, "SPICE2, A Computer Program to Simulate Semiconductor Circuits," Technical Report ERL Memorandum ERL-M520, U. of California, Electronics Research Laboratory, Berkeley, California, May 1975. PhD Dissertation, Dept. of EECS.

[485] L. W. Nagel and D. O. Pederson, "SPICE – Simulation Program with Integrated Circuit Emphasis," Memo ERL-M382, EECS Dept., University of California, Berkeley, Apr. 1973.

[486] L. W. Nagel and R. Rohrer, "Computer Analysis of Nonlinear Circuits Excluding Radiation (CANCER)," *IEEE Journal of Solid-State Circuits*, vol. 6, no. 4, pp. 166–182, Aug. 1971.

[487] N. Nagi, A. Chatterjee, and J. A. Abraham, "DRAFTS: Discretized Analog Circuit Fault Simulator," in *Proc. of the 30th Design Automation Conf.*, June 1993, pp. 509–514.

[488] N. Nagi, A. Chatterjee, A. Balivada, and J. A. Abraham, "Fault-Based Automatic Test Generator for Linear Analog Circuits," in *Proc. of the International Conf. on Computer-Aided Design*, Nov. 1993, pp. 88–91.

[489] N. Nagi, A. Chatterjee, A. Balivada, and J. A. Abraham, "Efficient Multisine Testing of Analog Circuits," in *Proc. of the 8th International Conf. on VLSI Design*, Jan. 1995, pp. 234–238.

[490] R. Nair, "Comments on 'An Optimal Algorithm for Testing Stuck-at Faults in Random-Access Memories'," *IEEE Trans. on Computers*, vol. C-28, no. 3, pp. 258–261, Mar. 1979.

[491] R. Nair, S. M. Thatte, and J. A. Abraham, "Efficient Algorithms for Testing Semiconductor Random-Access Memories," *IEEE Trans. on Computers*, vol. C-28, no. 3, pp. 572–576, Mar. 1978.

[492] S. Naito and M. Tsunoyama, "Fault Detection for Sequential Machines by Transition-Tours," in *Proc. of the 11th International Fault-Tolerant Computing Symp.*, June 1981, pp. 238–243.

[493] G. F. Nelson and W. F. Boggs, "Parametric Tests Meet The Challenge of High-Density ICs," *Electronics*, vol. 48, no. 5, pp. 108–111, Dec. 1975.

[494] M. Nicolaidis, "An Efficient Built-In Self-Test Scheme for Functional Test of Embedded RAMs," in *Proc. of the IEEE Fault Tolerant Computer Systems Conf.*, (University of Michigan, Ann Arbor, MI), 1985, pp. 118–123.

[495] M. Nicolaidis, "Transparent BIST for RAMs," in *Proc. of the International Test Conf.*, Sept. 1992, pp. 598–607.

[496] T. M. Niermann, W.-T. Cheng, and J. H. Patel, "PROOFS: A Fast, Memory-Efficient Sequential Circuit Fault Simulator," *IEEE Trans. on Computer-Aided Design*, vol. 11, no. 2, pp. 198–207, Feb. 1992.

[497] T. M. Niermann and J. H. Patel, "HITEC: A Test Generation Package for Sequential Circuits," in *Proc. of the European Design Automation Conf.*, Feb. 1991, pp. 214–218.

[498] P. Nigh, *Built-In Current Testing*. PhD thesis, ECE Dept., Carnegie Mellon U., Pittsburgh, Pennsylvania, 1990.

[499] P. Nigh and W. Maly, "A Self-Testing ALU Using Built-In Current Sensing," in *Proc. of the Custom Integrated Circuits Conf.*, May 1989, pp. 22.1.1–22.1.4.

[500] P. Nigh and W. Maly, "Test Generation for Current Testing," *IEEE Design & Test of Computers*, vol. 7, no. 1, pp. 26–38, Feb. 1990.

[501] P. Nigh, W. Needham, K. Butler, P. Maxwell, and R. Aitken, "An Experimental Study Comparing the Relative Effectiveness of Functional, Scan, I_{DDQ}, and Delay-Fault Testing," in *Proc. of the 15th VLSI Test Symp.*, April-May 1997, pp. 459–464.

[502] P. Nigh, W. Needham, K. Butler, P. Maxwell, R. Aitken, and W. Maly, "So What Is an Optimal Test Mix? A Discussion of the SEMATECH Methods Experiment," in *Proc. of the International Test Conf.*, Nov. 1997, pp. 1037–1038.

[503] P. Nigh, D. Vallett, A. Patel, J. Wright, F. Motika, D. Forlenza, R. Kurtulik, and W. Chong, "Failure Analysis of Timing and IDDq-only Failures from the SEMATECH Test Methods Experiment," in *Proc. of the International Test Conf.*, Sept. 1999, pp. 1152–1161.

[504] A. Osseiran, editor, *Analog and Mixed-Signal Boundary-Scan A Guide to the IEEE 1149.4 Test Standard*. Boston: Kluwer Academic Publishers, 1999.

[505] C.-Y. Pan and K.-T. Cheng, "Pseudo-Random Testing and Signature Analysis for Mixed-Signal Circuits," in *Proc. of the International Conf. on Computer-Aided Design*, Nov. 1995, pp. 102–107.

[506] C. A. Papachristou and N. B. Saghal, "An Improved Method for Detecting Functional Faults in Random Access Memories," *IEEE Trans. on Computers*, vol. C-34, no. 3, pp. 110–116, Mar. 1985.

[507] E. S. Park and M. R. Mercer, "An Efficient Delay Test Generation System for Combinational Logic Circuits," *IEEE Trans. on Computer-Aided Design*, vol. 11, no. 7, pp. 926–938, July 1992.

[508] K. P. Parker, "Adaptive Random Test Generation," *Journal of Design Automation and Fault-Tolerant Computing*, vol. 1, no. 1, pp. 62–83, Oct. 1976.

[509] K. P. Parker, *Integrating Design and Test: Using CAE Tools for ATE Programming*. Los Alamitos, California: IEEE Computer Society Press, 1987.

[510] K. P. Parker, "Observations on the 1149.x Family of Standards," in *Proc. of the International Test Conf.*, Oct. 1994, p. 1023.

[511] K. P. Parker, *The Boundary-Scan Handbook*. Boston: Kluwer Academic Publishers, second edition, 1998.

[512] K. P. Parker and E. J. McCluskey, "Probabilistic Treatment of General Combinational Networks," *IEEE Trans. on Computers*, vol. C-24, no. 6, pp. 668–670, June 1975.

[513] K. P. Parker, J. E. McDermid, and S. Oresjo, "Structure and Metrology for an Analog Testability Bus," in *Proc. of the International Test Conf.*, Oct. 1993, pp. 309–320.

[514] C. G. Parodi, V. D. Agrawal, M. L. Bushnell, and S. Wu, "A Non-Enumerative Path Delay Fault Simulator for Sequential Circuits," in *Proc. of the International Test Conf.*, Oct. 1998, pp. 934–943.

[515] G. Parthasarathy and M. L. Bushnell, "Simultaneous Delay-Fault Built-In Self-Test and Partial-Scan Insertion." U.S. Provisional Patent App. # 98-0081, Filed March. 2, 1998.

[516] G. Parthasarathy and M. L. Bushnell, "Towards Simultaneous Delay-Fault Built-In Self-Test and Partial-Scan Insertion," in *Proc. of the 16th VLSI Test Symp.*, Apr. 1998, pp. 210–217.

[517] S. Pateras and J. Rajski, "Cube-Contained Random Patterns and their Application to the Complete Testing of Synthesized Multi-Level Circuits," in *Proc. of the International Test Conf.*, Oct. 1991, pp. 473–482.

[518] A. Pawla and R. A. Rohrer, "Band-Faults: Efficient Approximations to Fault Bands for the Simulation before Fault Diagnosis of Linear Circuits," *IEEE Trans. on Circuits and Systems*, vol. 29, no. 2, pp. 81–88, Feb. 1982.

[519] R. J. Perry, "I_{DDQ} Testing in CMOS Digital ASIC's," *Journal of Electronic Testing: Theory and Applications*, vol. 3, no. 4, pp. 317–325, Dec. 1992.

[520] R. J. Perry, "I_{DDQ} Testing in CMOS Digital ASIC's – Putting it All Together," in *Proc. of the International Test Conf.*, Sept. 1992, pp. 151–157.

[521] W. W. Peterson and E. J. Weldon, Jr., *Error-Correcting Codes*. New York: John Wiley & Sons, Inc., 1972.

[522] J. S. Pittman and W. C. Bruce, "Test Logic Economic Considerations in a Commercial VLSI Chip Environment," in *Proc. of the International Test Conf.*, Oct. 1984, pp. 31–39.

[523] J. F. Poage, "Derivation of Optimum Tests to Detect Faults in Combinational Circuits," in *Proc. of the Symp. on Mathematical Theory of Automata*, (New York), Polytechnic Press, Apr. 1963, pp. 483–528.

[524] I. Pomeranz, L. N. Reddy, and S. M. Reddy, "COMPACTEST: A Method to Generate Compact Test Sets for Combinational Circuits," *IEEE Trans. on Computer-Aided Design*, vol. 12, no. 7, pp. 1040–1049, July 1993.

[525] I. Pomeranz and S. M. Reddy, "The Multiple Observation Time Test Strategy," *IEEE Trans. on Computers*, vol. 41, no. 5, pp. 627–637, May 1992.

[526] I. Pomeranz and S. M. Reddy, "An Efficient Nonenumerative Method to Estimate the Path Delay Fault Coverage in Combinational Circuits," *IEEE Trans. on Computer-Aided Design*, vol. 13, no. 2, pp. 240–250, Feb. 1994.

[527] I. Pomeranz and S. M. Reddy, "On Identifying Undetectable and Redundant Faults in Synchronous Sequential Circuits," in *Proc. of the 12th VLSI Test Symp.*, Apr. 1994, pp. 8–14.

[528] D. K. Pradhan and K. Son, "The Effect of Untestable Faults in PLAs and a Design for Testability," in *Proc. of the International Test Conf.*, Nov. 1980, pp. 359–367.

[529] R. P. Prasad, *Surface Mount Technology Principles and Practice*. New York: Van Nostrand Reinhold, 1989.

[530] G. R. Putzolu and T. P. Roth, "A Heuristic Algorithm for the Testing of Asynchronous Circuits," *IEEE Trans. on Computers*, vol. C-20, no. 6, pp. 639–647, June 1971.

[531] J. Rajski and H. Cox, "A Method to Calculate Necessary Assignments in Algorithmic Test Pattern Generation," in *Proc. of the International Test Conf.*, Aug. 1990, pp. 25–34.

[532] J. Rajski and J. Tyszer, *Arithmetic Built-In Self-Test for Embedded Systems*. Upper Saddle River, New Jersey: Prentice-Hall, 1998.

[533] R. Ramadoss and M. Bushnell, "Test Generation for Mixed-Signal Devices using Signal Flow Graphs," in *Proc. of the 9th International Conf. on VLSI Design*, Jan. 1996, pp. 242–248.

[534] R. Ramadoss and M. Bushnell, "Test Generation for Mixed-Signal Devices using Signal Flow Graphs," *Journal of Electronic Testing: Theory and Applications*, vol. 14, no. 3, pp. 189–205, June 1999.

[535] M. J. Raposa, "Dual-Port Static RAM Testing," in *Proc. of the International Test Conf.*, Sept. 1988, pp. 362–368.

[536] I. M. Ratiu, "VICTOR: A Fast VLSI Testability Analysis Program," in *Proc. of the International Test Conf.*, Nov. 1982, pp. 397–401.

[537] B. Rayner, "Market Driven Quality: IBM's Six Sigma Crusade," *Electronic Business*, vol. 16, no. 19, pp. 68–76, Oct. 1990.

[538] M. K. Reddy, S. M. Reddy, and P. Agrawal, "Transistor Level Test Generation for MOS Circuits," in *Proc. of the 22nd Design Automation Conf.*, June 1985, pp. 825–828.

[539] S. M. Reddy, C. J. Lin, and S. Patil, "An Automatic Test Pattern Generator for the Detection of Path Delay Faults," in *Proc. of the International Conf. on Computer-Aided Design*, Nov. 1987, pp. 284–287.

[540] S. M. Reddy, M. K. Reddy, and V. D. Agrawal, "Robust Tests for Stuck-Open Faults in CMOS Combinational Logic Circuits," in *Proc. of the 14th International Fault-Tolerant Computing Symp.*, June 1984, pp. 44–49.

[541] M. Renovell and G. Cambon, "Topology Dependence of Floating Gate Faults in MOS Circuits," *IEEE Electronics Letters*, vol. 22, no. 3, pp. 152–153, Jan. 1986.

[542] A. Richardson, A. Bratt, I. Baturone, and J. L. Huertas, "The Application of I_{DDX} Test Strategies in Analogue and Mixed Signal ICs," in *Proc. of the IEEE International Mixed-Signal Test Workshop*, June 1995, pp. 206–211.

[543] M. J. Riezenman, "Wanlass's CMOS Circuit," *IEEE Spectrum*, vol. 28, no. 5, p. 44, May 1991.

[544] J. Rius and J. Figueras, "Proportional BIC Sensor for Current Testing," *Journal of Electronic Testing: Theory and Applications*, vol. 3, no. 4, pp. 387–396, Dec. 1992.

[545] R. Rodriguez-Montanes, J. A. Segura, V. H. Champac, J. Figueras, and J. A. Rubio, "Current vs. Logic Testing of Gate Oxide Short, Floating Gate and Bridging Failures in CMOS," in *Proc. of the International Test Conf.*, Oct. 1991, pp. 510–519.

[546] A. Rogers, *Statistical Analysis of Spatial Dispersions*. London, United Kingdom: Pion Limited, 1974.

[547] A. L. Rosenblum and M. S. Ghausi, "Multiparameter Sensitivity in Active RC Networks," *IEEE Trans. on Circuit Theory*, vol. CT-18, no. 6, pp. 592–599, Nov. 1971.

[548] E. Rosenfeld, "Accuracy and Repeatability with DSP Test Methods," in *Proc. of the International Test Conf.*, Sept. 1986, pp. 788–795.

[549] E. Rosenfeld, "Timing Generation for DSP Testing," in *Proc. of the International Test Conf.*, Sept. 1988, pp. 755–763.

[550] J. P. Roth, "Diagnosis of Automata Failures: A Calculus and a Method," *IBM Journal of Research and Development*, vol. 10, no. 4, pp. 278–291, July 1966.

[551] J. P. Roth, W. G. Bouricius, and P. R. Schneider, "Programmed Algorithms to Compute Tests to Detect and Distinguish Between Failures in Logic Circuits," *IEEE Trans. on Electronic Computers*, vol. EC-16, no. 5, pp. 567–580, Oct. 1967.

[552] J. S. Roychowdhury and R. C. Melville, "Homotopy Techniques for Obtaining a DC Solution of Large-Scale MOS Circuits," in *Proc. of the 33rd Design Automation Conf.*, June 1996, pp. 286–291.

[553] E. M. Rudnick, J. H. Patel, G. S. Greenstein, and T. M. Niermann, "A Genetic Algorithm Framework for Test Generation," *IEEE Trans. on Computer-Aided Design*, vol. 16, no. 9, pp. 1034–1044, Sept. 1997.

[554] R. Russell, "A Method of Extending an 1149.1 Bus for Mixed-Signal Testing," in *Proc. of the International Test Conf.*, Oct. 1996, pp. 410–416.

[555] R. A. Rutman, "Fault-Detection Test Generation for Sequential Logic by Heuristic Tree Search," Technical Report TP-72-11-4, U. of Southern California, Dept. of EE-Systems, Los Angeles, California, Feb. 1972.

[556] D. G. Saab, Y. G. Saab, and J. A. Abraham, "Automatic Test Vector Cultivation for Sequential VLSI Circuits Using Genetic Algorithms," *IEEE Trans. on Computer-Aided Design*, vol. 15, no. 10, pp. 1278–1285, Oct. 1996.

[557] M. Sachdev, *Defect Oriented Testing for CMOS Analog and Digital Circuits*. Boston: Kluwer Academic Publishers, 1998.

[558] K. K. Saluja and K. Kinoshita, "Test Pattern Generation for API Faults in RAM," *IEEE Trans. on Computers*, vol. C-34, no. 3, pp. 284–287, Mar. 1985.

[559] K. K. Saluja, S. H. Sng, and K. Kinoshita, "Built-In Self-Testing RAM: A Practical Alternative," *IEEE Design & Test of Computers*, vol. 4, no. 1, pp. 42–51, Feb. 1987.

[560] P. A. Samuelson, *Economics, An Introductory Analysis*. New York: McGraw-Hill, 1976.

[561] Y. Savaria, M. Yousef, B. Kaminska, and M. Koudil, "Automatic Test Point Insertion for Pseudo-Random Testing," in *Proc. of the International Symp. on Circuits and Systems*, June 1991, pp. 1960–1963.

[562] J. Savir, "Syndrome-Testable Design of Combinational Circuits," *IEEE Trans. on Computers*, vol. C-29, no. 6, pp. 442–451, June 1980.

[563] J. Savir, "Good Controllability and Good Observability do not Guarantee Good Testability," *IEEE Trans. on Computers*, vol. C-32, pp. 1198–1200, Dec. 1983.

[564] J. Savir, "Skewed-Load Transition Test: Part I, Calculus," in *Proc. of the International Test Conf.*, Oct. 1992, pp. 705–713.

[565] J. Savir, "Skewed-Load Transition Test: Part II, Coverage," in *Proc. of the International Test Conf.*, Oct. 1992, pp. 714–722.

[566] J. Savir, "On Broad-Side Delay Testing," in *Proc. of the 12th VLSI Test Symp.*, Apr. 1994, pp. 284–290.

[567] J. Savir and P. H. Bardell, "On Random Pattern Test Length," *IEEE Trans. on Computers*, vol. C-33, no. 6, pp. 467–474, June 1984.

[568] J. Savir, G. S. Ditlow, and P. H. Bardell, "Random Pattern Testability," *IEEE Trans. on Computers*, vol. C-33, no. 1, pp. 79–90, Jan. 1984.

[569] D. R. Schertz and G. Metze, "A New Representation for Faults in Combinational Digital Circuits," *IEEE Trans. on Computers*, vol. C-1, no. 8, pp. 858–866, Aug. 1972.

[570] P. R. Schneider, "On the Necessity to Examine D-Chains in Diagnostic Test Generation – An Example," *IBM Journal of Research and Development*, vol. 10, no. 1, p. 114, Jan. 1967.

[571] H. D. Schnurmann, E. Lindbloom, and R. G. Carpenter, "The Weighted Random Test Pattern Generator," *IEEE Trans. on Computers*, vol. C-24, no. 7, pp. 695–700, July 1975.

[572] L. Schrage, *LINDO User's Manual*, 4th edition. Linear and Quadratic Programming with LINDO.

[573] D. M. Schuler, E. G. Ulrich, T. E. Baker, and S. P. Bryant, "Random Test Generation Using Concurrent Logic Simulation," in *Proc. of the 12th Design Automation Conf.*, June 1975, pp. 261–267.

[574] M. H. Schulz and E. Auth, "Advanced Automatic Test Pattern Generation and Redundancy Identification Techniques," in *Proc. of the International Fault-Tolerant Computing Symp.*, June 1988, pp. 30–35.

[575] M. H. Schulz and E. Auth, "ESSENTIAL: An Efficient Self-Learning Test Pattern Generation Algorithm for Sequential Circuits," in *Proc. of the International Test Conf.*, Aug. 1989, pp. 28–37.

[576] M. H. Schulz and E. Auth, "Improved Deterministic Test Pattern Generation with Applications to Redundancy Identification," *IEEE Trans. on Computer-Aided Design*, vol. 8, no. 7, pp. 811–816, July 1989.

[577] M. H. Schulz, E. Trischler, and T. M. Serfert, "SOCRATES: A Highly Efficient Automatic Test Pattern Generation System," *IEEE Trans. on Computer-Aided Design*, vol. CAD-7, no. 1, pp. 126–137, Jan. 1988.

[578] M. D. Schuster and R. E. Bryant, "Concurrent Fault Simulation of Digital MOS Circuits," in P. Penfield, Jr., editor, *Proc. of the Conf. on Advanced Research in VLSI*, Jan. 1984, pp. 129–138.

[579] J. Segura and A. Rubio, "GOS Defects in SRAM: Fault Modeling and Testing Possibilites," in *IEEE Int'l. Workshop on Memory Technology, Design, and Testing*, Aug. 1994, pp. 66–71.

[580] J. Segura and A. Rubio, "A Detailed Analysis of CMOS SRAMs with Gate Oxide Short Defects," *IEEE Journal of Solid-State Circuits*, vol. 32, no. 10, pp. 1543–1550, Oct. 1997.

[581] B. H. Seiss, P. M. Trouborst, and M. H. Schulz, "Test Point Insertion for Scan-Based BIST," in *Proc. of the European Test Conf.*, Apr. 1991, pp. 253–262.

[582] F. F. Sellers, M. Y. Hsiao, and L. W. Bearnson, "Analyzing Errors with the Boolean Difference," *IEEE Trans. on Computers*, vol. C-17, no. 7, pp. 676–683, July 1968.

[583] F. F. Sellers Jr., M.-Y. Hsiao, and L. W. Bearnson, *Error Detecting Logic for Digital Computers*. New York: McGraw-Hill, 1968.

[584] Semiconductor Industry Assoc., *The National Technology Roadmap for Semiconductors*. 181 Metro Drive, Ste. 450, San Jose, California 95110: Semiconductor Industry Association, 1997.

[585] M. Serra, T. Slater, J. C. Muzio, and D. M. Miller, "The Analysis of One-Dimensional Linear Cellular Automata and their Aliasing Properties," *IEEE Trans. on Computer-Aided Design*, vol. CAD-9, no. 7, pp. 767–778, July 1990.

[586] S. Seshu, "On an Improved Diagnosis Program," *IEEE Trans. on Electronic Computers*, vol. EC-14, no. 1, pp. 76–79, Feb. 1965.

[587] S. Seshu and D. N. Freeman, "The Diagnosis of Asynchronous Sequential Switching Systems," *IRE Trans. on Electronic Computers*, vol. EC-11, Aug. 1962.

[588] S. C. Seth and V. D. Agrawal, "Characterizing the LSI Yield from Chip Test Data," in *Proc. of the International Conf. on Circuits and Computers*, Sept. 1982, pp. 556–559.

[589] S. C. Seth and V. D. Agrawal, "Characterizing the LSI Yield from Wafer Test Data," *IEEE Trans. on Computer-Aided Design*, vol. CAD-3, no. 2, pp. 123–126, Apr. 1984.

[590] S. C. Seth and V. D. Agrawal, "A New Model for Computation of Probabilistic Testability in Combinational Circuits," *Integration, the VLSI Journal*, vol. 7, no. 1, pp. 49–75, Apr. 1989.

[591] S. C. Seth, V. D. Agrawal, and H. Farhat, "A Statistical Theory of Digital Circuit Testability," *IEEE Trans. on Computers*, vol. C-39, no. 4, pp. 582–586, Apr. 1990.

[592] S. C. Seth, L. Pan, and V. D. Agrawal, "PREDICT – Probabilistic Estimation of Digital Circuit Testability," in *Proc. of the International Fault-Tolerant Computing Symp.*, June 1985, pp. 220–225.

[593] I. P. Shaik and M. L. Bushnell, "A Graph Approach to DFT Hardware Placement for Robust Delay Fault BIST," in *Proc. of the 8th International Conf. on VLSI Design*, Jan. 1995, pp. 177–182.

[594] I. P. Shaik and M. L. Bushnell, "Circuit Design for Low Overhead Delay Fault BIST Using Constrained Quadratic 0-1 Programming," in *Proc. of the 13th VLSI Test Symp.*, April-May 1995, pp. 393–399.

[595] C. E. Shannon, "Communication in the Presence of Noise," *Proc. of the Institute of Radio Engineers*, vol. 37, no. 1, pp. 10–21, July 1949.

[596] J. P. Shen, W. Maly, and F. J. Ferguson, "Inductive Fault Analysis of MOS Integrated Circuits," *IEEE Design & Test of Computers*, vol. 2, no. 6, pp. 13–26, Dec. 1985.

[597] J. W. Sheppard and W. R. Simpson, *Research Perspectives and Case Studies in System Test and Diagnosis*. Boston: Kluwer Academic Publishers, 1998.

[598] N. A. Sherwani, *Algorithms for VLSI Physical Design Automation*. Boston: Kluwer Academic Publishers, 1993.

[599] N. A. Sherwani, S. Bhingarde, and A. Panyam, *Routing in the Third Dimension*. New York: IEEE Press, 1995.

[600] C.-J. Shi, *Analog and Mixed-Signal Test*, pp. 55–92. Upper Saddle River, New Jersey: Prentice-Hall, 1998. Editor: B. Vinnakota.

[601] C.-J. Shi and M. W. Tian, "Automated Test Generation for Linear(ized) Analog Circuits under Parameter Variations," in *Proc. of the Asian Pacific – Design Automation Conf. '98*, Feb. 1998, pp. 501–506.

[602] C.-J. Shi and M. W. Tian, "Simulation and Sensitivity of Linear Analog Circuits under Parameter Variations by Robust Interval Analysis," *ACM Transactions on Design Automation of Electronic Systems*, vol. 4, no. 5, pp. 280–312, July 1999.

[603] J. Sienicki, *Algorithms and Models for Distributed Test Generation*. PhD thesis, ECE Department, Rutgers University, New Brunswick, New Jersey, Oct. 1995.

[604] D. P. Siewiorek and R. S. Schwartz, *The Theory and Practice of Reliable System Design*. Bedford, Massachusetts: Digital Press, Digital Equipment Corp., 1982.

[605] J. P. M. Silva and K. A. Sakallah, "Grasp – A New Search Algorithm for Satisfiability," in *Proc. of the International Conf. on Computer-Aided Design*, Nov. 1996, pp. 220–227.

[606] W. R. Simpson and J. W. Sheppard, *System Test and Diagnosis*. Boston: Kluwer Academic Publishers, 1994.

[607] G. Singer, "Current Trends and Future Directions in Test and DFT," in *Proc. of the 15th VLSI Test Symp.*, April-May 1997, p. xxx. *Keynote Address*.

[608] M. Sivaraman and A. J. Strojwas, *A Unified Approach for Timing Verification and Delay Fault Testing*. Boston: Kluwer Academic Publisher, 1998.

[609] M. Slamani and B. Kaminska, "Analog Circuit Fault Diagnosis Based on Sensitivity Computation and Fault Testing," *IEEE Design & Test of Computers*, vol. 9, no. 1, pp. 30–39, Mar. 1992.

[610] M. Slamani and B. Kaminska, "Testing Analog Circuits by Sensitivity Computation," in *Proc. of the European Design Automation Conf.*, Feb. 1992, pp. 532–537.

[611] G. L. Smith, "Model for Delay Faults Based Upon Paths," in *Proc. of the International Test Conf.*, Nov. 1985, pp. 342–349.

[612] J. E. Smith, "Measure of the Effectiveness of Fault Signature Analysis," *IEEE Trans. on Computers*, vol. C-29, pp. 510–514, June 1980.

[613] T. J. Snethen, "Simulator-Oriented Fault Test Generator," in *Proc. of the 14th Design Automation Conf.*, June 1977, pp. 88–93.

[614] J. M. Soden, R. R. Fritzemeier, and C. F. Hawkins, "Zero-Defect or Zero Stuck-At Faults – CMOS IC Process Improvement with I_{DDQ}," in *Proc. of the International Test Conf.*, Sept. 1990, pp. 255–256.

[615] J. M. Soden and C. F. Hawkins, "Reliability of CMOS ICs with Gate Oxide Shorts," *Semiconductor International*, vol. 10, no. 6, pp. 240–245, May 1987.

[616] J. M. Soden and C. F. Hawkins, "Electrical Properties and Detection Methods for CMOS IC Defects," in *Proc. of the European Test Conf.*, Apr. 1989, pp. 159–167.

[617] J. M. Soden, C. F. Hawkins, R. K. Gulati, and W. Mao, "IDDQ Testing: A Review," *Journal of Electronic Testing: Theory and Applications*, vol. 3, no. 4, pp. 5–17, Dec. 1992.

[618] J. M. Soden, R. K. Treece, M. R. Taylor, and C. F. Hawkins, "CMOS IC Stuck-Open Fault Electrical Effects and Design Considerations," in *Proc. of the International Test Conf.*, Aug. 1989, pp. 423–430.

[619] M. Soma, "A Design-for-Test Methodology for Active Analog Filters," in *Proc. of the International Test Conf.*, Sept. 1990, pp. 183–192.

[620] M. Soma, "Fault Coverage of DC Parametric Tests for Embedded Analog Amplifiers," in *Proc. of the International Test Conf.*, Oct. 1993, pp. 566–573.

[621] M. Soma, "Automatic Test Generation Algorithms for Analogue Circuits," *IEE Proceedings G: Circuits and Devices*, vol. 143, no. 6, pp. 366–373, Dec. 1996.

[622] M. Soma and V. Kolarik, "A Design-for-Test Technique for Switched-Capacitor Filters," in *Proc. of the 12th VLSI Test Symp.*, Apr. 1994, pp. 42–47.

[623] S. S. Somayajula, E. Sanchez-Sinencio, and J. P. de Gyvez, "A Power Supply Ramping and Current Measurement Based Technique for Analog Fault Diagnosis," in *Proc. of the 12th VLSI Test Symp.*, Apr. 1994, pp. 234–239.

[624] S. S. Somayajula, E. Sanchez-Sinencio, and J. P. de Gyvez, "Analog Fault Diagnosis Based on Ramping Power Supply Current Signature Clusters," *IEEE Trans. on Circuits and Systems*, vol. 43, no. 10, pp. 703–712, Oct. 1996. Part II.

[625] K. Son, "Fault Simulation with the Parallel Value List Algorithm," *VLSI Systems Design*, vol. 6, no. 12, pp. 36–43, Dec. 1985.

[626] T. M. Souders and D. R. Flach, "An NBS Calibration Service for A/D and D/A Converters," in *Proc. of the International Test Conf.*, Oct. 1981, pp. 290–303. Electrosystems Division, National Institute of Standards, Washington, DC 20234.

[627] T. M. Souders and G. N. Stenbakken, "A Comprehensive Approach for Modeling and Testing Analog and Mixed-Signal Devices," in *Proc. of the International Test Conf.*, Sept. 1990, pp. 169–176.

[628] T. M. Souders and G. N. Stenbakken, "Cutting the High Cost of Testing," *IEEE Spectrum*, vol. 28, no. 3, pp. 48–51, Mar. 1991.

[629] S. J. Spinks and I. M. Bell, "A Comparison of Relative Accuracy of Fault Coverage Analysis Techniques Based on Analogue Fault Simulation," in *Proc. of the IEEE International Mixed-Signal Test Workshop*, 1996, pp. 17–22.

[630] T. Sridhar, D. S. Ho, T. J. Powell, and S. M. Thatte, "Analysis and Simulation of Parallel Signature Analyzers," in *Proc. of the International Test Conf.*, Nov. 1982, pp. 656–661.

[631] M. Srinivas and L. M. Patnaik, "A Simulation-Based Test Generation Scheme Using Genetic Algorithms," in *Proc. of the 6th International Conf. on VLSI Design*, Jan. 1993, pp. 132–135.

[632] M. K. Srinivas, M. L. Bushnell, and V. D. Agrawal, "Flags and Algebra for Sequential Circuit VNR Path Delay Fault Test Generation," in *Proc. of the 10th International Conf. on VLSI Design*, Jan. 1997, pp. 88–94.

[633] W. Stahnke, "Primitive Binary Polynomials," *Mathematical Computation*, vol. 27, no. 124, pp. 977–980, 1973.

[634] K. Stalnaker, "Practical Test Methods for Verification of the EDRAM," in *Proc. of the International Test Conf.*, Oct. 1994, p. 362.

[635] T. Stanion and D. Bhattacharya, "TSUNAMI: A Path Oriented Scheme for Algebraic Test Generation," in *Proc. of the International Fault-Tolerant Computing Symp.*, June 1991, pp. 36–43.

[636] C. H. Stapper, "On Yield, Fault Distributions, and Clustering of Particles," *IBM Journal of Research and Development*, vol. 30, no. 3, pp. 326–338, May 1986.

[637] G. N. Stenbakken and T. M. Souders, "Linear Error Modeling of Analog and Mixed-Signal Devices," in *Proc. of the International Test Conf.*, Oct. 1991, pp. 573–581.

[638] P. Stephan, R. K. Brayton, and A. L. Sangiovanni-Vincentelli, "Combinational Test Generation Using Satisfiability," *IEEE Trans. on Computer-Aided Design*, vol. 15, no. 9, pp. 1167–1176, Sept. 1996.

[639] A. K. Stevens, *Introduction to Component Testing*. Reading, Massachusetts: Addison-Wesley, 1986.

[640] S. N. Stevens and P.-M. Lin, "Analysis of Piecewise-Linear Resistive Networks Using Complementary Pivot Theory," *IEEE Trans. on Circuits and Systems*, vol. 28, no. 5, pp. 429–441, May 1981.

[641] S. Stoica, "Generating Functional Design Verification Tests," *IEEE Design & Test of Computers*, vol. 16, no. 3, pp. 53–63, July-Sept. 1999.

[642] A. C. Stover, *ATE: Automatic Test Equipment*. New York: McGraw-Hill, 1984.

[643] S. Su and R. Z. Makki, "Testing of Static Random-Access Memories by Monitoring Dynamic Power Supply Current," *Journal of Electronic Testing: Theory and Applications*, vol. 3, no. 3, pp. 265–278, Aug. 1992.

[644] D. S. Suk and S. M. Reddy, "An Algorithm to Detect a Class of Pattern Sensitive Faults in Semiconductor Random Access Memories," in *Proc. of the International Fault-Tolerant Computing Symp.*, 1979, pp. 219–225.

[645] D. S. Suk and S. M. Reddy, "Test Procedures for a Class of Pattern-Sensitive Faults in Semiconductor Random-Access Memories," *IEEE Trans. on Computers*, vol. C-29, no. 6, pp. 419–429, June 1980.

[646] D. S. Suk and S. M. Reddy, "A March Test for Functional Faults in Semiconductor Random-Access Memories," *IEEE Trans. on Computers*, vol. C-30, no. 12, pp. 982–985, Dec. 1981.

[647] T. G. Szymanski, "LEADOUT: A Static Timing Analyzer for MOS Circuits," in *Proc. of the International Conf. on Computer-Aided Design*, Nov. 1986, pp. 130–133.

[648] P. Tafertshofer, A. Ganz, and M. Henftling, "A SAT-Based Implication Engine for Efficient ATPG, Equivalence Checking, and Optimization of Netlists," in *Proc. of the International Conf. on Computer-Aided Design*, Nov. 1997, pp. 648–655.

[649] Y. Takamatsu and K. Kinoshita, "Extended Selection of Switching Target Faults in CONT Algorithm for Test Generation," *Journal of Electronic Testing: Theory and Applications*, vol. 1, no. 3, pp. 183–189, Oct. 1990.

[650] N. Tamarapalli and J. Rajski, "Constructive Multi-Phase Test Point Insertion for Scan-Based BIST," in *Proc. of the International Test Conf.*, Oct. 1996, pp. 649–658.

[651] S. Tani, M. Teramoto, T. Fukazawa, and K. Matsuhiro, "Efficient Path Selection for Delay Testing Based on Path Clustering," *Journal of Electronic Testing: Theory and Applications*, vol. 15, no. 1/2, pp. 75–85, Aug. 1999.

[652] Y. Taur, "The Incredible Shrinking Transistor," *IEEE Spectrum*, vol. 36, no. 7, pp. 25–29, July 1999.

[653] G. C. Temes, "Efficient Methods of Fault Simulation," in *Proc. of the 20th Midwest Symp. on Circuits and Systems*, Aug. 1977, pp. 191–194.

[654] N. Tendolkar, R. Molyneaux, C. Pyron, and R. Raina, "At-Speed Testing of Delay Faults for Motorola's MPC7400, a PowerPC Microprocessor," in *Proc. of the 18th VLSI Test Symp.*, Apr.-May 2000, pp. 3–8.

[655] P. A. Thaker, *Register-Transfer Level Fault Modeling and Test Evaluation Technique for VLSI Circuits*. PhD thesis, George Washington University, Washington, D.C., May 2000.

[656] P. A. Thaker, V. D. Agrawal, and M. E. Zaghloul, "Validation Vector Grade (VVG): A New Coverage Metric for Validation and Test," in *Proc. of the 17th VLSI Test Symp.*, Apr. 1999, pp. 182–188.

[657] P. A. Thaker, V. D. Agrawal, and M. E. Zaghloul, "Register-Transfer Level Fault Modeling and Test Evaluation Technique for VLSI Circuits," in *Proc. of the International Test Conf.*, Oct. 2000.

[658] C. W. Thatcher and R. E. Tulloss, "Towards a Test Standard for Board and System Level Mixed-Signal Interconnects," in *Proc. of the International Test Conf.*, Oct. 1993, pp. 300–308.

[659] S. M. Thatte and J. A. Abraham, "Testing of Semiconductor Random Access Memories," in *Proc. of the International Fault-Tolerant Computing Symp.*, June 1977, pp. 81–87.

[660] S. M. Thatte and J. A. Abraham, "Test Generation for Microprocessors," *IEEE Trans. on Computers*, vol. C-29, no. 6, pp. 429–441, June 1980.

[661] C. Thibeault, "Detection and Location of Faults and Defects Using Digital Signal Processing," in *Proc. of the 13th VLSI Test Symp.*, Apr. 1995, pp. 262–267.

[662] C. Thibeault, "A Novel Probabilistic Approach for IC Diagnosis Based on Differential Quiescent Current Signatures," in *Proc. of the 15th VLSI Test Symp.*, April-May 1997, pp. 80–85.

[663] C. Thibeault, "Increasing Current Testing Resolution," in *Proc. of the IEEE International Symp. on Defect and Fault Tolerance in VLSI Systems*, 1998, pp. 126–134.

[664] C. Thibeault, "A Histogram Based Procedure for Current Testing of Active Defects," in *Proc. of the International Test Conf.*, Sept. 1999, pp. 714–723.

[665] C. Thibeault, "On the Comparison of ΔI_{DDQ} and I_{DDQ} Testing," in *Proc. of the 17th VLSI Test Symp.*, Apr. 1999, pp. 143–150.

[666] C. Thibeault and L. Boisvert, "Diagnosis Method Based on Delta I_{DDQ} Probabilistic Signatures: Experimental Results," in *Proc. of the International Test Conf.*, Oct. 1998, pp. 1019–1026.

[667] D. E. Thomas and P. R. Moorby, *The Verilog Hardware Description Language*. Boston: Kluwer Academic Publishers, second edition, 1995.

[668] E. W. Thompson and S. A. Szygenda, "Digital Logic Simulation in a Time-Based Table-Driven Environment; Part 2. Parallel Fault Simulation," *Computer*, vol. 8, no. 3, pp. 38–49, Mar. 1975.

[669] M. W. Tian and C.-J. Shi, "Rapid Frequency-Domain Analog Fault Simulation under Parameter Tolerances," in *Proc. of the 34th Design Automation Conf.*, June 1997, pp. 275–280.

[670] M. W. Tian and C.-J. Shi, "Efficient DC Fault Simulation of Nonlinear Analog Circuits," in *Proc. of the Design Automation and Test in Europe Conf.*, Feb. 1998, pp. 899–904.

[671] M. W. Tian and C.-J. Shi, "Nonlinear DC-Fault Simulation by One-Step Relaxation – Linear Circuit Models are Sufficient for Nonlinear DC-Fault Simulation," in *Proc. of the 16th VLSI Test Symp.*, Apr. 1998, pp. 126–131.

[672] Y. Tokunaga and J. Frosien, "High Performance Electron Beam Tester for Voltage Measurement on Unpassivated and Passivated Devices," in *Proc. of the International Test Conf.*, Aug. 1989, pp. 917–922.

[673] K. A. E. Totton, "Review of Built-In Self-Test Methodologies for Gate Arrays," *IEE Proceedings*, vol. 132, no. 2, pp. 121–129, March/April 1985. Parts E&I.

[674] N. A. Touba and E. J. McCluskey, "Automated Logic Synthesis of Random Pattern Testable Circuits," in *Proc. of the International Test Conf.*, Oct. 1994, pp. 174–183.

[675] N. A. Touba and E. J. McCluskey, "Transformed Pseudo-Random Patterns for BIST," in *Proc. of the 13th VLSI Test Symp.*, April-May 1995, pp. 410–416.

[676] N. A. Touba and E. J. McCluskey, "Test Point Insertion Based on Path Tracing," in *Proc. of the 14th VLSI Test Symp.*, April-May 1996, pp. 2–8.

[677] K. S. Trivedi, *Probability and Statistics with Reliability, Queuing and Computer Science Applications*. Englewood-Cliffs, New Jersey: Prentice-Hall, 1982.

[678] K.-H. Tsai, S. Hellebrand, J. Rajski, and M. Marek-Sadowska, "STARBIST: Scan Autocorrelated Random Pattern Generation," in *Proc. of the 34th Design Automation Conf.*, June 1997, pp. 472–477.

[679] S.-J. Tsai, "Test Vector Generation for Linear Analog Devices," in *Proc. of the International Test Conf.*, Oct. 1991, pp. 592–597.

[680] R. R. Tummala and E. J. Rymaszewski, editors, *Microelectronics Packaging Handbook*. New York: Van Nostrand Reinhold, 1989. Page 677.

[681] J. G. Udell Jr., "Test Set Generation for Pseudo-Exhaustive BIST," in *Proc. of the International Conf. on Computer-Aided Design*, Nov. 1986, pp. 52–55.

[682] J. G. Udell Jr. and E. J. McCluskey, "Partial Hardware Partitioning: A New Pseudo-Exhaustive Test Implementation," in *Proc. of the International Test Conf.*, Sept. 1988, p. 1000.

[683] E. G. Ulrich, "Exclusive Simulation of Activity in Digital Networks," *Communications of the ACM*, vol. 12, no. 2, pp. 102–110, Feb. 1969.

[684] E. G. Ulrich, V. D. Agrawal, and J. H. Arabian, *Concurrent and Comparative Discrete Event Simulation*. Boston: Kluwer Academic Publishers, 1994.

[685] E. G. Ulrich and T. Baker, "Concurrent Simulation of Nearly Identical Digital Networks," *Computer*, vol. 7, pp. 39–44, Apr. 1974.

[686] S. H. Unger, *Asynchronous Sequential Switching Circuits*. New York: Wiley-Interscience, 1969.

[687] D. van Campenhout, H. Al-Asaad, J. P. Hayes, T. Mudge, and R. B. Brown, "High-Level Design Verification of Microprocessors via Error Modeling," *ACM Trans. on Design Automation of Electronic Systems (TODAES)*, vol. 3, pp. 581–599, Oct. 1998.

[688] A. J. van de Goor, *Testing Semiconductor Memories: Theory and Practice*. Chichester, UK: John Wiley & Sons, Inc., 1991.

[689] A. J. van de Goor and C. A. Verruijt, "An Overview of Deterministic Functional RAM Chip Testing," *ACM Computing Surveys*, vol. 22, no. 1, pp. 5–33, Mar. 1990.

[690] J. T. van der Linden, *Automatic Test Pattern Generation of Three-State Circuits*. PhD thesis, Technical University of Delft, Delft, The Netherlands, May 1996.

[691] J. van Sas, F. Catthoor, and H. D. Man, "Cellular Automata-Based Self-Test for Programmable Data Paths," in *Proc. of the International Test Conf.*, Sept. 1990, pp. 769–778.

[692] L. Vandenberghe, B. L. D. Moor, and J. Vandewalle, "The Generalized Linear Complementarity Problem Applied to the Complete Analysis of Resistive Piecewise-Linear Circuits," *IEEE Trans. on Circuits and Systems*, vol. 367, no. 11, pp. 1382–1391, Nov. 1989.

[693] D. Vazquez, J.-L. Huertas, and A. Rueda, "Reducing the Impact of DFT on the Performance of Analog Integrated Circuits: Improved sw-OPAMP Design," in *Proc. of the 14th VLSI Test Symp.*, April-May 1996, pp. 42–47.

[694] D. Vazquez, A. Rueda, and J.-L. Huertas, "A New Strategy for Testing Analog Filters," in *Proc. of the 12th VLSI Test Symp.*, Apr. 1994, pp. 36–41.

[695] D. Vazquez, A. Rueda, and J.-L. Huertas, "High Q Bandpass SC Filter with Enhanced Testability," *IEEE Journal of Solid-State Circuits*, vol. 33, no. 7, pp. 976–986, July 1998.

[696] D. Vazquez, A. Rueda, J.-L. Huertas, and A. M. D. Richardson, "Practical DFT Strategy for Fault Diagnosis in Active Analog Filters," *IEEE Electronics Letters*, vol. 31, no. 15, pp. 1221–1222, July 1995.

[697] P. K. Veenstra, F. P. M. Beenker, and J. J. M. Koomen, "Testing of Random Access Memories: Theory and Practice," *IEE Proceedings G*, vol. 135, no. 1, pp. 24–28, Feb. 1988.

[698] J. Villoldo, P. Agrawal, and V. D. Agrawal, "STAFAN Algorithms for MOS Circuits," in *Proc. of the International Conf. on Computer Design*, Oct. 1991, pp. 56–59.

[699] B. Vinnakota, editor, *Analog and Mixed-Signal Test*. Upper Saddle River, New Jersey: Prentice-Hall, 1998.

[700] R. S. Vogelsong, "Trade-offs in Analog Behavioral Model Development: Managing Accuracy and Efficiency," in *Proc. of the IEEE/VIUF International Workshop on Behavioral Modeling and Simulation*, (Washington, D.C.), Oct. 1997, pp. 33–40.

[701] R. L. Wadsack, "Fault Modeling and Logic Simulation of CMOS and MOS Integrated Circuits," *Bell System Technical Journal*, vol. 57, no. 5, pp. 1449–1474, May-June 1978.

[702] K. D. Wagner, C. K. Chin, and E. J. McCluskey, "Pseudorandom Testing," *IEEE Trans. on Computers*, vol. C-36, no. 3, pp. 332–343, Mar. 1987.

[703] K. D. Wagner and T. W. Williams, "Design for Testability of Mixed Signal Integrated Circuits," in *Proc. of the International Test Conf.*, Sept. 1988, pp. 823–828.

[704] J. A. Waicukauski, E. B. Eichelberger, D. O. Forlenza, E. Lindbloom, and T. McCarthy, "Fast Simulation for Structured VLSI," *VLSI Systems Design*, vol. 6, no. 12, pp. 20–32, Dec. 1985.

[705] J. A. Waicukauski and E. Lindbloom, "Fault Detection Effectiveness of Weighted Random Patterns," in *Proc. of the International Test Conf.*, Sept. 1988, pp. 245–255.

[706] J. A. Waicukauski, E. Lindbloom, E. B. Eichelberger, and O. P. Forlenza, "WRP: A Method for Generating Weighted Random Test Patterns," *IBM Journal of Research and Development*, vol. 33, no. 2, pp. 149–161, Mar. 1989.

[707] J. A. Waicukauski, E. Lindbloom, B. K. Rosen, and V. S. Iyengar, "Transition Fault Simulation," *IEEE Design & Test of Computers*, vol. 4, no. 2, pp. 32–38, Apr. 1987.

[708] J. A. Waicukauski, P. A. Shupe, D. J. Giramma, and A. Matin, "ATPG for Ultra-Large Structured Designs," in *Proc. of the International Test Conf.*, Sept. 1990, pp. 44–51.

[709] A. Wald, *Sequential Analysis*. New York: Dover Publications, Inc., 1973.

[710] D. M. H. Walker, *Yield Simulation for Integrated Circuits*. Boston: Kluwer Academic Publishers, 1987.

[711] H. Walker and S. Director, "VLASIC: A Catastrophic Fault Yield Simulator for Integrated Circuits," *IEEE Trans. on Computer-Aided Design*, vol. CAD-5, no. 4, pp. 463–466, Oct. 1986.

[712] K. M. Wallquist, A. W. Righter, and C. F. Hawkins, "Implementation of a Voltage Decay Method for I_{DDQ} Measurement on the HP 82000," June 1992. Hewlett-Packard User Group Meeting.

[713] L.-C. Wang and M. S. Abadir, "Test Generation Based on High-Level Assertion Specification for PowerPC^{TM} Microprocessor Embedded Arrays," *Journal of Electronic Testing: Theory and Applications*, vol. 13, pp. 121–135, Oct. 1998.

[714] F. M. Wanlass and C. T. Sah, "Nanowatt Logic Using Field-Effect Metal-Oxide Semiconductor Triodes," in *Proc. of the Solid State Circuits Conf.*, Feb. 1963, pp. 32–33.

[715] R. C. Weast, editor, *Handbook of Chemistry and Physics*. Cleveland, OH: Chemical Rubber Co., 49th edition, 1968-69.

[716] E. Wehrhahn, "Hierarchical Circuit Analysis," in *Proc. of the International Symp. on Circuits and Systems*, May 1989, pp. 701–704.

[717] R. S. Wei and A. Sangiovanni-Vincentelli, "PROTEUS: A Logic Verification System for Combinational Circuits," in *Proc. of the International Test Conf.*, Oct. 1986, pp. 350–359.

[718] N. H. E. Weste and K. Eshraghian, *Principles of CMOS VLSI Design: A Systems Perspective*. Reading, Massachusetts: Addison-Wesley, second edition, 1993.

[719] L. Whetsel, "Proposal to Simplify Development of a Mixed-Signal Test Standard," in *Proc. of the International Test Conf.*, Oct. 1996, pp. 400–409.

[720] M. V. Wilkes, *A Short Introduction to Numerical Analysis*. London, UK: Cambridge University Press, 1966.

[721] B. R. Wilkins, *Testing Digital Circuits, An Introduction*. Berkshire, UK: Van Nostrand Reinhold, 1986.

[722] M. J. Y. Willaims and J. B. Angell, "Enhancing Testability of Large-Scale Integrated Circuits via Test Points and Additional Logic," *IEEE Trans. on Computers*, vol. C-22, no. 1, pp. 46–60, jan 1973.

[723] T. W. Williams, "Test Length in Self-Testing Environment," *IEEE Design & Test of Computers*, vol. 2, pp. 59–63, Apr. 1985.

[724] T. W. Williams, editor, *VLSI Testing*. Amsterdam, The Netherlands: North-Holland, 1986.

[725] T. W. Williams and N. C. Brown, "Defect Level as a Function of Fault Coverage," *IEEE Trans. on Computers*, vol. C-30, no. 12, pp. 987–988, Dec. 1981.

[726] T. W. Williams, W. Daehn, M. Gruetzner, and C. W. Starke, "Comparison of Aliasing Errors for Primitive and Non-Primitive Polynomials," in *Proc. of the International Test Conf.*, Sept. 1986, pp. 282–288.

[727] T. W. Williams, W. Daehn, M. Gruetzner, and C. W. Starke, "Aliasing Errors in Signature Analysis Registers," *IEEE Design & Test of Computers*, vol. 4, no. 2, pp. 39–45, Apr. 1987.

[728] T. W. Williams, W. Daehn, M. Gruetzner, and C. W. Starke, "Aliasing Errors with Primitive and Non-Primitive Polynomials," in *Proc. of the International Test Conf.*, Sept. 1987, pp. 637–644.

[729] T. W. Williams, W. Daehn, M. Gruetzner, and C. W. Starke, "Bounds on Aliasing Errors in Linear Feedback Shift Registers," in *Proc. of the CompEuro Conf.*, May 1987, pp. 373–377.

[730] T. W. Williams, W. Daehn, M. Gruetzner, and C. W. Starke, "Bounds and Analysis of Aliasing Errors in Linear-Feedback Shift-Registers," *IEEE Trans. on Computer-Aided Design*, vol. CAD-7, no. 1, pp. 75–83, Jan. 1988.

[731] T. W. Williams, R. H. Dennard, B. Kapur, M. R. Mercer, and W. Maly, "I_{DDQ} Test: Sensitivity Analysis of Scaling," in *Proc. of the International Test Conf.*, Oct. 1996, pp. 786–792.

[732] T. W. Williams and R. Mercer, "Techniques for Designing More Testable Logic Networks," June 1990. Tutorial presented at the *1990 ACM/IEEE 27th* Design Automation Conf. .

[733] T. W. Williams and K. Parker, "Design for Testability – A Survey," *IEEE Trans. on Computers*, vol. C-31, no. 1, pp. 2–15, Jan. 1982.

[734] T. W. Williams and K. P. Parker, "Design for Testability – A Survey," *Proc. of the IEEE*, vol. 71, no. 1, pp. 98–112, Jan. 1983.

[735] T. W. Williams, R. C. Walther, P. S. Bottorff, and S. D. Gupta, "Experiment to Investigate Self Testing Techniques in VLSI," *IEE Proceedings*, vol. 132, no. 3, pp. 105–107, June 1985. Part G.

[736] S. Winegarden and D. Pannell, "Paragons for Memory Test," in *Proc. of the International Test Conf.*, Oct. 1981, pp. 44–48.

[737] D. M. Wolf and S. R. Sanders, "Multiparameter Homotopy Methods for Finding DC Operating Points of Nonlinear Circuits," *IEEE Trans. on Circuits and Systems*, pp. 824–838, 1996. Part I.

[738] S. Wolfram, "Statistical Mechanics of Cellular Automata," *Review of Modern Physics*, vol. 55, no. 3, pp. 601–644, 1983.

[739] H.-J. Wunderlich, "On Computing Optimized Input Probabilities for Random Tests," in *Proc. of the 24th Design Automation Conf.*, June-July 1987, pp. 392–398.

[740] H.-J. Wunderlich, "Multiple Distribution for Biased Random Test Patterns," in *Proc. of the International Test Conf.*, Sept. 1988, pp. 236–244.

[741] H.-J. Wunderlich and G. Kiefer, "Bit-Flipping BIST," in *Proc. of the International Conf. on Computer-Aided Design*, Nov. 1996, pp. 337–343.

[742] A. Yamazaki, K. Dosaka, T. Ogawa, M. Kuroiwa, H. Fukuda, G. Johnson, and M. Kumanoga, "Concurrent Operating CDRAM for Low Cost Multi-Media," in *Symp. on VLSI Circuits, Digest of Technical Papers*, (Kyoto, Japan), IEEE, May 1993, pp. 61–62.

[743] M. Yannakakis, "Personal communication," 1998.

[744] Y. Zorian, editor, *Multi-Chip Module Test Strategies*. Boston: Kluwer Academic Publishers, 1997.

[745] Y. Zorian, "Testing the Monster Chip," *IEEE Spectrum*, vol. 36, no. 7, pp. 54–60, July 1999.

[746] Y. Zorian, E. J. Marinissen, and S. Dey, "Testing Embedded-Core Based System Chips," in *Proc. of the International Test Conf.*, Oct. 1998, pp. 130–143.

INDEX

(n,k) cyclic code, 616
90° phase lag component, 345
A, 258
AC, 243
B, 258
C, 258
$C0$, 130
$C1$, 130
$CC0$, 131
$CC1$, 131
D, 160
$DC0$, 242
$DC1$, 242
D_{AT1}, 581
D_{AT2}, 581
F_s, 337
F_t, 337
I_{DDQ} current limit, 452
I_{DDQ} decision threshold, 455, 456
I_{DDQ} fault, 66
I_{DDQ} testing, 14, 439
I_{DDQ} testing time, 453
I_{DDT} test, 258
I_{defmin}, 459
LU factorization, 396
M, 258
M/N, 338
M/N synchronization, 319
N, 258
N_{max}, 459
$O(n)$ complexity, 263
P, 337
PC, 243
R, 342
$SC0$, 131
$SC1$, 131
SO, 131
$S_{x_i}^{T_j}$, 398
T_S, 504
T_j, 399
T_{aps}, 306
V_G, 583
V_H, 583, 585
V_L, 583, 585
V_L switch, 591
V_{TH}, 581
ΔI_{DDQ} testing, 456
Δ, 316, 339
$\Downarrow$, 262

$\Rightarrow$, 263
$\Uparrow$, 262
$\alpha 0$, 179
$\alpha 1$, 179
α particle, 259, 261
$\beta 0$, 179
$\beta 1$, 179
β, 49
$\downarrow$, 262
$\forall$, 263
λ, 178
μ, 178
μ law ADC, 338
μ law encoder, 361, 367
$\bar{D}$, 160
ϕ, 178
ψ, 178
$\rho_{x_i}^{T_j}$, 398
$\rightarrow$, 262
$\sin(x)/x$ distortion, 354
$\sin(x)/x$, 320
$\uparrow$, 262
$\updownarrow$, 262
$\rightarrow$, 262
d_{seq}, 226
k, 258
n, 258
t-WSF, 536
0-controllability, 130
0-observability, 131
0TLP, 368
1-controllability, 130
1-observability, 131
13-valued algebra, 433
2-SAT, 164
2-coupling fault, 268
3-SAT, 164
41-valued algebra, 434

A law encoder, 361, 367
A/D converter, 322, 338
AB1, 580, 583, 591
AB1N, 589
AB2, 580, 583, 591
AB2N, 589
Abadir, 263, 627
ABF, 258, 271
ABM, 580, 583
ABna, 590

ABnb, 590
ABnN, 589
Abraham, 266, 396
Abramovici, 78, 171, 206, 621
absolute dominator, 197
AC, 36
AC parametric test, 30, 32
AC voltmeter, 346
acceptance test, 17, 48, 596
access time test, 303
activation cost, 243
activation energy, 261
active and passive neighborhood pattern, 288
active and passive neighborhood pattern sensitive fault, 258
active area, 476
active gate, 103
active neighborhood pattern, 288
active neighborhood pattern sensitive fault, 258, 272
activity list, 103
ad-hoc DFT, 466
adaptive GA, 247
ADC, 322, 338
address decoder, 264
address decoder fault, 63, 258, 273
address descrambler, 26
address failure memory, 26
address formatting, 306
address generator, 530
address set-up time sensitivity test, 303
address stepper, 533
Advantest Model T6682 tester, 24
AF, 258, 273
Agarwal, 532
age defect, 58
aging factor, 37
Agrawal, P., 53, 160, 239, 502
Agrawal, V. D., 53, 65, 129–131, 160, 239, 477, 479, 502, 621, 624, 628
AI, 18
air pollution, 259
Aitken, 441, 447
Akers, 158, 206
alias image, 352
aliasing, 512, 514, 518
aliasing probability, 518
all instructions with random data fault model, 597
alloying, 261
alternate assignment, 162, 174
Ambler, 53, 627
Amerasekera, 627
analog ATPG, 411
analog automatic test-pattern generation, 397
analog backtrace, 407
analog bipartite graph, 400, 401, 403
analog boundary module, 580, 583
analog circuit connectivity matrix, 399

analog circuit design for testability, 401, 404, 413
analog circuit element, 399
analog circuit graph, 399
analog circuit parameter, 399
analog circuit performance, 399, 401
analog component deviation, 398
analog double-fault model, 403, 405, 406
analog element observability, 404
analog element tolerance, 401
analog element tolerance box, 400, 403
analog element variation, 399
analog fault coverage, 398, 401
analog fault observation, 401
analog fault ordering, 393
analog fault simulation, 390
analog multiple fault model, 398
analog output parameter, 398, 400
analog output parameter measurement, 399
analog reverse simulation, 407, 410
analog signal flow graph inversion, 408
analog single-fault model, 405
analog switch resistance, 591
analog switch sizing, 591
analog test access port, 579
analog test bus, 576
analog test bus chaining, 579
analog test bus switching pattern, 581
analog testability analysis, 401
analog tolerance computation, 398
analog triple-fault model, 405, 406
AND bridging fault, 61, 258, 271
ANP, 288
ANPSF, 258, 272, 286
aperture uncertainty, 352
APNP, 288
APNPSF, 258
appearance fault, 61, 65
application specific integrated circuit, 8
Arabian, 622, 628
arbitration priority set-up time, 306
arbitration test, 305
architectural design, 7
area overhead, 475, 493
Armstrong, 109, 176
array level, 261
artificial intelligence, 18
ASIC, 8
assertion fault, 60
asymmetric coupling fault, 270
asynchronous clear, 230
asynchronous loop, 233
asynchronous preset, 230
at-speed test, 9, 42, 490, 611
at-speed testing, 435
AT1, 580
AT1N, 589
AT2, 580
AT2N, 589
AT&T standard telecom test, 359

INDEX

AT&T standards, 372
ATE, 10, 18, 22, 24
ATE algorithmic pattern generator, 26
ATE noise, 380
ATE scan pattern generator, 26
ATE sequential pattern generator, 26
ATPG, 155
attenuation distortion test, 373
audio front end, 27
auditing, 8
autocorrelation, 343
autocorrelation property, 504
automatic test equipment, 10, 18, 22, 24, 364
automatic test-pattern generation, 155
automatic test-pattern generation computational complexity, 166
average costs, 36
average diagnostic length, 602
average product, 39

B1 cell, 583
B2 cell, 583
BACK algorithm, 219
back substitution, 396
backdriving, 492
backtrace, 175, 219
backtracing, 150, 187
backtrack, 162, 174
backup, 162
backward implication, 173
backward time expansion, 223
backwards logic simulation, 163
bad-gate, 113
band fault simulation, 397
band-pass case, 353
Bardell, 506, 619, 624
bare-board interconnect test, 21
base cell, 272
bathtub curve, 37
Bayes' rule, 5
BDD, 158
Bearnson, 625
bed-of-nails fixture, 491
bed-of-nails tester, 21, 549
Beenker, 624
behavioral level, 60, 91, 92, 148
behavioral level testability, 148
behavioral model, 264
behavioral synthesis, 148
behavioral view, 389
benefit-cost analysis, 35, 492
benefit-cost ratio, 42
benefits, 41
Bening, 625
Bennetts, 610, 623, 624
Bergeron, 626
BF, 258, 271
Bhattacharya, 204, 625
Bhavsar, 477
BIC sensor, 458

BIDIR cell, 571
bidirectional behavior, 167
bidirectional fault, 271
bidirectional input, 448
BILBO, 495, 497
bilinear transform, 397
binary counter, 499
binary decision diagram, 158
binary tree, 158
binding, 149
binomial coefficient, 122
BIST, 36, 43, 489
BIST controller, 533, 539
BIST initialization, 527
black-box model, 264
Bleeker, 623
board inductance, 551
board level, 261
Boltzmann voltage noise, 358
bonding deterioration, 260
Boolean contrapositive, 198
Boolean difference, 161
Boolean false expression, 165
Boolean false function, 203
Boolean partial derivative, 161
Boolean polynomial, 615
Boolean satisfiability, 164
Bose, 424, 434
bottom-up construction, 595
boundary register, 553, 581
boundary register cell, 553
boundary scan, 43, 527
boundary scan data inversion rules, 566
boundary scan description language, 569
boundary scan hold register, 565
boundary scan propagation delay measurement, 567
boundary scan setup and hold time measurement, 567
boundary scan standard, 549
boundary-scan master, 435
branch coverage, 60, 597
branch fault, 61
branch-and-bound search, 166, 176
breadboard, 83
breadth-first search, 195
Breuer, 238, 543, 621, 625
Brglez, 131
brick wall filter, 381
bridging fault, 14, 61, 65, 178, 258, 271, 443
bridging resistance, 444
broad-side delay test, 431
broadband noise, 381
Brooks, 595
brown noise, 381
brown-out, 14
Bryant, 69, 92
BSDL, 569
BSDL entity description, 570
BSM2, 435

buffer pin, 571
built-in current sensor, 458
built-in current testing, 458
built-in logic block observer, 495, 497, 519
built-in self-test, 36, 43, 466, 489
burn-in, 20, 44
burn-in coverage, 439
burst mode, 318
bus driver conflict, 552
bus fault, 61
bus-oriented BIST, 498
Bushnell, 199, 406, 628
Butler, 628
BYPASS instruction, 564, 585
bypass register, 553, 562, 564

C cell, 583
C-message filter, 373
C-message weighting curve, 375
CA, 511
Ca, 581
cache block, 298
cache DRAM, 254, 296
cache miss test, 298
calculus of redundant faults, 200
calibrate cell, 581
calibration, 313, 590
calibration error, 360
Campbell, 627
Capture-IR state, 571
Case, 68, 78, 125
CATAPULT, 204
catastrophic analog fault, 387, 406
catastrophic analog fault test, 391, 398, 399
catastrophic fault, 314
CCITT, 359, 372
CDRAM, 254
CE scaling, 14
cell, 267
cell coupling fault, 63
cell row, 475
cellular automaton, 499, 511
center frequency, 380
central office, 366
CF, 258, 266, 268
CFdyn, 258, 269, 270
CFid, 258, 270
CFin, 258, 269
CFS, 238
Chakraborty, K., 626
Chakraborty, T. J., 432
Chakradhar, 164, 201, 203, 628
Chakravarty, 626
Chang, 465, 625
channel, 26
channel-connected components, 93, 95
characteristic polynomial, 505, 507, 508, 516, 519
characterization test, 18, 375, 390, 591
Chatterjee, 396

Chattopadhyay, 624
checkpoint fault, 78
checkpoint theorem, 78
checksum test, 300
chemical sensor, 613
Chen, 628
Cheng, K.-T., 239, 479, 481, 624, 626, 628
Cheng, W.-T., 117, 219, 221
chip level, 261
Ciciani, 628
circuit element, 398
circuit level, 91, 93
circuit-under-test, 24, 495
circular fault masking, 64
circular self-test path BIST, 525
CLAMP instruction, 562, 587
Clarke, 85, 625
clipping level, 325
clock, 212
CLOCK cell, 571
clock divider, 320
clock fault, 214, 231
clock rate, 10
clock skew, 429
ClockDR signal, 559, 560
closest neighbor scan wiring, 478
clustered defects, 45
clustering parameter, 45, 49
CMOS stuck-open fault, 274
Co, 581
CO (central office), 366
CO (combinational observability), 131
code format, 323
code histogram, 326
code polynomial, 616
code word, 616
CODEC, 366
CODEC test, 375
coder-decoder, 366
coherence, 320
coherence requirement, 339
coherent correlation, 342
coherent filtering, 342
coherent measurement, 337
coherent multi-tone testing, 356
coherent sampling, 339
coherent testing, 339
collapse ratio, 74
column decoder, 266
combinational 0-controllability, 131
combinational 1-controllability, 131
combinational fault simulation complexity, 166
combinational loop, 233
combinational observability, 131
Combinational robust delay test, 433
common-mode voltage, 592
COMPACTEST, 206
compaction, 513
companion matrix, 505, 507

INDEX 675

comparator, 380
compiled-code simulator, 102
complementarity pivoting, 391, 392
complementarity variable, 392
complementary pivot method, 391
complete multiple backtrace, 200
completeness, 159
component boundary, 451
component level, 60
compression, 513
computed distortion, 376
concurrent BIST, 530
concurrent fault detection phase, 240
concurrent fault simulator, 113, 238
concurrent operation test, 298
concurrent testing, 495
conductance fault model, 446
conducting path, 445
cone segmentation, 500
conflict, 162
consistency procedure, 178
constant electric-field scaling, 14
Constructive Dilemma, 198
CONT-2, 239
contact test, 30
contactor, 27
contention circuit, 305
contention test, 306
CONTEST, 239
CONTROL cell, 571
control cell, 581
control point, 529
control-and-observe boundary scan cell, 565
controllability, 129
controllability iteration, 142
controlling event, 423
CONTROLR cell, 571
convergence, 114
converter gain, 323
COP, 131
core, 529, 605, 606
core isolation, 591
core-disconnect state, 585
core-disconnect switch, 587
core-test language, 607
correlated delay defects, 435
correlator, 343
corrosion, 259, 260
cosmic ray, 261
cost, 36
cost function, 240, 241, 402
count function, 537
coupled cell, 269, 270
coupling capacitance, 592
coupling cell, 269
coupling fault, 256, 258, 268
CRC, 300, 514
CREATOR, 116
CRIS, 247
critical area, 283

critical data pattern, 84
critical path, 418
critical path length, 283
critical path tracing, 125, 206
critical resistance, 444
critical timing path, 434
cross-correlation, 343
cross-point, 61
cross-point fault, 61, 65
crossover, 247
Crouch, 624
CSTP, 525
CTL, 607
cube intersection, 177
current noise, 380
current testing, 14, 440
current transmission, 591
current vector, 240
CUT, 24, 495
cutoff frequency, 381
cutting algorithm, 130
cycle breaking, 481
cycle-free circuit test complexity, 227
cycle-free circuits, 225
cyclic redundancy check code, 514
cyclic redundancy code, 300, 615
cyclic redundancy code test, 300
cyclic s-graph, 229

D cell, 583
D flip-flop, 212
D-ALG, 176
D-Algorithm, 176, 180
D-chain, 180
D-cube, 177
D-drive, 178
D-frontier, 163, 174
D-intersection, 177, 181
D/A converter, 322
D1, 581
D2, 581
DAC, 322
DAC glitch area, 323, 352
DAC settling time, 323
DAG, 226, 427, 480
Das Gupta, 483
data flow graph, 148
data generator, 530, 533
data path synthesis rule SR1, 148
data path synthesis rule SR2, 148
data polynomial, 616
data retention fault, 258
data transfer test, 298
Data1 cell, 581
Data2 cell, 581
David, 628
Davis, 627
dB, 325, 368
DBM, 580
dBm0, 368

DC analog fault simulation, 391, 392
DC bin, 350, 357
DC offset error, 323
DC parametric test, 30
DC testing, 435
DCC, 242
de-correlator, 522
de-glitching of DAC, 318
decibel, 325, 368
decision point, 178
deductive fault simulator, 109
deep submicron device, 613
defect, 45, 58, 61
defect coverage, 16
defect density, 45
defect frequency, 58
defect level, 36, 47
defect per million, 439
defect-oriented fault, 61, 65
defect-oriented testing, 54
deGyvez, 627
delay element, 295
delay fault, 61, 418, 444
delay fault BIST, 540
delay fault testing, 483
delay size, 61
delay test, 87
delay test with scan, 431
delay testing, 12
delay-fault BIST, 435
deleted neighborhood, 272
Demidenko, 624
Deming, 17
derived measurement, 381
design debug test, 18
design for testability, 4, 35, 466, 491, 608
design level, 60
design verification, 60, 83
detection probability, 505
Devadas, 628
device ID register, 553
device identification register, 563
device specification document, 22
device-under-test, 24
DFG, 148
DFT, 35, 341, 346, 348, 466, 491
diagnosis, 6, 47, 58, 63, 89, 106, 124, 257, 489, 496
diagnostic resolution, 489, 596, 598, 610
diagnostic run time, 540
diagnostic test, 390, 423, 596, 598
diagnostic tree, 598, 600
dielectric permeability, 12
dielectric permittivity, 12
difference state, 221
differential I_{DDQ} testing, 456
differential ATAP port, 589
differential fault simulation, 117
differential interconnect, 577
differential linearity error function, 325

differential non-linearity, 325
differential phase, 362
differential sensitivity, 398
digital anti-aliasing filter, 356
digital boundary module, 580
digital reconstruction filter, 351
digital signal processor, 15
digital waveform capture memory, 27
digitizer, 27, 591
digitizer uncertainty, 361
digitizing threshold, 591
Dirac delta function, 505
directed acyclic graph, 226, 427, 480
directed simulation-based search approach, 246
Director, 627
disappearance fault, 61, 65
discontinuous amplitude quantization function, 360
discontinuous function, 360
discontinuous time sampling function, 360
discrete event simulation, 102
discrete Fourier transform, 341, 346, 348
distortion power, 325
distortion power due to dynamic non-linearity, 376
dither, 314
divergence, 114
divided bit-line RAM, 301
DLE, 325
DNL, 325
dominance fault collapsing, 76
dominator, 197
DRAM, 253
DRF, 258
drivability, 220
driven guard wire, 592
DSC, 242
DSP, 15
DSP automatic test equipment, 346
DSP core, 606
DSP testing, 598
DSP-based analog testing, 388
dual in-line package, 549
dual transmission line, 299
duality property, 289
DUT, 24
dynamic burn-in, 44
dynamic combinational controllability, 242
dynamic compaction, 205
dynamic coupling fault, 258, 269, 270
dynamic dominator, 199
dynamic learning, 197
dynamic neighborhood pattern sensitive fault, 272
dynamic offset, 357
dynamic overload, 362
dynamic programming, 199
dynamic RAM, 253
dynamic sequential controllability, 242

INDEX

dynamic testability measures, 242

E-beam testing, 157
E-frontier, 199
EBT, 219
ECAT, 186
economic efficiency, 39
economy, 6
EEPROM, 254, 563
Eichelberger, 160, 467, 477, 502, 624
Elcherif, 391
Eldred, 59, 155
electrical fault, 30
electromagnetic interference, 12, 260
electromigration, 260
electron beam tester, 18
electron beam testing, 129
electronic bed-of-nails, 491
electronically erasable programmable ROM, 254
electrostatic discharge protector, 583
element coverage, 400
element node, 399
embedded-ATE, 10
embedded-ATE test, 611
EMI, 12, 260
emulated instrument, 336
energy function, 165
engineering economics, 35
enhanced-scan delay testing, 430
enterprise, 38
envelope delay distortion, 323
EPROM, 254
equivalence collapsed set, 74
equivalence collapsing, 75
equivalence state, 221
equivalent fault, 599
equivalent input noise, 380
equivalent state hashing, 199
erasable programmable read-only memory, 254
error, 58, 259
error bound, 121
error correction and translation circuit, 186
error detection probability, 518
error rate statistics, 610
error seeding, 610
error tolerance, 316
error vector, 507
error-correcting code, 489
escape probability, 16
ESD damage, 439, 583
ESSENTIAL, 221
essential prime implicant, 177
EST, 199
Eulerian sequence, 288
evaluation frontier, 199
event, 103
event digitizer, 379
event scheduling, 103

event-driven simulator, 103
ex situ testing, 501
excess quantization distortion, 372
exclusive-or operator, 617
exhaustive pattern generator, 519
exhaustive test generation, 160
exhaustive testing, 495, 499
expert system, 18
explicit clock model, 231
explicit digitization, 377
exponential number of paths, 427
extended backtrace, 219
extended interconnect, 577
external exclusive-OR LFSR, 503
external frequency interference, 358
EXTEST instruction, 560, 567, 585
extra cross-point, 61
extrinsic error, 370

fabrication, 8
failure analysis, 457
failure mode analysis, 8, 23, 58
failure rate, 439
fallout rate, 50
false path, 426
false-path, 436
FAN, 192
fanout branch, 133, 162
fanout stem, 133, 162
fast Fourier transform, 341, 346, 348
fast modeling clock phase, 236
FASTEST, 221
fault, 48, 58, 259
fault 1, 273
fault 2, 273
fault 3, 273
fault 4, 274
fault collapsing, 71, 74, 391
fault cone, 172
fault coverage, 36, 88, 204
fault coverage estimation, 150, 505
fault density, 49
fault detection probability, 131, 529
fault detection probability distribution, 506
fault detection time-frame, 222
fault diagnosis, 89, 596
fault dictionary, 598
fault distribution, 53
fault dominance, 76
fault dropping, 204
fault efficiency, 204
fault equivalence, 73
fault equivalence set, 74
fault excitation, 162
fault frequency, 281, 283
fault hierarchy, 278
fault isolation, 596
fault linkage, 278
fault list, 109
fault masking, 64, 169, 278

fault model, 4, 59, 257
fault propagation, 162
fault sampling, 121, 506
fault sensitization, 162
fault simulator, 88
fault tree, 600
fault-event, 113
fault-list, 113
fault-parallelism, 248
faulty column map, 257
faulty row map, 257
FC, 36
FDM, 352
feature scaling, 14
feedback bridging fault, 61, 64
feedback index, 235
feedback network, 503, 504, 523
feedback set, 235
feedback-free circuit, 226
FFT, 341, 346, 348
FFT leakage, 358
fictitious delay element, 235
field, 495
field reject rate, 48
field repair, 495
field-programmable gate array, 563
filter settling time, 317, 337, 342, 370
final objective, 194
firm core, 606
first-order partial derivative of network function, 398
first-order sensitivity, 398
fitness function, 247
five-valued algebra, 159
fixed costs, 35, 36
fixturing setup time, 524
flat description, 94
flicker noise, 380
flip-chip technology, 13
flip-flop distance contribution, 241
flip-flop initialization, 466
flip-flop reset hardware, 523
floating gate defect, 441
floating node, 98
floating pin technique, 452
floating point ADC, 338
floating state, 98
FMA, 8, 23, 58
FMCK, 236
FMOSSIM, 116
formal verification method, 85
forward implication, 172, 189
forward substitution, 396
Foster, 625
Fourier, 341
Fourier voltmeter, 343, 345
Fourier's first principle, 343
Fourier's second principle, 344
FPGA, 563
frame processor, 26

freak failures, 20
Freeman, 109, 156, 238
frequency bin, 341, 349
frequency division multiplexing, 314, 352
frequency leakage, 349
frequency modulation, 353
frequency-shift keyed signal, 320
Friedman, 621, 625
Frohwerk, 489, 513, 514
FSK, 320
FSR, 323
Fujiwara, 65, 176, 192, 193, 621
full channel test, 366
full scale range, 323
full-scan, 467
full-scan design, 157
functional fault, 61
functional level, 60, 91, 92, 389
functional model, 264
functional test, 21, 30, 59, 78, 156, 312, 386, 389, 596, 597
functional tests for ATE, 11
functionally sensitizable PDF, 426
FVM, 345

GA population, 247
gain error, 323
gain tracking test, 369, 372, 375
Galey, 78, 156
Galois field, 504
Gamma function, 46
Gandhi, 3
gate overhead, 474
gate oxide short, 441, 444, 446, 448
gate-delay fault, 61
GATEST, 247
GATTO, 247
Gaussian probability density, 122
generator polynomial, 616
genetic algorithm, 246
geometrical model, 264
Ghosh, 628
glitch area, 314
glitch code, 327
global testing problem, 172
Gmin stepping, 393
Goel, 158, 176, 186, 187, 205, 219
Goessel, 625
Goldberg, 246
Goldstein, 130, 175
Golomb, 489, 615, 624
good machine signature, 513
good-event, 113
good-gate, 113
GPIB, 28
Graf, 625
graph theory flow problem, 400
graph transitive closure, 201
GRASP, 203
Gray code pattern generator, 543

INDEX 679

gray-box model, 264
gross-defects, 435
ground loop, 260
growth fault, 61, 65
Grumberg, 625
guard pin, 592
guard wire, 592
guardbanding, 21
Gulati, 444, 448, 449, 626

Hachtel, 625
half-channel test, 366, 369
Hamida, 398, 402
Hamiltonian sequence, 287
Hamming distance, 252, 600, 617
hard core, 606
hard fault, 387
hard-core, 465, 498
hardware description language, 91, 569
hardware diffractor, 512
hardware partitioning, 501
hardware sharing, 149
hardware test, 59
hardware-software co-design, 597
Harjani, 385
harmonic analog test, 398
harmonic distortion test, 374
Hawkins, 626
Hayes, 287, 514, 625
hazard, 540
hazard elimination, 169
hazard-free stuck-fault test, 69
HDL, 91
headline, 193
headline objective, 194
Healy, 622
Henftling, 203
heterodyning, 315, 352
heuristic, 176
heuristic algorithm, 176
hierarchical decomposition, 491
hierarchical sensitivity analysis, 399
hierarchical test, 525
hierarchical test generation, 206
high impedance state, 98
high level, 60
high-speed operation test, 298
high-speed reload server, 26
high-speed undersampling, 341
HIGHZ instruction, 563, 585, 587
HILO, 109
HITEC, 221, 224
Hnatek, 623
hold time test, 33
Holland, 246
Holzman, 625
homotopy/simulation continuation, 393
Hopfield model, 165
Householder, 393–396
Householder's formula, 395

Hsiao, 247, 625
Huang, 626
humidity, 260
hybrid cellular automaton, 511
hybrid pattern generator, 540, 543
hyperactive fault, 62
hypergeometric probability density function, 122
hysteresis, 326

IBM TestBench, 248
ICT, 491, 605
IDCODE instruction, 562, 585
ideal converter transfer function, 322
ideal white noise measurement, 381
idempotent coupling fault, 258, 270
idle channel noise, 374
IEEE 1149.1 standard, 549
IEEE 1149.4 standard, 575
IEEE 1450 standard, 607
IEEE P1500 standard, 607
IFA, 294
IFA-13 and Delay test, 296
IFA-13 test, 295, 296
IFA-6 test, 295
IFA-9 and Delay test, 296
IFA-9 test, 295
ILE, 326
IM, 315, 323, 349, 356
image number, 353
immediate assignment of uniquely-determined signals, 193
impedance matching, 578
implication arc, 202
implication graph, 165, 201
implication graph AND node, 203
implication procedure, 178
implication stack, 173
implicit clock model, 231
implicit clock rate, 340
implicit digitization, 377
improved unique sensitization procedure, 198
in pin, 571
in situ testing, 501
in-band test, 363
in-circuit test, 21, 491, 549, 575, 605
in-phase component, 344
incoming inspection, 17
incoming inspection test, 20
incompatible cubes, 177
incorrect component orientation, 575
increasing returns to scale, 41
incremental sensitivity, 398
independent analog parameter, 404
independent faults, 206
indistinguishability condition, 73
inductance, 12
inductive fault analysis, 258, 264, 294
inertial delay, 99
infant mortality failures, 20

infinitely sharp band-pass filter, 346
information content, 339, 341
information loss, 322, 513
information redundancy, 259
initial condition, 518
initial state, 223
initialization, 212
initialization fault, 62, 70, 217
initialization phase, 240
initialization vectors, 239
INL, 326, 364
inout pin, 571
INPUT cell, 571
input leakage, 302
input MUX, 494, 496, 512, 526
Instruction 1, 198
Instruction 2, 198
instruction fault, 62, 597
instruction register, 553, 571
instruction register OPCODE, 571
integer-ratio synchronization, 319
integral linearity error function, 326
integral non-linearity, 326
integral non-linearity error, 364
integrated services data network, 540
integration interval, 337
intellectual property, 606
inter-gate bridge, 447
interconnect analog test, 586
interconnect bridging, 441
interconnect test, 591
interdiffusion, 260
intermittent fault, 62, 65, 259
intermodulation component, 356
intermodulation distortion, 315, 323, 356, 359
intermodulation distortion test, 374
intermodulation product, 349, 361
INTERNAL cell, 571
internal exclusive-OR LFSR, 507
internal memory state, 212
International Telegraph and Telephone Consultative Committee, 359
interrupt test, 305
interval counter, 380
INTEST instruction, 560, 588, 589
intra-gate bridge, 447
intrinsic error, 370
intrinsic parameter, 315, 323
inversion coupling fault, 258, 269
inverted signal flow graph, 407, 412
invertible function, 513
ionic contamination, 261
IP, 606
IP core, 606
irreducible polynomial, 495, 616
irreducible tones, 356
irredundant test, 263
ISDN, 540
isolation transistor, 585
Iyengar, 61

Jacobian matrix, 393
Jain, 131, 534
Jalote, 625
JEDEC, 562
Jha, 69, 148, 625
jitter, 315
jitter noise, 376
jitter testing, 299
jitter-induced sampling error, 358
Johnson, 625
Johnson counter pattern generator, 543
Joint Electron Device Engineering Council, 562
Joint Test Action Group, 549
JTAG, 549
JTAG bi-directional pin, 565
JTAG boundary scan, 496

k-coupling fault, 269, 272
Kaminska, 206, 398, 402, 622
Khare, 627
Kime, 602, 610
Kinoshita, 239, 536, 537
KMS algorithm, 437
Kohavi, 621
Koren, 625
Krstić, 624
ks/s, 315
Kundu, 625
Kunz, 200, 626
Kurshan, 626

Lala, 625
latch-up, 459
lattice, 261
law of diminishing returns, 40
layout optimization, 434
leaf-macro, 610
leakage, 591
leakage current, 302
leakage fault, 444, 446, 448
leakage fault table, 450
leakage test, 302
learning procedure, 197
Lee, C. Y., 158
Lee, M. T. C., 626
legacy core, 606
level numbering, 180
level order, 132, 135
level ordering algorithm, 136
level-sensitive scan design, 50, 477, 524, 528
levelization, 102
levelization procedure, 226
Levi, 439
LFSR, 300, 495, 498, 503
LFSR cycle length, 505
LFSR up/down counter, 531, 534
LFSR with all-zero pattern, 530
lightening arrester, 15
Lin, 391, 422

INDEX

Lin-Reddy algebra, 424
Lindbloom, 624
line, 267
line delay test, 428
line justification, 162
line-delay fault, 61, 62
linear analog fault simulation, 395
linear depreciation, 38, 54
linear feedback shift register, 300, 495, 498, 503, 519
linear phase shifter, 512
linear quantization error, 364
linear system, 516
linkage pin, 571
linked fault, 278
Linn, 626
logic analyzer software, 28
logic cone, 160
logic design, 7
logic error, 259
logic level, 91, 92, 132, 148
logic transformation fault, 180
logical fault, 63, 263
Lombardi, 621
long-run production, 38
loop, 233
loopback circuit, 527
loopback test, 540, 581
loss variability test, 372
low-pass case, 353
lowest replaceable unit, 596
LRU, 596
LSSD, 50, 477, 524, 528
LTX Fusion tester, 28

M1, 581
M2, 581
macro test, 609, 610
macroeconomics, 36
Mahoney, 309, 622
maintenance log, 601
maintenance test, 48
Malaiya, 61, 446, 626
Maly, 627
mandatory assignment, 197
Manning, 465, 625
manufacturing test, 17, 59
Mao, 444, 448, 449
MARCH A test, 281, 285
MARCH B test, 281, 285
MARCH C and Delay test, 296
MARCH C test, 281, 296, 532, 539
MARCH C− test, 285, 304
march element, 263
MARCH G test, 295
march test, 263
MARCH X test, 285
MARCH Y test, 285
MARCHING 1/0 test, 281
marginal product, 39

Marlett, 219
MARS hardware accelerator, 116
mass production, 41
master, 212
material defect, 58
material grains, 260
matrix period, 504
MATS test, 281, 285
MATS+ and Delay test, 296
MATS+ test, 263, 285, 296, 533, 536
MATS++ test, 285, 304
matured process, 46
Maunder, 623
maximal element tolerance, 401
maximal length LFSR, 503
maximum clock rate, 323
maximum recursion depth, 201
maximum relative variation, 403
Maxwell, 441, 447
Mazumder, 247, 626, 628
McAnney, 624
McCluskey, 54, 130, 502, 517, 528, 621
MCM, 605
McMillan, 626
MDCCS, 249
Mealy machine, 271
mean, 360
measured distortion, 376
measurement, 315
measurement error, 315, 386, 577
measurement tolerance, 407
measurement uncertainty, 360
memory address scrambling, 534
memory BIST, 529, 543
memory cell, 258
memory parametric testing, 301
memory system bus burnout, 567
memory width, 258
MEMS, 259, 310
MEMS part, 613
Menon, 78, 111, 112, 621
Mercer, 197, 204, 428, 628
metal migration, 14
Metze, 465, 625
MFVS, 481
micro electro-mechanical system, 259, 310
micro-strip probe card, 299
microeconomics, 35
microprocessor clock rate, 9
microprocessor core, 606
microprocessor test, 10, 60, 62, 157, 158, 248, 598
Miczo, 621
Miller capacitance, 591
minimal element tolerance, 401
minimum feedback vertex set, 481
minimum relative variation, 403
minmax-delay simulator, 100
misloaded component, 577
MISR, 300, 499, 516

INDEX

MISR aliasing, 519
missing code, 326
missing cross-point, 61
mixed-signal circuits, 578
MLT, 115
MNA, 395
model A, 433
model B, 433
model C, 433
model checking, 85
modified nodal analysis, 395
modular LFSR, 507
modulo-2 arithmetic, 504, 615, 618
Modus Ponens, 198
Modus Tollens, 198
monic polynomial, 509
Monte-Carlo analysis, 265
Monte-Carlo simulation, 397
Moore, 12, 168
Moore's Law, 12, 168
MOTIS, 116
MOZART, 116, 246
multi-chip module, 605
multi-domain concurrent and comparative simulation, 249
multi-list traversal, 115
multi-phase test point, 529
multi-site testing, 29
multi-tone measurement, 358
multi-tone testing, 315, 359
multi-valued algebra, 424
multiple analog fault model, 402
multiple backtrace, 194
multiple fault, 63
multiple fault analog test, 412
multiple fault model, 256
multiple parametric fault, 388
multiple path sensitization, 162
multiple scan register, 474
multiple signature checking, 517
multiple-delay simulator, 100
multiple-input signature register, 300, 499, 516
multiple-observation test, 218
multiple-path fault sensitization, 159
multiple-use drive enable signals, 565
multiple-weight set, 510
multiply-testable path-delay fault, 63
multiply-testable PDF, 426
mutation, 247
Muth, 159, 217, 218, 245
mutual comparator, 531, 534

Nadeau-Dostie, 532, 606, 607, 623
nail, 549
Nandi, 624
nano electronics, 613
narrow-band tone set, 362
near-Gaussian amplitude distribution, 362
Needham, 624

negative binomial probability density function, 45, 49
negative WSF, 536
neighborhood, 258, 272
neighborhood pattern sensitive fault, 258, 272
neighborhood pattern sensitive fault RAM BIST, 538
neighborhood size, 258
NEMESIS, 203
netlist, 70, 84
network transfer function, 398
neural net ATPG, 201
neural network, 165
new generation, 247
Newton, 628
Newton's method, 412
next state, 212
Nicolaidis, 534, 539, 625
nine-valued algebra, 159, 218
noise, 315
noise power, 381
noise power ratio, 323
noise referred to input, 380
noise rejection test, 589
noise uncertainty, 381
nominal component value, 398
non-Boolean primitive, 206
non-classical fault, 64
non-coherent sampling, 350
non-coherent waveform, 350
non-concurrent BIST, 530
non-concurrent testing, 495
non-deterministic device, 315
non-feedback bridging fault, 61
non-harmonic bin, 357
non-linear distortion, 357
non-linear distortion measurement, 357
non-linear system, 525
non-linearity error, 323
non-permanent fault, 259
non-primitive polynomial, 518
non-robust path-delay test, 420
non-separable code, 617
non-static analog test, 591
Norby, 78, 156
normal probability density, 122
normal system function, 498
normalized correlation, 343
NORTEL standards, 372
NP-Completeness, 166
NPR, 323
NPSF, 258, 272
NPSF fault detection algorithm, 290
NPSF fault location algorithm, 290
NS, 212
number of equivalent bits, 364
numerical differentiation formula, 409, 411
Nyquist, 340
Nyquist frequency, 352
Nyquist frequency bin, 341

INDEX

683

Nyquist interval, 352
Nyquist limit, 341, 353
Nyquist region, 352

OBF, 258, 271
objective, 175, 187
observability, 129
observation point, 404, 529
observe-only boundary scan cell, 564
OBSERVE_ONLY cell, 571
octave, 361
off-path signal, 420
off-set, 222
on-path signal, 420
on-set, 222
one-step relaxation, 392
ones counting, 513
opto-electronic devices, 613
opto-electronic interconnect, 613
OR bridging fault, 61, 258, 271
orthogonal tones, 356
orthogonality, 348
oscillation fault, 64, 108
oscilloscope software, 28
Osseiran, 622
out pin, 571
out-of-band measurement uncertainty, 363
out-of-band test, 363
output drive current test, 32
output short current test, 31
OUTPUT2 cell, 571
OUTPUT3 cell, 571
over-voltage power supply, 20
oversampling filter, 352

P/AR, 359
package defect, 58
package handler, 27
package sealing, 260
Palchaudhuri, 624
parallel BIST, 538
parallel division synchronization, 340
parallel fault simulator, 107
parallel iterative simulator, 125
parallel memory BIST, 533
parallel shift register pattern generator, 522
parallel value list, 125
parallel-pattern single-fault propagation, 125, 502
parameter node, 399
parametric analog fault test, 399
parametric fault, 64, 315, 387
parametric fault test, 398
parametric measurement unit, 27, 30
parametric test, 11, 21
parasitic, 58, 98, 294, 386, 442, 581
PARIS, 125
parity checking, 513
parity test, 300
Parker, 130, 238, 502, 549, 552, 575, 623

partial hardware partitioning, 501
partial-scan, 479
partially detectable fault, 226
partially testable path, 426
partitioned test bus interface controller, 590
partitioning, 171, 490, 595, 608
pass/fail test, 315
passive neighborhood pattern, 288
passive neighborhood pattern sensitive fault, 258, 272, 287
Patel, 204
path classification, 428
path count, 162
path sensitization, 162
path variable, 203
path-delay fault, 61, 64, 420
path-delay fault BIST, 542
pattern generator, 492
pattern matching, 26
pattern memory, 26
pattern multiplexing, 24
pattern sensitive fault, 63, 65, 272
PBX, 366
PCB, 58, 65, 490
PCM, 366
PDF, 178
PE, 26
peak-to-average ratio test, 359
peak-to-peak composite waveform swing, 360
peak-to-RMS ratio, 362
Peled, 625
perfect test, 309
performance overhead, 477, 493
performance parameter, 316, 323
periodic spectrum, 351, 352
permanent fault, 65
Peterson, 489, 509
PGA, 322
PGF, 49
phase drift, 359
phase lead measurement, 359
phase meter, 346
phase shifter, 522
phase-locked loop, 299, 315, 319
phone ringing frequency, 366
physical design, 8
physical fault, 65, 263
PI, 12, 212
Pilarski, 624
pin, 26
pin electronics, 26
pin fault, 65
pin multiplexing, 24, 27
pin overhead, 493
pin protection diode, 30
pin solder fault, 575
pin-permission mode, 552
pinhole short, 441
pink noise, 316, 381
pipeline circuit, 226

PLA, 61, 65, 502
PLA fault, 65
PLL, 299, 315, 319
PMU, 27, 30
PNP, 288
PNPSF, 258, 272, 287
PO, 212
Poage, 156
PODEM, 186
pogo pin, 26
polar output, 346
polynomial divider, 515, 617
polynomial modulus arithmetic, 615
polynomial multiplier, 616
Pomeranz, 206, 427, 428
population size, 121
port X/Y separation, 305
positive WSF, 536
potentially detectable fault, 62, 66, 112, 218
power consumption during test, 436
power consumption test, 31
power density, 14
power supply fluctuation, 260
power-set, 201
power-to-bandwidth ratio, 381
power-up state, 218
PowerMill, 479
PPI, 135, 212
PPO, 135, 212
PPSFP, 125, 502
Pradhan, 200, 625, 627
precision measurement unit, 452
PREDICT, 130
PRELOAD instruction, 559, 567, 585
present state, 212
pressure, 260
primary input, 12, 212
primary output, 212
prime tones, 356
prime-ratio locking, 319
primitive band, 316, 341
primitive D-cube of failure, 178
primitive frequency, 316, 339
primitive frequency bin, 350
primitive period, 341
primitive polynomial, 300, 495, 509, 518, 619
primitive spacing, 339
primitive spectrum, 352
printed circuit board, 58, 65, 490
private branch exchange, 366
probability density function, 123
probability generating function, 49
probe card, 24
PROBE instruction, 587
probe membrane, 24
probe needle, 24
probe pad, 591
probe test, 20, 30
process defect, 58
process diagnosis, 8, 47

process monitor resistor, 591
process monitor transistor, 591
process simulation, 46
process yield, 44
product quality, 35
production, 38
production output, 38
production test, 19, 390
programmable anti-aliasing filter, 322
programmable logic array, 61, 65, 502
programmable reconstruction filter, 320
programmable-gain amplifier, 322
propagation cost, 243
propagation D-cube, 177
propagation delay, 99, 101
propagation delay test, 33
propagation profile, 529
prototyping turn, 489
PS, 212
pseudo LFSR, 540
pseudo-Boolean equation, 165
pseudo-exhaustive testing, 495, 499, 501
pseudo-primary input, 135, 212
pseudo-primary output, 135, 212
pseudo-random phase distribution, 362
pseudo-random testing, 496, 498
pseudo-stuck-at fault, 447
PSF, 272
pull-up resistance, 448
pulse code modulation, 366
pulse modulation, 353
pulse train matching, 26
PVL, 125

quadrature computation, 346
quadrature correlator, 345
quality, 6
quantization bin, 358
quantization distortion, 357, 358, 364
quantization distortion component, 349
quantization distortion measurement error, 363
quantization error, 316, 364
quantization noise, 376
quantization power, 357
quantization uncertainty, 381
quantum electronics, 613
quantum voltage, 316, 325, 326
quiescent current, 14, 69, 440
quiescent current monitor, 458
QUIETEST algorithm, 451

r, 262
r0, 262
r1, 262
race, 66, 214, 233
race fault, 66, 108, 112
radiation, 261
radiation coupling, 12
radiation noise induced errors, 12

INDEX 685

radio frequency circuit, 613
Rajski, 201, 506, 529, 624
Rajsuman, 625–627
RAM, 253
random defects, 45
random logic BIST, 495, 543
random noise, 349, 357, 386
random noise power, 376
random pattern generation, 502
random power, 357
random sampling, 339
random-access scan, 467, 484
random-first-detection variable, 506
random-pattern resistant fault, 512
range of measurement uncertainty, 361
RAS, 467, 484
rated-clock, 9
rated-clock delay test, 434
read-only memory, 254
real analog fault coverage, 404
real event, 423
receive filter, 355
receive memory, 318
reciprocal characteristic polynomial, 531
reconstruction, 352
reconstruction filter, 318
reconvergent fanout, 67, 130, 132
rectangular output, 346
recursive learning, 200
Reddy, 263, 422, 427, 428
reduced coupling analog switch, 592
reduced functional fault, 266, 267
redundancy identification, 168
redundant fault, 66, 70
redundant implicant, 169
redundant memory column, 257
redundant memory row, 257
reference quality voltage, 583
reference RAM, 380
refresh signal, 267
Reghbati, 263, 621
register-transfer level, 60, 92, 149
reject rate, 453
reject ratio, 48, 491
relative amplitude, 363
relative analog element deviation, 401
relative deviation, 401
relative prime numbers, 338
release time test, 33
reliability, 492
reliability reduction, 493
Rent's rule, 13, 609
repair, 47
repair using redundant cells, 21
requirements, 7
residue, 616
resistive bridge, 447
resistive short, 444
response compacter, 492
restricted SNPSF Test, 538

restricted stuck-at fault, 300
reverse pattern simulation, 205
reverse-order simulation, 90
RF circuit, 613
RFD, 506
ripple, 336
rise and fall time test, 33
RMS, 336, 346
RMS function, 381
RMS quantization uncertainty, 325
Roberts, 622
ROM, 254, 498
root mean square, 336, 346
root mean square function, 381
root node, 158
Ross, 78
Roth, 78, 116, 155, 156, 159, 176, 218, 626
routing channel, 475
row decoder, 266
Roychowdhury, 624
RPG, 502
RTI, 380
RTL, 60, 92, 149
Rudnick, 247, 628
rule 150, 511
rule 90, 511
rule of ten, 44, 495
RUNBIST instruction, 552, 562, 588
running time tests, 304

s-graph, 225, 480
S/N ratio, 360
s/s, 316
SA0, 267
SA1, 267
Sachdev, 627
SAF, 258, 266
Saluja, 536, 537
Sami, 621
sample, 121
sample coverage, 121
SAMPLE instruction, 559, 567, 585
sample set information, 377
sample size, 121
sampler, 27
sampling interval, 337
sampling jitter, 352
sampling point, 351
sampling rate, 337
sampling with replacement, 121
sampling without replacement, 121
Samuelson, 35
Sangiovanni-Vincentelli, 621
Sastry, 627
satisfiability expression, 165
SATURN, 248
Savir, 130, 506, 624
SB1 switch, 583
SB2 switch, 583
scan chain, 522

scan chain boundary, 448
scan clock frequency, 479
scan design, 43, 467
scan flip-flop, 468
scan flush test, 50
scan for delay test, 431
scan multiplexer skew, 479
scan overhead formula, 474
scan overheads, 474
scan power dissipation, 479
scan register, 467
scan shift delay test, 431
scan test length, 473
scan testability rules, 469
scan-functional delay test, 431
scan-hold flip-flop, 483
SCANIN, 468
scanning electron microscope, 18
SCANOUT, 468
scatter, 360
SCF, 258, 271
schedule slippage, 489
Schuler, 238
Schulz, 197, 198
SCIRTSS, 222
SCOAP, 130
scrambled address lines, 294
SD switch, 583
search space, 158
search-space parallelism, 249
second harmonic distortion, 356
second-order analog transfer function, 406
second-order harmonic, 361
seed value, 515
segment-delay fault, 61, 67
selection, 247
self-test control, 498
Sellers, 625
SEM, 18
semaphore, 306
semaphore test, 306
SEMATECH experiment, 50, 455
send memory, 318
sense amplifier, 266
sense amplifier recovery fault test, 304
sensitivity analysis, 401
sensitivity matrix, 399
sensitivity-based analog test generation, 398
sensitization value set, 236
sensitized path segmentation, 501
sequential 0-controllability, 131, 140
sequential 1-controllability, 131, 140
sequential depth, 148, 226
sequential fault simulation complexity, 166
sequential observability, 131, 140
sequential robust test, 433
sequential sampling, 125
serial fault simulation, 106
service interruption, 494
Seshu, 109, 156, 238

SEST, 221
set-up time test, 33
Seth, 53, 65, 69, 129–131, 505, 621
Seth and Agrawal model, 53
settling time, 316, 448
SFG, 406
SG switch, 583
SH switch, 583
Shannon, 353
Shannon's expansion theorem, 161
Shannon's sampling theorem, 353
Sharma, 626
Sheppard, 627
Shi, 392–395
shift register sequence, 489
shift test, 471
Shift-DR state, 562
shift-induced bit correlation, 511
Shmoo plot, 19, 28
shop replaceable unit, 596
short resistance, 446
short-run production, 38
shot noise, 380
shrinkage fault, 61, 65
SIC, 436, 542
SIC test, 425
signal cycle, 337
signal flow graph, 389, 396, 406
signal flow graph self-loop, 409
signal information preservation, 353
signal lag, 342
signal lead, 343
signal-to-distortion test, 373
signal-to-noise ratio, 360
signal-to-quantization noise, 319
signal/total distortion test, 375
signature, 489, 496, 513, 514
signature analysis, 513
silicon on insulator, 256
simple interconnect, 577
simplex method, 398, 401, 402
Simpson, 627
simulation, 101
simulation level, 91
simulation-based test generator, 214
simulator, 83
sinc correction, 320
sinc distortion, 354
sine wave generator, 27
Singh, 628
single cell stuck-at fault, 63
single fault detection phase, 242
single input change test, 425
single parametric analog fault, 387, 401
single stuck-at fault, 71
single-clock scan, 471
single-input changing pattern, 542
single-input-change, 436
single-tone testing, 316
single-tone uncertainty, 363

INDEX

single-weight set, 510
singly-testable path-delay fault, 420
singly-untestable path, 426
singular cover, 177
sink node, 407
Sivaraman, 624
six sigma, 439
six sigma quality, 491
skewed-load delay test, 431
SL switch, 583
slave, 212
sleeping sickness, 295
slew limiting, 362
SLIC, 366
slow clock test, 11
slow-clock delay testing, 432
Smith, 422
SMT, 491, 551, 605
Snethen, 219
SNPSF, 258, 273, 287, 536
SOAF, 258
SOC, 15, 386, 605, 611, 613
SOCRATES, 197, 204
soft core, 606
soft fault, 387
soft fault test, 398
SOFTG, 219
SOI, 256
solder bump, 549
Soma, 386, 622
Somenzi, 625
Souders, 335
source event, 102
source node, 407
source stepping, 393
spare column, 257
spare row, 257
sparkle code, 327
sparse sampling, 381
specialization, 41
specification, 4, 6, 7
specification testing, 389
specification-based test, 312
spectra, 341
spectral component, 342
spectral image, 354
spectral line power, 349
spectral mirror image, 351
speed binning, 21
spike suppression, 127
SPLIT data structure, 221
spot delay defect, 435
SRAM, 253
SRU, 596
SSQ, 349, 381
stacked capacitor, 256
STAFAN, 125
Stahnke, 619
staircasing, 320
STALLION, 222

standard cell, 475
standard cell library, 94
standard deviation, 360
standard LFSR, 503
standard test interface language, 607
standby current test, 305
STAR-BIST, 512
star-fault, 64, 66
state coupling fault, 258, 271
state observation, 212
state transition graph, 222
statement coverage, 60, 597
static analog test, 398
static analysis, 129
static burn-in, 44
static compaction, 205
static differential linearity, 323
static electrical discharge, 260
static integral linearity, 323
static learning, 197
static neighborhood pattern sensitive fault, 258, 273, 536
static path sensitization, 420
static RAM, 253
static testing, 435
static timing analysis, 434
statistical fault analysis, 125
statistical sampling, 381
statistically independent signals, 348
statistically orthogonal signals, 348
steady state, 14
STEED, 222
STG, 222
STIL, 607
Stoffel, 626
STRATEGATE, 247
stratified sampling, 125
strobe time, 26
Strojwas, 624, 627
Stroud, 534
structural analog circuit test, 406
structural fault, 67
structural test, 59, 78, 155, 156, 386
structural view, 389
structure graph, 480
structured DFT, 467
stuck-at fault, 65, 67, 156, 258, 267
stuck-closed fault, 446
stuck-on fault, 14, 68
stuck-open address decoder fault, 258
stuck-open fault, 67, 444
stuck-open transistor, 448
stuck-short fault, 67, 70
STUMPS, 522
sub-threshold conduction, 440, 455
sub-threshold conduction leakage, 459
subscriber loop, 366
subscriber loop interface circuit, 366
substrate coupling noise, 380
Suk, 263

sum of squares, 349, 381
supergate, 130
superlinear speed up, 41, 249
superposition error, 326
superposition principle, 516
surface-mount technology, 491, 551, 605
swept frequency measurement, 358
switch level, 91, 92
switching delay, 99
switching relay, 318
symbolic simulation, 97
symmetric coupling fault, 270
symmetrical quantization, 357
synchronous circuit, 212
synchronous interference, 360
syndrome, 604
syndrome polynomial, 617
syndrome testing, 513
Synopsys, 171
synthesis, 84
system, 259, 595
system clock, 523
system clock phase, 235, 236
system logic, 552
system test, 48
system voltmeter, 587
system-on-a-chip, 15, 386, 575, 605, 611, 613

t-diagnosability, 610
Tafertshofer, 167, 203
TAM, 608
tap coefficient, 504
TAP controller, 555
TAP controller power-up reset, 557
TAP controller timing, 557
TBIC, 581
TC, 36
TCK, 553, 566, 580
TDANPSF1G test, 291
TDI, 553, 580
TDO, 553, 580
TDR, 299
TDSNPSF1G test, 291
technological efficiency, 35, 38
technology-dependent fault, 60
TEGAS, 109
TEGUS, 203
temperature, 260
termination resistor, 586
test access mechanism, 608
test bus interface circuit, 580
test complexity, 13
test controller, 496
test cube, 180
test description language, 28
test economics, 35
test escape, 456
test generation algebra, 159
test generation complexity, 166
test generation system, 204

test head, 24, 26, 29
test invalidation, 422
test invalidation problem, 64, 542
test length estimation, 505
test pattern diffraction, 512
test period, 337
test plan, 22
test point, 404, 491
test point insertion, 528
test program, 18, 22, 30
test program generation system, 22
test scheduling, 610
test set length prediction, 146
test sink, 608
test site characterization, 20
test source, 608
test syndrome, 599
test vectors, 18, 22
test-collar, 606
TEST-DETECT procedure, 116
Test-Logic-Reset state, 567, 571
test-pattern augmentation, 512
test-pattern compaction, 512
test-pattern generator, 496
test-per-clock system, 521
test-per-scan system, 521
test-wrapper, 606
testability analysis, 129
testability index, 148
testable path, 426
testbench, 9, 92, 102
tester load board, 452
tester-independent program, 23
testing, 8
testing burnout, 14
testing cost, 10, 35
testing epoch, 516
TF, 258, 266, 268
Thadikaran, 626
Thatte, 266
THD, 358
thermal noise, 380
thermal verification, 614
thermometer code, 327
Thibeault, 456, 457
third harmonic distortion, 356
three-satisfiability, 164
three-sigma range, 123
three-state truth table, 97
threshold test, 32
threshold voltage, 14
tiling method, 289
time difference, 380
time domain analog measurement, 398
time domain reflectometer, 299
time measurement unit, 379
time slot, 105
time to market, 35
time wheel, 105
time-frame, 216

INDEX

time-frame bound, 225
time-frame expansion, 214, 215
timed integrator, 336
timing analysis, 434
timing design of scan path, 479
timing level, 91, 92
Timoc, 624
TLAPNPSF1G test, 290–292
TLAPNPSF1T test, 291
TLAPNPSF2T test, 291
TLSNPSF1G test, 281, 291
TLSNPSF1T test, 291, 292
TLSNPSF2T test, 291
TMS, 553, 580
TMS1, 556
TMS2, 556
TMU, 379
toggle coverage, 61
tolerances, 386
tone, 316
tone frequency, 337
tone pruning, 361
top-down design, 595
topological analysis, 129
TOPS, 197
total bus fault, 61
total cost, 36, 39
total harmonic distortion, 358
TPG, 22, 496
TRAN, 201, 203
transfer map, 323
transient fault, 259
transient region, 99
transistor fault, 70
transistor trans-conductance, 387
transition count, 514
transition count test, 489, 514
transition counter, 513, 519
transition fault, 258, 268, 428
transition tour, 87
transition-delay fault, 61, 70
transitive closure, 166
transmission parameter, 316, 323
transmission test, 317, 372
transmit RAM, 371
transparent test, 530, 539
transport delay, 99
tree, 67
tree leaf, 158
tri-state behavior, 167
tri-state logic, 206
trial vector, 240
tristated leakage, 302
tristated output leakage current, 302
TRST, 553, 580
true power, 348
true process yield, 50
true-value simulator, 84
trunk signaling test, 359
truth expression, 165

Tsui, 624
TSUNAMI, 204
Tulloss, 623
Turino, 624
two-clock scan, 471
two-coupling fault, 268
two-group method, 289, 536
two-satisfiability, 164
type 1 LFSR, 503
type 2 LFSR, 507
Type-1 neighborhood, 272, 536
Type-2 neighborhood, 272, 538
Tyszer, 624

Ulrich, 115, 628
undersampling, 352
unfactorable tones, 356
unidirectional stuck-at fault, 300
uniform sampling, 353
uninitialized memory state, 215
unique sensitization, 193
unit test period, 316, 337, 341
unit-delay simulator, 100
universal rule for non-coherent sampling, 353
unlinked fault, 278
unnecessary hardware, 168
unpowered open, 441
unpowered testing, 578
unrolling, 215
untestable fault, 66, 70, 158
UpdateDR signal, 559
USE statement, 571
user-defined instruction, 553
USERCODE instruction, 563, 585
UTP, 337, 341
Uyar, 626

validatable non-robust test, 426
van de Goor, 253, 263, 266, 274, 278, 530, 534, 537, 626
variable costs, 35, 36
variable-clock delay testing, 432
variance, 123
VC, 36
VCO, 337
vector bus architecture, 318
vector compaction, 9, 87
vector dot product, 342, 347
vector editor, 23
verification, 8
verification testing, 14, 17, 18, 500
verification vectors, 87
Verilog, 91, 490
VHDL, 91, 490, 569
VHDL package, 570
VHDL package body, 570
via resistance, 444
vibration, 260
Vinnakota, 385, 622
virtual decision level, 325

virtual edge, 325, 368
virtual memory, 253
virtual test, 28, 314
VLASIC, 294
VNR test, 426
voltage bump, 301
voltage bump test, 301
voltage compliance limit, 587
voltage drop device, 458
voltage noise, 380
voltage testing, 155
voltage transmission, 591
voltage-controlled oscillator, 337

w, 262
w0, 262
w1, 262
Wadsack, 68
wafer prober, 27
wafer sort test, 20, 30
wafer yield, 45
Waicukauski, 160, 624
Walker, 628
waveform correlation, 342
waveform digitizer, 318, 322
waveform generator, 27
waveform information content, 355
waveform synthesizer, 318, 320
weak fault, 445, 448, 449
weak fault table, 450
weight select line, 511
weight-sensitive fault, 536
weighted pseudo-random pattern generation, 502, 510
weighted random patterns, 160
Weldon, 489, 509
white noise, 316, 381
Wilkes, 57
Wilkins, 621
Williams and Brown model, 53, 55
Williams, M. J. Y., 467
Williams, T. W., 53, 55, 467, 483, 506, 518, 622, 624
wire-wrap board, 83
write driver, 266
write recovery fault, 531
write recovery fault test, 304
wrong chip insertion, 575
WSF, 536
Wunderlich, 161, 483

X-list propagation, 610
X-PATH-CHECK, 187

Yarmolik, 623, 624
yield, 19, 36, 45, 46, 53, 492
yield loss, 6, 456, 493
Yu, 117

Z-domain, 397

zero defects, 8, 439
zero transmission level point, 368
zero-delay simulator, 100
zero-order-hold sampling, 355
zero-width samples, 354
zone number, 353
Zorian, 623, 625